T0224663

TOWARDS HIGHER MATHEMATICS:
A COMPANION

Containing a large and varied set of problems, this rich resource will allow students to stretch their mathematical abilities beyond the school syllabus, and bridge the gap to university-level mathematics. Many proofs are provided to better equip students for the transition to university.

The author covers substantial extension material using the language of sixth form mathematics, thus enabling students to understand the more complex material. Exercises are carefully chosen to introduce students to some central ideas, without building up large amounts of abstract technology. There are over 1500 carefully graded exercises, with hints included in the text, and solutions available online. Historical and contextual asides highlight each area of mathematics and show how it has developed over time.

RICHARD EARL is currently Director of Undergraduate Studies in the Mathematical Institute, Oxford, and a Tutor in Mathematics at Worcester College. From 2003 to 2013, he was Admissions Coordinator and Schools Liaison Officer in the department and has more than a decade of experience setting the MAT (Oxford's Mathematics Admissions Test). He has won several teaching awards within the university for his teaching and lecturing. This book grew out of a residential week he ran for several years at the University of Oxford for new mathematics undergraduates who had not had the chance to study Further Mathematics at A-level.

TOWARDS HIGHER MATHEMATICS: A COMPANION

RICHARD EARL
University of Oxford

CAMBRIDGE
UNIVERSITY PRESS

CAMBRIDGE
UNIVERSITY PRESS

University Printing House, Cambridge CB2 8BS, United Kingdom

One Liberty Plaza, 20th Floor, New York, NY 10006, USA

477 Williamstown Road, Port Melbourne, VIC 3207, Australia

4843/24, 2nd Floor, Ansari Road, Daryaganj, Delhi – 110002, India

79 Anson Road, #06-04/06, Singapore 079906

Cambridge University Press is part of the University of Cambridge.

It furthers the University's mission by disseminating knowledge in the pursuit of education, learning, and research at the highest international levels of excellence.

www.cambridge.org
Information on this title: www.cambridge.org/9781107162389
DOI: 10.1017/9781316677193

First published 2017

A catalogue record for this publication is available from the British Library

ISBN 978-1-107-16238-9 Hardback
ISBN 978-1-316-61483-9 Paperback

In memory of Jazz.

Never the most focused of students, but your tender presence in tutorials
and around college is sorely missed.

Contents

Contents

Glossary

Relating to Sets

The set of **real numbers** is denoted as $\mathbb{R}$.

A **natural number** is a non-negative whole number. The set of natural numbers is denoted $\mathbb{N}$. Note that by this definition 0 is a natural number; some texts consider only positive whole numbers to be natural.

An **integer** is a whole number. The set of integers is denoted $\mathbb{Z}$. The letter zed is used from the German word *Zahlen* meaning 'numbers'.

A **rational number** is a real number which can be written as the ratio of two integers. The set of rational numbers is denoted $\mathbb{Q}$. A real number which is not rational is said to be **irrational**.

If S is a set, we write $x \in S$ to denote that x is an **element** of S. So $\sqrt{2} \in \mathbb{R}$ but $\sqrt{2} \notin \mathbb{Q}$, as $\sqrt{2}$ is real but irrational.

If S is a set, the notation $\{x \in S : P(x)\}$ denotes the **subset** of S consisting of those elements x that satisfy some property P. So, for example, $\{x \in \mathbb{R} : x > 0\}$ denotes the set of positive real numbers.

Two sets A and B are said to be **disjoint** if there is no element common to both sets. Equivalently this means that the intersection $A \cap B$ is empty.

A **partition** of a set X is a collection of disjoint non-empty subsets of X whose union equals X. So every element of X is an element of one and only one subset from the partition.

Relating to Logic

Given statements P and Q, we say that P **implies** Q, and write $P \Longrightarrow Q$, if whenever P is true then Q is true. The **converse** implication is $Q \Longrightarrow P$. So, for example, the statement '$x > 0$' implies '$x^2 > 0$'. In this case the converse is false: '$x^2 > 0$' does not imply '$x > 0$', as $(-1)^2 > 0$.

If statement P implies statement Q, we say that Q is **necessary** for P and P is **sufficient** for Q. For example, '$x = 2$' implies 'x is even'. It is necessary for x to be even for x to equal 2; it is sufficient for x to equal 2 for x to be even. It is not necessary for x to equal 2 for x to be even, and it is not sufficient for x to be even for x to equal 2; these last two facts show the converse is false: 'x is even' does not imply '$x = 2$'.

Given two statements P and Q such that $P \Longrightarrow Q$ and $Q \Longrightarrow P$, then we say P is true **if and only if** Q is true and write $P \Longleftrightarrow Q$. We also say that P is **necessary and sufficient** for Q.

Relating to Functions

Given sets X and Y, a **function** or **map** f from X to Y is a rule which assigns an element $f(x)$ of Y to each x in X. This is commonly denoted as $f: X \to Y$.

Given a function f from X to Y, we refer to X as the **domain** and Y as the **codomain**. The **image** or **range** of f is the subset $\{f(x) : x \in X\}$ of Y. So given the function $f(x) = x^2$ from $\mathbb{R}$ to $\mathbb{R}$, its image is the set of non-negative real numbers. A **bijection** f from a set S to a set T is a function satisfying:

(i) if $f(s_1) = f(s_2)$ where s_1, s_2 are in S, then $s_1 = s_2$; that is, f is **injective** or **1-1**.
(ii) if t is in T then there is s in S such that $f(s) = t$; that is, f is **surjective** or **onto**.

For example, as functions from $\mathbb{R}$ to $\mathbb{R}$, $2x + 1$ is bijective, e^x is only 1–1, $x^3 - x$ is only onto and $\sin x$ is neither.

A function f from a set S to a set T is said to be **invertible** if there is a map g from T to S such that $g(f(s)) = s$ for all s in S and $f(g(t)) = t$ for all t in T. We refer to g as the **inverse** of f and write $g = f^{-1}$. It is a fact that a function is invertible if and only if it is bijective.

Given a function f from a set S to a set T, and if R is a subset of S, then the **restriction** of f to R, denoted $f|_R$, is the map from R to T defined by $f|_R(r) = f(r)$ for all r in R.

Let R, S, T be sets and f a map from R to S and g a map from S to T. Then the **composition** $g \circ f$ from R to T is defined as

$$(g \circ f)(r) = g(f(r)) \qquad \text{for all } r \text{ in } R.$$

Let f be a real-valued function, defined on some subset S of $\mathbb{R}$. We say f is **increasing** if $f(x) \leqslant f(y)$ whenever $x \leqslant y$ and is **strictly increasing** if $f(x) < f(y)$ whenever $x < y$. Similarly we say f is **decreasing** if $f(x) \geqslant f(y)$ whenever $x \leqslant y$ and is **strictly decreasing** if $f(x) > f(y)$ whenever $x < y$.

Let f be a real-valued function, defined on some subset S of $\mathbb{R}$. We say that f is **bounded** if there is a positive number M such that $|f(x)| < M$ for all x in S. A function which is not bounded is said to be **unbounded**. For example, $\sin x$ and e^{-x^2} are bounded on $\mathbb{R}$, whilst x^{-1} is unbounded on the interval $0 < x < 1$.

A **sequence** (x_n) is an infinite ordered list of real (or complex) numbers. We denote the nth term of the sequence as x_n. Sequences typically begin from $n = 0$ or $n = 1$.

A real sequence (x_n) is **increasing** if $x_m \leqslant x_n$ when $m \leqslant n$ and **strictly increasing** if $x_m < x_n$ when $m < n$. Likewise (x_n) is **decreasing** if $x_m \geqslant x_n$ when $m \leqslant n$ and **strictly decreasing** if $x_m > x_n$ when $m < n$.

A real (or complex) sequence (x_n) is said to be **bounded** if there is a positive number M such that $|x_n| < M$ for all n and is otherwise said to be **unbounded**. For example (2^{-n}) is bounded while (n^2) is unbounded.

Miscellaneous

An **equivalence relation** on a set S is a binary relation $\sim$ on S, binary here meaning $\sim$ compares two elements of S and says whether the relation is true or not for that pair. Further $\sim$ must satisfy (i) $s \sim s$ for all s in S, (ii) if $s \sim t$ then $t \sim s$, (iii) if $s \sim t$ and $t \sim u$ then $s \sim u$. So similarity is an equivalence relation on the set of triangles in the xy-plane. However, $\leqslant$ is not an equivalence relation on $\mathbb{R}$, as it does not satisfy condition (ii); for example, '$3 \leqslant 4$' is true but '$4 \leqslant 3$' is false.

A **group** $(G, *)$ is a set G together with a binary operation $*$ on G, binary here meaning that $*$ takes two inputs g_1, g_2 from G and returns an output $g_1 * g_2$ in G. Further:

(i) the operation needs to be **associative**, that is, $(g_1 * g_2) * g_3 = g_1 * (g_2 * g_3)$ for all g_1, g_2, g_3 in G;
(ii) there needs to be an identity element e satisfying $e * g = g = g * e$ for all g;
(iii) for each element there needs to be an inverse g^{-1} such that $g * g^{-1} = e = g^{-1} * g$.

There are many examples of groups in mathematics including the real numbers under addition, the non-zero rational numbers under multiplication, and the bijections from a set to itself under composition.

Two positive integers m, n are said to be **coprime** if the only positive integer which divides both is 1.

Given a natural number n, its **factorial** $n!$ is defined inductively by $0! = 1$ and $n! = n \times (n-1)!$ for $n \geqslant 1$.

Two random variables X and Y are said to be **independent** if for all x, y we have

$$P(X = x \text{ and } Y = y) = P(X = x) \times P(Y = y).$$

For example, two rolls of a fair die are independent, but the sum of the rolls is not independent of the first roll.

Symbols

General Notation

Notation	Meaning	Page	Notation	Meaning	Page
$\mathbb{R}$	the real numbers	p.ix	$\binom{n}{k}$	binomial coefficient	p.85
$\mathbb{Q}$	the rational numbers	p.ix	$\binom{n}{i,j,k}$	trinomial coefficient	p.89
$\mathbb{Z}$	the integers	p.ix	sinh	hyperbolic sine	p.364
$\mathbb{N}$	the natural numbers	p.ix	cosh	hyperbolic cosine	p.364
$\mathbb{C}$	the complex numbers	p.9	tanh	hyperbolic tangent	p.364
$n!$	factorial	p.xi	$[a,b]$	interval $a \leqslant x \leqslant b$	p.341
$\lfloor x \rfloor$	integer part (or floor)	p.78	(a,b)	interval $a < x < b$	p.341

Relating to Complex Numbers

Notation	Meaning	Page	Notation	Meaning	Page
i	$\sqrt{-1}$	p.9	$\lvert z \rvert$	modulus of z	p.14
$\mathrm{Re}\,z$	real part of z	p.9	$\arg z$	argument of z	p.15
$\mathrm{Im}\,z$	imaginary part of z	p.9	$\mathrm{cis}\,\theta$	$\cos\theta + i\sin\theta$	p.15
$\bar{z}$	conjugate of z	p.11	$\sqrt{z}$	square roots of z	p.23

Relating to Vectors and Matrices

Notation	Meaning	Page	Notation	Meaning	Page
$\mathbb{R}^n$	space of $1 \times n$ row vectors	p.129	$\mathbf{e}_i$	standard basis vector in $\mathbb{R}^n$	p.131
$\mathbb{R}_n$	space of $n \times 1$ column vectors	p.129	$\mathbf{i},\mathbf{j},\mathbf{k}$	standard basis in $\mathbb{R}^3$	p.131
M_{mn}	set of $m \times n$ real matrices	p.147	E_{ij}	standard basis vector in M_{mn}	p.154

Notation	Meaning	Page	Notation	Meaning	Page
Row(A)	rowspace of A	p.181	$A = (a_{ij})$	the (i,j)th entry of A is a_{ij}	p.146
Col(A)	column space of A	p.204	$[A]_{ij}$	(i,j)th entry of A	p.146
Null(A)	the null space of A	p.204	A^T	the transpose of A	p.153
dim	dimension	p.199	A^{-1}	the inverse of A	p.165
rank(A)	rank of A	p.184	$\mathbf{x} \cdot \mathbf{y}$	scalar or dot product	p.135
RRE(A)	reduced row echelon form of A	p.182	$\mathbf{x} \wedge \mathbf{y}$	vector or cross product	p.248
I_n	$n \times n$ identity matrix	p.150	$\oplus$	direct sum	p.270
0_{mn}	$m \times n$ zero matrix	p.147	$\det A$	determinant of A	p.223
δ_{mn}	Kronecker delta	p.150	R_θ	anticlockwise rotation by θ	p.212
S_{ij}	ERO swapping ith and jth rows	p.172	S_θ	reflection in $y = x\tan\theta$	p.212
$M_i(\lambda)$	ERO multiplying ith row by λ	p.172	$_W T_V$	matrix of T wrt bases $\mathcal{V}$ and $\mathcal{W}$	p.284
$A_{ij}(\lambda)$	ERO adding λ(row i) to row j	p.172	$J(\lambda, r)$	Jordan block	p.160
$\langle S \rangle$	the span of S	p.196	$Z(\mathbf{v}, A)$	cyclic subspace	p.335
$(\mathbf{c}_1 \mid \cdots \mid \mathbf{c}_n)$	matrix with columns $\mathbf{c}_1, \ldots, \mathbf{c}_n$	p.196	$C(f)$	companion matrix	p.335
$(\mathbf{r}_1 / \cdots / \mathbf{r}_m)$	matrix with rows $\mathbf{r}_1, \ldots, \mathbf{r}_m$	p.196	K_n	complete graph	p.275
diag($a, \ldots, z$)	diagonal matrix	p.151	$K_{m,n}$	complete bipartite graph	p.275
diag(A, B)	matrix with blocks A, B	p.151			

Relating to Special Functions

Notation	Meaning	Page	Notation	Meaning	Page
B_n	Bernoulli numbers	p.123	$\operatorname{erf} x$	error function	p.368
C_n	Catalan numbers	p.118	$\operatorname{li}(x)$	logarithmic integral	p.393
F_n	Fibonacci numbers	p.100	$B_n(x)$	Bernoulli polynomials	p.123
H_n	harmonic numbers	p.424	$H_n(x)$	Hermite polynomials	p.123
L_n	Lucas numbers	p.104	$P_n(x)$	Legendre polynomials	p.124
$\Gamma(a)$	gamma function	p.380	$T_n(x), U_n(x)$	Chebyshev polynomials	p.125
$B(a,b)$	beta function	p.377	$L_n(x)$	Laguerre polynomials	p.125
$\psi(a)$	digamma function	p.424	γ	Euler's constant	p.369
$E(k), K(k)$	elliptic integrals	p.419	G	Catalan's constant	p.381

Relating to Integration and Differential Equations

Notation	Meaning	Page	Notation	Meaning	Page
$\mathbf{1}_I(x)$	indicator function	p.342	$\bar{f}(s)$	Laplace transform of $f(x)$	p.467
$[a,b]$	$\{x : a \leqslant x \leqslant b\}$	p.342	$\mathcal{L}$	Laplace transform	p.467
(a,b)	$\{x : a < x < b\}$	p.342	$H(x)$	Heaviside function	p.468
$[a,b)$	$\{x : a \leqslant x < b\}$	p.342	$\delta(x)$	Dirac delta function	p.477
$(a,b]$	$\{x : a < x \leqslant b\}$	p.342			

Abbreviations

Notation	Meaning	Page	Notation	Meaning	Page
LHS	left-hand side		RHS	right-hand side	
RRE	row-reduced echelon form	p.177	DE	differential equation	p.426
ERO	elementary row operation	p.172	ODE	ordinary differential equation	p.426
ECO	elementary column operation	p.192	PDE	partial differential equation	p.432

Notation	Meaning	Page	Notation	Meaning	Page
FTC	fundamental theorem of calculus	p.354	SHM	simple harmonic motion	p.450
IBP	integration by parts	p.372	pdf	probability density function	p.413
			cdf	cumulative distribution function	p.413

The Greek Alphabet

A,α	alpha	H,η	eta	N,ν	nu	T,τ	tau
B,β	beta	Θ,θ	theta	Ξ,ξ	xi	Y,υ	upsilon
Γ,γ	gamma	I,ι	iota	O,o	omicron	Φ,ϕ	phi
Δ,δ	delta	K,κ	kappa	Π,π	pi	X,χ	chi
E,ϵ	epsilon	Λ,λ	lambda	P,ρ	rho	Ψ,ψ	psi
Z,ζ	zeta	M,μ	mu	Σ,σ,ς	sigma	Ω,ω	omega

Introduction

I.1 Mathematics in Higher Education

Monday – tried to prove theorem,
Tuesday – tried to prove theorem,
Wednesday – tried to prove theorem,
Thursday – tried to prove theorem;
Friday – theorem false.

Julia Robinson to the Berkeley Personnel Office. (Photo reproduced with permission from Dan Reid)

So where is mathematics going to take you in the next few years? This is an entirely reasonable question if you're thinking about committing to a degree in maths or the physical sciences in the short-term future. This might also be a good time to ask yourself why you enjoy maths, or what it is about the subject that generates that enjoyment.

Mathematics within higher education (whether studied solely, or as a joint degree, or as a means to investigate some cognate discipline) will be more varied in its theory and applications than the subject you have so far met at school or college. Much of it will be more challenging, possibly more so initially, but challenging aspects will remain. Further, the expected mode of learning, and the range of abilities of the students around you, could well be very different from your current experiences. All of which may sound a little daunting.

You can imagine that the increased variety now offered by the subject is largely welcomed by students. There is no particular downside to this, save that some students may find themselves spoilt for choice when it comes to selecting options. The increased challenge in the content, and the emphasis on more independent learning, need considerable effort to become habit. But these are not challenges unique to mathematics, and are largely the point of higher education.

Julia Robinson is pictured at the start of this introduction. She was the first female president of the AMS – the American Mathematical Society – and made significant contributions to number theory, though any reading of her biography (Reid [25]) will make clear her modesty, awkwardness in the limelight, and desire to be remembered for her mathematical contributions rather than record-making firsts. The quote accompanying her photo above is a response she made to the personnel department at Berkeley when asked to describe her average week and I include it here to encourage reflection on what sort of week Julia was having as she wrote that.

A life spent researching mathematics is at some orders of magnitude removed from the undergraduate experience; however, they do have some flavours in common. The undergraduate mathematician is both aware of a lot more mathematics than previously, but also aware of how much more maths there is that s/he still hasn't met; whilst exams may come with a syllabus, s/he has seen that mathematics doesn't, that there is great interconnectedness within the subject and that it can be relatively easy to ask questions that are difficult or even have no currently known answer. I would suggest Julia's quotation was in no way intended to reflect negatively on how her week had just gone, nor of the nature of a career in research generally. Growth and learning come with the price of occasional failure; there is some small merit and reassurance in tackling problems set on material with which you feel comfortable, but the greater merit comes in pushing the envelope in more widely applying that knowledge and/or seeking a richer coherence of understanding. So after five days of deep thinking Julia might not have had something particularly tangible from her efforts (such as a research paper) but she knew her original sense of a theorem wasn't true; she likely had a very good sense of why it wasn't true (rather than a solitary counter-example); she would have a more coherent sense of how the internal logic of the problem all tied together; and hopefully she would have a better sense of what she should be seeking to prove and perhaps a sense of how she might go about that. In the words of Beckett: "Ever tried. Ever failed. No matter. Try again. Fail again. Fail better."

Reassuringly for undergraduates, problems of such difficulty won't appear on their regular exercise sheets; rather the exercises will have been selected for being appropriate to the students' current understanding and development, and relevant to current lecture material. It may be that some problems will need several attempts to complete, and that one or two will not get completed, or only partially resolved, before the submission deadline. The student will also have the opportunity at some point soon afterwards to find out the solutions and to discuss the exercises and any misunderstandings further.

What the above does not make explicit is the importance of your *personal* relationship with mathematics. Everyone studying mathematics thinks about the subject in a somewhat different way, and the process of internalizing, for the first time, its ideas and theory can be even more personal; the order in which things click for you, the recognition of the importance of a definition, or appreciation for the subtle rigour of a proof, and the previous examples, visualization, logic with which you are currently comfortable and on which new material builds, mean that you are making your way through mathematics in a highly individual way. Moreover, your personal appreciation of mathematics – its structures, techniques and logic – cannot properly grow without your doing and thinking about mathematics. There will be occasions when you are just told something and it simply makes sense, but by and large understanding will follow as a consequence of time and thought put into working through exercises and juggling relevant theory. Indeed, even the occasion where you are simply given an answer that makes sense may well be a consequence of the effort put into isolating just what question it is that you wanted answering.

Of course there is a commonality in the language of mathematics – one of the main purposes of a maths degree is also to train students in presentation and argumentation, so that they can irrefutably make clear their work to others – but mathematicians do not understand their subject as strings of discrete, logical steps and rather recognize the intent and direction of a proof or method, likely with a picture or suitably general instance of the result in mind. Being comfortable taking on new material, and aligning/comparing/contrasting that new material with previously understood similar theory, and adding it to the growing body of work you understand, will always remain highly individual – *your* way of thinking about mathematics. This can sometimes become very apparent when two different mathematicians seek to answer a layperson's question; the examples or emphases the mathematicians first draw on to explain themselves can often be very different, reflecting their individual views of how the topic fits into their larger understanding or the primary examples or results they have in mind that best embody an idea.

But your relationship with mathematics will also be personal in the sense of what you hope to gain from mathematics. Here and there in this text are brief biographies of significant mathematicians, such as Euler, Noether, Riemann; however it remains the case for almost all of us that we won't ever make such lasting contributions. What enjoyment and worth comes then from studying mathematics? A training in mathematics will certainly serve someone well in terms of employment, and fairly diverse employment at that, but I hope you are able to find ways to engage with mathematics much more variously than employment concerns or a desire to produce significant original research – just as students of English literature or art are unlikely to see themselves as the next Shakespeare or Picasso, but still might enjoy their subject beyond the bounds of their degrees, in recreational reading, writing, painting or sculpture, or welcoming the sharper analysis such a degree would lend to appreciating literature, film, theatre. Mathematics is one of humankind's most successful means of understanding the world around us and has pattern and complexity enough of its own to delight and fascinate. The analytical and logical training alone in a mathematics degree provides a potent mindset and a varied technical toolkit for seeking to understand new problems, physical and philosophical. More optimistically, I hope that mathematics proves to be something enjoyable for you at undergraduate level, and an enjoyment that continues with you in one manner or another beyond university.

Acknowledgements

I would like to thank the following for their helpful comments on earlier drafts of the text: Sam Barnett, Rebecca Cotton-Barratt, Janet Dyson, Will Fourie, Martin Galpin, Andy Krasun, Simon Morris, Vicky Neale, Paul Williams. I would especially like to thank Andy, who also annually supported the original bridging course.

I.2 Using These Notes (Important to Read)

This book grew out of some 'Bridging Notes' written for an annual, week-long, pre-term course at Oxford University; the course was for first-year mathematics students who had had no or limited chance to previously study Further Mathematics. The focus of those notes was on the missing Further Mathematics material most relevant to the Oxford mathematics undergraduate degree, such as complex numbers, matrices, induction and differential equations. Their aim was to help students with the transition to university mathematics, largely from a point of view of quickly bridging any gap in important techniques and knowledge that the students had missed.

This still remains one of the aims of this volume, but I also hope to allow (willing) readers to progress further and also provide those readers with something of an impression of what university maths entails. This includes further content for the material, more emphasis on proofs and understanding, rather than solely on method and calculation, and extension exercises and sections ranging beyond any Further Mathematics syllabus. So the text still includes that same bridging material but also addresses somewhat the changes in style and emphasis that come with the transition to higher education.

It is envisaged that there is a range of students who will find this book of use; likewise it is envisaged that some students will be self-teaching whilst others may have support from teachers at school, college or university. It is almost certain that the 'current you' will find certain sections or types of exercise of more use than others; the 'future you' will no doubt have moved on to other topics or more complicated exercises and so find some of the advanced material more of use. With this in mind, this volume is set up so that it should prove useful both for students meeting a topic for the first time and for those looking for a deeper understanding of the material, or to cover further topics beyond the typical sixth form syllabus. Some of the more advanced material should prove useful for a first-year university student of mathematics or the physical sciences. Consequently you will need, to some extent, to navigate through the book, making use of the guidance and gradings given throughout; this is certainly not a text that is to be read from beginning to end and the chapters are largely independent of one another.

The six chapters are subdivided into sections. Some sections appear with an asterisk (*). This signifies that the material of that chapter/section is beyond the typical Further Mathematics syllabus or what would normally be considered bridging material. First-time readers, or those looking simply to familiarize themselves with Further Mathematics material, may choose to omit these sections. This can be safely done, as later 'unasterisked' sections make no assumptions of familiarity with asterisked material.

The only significant digression from the above is in the material on linear algebra and matrices. Much of the chapter *Vectors and Matrices* contains material that is in Further Mathematics, though perhaps more generally introduced here. However, the material defining dimension and the following chapter, *More on Matrices*, go considerably beyond this. This is a deliberate effort to concretely introduce more advanced material without the abstraction of vector spaces. Such abstraction has an important place in university mathematics, but appreciation of this often takes time and I feel is best garnered at university, in a supportive teaching environment. Nonetheless, much of the purpose and power of this linear algebra can be demonstrated without needing abstraction; in particular, dimension can be rigorously introduced via more concrete theory developed to deal with linear systems and elementary row operations. This material also provides fertile terrain for introducing examples of proofs, occasionally of some sophistication.

There are exercises at the end of each section, numbered alongside a # symbol. If you are meeting a topic for the first time (rather than simply refreshing yourself with the material), then you are strongly advised to attempt a number of the exercises before progressing to the next section. In the margins alongside the text are occasional lists of exercises most suited to the current material. If you wish to familiarize yourself with the immediate material, or feel rather confused by something in a given section, you would be best off first trying some of that section's exercises before reading on further. This is particularly advisable for the longer sections. To help guide you further, the exercises are graded as follows.

- **A or a**: These are very routine questions, which typically require only an application of a definition without much or any further thought. They usually involve simple calculations to get you used to a definition or involve verifying simple parts of proofs that have been left undone.
- **B or b**: These are intended still to be relatively straightforward, but will usually require some understanding of the material to complete without being particularly difficult. For those meeting the section's material for the first time, these exercises probably make the best test of whether you have sufficiently understood the topics, and first-time readers are advised to attempt most of these exercises.
- **C or c**: These are more exercises of the difficulty of **A/a** and **B/b** questions. If you have completed most of the grade **B/b** questions then these questions can be skipped. However, if you are seeking further exercises for practice or if you have struggled somewhat with a particular section's **B/b** questions, these **C/c** questions should prove helpful.
- **D or d**: These are rather more difficult questions, that are often less structured, require a fuller understanding of the material to complete and/or are more challenging in the technical calculations they involve. They are loosely comparable with the style of MAT, AEA or STEP questions and you certainly shouldn't be concerned if these prove more testing or if you have to try several approaches before completing these. Some of these exercises are parts of past first-year Oxford mathematics undergraduate exams.
- **E**: these are deliberately challenging exercises, sometimes completely unstructured and will likely be technically involved and/or conceptually difficult.
- † There are short one-line hints, at the end of the text before the index, for questions appearing with this symbol.

At the end of each chapter is a section entitled *Further Exercises*; these sections are always asterisked. The exercises are intended as tasters, often introducing readers to new ways in which the chapter's material reappears again at university or covering further topics that could not be reasonably considered core enough to appear in the main chapter. There are often supporting passages explaining something of the history or motivation behind the problems. Many of these exercises are consequently somewhat ambitious (especially for first-time readers), but may well prove of interest for those with some prior appreciation of the chapter's material or when revisiting the chapter a second time.

Answers, sketched solutions or whole solutions are available to almost all questions at

http://people.maths.ox.ac.uk/earl/

- There are completely argued solutions to questions graded **A**, **B**, **C**, **D** and **E**.
- There are typically answers or sketched solutions to questions graded **a**, **b**, **c** and **d**. In some few cases – usually where the answer was already given in the question – no further answer or solution exists.
- All grade **E** exercises and all *further exercises* have solutions provided.

The solutions are provided by way of support and reassurance, especially for those self-teaching material from this text. It is important though that they don't become a crutch. It can be all too easy to look at a solution, and even understand the solution, and think "I could have done that if only I'd tried that approach..." without properly getting to grips with the associated ideas and techniques. If an exercise is problematic then you might instead try other exercises first and return later to the troublesome exercise; likewise, if an exercise's hint does not prompt, with a little work, any new avenues of thought, then I would encourage you to move on to other exercises, rather than immediately referring to the solution.

In a similar vein, when reading the text do not let the mathematics wash over you without properly registering – do not be a *passive* reader! It is all too easy to begin reading mathematical language as if it were prose without taking on board its meaning. The marginal recommended exercises should help break this up somewhat, but you are also strongly encouraged to take some time questioning the logic of the mathematics you are reading, reflecting on quite why it is true and doing for yourself any algebraic manipulation that is omitted. On that point, it would be a good idea to have blank paper with you whenever reading the text so that you can jot down any such calculations and/or thoughts.

The Further Exercises are deliberately wide-ranging and commonly difficult. They can include elements of material that mathematics undergraduates might not see till the second or third year of a degree – where, of course, that material will be covered more broadly, rigorously and often more abstractly. However, their range is not meant as a challenge – no beginning undergraduate would have command of all the content these exercises cover – and rather their breadth is intended to offer a wide menu of possible further interest to students which can be selected from according to taste. In a similar manner, the difficulty of the **E** exercises is not meant as a general challenge, and most mathematics undergraduates would either find them difficult or best approached using more advanced ideas and techniques they have met at university. Any reader genuinely looking for questions at this level, and with the time to solve them, should find them useful; others may choose to give them a wide berth; some others may find it useful to note the results these exercises contain, and selectively review the solutions if still interested. The solutions all rely on mathematical knowledge from Mathematics and Further Mathematics A-levels or else introduced within this book.

In summary then, *this text is not at all intended to be read from beginning to end.* There is a certain onus on you to decide what you want from the text, which will change with time, and to use it to your ends. With the help of the asterisks and the exercise grades you will hopefully find a profitable way through the material. Further, the chapters are largely independent of one another (with the exception of Chapter 4 building on Chapter 3), so you might reasonably start with almost any chapter. Depending on previous familiarity with some material you might find **D/d** exercises approachable in one chapter and **B/b** exercises much more appropriate when meeting totally new material. So don't be despondent if some exercises prove hard, and look to raise your game if you start finding some grade of exercise too routine. Most importantly, the exercises – and seriously attempting them – form a key part of making the most of this text. The sections are relatively brief, with crucial theory and examples discussed, but not usually more than one example of any given type, and only by attempting a range of the exercises and reflecting on their solution will you arrive at a more coherent understanding of the material's ideas and techniques.

1

Complex Numbers

1.1 The Need for Complex Numbers

It is well known that the two roots of the quadratic equation $ax^2 + bx + c = 0$ (a, b, c real and $a \neq 0$) are

$$x = \frac{-b \pm \sqrt{b^2 - 4ac}}{2a}, \tag{1.1}$$

and mathematicians have been solving quadratic equations since the time of the Babylonians. When the *discriminant* $b^2 - 4ac$ is positive then these two roots are real and distinct; graphically they are where the curve $y = ax^2 + bx + c$ cuts the x-axis. When $b^2 - 4ac = 0$ then we have one real root and the curve just touches the x-axis here. But what happens when $b^2 - 4ac < 0$? Then there are no real solutions to the equation, as no real number squares to give the negative $b^2 - 4ac$. From the graphical point of view the curve $y = ax^2 + bx + c$ lies entirely above or below the x-axis (Figure 1.1).

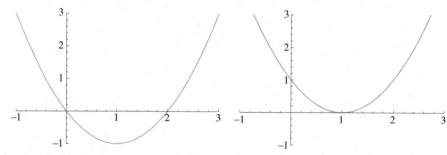

Distinct real roots   Repeated real root

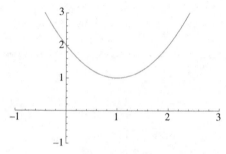

Complex roots

Figure 1.1

It is only comparatively recently that mathematicians have been comfortable with the roots in (1.1) when $b^2 - 4ac < 0$. During the Renaissance the quadratic would have been considered unsolvable or its roots would have been called *imaginary*. But if we imagine $\sqrt{-1}$ to exist, and that it behaves (adds and multiplies) much the same as other numbers, then the two roots in (1.1) can be written in the form

$$x = A \pm B\sqrt{-1} \qquad (1.2)$$

#1
#2
where $A = -b/(2a)$ and $B = \sqrt{4ac - b^2}/(2a)$ are real numbers. But what meaning can such roots have? It was this philosophical point which pre-occupied mathematicians until the start of the nineteenth century, when these 'imaginary' numbers started proving so useful (especially in the work of Cauchy and Gauss) that the philosophical concerns ultimately became side issues. (See Derbyshire [9] for an approachable account of the eventual acceptance of complex numbers.)

Notation 1.1 We shall, from now on, write

$$i = \sqrt{-1}$$

though many books, particularly those written for engineers and physicists, use j instead. The notation i was first introduced by the Swiss mathematician Leonhard Euler in 1777. (See p.18 for a brief biography.)

Definition 1.2 A **complex number** is a number of the form $a + bi$, where a and b are real numbers. We will usually denote a complex number with a single letter such as z or w.

If $z = a + bi$, where a and b are real, then a is known as the **real part** of z and b as the **imaginary part**. We write

$$a = \operatorname{Re} z \quad \text{and} \quad b = \operatorname{Im} z.$$

When we write 'let $z = a + bi$' we will implicitly assume that a and b are real so that $a = \operatorname{Re} z$ and $b = \operatorname{Im} z$.

- Note that real numbers are complex numbers; a real is just a complex number with an imaginary part of zero.

Notation 1.3 We write $\mathbb{C}$ for the set of all complex numbers.

One of the first major results concerning complex numbers, and which conclusively demonstrated their usefulness, was proved by Gauss in 1799 in his doctoral thesis. From the quadratic formula (1.1) we know that all quadratic equations can be solved using complex numbers – what Gauss was the first to prove was the much more general result:

Theorem 1.4 *(Fundamental Theorem of Algebra)* Let $p(z) = a_0 + a_1 z + \cdots + a_n z^n$ be *a polynomial of degree $n \geqslant 1$ with real (or complex) coefficients a_k. Then the roots of the equation $p(z) = 0$ are complex. That is, there are n (not necessarily distinct) complex numbers $\gamma_1, \ldots, \gamma_n$ such that*

$$a_0 + a_1 z + \cdots + a_n z^n = a_n(z - \gamma_1)(z - \gamma_2) \cdots (z - \gamma_n).$$

In particular, the theorem shows that a degree n polynomial has, counting repetitions, n roots in $\mathbb{C}$.

The proof of this theorem is far beyond the scope of this text. Note that the theorem only guarantees the *existence* of the roots of a polynomial somewhere in $\mathbb{C}$, unlike the quadratic formula, which determines the roots exactly. The theorem gives no hints as to where in $\mathbb{C}$ these roots are to be found.

Exercises

#1a Which of the following quadratic equations require the use of complex numbers to solve them?

$$3x^2 + 2x - 1 = 0; \quad 2x^2 - 6x + 9 = 0; \quad -4x^2 + 7x - 9 = 0.$$

#2a Find the four solutions of the equation $(2 + x - x^2)^2 = 16$.

#3C For what values of c does the equation $x^4 = (x - c)^2$ have four real solutions?

#4B On separate axes, sketch the graphs of the following cubics, being sure to carefully label any turning points. In each case, state how many of the cubic's roots are real.

$$y_1(x) = x^3 - x^2 - x + 1; \quad y_2(x) = 3x^3 + 5x^2 + x + 1; \quad y_3(x) = -2x^3 + x^2 - x + 1.$$

#5c† Let c be a real number and $p(x) = 3x^4 - 20x^3 + 48x^2 - 48x$. How many real roots (including multiplicities) does the equation $p(x) = c$ have? You will need to consider various different cases for c.

#6D Show, for all values of x and a, that

$$3x^4 + (8a - 4)x^3 + (6a^2 - 12a + 6)x^2 - (12a^2 + 12)x + (7a^2 + 6a + 9) > 0.$$

What is the smallest value taken by the above expression as x and a vary?

#7d† How many real roots (including multiplicities) have the following polynomial equations?

$$(x^2 + 1)^{10} = (2x - x^2 - 2)^7; \qquad (x^2 + 2x + 3)^8 = (2 + 2x - x^2)^5;$$
$$(x^2 + 1)^5 = (1 - 2x^4)^7.$$

#8B Let a, b, c be positive reals. By expanding $(a + b + c)(a^2 + b^2 + c^2 - ab - bc - ca)$ and considering the second factor as a quadratic in a, show that

$$a^3 + b^3 + c^3 \geqslant 3abc \quad \text{if } a, b, c > 0.$$

When is there equality in the above inequality?

1.2 Their Algebra

We add, subtract, multiply and divide complex numbers in obvious sensible ways. To **add/subtract** complex numbers, we just add/subtract their real and imaginary parts:

$$(a + bi) + (c + di) = (a + c) + (b + d)i; \qquad (a + bi) - (c + di) = (a - c) + (b - d)i. \quad (1.3)$$

Note that these equations are entirely comparable to adding or subtracting two vectors in the xy-plane. Unlike with vectors, we can also **multiply** complex numbers by expanding the brackets in the usual fashion and remembering that $i^2 = -1$:

$$(a+bi)(c+di) = ac + bci + adi + bdi^2 = (ac - bd) + (ad + bc)i. \qquad (1.4)$$

To **divide** complex numbers, we first note that $(c+di)(c-di) = c^2 + d^2$ is real. So

$$\frac{a+bi}{c+di} = \frac{a+bi}{c+di} \times \frac{c-di}{c-di} = \left(\frac{ac+bd}{c^2+d^2}\right) + \left(\frac{bc-ad}{c^2+d^2}\right)i. \qquad (1.5)$$

Example 1.5 Calculate, in the form $a+bi$, the following complex numbers:

$$(1+3i) + (2-6i), \quad (1+3i) - (2-6i), \quad (1+3i)(2-6i), \quad (1+3i)/(2-6i).$$

Solution

$$(1+3i) + (2-6i) = (1+2) + (3+(-6))i = 3 - 3i;$$
$$(1+3i) - (2-6i) = (1-2) + (3-(-6))i = -1 + 9i;$$
$$(1+3i)(2-6i) = 2 + 6i - 6i - 18i^2 = 2 + 18 = 20.$$

#9
#12
Division takes a little more care, and we need to remember to multiply both the numerator and the denominator by $2+6i$.

$$\frac{1+3i}{2-6i} = \frac{(1+3i)(2+6i)}{(2-6i)(2+6i)} = \frac{2+6i+6i+18i^2}{2^2+6^2} = \frac{-16+12i}{40} = -\frac{2}{5} + \frac{3}{10}i. \qquad \square$$

Remark 1.6 Division of complex numbers is very similar to the method of rationalizing a surd. Recall that to write a quotient such as $(2+3\sqrt{2})/(1+2\sqrt{2})$ in the form $q_1 + q_2\sqrt{2}$, where q_1 and q_2 are rational numbers, we multiply the numerator and the denominator by $1 - 2\sqrt{2}$ to get

$$\frac{2+3\sqrt{2}}{1+2\sqrt{2}} = \frac{2+3\sqrt{2}}{1+2\sqrt{2}} \times \frac{1-2\sqrt{2}}{1-2\sqrt{2}} = \frac{2+3\sqrt{2}-4\sqrt{2}-12}{1-8} = \frac{10}{7} + \frac{1}{7}\sqrt{2}. \qquad \blacksquare$$

The number $c - di$, used in (1.5), as relating to $c + di$, has a special name and some useful properties (Proposition 1.16).

Definition 1.7 Let $z = a + bi$. The **conjugate** of z is the number $a - bi$ and is denoted as $\bar{z}$ (or sometimes as z^*).

- Note z is real if and only if $z = \bar{z}$. Also

$$\operatorname{Re} z = \frac{z+\bar{z}}{2}; \qquad \operatorname{Im} z = \frac{z-\bar{z}}{2i}. \qquad (1.6)$$

- Note that two complex numbers are equal if and only if their real and imaginary parts correspond. That is,

$$z = w \quad \text{if and only if} \quad \operatorname{Re} z = \operatorname{Re} w \quad \text{and} \quad \operatorname{Im} z = \operatorname{Im} w.$$

This is called **comparing real and imaginary parts**.

- Note from (1.2) that when the *real* quadratic equation $ax^2 + bx + c = 0$ has complex roots then these roots are conjugates of one another. More generally, if z_0 is a root of the polynomial $a_n z^n + a_{n-1} z^{n-1} + \cdots + a_0 = 0$, where the coefficients a_k are real, then its conjugate $\overline{z_0}$ is also a root (Corollary 1.17).

We present the following problem because it highlights a potential early misconception involving complex numbers – if we need a new number i as the square root of -1, then shouldn't we need another one for the square root of i? However, $z^2 = i$ is just another polynomial equation, with complex coefficients, and two roots in $\mathbb{C}$ are guaranteed by the fundamental theorem of algebra. They are also quite straightforward to calculate.

Example 1.8 Find all those complex numbers z that satisfy $z^2 = i$.

Solution Suppose that $z^2 = i$ and $z = a + bi$. Then $i = (a + bi)^2 = (a^2 - b^2) + 2abi$. Comparing real and imaginary parts, we obtain two simultaneous equations

$$a^2 - b^2 = 0 \quad \text{and} \quad 2ab = 1.$$

So $b = \pm a$ from the first equation. Substituting $b = a$ into the second gives $a = b = 1/\sqrt{2}$ or $a = b = -1/\sqrt{2}$. Substituting $b = -a$ into the second equation gives $-2a^2 = 1$, which has no real solution a. So the two z which satisfy $z^2 = i$, i.e. the two square roots of i, are $(1 + i)/\sqrt{2}$ and $(-1 - i)/\sqrt{2}$. □

#10
#13
#16
#20

Notice, as with the square roots of real numbers, that the two square roots of i are negatives of one another, and this is generally the case (#20). However, it's typically the case that neither of the two square roots is more preferential than the other. So we reserve the notation $\sqrt{z}$, denoting a preferred choice of square root of z, for a positive number z and define $\sqrt{z}$ to be the positive root of z in such a case. (See Definition 1.23 et seq.) More generally we will use the notation $\sqrt{z}$ to represent both of the possible square roots of z.

Example 1.9 Use the quadratic formula to solve $z^2 - (3 + i)z + (2 + i) = 0$.

Solution We have $a = 1$, $b = -3 - i$ and $c = 2 + i$ (in the notation of the quadratic formula). So

$$b^2 - 4ac = (-3 - i)^2 - 4 \times 1 \times (2 + i) = 9 - 1 + 6i - 8 - 4i = 2i.$$

Knowing that $\sqrt{i} = \pm(1 + i)/\sqrt{2}$ from Example 1.8, we see that the two square roots of $2i$ are $\pm(1 + i)$. So

#21
#23

$$z = \frac{-b \pm \sqrt{b^2 - 4ac}}{2a} = \frac{(3 + i) \pm \sqrt{2i}}{2} = \frac{(3 + i) \pm (1 + i)}{2} = \frac{4 + 2i}{2} \quad \text{or} \quad \frac{2}{2} = 2 + i \text{ or } 1.$$

Note that the two roots are not conjugates of one another – this need not be the case when the coefficients a, b, c are not all real. □

Exercises

#9 a Put each of the following numbers into the form $a + bi$.

$$(3 + 2i) + (2 - 7i), \qquad (1 + 2i)(3 - i), \qquad (1 + 2i)/(3 - i), \qquad (1 + i)^4.$$

#10a Let $z_1 = 1 + i$ and let $z_2 = 2 - 3i$. Put each of the following into the form $a + bi$.

$$z_1 + z_2, \quad z_1 - \overline{z_2}, \quad z_1 z_2, \quad \overline{z_1}\, \overline{z_2}, \quad z_1/z_2.$$

#11C Verify the identities in equation (1.6).

#12A Show that multiplication of complex numbers is **commutative** and **associative**. That is, show for any complex numbers z_1, z_2, z_3, that $z_1 z_2 = z_2 z_1$ and $z_1(z_2 z_3) = (z_1 z_2)z_3$.

#13B Show that if $zw = 0$, where z, w are two complex numbers, then $z = 0$ or $w = 0$ (or both).

#14B† The order relation $\leqslant$ for the real numbers has the following properties:

(A) If $0 \leqslant x$ and $0 \leqslant y$ then $0 \leqslant x + y$ and $0 \leqslant xy$.
(B) For any x precisely one of the following holds: $0 \leqslant x$ or $0 \leqslant -x$ or $0 = x$.

Show that no such relation exists for the complex numbers.

#15B If the quadratic equation $z^2 + az + b = 0$ has conjugate complex roots α and $\overline{\alpha}$, show that a and b are real.

#16B† Find all real solutions x, y, if any, of the following equations:

$$\frac{x + yi}{x - yi} = i; \qquad \frac{x + yi}{x - yi} = 2; \qquad (x + yi)^2 = (x - yi)^2; \qquad (x + yi)^4 = 1.$$

#17c† Find the three cube roots of i.

#18b Find all solutions z to the equation $z\overline{z} + 2z + 3\overline{z} = 2i$.

#19c† For what real numbers A, B, C does the equation $z\overline{z} + Az + B\overline{z} = Ci$ have a unique solution z?

#20B Show that given any non-zero complex number z_0, there exists a complex number w such that $w^2 = z_0$. Show that the two solutions of $z^2 = z_0$ are $z = w$ and $z = -w$.

#21b Find the two square roots of $-5 - 12i$. Hence solve $z^2 - (4 + i)z + (5 + 5i) = 0$.

#22c† Find the four roots of $z^4 + 6z^2 + 25 = 0$.

#23B† Show that $3/2 + 5i$ solves $16z^3 - 76z^2 + 520z - 763 = 0$ and hence find all three roots.

#24c Show that $1 + i$ is a root of $z^3 + z^2 + (5 - 7i)z - (10 + 2i) = 0$ and hence find the other two roots.

1.3 The Argand Diagram

The real numbers are often represented on the **real line**, each point of which corresponds to a unique real number. This number increases as we move from left to right along the line. The complex numbers, having two components, their real and imaginary parts, can be represented on a plane; indeed $\mathbb{C}$ is sometimes referred to as the **complex plane**, but more commonly, when we represent $\mathbb{C}$ in this manner, we call it the **Argand diagram** [1].

In the Argand diagram the point (a,b) represents the complex number $a + bi$ so that the x-axis contains all the real numbers, and so is termed the **real axis**, and the y-axis contains all those complex numbers which are purely imaginary (i.e. have no real part) and so is referred to as the **imaginary axis** (Figure 1.2).

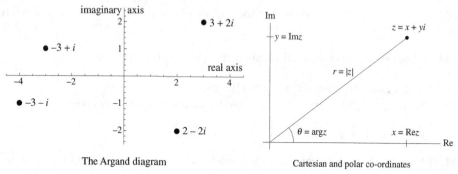

The Argand diagram   Cartesian and polar co-ordinates

Figure 1.2

Remark 1.10 We can think of $z_0 = a + bi$ as a point in an Argand diagram, but it is often useful to think of it as a vector as well. Adding z_0 to another complex number translates that number by the vector (a,b). That is, the map $z \mapsto z + z_0$ represents a translation a units to the right and b units up in the complex plane. Note also that the conjugate $\bar{z}$ of a point z is its mirror image in the real axis. So the map $z \mapsto \bar{z}$ represents reflection in the real axis. We shall discuss in more detail the geometry of the Argand diagram in §1.6.   ∎

#25   A complex number z in the complex plane can be represented by Cartesian co-ordinates,
#26   its real and imaginary parts, x and y, but equally useful is the representation of z by **polar**
#44   **co-ordinates**, r and θ. If we let r be the distance of z from the origin and, if $z \neq 0$, we let θ be the angle that the line connecting z to the origin makes with the positive real axis (see Figure 1.2, right), then we can write

$$z = x + yi = r\cos\theta + (r\sin\theta)i. \tag{1.7}$$

The relations between z's Cartesian and polar co-ordinates are simple:

$$x = r\cos\theta \quad \text{and} \quad y = r\sin\theta;$$

$$r = \sqrt{x^2 + y^2} \quad \text{and} \quad \tan\theta = y/x.$$

Definition 1.11 The number r is called the **modulus** of z and is written $|z|$. If $z = x + yi$ then $|z| = \sqrt{x^2 + y^2}$.

---

[1] After the Swiss mathematician Jean-Robert Argand (1768–1822), although the Norwegian mathematician Caspar Wessel (1745–1818) had previously had the same idea for representing complex numbers, but his work went unnoticed.

• Note that $|z| \geqslant 0$ for all z and that if $|z| = 0$ then $z = 0$. Note also

$$|\operatorname{Re} z| \leqslant |z| \quad \text{and} \quad |\operatorname{Im} z| \leqslant |z|. \tag{1.8}$$

Definition 1.12 The number θ is called the **argument** of z and is written $\arg z$. If $z = x + yi$ then

$$\sin \arg z = \frac{y}{\sqrt{x^2 + y^2}}; \qquad \cos \arg z = \frac{x}{\sqrt{x^2 + y^2}}; \qquad \tan \arg z = \frac{y}{x}.$$

Note that the argument of 0 is undefined. Note also that $\arg z$ is defined only up to multiples of 2π. For example, the argument of $1 + i$ could be $\pi/4$ or $9\pi/4$ or $-7\pi/4$, etc. Here $\pi/4$ would be the preferred choice, as for definiteness we shall take the **principal values for argument** to be in the range $0 \leqslant \theta < 2\pi$.

Notation 1.13 For ease of notation we shall write $\operatorname{cis} \theta$ for $\cos \theta + i \sin \theta$, so that complex numbers in polar form as in (1.7) will now be written $z = r \operatorname{cis} \theta$.

Remark 1.14 Euler's famous result (see p.27 for more) which shows $e^{i\theta} = \cos \theta + i \sin \theta$ is not discussed in detail in this volume, but rather is left until the second volume as the theorem requires a proper understanding of power series and the complex exponential. This remark is purely to inform readers who may have seen complex numbers written in polar form as $re^{i\theta}$ in other texts. For our intents and purposes, regarding the algebra and geometry of the complex numbers here, the notations $re^{i\theta}$ and $r \operatorname{cis} \theta$ are interchangeable without causing confusion or any lack of insight. ∎

Proposition 1.15 *Let α and β be real numbers. Then*

$$\operatorname{cis}(\alpha + \beta) = \operatorname{cis}(\alpha) \operatorname{cis}(\beta); \qquad \overline{\operatorname{cis} \alpha} = \operatorname{cis}(-\alpha) = (\operatorname{cis} \alpha)^{-1}.$$

Proof Recalling the formulae for $\cos(\alpha + \beta)$ and $\sin(\alpha + \beta)$ we have

$$\operatorname{cis}(\alpha) \operatorname{cis}(\beta) = (\cos \alpha + i \sin \alpha)(\cos \beta + i \sin \beta)$$
$$= (\cos \alpha \cos \beta - \sin \alpha \sin \beta) + i(\sin \alpha \cos \beta + \cos \alpha \sin \beta)$$
$$= \cos(\alpha + \beta) + i \sin(\alpha + \beta) = \operatorname{cis}(\alpha + \beta).$$

From this we then have $\operatorname{cis}(-\alpha) \operatorname{cis} \alpha = \operatorname{cis}(-\alpha + \alpha) = \operatorname{cis}(0) = 1$, and finally

$$\overline{\operatorname{cis} \alpha} = \overline{\cos \alpha + i \sin \alpha} = \cos \alpha - i \sin \alpha = \cos(-\alpha) + i \sin(-\alpha) = \operatorname{cis}(-\alpha). \qquad \square$$

We now prove some important formulae about properties of the modulus, argument and conjugation.

#32
#42
#47

Proposition 1.16 *The following identities and inequalities hold for complex numbers z, w:*

$$|zw| = |z||w| \tag{1.9}$$

$$|z/w| = |z|/|w| \text{ if } w \neq 0 \tag{1.10}$$

$$\arg(zw) = \arg z + \arg w \text{ if } z, w \neq 0 \tag{1.11}$$

$$\arg(z/w) = \arg z - \arg w \text{ if } z, w \neq 0 \tag{1.12}$$

$$\arg \bar{z} = -\arg z \; if \; z \neq 0 \tag{1.13}$$

$$z\bar{z} = |z|^2 \tag{1.14}$$

$$\overline{z \pm w} = \bar{z} \pm \bar{w} \tag{1.15}$$

$$\overline{zw} = \bar{z}\bar{w} \tag{1.16}$$

$$\overline{z/w} = \bar{z}/\bar{w} \; if \; w \neq 0 \tag{1.17}$$

$$|\bar{z}| = |z| \tag{1.18}$$

$$|z + w| \leqslant |z| + |w| \tag{1.19}$$

$$||z| - |w|| \leqslant |z - w| \tag{1.20}$$

*(1.19) is known as the **triangle inequality**.*

Proof We prove here a selection of these identities. The remainder are left to #39.

- Identity (1.9): $|zw| = |z| \, |w|$. If $z = a + bi$, $w = c + di$, then $zw = (ac - bd) + (bc + ad)i$ so that

$$|zw| = \sqrt{(ac-bd)^2 + (bc+ad)^2} = \sqrt{a^2c^2 + b^2d^2 + b^2c^2 + a^2d^2}$$
$$= \sqrt{(a^2+b^2)(c^2+d^2)} = \sqrt{a^2+b^2}\sqrt{c^2+d^2} = |z|\,|w|.$$

- Identity (1.11): $\arg(zw) = \arg z + \arg w$. Let $z = r\operatorname{cis}\theta$ and $w = R\operatorname{cis}\Theta$. Then

$$zw = (r\operatorname{cis}\theta)(R\operatorname{cis}\Theta) = rR\operatorname{cis}\theta\operatorname{cis}\Theta = rR\operatorname{cis}(\theta + \Theta),$$

by Proposition 1.15. We can read off that $|zw| = rR = |z|\,|w|$, which is a second proof of (1.9), and also that

$$\arg(zw) = \theta + \Theta = \arg z + \arg w \qquad \text{up to multiples of } 2\pi.$$

- Identity (1.16): $\overline{zw} = \bar{z}\,\bar{w}$. With $z = a + bi$, $w = c + di$ then

$$\overline{zw} = \overline{(ac-bd) + (bc+ad)i} = (ac-bd) - (bc+ad)i = (a-bi)(c-di) = \bar{z}\bar{w}.$$

- Identity (1.19): the triangle inequality $|z + w| \leqslant |z| + |w|$.

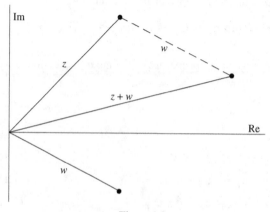

Figure 1.3

A geometric proof of this is simple, explains the inequality's name, and is represented diagrammatically in Figure 1.3.

Note that the shortest distance between 0 and $z + w$ is $|z + w|$. This is the length of one side of the triangle with vertices $0, z, z+w$ and so is shorter in length than the path which goes straight from 0 to z and then straight on to $z + w$. The total length of this second path is $|z| + |w|$.

For an algebraic proof, we note for complex numbers z, w using (1.6), (1.8), (1.9) and (1.18) that

$$\frac{z\overline{w} + \overline{z}w}{2} = \text{Re}(z\overline{w}) \leqslant |z\overline{w}| = |z|\,|\overline{w}| = |z|\,|w|. \tag{1.21}$$

Hence, using (1.14), (1.15) and (1.21) we have

$$|z+w|^2 = (z+w)\overline{(z+w)} = (z+w)(\overline{z}+\overline{w}) = z\overline{z} + z\overline{w} + \overline{z}w + w\overline{w}$$
$$\leqslant |z|^2 + 2|z|\,|w| + |w|^2 = (|z|+|w|)^2,$$

to give the required result. There is equality only if $z\overline{w} \geqslant 0$ so that $\text{Re}(z\overline{w}) = |z\overline{w}|$; if z and w are non-zero this is equivalent to requiring $z = \mu w$ for some $\mu > 0$, meaning that $0, z, w$ are on the same half-line from 0. $\qquad\square$

Corollary 1.17 (*Conjugate Pairs*) *Suppose that z_0 is a root of the degree n polynomial $p(z) = a_n z^n + \cdots + a_1 z + a_0$ where the coefficients a_k are all real. Then the conjugate $\overline{z_0}$ is also a root of $p(z)$. Consequently*

$$p(z) = a_n(z - \alpha_1) \cdots (z - \alpha_r) q_1(z) \cdots q_s(z)$$

where $\alpha_1, \ldots, \alpha_r$ are the real roots of $p(z)$ and $q_1(z), \ldots, q_s(z)$ are real quadratic polynomials with conjugate complex roots. In particular, an odd-degree real polynomial has at least one real root.

#37 **Proof** Note

$$p(\overline{z_0}) = a_n(\overline{z_0})^n + a_{n-1}(\overline{z_0})^{n-1} + \cdots + a_1\overline{z_0} + a_0$$
$$= \overline{a_n}(\overline{z_0})^n + \overline{a_{n-1}}(\overline{z_0})^{n-1} + \cdots + \overline{a_1 z_0} + \overline{a_0} \qquad \text{[as the coefficients } a_k \text{ are real]}$$
$$= \overline{a_n z_0^n + a_{n-1} z_0^{n-1} + \cdots + a_1 z_0 + a_0} \qquad \text{[using (1.15) and (1.16)]}$$
$$= \overline{p(z_0)} = \overline{0} = 0.$$

The fundamental theorem of algebra tells us a polynomial's roots can be found amongst the complex numbers. So the roots of $p(z)$ are either real, call these $\alpha_1, \ldots, \alpha_r$, or come in conjugate complex pairs $\beta_1, \overline{\beta_1}, \ldots, \beta_s, \overline{\beta_s}$. Now

$$(z - \beta_k)(z - \overline{\beta_k}) = z^2 - (\beta_k + \overline{\beta_k})z + \beta_k\overline{\beta_k} = z^2 - (2\,\text{Re}\,\beta_k)z + |\beta_k|^2.$$

If we denote this real quadratic as $q_k(z)$ then $p(z) = a_n(z - \alpha_1) \cdots (z - \alpha_r) q_1(z) \cdots q_s(z)$, where $n = r + 2s$ from equating the degrees of the polynomials. Note that if n is odd then $r \geqslant 1$ and so $p(z)$ has at least one real root. $\qquad\square$

Exercises

#25 a Find the modulus and argument of each of the following numbers:

$$1 + \sqrt{3}i, \qquad (2+i)(3-i), \qquad (1+i)^5, \qquad (1+2i)^3/(2-i)^3.$$

#26 B† Given α in the range $0 < \alpha < \pi/2$, find the modulus and argument of the following numbers:

$$\cos\alpha - i\sin\alpha, \quad \sin\alpha - i\cos\alpha, \quad 1 + i\tan\alpha, \quad 1 + \cos\alpha + i\sin\alpha.$$

Leonhard Euler (1707–1783) (pronounced 'oil-er') was a prolific Swiss mathematician (over 800 papers bear his name) and indisputably the greatest mathematician of the eighteenth century. He made major contributions in many areas of mathematics, but especially in the study of infinite series, and in particular he determined the sum of the series $1 + \frac{1}{4} + \frac{1}{9} + \frac{1}{16} + \cdots$ to be $\frac{\pi^2}{6}$ (the so-called *Basel problem*). Euler's identity, $e^{i\theta} = \cos\theta + i\sin\theta$, and in particular $e^{i\pi} = -1$, can be proved using infinite series for the exponential and trigonometric functions (p.27). But Euler's name appears widely throughout pure and applied mathematics: Euler's equation is fundamental in the calculus of variations (which involves problems such as showing the shortest curve between two points is a straight line); Euler's theorem in number theory is a generalization of Fermat's little theorem; three important centres of a triangle lie on the Euler line (#548).

Euler also produced some of the first *topological* results: he famously showed that it was impossible to traverse the seven bridges in Königsberg without repetition (#1015), in fact determining more generally which networks (or *graphs*) can be traversed. He also showed for a (convex) polyhedron that $V - E + F = 2$, where V, E, F respectively denote the number of vertices (corners), edges, faces that the polyhedron has (p.274). Both these results depend on shape (e.g. how the points are connected) rather than geometry (e.g. the lengths of the edges).

#27A Revisit #13, now making use of properties of the modulus.

#28a Let ω be a cube root of unity (i.e. $\omega^3 = 1$) such that $\omega \neq 1$. Show that $1 + \omega + \omega^2 = 0$. Hence determine the three cube roots of unity in the form $a + bi$.

#29B† Show, for any complex numbers z, w, that $|1 - \bar{z}w|^2 - |z - w|^2 = (1 - |z|^2)(1 - |w|^2)$. Deduce that

$$\left| \frac{z - w}{1 - \bar{z}w} \right| = 1 \qquad \text{if} \quad |w| = 1 \quad \text{and} \quad z \neq w.$$

#30C† Show, for any distinct complex numbers z, w, that

$$\mathrm{Re}\left(\frac{z + w}{z - w} \right) = \frac{|z|^2 - |w|^2}{|z - w|^2}.$$

For given $w \neq 0$, describe the locus of points z that satisfy $\mathrm{Re}\left((z + w)/(z - w)\right) = 1$.

#31 C† Let a, b, c be complex numbers such that $|a| = |b| = |c|$ and $a + b + c = 0$. Show that $a^2 + b^2 + c^2 = 0$.

#32 B Let $z_1 = 1 + i$, $z_2 = \sqrt{3} + i$. Find the real and imaginary parts, modulus and argument of z_1/z_2. Deduce

$$\cos \frac{\pi}{12} = \frac{\sqrt{3} + 1}{2\sqrt{2}} \quad \text{and} \quad \sin \frac{\pi}{12} = \frac{\sqrt{3} - 1}{2\sqrt{2}}.$$

Use the $\cos 2\theta$ formula to show that $\cos(\pi/12) = \frac{1}{2}\sqrt{2 + \sqrt{3}}$. Verify that these two answers for $\cos(\pi/12)$ agree.

#33 D† Let $z \neq 0$. Show there is an integer $1 \leqslant n \leqslant 628$ such that $-0.01 \leqslant \arg(z^n) \leqslant 0.01$.

#34 B Find all solutions, if any, to the following equations:

(i) $|z| = z$;   (ii) $z^2 + |z| = 1$;   (iii) $z + 2\bar{z} = 3$;   (iv) $|z|^4 + \bar{z} = z$.

#35 c† Find all solutions, if any, to the following equations:

(i) $z^2 + 2\bar{z} = 2$;   (ii) $z^2 + i\bar{z}^2 = 1$;   (iii) $z^4 = 8\bar{z}$;   (iv) $z^2 - \bar{z}^2 = 1$.

#36 c Suppose that the roots of the polynomial $p(z) = z^n + a_{n-1}z^{n-1} + \cdots + a_0$ are either real or come in conjugate pairs. Show that coefficients $a_{n-1}, \ldots, a_0$ are all real numbers.

#37 b† Show that $2 + 3i$ solves the quartic $z^4 - 4z^3 + 17z^2 - 16z + 52 = 0$ and hence find the other three roots.

#38 B† By a half-plane of $\mathbb{C}$ we mean the set of points that lie to one side of a line. Prove that every half-plane can be written in the form $\text{Im}((z - a)/b) > 0$ for complex numbers a, b where $b \neq 0$.

#39 C Prove the remaining identities from Proposition 1.16.

#40 B† Let $P(z)$ be a polynomial with (possibly repeated) roots $\alpha_1, \alpha_2, \ldots, \alpha_k$. Show that

$$\frac{P'(z)}{P(z)} = \frac{1}{z - \alpha_1} + \frac{1}{z - \alpha_2} + \cdots + \frac{1}{z - \alpha_k}.$$

Deduce that if $\text{Im}\,\alpha_i > 0$ for each i, then $\text{Im}\,\beta > 0$ for any root β of $P'(z)$. Hence prove **Lucas' theorem**[2], which states that if all the roots of a polynomial $P(z)$ lie in a half-plane then so do all the roots of $P'(z)$.

#41 C† Use Lucas' theorem to show that if all the roots α of a polynomial $P(z)$ satisfy $|\alpha| < R$ then all the roots β of $P'(z)$ satisfy $|\beta| < R$.

#42 B† Let α be a complex number and $k > 0$. Show the equation $|z - \alpha| + |z + \alpha| = k$ has solutions in z if and only if $2|\alpha| \leqslant k$. (See also Example 1.36.)

---

[2] Proven in 1879 by Felix Lucas (rather than Édouard Lucas). The theorem is also sometimes referred to as the Gauss–Lucas theorem.

#43c Show that the locus of $|z - a| - |z - b| = r$, where a, b are distinct complex numbers and r is real, is empty if $|r| > |a - b|$. What is the locus if $r = \pm|a - b|$? (See also Example 1.36 and Proposition 4.50.)

#44B† On separate Argand diagrams sketch the following sets:

(i) $|z| < 1$;   (ii) $\mathrm{Re}\, z = 3$;   (iii) $|z - 1| = |z + i|$;
(iv) $\arg(z - i) = \pi/4$; (v) $-\pi/4 < \arg z < \pi/4$; (vi) $\mathrm{Re}\,(z + 1) = |z - 1|$;
(vii) $|z - 3 - 4i| = 5$;   (viii) $\mathrm{Re}\,((1 + i)z) = 1$;   (ix) $\mathrm{Im}\,(z^3) > 0$.

#45D† Let $k > 0$. Sketch the curve C_k with equation $|z + 1/z| = k$. What are the extreme values of $|z|$ on C_k?

#46d With notation as in #45, show that the curve C_2 consists of precisely two circles.

#47A† Multiplication by i takes the point $x + yi$ to the point $-y + xi$. What transformation of the Argand diagram does this represent? What is the effect of multiplying a complex number by $(1 + i)/\sqrt{2}$?

#48C Let $A = 1 + i$ and $B = 1 - i$. Find the two numbers C and D such that ABC and ABD are equilateral triangles in the Argand diagram. Show that if $C < D$ then

$$A + \omega C + \omega^2 B = 0 = A + \omega B + \omega^2 D,$$

where $\omega = \mathrm{cis}(2\pi/3)$.

#49d(i) Let A, B, C be complex numbers with $|A| \neq |B|$. Consider the equation

$$Az + B\bar{z} = C \tag{1.22}$$

in z. Show that (1.22) has a unique solution.
(ii) Suppose now that $|A| = |B| \neq 0$. Show that (1.22) has no solutions if $C\bar{A} \neq \bar{C}B$ and has infinitely many solutions if $C\bar{A} = \bar{C}B$. In the latter case describe geometrically the solutions as a subset of the Argand diagram.

1.4 Roots of Unity

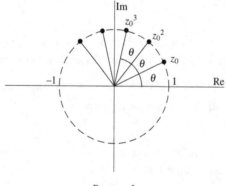

Powers of z_0

Figure 1.4

Consider the complex number $z_0 = \mathrm{cis}\,\theta$ where θ is some real number in the range $0 \leqslant \theta < 2\pi$. The modulus of z_0 is then 1 and the argument of z_0 is θ (Figure 1.4).

In Proposition 1.16 we proved that $|zw| = |z|\,|w|$ and for $z, w \neq 0$ that $\arg(zw) = \arg z + \arg w$. Hence

$$|z^n| = |z|^n \text{ and } \arg(z^n) = n \arg z$$

for any integer n and $z \neq 0$. So the modulus of $(z_0)^n$ is $1^n = 1$, and the argument of $(z_0)^n$ is $n\theta$ up to multiples of 2π. Putting this another way, we have the following famous theorem due to De Moivre.

Theorem 1.18 *(De Moivre's Theorem [3]) For a real number θ and integer n we have that*

$$\cos n\theta + i\sin n\theta = (\cos\theta + i\sin\theta)^n,$$

or more succinctly, $\mathrm{cis}(n\theta) = (\mathrm{cis}\,\theta)^n.$

Example 1.19 Find expressions for $\cos 4\theta$ and $\sin 4\theta$.

#74
#75
Solution Writing c for $\cos\theta$ and s for $\sin\theta$ we have

$$\cos 4\theta + i\sin 4\theta = (c+is)^4 = c^4 + 4ic^3s - 6c^2s^2 - 4ics^3 + s^4.$$

Comparing real and imaginary parts we have

$$\cos 4\theta = c^4 - 6c^2(1-c^2) + (1-c^2)^2 = 8c^4 - 8c^2 + 1;$$
$$\sin 4\theta = 4c^3s - 4c(1-c^2)s = 8c^3s - 4cs. \qquad \square$$

We now apply the ideas from the proof of De Moivre's theorem to the following problem, that of

finding the roots of $z^n = 1$ where $n \geqslant 1$ is an integer.

We know from the fundamental theorem of algebra that there are (counting multiplicities) n solutions – these are known as the *n*th **roots of unity**. Let's first solve directly $z^n = 1$ for $n = 2, 3, 4$.

- $n = 2$: we have $0 = z^2 - 1 = (z-1)(z+1)$ and so the square roots of 1 are ± 1.
- $n = 3$: $0 = z^3 - 1 = (z-1)(z^2 + z + 1)$. The cube roots of unity are then $(-1 \pm i\sqrt{3})/2$ and 1.
- $n = 4$: $0 = z^4 - 1 = (z^2 - 1)(z^2 + 1)$, so that the fourth roots of 1 are ± 1 and $\pm i$.

Plotting these roots on Argand diagrams, as in Figure 1.5, we can see a pattern developing

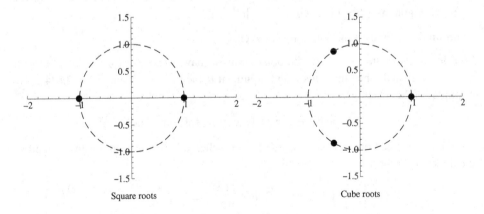

Square roots   Cube roots

---

3 Abraham De Moivre (1667–1754), a French Protestant who fled religious persecution in France to move to England, is best remembered for this formula but he also made important contributions in probability which appeared in his *The Doctrine of Chances* (1718).

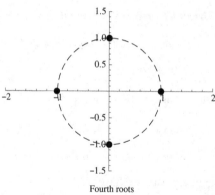

Fourth roots

Figure 1.5

Returning to the general case, suppose that

$$z = r\operatorname{cis}\theta \qquad \text{and satisfies} \qquad z^n = 1.$$

Then z^n has modulus r^n and has argument $n\theta$, whilst 1 has modulus 1 and argument 0. Comparing their moduli we see $r^n = 1$ and hence $r = 1$ (as $r > 0$). Comparing arguments we see $n\theta = 0$ up to multiples of 2π. That is, $n\theta = 2k\pi$ for some integer k, giving $\theta = 2k\pi/n$. So if $z^n = 1$ then z has the form $z = \operatorname{cis}(2k\pi/n)$ where k is an integer. At first glance there seems to be an infinite number of roots but, as cosine and sine have period 2π, then $\operatorname{cis}(2k\pi/n)$ repeats with period n. Hence we have shown:

Proposition 1.20 (*Roots of Unity*) *The nth roots of unity, that is the solutions of the equation* $z^n = 1$, *are*

$$z = \operatorname{cis}(2k\pi/n) \qquad \text{where } k = 0, 1, 2, \ldots, n-1.$$

- When plotted these nth roots of unity form a regular n-gon inscribed within the unit circle with a vertex at 1. More generally, for $c \neq 0$, the n solutions of $z^n = c$ make a regular n-gon inscribed in the circle $|z| = |c|^{1/n}$.

Example 1.21 Find all the solutions of the cubic $z^3 = -2 + 2i$.

Solution If we write $-2 + 2i$ in its polar form we have $-2 + 2i = \sqrt{8}\operatorname{cis}(3\pi/4)$. So if $z^3 = -2 + 2i$ and z has modulus r and argument θ then $r^3 = \sqrt{8}$ and $3\theta = 3\pi/4$ up to multiples of 2π, which gives

$$r = \sqrt{2} \quad \text{and} \quad \theta = \pi/4 + 2k\pi/3 \quad \text{for some integer } k.$$

As before, we need only consider $k = 0, 1, 2$ (as other values of k lead to repeats) and so the three cube roots are

$$\sqrt{2}\operatorname{cis}\left(\frac{\pi}{4}\right) = 1 + i, \qquad \sqrt{2}\operatorname{cis}\left(\frac{11\pi}{12}\right) = -\frac{(1+\sqrt{3})}{2} + \frac{(\sqrt{3}-1)}{2}i,$$

$$\sqrt{2}\operatorname{cis}\left(\frac{19\pi}{12}\right) = \frac{(-1+\sqrt{3})}{2} - \frac{(1+\sqrt{3})}{2}i. \qquad \square$$

#51
#53
#57
#58
#66

• Note that the above method generalizes naturally to solving any equation of the form $z^n = c$, where c is a complex number and $n \geqslant 1$ is an integer (#53, #54).

There are many mathematical fallacies associated with roots and complex numbers, but these can be easily avoided if care is taken. A classical example of such a fallacy is the following.

Example 1.22 Clearly

$$\frac{-1}{1} = \frac{1}{-1} \qquad \text{so that} \qquad \sqrt{\frac{-1}{1}} = \sqrt{\frac{1}{-1}}$$

and as $\sqrt{a/b} = \sqrt{a}/\sqrt{b}$ then

$$\frac{i}{1} = \frac{\sqrt{-1}}{\sqrt{1}} = \frac{\sqrt{1}}{\sqrt{-1}} = \frac{1}{i}.$$

This rearranges to give $i^2 = 1$, which is plainly false as $i^2 = -1$.

This fallacy comes about in the use of the 'rule' $\sqrt{a/b} = \sqrt{a}/\sqrt{b}$ and not appreciating what the root sign $\sqrt{}$ can mean.

Definition 1.23 Given a positive real number x, then $\sqrt{x}$ denotes the positive square root of x. More generally given a complex number z there is no preferred square root of z. Instead we define

$$\sqrt{z} = \{w \in \mathbb{C} : w^2 = z\}.$$

• For non-zero z there are two square roots, which are negatives of one another (#20, #50).

Definition 1.24 Given a complex number z and rational number $q = m/n$ in lowest form, then we may likewise define

$$z^q = \{w \in \mathbb{C} : w^n = z^m\}.$$

• If $q = m/n$ in lowest form and $z \neq 0$ then z^q is a set containing n elements; if Z is a particular element in z^q then $z^q = \{Z, \omega Z, \omega^2 Z, \ldots, \omega^{n-1} Z\}$ where $\omega = \text{cis}(2\pi/n)$ is an nth root of unity (#71).
• Sense can be made of z^q when q is an irrational real or even a complex number, but this requires knowledge of the complex exponential and logarithm. This is discussed in the second volume.

#71

The following two rules about powers hold for positive real numbers.

$$r^p r^q = r^{p+q} \quad \text{where } p, q, r \text{ are real and } r > 0; \tag{1.23}$$

$$(rs)^p = r^p s^p \quad \text{where } p, r, s \text{ are real and } r, s > 0. \tag{1.24}$$

It is reasonable to ask whether the two rules (1.23) and (1.24) make any sense in the context of Definition 1.24. As z^q, in that definition, is a set then we first need to consider how to multiply sets.

Definition 1.25 If S and T are subsets of $\mathbb{C}$ then we define

$$ST = \{st : s \in S, t \in T\}. \tag{1.25}$$

That is, to calculate the product ST we multiply each element of S by each element of T.

Example 1.26 Let z, w be non-zero complex numbers and $Z^2 = z$, $W^2 = w$. Then

$$\sqrt{z} = \{Z, -Z\} \quad \text{and} \quad \sqrt{w} = \{W, -W\}.$$

Multiplying $\sqrt{z}$ and $\sqrt{w}$ according to (1.25) then we have four products to consider:

$$ZW, \quad (-Z)W = -ZW, \quad Z(-W) = -ZW, \quad (-Z)(-W) = ZW.$$

Hence $\sqrt{z}\sqrt{w} = \{ZW, -ZW\}$. As $(ZW)^2 = Z^2 W^2 = zw$ then $\sqrt{zw} = \{ZW, -ZW\}$. Thus we've shown

$$\sqrt{zw} = \sqrt{z}\sqrt{w} \qquad \text{for } z, w \neq 0, \tag{1.26}$$

in the sense of Definitions 1.24 and 1.25. It is an easy check that (1.26) holds also if either or both of z, w are zero.

More generally (1.24) holds true in the form $(zw)^p = z^p w^p$ for non-zero complex z, w and rational p in the context of Definitions 1.24 and 1.25 (#72). However, (1.23) does not generally hold, as we see now.

Example 1.27 Take $z = -1$ and $p = q = 1/2$. Then

$$(-1)^{1/2} = \left\{ w \in \mathbb{C} : w^2 = -1 \right\} = \{i, -i\}.$$

There are four products in $(-1)^{1/2}(-1)^{1/2}$ namely

$$ii = -1, \quad i(-i) = 1, \quad (-i)i = 1, \quad (-i)(-i) = -1,$$

showing $(-1)^{1/2}(-1)^{1/2} = \{1, -1\}$. However, $(-1)^{1/2+1/2} = (-1)^1 = \{-1\}$.

Exercises

#50 A Prove, using modulus and argument, that every non-zero complex number has two square roots, and that these are negatives of one another. (See also #20.)

#51 A† Let $\omega^3 = 1$, $\omega \neq 1$. Show that $z^3 - a^3 = (z - a)(z - \omega a)(z - \omega^2 a)$ for any complex numbers z and a.

#52 C Let $\omega^3 = 1$, $\omega \neq 1$. Show for any complex numbers a, b, c that

$$a^3 + b^3 + c^3 - 3abc = (a + b + c)(a + \omega b + \omega^2 c)(a + \omega^2 b + \omega c).$$

Hence find all real numbers a, b, c which satisfy $a^3 + b^3 + c^3 = 3abc$.

#53 a Find the fifth roots of $2 + 3i$, giving real and imaginary parts to five decimal places.

#54 c Find the sixth roots of $3 - 7i$, giving real and imaginary parts to five decimal places.

#55 C Verify the expressions for $\sqrt{2}\text{cis}\,(11\pi/12)$ and $\sqrt{2}\text{cis}\,(19\pi/12)$, as given in Example 1.21, are indeed correct.

#56c An nth root of unity z is said to be **primitive** if $r = n$ is the smallest positive integer such that $z^r = 1$. What are the primitive fifth roots of unity? What are the primitive sixth roots of unity?

#57B† Write down the seven roots of $z^7 + 1 = 0$. By considering the coefficient of z^6 in the factorization of $z^7 + 1$, show that $\cos(\pi/7) + \cos(3\pi/7) + \cos(5\pi/7) = 1/2$.

#58B† Let $\zeta = \text{cis}(2\pi/5)$. Show that $1 + \zeta + \zeta^2 + \zeta^3 + \zeta^4 = 0$. Show further that

$$(z - \zeta - \zeta^4)(z - \zeta^2 - \zeta^3) = z^2 + z - 1$$

and deduce that $\cos(2\pi/5) = (\sqrt{5} - 1)/4$.

#59c† Let $\zeta = \text{cis}(\pi/7)$. Simplify the expression $(\zeta - \zeta^6)(\zeta^3 - \zeta^4)(\zeta^5 - \zeta^2)$. Hence show that

$$\cos(\pi/7)\cos(3\pi/7)\cos(5\pi/7) = -1/8. \qquad (1.27)$$

#60c Find all the roots of the equation $z^8 = -1$. Hence factorize $z^8 + 1$ as the product of four real quadratics.

#61D† Given $\cos 7\theta = 64\cos^7\theta - 112\cos^5\theta + 56\cos^3\theta - 7\cos\theta$, rederive the result of #59.

#62C Apply Lucas' theorem (#40 and #41) to the polynomial $(z^3 + 1)(z^4 + 1)(2z - 1)$ to show that all the roots of $16z^7 - 7z^6 + 10z^4 + 4z^3 - 3z^2 + 2$ have modulus less than or equal to 1.

#63b† Find all roots of

(i) $1 + z^2 + z^4 + z^6 = 0$;   (ii) $1 + z^3 + z^6 = 0$;   (iii) $1 + z^2 + z^6 + z^8 + z^{12} + z^{14} = 0$.

#64D† Find all roots of the following equations:

(i) $(1 + z)^5 - z^5 = 0$;   (ii) $(z + 1)^9 + (z - 1)^9 = 0$;   (iii) $(2z + 1)^6 + (z^2 - 1)^3 = 0$.

#65D(i) Write down the $n - 1$ roots of the polynomial $[(z + 1)^n - 1]/z$. What is the polynomial's constant term?
(ii) There are n points equally spaced around the circumference of a unit circle. Prove that the product of the distances from one of these points to each of the others is n.

#66B† Show, for any complex number z and positive integer n, that

$$z^{2n} - 1 = (z^2 - 1)\prod_{k=1}^{n-1}\left\{z^2 - 2z\cos\left(\frac{k\pi}{n}\right) + 1\right\}. \qquad (1.28)$$

Deduce for any real θ that

$$\sin n\theta = 2^{n-1}\sin\theta\prod_{k=1}^{n-1}\left\{\cos\theta - \cos\left(\frac{k\pi}{n}\right)\right\}.$$

#67d† Let n be a positive integer. Determine

(i) $\displaystyle\prod_{k=1}^{2n}\cos\left(\frac{k\pi}{2n+1}\right)$,   (ii) $\displaystyle\prod_{k=1}^{n}\cos\left(\frac{k\pi}{2n+1}\right)$,   (iii) $\displaystyle\prod_{k=1}^{n}\cos\left(\frac{(2k-1)\pi}{2n+1}\right)$.

#68D† Let n be a positive integer. Show that

$$\prod_{k=1}^{n-1}\cos\left(\frac{k\pi}{2n}\right) = 2^{1-n}\sqrt{n}.$$

#69C Determine, using the results of #67 and #68, the products

$$\prod_{k=0}^{n-1}\cos\left(\frac{(3k+1)\pi}{3n}\right) \quad\text{and}\quad \prod_{k=0}^{n-1}\cos\left(\frac{(3k+2)\pi}{3n}\right).$$

#70D Let n be a positive integer. By considering the factorizations of $z^{4n}+1$ and $z^{4n+2}+1$, show that

$$\prod_{k=1}^{n}\cos\left(\frac{(2k-1)\pi}{4n}\right) = \frac{\sqrt{2}}{2^n} \quad\text{and}\quad \prod_{k=1}^{n}\cos\left(\frac{(2k-1)\pi}{4n+2}\right) = \frac{\sqrt{2n+1}}{2^n}.$$

#71B Let $z = r\operatorname{cis}\theta$ be a non-zero complex number and $q = m/n$ be a rational in its lowest form.

(i) Show that $z^q = \{w \in \mathbb{C}: w^n = z^m\}$ contains n elements and find them.
(ii) Show further, that if Z is in z^q, then

$$z^q = \left\{Z, \omega Z, \omega^2 Z, \ldots, \omega^{n-1}Z\right\} \qquad\text{where}\qquad \omega = \operatorname{cis}(2\pi/n).$$

(iii) Let k be an integer. Show, in the sense of Definition 1.24, that $z^k = \{z^k\}$.

#72C Let $z, w \neq 0$ be complex and p rational. Show that $(zw)^p = z^p w^p$ in the context of Definitions 1.24 and 1.25.

#73C† Let S denote the 'cut-plane' of $\mathbb{C}$ without the non-positive real numbers. Each z in S can be written uniquely in the form $z = r\operatorname{cis}\theta$, where $-\pi < \theta < \pi$ and $r > 0$; for z in S we define $f(z) = \sqrt{r}\operatorname{cis}(\theta/2)$.

(i) Show that $\sqrt{z} = \{f(z), -f(z)\}$.
(ii) Let $\varepsilon > 0$ be small, so small that ε^2 may be considered negligible. Determine the values of r and θ associated with $-1 + \varepsilon i$ and $-1 - \varepsilon i$. Show that

$$f(-1+\varepsilon i) \approx i + \varepsilon/2 \qquad\text{and}\qquad f(-1-\varepsilon i) \approx -i - \varepsilon/2.$$

This means that there is a sign change in the values of $f(z)$ across the negative real axis. $f(z)$ is called a **branch** of $\sqrt{z}$, essentially a choice of principal values for $\sqrt{z}$ on the cut plane. The only other branch on this cut plane is $-f(z)$.

#74D† Determine $\cos 5\theta$ in terms of powers of $\cos\theta$. Factorize the quintic

$$16z^5 - 20z^3 + 5z + 1$$

and deduce that $\cos(\pi/5) = (1+\sqrt{5})/4$. Find a similar expression for $\cos(3\pi/5)$.

#75 B(i) Let $z = \operatorname{cis} \theta$ and let n be an integer. Show $2\cos\theta = z + z^{-1}$ and $2i\sin\theta = z - z^{-1}$.
(ii) Using De Moivre's theorem, show that $2\cos n\theta = z^n + z^{-n}$ and $2i\sin n\theta = z^n - z^{-n}$.
(iii) Deduce that $16\cos^5\theta = \cos 5\theta + 5\cos 3\theta + 10\cos\theta$ and evaluate $\int_0^{\pi/2}\cos^5\theta\, d\theta$.

#76 c Use De Moivre's theorem to write $\cos 6\theta$ in terms of powers of $\cos\theta$ and $\sin 5\theta$ in terms of powers of $\sin\theta$.

#77 D† Use De Moivre's theorem to show, for any real value x and integer $n \geqslant 0$, that

$$\sum_{k=0}^{n}\cos kx = \frac{\sin\left((n+1)x/2\right)\cos\left(nx/2\right)}{\sin\left(x/2\right)},$$

and find a similar formula for $\sum_{k=1}^{n}\sin kx$.

#78 C By differentiating the previous result, determine $\sum_{k=1}^{n}k\sin kx$.

#79 D† Show, for any real value x, that

$$\sum_{k=0}^{\infty}\frac{\cos kx}{2^k} = \frac{4 - 2\cos x}{5 - 4\cos x}.$$

#80 d Express $\tan 7\theta$ in terms of $\tan\theta$ and its powers. Hence solve $x^6 - 21x^4 + 35x^2 - 7 = 0$.

Euler's Identity (1748) is the top identity to the right. It is true for all complex numbers z and is usually proved using power series. Of course, we first need definitions for what are meant by the exponential, cosine and sine functions of a complex number and these are given by the power series opposite. For any complex number z, and for any of those three series, as we add more and more terms in the series their sum begins to converge to a certain complex number. In the *limit* (i.e. when all the terms have been summed) we have our answer. (The technical details behind this statement are largely left to the second volume – those interested should go to p.358). Assuming sensible properties of infinite sums, guided proofs of the identity appear in #81 and #82. There are also technical issues with what it means to raise one complex number to the power of another. For this reason you may commonly see $\exp(iz)$ written instead of e^{iz}. For *real* values of z, the exponential series (second row) agrees with our usual notion of e^z.	$e^{iz} = \cos z + i\sin z.$
	$e^z = 1 + z + \dfrac{z^2}{2!} + \dfrac{z^3}{3!} + \cdots$
	$\cos z = 1 - \dfrac{z^2}{2!} + \dfrac{z^4}{4!} - \cdots$
	$\sin z = z - \dfrac{z^3}{3!} + \dfrac{z^5}{5!} - \cdots$
	$e^{\pi i} = -1$

#81 C By substituting iz into the exponential series above, prove Euler's identity.

#82 C(i) Assuming that term-by-term differentiation of the three power series is permitted, show that

$$\exp' = \exp, \qquad \sin' = \cos, \qquad \cos' = -\sin.$$

(ii) The exponential function is uniquely characterized by the properties: $\exp' = \exp$ and $\exp(0) = 1$. Deduce Euler's identity from this characterization.

1.5 Solving Cubic Equations*

A cubic equation is one of the form $Az^3 + Bz^2 + Cz + D = 0$, where $A \neq 0$. By dividing through by A if necessary we can assume $A = 1$. Further, we can make a substitution to simplify a cubic equation further.

Proposition 1.28 *The substitution $z = Z - a/3$ turns the equation $z^3 + az^2 + bz + c = 0$ into one of the form*

$$Z^3 + mZ + n = 0.$$

Proof With the substitution $z = Z - a/3$, the given cubic equation becomes

$$(Z - a/3)^3 + a(Z - a/3)^2 + b(Z - a/3) + c = 0.$$

The contributions to the coefficient of Z^2 add as $3(-a/3) + a + 0 + 0 = 0$ and so the new equation has the required form.   □

This leaves us in a position to introduce Cardano's method for solving such cubics.

Algorithm 1.29 *(Cardano's Method for Solving Cubics) Consider the cubic equation*

$$z^3 + mz + n = 0, \tag{1.29}$$

where m and n are real numbers. Let D be such that $D^2 = m^3/27 + n^2/4$. (As D^2 is real then D is a real or purely imaginary number.) We then define t and u by

$$t = -n/2 + D \qquad and \qquad u = n/2 + D,$$

and let T and U respectively be cube roots of t and u. It follows that tu is real, and if T and U are chosen appropriately, so that $TU = m/3$, then $z = T - U$ is a solution of the original cubic equation.

Remark 1.30 The **discriminant** of a cubic is $\Delta = (\alpha - \beta)^2 (\beta - \gamma)^2 (\gamma - \alpha)^2$, where α, β, γ are the roots of the cubic. We will see in #843 that $\Delta = -4m^3 - 27n^2$ for (1.29). So $D^2 = -\Delta/108$. For $m \leqslant 0$, we see in #90 that (1.29) has three, two or one real root when $\Delta < 0$, $\Delta = 0$ or $\Delta > 0$.   ■

Proof Putting $z = T - U$ into the LHS of (1.29) we get

$$
\begin{aligned}
\text{LHS} &= (T - U)^3 + m(T - U) + n \\
&= T^3 - 3UT^2 + 3U^2T - U^3 + mT - mU + n \\
&= t - u + (m - 3UT)(T - U) + n \\
&= (-n/2 + D) - (n/2 + D) + (m - 3UT)(T - U) + n \\
&= (m - 3TU)(T - U).
\end{aligned}
$$

Now

$$ (TU)^3 = tu = (-n/2 + D)(n/2 + D) = -n^2/4 + D^2 = m^3/27 = (m/3)^3. $$

The cube roots of $tu = (m/3)^3$ are $m/3$, $\omega m/3$ and $\omega^2 m/3$, where $\omega \neq 1$ is a cube root of unity. If T is a cube root of t and U_0 is a cube root of u, then TU_0 is a cube root of tu. The other cubes roots of u are ωU_0 and $\omega^2 U_0$; if we choose U appropriately from $U_0, \omega U_0, \omega^2 U_0$ we have $TU = m/3$. Thus, with these T and U, we've shown $z = T - U$ satisfies $z^3 + mz + n = (m - 3TU)(T - U) = 0$. Note that if T and U are chosen carefully (so that $TU = m/3$) then the other two roots are

$$ \omega T - \omega^2 U, \qquad \omega^2 T - \omega U $$

as $\omega T, \omega^2 T$ and $\omega U, \omega^2 U$ are respectively cube roots of t and u and we can further note $(\omega T)(\omega^2 U) = (\omega^2 T)(\omega U) = TU = m/3$. In fact a highly non-memorable formula for the three roots z_0, z_1, z_2 of the cubic is

$$ z_k = \omega^k \sqrt[3]{-n/2 + \sqrt{n^2/4 + m^3/27}} + \omega^{2k} \sqrt[3]{-n/2 - \sqrt{n^2/4 + m^3/27}} \qquad (1.30) $$

$\square$

Example 1.31 Find the three roots of $z^3 - 12z + 8 = 0$.

Solution We have $m = -12$ and $n = 8$. As $D^2 = m^3/27 + n^2/4 = -48$, then we can take $D = 4\sqrt{3}i$. We set

$$ t = -n/2 + D = -4 + 4\sqrt{3}i; \qquad u = n/2 + D = 4 + 4\sqrt{3}i. $$

Now T is a cube root of t. As $|t| = 8$ and $\arg t = 2\pi/3$, then t's three cube roots are

$$ T_1 = 2\operatorname{cis}(2\pi/9); \qquad T_2 = 2\operatorname{cis}(8\pi/9); \qquad T_3 = 2\operatorname{cis}(14\pi/9). $$

Similarly $|u| = 8$ and $\arg u = \pi/3$, giving

$$ U_1 = 2\operatorname{cis}(7\pi/9); \qquad U_2 = 2\operatorname{cis}(\pi/9); \qquad U_3 = 2\operatorname{cis}(13\pi/9) $$

respectively, where the U_i are chosen so that $T_i U_i = m/3 = -4$. Hence the three roots of $z^3 - 12z + 8 = 0$ are

$$
\begin{aligned}
T_1 - U_1 &= 2(\operatorname{cis}(2\pi/9) - \operatorname{cis}(7\pi/9)) = 4\cos(2\pi/9); \\
T_2 - U_2 &= 2(\operatorname{cis}(8\pi/9) - \operatorname{cis}(\pi/9)) = 4\cos(8\pi/9); \\
T_3 - U_3 &= 2(\operatorname{cis}(14\pi/9) - \operatorname{cis}(13\pi/9)) = 4\cos(14\pi/9).
\end{aligned}
$$

Numerically these roots are 3.06418, -3.75877 and 0.69459 to five decimal places. Note that the three roots are all real even though none of t, u, T_i, U_i is real. $\square$

Girolamo Cardano

Solving the Cubic The story of how the cubic equation came to be solved is a colourful one, but is also a window on a time when the habits of the mathematical community were very different from those of today. The three roots of the cubic equation $z^3 + mz + n = 0$ appear in (1.30), where k equals 0, 1 or 2 and ω is a cube root of unity other than 1. The formula is considerably more complicated than the quadratic formula. A further complication, for the time, came with the unease sixteenth-century mathematicians had with negative numbers, let alone complex ones. So a cubic equation of the form $z^3 + mz = n$ would only have been considered meaningful if m and n were positive and only positive roots would have been of interest; an equation of the form $z^3 = mz + n$

(with m and n again positive) would have been considered an entirely different type of cubic equation. The first person to solve equations of the form $z^3 + mz = n$ was Scipione Del Ferro (1456–1526) who was a mathematician at Bologna University. Before he died, Del Ferro shared, in secret, his solution with a student, Fiore. Armed with this knowledge, of which he thought himself to be the sole possessor, Fiore rather fancied that he could make a reputation for himself as a teacher of mathematics and challenged the Venetian mathematician Nicolo Tartaglia (1499–1557) to a competition to solve cubic equations which they would set one another. Unfortunately for Fiore, Tartaglia was more than up to the task, as he proved both able to solve equations of the form $z^3 + mz = n$ but could also pose and solve equations of the form $z^3 = mz + n$ which were beyond Fiore's ken. Girolamo Cardano (1501–1576, pictured left), heard of Tartaglia's victory and – after some considerable effort on his part – eventually convinced Tartaglia to share his methods, but only after Cardano had made an oath to also keep them secret, which Cardano largely did. But the situation became yet more confused when Cardano shared the solution of the cubic with his student and secretary Lodovico Ferrari (1522–1565), who subsequently used it to solve quartic (degree four) equations: both the solutions to the cubic and quartic were now held in secrecy by Cardano's oath to Tartaglia. The situation eventually resolved itself when Cardano heard that Fiore had gained his solution from Del Ferro. Cardano then travelled to Bologna to see Del Ferro's papers and was able to see that his original method had been the same as Tartaglia's. Feeling unburdened of the responsibilities of his oath, Cardano published the solutions to the cubic and quartic in his *Ars Magna* (1545), citing priority with Del Ferro and citing Tartaglia as an independent later solver of the cubic. Despite Cardano acknowledging the two as the original solvers, the method of solution is usually referred to as Cardano's method. For more historical detail see Derbyshire [9, pp.72–80], Hellman [19, chapter 1].

The method, then, for solving cubics is not a particularly elaborate one: a substitution of the form $z = Z - k$ with k chosen to eliminate the Z^2-term, followed by an application of Algorithm 1.29. There are cubic formulae, akin to the quadratic formula, which give the three roots of a general cubic equation in terms of its coefficients. But they are hideous compared with (1.1) and not at all memorable. The next example shows how messy the formulae for the roots can become even with small integer values for the coefficients.

Example 1.32 Find the three roots of $2z^3 + 5z^2 - 7z + 3 = 0$.

Solution Dividing the equation by 2, it becomes $z^3 + (5/2)z^2 - (7/2)z + 3/2 = 0$, and making the substitution from Proposition 1.28, namely $z = Z - 5/6$, we obtain with some simplifying

$$Z^3 - (67/12)Z + (301/54) = 0. \tag{1.31}$$

So $m = -67/12$ and $n = 301/54$ and we introduce

$$D^2 = \frac{m^3}{27} + \frac{n^2}{4} = \frac{1}{27}\left(\frac{-67}{12}\right)^3 + \frac{1}{4}\left(\frac{301}{54}\right)^2 = \frac{-300763}{46656} + \frac{90601}{11664} = \frac{761}{576},$$

and we may take $D = \sqrt{761/576} = \frac{1}{24}\sqrt{761}$. Then

$$t = -\frac{n}{2} + D = -\frac{301}{108} + \frac{\sqrt{761}}{24}; \qquad u = \frac{n}{2} + D = \frac{301}{108} + \frac{\sqrt{761}}{24},$$

and we can take the real cube roots

$$T = \sqrt[3]{\frac{\sqrt{761}}{24} - \frac{301}{108}}; \qquad U = \sqrt[3]{\frac{\sqrt{761}}{24} + \frac{301}{108}}.$$

Then one root of (1.31) is $Z_1 = T - U$ and the other two roots are $Z_2 = \omega T - \omega^2 U$ and $Z_3 = \omega^2 T - \omega U$ where $\omega \neq 1$ is a cube root of unity. Finally, then, the three roots of the original equation are

$$z_1 = -\frac{5}{6} + \sqrt[3]{\frac{\sqrt{761}}{24} - \frac{301}{108}} - \sqrt[3]{\frac{\sqrt{761}}{24} + \frac{301}{108}} \approx -3.59099;$$

$$z_2 = -\frac{5}{6} + \omega\sqrt[3]{\frac{\sqrt{761}}{24} - \frac{301}{108}} - \omega^2\sqrt[3]{\frac{\sqrt{761}}{24} + \frac{301}{108}} \approx 0.54549 + 0.34663i;$$

$$z_3 = -\frac{5}{6} + \omega^2\sqrt[3]{\frac{\sqrt{761}}{24} - \frac{301}{108}} - \omega\sqrt[3]{\frac{\sqrt{761}}{24} + \frac{301}{108}} \approx 0.54549 - 0.34663i,$$

with the real and imaginary parts of the roots given to five decimal places. □

Solving quartics (degree four polynomial equations) requires only a little further insight once we can solve cubics. However, the method leads to still more involved expressions for the roots; details are left to #95 and #96.

Exercises

#83B† Solve $x^3 = 6x + 6$ and hence solve $8x^3 - 12x^2 - 42x + 71 = 0$.

#84c† By making a different substitution, from that of #83, relate the roots of the cubic $8x^3 - 12x^2 - 6x - 1$ to those of $x^3 - 6x - 6$.

#85B† (See also #8 and #52.) Consider the identity

$$x^3 + y^3 + z^3 - 3xyz = (x + y + z)(x^2 + y^2 + z^2 - xy - yz - zx).$$

By treating z as a variable, explain how this identity can be used to solve the cubic equation $z^3 + mz + n = 0$ and relate this method to Cardano's method (Algorithm 1.29).

#86B Show that $\cos 3\theta = 4\cos^3 \theta - 3\cos\theta$. By setting $z = 4\cos\theta$, rederive the roots of $z^3 - 12z + 8 = 0$.

#87B (Viète's substitution[4]) Let p, q be complex numbers. Show that the substitution $z = w - p/(3w)$ turns the equation $z^3 + pz = q$ into the quadratic $(w^3)^2 - q(w^3) - p^3/27 = 0$ in w^3.

#88b Use Viète's substitution to solve the cubic $z^3 - 12z + 8 = 0$ from Example 1.31.

#89C Show that $z = \cos\alpha$ is a root of the cubic $4z^3 - 3z - \cos 3\alpha$ and determine the remaining two roots.

#90b† Let p and q be real numbers with $p \leqslant 0$. Find the co-ordinates of the turning points of the cubic $y = x^3 + px + q$. Show that the cubic equation $x^3 + px + q = 0$ has three real roots, with two or more repeated, precisely when

$$4p^3 + 27q^2 = 0.$$

Under what conditions on p and q does $x^3 + px + q = 0$ have (i) three distinct real roots or (ii) just one real root? How many real roots does the equation $x^3 + px + q = 0$ have when $p > 0$?

#91b† Under what conditions on a and b is the line $y = ax + b$ a tangent to the curve $y = x^3 - x$?

#92a Let α, β be the roots of $ax^2 + bx + c = 0$. Write the quadratic's discriminant $(\alpha - \beta)^2$ in terms of a, b, c.

#93C† Show that if the equation $z^3 - mz + n = 0$ has three real roots then these can be determined by a substitution $z = k\cos\theta$ for an appropriate value of k.

---

[4] The French mathematician Francois Viète (1540–1603) was arguably the most important European mathematician of the sixteenth century. Whilst his name is associated with various pieces of mathematics (e.g., the above substitution and the formulas connecting the roots of a polynomial and its coefficients) his main influence was in the improvements he made to algebraic notation in *In Artem Analyticam Isagoge* (1591), particularly in the use of letters for variables. His use of vowels for unknowns and consonants for parameters was replaced later by Descartes, who used letters at the end of the alphabet for unknowns and those at the start for constants, a convention still adhered to today.

Evariste Galois

The Quintic Despite the successes in the sixteenth century in solving cubics and quartics, more than two centuries passed before progress was made with the quintic. During that time, perhaps the most insightful view of the mathematics to unfold came in 1770 when Joseph-Louis Lagrange (1736–1813) rederived the solutions for quadratic, cubic and quartic equations through an analysis of permutations of their roots; at that time though it seemed that the mathematical community largely expected quintic equations to be likewise solvable and Paolo Ruffini (1765–1822), Niels Henrik Abel (1802–1829) and Evariste Galois (1811–1832) encountered various difficulties convincing contemporaries of the insolvability of the general quintic. Insolvability, here, means that there is no formula for the solutions of a general quintic polynomial that involves only addition, subtraction, multiplication, division and taking roots, as is the case for polynomials of degree four or less; that is to say, the general quintic is not *solvable by radicals*. Ruffini gave an incomplete proof of this fact in 1799 but, despite writing to Lagrange several times, his work failed to interest other mathematicians even to the extent of the gap in the proof being pointed out. Ruffini's work contains many important ideas about *permutation groups* and it may have been the novelty of his work that led to it going unheeded. It was to be Abel, instead, who provided the first complete proof in 1824. But it was Galois, around 1830, who saw deepest into the problem. Abel had shown that quintics are insolvable by radicals via a single formula, though some (e.g. $x^5 = 0$) clearly are. Galois' genius was to give a criterion showing precisely which polynomials (of any degree) are solvable by radicals. He associated with a polynomial an algebraic structure now known as its *Galois group*.

If the roots of the polynomial could be extracted through a succession of everyday algebra and taking roots, then its Galois group could correspondingly be 'built up' in a certain technical sense – the group would be what is now known as *solvable*. And the Galois group of the general quintic is not solvable. Galois' legacy in mathematics is enormous and many important ideas relating to *groups* and *fields* (p.160) – two fundamental structures in modern algebra – can be traced back to his work; in fact finite fields are commonly called *Galois fields* after him. However, Galois, too, found it difficult to share his ideas during his lifetime despite submitting his work to the Paris Academy three times. It was only thanks to Joseph Liouville (1809–1882) that Galois' work posthumously reached a wider audience, though this wasn't until 1846. See Livio [22] for a popular account of the above.

#94d Make a substitution $X = x - \alpha$, for a certain choice of α, to transform the equation $X^3 + aX^2 + bX + c = 0$ into one of the form $x^3 + px + q = 0$. Hence find conditions involving a, b, c, under which the original cubic in X has (i) three distinct real roots, (ii) three real roots involving repetitions, (iii) just one real root.

#95D(i) Show a substitution $z = Z - k$ can convert the equation $z^4 + az^3 + bz^2 + cz + d = 0$ into one of the form $Z^4 + LZ^2 + MZ + N = 0$. Find expressions for k, L, M, N in terms of a, b, c, d.

(ii) Let $p(Z) = Z^4 + LZ^2 + MZ + N$. Show that there exist complex numbers A, B, C such that

$$p(Z) + (AZ + B)^2 = (Z^2 + C)^2 \quad \text{for all } Z,$$

where $B = -M/(2A)$, $C = (A^2 + L)/2$ and A^2 is a non-zero root of the equation

$$x^3 + 2Lx^2 + (L^2 - 4N)x - M^2 = 0. \tag{1.32}$$

(iii) Write down the four roots of $p(Z)$ in terms of A, B, C.

#96D(i) Show with a suitable substitution that the equation $3z^4 - 2z^3 + 5z^2 + 2z - 1 = 0$ becomes

$$Z^4 + (3/2)Z^2 + (32/27)Z - (77/432) = 0. \tag{1.33}$$

(ii) Use Algorithm 1.29 to find a non-zero solution of $x^3 + 3x^2 + (80/27)x - 1024/729 = 0$.
(iii) Using the method of #95(ii), show that we can now add a term $(AZ + B)^2$ to both sides of (1.33) in such a way that it becomes $(Z^2 + C)^2 = (AZ + B)^2$. Show that

$$A = 0.584877\ldots, \qquad B = -1.013191\ldots, \qquad C = 0.921040\ldots.$$

(iv) Deduce that the four roots of the original quartic, to five decimal places, are

$$z_1 = 0.45911 + 1.35967i, \qquad z_2 = 0.45911 - 1.35967i,$$

$$z_3 = 0.29574, \qquad z_4 = -0.54728.$$

1.6 Their Geometry

Using the modulus and argument functions, we can measure distances and angles in the Argand diagram.

Definition 1.33 Given two complex numbers $z = z_1 + z_2 i$ and $w = w_1 + w_2 i$ then the **distance** between them is $|z - w|$. This follows from Pythagoras' theorem as

$$\sqrt{(z_1 - w_1)^2 + (z_2 - w_2)^2} = |(z_1 - w_1) + (z_2 - w_2)i| = |z - w|.$$

Definition 1.34 Given three complex numbers a, b, c then the **angle** $\angle abc$ equals

$$\arg(c - b) - \arg(a - b) = \arg\left(\frac{c - b}{a - b}\right).$$

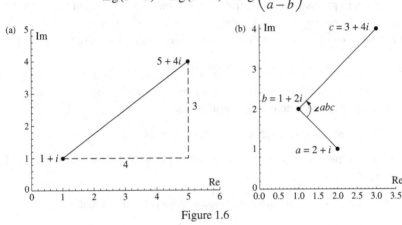

Figure 1.6

In Figure 1.6a we see that the distance from $1 + i$ and $5 + 4i$ equals

$$\sqrt{(5 - 1)^2 + (4 - 1)^2} = 5.$$

In Figure 1.6b $a = 2 + i$, $b = 1 + 2i$ and $c = 3 + 4i$. The angle at b is

$$\arg\left(\frac{(3 + 4i) - (1 + 2i)}{(2 + i) - (1 + 2i)}\right) = \arg\left(\frac{2 + 2i}{1 - i}\right) = \arg(2i) = \frac{\pi}{2}.$$

- Notice that Definition 1.34 gives a signed angle in the sense that it is measured in an anticlockwise fashion from the segment ba round to the segment bc, and the above formula for the angle at b would give negative the previous result if the roles of a and c were swapped. Note also that this signed angle is defined only up to multiples of 2π.

Example 1.35 Find the smaller angle $\angle cab$ where $a = 1 + i$, $b = 3 + 2i$ and $c = 4 - 3i$.

Solution The angle $\angle cab$ is given by

$$\arg\left(\frac{b - a}{c - a}\right) = \arg\left(\frac{2 + i}{3 - 4i}\right) = \arg\left(\frac{2 + 11i}{25}\right).$$

So $\tan \angle cab = 11/2$ with the number $(2 + 11i)/25$ clearly in the first quadrant. (There is a solution to $\tan\theta = 11/2$ in the opposite quadrant.) Hence $\angle cab = \tan^{-1}(11/2) \approx 1.39$ radians. □

#97
#98
#99
#100
#102

We can use these definitions of distance and angle to describe various regions of the plane. (See also #44.)

Example 1.36 Describe the following regions of the Argand diagram.

- $|z - a| = r$, where a is a complex number and $r > 0$.

This is clearly the locus of points at distance r from a, i.e. the circle with centre a and radius r. The regions $|z - a| < r$ and $|z - a| > r$ are then the interior and exterior of the circle respectively.

- $|z - a| = |z - b|$, where a, b are distinct complex numbers.

This is the set of points equidistant from a and b and so is the perpendicular bisector of the line segment connecting a and b. The regions $|z - a| < |z - b|$ and $|z - a| > |z - b|$ are then the half-planes on either side of this line and which respectively contain a and b.

- $\arg(z - a) = \theta$, where a is a complex number and $0 \leqslant \theta < 2\pi$.

This is a half-line emanating from the point a and making an angle θ with the positive real axis. Note that it doesn't include the point a itself.

- $|z - a| + |z - b| = r$ where a, b are complex numbers and $r > 0$.

This is an **ellipse** with foci a and b provided $r > |a - b|$ (see Figure 1.7a and Proposition 4.50).

- $|z - a| - |z - b| = r$ where a, b are complex numbers and r is real.

If $|r| < |a - b|$ then this is one branch of a **hyperbola** with foci a and b. If $r > 0$ then this is a branch closer to b and if $r < 0$ this is a branch closer to a (see Figure 1.7b, Proposition 4.50 and #43).

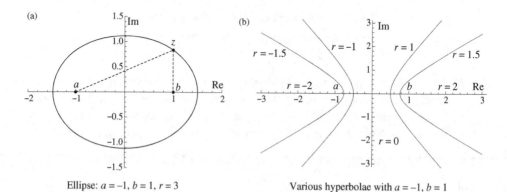

(a) Ellipse: $a = -1$, $b = 1$, $r = 3$

(b) Various hyperbolae with $a = -1$, $b = 1$

Figure 1.7

Example 1.37 Describe the set of points z which satisfy the equation

$$\arg\left(\frac{z+1}{z-1}\right) = \frac{\pi}{2}. \tag{1.34}$$

Solution *Method One: parametrization.* We can rewrite (1.34) as

$$1 + \frac{2}{z-1} = \frac{z+1}{z-1} = it \quad \text{where } t > 0,$$

as these points it are precisely those with argument $\pi/2$. Solving for z we have

$$z = \frac{1+ti}{-1+ti} = \left(\frac{t^2-1}{t^2+1}\right) + i\left(\frac{-2t}{1+t^2}\right). \tag{1.35}$$

It is quite easy to spot, from the first expression for z in (1.35), that the denominator and numerator have the same modulus and so $|z| = 1$. Those with knowledge of the half-angle tangent formulae (5.27) may also have spotted them in (1.35).

So z must lie on the unit circle $|z| = 1$. We see that the real part $x(t)$ in (1.35) varies across the range $-1 < x(t) < 1$ as t varies across the positive numbers, but the imaginary part $y(t)$ is always negative. Thus (1.34) is the equation of the lower semicircle from $|z| = 1$, not including the points -1 and 1.

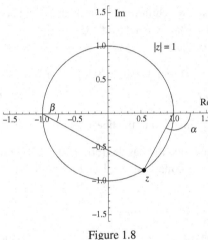

Figure 1.8

Method Two: a geometric approach.
In Figure 1.8 we have $\arg(z-1) = -\alpha$ and $\arg(z+1) = -\beta$, where α, β represent the actual magnitudes of the angles irrespective of their sense. By (1.12) we have

$$\arg((z+1)/(z-1)) = \pi/2$$
$$\Longleftrightarrow \arg(z+1) - \arg(z-1) = \pi/2$$
$$\Longleftrightarrow -\beta - (-\alpha) = \pi/2$$
$$\Longleftrightarrow \beta + (\pi - \alpha) = \pi/2$$

i.e. when the angle at z is a right angle, which holds when -1, 1 is the diameter of the circle through -1, 1, z. This is Thales' theorem (Theorem 1.48).

A similar application of Thales' theorem shows that $\arg((z+1)/(z-1)) = -\pi/2$ on the upper semicircle. □

An important class of maps of the complex plane is that of the *isometries*, i.e. the distance-preserving maps.

Definition 1.38 We say a map f from $\mathbb{C}$ to $\mathbb{C}$ is an **isometry** if it preserves distance, that is

$$|f(z) - f(w)| = |z - w| \quad \text{for any complex } z, w.$$

Example 1.39 Of particular note are the following three isometries:

- $z \mapsto (\text{cis}\,\theta)z$ which is rotation anticlockwise by θ about 0;
  in particular $z \mapsto iz$ is rotation by a right angle anticlockwise about 0.
- $z \mapsto z + k$ which is translation by $\text{Re}\,k$ to the right, and $\text{Im}\,k$ up.
- $z \mapsto \bar{z}$ which is reflection in the real axis.

Proposition 1.40 *(Isometries of* $\mathbb{C}$*) (See §4.2 for further details if necessary.) Let f be an isometry of* $\mathbb{C}$*.*

If f is **orientation-preserving** *then f has the form* $f(z) = az + b$ *for some complex a,b with* $|a| = 1$*. These maps are all* **rotations**. *If f reverses* **orientation** *then f has the form* $f(z) = a\bar{z} + b$ *for some complex a,b with* $|a| = 1$*. These include, but are not limited to, reflections (#107 and #110).*

Proof This is simply a rephrasing of Theorem 4.40 and Example 4.18, which tell us that every orientation-preserving isometry of the *xy*-plane can be written in matrix form as

$$\begin{pmatrix} x \\ y \end{pmatrix} \mapsto \begin{pmatrix} \cos\theta & -\sin\theta \\ \sin\theta & \cos\theta \end{pmatrix} \begin{pmatrix} x \\ y \end{pmatrix} + \begin{pmatrix} b_1 \\ b_2 \end{pmatrix}$$

for some real θ, b_1, b_2. Or equivalently, written in terms of the individual co-ordinates this means

$$(x,y) \mapsto (x\cos\theta - y\sin\theta + b_1, x\sin\theta + y\cos\theta + b_2).$$

If we write $z = x + yi$, $b = b_1 + b_2 i$, and $a = \text{cis}\,\theta$, which has modulus 1, then the above equation reads as $z \mapsto az + b$. That this is indeed a rotation is left to #135.

Similarly orientation-reversing isometries are of the form

$$(x,y) \mapsto (x\cos\theta + y\sin\theta + b_1, -x\sin\theta + y\cos\theta + b_2),$$

which reads as $z \mapsto a\bar{z} + b$ for the same choices of z, a, b.   □

Example 1.41 Express in the form $f(z) = a\bar{z} + b$ reflection in the line $x + y = 1$.

Solution Knowing from Proposition 1.40 that the reflection has the form $f(z) = a\bar{z} + b$ we can find a and b by considering where two points are mapped. As 1 and i both lie on the line of reflection then they are both fixed. So

$$1 = a\bar{1} + b = a1 + b; \qquad i = a\bar{i} + b = -ai + b.$$

Substituting $b = 1 - a$ into the second equation we find $a = -i$ and $b = 1 + i$. Hence $f(z) = -i\bar{z} + 1 + i$.   □

There are other important maps of the complex plane which aren't isometries. Here are some examples which involve determining the images of certain regions. The methods used involve either relying on the fact that the map in question has an inverse or parametrizing the region with Cartesian or polar co-ordinates.

Example 1.42 In the two cases below, find the image in $\mathbb{C}$ of the given subset under the given map.

- A is the region $\text{Im}\,z > \text{Re}\,z > 0$ and $f(z) = z^2$.

A general point in A can be written in the form $z = r\text{cis}\theta$ where $\pi/4 < \theta < \pi/2$ and $r > 0$. De Moivre's theorem gives us that $z^2 = r^2\text{cis}2\theta$ which has an argument of 2θ in the range $\pi/2 < 2\theta < \pi$ and independently a modulus of r^2 in the range $r^2 > 0$.

So $f(A)$ is the second quadrant of Figure 1.9.

(margin notes:) #104 #107   #112 #113

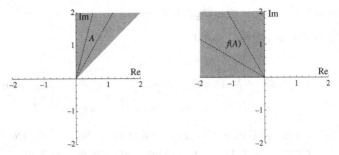

Figure 1.9

- B is the unit disc $|z| < 1$ and $g(z) = \frac{1+z}{1-z}$.

  The map g maps from the set $\{z \in \mathbb{C}: z \neq 1\}$ to $\{z \in \mathbb{C}: z \neq -1\}$, with an inverse $g^{-1}(z) = (z-1)/(z+1)$. To see this we note

  $$w = (1+z)/(1-z) \Leftrightarrow w - wz = 1 + z \Leftrightarrow w - 1 = z(1+w) \Leftrightarrow z = (w-1)/(w+1).$$

  Note z is in $g(B)$ $\Leftrightarrow g^{-1}(z)$ is in B $\Leftrightarrow |(z-1)/(z+1)| < 1$ $\Leftrightarrow |z-1| < |z+1|$,

  which is the half-plane of points closer to 1 than -1, (Example 1.36) or equivalently the half-plane $\operatorname{Re} z > 0$ (see Figure 1.10).

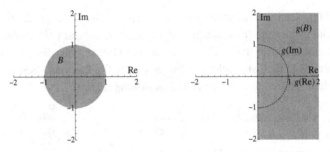

Figure 1.10

There is a useful general equation which encompasses both circles and lines. (This fact is less surprising with an appreciation of the extended complex plane, circlines and the Riemann sphere; see #187 and #188.)

Proposition 1.43 *(Circles and Lines in $\mathbb{C}$) Let A and C be real and B complex, with A, B not both zero. Then*

$$A z \bar{z} + B \bar{z} + \bar{B} z + C = 0, \tag{1.36}$$

represents:

(a) a line in direction iB when $A = 0$;

(b) a circle, if $A \neq 0$ and $|B|^2 \geqslant AC$, with centre $-B/A$ and radius $|A|^{-1} \sqrt{|B|^2 - AC}$;

and otherwise has no solutions. Moreover, every circle and line can be represented in the form of (1.36).

Proof If $A \neq 0$ then we can rearrange (1.36) as

$$z\bar{z} + (B/A)\bar{z} + (\bar{B}/A)z + C/A = 0,$$

$$(z + B/A)(\bar{z} + \bar{B}/A) = B\bar{B}/A^2 - C/A,$$

$$(z + B/A)\overline{(z + B/A)} = B\bar{B}/A^2 - C/A, \qquad \text{[as } A \text{ is real and using (1.15)]}$$

$$|z + B/A|^2 = (|B|^2 - AC)/A^2, \qquad \text{[using (1.14)]}.$$

If $|B|^2 \geqslant AC$ then this is a circle with centre $-B/A$ and radius $|A|^{-1}\sqrt{|B|^2 - AC}$ and otherwise there are no solutions to (1.36). Conversely, note that the equation of a general circle is $|z - a| = r$, where a is a complex number and $r \geqslant 0$. This can be rearranged using (1.14) as

$$z\bar{z} - a\bar{z} - \bar{a}z + (|a|^2 - r^2) = 0,$$

which is in the form of (1.36) with $A = 1$, $B = -a$ and $C = |a|^2 - r^2$.

If $A = 0$ then we have the equation $B\bar{z} + \bar{B}z + C = 0$. If we write $B = u + iv$ and $z = x + yi$ then this equation becomes

$$2ux + 2vy + C = 0, \tag{1.37}$$

which is the equation of a line. Moreover we see that every line appears in this form by choosing u, v, C appropriately. The line (1.37) is parallel to the vector $(v, -u)$ or equivalently $v - ui = i(u + iv) = iB$. $\qquad\qquad\square$

We end this section with a theorem due to Apollonius of Perga (c. 262–190 BC). Apollonius wrote *Conics,* a series of eight books which is second only to Euclid's *Elements* in its importance amongst the works of the ancient Greeks. The theorem shows that certain loci are circles. In fact, all circles can be represented in this way – see #127. In this form a circle is often referred to as a **Circle of Apollonius** (Figure 1.11).

Theorem 1.44 (Apollonius' Theorem) *Let $k > 0$ with $k \neq 1$, and let a, b be distinct complex numbers. Then the locus of points satisfying the equation*

$$|z - a| = k|z - b| \tag{1.38}$$

is a circle with centre c and radius r where

$$c = \frac{k^2 b - a}{k^2 - 1}, \qquad r = \frac{k|a - b|}{|k^2 - 1|}.$$

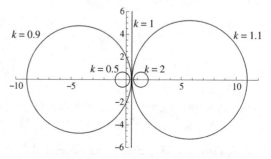

Circles of Apollonius with $a = 0$, $b = 1$ and different k

Figure 1.11

*Further, a and b are **inverse points** – i.e. a, b, c are collinear, with a, b on the same side of c, and $|c - a||c - b| = r^2$.*

#118 **Proof** Squaring (1.38) and using (1.14) and (1.15) we have $(z - a)(\bar{z} - \bar{a}) = k^2$
#121 $(z - b)(\bar{z} - \bar{b})$ which rearranges to
#124
#127
$$(k^2 - 1)z\bar{z} + (\bar{a} - k^2\bar{b})z + (a - k^2 b)\bar{z} = a\bar{a} - k^2 b\bar{b}$$

which rearranges to

$$\left| z - \left(\frac{k^2 b - a}{k^2 - 1} \right) \right|^2 = \frac{a\bar{a} - k^2 b\bar{b}}{k^2 - 1} + \frac{\left| k^2 b - a \right|^2}{(k^2 - 1)^2} = \frac{k^2 a\bar{a} + k^2 b\bar{b} - k^2 a\bar{b} - k^2 \bar{a}b}{(k^2 - 1)^2} = \frac{k^2 |a - b|^2}{(k^2 - 1)^2}.$$

This is the circle with centre c and radius r as given previously. That a and b are inverse points is left to #127. □

Exercises

#97a Which pair of the points $2 + i$, $3 + 6i$, $1 - 3i$, $-3 + 4i$ is closest? Which two are farthest apart?

#98a Find the lengths and interior angles of the triangle with vertices $1 + i$, $3 + 2i$, $2 + 4i$.

#99A Let $a = 2$, $b = 1 + i$, $c = 3 + ti$. Find the real value of t such that $\angle abc = \pi/3$. As t becomes large and positive, what angle does $\angle abc$ approximate?

#100B† Let a, b be non-zero complex numbers. Show $|\lambda a + \mu b| \geqslant |\lambda a|$ for all real λ, μ if and only if $\angle a0b$ is a right angle.

#101b Using either of the two methods of Example 1.37, describe the set of points satisfying $\arg(1 - 1/z) = \pi/3$.

#102D† Let z_1, z_2, z_3 be distinct complex numbers with $z_1^2 + z_2^2 + z_3^2 = z_1 z_2 + z_2 z_3 + z_3 z_1$. Show that $\triangle z_1 z_2 z_3$ is an equilateral triangle.

#103C Let a, b be complex numbers. Show directly that the map $f(z) = az + b$ preserves angles, and that f preserves distances if and only if $|a| = 1$.

#104b Write in the form $z \mapsto az + b$ the rotation through $\pi/3$ radians anticlockwise about the point $2 + i$. Write in the form $z \mapsto a\bar{z} + b$ the reflection in the line $3x + 2y = 6$.

#105c† Find two reflections h and k such that $k(h(z)) = iz$ for all z.

#106C Let $f(z) = iz + 3 - i$. Find a map $g(z) = az + b$ where $|a| = 1$ such that $g(g(z)) = f(z)$. How many such maps g are there? Geometrically what transformations do these maps g and the map f represent?

#107B† Given that the map $z \mapsto a\bar{z} + b$ is a reflection, show that $|a| = 1$ and that $a\bar{b} + b = 0$.

#108D† Let d be a real number and a be a complex number with $|a| = 1$. Show that the map $z \mapsto -a^2\bar{z} + 2da$ represents the reflection in the line through da and parallel to ia. Why does this prove the converse of #107?

#109C Show that every orientation-reversing isometry of $\mathbb{C}$ can be written as the composition of a reflection followed by a rotation about the origin.

#110D† Show that every orientation-reversing isometry of $\mathbb{C}$ is a **glide reflection**, i.e. it can be written as the composition of a reflection followed by a translation parallel to the line of reflection.

#111 C Let z_1, z_2, z_3 and w_1, w_2, w_3 be two triples of distinct complex numbers satisfying $|z_i - z_j| = |w_i - w_j|$ for each i, j. Show there is a unique isometry f of $\mathbb{C}$ such that $f(z_i) = w_i$ for $i = 1, 2, 3$.

#112 B† For $i = 1, 2, 3, 4$, find the image of the region A_i under the given map f_i.

$$A_1 = \{z \in \mathbb{C} : |z| < 1\}, \quad f_1(z) = z^3;$$
$$A_2 = \{z \in \mathbb{C} : \mathrm{Im}\, z > 0\}, \quad f_2(z) = 1/z;$$
$$A_3 = \{z \in \mathbb{C} : \mathrm{Im}\, z > 0\}, \quad f_3(z) = 2z + 1;$$
$$A_4 = \{z \in \mathbb{C} : |z - 1| < 1\}, \quad f_4(z) = z^2 - 2z + 2.$$

#113 B Let t be a real number. Find expressions for $x = \mathrm{Re}\,(2 + ti)^{-1}$ and $y = \mathrm{Im}\,(2 + ti)^{-1}$. Find an equation relating x and y by eliminating t. Deduce that the image of the line $\mathrm{Re}\, z = 2$ under the map $z \mapsto 1/z$ is contained in a circle. Does the image of the line comprise all of the circle?

#114 c Using the method of #113, find the image of the line $\mathrm{Re}\, z = 2$ under the maps $z \mapsto iz$; $z \mapsto z^2$; $z \mapsto (z - 1)^{-1}$.

#115 c Sketch the image of the square $\{z \in \mathbb{C} : 0 \leqslant \mathrm{Re}\, z \leqslant 1, 0 \leqslant \mathrm{Im}\, z \leqslant 1\}$ under the map $z \mapsto z^2$. Find the equations of the three curves which bound the image. Verify that these three curves meet at right angles.

#116 C Determine the images $f_i(A_i)$ of the following regions A_i in $\mathbb{C}$ under the given maps f_i:

(i) $A_1 = \{z \in \mathbb{C} : |z - i| = |z|\}; f_1(z) = (1 + i)z$.
(ii) $A_2 = \{z \in \mathbb{C} : \mathrm{Re}\, z = 1\}; f_2(z) = z/(z - 1)$.
(iii) $A_3 = \{z \in \mathbb{C} : |z - 1| + |z - i| = 2\}; f_3(z) = -i\bar{z} + (1 + i)$. [Hint: see Example 1.41.]

#117 D† Show that the image of the circle $|z - 1| = 1$ under the map $z \mapsto z^2$ is the **cardioid** with equation in polar co-ordinates $r = 2(1 + \cos\theta)$. Sketch the original curve and its image.

#118 b Determine the equation of the following circles and lines in the form of (1.36).

(i) The circle with centre $3 + 4i$ and radius 5.
(ii) The circle which passes through $1, 3$ and i.
(iii) The line through $1 + 3i$ and $2 - i$.
(iv) The line through 2, making an angle θ with the real axis.

#119 c A circle or line S has equation (1.36). Determine the image of S under the maps (i) $f(z) = 1/z$, (ii) $g(z) = cz$, (iii) $h(z) = z + k$, where c, k are complex numbers and $c \neq 0$, putting all your answers in the form of (1.36).

#120 C† Let z be a complex number with $z \neq 0, 1$. Show, by direct calculation, there are complex numbers A, B, C such that $0, 1, z$ satisfy (1.36). Show further that these A, B, C are unique up to multiplication by a non-zero real number. Deduce that given three distinct points z_1, z_2, z_3 in $\mathbb{C}$ there are (up to multiplication by a non-zero real number) unique A, B, C such that z_1, z_2, z_3 satisfy (1.36).

#121b Let a be a complex number with $|a| > 1$. Show that image of the unit circle $\{z \in \mathbb{C}: |z| = 1\}$ under the map $z \mapsto (z-a)^{-1}$ is a circle. Find its centre and radius.

#122C Let C be the circle, centre c, radius r. Show that α, β are inverse points of C if and only if $(c-\alpha)(\bar{c}-\bar{\beta}) = r^2$.

#123c Given distinct complex numbers a, b, for what different values of k are the circles $|z-a| = k|z-b|$ of the same area?

#124a Find an equation of the form $|z-a| = k|z|$ for the circle with radius 1 and centre at 2.

#125C† Let C_1 be the circle with centre 0 and radius 1, and let C_2 be the circle with centre $c > 0$ and radius R. Under what conditions are there (real) α, β which are inverse points to both C_1 and C_2. Geometrically, what do these conditions require of C_2?

#126C Find the common inverse points α and β of the circles with equations $|z| = 1$ and $|z - 3i| = 1$. Then give equations for the two circles in the form $|z-\alpha| = k|z-\beta|$, with two different values of k.

#127B† Let a, b be distinct complex numbers and $0 < k < 1$. Consider the circle of Apollonius defined by (1.38).

  (i) Show that a lies inside the circle and that b lies outside the circle. Show further that a and b are inverse points.
  (ii) Conversely, given any circle C, and any point a within the circle other than the centre, show that there exists a unique complex number b and $0 < k < 1$ such that (1.38) is the equation of C.

#128D† Let $0 < k < 1$. Determine those complex numbers w such that $\text{Im}\,w > 0$ and $|w| < 1$ and such that the circle through $0, k, w$ lies entirely in the disc $|z| < 1$. Sketch the set of such w.

#129B(i) Let C denote the circle with centre a and radius $r > 0$. **Inversion in the circle** C is the map which takes a point z to the point w in such a way that a, z, w are collinear (in that order) and $|z-a||w-a| = r^2$. Find an expression for w in terms of $\bar{z}$.
(ii) Show that inversion in the circle $z\bar{z} + \bar{A}z + A\bar{z} + C = 0$, $|A|^2 > C$, is given by the map $I(z) = -(A\bar{z} + C)/(\bar{z} + \bar{A})$.

#130C† Write down the formula $I_\varepsilon(z)$ for inversion in the circle $\varepsilon z\bar{z} + \bar{A}z + A\bar{z} + C = 0$, where $\varepsilon > 0$. Show that as ε becomes small, $I_\varepsilon(z)$ approximates to $I_0(z) = -(A\bar{z} + C)/\bar{A}$ and that $I_0(z)$ represents reflection in a line.

Remark 1.45 Consequently, by the expression **inversion in a circline**, we shall mean inversion in a circle or reflection in a line. We see from #129 and #130 that inversion in the circline $Az\bar{z} + \bar{B}z + B\bar{z} + C = 0$ is given by

$$I(z) = -\left(\frac{B\bar{z} + C}{A\bar{z} + \bar{B}}\right). \qquad \blacksquare$$

#131 c† Given $-1 < c < 1$ we define $f(z) = z + c/z$. Show that f maps the unit circle $|z| = 1$ to an ellipse.

#132 D† Show that the **Joukowski**[5] **map** $f(z) = z + 1/z$ is an invertible map from $\{z \colon |z| > 1\}$ to the complement of the interval $\{z \colon -2 \leqslant \operatorname{Re} z \leqslant 2, \operatorname{Im} z = 0\}$.

#133 c Let w be a complex number with $\operatorname{Re} w > 0$. With the help of a computer, and for several different values of w, sketch the images under the Joukowski map $f(z) = z + 1/z$ of the circles with centre $-w$ and radius $|1 + w|$. (It is because of the shape of these images that the Joukowski map was historically used to study aerofoils.)

1.7 Some Geometric Theory

Typically the hypotheses of a geometric theorem (i.e. the assumptions in the theorem's set-up) describe a situation something along the lines of 'A triangle has vertices $A, B, C, \ldots$' or 'A circle with diameter $PQ \ldots$'. The theorem makes no mention of Cartesian co-ordinates or of complex numbers and takes place in a generally featureless *Euclidean plane*, devoid of any origin or axes. The geometric theorems we will prove each have a certain level of generality, but proofs of them can be made much easier with a sensible choice of co-ordinates which reflects the theorem's set-up. We have certain degrees of freedom in the choices we make: where we choose to place the origin; the directions in which our perpendicular axes point; in our choice of unit length.

For example, if faced with proving a theorem involving a triangle ABC then we can place our origin at A so that the point is represented by the complex number 0; we can further direct the positive real axis along AB and take AB as our unit length so that B is represented by the complex number 1. But at this point we have largely used up our degrees of freedom and we will need to represent C by a general complex number z (see Figure 1.12). A fourth point D would similarly have to be assigned a general complex co-ordinate, w say, unless further specific facts were given.

- **With the introduction of complex co-ordinates, geometric theorems in the plane become algebraic identities in a handful of complex variables.**

One obvious concern is whether introducing co-ordinates is valid; are our proofs independent on our choice of origin, axes and unit length? Well, the answer is positive but unfortunately a rigorous and comprehensive treatment would fill a book in itself. The axioms of Euclidean geometry and the theorems validating the use of co-ordinates make a substantial piece of theory – see, for example, Roe [26] or Greenberg [15]. Given the power of co-ordinates to turn a geometric problem into an algebraic one, something which has become very natural to school and undergraduate mathematicians alike, it is quite amazing

---

[5] Nikolai Joukowski (1847–1921), sometimes referred to as 'The Father of Russian Aviation', was a remarkably prolific mathematician who produced over 200 papers on mechanics and fluid dynamics, especially aerodynamics. In 1906 he calculated the lift on an aerofoil, a result now known as the Kutta–Joukowski theorem (jointly with Martin Wilhelm Kutta (1867–1944)). This theorem states that the lift per unit length on an aerofoil equals the product of the fluid's density, its velocity and the circulation around the aerofoil.

that the ancient Greek mathematicians (with the exception of Apollonius) developed so much geometry without any use of co-ordinates.

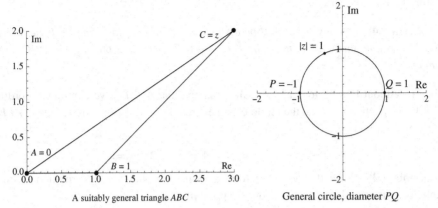

A suitably general triangle *ABC*   General circle, diameter *PQ*

Figure 1.12

#136
#142
#143

We now prove a selection of basic geometric facts. Here is a reminder of some identities which will prove useful.

$$\operatorname{Re}z = \frac{z+\bar{z}}{2}; \qquad z\bar{z} = |z|^2; \qquad \cos\arg z = \frac{\operatorname{Re}z}{|z|}; \qquad \sin\arg z = \frac{\operatorname{Im}z}{|z|}. \qquad (1.39)$$

Theorem 1.46 *(**Cosine Rule**) Let ABC be a triangle with angles $\hat{A}, \hat{B}, \hat{C}$. Then*

$$|BC|^2 = |AB|^2 + |AC|^2 - 2|AB|\,|AC|\cos\hat{A}. \qquad (1.40)$$

Proof We can introduce co-ordinates in the plane so that A is at the origin and B is at 1. Let C be at the point z. So in terms of our co-ordinates: $|AB| = 1$, $|BC| = |z-1|$, $|AC| = |z|$, $\hat{A} = \arg z$. Hence, by (1.39),

$$\begin{aligned}
\text{RHS of } (1.40) &= 1 + |z|^2 - 2|z|\cos\arg z = 1 + z\bar{z} - 2|z| \times \tfrac{\operatorname{Re}z}{|z|} \\
&= 1 + z\bar{z} - 2\times(z+\bar{z})/2 \quad = 1 + z\bar{z} - z - \bar{z} \\
&= (z-1)(\bar{z}-1) \qquad\qquad = \text{LHS of } (1.40). \qquad \Box
\end{aligned}$$

The sine rule also follows quickly.

Theorem 1.47 *(**Sine Rule**) Let ABC be a triangle with angles $\hat{A}, \hat{B}, \hat{C}$. Then*

$$\frac{|AB|}{\sin\hat{C}} = \frac{|BC|}{\sin\hat{A}} = \frac{|CA|}{\sin\hat{B}}.$$

Proof With A, B, C represented by complex numbers $0, 1, z$ respectively we have

$$\frac{|CA|}{\sin\hat{B}} = \frac{|z|\,|z-1|}{\operatorname{Im}(z-1)} = \frac{|z|\,|z-1|}{\operatorname{Im}z} = \frac{|BC|}{\sin\hat{A}}.$$

The remaining equality follows by symmetry. $\qquad\qquad\qquad\qquad\qquad\qquad\qquad\Box$

Whilst a simple enough theorem to prove in a geometric manner, Thales' theorem is arrived at nicely with the use of complex numbers, and the theorem's converse also comes naturally using algebra.

Theorem 1.48 *(Thales [6]) Let A, B be distinct points in the plane and let C be the circle with AB as a diameter. Then the point P lies on C if and only if the angle $\angle APB$ is a right angle.*

Proof Without loss of generality we may assume that A, B, P have complex co-ordinates $-1, 1, z$ respectively so that the circle C is the circle $|z| = 1$. By the cosine rule, $\angle APB$ is a right angle if and only if

$$2^2 = |z - 1|^2 + |z + 1|^2 \quad \Longleftrightarrow \quad 4 = 2z\bar{z} + 2 \quad \Longleftrightarrow \quad |z| = 1. \qquad \square$$

Example 1.49 Write down, in the form $z \mapsto az + b$ with $|a| = 1$, rotation through θ anticlockwise (see Figure 1.13) about the point p.

#134
#135

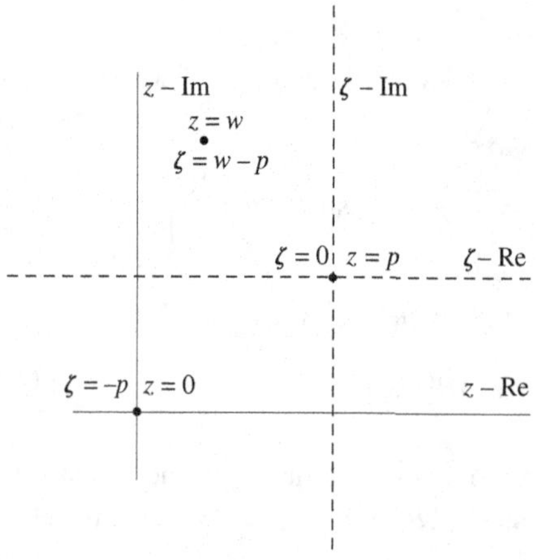

z and ζ co-ordinates

Figure 1.13

This example is straightforward enough to solve by means of a method akin to Example 1.41 but is included here so that we can solve it by means of a **change of co-ordinates**. We shall use two sets of co-ordinates, the given z-co-ordinate with its origin at 0 and a second co-ordinate system translated from z so that its origin is the point with z-co-ordinate p. Let's call this second co-ordinate ζ. Note that if a point has z-co-ordinate a then the same point has ζ-co-ordinate $a - p$. If we begin with a point with z-co-ordinate z_0 then the same point has ζ-co-ordinate $z_0 - p$. The transformation we are considering is rotation by θ about ζ's origin and so is given by $\zeta \mapsto \zeta \text{cis}\theta$ in terms of the ζ-co-ordinate (see Example 1.39). Hence, after rotating, our initial point has moved to the point with ζ-co-ordinate $\text{cis}\theta(z_0 - p)$. Finally we need to determine the z-co-ordinate of this image point which is $\text{cis}\theta(z_0 - p) + p$. So this rotation, in terms of the z-co-ordinate, is the map $z \mapsto \text{cis}\theta(z - p) + p$. $\qquad \square$

Example 1.50 The triangle abc in $\mathbb{C}$ (with the vertices taken in anticlockwise order) is equilateral if and only if $a + \omega b + \omega^2 c = 0$ where $\omega = \text{cis}(2\pi/3)$ is a cube root of unity.

---

[6] After the Greek mathematician and philosopher Thales (c.624BC–c.547BC). Thales might reasonably be considered the first Western philosopher – in that he sought to explain phenomena without reference to mythology – though it is unclear to what extent he was able to prove the various theorems of geometry that are attributed to him. At that time it is more likely that these results were appreciated as useful rules of thumb employed by engineers.

#139 Solution From #28 we have that $1+\omega+\omega^2 = 0$. The triangle abc is equilateral if and only if $c-b$ is the side $b-a$ rotated through $2\pi/3$ anticlockwise (see Figure 1.14) – i.e. if and only if

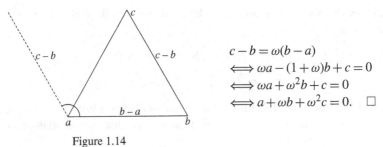

$$c-b = \omega(b-a)$$
$$\Longleftrightarrow \omega a - (1+\omega)b + c = 0$$
$$\Longleftrightarrow \omega a + \omega^2 b + c = 0$$
$$\Longleftrightarrow a + \omega b + \omega^2 c = 0. \quad \square$$

Figure 1.14

Exercises

#134A Rederive the answer of Example 1.49 using the method of Example 1.41.

#135b† Show that every orientation-preserving isometry of $\mathbb{C}$ is a rotation or translation. What is the centre of rotation of the map $z \mapsto az+b$ where $|a|=1, a \neq 1$? What is the angle of rotation?

#136A Let p,q,r be three distinct points in the complex plane. Show that r is collinear with p and q if and only if $(r-p)/(q-p)$ is real. Hence show that the equation of the line connecting p and q is

$$(\bar{q}-\bar{p})z + (p-q)\bar{z} = p\bar{q} - q\bar{p}.$$

#137c(i) Let a,b,c be three distinct complex numbers representing points in $\mathbb{C}$ which are not collinear. Show every complex number z can be uniquely written as $z = \alpha a + \beta b + \gamma c$ where α, β, γ are real and $\alpha + \beta + \gamma = 1$.
(ii) Sketch the lines $\alpha = 0$, $\beta = 0$ and $\gamma = 0$. Sketch the three regions where

$$\alpha > 0 \text{ and } \beta > 0 \text{ and } \gamma > 0; \quad \alpha > 0 \text{ and } \beta > 0 \text{ and } \gamma < 0; \quad \alpha > 0 \text{ and } \beta < 0 \text{ and } \gamma < 0.$$

The co-ordinates α, β, γ are called **barycentric co-ordinates** and were first introduced by Möbius.

#138D†(i) Let O, A, B, C be four points in the plane such that the triangle CAB contains the triangle OAB. Show that $|BC| + |CA| \geqslant |BO| + |OA|$.
(ii) Say now that a polygon contains a triangle and has one side in common with it. Show that the perimeter of the polygon exceeds that of the triangle.
(iii) Prove more generally that if a polygon contains a triangle then the polygon's perimeter exceeds the triangle's.

#139C†(i) Given a triangle ABC in the complex plane whose vertices are represented by the complex numbers a, b, c respectively, write down the complex numbers which represent the midpoints of AB, BC, CA.
(ii) A **median** of a triangle is a line connecting a vertex to the midpoint of its opposite side. Show that the three medians of ABC meet at the triangle's **centroid**, that is, the point represented by $(a+b+c)/3$.

#140 B† (Napoleon's Theorem) Let ABC be an arbitrary triangle. Place three equilateral triangles ADB, BEC, CFA, one on each edge and pointing outwards. Show that the centroids of these three new triangles define a fourth equilateral triangle.

#141 c Prove that the midpoints of the sides of an arbitrary quadrilateral are the vertices of a parallelogram.

#142 a Let z_1 and z_2 be two complex numbers. Show that

$$|z_1 - z_2|^2 + |z_1 + z_2|^2 = 2(|z_1|^2 + |z_2|^2). \tag{1.41}$$

Remark 1.51 Equation (1.41) is called the **parallelogram law**. If we consider the parallelogram with vertices $0, z_1, z_1 + z_2, z_2$ then we see that the parallelogram law states that

'in a parallelogram, the sum of the squares of the diagonals equals the sum of the squares of the four sides'. (1.42)

∎

#143 b Consider a quadrilateral $OABC$ in the complex plane whose vertices are at the complex numbers $0, a, b, c$. Show that the equation

$$|b|^2 + |a - c|^2 = |a|^2 + |c|^2 + |a - b|^2 + |b - c|^2$$

can be rearranged as $|b - a - c|^2 = 0$. Hence show that the only quadrilaterals to satisfy (1.42) are parallelograms.

#144 C† Let ABC be a triangle whose vertices (counted anticlockwise) are represented by the complex numbers a, b, c. Show that the triangle has area $\mathrm{Im}\,(\bar{a}b + \bar{b}c + \bar{c}a)/2$.

#145 D† Let ABC be an equilateral triangle and D a point inside the triangle. Let E, F, G be the feet of the altitudes from D to the sides AB, BC and CA respectively. Show that

$$\mathrm{Area}(ADE) + \mathrm{Area}(BDF) + \mathrm{Area}(CDG) = \mathrm{Area}(BDE) + \mathrm{Area}(CDF) + \mathrm{Area}(ADG).$$

#146 C Let ABC be a triangle and with squares $ABDE, ACFG$ constructed on the outside of its edges AB, AC. Let P be the midpoint of BC and let Q be the foot of the altitude from A to the side BC. Show that: (i) EG is perpendicular to AP. (ii) $|EG| = 2|AP|$. (iii) The line AQ bisects EG.

#147 D† Let a, b be complex numbers such that $0, a, b$ aren't collinear in $\mathbb{C}$. Find the centre c and radius r of the circle which passes through each of $0, a, b$. By changing origin, find the centre and radius of the circle which passes through three non-collinear points A, B, C in $\mathbb{C}$. This circle is called the **circumcircle** of the triangle ABC and its centre the **circumcentre** of the triangle.

#148 b† Show that the map $f(z) = 1/z$

(i) maps a circle not through the origin to a circle not through the origin;
(ii) maps non-zero points on a circle through the origin to a line not through the origin;
(iii) maps non-zero points on a line through the origin to non-zero points on a line through the origin.

#149D† (i) In Figure 1.15a the lines L_1 and L_2 are parallel; the circle C_1 is tangential to L_1 at P; the circle C_2 is tangential to L_2 at R; the circles C_1 and C_2 are tangential at Q. Show that P, Q, R are collinear.

(ii) In Figure 1.15b are four circles externally arranged to one another such that C_1 is tangential to C_2 which is tangential to C_3 which is tangential to C_4 which in turn is tangential to C_1. Show that the four points where the circles are tangential lie on a circle.

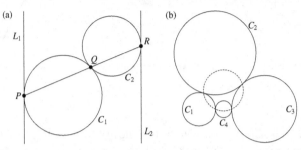

Figure 1.15

#150B† Let A, B, C, D be distinct points in the plane. By taking A as the origin, and using the properties of the map f proved in #148, show that

$$|AB|\,|CD| + |AD|\,|BC| \geqslant |AC|\,|BD| \tag{1.43}$$

with equality if and only if A, B, C, D are **concyclic** (i.e. lie on a circle) or collinear in the given order (clockwise or anticlockwise). This is known as **Ptolemy's theorem.** [7]

#151c For any $a \neq 0$, use Ptolemy's theorem to show that $a, -\bar{a}, a^{-1}, -\bar{a}^{-1}, 1, -1$ are concyclic.

#152D† Use Ptolemy's theorem to show that four points A, B, C, D (in that order) are concyclic or collinear if and only if $\angle ADC + \angle ABC = \pi$.

#153D† Let G denote the set of maps from $\mathbb{C}$ to $\mathbb{C}$ of the form $z \mapsto az + b$ where a, b are complex and $a \neq 0$.

(i) Let w_1, w_2, w_3 and z_1, z_2, z_3 be two triples of distinct points in $\mathbb{C}$. Suppose that there exists f in G such that $f(z_i) = w_i$ for $i = 1, 2, 3$; show that the triangles $\triangle w_1 w_2 w_3$ and $\triangle z_1 z_2 z_3$ are similar. Is the converse true?

(ii) Let C_1, C_2 be circles with centres a_1, a_2 and radii r_1, r_2 respectively where $r_1 \neq r_2$. Determine which maps f in G satisfy $f(C_1) = C_2$. Show that the fixed points of these maps comprise a circle (called the **circle of similitude**).

(iii) Show that there are two points p, q and real numbers λ, μ such that the maps

$$z \mapsto \lambda(z - p) + p,$$
$$z \mapsto \mu(z - q) + q$$

map C_1 to C_2. The points p, q are called **centres of similarity**. Show that p and q lie at opposite ends of a diameter of the circle of similitude (see Figure 1.16).

---

[7] Named after the Greek mathematician and astronomer, Claudius Ptolemy (c.85–c.165).

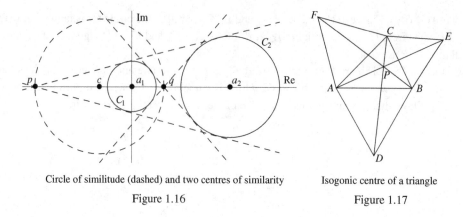

Circle of similitude (dashed) and two centres of similarity   Isogonic centre of a triangle

Figure 1.16   Figure 1.17

#154E† Let $\omega = \text{cis}\,(2\pi/3)$. Consider the triangle in $\mathbb{C}$ with vertices $A = 0$, $B = 1$ and $C = z$. Three equilateral triangles ADB, BEC, CFA, one on each edge and pointing outwards are constructed. Show that

  (i) The line segments AE, BF, CD are all of the same length.

 (ii) These three line segments are all at $60°$ to one another.

(iii) These three line segments are concurrent at

$$P = \frac{z - \omega^2 \bar{z}}{(1 - \omega)(1 + \omega \bar{z})}.$$
(1.44)

(iv) P lies on the circumcircles of ADB, BEC, CFA. The point P is called the **first isogonic centre** (Figure 1.17).

#155c(i) Show that the first isogonic centre and centroid coincide when the triangle ABC is equilateral.

(ii) What happens to the first isogonic centre if C moves towards A or B?

(iii) Show the converse of (i), namely that if the first isogonic centre and the centroid coincide then ABC is equilateral.

#156c† Given distinct complex numbers a, b, c find values of A, B, z such that $w \mapsto Aw + B$ maps $0, 1, z$ to a, b, c respectively. Hence use formula (1.44) to find the first isogonic centre of the triangle with vertices $2, 2 + 2i, i$.

#157C With notation as in #154, describe the set of values z such that (i) P is at 0; (ii) P is at 1; (iii) P is at z. What is the angle at 0 in case (i), at 1 in case (ii), at z in case (iii)?

#158D† With notation as in #154, determine for which z it follows that P lies on the real axis. Deduce that the only real values P can take are 0 and 1.

#159E† With notation as in #154, suppose that $\text{Im}\,z > 0$. Under what conditions on z is $\text{Im}\,P > 0$? Deduce that the first isogonic centre lies within the triangle if and only if the triangle's three angles are all less than $2\pi/3$.

#160 E Let ABC be a triangle with none of its three angles exceeding $2\pi/3$, so that its first isogonic point P lies inside the triangle, For an arbitrary point Q inside the triangle, we define

$$d(Q) = |AQ| + |BQ| + |CQ|.$$

The aim of this question is to determine when $d(Q)$ is minimal.

(i) Without loss of generality let $A = 0, B = 1, C = z$ and $Q = q$. Let $\omega = \operatorname{cis}(2\pi/3)$. Show that

$$d(Q) = |z - q| + |q - (-\omega q)| + |-\omega q - (-\omega)|.$$

(ii) Deduce that $d(Q)$ would be minimal if $z, q, -\omega q$ and $-\omega$ were to lie in a line in that order and show that this is indeed the case when $Q = P$.

(iii) Show that $Q = P$ is the unique minimum for $d(Q)$.

Definition 1.52 The **Fermat point** of the triangle ABC is the point Q such that $d(Q)$ is minimal.

#161 E† (i) Show that the Fermat point can never lie outside a triangle.

(ii) Show that if the angle at A exceeds $2\pi/3$ then the Fermat point is A.

1.8 Further Exercises*

Exercises on Cyclotomic Polynomials

Definition 1.53 Let n be a positive integer. The nth **cyclotomic polynomial** $\Phi_n(x)$ is defined as

$$\Phi_n(x) = \prod_{\omega \text{ primitive}} (x - \omega) = \prod_{\substack{0 < k < n \\ k, n \text{ coprime}}} (x - \operatorname{cis}(2k\pi/n)),$$

where the first product is taken over all primitive nth roots of unity ω (#56) or equivalently over all k in the range $1 \leqslant k < n$ where k and n are coprime.

#162 B Show that the first six cyclotomic polynomials, $\Phi_1(x), \ldots, \Phi_6(x)$ are

$$x - 1, \quad x + 1, \quad x^2 + x + 1, \quad x^2 + 1, \quad x^4 + x^3 + x^2 + x + 1, \quad x^2 - x + 1.$$

#163 B† Let p be a prime and n a positive integer. Show that

$$\Phi_p(x) = x^{p-1} + x^{p-2} + \cdots + x + 1 \quad \text{and} \quad \Phi_{p^n}(x) = \Phi_p\left(x^{p^{n-1}}\right).$$

#164 D† Let n be a positive integer. Show that

$$x^n - 1 = \prod_{d|n} \Phi_d(x) \qquad \text{and} \qquad x^n + 1 = \prod_{d|2n,\, d\nmid n} \Phi_d(x),$$

where the first product is taken over all factors d of n and the second product is taken over all factors d of $2n$ which aren't factors of n.

#165 D† Show that $\Phi_{4n}(x) = \Phi_{2n}(x^2)$ and $\Phi_{4n+2}(x) = \Phi_{2n+1}(-x)$, where n is a positive integer.

#166 C Show, for $n > 1$, that Φ_n is palindromic, that is $\Phi_n(x) = x^m \Phi_n(x^{-1})$ where m is the degree of $\Phi_n(x)$.

#167 D† Determine $\Phi_{15}(x)$ and deduce that

$$\cos\left(\frac{\pi}{15}\right) \cos\left(\frac{2\pi}{15}\right) \cos\left(\frac{4\pi}{15}\right) \cos\left(\frac{7\pi}{15}\right) = \frac{1}{16}.$$

#168 C Show that the coefficients of a cyclotomic polynomial are integers. [8]

#169 D Let $n \geq 2$. Show that $\Phi_n(1) = p$ if $n = p^k$ for some prime p and otherwise that $\Phi_n(1) = 1$.

Exercises on Quaternions

#170 B A **quaternion** is a number of the form $a + bi + cj + dk$ where a, b, c, d are real numbers and i, j, k satisfy

$$i^2 = j^2 = k^2 = ijk = -1. \tag{1.45}$$

Quaternions then add and multiply as one would expect (associatively and distributively), with real scalars commuting.

(i)   Show that $ij = k$ and that $ij = -ji$. (Hence quaternion multiplication is not commutative.)

(ii) Show that $(a + bi + cj + dk)(a - bi - cj - dk)$ is real and non-negative.

(iii) Find the multiplicative inverse of $2 + i + k$.

(iv) How many solutions are there amongst the quaternions to the equation $z^2 = -1$?

Exercises on Complex Dynamics

#171 D† (i) Show for any $m \geq 2$ that the remainders, when the powers of 2 are divided by m, eventually become periodic. For example, when the powers of 2 are divided by 24 we get the remainders $2, 4, 8, 16, 8, 16, 8, 16, \ldots$.

(ii) What is the last digit in 2^{2017}?

---

[8] On the basis of the examples above you may be tempted to believe these coefficients are always $0, 1$ or -1. This is not true, though Φ_{105} is the first counter-example to this.

Hamilton plaque (Photograph by Patrick E. Dorey)

Division Algebras The quaternions were discovered by the Irish mathematician William Rowan Hamilton (1805–1865) in 1843. In his honour, the set of quaternions is denoted $\mathbb{H}$. He had originally been seeking a three-dimensional analogue of the complex numbers to describe rotations in three dimensions, the way complex numbers can in the plane. The appreciation that he would need a four-dimensional algebra, and its rules of multiplication, came to him as he walked along the canal by Broom Bridge in Dublin and he famously carved the equations in (1.45) into a stone of the bridge.

The real numbers, complex numbers and quaternions are all examples of real *division algebras*. In fact, as Frobenius showed in 1878, they are the only examples of associative real division algebras. A real division algebra has, loosely put, a notion of addition, multiplication, scalar multiplication (multiplication by reals) and division by non-zero elements which obey self-evident rules (*axioms*) as to how these operations interact. It might further be assumed that multiplication is commutative and/or associative. $\mathbb{R}$ is (essentially) the only one-dimensional real division algebra, but there are infinitely many in two dimensions – for example, the rule $x \times y = x\bar{y}$ makes $\mathbb{C}$ into a non-commutative, non-associative division algebra – and their classification was determined only in the 1980s. It is more generally known that any real division algebra must have dimension 1, 2, 4 or 8 (proved by Bott and Milnor, and also by Kervaire, in 1958). (See #1226 for the use of quaternions to describe rotations in three dimensions.)

Cobwebbing If you have ever idly, repeatedly, pressed a function key on your calculator, you might have produced a sequence of numbers like the following:

$$0, \ 1, \ 0.540302, \ 0.857553, \ 0.65429, \ 0.79348, \ 0.701369, \ 0.76396, \ 0.722102, \ldots$$

which is the sequence you arrive at by repeatedly applying 'cos' (in radians mode) having started with 0. The numbers oscillate back and forward somewhat but eventually settle down to a number $\alpha = 0.739085\ldots$. What you have found, in this case, is a solution of the equation $x = \cos x$.

This is an instance of a method called *cobwebbing* for finding roots to equations of the form $f(x) = x$, so-called *fixed points* of f. The reason for the name becomes more

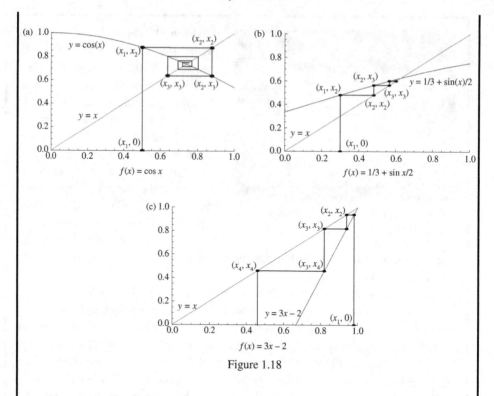

Figure 1.18

apparent when we analyze Figure 1.18a. Sketched on the axes are $y = x$ and $y = \cos x$. The point where these graphs intersect is (α, α). On this sketch our initial term x_1 equals 0.5. Our second term is $x_2 = \cos x_1$. We can appreciate where this is on the graph by going up from the point $(x_1, 0)$ to the $y = \cos x$ graph. We are then at the point (x_1, x_2). Going across from this point to the $y = x$ line takes us to (x_2, x_2). From here if we go down to the cosine graph we come to (x_2, x_3) and going across to the line takes us to (x_3, x_3). If we continue in this fashion we produce a sequence (x_n) whose odd terms are a little less than α, whose even terms a little greater, but importantly they are closing in on α as n increases. It's not hard to appreciate from the sketch that there was nothing particularly important about the starting value $x_1 = 0.5$ save that it was reasonably close to α. A similar pattern would emerge from any other suitably close x_1.

If we look to the second and third sketches we see different behaviours. In the second sketch the sequence increases to the fixed point from below (or decreases down to it if the initial term is an overestimate). In the third sketch the sequence moves away from the fixed point. What is important in each case is the gradient of $f(x)$ at the fixed point. If this gradient is in the range $-1 < f'(\alpha) < 0$ then the oscillatory behaviour of the first sketch will arise. If $0 < f'(\alpha) < 1$ then the wholly increasing (or decreasing) behaviour of Figure 1.18b will occur. When $|f'(\alpha)| < 1$ we say that α is *attracting*. Finally if $|f'(\alpha)| > 1$ then the fixed point is *repelling* and the sequence will move away from α (#175).

#172D What is the next to last digit in 2^{2017}?

#173D† Let $P(z) = z^2$ and for each possible complex number z_0 consider the sequence

$$z_0, P(z_0), P(P(z_0)), P(P(P(z_0))), \ldots \tag{1.46}$$

formed by repeated application of the function P.

(i) For what values of z_0 does the sequence (1.46) become unbounded?

(ii) For what values of z_0 does the sequence (1.46) converge?

(iii) Give an example of a value of z_0, for which the sequence (1.46) becomes periodic (without converging).

(iv) Give an example of a value of z_0 for which the sequence does not converge and is not periodic.

#174B (Knowledge of induction may prove helpful.) Given a complex number c, we recursively define a sequence $z_0(c), z_1(c), z_2(c), \ldots$ by setting

$$z_0(c) = 0 \qquad z_{n+1}(c) = z_n(c)^2 + c.$$

We say that c lies in the **Mandelbrot set** (see p.58 for a plot) if the sequence $z_0(c), z_1(c), z_2(c), \ldots$ is bounded.

(i) Show that $-2, -1, 0, i$ each lie in the Mandelbrot set but that 1 lies outside the set.

(ii) Show that if c lies in the Mandelbrot set then so does $\bar{c}$.

(iii) Show that if $|c| = 2 + \varepsilon$ where $\varepsilon > 0$ then $|z_n(c)| \geqslant 2 + n\varepsilon$ for $n \geqslant 1$. Deduce that the Mandelbrot set lies entirely within the disc $|z| \leqslant 2$.

(iv) Show that if $|c| < 2$ and $|z_n(c)| > 2$ for some n then c is not in the Mandelbrot set. Deduce that neither $1 + i$ nor $-1 + i$ lies in the Mandelbrot set.

#175B† Say that α is a solution of $f(x) = x$. If x_1 is close to α explain why

$$|x_2 - \alpha| \approx |f'(\alpha)| \times |x_1 - \alpha| \qquad \text{where } x_2 = f(x_1).$$

Verify for the fixed points α in the sketches above that $-1 < f'(\alpha) < 0$, $0 < f'(\alpha) < 1$, $|f'(\alpha)| > 1$ respectively.

#176B Let α denote the solution of $\cos x = x$ from the first sketch. Verify (with a calculator) that $\alpha = 0.739085$ to 6 decimal places.

#177C Let $f(x) = 2x^2 + x^3$ and $g(x) = 3x^3 + 2x^4$. Note that both functions have a fixed point $\alpha = 0$ and that $f'(\alpha) = 0 = g'(\alpha)$. For each function, describe the behaviour of the sequence x_n generated by cobwebbing when the initial estimate x_1 is close to 0. Show, for suitably small x_n, that

$$\text{for } f: \quad x_{n+1} - \alpha \approx 2(x_n - \alpha)^2;$$

$$\text{for } g: \quad x_{n+1} - \alpha \approx 3(x_n - \alpha)^3.$$

#178D Let $f(x) = e^x - 2$. Show that f has two fixed points, one attracting, one repelling. For each possible initial value x_0, describe the behaviour of the sequence $x_{n+1} = f(x_n)$.

#179 D† For each possible initial value x_0, describe qualitatively the behaviour of the sequence $x_{n+1} = 2\sin x_n$.

#180 B(i) Let $f(x) = x^3 - 3x^2 + 1$. Show that f has three real fixed points $\alpha < \beta < \gamma$ that are each repelling.

(ii) Show that $g(x) = 3 + x^{-1} - x^{-2}$ has the same three fixed points as f. Determine which fixed points of g are attracting, which repelling.

(iii) Show $h(x) = \sqrt{(x^3 - x + 1)/3}$ has β as an attracting fixed point and $-h(x)$ has α as an attracting fixed point.

#181 B† Sketch graphs of $y = x^2 + 1/4$ and $y = x$ on the same xy-axes. Show, using cobwebbing, that $1/4$ lies in the Mandelbrot set. Deduce that if $|c| \leqslant 1/4$ then c lies in the Mandelbrot set.

#182 D Let c be a real number. With separate axes for each of the following cases, sketch the graphs of $y = x^2 + c$ and $y = x$: (i) $c > 1/4$; (ii) $0 \leqslant c < 1/4$; (iii) $-3/4 < c < 0$; (iv) $-5/4 < c \leqslant -3/4$; (v) $-2 \leqslant c \leqslant -5/4$; (vi) $c < -2$. In each case use cobwebbing to qualitatively describe the behaviour of the sequence $z_n(c)$ as n increases. Deduce that the intersection of the Mandelbrot set and the real line is the interval $-2 \leqslant c \leqslant 1/4$.

#183 D† Let $P_c(z) = z^2 + c$. Show that P_c has an attracting fixed point if and only if $c = (w/2)(1 - w/2)$ for some $|w| < 1$ and that the attracting fixed point is $\alpha = w/2$. Such c comprise the main **cardioid** of the Mandelbrot set. Show that this cardioid can be parametrized as

$$x = \frac{\cos t}{2} - \frac{\cos 2t}{4}, \qquad y = \frac{\sin t}{2} - \frac{\sin 2t}{2}, \qquad (0 \leqslant t < 2\pi).$$

By changing origin, and using polar co-ordinates, show that the cardioid can be described by equation $r = a(1 - \cos\theta)$ for some a.

Definition 1.54 We say distinct points α, β are a **period 2 orbit** of f if $f(\alpha) = \beta$ and $f(\beta) = \alpha$, and call α, β **periodic points of period 2**. We say that α is an **attracting periodic point of period 2** if $|(f \circ f)'(\alpha)| < 1$.

#184 D† Show, with the aid of sketches, that $f(x) = 2\sin x$ does not have any period 2 orbits but that $g(x) = 3\sin x$ does have. Is the latter orbit attracting?

#185 C† Assuming $c \neq -3/4$, show that P_c has precisely one period 2 orbit. For what c is 0 in this orbit?

#186 D Show that P_c has attracting periodic points of period 2 if and only if c lies in the disc $|c + 1| < 1/4$.

Remark 1.55 It follows that points of the cardioid and disc found in #183 and #186 are themselves in the Mandelbrot set according to its definition in #174. This is not easy to demonstrate; for completeness a reference is included here, namely Devaney [8, Theorem 4.6, p.281], but the result is technical and beyond the scope of this chapter. ∎

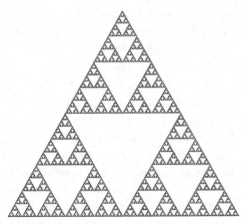

Figure 1.19

Fractals If you think about it you may realise that 'dimension' is a word that you have commonly been using in mathematics without it ever having been properly defined for you. What does it mean to say something is two- or three-dimensional? Not unreasonably you may consider an object (like a sphere) to be two-dimensional because it takes two co-ordinates to determine a point (e.g. longitude and latitude). However, it became clear in the nineteenth century (especially with the discovery of *space-filling curves*) that a rigorous definition was needed. There are several definitions for *topological dimension* which take integer values but intriguingly *Hausdorff dimension* (after the German mathematician Felix Hausdorff (1868–1942)) can take non-integer values.

One property of dimension is the following: if you scale a square's sides by k then the area goes up by a factor of k^2; likewise for a cube the volume goes up by a factor of k^3. Figure 1.19 is a picture of the Sierpinski triangle (named after the Polish mathematician Waclaw Sierpinski (1882–1969)); it is formed from an equilateral triangle by removing an equilateral triangle from the middle to leave three smaller versions and then taking the middle out of each of these three triangles and so on and so on... If instead the Sierpinski triangle had been made from a triangle twice as big, we would have produced three versions of the original. Thinking along these lines suggests that the dimension d of the Sierpinski triangle should satisfy $2^d = 3$; i.e. that $d = \log_2 3$ which is a number strictly between 1 and 2. This is, in fact, the correct value for the Hausdorff dimension of the Sierpinski triangle. The Cantor set (#1522) is also a fractal and can similarly be shown to have dimension $d = \log_3 2$.

Figure 1.20

One definition of a *fractal* is an object with a greater Hausdorff dimension than its topological dimension. (There are certain problems with this definition as it isn't satisfied by certain sets with fractal-like properties.) Fractals have been known about for a long time but interest in them accelerated with the advent of fast computers and also because of the popularising work of Benoit Mandelbrot (1924–2010) who first studied his eponymous set (see #174). Much work had already been done on the dynamics associated with repeatedly applying a polynomial by the French mathematician, Gaston Julia (1892–1978). With c now fixed, if $P_c(z)$ is repeatedly applied to an initial value z_0 then the sequence z_n formed may behave very differently depending on the value of z_0, remaining bounded or becoming unbounded.

The *Julia set* of $P_c(z)$ is the boundary set between these differently behaving z_0. Typically the Julia set is intricate, fractal and exhibits *self-similarity* (as in the way, with the Sierpinski triangle, exact copies of the original can be found within at all levels of magnification). An example of a Julia set is given in Figure 1.20.

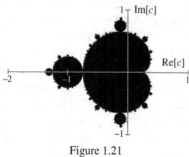

Figure 1.21

The Mandelbrot set (Figure 1.21) is a set of different c-values that classifies the different possible Julia sets of P_c as c varies. For c in the main cardioid of the Mandelbrot set, the Julia set of P_c is vaguely circular. For values of c in the other components of the Mandelbrot set the Julia set becomes more complicated but remains of the same type whilst in the same component, and will always be *connected*, meaning essentially that the set is in one piece.

A good introduction to dimension theory can be found in Edgar [11]. Many detailed images of the Mandelbrot set, showing its fine detail to incredible levels of magnification, can be found on the internet, showing the different ways to colour the Mandelbrot set. Rather than simply colouring a point c one of two colours, depending on the sequence $z_n(c)$ remaining bounded, the c outside the Mandelbrot set can be assigned a variety of different colours on the basis of how quickly $z_n(c)$ tends to infinity.

Exercises on Stereographic Projection

#187 B The sphere S, with centre $(0,0,0)$ and radius one, has equation $x^2 + y^2 + z^2 = 1$.

Thinking of $\mathbb{C}$ as the xy-plane, every complex number $P = X + Yi$ can be identified with a point Q on S by drawing a line from $(X, Y, 0)$ in the xy-plane to the sphere's north pole $N = (0, 0, 1)$; this line intersects the sphere at two points Q and N. We define a map f from $\mathbb{C}$ to the sphere S by setting $f(P) = Q$ (see Figure 1.22).

(i) Show that $f(X + Yi)$ equals

$$\left(\frac{2X}{1 + X^2 + Y^2}, \frac{2Y}{1 + X^2 + Y^2}, \frac{X^2 + Y^2 - 1}{1 + X^2 + Y^2} \right).$$

(ii) Show that the image of f equals the sphere S except for N.

(iii) The inverse map $\pi = f^{-1}$ is called **stereographic projection**. Show, for $(x, y, z) \neq N$, that

$$\pi(x, y, z) = \frac{x + yi}{1 - z}.$$

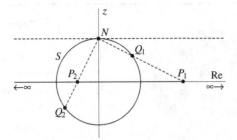

Stereographic projection, cross section

Figure 1.22

- Note that π maps points near N to complex numbers with large moduli. It is consequently natural to identify a single 'infinite' point, written ∞, with the north pole N.
- The set $\mathbb{C} \cup \{\infty\}$ is called the **extended complex plane** and this is denoted $\tilde{\mathbb{C}}$.
- If we identify $\tilde{\mathbb{C}}$ via stereographic projection with the sphere S, then S is known as the **Riemann sphere**. (See p.345 for biographical notes about Riemann.)
- Under stereographic projection, lines in $\mathbb{C}$ correspond to circles on S which pass through N, and circles in $\mathbb{C}$ correspond to circles on S which don't pass through N. For this reason the term **circline** is often used to denote a circle or a line in $\tilde{\mathbb{C}}$, and a line should be viewed as a special type of circle which passes through ∞. Details of this are in the following exercise.

#188D†(i) Consider the plane Π with equation $Ax + By + Cz = D$ (see Proposition 3.26 for further details if necessary). Show that Π intersects the Riemann sphere S in a circle if $A^2 + B^2 + C^2 > D^2$.
(ii) Given $A^2 + B^2 + C^2 > D^2$, show that stereographic projection maps this circle to the curve in the XY-plane with equation

$$(C - D)(X^2 + Y^2) + 2AX + 2BY - C - D = 0. \tag{1.47}$$

(iii) Show that if $C \neq D$ then (1.47) is the equation of a circle. Show further that every circle can be represented in this way for appropriate choices of A, B, C, D.
(iv) Show that if $C = D$ then (1.47) is the equation of a line and that every line can be represented in this way. Note that $C = D$ if and only if $N = (0, 0, 1)$ lies in the plane Π.

#189C Show that the **great circles** on S (i.e. the intersections of S with planes through the origin) correspond to circlines of the form $\alpha(z\bar{z} - 1) + \bar{\beta}z + \beta\bar{z} = 0$, where α is real, β is complex, and $z = X + Yi$.

#190B Show that if the point M on the Riemann sphere S corresponds under stereographic projection to the complex number z then the **antipodal point** $-M$ corresponds to $-1/\bar{z}$.

Augustin-Louis Cauchy (Bettmann/Getty Images)

Complex Analysis is a central topic in mathematics and widely regarded as both one of the most aesthetic and harmonious but also highly applicable. Most undergraduates meet analysis for the first time at university in real analysis courses (pure calculus essentially) containing results they might have otherwise considered obvious, or 'known' from school, treated in a seemingly pedantic manner and with a focus on the pathological. By and large, a first course in real analysis contains more you-can't-do-that moments than most undergraduates welcome. By comparison, complex analysis is the can-do sibling of real analysis. The main subject of study is the set of *holomorphic functions*, essentially consisting of those complex-valued functions that have a derivative. It turns out that being differentiable in this complex sense is a more demanding requirement; consequently it is possible to build up a richer theory about these functions – for example, holomorphic functions have in fact derivatives of all orders and are defined locally by some Taylor series. But the set of holomorphic functions is still wide enough to include all the important functions you are likely to be interested in. Complex analysis, though, is more than simply another version of real analysis in which the results are cleaner and more positive. The very nature of the subject is different and has a much more *topological* flavour.

The first major result of complex analysis is *Cauchy's theorem*, after Augustin-Louis Cauchy (1789–1857). In fact you will see Cauchy's name populating much of a complex analysis course as he almost single-handedly developed the subject. This theorem states that the integral around any closed curve Γ (essentially a loop) in the complex plane of a holomorphic function $f(z)$ is zero. If the integrand isn't holomorphic everywhere – like $1/z$, where only $z = 0$ is a problem – then the integral now only depends on whether the loop encloses 0. If not, the answer is still zero; if it wraps around the origin once anticlockwise the answer is $2\pi i$; twice and the answer is $4\pi i$.

You might reasonably raise the issue that a real integral often represents something, well, real – for example area or arc-length: what meaning can a complex answer have? But by taking the real or imaginary parts of such an answer, real integrals can be determined, often ones that are difficult or impossible to calculate by real methods alone.

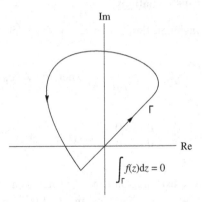

Figure 1.23

By such (relatively standard) means a first course in complex analysis would likely have you determining, without much calculation, integrals and infinite sums such as

$$\int_0^\infty \frac{\sin x}{x}\,dx = \frac{\pi}{2}; \qquad \int_0^{2\pi} \frac{dt}{2+\sin t} = \frac{2\pi}{\sqrt{3}};$$

$$\int_0^\infty \frac{dx}{1+x^n} = \frac{\pi}{n}\csc\left(\frac{\pi}{n}\right); \qquad \sum_1^\infty \frac{1}{n^4} = \frac{\pi^4}{90}.$$

It is usually via complex analysis that one first sees a proof of the fundamental theorem of algebra (Theorem 1.4). Complex analysis is also widely applicable. In pure mathematics it leads naturally into the study of Riemann surfaces (for example, $x^2 + y^2 = 1$, rather than defining a curve in the real plane $\mathbb{R}^2$, defines a two-dimensional surface in four-dimensional $\mathbb{C}^2$) and aspects of algebraic geometry; also it is important within analytic number theory via Riemann's zeta function (p.345). In both pure and applied analysis there are applications, with complex methods being widely used in the solution of partial differential equations. And the links between Laplace's equation and holomorphic functions make complex analysis very useful in the study of ideal (inviscid) fluid flow in two dimensions.

#191D Let p be a point on the real axis in $\mathbb{C}$ and let $0 \leqslant \theta < 2\pi$. Consider the two lines through p parametrized by $z_1(t) = p+t$, $z_2(t) = p+t\,\mathrm{cis}\,\theta$, where t is real.

(i) Find the images $\mathbf{r}_1(t) = f(z_1(t))$ and $\mathbf{r}_2(t) = f(z_2(t))$ on the Riemann sphere which correspond to $z_1(t)$ and $z_2(t)$.

(ii) Show that the angle between the tangent vectors $\mathbf{r}_1'(0)$ and $\mathbf{r}_2'(0)$ is θ. [See §3.2 for details on the scalar product if necessary.] This shows that stereographic projection is **conformal** (i.e. angle-preserving).

Exercises on Möbius Transformations

#192 B† A **Möbius**[9] **transformation** is a map f of the extended complex plane $\tilde{\mathbb{C}}$ of the form

$$f(z) = \frac{az+b}{cz+d} \qquad \text{where } ad \neq bc. \tag{1.48}$$

We define

$$f(\infty) = \begin{cases} \frac{a}{c} & \text{if } c \neq 0, \\ \infty & \text{if } c = 0, \end{cases} \qquad f(-d/c) = \infty \text{ if } c \neq 0.$$

(i) Show that a Möbius transformation can be written as a composition of **translations** (maps of the form $z \mapsto z + k$), **dilations** (maps of the form $z \mapsto kz$ where $k \neq 0$) and **inversion** (the map $z \mapsto 1/z$).

(ii) Deduce that a Möbius transformation maps a circline to a circline. [Hint: use the results of #119.]

(iii) Show[10] that if f and g are Möbius transformations then so are the composition $g \circ f$ and the inverse f^{-1}.

#193 B† Show that every Möbius transformation, other than the identity map, fixes one or two points in $\tilde{\mathbb{C}}$. Under what conditions on a, b, c, d does (1.48) fix one or two points?

#194 D Suppose that the Möbius transformation f has two fixed points α, β in $\mathbb{C}$. Determine $f'(\alpha)f'(\beta)$ and deduce that if one fixed point is attracting (i.e. $|f'(\alpha)| < 1$) then the other is repelling (i.e. $|f'(\beta)| > 1$).

#195 D† (See also #120.) (i) Let z_1, z_2, z_3 be three distinct points in $\tilde{\mathbb{C}}$. Find a Möbius transformation f such that

$$f(z_1) = 0, \qquad f(z_2) = 1, \qquad f(z_3) = \infty.$$

(ii) Let w_1, w_2, w_3 be three distinct points in $\tilde{\mathbb{C}}$. Show that there is a unique Möbius transformation g such that $g(z_i) = w_i$ for $i = 1, 2, 3$. Deduce that, given three distinct points in $\tilde{\mathbb{C}}$, there exists a unique circline passing through the points.

#196 D Consider the curves $\gamma(t) = 1 + t$ and $\Gamma(t) = 1 + t\operatorname{cis}\theta$ which intersect in angle θ at the point 1 in $\mathbb{C}$. Show that, under each of the maps

$$z \mapsto z + k; \qquad z \mapsto kz \ (k \neq 0); \qquad z \mapsto 1/z$$

the angle between the curves' images remains θ. Deduce that Möbius transformations are conformal.

---

[9] August Ferdinand Möbius (1790–1868) is most widely remembered today for the Möbius strip, a closed loop with one twist so as to make a non-orientable surface (i.e. a surface with just one side). However, Möbius made contributions in many areas of mathematics and astronomy, including introducing homogeneous co-ordinates in projective geometry and the so-called Möbius function in number theory.

[10] The Möbius transformations form a group, denoted $PGL(2, \mathbb{C})$, under composition; this follows from (iii), that the identity is a Möbius transformation and the associativity of composition.

#197D† In each of the following cases find a Möbius transformation which

(i) maps the unit circle $|z| = 1$ to the real axis and the real axis to the imaginary axis;

(ii) maps the points $1, 2, 3$ to $2, 3, 1$ respectively;

(iii) maps the real axis to the circle $|z - 1| = \sqrt{2}$ and the imaginary axis to the circle $|z + 1| = \sqrt{2}$;

(iv) maps the real axis to the circle $|z| = 1$ and the line $\operatorname{Im} z = 1$ to the circle $|z - 2| = 1$.

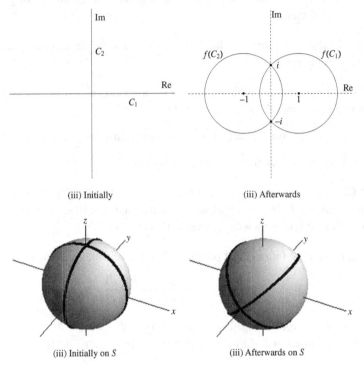

(iii) Initially   (iii) Afterwards

(iii) Initially on S (iii) Afterwards on S

Figure 1.24

Remark 1.56 Whilst the 'before' and 'after' diagrams from #197 (Figure 1.24) look very different when sketched in the Argand diagram, their similarities are much more apparent on the Riemann sphere. Take, for example, part (iii). In the Argand diagram we are initially presented with two lines crossing at the origin whilst we are directed to map these to two circles which intersect twice. Thought of as two circlines, those two lines meet a second time at ∞. Under the inverse of stereographic projection, these diagrams appear much more alike on the Riemann sphere. The first diagram corresponds to two circles meeting at right angles at $(0, 0, \pm 1)$ and the second corresponds to two circles meeting at right angles at $(0, \pm 1, 0)$ and are clearly connected by a rotation of the Riemann sphere. Though we shall not prove the result here (see Berger [3, Theorem 18.10.4]), the conformal (i.e. angle-preserving) mappings of the Riemann sphere are precisely those maps which correspond to the Möbius transformations once we have identified the Riemann sphere with the extended complex plane. Most such conformal mappings of the sphere are not simply rotations (#202). ∎

#198 D Find all the Möbius transformations which map the unit circle $|z| = 1$ to the real axis and the real axis to the imaginary axis.

#199 D† Let A, B be complex numbers, not both zero. Show that the Möbius transformation

$$T(z) = \frac{Az + B}{-\bar{B}z + \bar{A}} \tag{1.49}$$

maps antipodal points to antipodal points. Deduce that T takes images of great circles to images of great circles.

#200 B Let T_1 and T_2 be maps of the form (1.49). Show that both $T_2 \circ T_1$ and $(T_1)^{-1}$ can be put in the same form.

#201 E† Given any map ρ of S, then there is a corresponding map $\tilde{\rho} = \pi \circ \rho \circ \pi^{-1}$ of $\tilde{\mathbb{C}}$. Rotation of S by θ about the x-axis is given by

$$R(x, y, z) = (x, y\cos\theta - z\sin\theta, y\sin\theta + z\cos\theta).$$

Show that $\tilde{R}(w) = (w + it)/(itw + 1)$ where $t = \tan(\theta/2)$.

#202 E† Given every rotation of S can be written as $R_1 \circ R \circ R_2$, where R is a rotation about the x-axis and R_1, R_2 are rotations about the z-axis (#1103), show the rotations of S correspond to the Möbius transformations in (1.49).

#203 B Let a, b, c, d be four distinct points in $\mathbb{C}$. We define their **cross-ratio** as

$$(a, b; c, d) = \frac{(a - c)(b - d)}{(a - d)(b - c)}.$$

(i) Show that $(a, b; c, d) = (b, a; d, c) = (c, d; a, b) = (d, c; b, a)$.
(ii) Show further that $(a, b; d, c) = (a, b; c, d)^{-1}$ and $(a, c; b, d) = 1 - (a, b; c, d)$.

#204 D†(i) Let a, b, c, d be four distinct points in $\mathbb{C}$. Show that a, b, c, d lie on a circline if and only if the cross-ratio $(a, b; c, d)$ is real. How would you make sense of the cross-ratio and this result if $a = \infty$?
(ii) Show that if f is a Möbius transformation then $(f(a), f(b); f(c), f(d)) = (a, b; c, d)$.
(iii) Given two quartets of distinct points z_1, z_2, z_3, z_4 and w_1, w_2, w_3, w_4 show that there is a Möbius transformation which maps each z_i to w_i respectively if and only if $(z_1, z_2; z_3, z_4) = (w_1, w_2; w_3, w_4)$.

#205 D† (For those with some knowledge of bijections.) Calculate the cross-ratio $(a, b; c, d)$ when $a = \infty$, $b = 1$ and $c = 0$. Deduce that if g is a bijection of $\tilde{\mathbb{C}}$ which preserves cross-ratios, then g is a Möbius transformation.

#206 E† Let A, B be real with $A \neq \pm B$. Show that $T(z) = (Az + B)/(Bz + A)$ fixes the unit circle $|z| = 1$. Deduce that, given any two non-intersecting circles, there is a Möbius transformation which maps them to concentric circles.

#207D Let f be a Möbius transformation.

(i) Suppose that f fixes only ∞. Show that f is a translation.

(ii) Suppose that f fixes 0 and ∞. Show that f is a dilation.

(iii) Show that if f fixes precisely one point then there is a Möbius transformation g such that $g^{-1}fg$ is a translation.

(iv) Show that if f fixes precisely two points there is a Möbius transformation g such that $g^{-1}fg$ is a dilation.

#208D† Show that two non-identity Möbius transformations commute if and only if they have the same fixed points.

#209B Show that inversion in a circline is the composition of complex conjugation with a Möbius transformation. Deduce that inversion in a circline is conformal. Deduce also that a composition of two inversions in circlines is a Möbius transformation.

#210D†(i) Show that any translation can be written as the composition of two inversions in circlines.

(ii) Let $a > 0$. Show that the dilation $z \mapsto az$ can be written as the composition of two inversions in circlines.

(iii) Let a be a complex number with $|a| = 1$. Show that the dilation $z \mapsto az$ can be written as the composition of two inversions in circlines.

#211D Show that the only dilations which can be written as the composition of two inversions are the ones described in #210(ii) and (iii).

#212D† If I is inversion in a circline and f is a Möbius transformation, show $f^{-1}If$ is also inversion in a circline.

Exercises on the Hyperbolic Plane

#213B Show that the maps

$$f_{a,\theta}(z) = \frac{\mathrm{cis}\,\theta\,(z-a)}{1-\bar{a}z} \quad \text{where } \theta \text{ is real and } |a| < 1,$$

are Möbius transformations which map the unit circle $|z| = 1$ to itself and a to 0.

#214B Given $f_{a,\theta}$ and $f_{b,\alpha}$ as in the notation of #213 show that both the inverse $f_{a,\theta}^{-1}$ and composition $f_{a,\theta} \circ f_{b,\alpha}$ can be written in the same form.

#215D† Show, further to #213, that any Möbius transformation which maps the unit circle $|z| = 1$ to itself, and the interior to the interior, has the form of $f_{a,\theta}(z)$ for some θ and a.

#216D Show the circline $Az\bar{z} + \bar{B}z + B\bar{z} + C = 0$ intersects the circle $|z| = 1$ at right angles if and only if $A = C$.

#217D† Show, given two points in the hyperbolic plane, that there is a unique hyperbolic line connecting them.

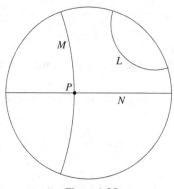

Figure 1.25

Hyperbolic Geometry The Möbius transformations in #213–#215 are important, being the
isometries of the unit disc model $D = \{z : |z| < 1\}$
for the hyperbolic plane due to Poincaré. Angles
in D are measured as usual but the distance
between two points, represented by complex
numbers a and b is given by the formula $d_D(a,b)$
at the end of this box. The hyperbolic plane is
an example of a *Non-Euclidean Geometry*, that
is a geometry which meets all of Euclid's postulates except the *parallel postulate*: this requires
that given any line L and any point P not on L,
there is a unique line through P which doesn't meet L, such a line being called a *parallel*. For two millennia it was considered that Euclid's parallel postulate should be
deducible from his other postulates. It was only in the nineteenth century that mathematicians seriously started considering the possibility that perhaps this wasn't the
case.

Nikolai Lobachevsky (1792–1856) and Janos Bolyai (1802–1860) were among the
first to consider the consequences for geometry if the parallel postulate were not true;
later models due to Eugenio Beltrami (1735–1900) and Henri Poincaré (1854–1912)
showed that non-Euclidean geometry is in fact just as consistent as Euclidean geometry. In the unit disc model D, lines are either diameters of the disc or the arcs of circles
which cut the circumference at right angles. (These 'lines' are still the *geodesics* in
the model; that is the shortest curve between two points in D is still an arc or segment
of a hyperbolic line.) We see that P is a point not on L and that M and N are two
different parallels to L through P – in fact there are infinitely many such parallels.
The geometry of the hyperbolic plane is very different to that of the Euclidean plane.
For example, the sum of a triangle's angles α, β, γ is always less than π radians and
its area is $\pi - \alpha - \beta - \gamma$; in fact similar triangles are automatically congruent as well.
Also through three distinct non-collinear points there need not be a circle (#219). The
circumference of a circle with radius r is $2\pi \sinh r$ and its area is $4\pi \sinh^2 (r/2)$; note
that these formulae approximate to the usual $2\pi r$ and πr^2 for small r. Similarly there
are cosine and sine rules in hyperbolic geometry which approximate to the Euclidean
versions when the triangles involved are small (#221, #222).

$$d_D(a,b) = 2\tanh^{-1} \left| \frac{b-a}{1-\bar{a}b} \right| = \ln \left\{ \frac{|1 - \bar{a}b| + |b - a|}{|1 - \bar{a}b| - |b - a|} \right\}.$$

#218, #219, #220, #221, #222 require knowledge of hyperbolic functions. For further
details see p.364.

#218D Given $f_{a,\theta}$ as in #213 and two points p, q in the hyperbolic plane, show that
$d_D(f_{a,\theta}(p), f_{a,\theta}(q)) = d_D(p,q)$. That is $f_{a,\theta}$ is an isometry of the hyperbolic plane.

#219E† Show that the points of a hyperbolic circle make up a Euclidean circle though the hyperbolic centre may differ from the Euclidean one. Find three non-collinear points in the hyperbolic plane which do not lie on a hyperbolic circle – something which is impossible to do in Euclidean geometry.

#220D The Möbius transformation $g(z) = (z - i)/(z + i)$ is a bijection from the upper half-plane $H = \{z: \mathrm{Im}\,z > 0\}$ to the unit disc D. If distance and angle are measured in H so as to make g an isometry between D and H, show that (i) the lines in H are either half-lines perpendicular to the real axis or semicircles which cut the real axis in right angles and (ii) the distance between two points p, q in H is given by

$$d_H(p,q) = 2\tanh^{-1}\left|\frac{p-q}{p-\bar{q}}\right| = \ln\left\{\frac{|p-q|+|p-\bar{q}|}{|p-q|-|p-\bar{q}|}\right\}.$$

#221E† Let ABC be a hyperbolic triangle. Prove the hyperbolic version of the cosine rule, namely

$$\cosh c = \cosh a \cosh b - \sinh a \sinh b \cos \hat{C},$$

where a, b, c denote the (hyperbolic) lengths of BC, CA, AB. Deduce from this cosine rule the triangle inequality for d_D, that is $d_D(p,q) + d_D(q,r) \geqslant d_D(p,r)$ for p, q, r in D. Show also that if a, b, c are small enough that third and higher degree powers of a, b, c are negligible, then the hyperbolic cosine rule approximates to the usual Euclidean cosine rule.

#222D† Let ABC be a hyperbolic triangle which has a right angle at C. With notation as in #221, prove the hyperbolic sine rule, namely $\sin\hat{A}/\sinh a = \sin\hat{B}/\sinh b = \sin\hat{C}/\sinh c$.

#223E For the hyperbolic triangle described in #221, prove the dual of the hyperbolic cosine rule, namely that

$$\cos\hat{A} = \cosh a \sin\hat{B}\sin\hat{C} - \cos\hat{B}\cos\hat{C}.$$

Deduce that $\hat{A} + \hat{B} + \hat{C} < \pi$. (Though we shall not prove it, the area of a hyperbolic triangle equals $\pi - \hat{A} - \hat{B} - \hat{C}$.)

2

Induction

2.1 Introduction

Mathematical statements can come in the form of a single proposition such as

$$3 < \pi \quad \text{or} \quad 0 < x < y \implies x^2 < y^2,$$

but often they come in a family of statements such as

A $2^x > 0$ for all real numbers x;

B $1 + 2 + \cdots + n = n(n+1)/2$ for $n = 1, 2, 3, \ldots$;

C $\displaystyle\int_0^\pi \sin^{2n} \theta \, d\theta = \frac{(2n)!}{(n!)^2} \frac{\pi}{2^{2n}}$ for $n = 0, 1, 2, 3, \ldots$;

D $2n$ can be written as the sum of two primes for all $n = 2, 3, 4, \ldots$.

    Induction (or more precisely *mathematical induction* [1]) is a particularly useful method of proof for dealing with families of statements, such as the last three statements above, which are indexed by the natural numbers, the integers or some subset of them. We shall prove both statements B and C using induction (Examples 2.1 and 2.8). Statements B and C can be approached with induction because, in each case, knowing that the nth statement is true helps enormously in showing that the $(n+1)$th statement is true – this is the crucial idea behind induction. Statement D, on the other hand, is a famous open problem known as *Goldbach's conjecture*. [2] If we let $D(n)$ be the statement that $2n$ can be written as the sum of two primes, then it is currently known that $D(n)$ is true for $n \leqslant 4 \times 10^{18}$ (see Silva, Herzog and Pardi [28]). What makes statement D different from B and C, and more intractable to induction, is that in trying to verify $D(n+1)$ we can't generally make much use of knowledge of $D(n)$ and so we can't build towards a proof. For example, we can verify $D(19)$ and $D(20)$ by noting that

$$38 = 7 + 31 = 19 + 19 \quad \text{and} \quad 40 = 3 + 37 = 11 + 29 = 17 + 23.$$

Here, knowing that 38 can be written as a sum of two primes is no great help in verifying that 40 can be, as none of the primes we might use for the latter was previously used in splitting 38.

---

[1] The term 'induction' also refers to inductive reasoning, where a universal conclusion is reached on the basis of a wealth of examples, even though the many examples don't guarantee a general truth. Mathematical induction is in fact a type of deductive reasoning.

[2] Christian Goldbach (1690–1764), who had been a professor of mathematics at St. Petersburg, made this conjecture in a letter to Euler in 1742 and it is still unproved.

By way of an example, we shall prove statement *B* by induction before giving a formal definition of just what induction is.

Example 2.1 For any integer $n \geqslant 1$,

$$1 + 2 + \cdots + n = n(n+1)/2. \tag{2.1}$$

(Whilst you may not explicitly have seen this formula before, it is just an instance of the formula for the sum of an arithmetic progression.)

#224
#227
#229

Solution For any integer $n \geqslant 1$, let $B(n)$ be the statement in (2.1). We shall prove two facts:

(a) $B(1)$ is true, and

(b) for any $n \geqslant 1$, if $B(n)$ is true then $B(n+1)$ is also true.

The first fact is the easy part, as we just need to note that

$$\text{LHS of } B(1) = 1 = 1 \times (1+1) \div 2 = \text{RHS of } B(1).$$

To verify (b) we need to prove *for each* $n \geqslant 1$ that $B(n+1)$ is true *assuming* $B(n)$ to be true. Now

$$\text{LHS of } B(n+1) = 1 + \cdots + n + (n+1).$$

But, assuming $B(n)$ to be true, we know the terms from 1 through to n add up to $n(n+1)/2$. *This is always a crucial part of a proof by induction, recognizing how knowledge of the nth case helps with the $(n+1)$th case.* So

$$\text{LHS of } B(n+1) = n(n+1)/2 + (n+1) = (n+1)(n/2+1) = (n+1)(n+2)/2$$
$$= \text{RHS of } B(n+1).$$

This verifies (b). □

Definition 2.2 The first statement to be proved ($B(1)$ in Example 2.1) is known as the **initial case** or **initial step.** Showing the $(n+1)$th statement follows from the nth statement is called the **inductive step**, and the assumption that the nth statement is true is called the **inductive hypothesis**.

Be sure that you understand Example 2.1: it contains the important steps common to any proof by induction.

• In particular, note in the final step that we retrieved our original formula of $n(n+1)/2$, but with $n+1$ now replacing n everywhere; *this was the expression that we always had to be working towards!*

By induction we now know that *B* is true, i.e. that $B(n)$ is true for any $n \geqslant 1$. How does this work? Well, suppose we wanted to be sure $B(3)$ is correct – we have just verified (amongst other things) the following three statements:

(i) If $B(1)$ is true.

(ii) If $B(1)$ is true then $B(2)$ is true.

(iii) If $B(2)$ is true then $B(3)$ is true.

Putting the three together, we then see that $B(3)$ is true. The first statement tells us that $B(1)$ is true and the second two are stepping stones, first to the truth about $B(2)$, and then on to proving $B(3)$. A similar chain of logic can be made to show that any $B(n)$ is true.

Formally, then, the principle of induction is as follows:

Theorem 2.3 (***The Principle of Induction***) *Let $P(n)$ be a family of statements indexed by the natural numbers $n = 0, 1, 2, \dots$. Suppose that*

- *(Initial Step) $P(0)$ is true.*
- *(Inductive Step) For any $n \geqslant 0$, if $P(n)$ is true then $P(n+1)$ is also true.*

Then $P(n)$ is true for all $n \geqslant 0$.

It is not hard to see how we might amend the hypotheses of Theorem 2.3 to show:

Corollary 2.4 *Let N be an integer and let $P(n)$ be a family of statements for $n \geqslant N$. Suppose that*

- *(Initial Step) $P(N)$ is true.*
- *(Inductive Step) For any $n \geqslant N$, if $P(n)$ is true then $P(n+1)$ is also true.*

Then $P(n)$ is true for all $n \geqslant N$.

In fact, we have already employed this corollary, with $N = 1$, to prove statement B.

Proof (of Theorem 2.3) Let S denote the set of all those natural numbers n for which $P(n)$ is *false*. We aim to show that S is empty, i.e. that no $P(n)$ is false, so that each $P(n)$ is true. Suppose, for a contradiction, that S is non-empty. Any non-empty set of natural numbers has a smallest element; denote the smallest element of S as m. As m is in S, then $P(m)$ is false; as $P(0)$ is true it can't be the case that $m = 0$ and so $m \geqslant 1$.

#233
#234

Consider $m - 1$. As $m \geqslant 1$ then $m - 1$ is a natural number. Further, as $m - 1$ is smaller than the smallest element of S, then $m - 1$ is not in S and hence $P(m-1)$ is true. But, as $P(m-1)$ is true, then the inductive step shows that $P(m)$ is also true. So $P(m)$ is both true and false, which is our required contradiction. If S being non-empty leads to a contradiction, the only alternative is that S is empty. That is, no $P(n)$ is false and so every $P(n)$ is true. □

This proof may appear a little abstract at first but the essence of it is simply this: if some statement on the list were false there would be a first such statement, which means the previous statement must be true but induction would then mean that our first false statement was in fact true.

Proof (of Corollary 2.4) The corollary follows by applying Theorem 2.3 to the statements $Q(n) = P(n+N)$, $n \geqslant 0$. □

Here is another version of induction, which is usually referred to as the *strong form of induction*. In some problems, the inductive step might depend on some earlier case but not necessarily the immediately preceding case – such a case is Example 2.13.

Theorem 2.5 (***Strong Form of Induction***) *Let $P(n)$ be a family of statements for $n \geqslant 0$. Suppose that*

- *(Initial Step) $P(0)$ is true.*
- *(Inductive Step) For any $n \geqslant 0$, if $P(0), P(1), \dots, P(n)$ are all true then so is $P(n+1)$.*

Then $P(n)$ is true for all $n \geqslant 0$.

#230 **Proof** For $n \geqslant 0$, let $Q(n)$ be the statement '$P(k)$ is true for $0 \leqslant k \leqslant n$'. The assumptions of Theorem 2.5 are equivalent to (i) $Q(0)$ (which is the same as $P(0)$) is true and (ii) if $Q(n)$ is true then $Q(n+1)$ is true. By induction (as stated in Theorem 2.3) we know that $Q(n)$ is true for all n. As $P(n)$ is a consequence of $Q(n)$, then $P(n)$ is true for all n. □

To reinforce the need for proof, and to show how patterns can at first glance deceive us, consider the following example. Take two points on the circumference of a circle and take a line joining them; this line then divides the circle's interior into two regions. If we take three points on the circumference then the lines joining them will divide the disc into four regions. Four points can result in a maximum of eight regions (see Figure 2.1). Surely, then, we can confidently predict that n points will maximally result in 2^{n-1} regions.

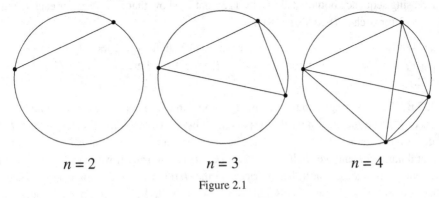

$n = 2$ $\qquad\qquad$ $n = 3$ $\qquad\qquad$ $n = 4$

Figure 2.1

Further investigation shows our conjecture to be true for $n = 5$ but, to our surprise, when we take six points on the circle, the maximum number of regions attained is 31, no matter how we choose the points. Indeed the maximum number of regions attained from n points on the circumference is given by the formula

$$\frac{1}{24}(n^4 - 6n^3 + 23n^2 - 18n + 24). \tag{2.2}$$

Our original guess was way out! This is known as *Moser's circle problem.*[3] (See #267 and #343).

Exercises

#224 A Let a and r be real numbers with $r \neq 1$. Prove, by induction, that

$$a + ar + ar^2 + \cdots + ar^{n-1} = a(r^n - 1)/(r - 1) \qquad \text{for } n \geqslant 1.$$

#225 a(i) Use the formula from statement B to show that the sum of an arithmetic progression with initial value a, common difference d and n terms is $n(2a + (n-1)d)/2$. (ii) Rederive the result of part (i) directly using induction.

#226 c Show by induction that the sum of an arithmetic progression is $n(a+l)/2$, where a and l are the first and last terms and n is the number of terms.

[3] After the Canadian mathematician Leo Moser (1921–1970).

#227a Prove **Bernoulli's inequality** [4]: $(1+x)^n \geqslant 1+nx$ for real $x \geqslant -1$ and integers $n \geqslant 1$.

#228a Assuming only the product rule, show, for $n \geqslant 1$, that the derivative of x^n equals nx^{n-1}.

#229B†(i) Prove by induction, for $n \geqslant 2$, that $\sum_{r=2}^{n} \frac{1}{r^2-1} = \frac{(n-1)(3n+2)}{4n(n+1)}$.
(ii) Now determine the sum in (i) directly using the identity $\frac{1}{r^2-1} = \frac{1}{2}\left\{\frac{1}{r-1} - \frac{1}{r+1}\right\}$.

#230A† Show that every integer $n \geqslant 1$ can be written as $n = 2^k l$, where k, l are natural numbers and l is odd.

#231C The sequence z_n is defined as $z_0 = 0$ and $z_{n+1} = (z_n)^2 + 1/4$. Show that z_n is an increasing sequence bounded above by $1/2$; that is, show that $z_{n+1} \geqslant z_n$ for each n and that $z_n \leqslant 1/2$ for each n. (See also #181.)

#232D† Let $-2 \leqslant c \leqslant 1/4$. The sequence $z_n(c)$ is defined recursively by $z_0(c) = 0$ and $z_{n+1}(c) = z_n(c)^2 + c$. Show that $|z_n(c)| \leqslant \beta$ for all n where $\beta = (1 + \sqrt{1-4c})/2$. (See #182.)

#233B What is wrong with the following 'proof' that all people are of the same height? [5]

'Let $P(n)$ be the statement that n persons must be of the same height. Clearly $P(1)$ is true, as a person is the same height as him-herself. Suppose now that $P(k)$ is true for some natural number k and we shall prove that $P(k+1)$ is also true. If we have $k+1$ people then we can invite one person to leave briefly so that k remain – as $P(k)$ is true we know that these people must all be equally tall. If we invite back the missing person and someone else leaves, then these k persons are also of the same height. Hence the $k+1$ persons are all of equal height and so $P(k+1)$ follows. By induction everyone is of the same height.'

#234B Below are several different families of statements $P(n)$, indexed by the natural numbers, which satisfy rules that are similar (but not identical) to the hypotheses required for induction. In each case, indicate for which n the truth of $P(n)$ must necessarily follow from the given rules.

(i) $P(0)$ is true; for $n \geqslant 0$, if $P(n)$ is true then $P(n+2)$ is true.
(ii) $P(1)$ is true; for $n \geqslant 0$, if $P(n)$ is true then $P(2n)$ is true.
(iii) For $n \geqslant 0$, if $P(n)$ is true then $P(n+3)$ is true.
(iv) $P(0)$ and $P(1)$ are true; for $n \geqslant 0$, if $P(n)$ is true then $P(n+2)$ is true.
(v) $P(0)$ and $P(1)$ are true; for $n \geqslant 0$, if $P(n)$ and $P(n+1)$ are true then $P(n+2)$ is true.
(vi) $P(0)$ is true; for $n \geqslant 0$, if $P(n)$ is true then $P(n+2)$ and $P(n+3)$ are both true.
(vii) $P(0)$ is true; for $n \geqslant 0$, if $P(n+1)$ is true then $P(n+2)$ is true.

#235B† Determine the first digits in the powers of 2^n for $0 \leqslant n \leqslant 9$. Show that this pattern recurs for $10 \leqslant n \leqslant 19$, for $20 \leqslant n \leqslant 29$ and for $30 \leqslant n \leqslant 39$. When does this pattern first break down? What is the reason behind the pattern emerging in the first place?

---

[4] Named after Jacob Bernoulli; see p.124 for a brief biography.
[5] The first example of this type of (incorrect) induction was due to the Hungarian mathematician George Polya (1887–1985).

Remark 2.6 In [16] and [17] Richard Guy humorously proposes two informal *Strong Laws of Small Numbers* (by way of contrast with the rigorous *Strong Law of Large Numbers*). His point is that there are few small numbers and many naturally occurring integer sequences. In these papers appear many examples where a familiar pattern seemingly emerges; the reader is invited to determine whether the sequence does indeed continue forever as expected or whether the initial familiarity is deceiving and the guessed pattern goes awry at some later stage. ∎

2.2 Examples

On a more positive note, many of the patterns found in mathematics won't trip us at some later stage and here are some further examples of proof by induction.

#237
#240
#244
#249

Example 2.7 Show that n lines in the plane, no two of which are parallel and no three meeting in a point, divide the plane into $n(n+1)/2+1$ regions.

Solution

When we have no lines in the plane then clearly we have just one region, as expected from putting $n=0$ into the formula $n(n+1)/2+1$.

Suppose now that we have n lines dividing the plane into $n(n+1)/2+1$ regions and we add a $(n+1)$th line. This extra line meets each of the previous n lines because, by assumption, it is parallel with none of them. Also, it meets each of these n lines in a distinct point, as we have assumed that no three lines are concurrent. These n points of intersection divide the new line into $n+1$ segments. For each of these $n+1$ segments there are now two regions, one on either side of the seg-

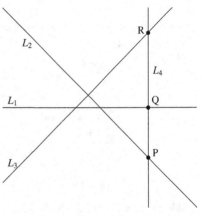

Figure 2.2

ment, where previously there had been only one region. So by adding this $(n+1)$th line we have created $n+1$ new regions. For example, in Figure 2.2, the four segments, 'below P', PQ, QR and 'above R' on the fourth line L_4, divide what were four regions previously into eight new ones. In total, then, the number of regions we now have is

$$(n(n+1)/2+1)+(n+1)=(n+1)(n+2)/2+1.$$

So the given formula is correct for the number of regions with $n+1$ lines; the result follows by induction. □

Example 2.8 Prove, for $n \geqslant 0$, that

$$\int_0^\pi \sin^{2n}\theta \, d\theta = \frac{(2n)!}{(n!)^2} \frac{\pi}{2^{2n}}. \tag{2.3}$$

#255
#262
#263

Solution Let's denote the integral on the LHS of equation (2.3) as I_n. The value of I_0 is easy to calculate, because the integrand is just 1, and so $I_0 = \pi$. We also see in (2.3) that $\text{RHS}(n=0) = (0!/(0!)^2)(\pi/2^0) = \pi$, verifying the initial case.

We now prove a *reduction formula* (see also §5.3) connecting I_n and I_{n+1}, so that we can use this in our induction.

$$I_{n+1} = \int_0^\pi \sin^{2(n+1)}\theta\, d\theta = \int_0^\pi \sin^{2n+1}\theta \times \sin\theta\, d\theta$$

$$= \left[\sin^{2n+1}\theta \times (-\cos\theta)\right]_0^\pi - \int_0^\pi (2n+1)\sin^{2n}\theta\cos\theta \times (-\cos\theta)\, d\theta$$

[using integration by parts]

$$= 0 + (2n+1)\int_0^\pi \sin^{2n}\theta(1-\sin^2\theta)\, d\theta \qquad [\text{using } \cos^2\theta = 1 - \sin^2\theta]$$

$$= (2n+1)\int_0^\pi (\sin^{2n}\theta - \sin^{2(n+1)}\theta)\, d\theta$$

$$= (2n+1)(I_n - I_{n+1}).$$

Rearranging gives the *reduction formula*

$$I_{n+1} = \left(\frac{2n+1}{2n+2}\right)I_n.$$

As our inductive hypothesis, suppose that (2.3) gives the right value of I_n for some natural number n. Then

$$I_{n+1} = \frac{2n+1}{2n+2} \times I_n \qquad\qquad\qquad [\text{by the reduction formula}]$$

$$= \frac{2n+1}{2(n+1)} \times \frac{(2n)!}{(n!)^2} \times \frac{\pi}{2^{2n}} \qquad\qquad [\text{by the inductive hypothesis}]$$

$$= \frac{2n+2}{2(n+1)} \times \frac{2n+1}{2(n+1)} \times \frac{(2n)!}{(n!)^2} \times \frac{\pi}{2^{2n}}$$

$$= \frac{(2n+2)!}{((n+1)!)^2} \times \frac{\pi}{2^{2(n+1)}},$$

which equals the RHS of (2.3) with n replaced by $n + 1$. The result follows by induction. □

Example 2.9 Show for $n, k \geqslant 1$, that

$$\sum_{r=1}^n r(r+1)(r+2)\cdots(r+k-1) = \frac{n(n+1)(n+2)\cdots(n+k)}{k+1}. \qquad (2.4)$$

#246
#261 **Remark 2.10** This problem differs from our earlier examples in that our family of statements now involves two variables n and k, rather than just the one variable. If we write $P(n,k)$ for the statement in (2.4) then we can use induction to prove all of the statements $P(n,k)$ in various ways:

- We could prove $P(1,1)$ and show how $P(n+1,k)$ and $P(n,k+1)$ both follow from $P(n,k)$ for $n,k \geqslant 1$.
- We could prove $P(1,k)$ *for all* $k \geqslant 1$ and show that knowing $P(n,k)$ *for all* k leads to the truth of $P(n+1,k)$ *for all* k – this reduces the problem to one application of induction in n to a family of statements at a time.

• We could prove $P(n,1)$ *for all* $n \geqslant 1$ and show how knowing $P(n,k)$ *for all* n leads to proving $P(n,k+1)$ *for all* n – in a similar fashion to the previous method, now inducting through k and treating n as arbitrary.

What these different approaches rely on is that all the possible pairs (n,k) are somehow linked to our initial pair (or pairs). Let

$$S = \{(n,k): n,k \geqslant 1\}$$

be the set of all possible pairs (n,k). The first method of proof uses the fact that the only subset T of S satisfying the properties

(i) $(1,1)$ is in T.
(ii) if (n,k) is in T then $(n,k+1)$ is in T.
(iii) if (n,k) is in T then $(n+1,k)$ is in T,

is S itself. The second and third bullet points above rely on the fact that the whole of S is the only subset having similar properties. (See #265.) ∎

Solution (of Example 2.9) In this case the second method of proof seems easiest; that is, we will prove that $P(1,k)$ holds for all $k \geqslant 1$ and show that assuming the statements $P(N,k)$, for a particular N and all k, is sufficient to prove the statements $P(N+1,k)$ for all k. First we note

$$\text{LHS of } P(1,k) = 1 \times 2 \times \cdots \times k \quad \text{and}$$

$$\text{RHS of } P(1,k) = \frac{1 \times 2 \times \cdots \times (k+1)}{k+1} = 1 \times 2 \times \cdots \times k$$

are equal, proving $P(1,k)$ for all $k \geqslant 1$. Now if $P(N,k)$ holds true, for particular N and all $k \geqslant 1$, we have

$$\text{LHS of } P(N+1,k) = \sum_{r=1}^{N+1} r(r+1)(r+2)\cdots(r+k-1)$$

$$= \frac{N(N+1)\ldots(N+k)}{k+1} + (N+1)(N+2)\cdots(N+k) \quad \text{[by hypothesis]}$$

$$= (N+1)(N+2)\cdots(N+k)\left(\frac{N}{k+1}+1\right)$$

$$= \frac{(N+1)(N+2)\cdots(N+k)(N+k+1)}{k+1}$$

$$= \text{RHS of } P(N+1,k),$$

proving $P(N+1,k)$ simultaneously for each k. This verifies all that is required for the second method. □

Let n be a positive integer and consider the set $\{1,2,\ldots,n\}$. By a **subset** of $\{1,2,\ldots,n\}$ we mean a collection of some, but not necessarily all or any, of the elements in $\{1,2,\ldots,n\}$.

Example 2.11 List the subsets of $\{1,2,3\}$.

Solution There are eight subsets of $\{1,2,3\}$, each listed below.

$$\{1,2,3\}, \quad \{1,2\}, \quad \{1,3\}, \quad \{2,3\}, \quad \{1\}, \quad \{2\}, \quad \{3\}, \quad \varnothing.$$

Most of them will be natural enough, though it may seem surprising that the **empty set**, ∅, the subset with no elements, is a permitted subset.

Note that there are 2^3 subsets of $\{1,2,3\}$. This isn't that surprising if we note that each subset of $\{1,2,3\}$ is entirely determined by how it would 'respond' to the following 'questionnaire':

Is 1 an element of you?   Is 2 an element of you?   Is 3 an element of you?

Each element can be independently in a subset or not, irrespective of what else the subset may contain. Three independent binary (i.e. yes or no) decisions lead to 2^3 possible sets of responses. Respectively, the eight subsets above correspond to the series of questionnaire responses

YYY,   YYN,   YNY,   NYY,   YNN,   NYN,   NNY,   NNN.   □

If we interpret the subsets of $\{1,2,\ldots,n\}$ as the possible responses to n yes-or-no questions then it is not surprising that there are 2^n subsets. Nonetheless we give a careful proof of this fact using induction.

Proposition 2.12 *(Subsets of a finite set)* Show that there are 2^n subsets of the set $\{1,2,\ldots,n\}$.

Proof The subsets of $\{1\}$ are ∅ and $\{1\}$, thus verifying the proposition for $n = 1$.

Suppose now that the proposition holds for a particular n and consider the subsets of $\{1,2,\ldots,n,n+1\}$. Such subsets come in two, mutually exclusive, varieties: they either contain the new element $n+1$ or they don't.

The subsets of $\{1,2,\ldots,n,n+1\}$ which *don't* include $n+1$ are precisely the subsets of $\{1,2,\ldots,n\}$ and by hypothesis there are 2^n of these. The subsets of $\{1,2,\ldots,n,n+1\}$ which *do* include $n+1$ are the previous 2^n subsets of $\{1,2,\ldots,n\}$ together with the new element $n+1$ included in each of them; including $n+1$ in these 2^n distinct subsets still leads to 2^n distinct subsets. So in all we have $2^n + 2^n = 2^{n+1}$ subsets of $\{1,2,\ldots,n,n+1\}$, completing the inductive step.   □

We end this section with one example which makes use of the strong form of induction (Theorem 2.5). Recall that a natural number $n \geqslant 2$ is called **prime** if the only natural numbers which divide it are 1 and n. (Note that 1 is not considered prime.) The list of prime numbers begins $2,3,5,7,11,13,\ldots$ and has been known to be infinite since the time of Euclid (see p.79). The prime numbers are, in a sense, the atoms of the natural numbers under multiplication, as every natural number $n \geqslant 2$ can be written as a product of primes in an essentially unique way – this fact is known as the **fundamental theorem of arithmetic**. Here we just prove the *existence* of such a product.

Example 2.13 Every natural number $n \geqslant 2$ can be written as a product of (one or more) prime numbers.

Proof Our initial step is taken care of as 2 is prime. As our inductive hypothesis, we assume that every number k in the range $2 \leqslant k \leqslant n$ can be written as a product of one or more prime numbers. We need to show $n+1$ is a product of prime numbers. If $n+1$ is prime then we are done, as it is then a product of just one prime number. If $n+1$ is not prime, then it has a factor m in the range $2 \leqslant m \leqslant n$ which divides $n+1$. Note also that $2 \leqslant (n+1)/m \leqslant n$ as m is at least 2. So, by our inductive hypothesis, we know that both m and $(n+1)/m$ are products of prime numbers and we can write

$$m = p_1 \times p_2 \times \cdots \times p_r, \qquad \frac{n+1}{m} = P_1 \times P_2 \times \cdots \times P_R,$$

where $p_1, \ldots, p_r$ and $P_1, \ldots, P_R$ are prime numbers. Finally we have that

$$n+1 = m \times \left(\frac{n+1}{m}\right) = p_1 \times p_2 \times \cdots \times p_r \times P_1 \times P_2 \times \cdots \times P_R,$$

showing $n+1$ to be a product of primes. The result follows using the strong form of induction. $\qquad\square$

Exercises

#236 A† Show, by induction, that $n^2 + n \geqslant 42$ when $n \leqslant -7$.

#237 A Prove that $(2n+1) + (2n+3) + \cdots + (4n-1) = 3n^2$ for $n \geqslant 1$, and then rederive this result using (2.1).

#238 a Prove for $n \geqslant 1$ that

$$\underbrace{\sqrt{2 + \sqrt{2 + \sqrt{2 + \cdots + \sqrt{2}}}}}_{n \text{ root signs}} = 2\cos\frac{\pi}{2^{n+1}}.$$

#239 B† Given n positive numbers $x_1, x_2, \ldots, x_n$ such that $x_1 + x_2 + \cdots + x_n \leqslant 1/3$, prove by induction that

$$(1 - x_1)(1 - x_2) \times \cdots \times (1 - x_n) \geqslant 2/3.$$

#240 B The nth **Fermat number**[6] is $F_n = 2^{2^n} + 1$. Show that $F_0 F_1 F_2 F_3 \cdots F_{n-1} = F_n - 2$.

#241 B Show, for $n \geqslant 1$, that $\sum_{r=1}^{n} \frac{1}{r^2} \leqslant 2 - \frac{1}{n}$.

#242 b† Prove, for $n \geqslant 1$, that

$$\sqrt{n} \leqslant \sum_{k=1}^{n} \frac{1}{\sqrt{k}} \leqslant 2\sqrt{n} - 1.$$

#243 b Let $S(n) = 0 + 1 + 4 + 9 + \cdots + n^2$ for $n \geqslant 0$. Show that there is a unique cubic $f(n)$ such that $f(n) = S(n)$ for $n = 0, 1, 2, 3$. Prove by induction that $f(n) = S(n)$ for $n \geqslant 0$.

---

[6] A number of the form $2^k + 1$ can be prime only if k is a power of 2. The French mathematician Pierre de Fermat (see p.93 for a biographical note) conjectured that the Fermat numbers are all prime. This is true for $0 \leqslant n \leqslant 4$ but Euler later showed that F_5 is divisible by 641 and in fact no further prime Fermat numbers have been found. A regular m-sided polygon is constructible using ruler and compass if and only $m = 2^a p_1 \cdots p_r$ where the p_i are distinct Fermat primes. So the equilateral triangle, square, pentagon, hexagon are all constructible. Gauss, in 1796, was the first to construct a 17-sided regular polygon.

#244 A Show that

$$n + 3\sum_{r=1}^{n} r + 3\sum_{r=1}^{n} r^2 = \sum_{r=1}^{n} \left\{ (r+1)^3 - r^3 \right\} = (n+1)^3 - 1.$$

Hence, using the formula from (2.1), find an expression for $\sum_{r=1}^{n} r^2$.

#245 b Extend the method of #244 to find expressions for $\sum_{r=1}^{n} r^3$ and $\sum_{r=1}^{n} r^4$.

#246 D† (See also #467.) Let k be a natural number. Deduce from Example 2.9 that

$$\sum_{r=1}^{n} r^k = \frac{n^{k+1}}{k+1} + E_k(n), \tag{2.5}$$

where $E_k(n)$ is a polynomial in n of degree at most k. What are $E_1(n)$ and $E_2(n)$?

Definition 2.14 Let x be a real number. The **integer part** or **floor** of x, denoted $\lfloor x \rfloor$, is the largest integer less than or equal to x. For example, $\lfloor \pi \rfloor = 3$, $\lfloor -\pi \rfloor = -4$, $\lfloor 3 \rfloor = 3$.

#247 D† (i) By considering areas under the graph of $y = \sqrt{x}$, prove $\sum_{k=1}^{n} \sqrt{k} \leqslant \frac{2}{3}(n+1)^{3/2}$ for $n \geqslant 1$.
(ii) Show that

$$\sum_{k=1}^{n} \lfloor \sqrt{k} \rfloor = (n+1)N - \frac{N(N+1)(2N+1)}{6} \qquad \text{where } N = \lfloor \sqrt{n} \rfloor.$$

(iii) Deduce that for large n

$$\sum_{k=1}^{n} \lfloor \sqrt{k} \rfloor \Big/ \sum_{k=1}^{n} \sqrt{k} \qquad \text{approximately equals 1.}$$

#248 D† Let x and b be integers with $x \geqslant 0$ and $b \geqslant 2$. Show that there are unique integers $a_0, a_1, a_2, \ldots$ with $0 \leqslant a_i < b$ such that

$$x = a_0 + a_1 b + a_2 b^2 + \cdots.$$

This is called the **expansion of x in base b**. When $b = 2$ this is **binary** and when $b = 10$ this is **decimal**.

Definition 2.15 Given n positive numbers $x_1, x_2, \ldots, x_n$ their **arithmetic mean** A, **geometric mean** G, **harmonic mean** H and **quadratic mean** (or **root mean square**) Q are defined by

$$A = \frac{x_1 + x_2 + \cdots + x_n}{n}; \qquad G = \sqrt[n]{x_1 x_2 \cdots x_n};$$

$$H = \frac{n}{\frac{1}{x_1} + \frac{1}{x_2} + \cdots + \frac{1}{x_n}}; \qquad Q = \sqrt{\frac{x_1^2 + \cdots + x_n^2}{n}}.$$

It is generally the case that $H \leqslant G \leqslant A \leqslant Q$ as shown in the following exercises.

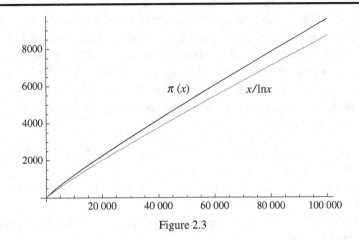

Figure 2.3

Prime Numbers The concept of a prime number is fundamental to much of mathematics, particularly in number theory (the study of the integers), but the notion of primality generalizes to many other important algebraic structures. Yet despite the elementary understanding needed to grasp what a prime number is, it is frustratingly easy to ask very difficult and even unanswered questions about the primes.

There are infinitely many prime numbers $2,3,5,7,11,\ldots$. You might wonder why 1 is not included on the list and the answer largely rests with the **fundamental theorem of arithmetic**. This says that any positive integer $n \geqslant 2$ can be written uniquely as a product of prime numbers. For example, $57 = 3 \times 19$. As multiplication commutes then $57 = 19 \times 3$ but we consider this to be essentially the same decomposition. If 1 were a prime number then 57 would have infinitely many different decompositions $1^k \times 3 \times 19$ where $k \geqslant 0$, as would any other n.

Assuming the fundamental theorem, it is relatively easy to show that there are infinitely many primes with a proof by contradiction. Suppose that there are only finitely many primes, say $p_1, p_2, \ldots, p_k$ and set $N = p_1 \times p_2 \times \cdots \times p_k + 1$. Note that when we divide N by any of the p_i there is a remainder of 1. By the fundamental theorem this means that either N is a new prime not on our list, or the prime factors of N are new primes not on our list. Either way, our assumption that we had listed all the primes was a false one and it must be the case that there aren't finitely many primes. This proof first appears in Book IX of Euclid's *Elements*, which dates from around 300 B.C..

Despite their seemingly sporadic nature there is some long-term regularity to the occurrence of prime numbers. This was first noted by Legendre and Gauss at the end of the eighteenth century. They estimated that $\pi(x)$, the number of primes less than x, is asymptotically $x/\ln x$. That is, $\pi(x) \ln x / x$ approximates to 1 as x becomes increasingly large. A sketch of these two graphs for $x < 10^5$ appears in Figure 2.3. This estimate was independently shown to be true in 1896 by Jacques Hadamard (1865–1963) and Charles Jean De La Vallée-Poussin (1866–1962). This result is known as the **prime number theorem**, which shows that, in the long term, primes occur less often with a certain regularity. Despite this, it is conjectured that there are infinitely many pairs of consecutive odd numbers that are prime, whilst at the same time it is easy to show that there are arbitrarily long gaps amongst the primes – e.g. for any $n \geqslant 2$, none of the numbers $n!+2, n!+3, \ldots, n!+n$ is prime.

The primes remain a source of fascinating and difficult mathematics. Here are some important known results:

- (Fermat, 1640) A prime which is 1 more than a multiple of 4 can be written as a sum of two squares.
- (Euler, 1737) The series $\sum 1/p$ diverges, where the sum is taken over all prime numbers p.
- (Dirichlet, 1837) Let a, b be coprime natural numbers. There are infinitely many primes of the form $ak+b$, $k \geqslant 1$.
- (Chebyshev 1848) Bertrand's postulate: For $n \geqslant 2$ there is a prime number p in the range $n < p < 2n$.
- (Brun, 1919) The series $\sum 1/p$ converges, where the sum is taken over all twin primes p such that $p+2$ is prime.
- (Chen, 1973) Every sufficiently large even number can be written as the sum of a prime and the product of at most two primes.
- (Green and Tao, 2004) There are arithmetic progressions amongst the prime numbers of any given length.
- (Zhang, 2013) There is an integer N such that infinitely many primes differ by less than N.

and here are some important open problems:

- (Goldbach's conjecture) Every even number $n \geqslant 4$ can be written as the sum of two prime numbers.
- (Twin prime conjecture) There are infinitely many pairs of consecutive odd numbers which are both prime.
- There are infinitely many Mersenne primes — primes of the form $M_p = 2^p - 1$ where p is prime.
- (Landau's problem) There are infinitely many primes of the form $n^2 + 1$.
- (Legendre's conjecture) There is always some prime number p within the interval $n^2 < p < (n+1)^2$ where $n \geqslant 1$.
- (Sophie Germain primes) Are there infinitely many primes p such that $2p+1$ is prime?

To give a sense of how a conjecture may well look highly plausible, only for a large counter-example to be found, consider Polya's conjecture. This states that for any given n, most numbers less than n (i.e. more than half of them) have an odd number of factors. The first n for which this is not true is $906,150,257$.

Carl Gauss (Photo by Fine Art Images/Heritage Images/Getty Images)

Unique Factorization Domains

Once properly understood, we see that a version of the fundamental theorem of arithmetic holds in many algebraic settings. It is likely you will have met *polynomial factorization*. The fundamental theorem of algebra guarantees that a polynomial can be factorized as a product of linear factors with complex coefficients, but to what extent must this factorization be unique? We can note, for example, that

$$2x^2 - 6x + 4 = (2x - 2)(x - 2) = (x - 1)(2x - 4).$$

Are these essentially distinct factorizations though? The factors are, after all, just $(x - 1)$ and $(x - 2)$ with an extra factor of 2 thrown in somewhere. So if we consider a polynomial and any non-zero constant multiple of it as essentially the same, it seems that we still may have unique factorization. Likewise, if we were to note in the integers that $6 = 2 \times 3 = (-2) \times -3$, then these can be seen as the same products if we consider a number and its negative as essentially the same factor. The non-constant polynomials in the first example, and ± 1 in the second, are so-called *units*. A unit is an element u such that $uv = 1$ for some element v. Both the complex polynomials and the integers are examples of *unique factorization domains* (UFDs). In a UFD any element can be written as a product of prime elements and such a factorization is unique up to the order of the factors and the positioning of some units.

How should we generalize our definition of what is meant by a *prime element*? We say, more generally, that p is prime if whenever p divides a product ab, then p must divide either a or b (or both). This replicates the atomic nature of prime integers – if p appears in a product, it must wholly appear in at least one of the factors. This new definition generalizes more readily to other algebraic settings, but with a little work it can be shown that this definition of a prime element concurs with our original idea of a prime number being divisible only by 1 and itself.

When working with complex polynomials, the prime elements are the linear factors. With real polynomials, the primes are the linear factors and quadratic factors with complex conjugate roots (Corollary 1.17). With the integers, the prime elements are the prime numbers (in the usual sense) and their negatives. There are other important examples of UFDs such as the *Gaussian integers*, complex numbers of the form $m + ni$ where m, n are integers, first studied by Carl Gauss (1777–1855). The units in this case are $1, -1, i, -i$. Such examples are important to *algebraic number theory*. But, importantly, some examples of this kind are not UFDs. If one considers complex numbers of the form $m + n\sqrt{-5}$, where m, n are integers, then $6 = 2 \times 3 = (1 + \sqrt{-5})(1 - \sqrt{-5})$ are two *essentially different* factorizations of 6. In 1847, Gabriel Lamé claimed to have proved Fermat's Last Theorem, which states that the equation $x^n + y^n = z^n$ has no solutions amongst the positive integers when $n > 2$. His method ultimately failed because he had assumed unique factorization properties for certain types of complex number. The German mathematician Eduard Kummer (1810–1893) had already appreciated that this was not generally the case. However, he was able to adapt Lamé's proof to the so-called *regular* primes and thus prove Fermat's Last Theorem for a wide range of exponents – for example, his work addressed all powers up to 100 except for 37, 59, 67, 74.

#249 B (AM–GM inequality) (i) Show directly for $n = 2$ that $A \geqslant G$ with equality if and only if $x_1 = x_2$.

(ii) Let $x_1, \ldots, x_{n+1}$ be $n + 1$ positive numbers with arithmetic mean μ and assume that $x_n < \mu < x_{n+1}$. Set

$$X_1 = x_n + x_{n+1} - \mu, \qquad X_2 = \mu.$$

Show that $X_1 X_2 > x_n x_{n+1}$ and that the numbers $x_1, x_2, \ldots, x_{n-1}, X_1$ have arithmetic mean μ. Deduce that $A_n \geqslant G_n$ for all n with equality if and only if all the x_i are equal.

#250 c (GM–HM inequality) Show, when $n = 2$, that $G^2 = AH$. More generally deduce, from the AM–GM inequality, that $G \geqslant H$ with equality if and only if the x_i are all equal.

#251 C† (QM–AM inequality) Show that $Q \geqslant A$ with equality if and only if all the x_i are equal.

#252 B† (MAT 2008 #2 adapted) (i) Find positive integers, x_1 and y_1, satisfying the equation $(x_1)^2 - 2(y_1)^2 = 1$.

(ii) Given integers a, b, we define two sequences $x_1, x_2, x_3, \ldots$ and $y_1, y_2, y_3, \ldots$ by setting

$$x_{n+1} = 3x_n + 4y_n, \qquad y_{n+1} = ax_n + by_n, \qquad \text{for } n \geqslant 1.$$

Find *positive* values for a, b such that $(x_{n+1})^2 - 2(y_{n+1})^2 = (x_n)^2 - 2(y_n)^2$. Deduce that there are infinitely many pairs of integer solutions x, y which solve the equation $x^2 - 2y^2 = 1$.

(iii) Find a pair of integers X, Y which satisfy $X^2 - 2Y^2 = 1$ such that $X > Y > 50$.

(iv) Using the values of a and b found in part (ii), what is the approximate value of x_n/y_n as n increases?

#253c†(i) Show, for any $n \geqslant 0$, that we may uniquely write $(1+\sqrt{2})^n = a_n + b_n\sqrt{2}$ where a_n and b_n are integers. Find formulae for a_{n+1} and b_{n+1} in terms of a_n and b_n.
(ii) Show that $(a_n)^2 - 2(b_n)^2 = (-1)^n$.
(iii) Using the fact that $(\sqrt{2}+1)(\sqrt{2}-1) = 1$, show that $(1+\sqrt{2})^{-n} = a_{-n} + b_{-n}\sqrt{2}$ where a_{-n} and b_{-n} are integers. Determine a_{-n} and b_{-n} in terms of a_n and b_n.

#254B† Let n be a positive integer. Prove that there are integers $a_n, a_{n-2}, a_{n-4}, \ldots$ such that

$$\cos n\theta = a_n \cos^n \theta + a_{n-2} \cos^{n-2} \theta + a_{n-4} \cos^{n-4} \theta + \cdots$$

and that $a_n = 2^{n-1}$.

#255B† Show, for integers $n \geqslant 1$, that

$$\sum_{k=1}^{n} \frac{1}{2^k} \tan\left(\frac{x}{2^k}\right) = \frac{1}{2^n} \cot\left(\frac{x}{2^n}\right) - \cot x.$$

#256b Show, for $n \geqslant 1$, that

$$\sum_{k=1}^{n} \cos(2k-1)x = \frac{\sin 2nx}{2\sin x}.$$

#257D (See also #77.) Show, for $n \geqslant 1$, that

$$\sum_{k=1}^{n} \sin kx = \frac{\sin\left\{\frac{1}{2}(n+1)x\right\} \sin\left\{\frac{1}{2}nx\right\}}{\sin\left\{\frac{1}{2}x\right\}}.$$

#258a† Show that the last two digits of 76^n are 76 for any integer $n \geqslant 1$.

#259b† Show that $3 \times 2^{2n} + 2 \times 3^{2n}$ is divisible by 5 for all natural numbers n.

#260b Show that $3^{3n} + 5^{4n+2}$ is divisible by 13 for all natural numbers n.

#261D†(i) Show that $7^{m+3} - 7^m$ and $11^{m+3} - 11^m$ are both divisible by 19 for all natural numbers m.
(ii) Calculate the remainder when $7^m - 11^n$ is divided by 19, for the cases $0 \leqslant m \leqslant 2$ and $0 \leqslant n \leqslant 2$.
(iii) Deduce that $7^m - 11^n$ is divisible by 19 precisely when $m+n$ is a multiple of 3.

#262b† By setting up a reduction formula between I_n and I_{n-2}, show that

$$I_n = \int_0^\pi \frac{\sin nx}{\sin x} \, dx$$

equals π when n is odd. What value does I_n take when n is even?

#263b† Let n be a natural number. Show that

$$I_{2n+1} = \int_0^{\pi/2} \cos^{2n+1} x \, dx = \frac{2^{2n}(n!)^2}{(2n+1)!}$$

and determine the corresponding formula for even powers.

- For further examples of reduction formulae, see §5.3.

#264 B Bertrand's postulate [7] states that

> for $n \geqslant 2$ there is a prime number p satisfying $n < p < 2n$.

Use Bertrand's postulate and the strong form of induction to show that every integer $n \geqslant 1$ can be written as a sum of distinct prime numbers. For the purposes of this question, you will need to regard 1 as a prime number. Deduce that for every integer $n \geqslant 2$, either n or $n - 1$ can be written as a sum of distinct prime numbers (in the usual sense). [8]

#265 C Carefully demonstrate the following to verify Remark 2.10. Let T be a set of ordered pairs (m, n) of natural numbers with any of the following three properties:

(i) $(0, 0)$ is in T and if (m, n) is in T then so are $(m, n + 1)$ and $(m + 1, n)$.
(ii) Each $(0, n)$ is in T and if (m, n) is in T for every n then $(m + 1, n)$ is in T for every n.
(iii) Each $(m, 0)$ is in T and if (m, n) is in T for every m then $(m, n + 1)$ is in T for every m.

Show that in each case T is the set of all ordered pairs of natural numbers.

2.3 The Binomial Theorem

All of you will have met the identity

$$(x + y)^2 = x^2 + 2xy + y^2$$

and may have met identities like

$$(x + y)^3 = x^3 + 3x^2 y + 3xy^2 + y^3.$$

It may even have been pointed out to you that these coefficients $1, 2, 1$ and $1, 3, 3, 1$ are simply the numbers that appear in **Pascal's triangle** [9] (see p.93 for a biographical note about Pascal) and that more generally the nth row (counting from $n = 0$) contains the coefficients in the expansion of $(x + y)^n$. Pascal's triangle is the infinite triangle of numbers that has 1s down both edges, and a number internal to some row of the triangle is calculated by adding the two numbers above it in the previous row. So the triangle grows as

---

[7] This postulate was made in 1845 by Joseph Bertrand (1822-1900), a professor of analysis at the École Polytechnique and later at the Collège de France. The postulate was first proved by Chebyshev (see p.278 for his biographical note) in 1850. A much simpler proof was given by Erdös in 1932.

[8] Hans-Egon Richert proved in 1949 the more difficult result that every $n \geqslant 7$ can be written as a sum of distinct primes.

[9] Despite the name, the numbers in Pascal's triangle (and indeed even their triangular representation) were known variously to earlier Greek, Indian, Persian, Chinese and other Western mathematicians, though Pascal did develop many uses and identities related to the triangle.

follows:

$$
\begin{array}{cccccccccccccc}
n=0 & & & & & & & 1 & & & & & & \\
n=1 & & & & & & 1 & & 1 & & & & & \\
n=2 & & & & & 1 & & 2 & & 1 & & & & \\
n=3 & & & & 1 & & 3 & & 3 & & 1 & & & \\
n=4 & & & 1 & & 4 & & 6 & & 4 & & 1 & & \\
n=5 & & 1 & & 5 & & 10 & & 10 & & 5 & & 1 & \\
n=6 & 1 & & 6 & & 15 & & 20 & & 15 & & 6 & & 1
\end{array}
$$

From the triangle we could, say, read off the identity

$$(x+y)^6 = x^6 + 6x^5y + 15x^4y^2 + 20x^3y^3 + 15x^2y^4 + 6xy^5 + y^6.$$

Of course we haven't *proved* this identity yet! These identities, for general n, are the subject of the *binomial theorem*. We introduce now the binomial coefficients; their connection with Pascal's triangle will become clear soon.

Definition 2.16 The (n,k)th **binomial coefficient** is the number

$$\binom{n}{k} = \frac{n!}{k!\,(n-k)!} \qquad \text{where} \qquad 0 \leqslant k \leqslant n.$$

$\binom{n}{k}$ is read as 'n **choose** k' and in some books is denoted as ^nC_k. By convention $\binom{n}{k} = 0$ if $k > n \geqslant 0$ or $n \geqslant 0 > k$.

#266
#268

- Note some basic identities concerning the binomial coefficients:

$$\binom{n}{k} = \binom{n}{n-k}; \qquad \binom{n}{0} = \binom{n}{n} = 1; \qquad \binom{n}{1} = \binom{n}{n-1} = n. \qquad (2.6)$$

The first identity reflects the fact that Pascal's triangle is left–right symmetrical, the second that the left and right edges are all 1s and the third that the next diagonals in from the edges progress $1, 2, 3, \ldots$.

The following lemma demonstrates that the binomial coefficients are precisely the numbers from Pascal's triangle.

Lemma 2.17 *Let* $1 \leqslant k \leqslant n$. *Then*

$$\binom{n}{k-1} + \binom{n}{k} = \binom{n+1}{k}.$$

Proof Putting the LHS over a common denominator we obtain

$$\frac{n!}{(k-1)!\,(n-k+1)!} + \frac{n!}{k!\,(n-k)!} = \frac{n!\,\{k+(n-k+1)\}}{k!\,(n-k+1)!} = \frac{n! \times (n+1)}{k!\,(n-k+1)!}$$

$$= \frac{(n+1)!}{k!\,(n+1-k)!} = \binom{n+1}{k}. \qquad \square$$

Corollary 2.18 *The kth number in the nth row of Pascal's triangle is $\binom{n}{k}$ (remembering to count from $n = 0$ and $k = 0$). In particular, the binomial coefficients are whole numbers.*

Proof We shall prove this by induction. Note that $\binom{0}{0} = 1$ gives the 1 at the apex of Pascal's triangle, proving the initial step. Suppose now that the numbers $\binom{n}{k}$, where $0 \leqslant k \leqslant n$, are the numbers that appear in the nth row of Pascal's triangle. The first and last entries of the $(n+1)$th row (associated with $k = 0$ and $k = n+1$) are

$$1 = \binom{n+1}{0} \quad \text{and} \quad 1 = \binom{n+1}{n+1}$$

as required. For $1 \leqslant k \leqslant n$, then the kth entry on the $(n+1)$th row is formed by adding the $(k-1)$th and kth entries from the nth row. By the inductive hypothesis and Lemma 2.17 their sum is

$$\binom{n}{k-1} + \binom{n}{k} = \binom{n+1}{k},$$

verifying that the $(n+1)$th row also consists of the correct binomial coefficients. The result follows by induction. □

Finally, we come to the binomial theorem.

Theorem 2.19 (*Binomial Theorem*) *Let n be a natural number and x, y be real numbers. Then*

$$(x+y)^n = \sum_{k=0}^{n} \binom{n}{k} x^k y^{n-k}.$$

#269
#271
#275
#276
#277
#295

Proof Let's check the binomial theorem first for $n = 0$. We can verify this by noting

$$\text{LHS} = (x+y)^0 = 1; \qquad \text{RHS} = \binom{0}{0} x^0 y^0 = 1.$$

For induction, we aim to show the theorem holds in the $(n+1)$th case assuming the nth case to be true. We have

$$\text{LHS} = (x+y)^{n+1} = (x+y)(x+y)^n = (x+y) \left(\sum_{k=0}^{n} \binom{n}{k} x^k y^{n-k} \right),$$

writing in our assumed expression for $(x+y)^n$. Expanding the brackets gives

$$\sum_{k=0}^{n} \binom{n}{k} x^{k+1} y^{n-k} + \sum_{k=0}^{n} \binom{n}{k} x^k y^{n+1-k}$$

$$= x^{n+1} + \sum_{k=0}^{n-1} \binom{n}{k} x^{k+1} y^{n-k} + \sum_{k=1}^{n} \binom{n}{k} x^k y^{n+1-k} + y^{n+1},$$

by taking out the last term from the first sum and the first term from the second sum. In the first sum we now make a change of variable; we set $k = l - 1$, noting that as k ranges over

$0, 1, \ldots, n-1$ then l ranges over $1, 2, \ldots, n$. So the above equals

$$x^{n+1} + \sum_{l=1}^{n} \binom{n}{l-1} x^l y^{n+1-l} + \sum_{k=1}^{n} \binom{n}{k} x^k y^{n+1-k} + y^{n+1}$$

$$= x^{n+1} + \sum_{k=1}^{n} \left\{ \binom{n}{k-1} + \binom{n}{k} \right\} x^k y^{n+1-k} + y^{n+1}$$

$$= x^{n+1} + \sum_{k=1}^{n} \binom{n+1}{k} x^k y^{n+1-k} + y^{n+1} = \sum_{k=0}^{n+1} \binom{n+1}{k} x^k y^{n+1-k}$$

by Lemma 2.17. The final expression is the RHS of the binomial theorem in the $(n+1)$th case as required. □

Example 2.20 Let n be a natural number. Show that

(a) $\quad \binom{n}{0} + \binom{n}{1} + \binom{n}{2} + \cdots + \binom{n}{n} = 2^n;$ \hfill (2.7)

(b) $\quad \binom{n}{0} + \binom{n}{2} + \binom{n}{4} + \cdots = 2^{n-1};$

(c) $\quad \binom{n}{1} + \binom{n}{3} + \binom{n}{5} + \cdots = 2^{n-1}.$

Note that the sums in (b) and (c) are not infinite as the binomial coefficients $\binom{n}{k}$ eventually become zero once $k > n$.

#307 Solution From the binomial theorem we have
#308
#309
#312
$$(1+x)^n = \binom{n}{0} + \binom{n}{1} x + \binom{n}{2} x^2 + \cdots + \binom{n}{n} x^n.$$

#313 If we set $x = 1$ we demonstrate (2.7). If we set $x = -1$ then we have
#314
#315
$$\binom{n}{0} - \binom{n}{1} + \binom{n}{2} - \cdots + (-1)^n \binom{n}{n} = (1-1)^n = 0. \hfill (2.8)$$

Equation $\frac{1}{2}(2.7) + \frac{1}{2}(2.8)$ gives (b) and equation $\frac{1}{2}(2.7) - \frac{1}{2}(2.8)$ gives (c). □

Our next example is due to Gottfried Leibniz (see p.429 for a brief biographical note).

Example 2.21 Use induction to prove **Leibniz's rule** for differentiation of a product – this says, for (sufficiently differentiable) functions $u(x)$ and $v(x)$ and $n \geqslant 1$, that

$$\frac{d^n}{dx^n}(u(x)v(x)) = \sum_{k=0}^{n} \binom{n}{k} \frac{d^k u}{dx^k} \frac{d^{n-k} v}{dx^{n-k}}. \hfill (2.9)$$

Solution Notice that, when $n = 1$, the identity (2.9) is just the product rule of differentiation, thus checking the initial step. Let's suppose that (2.9) holds true in the nth case. Then

$$\frac{d^{n+1}}{dx^{n+1}}(u(x)v(x)) = \frac{d}{dx}\left(\frac{d^n}{dx^n}(u(x)v(x)) \right) = \frac{d}{dx}\left(\sum_{k=0}^{n} \binom{n}{k} \frac{d^k u}{dx^k} \frac{d^{n-k} v}{dx^{n-k}} \right)$$

from our inductive hypothesis. Using the sum and product rules of differentiation the above equals

$$\left(\sum_{k=0}^{n}\binom{n}{k}\frac{d^{k+1}u}{dx^{k+1}}\frac{d^{n-k}v}{dx^{n-k}}\right)+\left(\sum_{k=0}^{n}\binom{n}{k}\frac{d^{k}u}{dx^{k}}\frac{d^{n+1-k}v}{dx^{n+1-k}}\right)$$

$$=\frac{d^{n+1}u}{dx^{n+1}}v+\left(\sum_{k=0}^{n-1}\binom{n}{k}\frac{d^{k+1}u}{dx^{k+1}}\frac{d^{n-k}v}{dx^{n-k}}\right)+\left(\sum_{k=1}^{n}\binom{n}{k}\frac{d^{k}u}{dx^{k}}\frac{d^{n+1-k}v}{dx^{n+1-k}}\right)+u\frac{d^{n+1}v}{dx^{n+1}}$$

$$=\frac{d^{n+1}u}{dx^{n+1}}v+\left(\sum_{k=1}^{n}\left\{\binom{n}{k-1}+\binom{n}{k}\right\}\frac{d^{k}u}{dx^{k}}\frac{d^{n+1-k}v}{dx^{n+1-k}}\right)+u\frac{d^{n+1}v}{dx^{n+1}}$$

by replacing k by $k-1$ in the first sum. From Lemma 2.17 we finally have

$$\frac{d^{n+1}u}{dx^{n+1}}v+\left(\sum_{k=1}^{n}\binom{n+1}{k}\frac{d^{k}u}{dx^{k}}\frac{d^{n+1-k}v}{dx^{n+1-k}}\right)+u\frac{d^{n+1}v}{dx^{n+1}}=\sum_{k=0}^{n+1}\binom{n+1}{k}\frac{d^{k}u}{dx^{k}}\frac{d^{n+1-k}v}{dx^{n+1-k}}$$

which demonstrates (2.9) for the $(n+1)$th case and hence the result follows by induction. ☐

Recall, from Definition 2.16, that $\binom{n}{k}=n!/\{k!(n-k)!\}$ is read as 'n **choose** k'; as we shall see, the reason for this is that there are $\binom{n}{k}$ ways of choosing k elements without repetition from the set $\{1,2,\ldots,n\}$, no interest being shown in the order that the k elements are chosen. By way of an example:

Example 2.22 Determine $\binom{5}{3}$ and list the subsets corresponding to this number.

Solution We have $\binom{5}{3}=\frac{5!}{3!2!}=\frac{120}{6\times2}=10$ and the corresponding subsets are

$$\{1,2,3\},\quad\{1,2,4\},\quad\{1,2,5\},\quad\{1,3,4\},\quad\{1,3,5\},$$
$$\{1,4,5\},\quad\{2,3,4\},\quad\{2,3,5\},\quad\{2,4,5\},\quad\{3,4,5\}.$$

Note that there are $6=3!$ orders of choosing the elements that lead to each subset. Any of the ordered choices

$$(1,2,3),\quad(1,3,2),\quad(2,1,3),\quad(2,3,1),\quad(3,1,2),\quad(3,2,1),$$

would have led to the same subset $\{1,2,3\}$. ☐

Proposition 2.23 *There are $\binom{n}{k}$ subsets of $\{1,2,\ldots,n\}$ with k elements.*

#284
#296
#338

Proof Let's think about how we might go about choosing k elements from n. For our 'first' element we can choose any of the n elements, but once this has been chosen it can't be put into the subset again. For our 'second' element, any of the remaining $n-1$ elements may be chosen, for our 'third' any of the $n-2$ that are left, and so on. So choosing a set of k elements from $\{1,2,\ldots,n\}$ in a particular order can be done in

$$n\times(n-1)\times(n-2)\times\cdots\times(n-k+1)=\frac{n!}{(n-k)!}\text{ ways.}$$

But there are lots of different orders of choosing that would have produced the same subset. Given a set of k elements there are $k!$ ways of ordering them (see #294) – that is to say, for each subset with k elements there are $k!$ different orders of choice that will each lead to that same subset. So the number $n!/(n-k)!$ is an 'overcount' by a factor of $k!$. Finally then, the number of subsets of size k equals $\binom{n}{k}$. $\qquad\square$

- In particular this means we can reinterpret the identity (2.7) as saying that the total number of subsets $\{1,2,3,\ldots,n\}$ equals 2^n (as demonstrated already in Proposition 2.12).

Remark 2.24 There is a **trinomial theorem** and further generalizations of the binomial theorem to greater numbers of variables. Given three real numbers x,y,z and a natural number n, we can apply the binomial theorem twice to show $(x+y+z)^n$ equals

$$\sum_{k=0}^{n}\frac{n!}{k!\,(n-k)!}x^k(y+z)^{n-k} = \sum_{k=0}^{n}\frac{n!}{k!\,(n-k)!}\sum_{l=0}^{n-k}\frac{(n-k)!}{l!\,(n-k-l)!}x^k y^l z^{n-k-l}$$

$$= \sum_{k=0}^{n}\sum_{l=0}^{n-k}\frac{n!}{k!\,l!\,(n-k-l)!}x^k y^l z^{n-k-l}.$$

#292
#301 This is a somewhat cumbersome expression; it's easier on the eye, and has a nicer symmetry, if we write $m=n-k-l$ and then we can rewrite the above as

$$(x+y+z)^n = \sum_{\substack{k+l+m=n \\ k,l,m \geqslant 0}}\frac{n!}{k!\,l!\,m!}x^k y^l z^m.$$

Given $k,l,m,n \geqslant 0$ such that $k+l+m=n$, the notation

$$\binom{n}{k,l,m} = \frac{n!}{k!\,l!\,m!}$$

is commonly used for this trinomial coefficient. Again it denotes the number of ways in which n objects can be split into a first subset of size k, a second subset of size l and a third subset of size m. $\qquad\blacksquare$

Remark 2.25 Pascal's tetrahedron corresponds to the trinomial theorem in the same way that his triangle corresponds to the binomial theorem. Each face of the tetrahedron contains a copy of Pascal's triangle, with entries in the interior of the tetrahedron being the sum of three entries in the floor above. The nth floor (counting from $n=0$) contains $(n+1)(n+2)/2$ entries (see #285), which sum to 3^n. $\qquad\blacksquare$

Example 2.26 Note that

$$\binom{4}{1,1,2} = \frac{4!}{1!\,1!\,2!} = 12.$$

This number corresponds to

$(\{1\},\{2\},\{3,4\}),\quad (\{2\},\{1\},\{3,4\}),\quad (\{1\},\{4\},\{2,3\}),\quad (\{4\},\{1\},\{2,3\}),$

$(\{1\},\{3\},\{2,4\}),\quad (\{3\},\{1\},\{2,4\}),\quad (\{2\},\{3\},\{1,4\}),\quad (\{3\},\{2\},\{3,4\}),$

$(\{2\},\{4\},\{1,3\}),\quad (\{4\},\{2\},\{1,3\}),\quad (\{3\},\{4\},\{1,2\}),\quad (\{4\},\{3\},\{1,2\}).$

Note this is twice the number of ways four elements can be partitioned into a pair and two singletons as each partition is double-counted; for example, $(\{1\},\{2\},\{3,4\})$ and $(\{2\},\{1\},\{3,4\})$ both lead to the same partition.

Exercises

#266a Verify the identities in (2.6).

#267a Show that the expression in (2.2) can be rewritten as $1 + \binom{n}{2} + \binom{n}{4}$.

#268A Show, for $n \geqslant k > 0$ that $(n/k)^k \leqslant \binom{n}{k} \leqslant n^k/k!$.

#269A† Use the binomial theorem to show, for $n \geqslant 1$, that $2 \leqslant \left(1 + \frac{1}{n}\right)^n < 3$.

#270c Find the number of pairs of positive integers x, y which solve the equation

$$x^3 + 6x^2y + 12y^2 + 8y^3 = 1,000,000.$$

#271B† Find all the roots of

$$z^3 + 21z^2 + 99z + 7 = 0 \qquad \text{and} \qquad z^4 + 12z^3 + 48z^2 + 72z + 32 = 0.$$

#272a Determine the full expansion of $(1 + x/2)^{10}$ and find the power of x which has the greatest coefficient.

#273B† Determine the power of x which has the greatest coefficient in the binomial expansion of $(1 + x/3)^{1000}$.

#274B Let $a > 0$ and n be a positive integer. The coefficients c_k are defined by the identity $(1 + ax)^n = \sum_{k=0}^{n} c_k x^n$. Show that $c_0 < c_1 < \cdots < c_{n-1} < c_n$ if and only if $n < a$.

#275b† Let $a > 0$ and let n be a positive integer. Show that the coefficients of x^k in $(1 + ax)^n$ increase up to a point and then subsequently decrease.

#276b Show that the greatest binomial coefficient on the $(2n)$th row of Pascal's triangle is $\binom{2n}{n}$ and the greatest binomial coefficients are $\binom{2n+1}{n} = \binom{2n+1}{n+1}$ on the $(2n+1)$th row.

#277b† What is the first line of Pascal's triangle containing a number greater than 10^6?

#278D† (See also #263 and #1414.) Show by induction that

$$\frac{2^{2n}}{2\sqrt{n}} \leqslant \binom{2n}{n} \leqslant \frac{2^{2n}}{\sqrt{3n+1}} \qquad \text{for } n \geqslant 1.$$

#279D† (**Wallis Product**) Let n be a natural number. Deduce from #263 that

$$\binom{2n}{n} \frac{\sqrt{\pi n}}{2^{2n}}$$

is approximately 1 for large values of n and that [10]

$$\frac{2}{\pi} = \frac{1}{2} \times \frac{3}{2} \times \frac{3}{4} \times \frac{5}{4} \times \frac{5}{6} \times \frac{7}{6} \times \frac{7}{8} \times \cdots.$$

#280D† Let $n \geqslant 2$ be an integer. Determine the finite product

$$\left(1 - \frac{1}{4}\right)\left(1 - \frac{1}{9}\right)\left(1 - \frac{1}{16}\right) \times \cdots \times \left(1 - \frac{1}{n^2}\right)$$

and the infinite products

$$\left(1 - \frac{1}{4}\right)\left(1 - \frac{1}{9}\right)\left(1 - \frac{1}{16}\right)\left(1 - \frac{1}{25}\right)\cdots \qquad \text{and}$$

$$\left(1 - \frac{1}{9}\right)\left(1 - \frac{1}{25}\right)\left(1 - \frac{1}{49}\right)\left(1 - \frac{1}{81}\right)\cdots.$$

#281b Deduce from #278 and #276, for $0 < k \leqslant n$, that $\binom{n}{k} \leqslant \frac{2^n}{\sqrt{n}}$.

#282D† Show that the number 3003 appears eight times in Pascal's triangle. [11]

#283C† Say k, l are natural numbers and l is odd. Show that $\binom{2^k l}{2^k}$ is odd.

#284B† Reinterpret Lemma 2.17 in terms of subsets of $\{1, 2, \ldots, n\}$ and of $\{1, 2, \ldots, n+1\}$.

#285c† Show that there are $(n+1)(n+2)/2$ different types of term $x^i y^j z^k$ in the expansion of $(x + y + z)^n$.

#286c Find the possible values of $k, l, m \geqslant 0$ with $k + l + m = 6$ such that $\binom{6}{k,l,m}$ is largest.

#287c Find the possible values of $k, l, m, n \geqslant 0$ with $k + l + m + n = 7$ such that $\binom{7}{k,l,m,n}$ is largest.

#288C† How many ways are there to partition $\{1, 2, 3, 4, 5, 6\}$ into three subsets of size 2? Why is this answer different from $\binom{6}{2,2,2}$?

#289c How many ways are there to partition $\{1, 2, 3, 4, 5, 6, 7\}$ into three subsets of size $2, 2, 3$? How does this answer relate to $\binom{7}{2,2,3}$?

#290B† How many anagrams are there of the words 'algebra', 'calculus', 'grassmannian'?

#291D† Let $n = 3N$ and $k, l, m \geqslant 0$ with $k + l + m = n$. Show $\binom{n}{k,l,m}$ is largest when $k = l = m$. When is the maximum achieved in the cases when $n = 3N + 1$ and $n = 3N + 2$?

---

[10] This infinite product was first evaluated by the English mathematician John Wallis (1616–1703) in his 1656 work *Arithmetica infinitorum*.

[11] 3003 is currently the only number known to appear eight or more times in Pascal's triangle. *Singmaster's conjecture*, made in 1971, by David Singmaster (b.1939), proposes that there is an upper bound to how many times a number can appear. He showed that there are infinitely many numbers that appear six times (see #363).

#292B If three fair dice are rolled what is the probability of getting a total of 16? A total of 15? How do these probabilities relate to coefficients in the expansion of

$$\left(\frac{x+x^2+x^3+x^4+x^5+x^6}{6}\right)^3 ?$$

#293b If four fair dice are rolled what is the probability of getting a total of 7? A total of 8?

#294B† Show that there are $n!$ ways of listing n elements where order matters.

#295b Deduce from #269 that $(n/3)^n < n! < (n/2)^n$ for $n \geqslant 6$.

#296A† Show by induction that $2^n \leqslant \binom{2n}{n}$ for $n \geqslant 0$. Rederive this result by considering subsets of $\{1, 2, \ldots, 2n\}$.

#297D† A poker hand is five cards from a standard pack [12] of 52 cards. So there are $\binom{52}{5}$ possible hands. Of these, how many give (i) four of a kind, (ii) a full house, (iii) a flush, (iv) a straight?

#298B Two gamblers agree to toss a fair coin 13 times. Each pays £1 to play the game with one gambler seeking heads, the other tails; all the money goes to the first gambler with 7 favourable tosses. However, the game is interrupted after 8 tosses, 5 of them heads, 3 of them tails. How should the money be split? Determine the answer by considering the 32 ways the remaining tosses might have fallen. Show also that this answer is the same when considering the probabilities of how the game might have ended if played to conclusion.

#299C Three gamblers A, B, C play a fair dice game, each paying £1 to play the game, and all the money goes to the winner. A die roll of 1 or 2 is favourable to A, 3 or 4 is favourable to B, 5 or 6 is favourable to C. The winner is the first player with 4 favourable rolls. However, the game is interrupted after five rolls of $4, 2, 1, 3, 2$. How should the money be divided amongst the players?

#300D† Generalize #298 to a situation where the coin is to be tossed $2n + 1$ times and the game is interrupted after h heads and t tails. You will need to leave your answer as a sum. Verify your answer agrees with the expected answers when $h = n$ or $t = n$.

#301A Find the coefficient of $x^2 y^3$ in the expansions of

$$(1+x+y)^3, \qquad (1+x+y)^6, \qquad (1+xy+x^2+y^2)^3, \qquad (xy+x+y)^6.$$

#302B Let m, n be natural numbers. Find the coefficient of $x^2 y^3$ in the expansions of

$$(2+3x+5y)^m, \qquad (1+x)^m(1+y)^n, \qquad (x+y^2)^m(y+x^2)^n, \qquad (1+xy)^m(x+y)^n.$$

#303B Use the method of #244 and the binomial theorem to find an alternative proof of (2.5).

---

[12] A standard pack of playing cards has 52 cards. These are split into four *suits* of 13 cards each – each suit includes the cards 1–13 (with 1 being called an *ace*, 11 a *jack*, 12 a *queen*, 13 a *king*). *Four of a kind* means that the hand has all four of one type (e.g. all four queens). A *full house* means the hand is three cards of one type and two of another (e.g. three 7s and two 4s). A *flush* means the hand is all of the same suit. A *straight* means the hand contains five consecutive cards (e.g. 3,4,5,6,7); for the purposes of a straight an ace can be high (above a king) or low (below a 2).

Pierre de Fermat

A Chance Correspondence Many of the key ideas of probability theory were laid down in 1654 in a brief but detailed correspondence between the French mathematicians **Pierre de Fermat** (1601–1665) and **Blaise Pascal** (1623–1662). Their letters had been prompted by the so-called *Problem of Points*: if a game of chance between two gamblers had to stop before completion, how should the winnings be divided up? By modern standards, this is a relatively trivial problem of probability, yet prior to 1654 many noteworthy mathematicians had declared the problem unsolvable or had determined wrong answers. That it took two great mathematicians to flesh out the details shows how far the seventeenth-century mindset needed to progress; at the time there was no precise notion of probability (a number between 0 and 1 signifying the chance of an event happening), let alone more complicated ideas of conditional probability and expected gain. See Devlin [10] for a highly readable account of the correspondence, other important contributors (Cardano, Jacob Bernoulli, De Moivre, Huygens, Bayes), and the huge impact on society from the introduction of risk management. To give a simple example of the Problem of Points, suppose that two gamblers agree to toss a fair coin five times, with gambler A winning if the majority gives heads, gambler B if the majority gives tails. Each gambler pays £1 to play, with all money being returned to the winner. But say the game is interrupted after three tosses with two heads, one tails having shown: how should the money be divided up?

Fermat argued (correctly) that there were four possible ways the remaining two coin tosses might land: HH, HT, TH, TT, (what would be called the *sample space* today). A wins in the first three eventualities and B only in the last. Consequently A should receive three quarters of the money (or £1.50) and B should receive one quarter (or £0.50). Even Pascal was a little confused by this reasoning – there were after all only three ways that the game would end H, TH, TT (for if the first coin is H then A wins and there is no further need to toss a second time). Anyone comfortable with probability would immediately counter that these three events don't have equal probability; there is a 0.5 chance of the H and 0.25 chance of TH so this still returns a probability of 0.75. Following on from Fermat's and Pascal's work, rapid progress was made in understanding the nuances of probabilistic thinking, though arguably the subject wasn't made properly rigorous until the work of the Russian mathematician **Andrey Kolmogorov** (1903–1987) in 1933 in laying down the axioms of a probability space.

Fermat was a titan of seventeenth-century mathematics. Besides in probability, he made enormous contributions to number theory, independently of Descartes invented co-ordinate geometry, and laid many important preliminary contributions paving the way to calculus (see p.429). He, more than any of his contemporaries, showed the power of algebraic analytical methods to old and new problems and 'in a very real sense, Fermat presided over the death of the classical Greek tradition in mathematics.' (Mahoney [23, p.365].)

The Binomial Distribution

Imagine a fair coin being tossed consecutively a number of times, say n times. There are 2^n possible outcomes and these are each equally likely, having probability $1/2^n$. The number of times N that heads appears is a number in the range $0 \leqslant N \leqslant n$. Note that $N = 0$ arises only when n consecutive tails appear – the probability of this happening is $1/2^n$. A single head ($N = 1$) can arise in n different ways (the head being one of n tosses and the remaining tosses being tails) and so has probability $n/2^n$. More generally we see that the probability of there being k heads, denoted $P(N = k)$, is given by

$$P(N = k) = \binom{n}{k} 2^{-n}.$$

If the coin is biased with probability p of a head and probabilty $q = 1 - p$ of tails, then the probability of k heads is given by

$$P(N = k) = \binom{n}{k} p^k q^{n-k}.$$

The bar charts opposite demonstrate how the probabilities change if 20 tosses are made, Figure 2.4a with an unbiased coin, and then coins that are increasingly likely to show tails (Figures 2.4b and 2.4c). The *expectation* of N (that is, the mean number of heads) is the sum

$$\mu = E(N) = \sum_{k=0}^{n} kP(N = k) = np.$$

(See #322(ii).) The *variance* of N is the expected value of $(N - \mu)^2$; the *standard deviation* σ is the square root of the variance. It is the case that

$$\sigma^2 = \text{Var}(N) = E((N - \mu)^2) = \sum_{k=0}^{n} (k - np)^2 P(N = k) = npq.$$

(See # 322(iii).) The variance is a measure of the spread of the possible values; note that it is maximal, for given n, when $p = 1/2$.

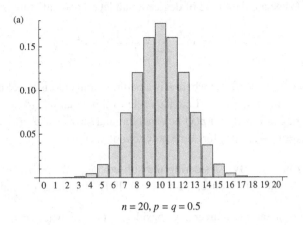

$n = 20, p = q = 0.5$

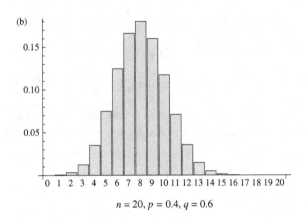

$n = 20, p = 0.4, q = 0.6$

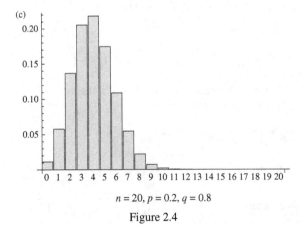

$n = 20, p = 0.2, q = 0.8$

Figure 2.4

#304 B† Let $p(x)$ be a polynomial of degree n and let a be a real number. Show that

$$p(x) = \sum_{k=0}^{n} \frac{p^{(k)}(a)}{k!}(x-a)^k.$$

#305 b† Let $p(x) = (x-a)^k q(x)$ where k is a natural number, a is a real number and $q(x)$ is a polynomial with $q(a) \neq 0$. Use Leibniz's rule to show that $p^{(r)}(a) = 0$ for $0 \leqslant r < k$ and that $p^{(k)}(a) \neq 0$. Conversely let $p(x)$ be a polynomial such that $p^{(r)}(a) = 0$ for $0 \leqslant r < k$. Show that $p(x) = (x-a)^k q(x)$ for some polynomial $q(x)$.

#306 B† Let n be odd. By considering the subsets of $\{1,2,3,\ldots n\}$, find a new proof of Example 2.20(b) and (c). Deduce, from this result, Example 2.20(b) and (c) for even n.

#307 A† Let n be a natural number. By considering $(1+i)^n$ where $i^2 = -1$, show that

$$\binom{n}{0} - \binom{n}{2} + \binom{n}{4} - \binom{n}{6} + \cdots = 2^{n/2} \cos \frac{n\pi}{4}.$$

What is the value of $\binom{n}{1} - \binom{n}{3} + \binom{n}{5} - \binom{n}{7} + \cdots$?

#308 a† Let n be a natural number. Determine $\binom{n}{0} + \binom{n}{4} + \binom{n}{8} + \binom{n}{12} + \cdots$.

#309 B Let n be a natural number and let $\omega = \text{cis}(2\pi/3)$, a cube root of unity. Simplify $1 + \omega^k + \omega^{2k}$ for the different cases $k = 3m, 3m+1, 3m+2$. By considering the binomial expansion of

$$(1+x)^n + (1+\omega x)^n + (1+\omega^2 x)^n$$

determine $\binom{n}{0} + \binom{n}{3} + \binom{n}{6} + \binom{n}{9} + \cdots$.

#310 d† Let n be a natural number and let $\omega = \text{cis}(2\pi/3)$, a cube root of unity. Find complex numbers a,b,c which satisfy the equations

$$a+b+c = 0; \qquad a+b\omega+c\omega^2 = 1; \qquad a+b\omega^2+c\omega = 0.$$

Hence determine $\binom{n}{1} + \binom{n}{4} + \binom{n}{7} + \binom{n}{10} + \cdots$ and $\binom{n}{2} + \binom{n}{5} + \binom{n}{8} + \binom{n}{11} + \cdots$.

#311 C Use the binomial theorem and De Moivre's theorem to rederive the solution of #254.

#312 b† Show that $\sum_{k=1}^{n} \binom{n}{k} \sin 2k\theta = 2^n \sin n\theta \cos^n \theta$.

#313 a† Use the identity $(1+x)^{2n} = (1+x)^n (1+x)^n$ to show that $\sum_{k=0}^{n} \binom{n}{k}^2 = \binom{2n}{n}$.

#314 B† For $n \geqslant 0$, determine $\sum_{k=0}^{n} (-1)^k \binom{n}{k}^2$.

#315 B(i) By differentiating the binomial expansion of $(1+x)^n$, show $\sum_{k=1}^{n} k\binom{n}{k} = n2^{n-1}$.
(ii) Similarly determine $\sum_{k=1}^{n} k^2 \binom{n}{k}$.

#316b By considering the identity $(1+x)^{m+n} = (1+x)^m(1+x)^n$, prove **Vandermonde's identity** [13]

$$\binom{m+n}{r} = \binom{m}{0}\binom{n}{r} + \binom{m}{1}\binom{n}{r-1} + \cdots + \binom{m}{r}\binom{n}{0} \quad \text{for } m,n,r \geqslant 0.$$

#317B† Let $n \geqslant 1$. By considering a certain product of polynomials, show that

$$\sum_{k=1}^{n} k\binom{n}{k}^2 = (2n-1)\binom{2n-2}{n-1}.$$

#318D† Let $n \geqslant 1$. Show that

$$\sum_{k=0}^{n} \binom{2n}{k,k,n-k,n-k} = \binom{2n}{n}^2.$$

#319D† Use #263 and the method of #75 to show, for a natural number n, that

$$(-1)^n \frac{2^{4n}(n!)^2}{(2n+1)!} = \sum_{k=0}^{n} \frac{(-1)^k}{2n-2k+1}\binom{2n+1}{k}.$$

#320D† Show that a_{n-2k}, as described in #254, equals

$$(-1)^k \frac{n(n-k-1)!}{k!(n-2k)!} 2^{n-2k-1} \quad \text{where} \quad 0 \leqslant 2k \leqslant n.$$

#321B† Show that

$$\sum_{k=0}^{n} \frac{1}{k+1}\binom{n}{k} = \frac{2^{n+1}-1}{n+1}.$$

#322B Let $0 \leqslant p \leqslant 1$, $q = 1-p$ and let s be a real number. Prove that

(i) $\displaystyle\sum_{k=0}^{n} \binom{n}{k} p^k q^{n-k} = 1;$ (ii) $\displaystyle\sum_{k=0}^{n} k\binom{n}{k} p^k q^{n-k} = np;$

(iii) $\displaystyle\sum_{k=0}^{n} (k-np)^2 \binom{n}{k} p^k q^{n-k} = npq;$ (iv) $\displaystyle\sum_{k=0}^{n} \binom{n}{k} p^k q^{n-k} s^k = (q+ps)^n.$

Results (i)–(iii) verify that the binomial distribution $B(n,p)$ is indeed a probability distribution and has expectation np and variance npq (see p.94). Result (iv) determines its probability generating function.

#323C† Let N be binomially distributed as $B(n,p)$ with mean $\mu = np$. The expected value of $|N - \mu|$ is

$$E(|N - \mu|) = \sum_{k=0}^{n} |k - np| \binom{n}{k} p^k q^{n-k}.$$

Distinguishing cases, calculate this sum when $n = 3$ and $0 \leqslant p \leqslant 1$. Sketch a graph of this value against p.

---

[13] After the French mathematician, Alexandre-Théophile Vandermonde (1735–1796), who also made contributions to the early history of group theory and whose name is also associated with an important determinant (see Example 3.165).

#324 C† Let N be binomially distributed as $B(n,p)$ with mean $\mu = np$. Show that

$$E(|N - \mu|)^2 \leqslant E((N - \mu)^2)$$

with equality only if $p = 0$ or 1.

#325 C Let $n = 10$ and $p = 1/2$. What is the standard deviation of $B(10, 1/2)$? If N has a $B(10, 1/2)$ distribution, what is the probability of N being within one standard deviation of the mean?

#326 c Use a computer to repeat #325 with $n = 200, 500, 1000, 5000$.

Definition 2.27 Let N have distribution $B(n,p)$. A **median** of N is a natural number m such that

$$P(N \leqslant m) \geqslant 1/2 \qquad \text{and} \qquad P(N \geqslant m) \geqslant 1/2.$$

A **mode** of N is a natural number m for which $P(N = m)$ is largest.

#327 C Let $n = 2$. For what values of p is the median of N equal to $0, 1$ or 2?

#328 C† Show that $B(n,p)$ has a unique mode unless $(n+1)p$ is a natural number, in which case there are two modes. Show that if there are two modes then the mean lies between them.

#329 C Assume now that there is a single mode. Under what circumstances does the mode of $B(n,p)$ exceed the mean? Give specific values of n and p to show that both cases can arise.

#330 C† A biased coin has probability p of landing heads (where $0 \leqslant p \leqslant 1$) and probability $1 - p$ of landing tails. If the coin is tossed n times what is the probability that an even number of heads will be tossed?

#331 A Let m, n be natural numbers. Deduce from Example 2.9 that

$$\sum_{k=0}^{n} \binom{m+k}{m} = \binom{n+m+1}{m+1}.$$

#332 B† Rederive the result of #331 combinatorially as follows. Consider words of length $m + n + 1$ using $m + 1$ copies of the letter x and n copies of the letter y, and separately determine this number by considering the number of letters y that the word starts with.

#333 D† Let n, r, s be natural numbers. Show that

$$\sum_{k=0}^{n} \binom{k}{r}\binom{n-k}{s} = \binom{n+1}{r+s+1}.$$

#334 c† Reinterpret the result of #333 by considering subsets of $\{0, 1, 2, \ldots n\}$ containing $r + s + 1$ elements with r elements smaller than k and s elements greater than k.

#335 C Let $n \geqslant r \geqslant 0$. Show $\sum_{k=r}^{n} \binom{n}{k}\binom{k}{r} = 2^{n-r}\binom{n}{r}$.

#336 D† Rederive the result of #335 by considering the number of ways of choosing a subset of $\{1, 2, \ldots, n\}$ of size k which itself contains a subset of size r.

Definition 2.28 A **composition** of a number n is a means of writing $n = n_1 + n_2 + \cdots + n_k$ where each n_i is a positive integer and order matters. So, for example, there are 8 compositions of 4 as

$$4 = 3+1 = 1+3 = 2+2 = 1+1+2 = 2+1+1 = 1+2+1 = 1+1+1+1.$$

For **partitions**, order does not matter and so there are 5 partitions of 4, namely

$$4 = 1+3 = 2+2 = 1+1+2 = 1+1+1+1$$

as (amongst others) the three compositions of 4 into $1, 1, 2$ are all considered the same partition.

#337B† Show that there are 2^{n-1} compositions of the number n.

#338B† Let n, k be positive integers. Show the number of solutions of $x_1 + x_2 + \cdots + x_k = n$ in integers $x_i \geqslant 0$ is $\binom{n+k-1}{k-1}$. How many solutions are there if we insist $x_i > 0$ for each i?

#339B Let n, k be positive integers. Show the number of solutions of $x_1 + x_2 + \cdots + x_k \leqslant n$ in integers $x_i \geqslant 0$ is $\binom{n+k}{k}$. Hence rederive the result of #331.

#340D† The triangle on the left below is formed by each number in a row being the *difference* of the numbers above it. How does it relate to Pascal's triangle? Determine precisely which rows are all 1s in this triangle.

```
            1                                      1
         1     1                               1   1   1
      1     0     1                          1   2   3   2   1
   1     1     1     1                      1   3   6   7   6   3   1
1     0     0     0     1                1   4  10  16  19  16  10  4   1
1  1     0     0     1     1          1   5  16  30  45  51  45  30  16  5   1
```

#341D The **trinomial triangle** [14] is defined in a similar fashion to Pascal's triangle, but an entry is defined as being the sum of the three entries above it. Its first few rows are given in the triangle on the right above. Show that:

(i) The numbers in the nth row are the coefficients in the expansion of $(1 + t + t^2)^n$.
(ii) The sum of the entries in the nth row, counting from $n = 0$, equals 3^n and that the alternating sum equals 1.

#342E† Show that, in any given row of Pascal's triangle, the number of odd numbers is a power of 2.

#343E Recall Moser's circle problem from p.71. Let n_k denote the number of different regions which are bounded by $k \geqslant 2$ sides (which may be line segments or arcs of the circle).

---

[14] The numbers in the trinomial triangle are sometimes referred to as *trinomial coefficients,* as are the coefficients that appear in the trinomial theorem and in Pascal's tetrahedron (Remarks 2.24 and 2.25), which makes for the possibility of confusion as these numbers are not at all the same.

(i) Determine n_2.

(ii) By considering the total number of edges of the internal regions, explain why

$$3n_3 + 4n_4 + 5n_5 + \cdots = 4\binom{n}{4} + n(n-2).$$

(iii) By considering the total angles of the internal polygons, show that

$$\pi(n_3 + 2n_4 + 3n_5 + \cdots) = 2\pi\binom{n}{4} + \pi(n-2).$$

(iv) Deduce that $n_2 + n_3 + n_4 + \cdots = 1 + \binom{n}{2} + \binom{n}{4}$.

2.4 Fibonacci Numbers*

Definition 2.29 The **Fibonacci numbers** F_n are defined recursively by

$$F_n = F_{n-1} + F_{n-2}, \qquad \text{for } n \geqslant 2, \tag{2.10}$$

with initial values $F_0 = 0$ and $F_1 = 1$.

#345
#347
#348
#352

So we see $F_2 = F_1 + F_0 = 1$, $F_3 = F_2 + F_1 = 1 + 1 = 2$, $F_4 = F_3 + F_2 = 2 + 1 = 3, \ldots$ and the Fibonacci sequence runs

$$0, 1, 1, 2, 3, 5, 8, 13, 21, 34, 55, 89, 144, \ldots.$$

The sequence continues to grow, always producing whole numbers and increasing by a factor of roughly 1.618 each time (see Proposition 2.30). We now use induction to prove some of the theory of Fibonacci numbers.

Proposition 2.30 *(Binet[15], 1843) For every integer $n \geqslant 0$,*

$$F_n = \frac{\alpha^n - \beta^n}{\sqrt{5}} \tag{2.11}$$

where $\alpha = (1 + \sqrt{5})/2$ and $\beta = (1 - \sqrt{5})/2$. In particular, F_n is the integer closest to $\alpha^n/\sqrt{5}$ and F_{n+1}/F_n approximates to $\alpha = 1.618\ldots$ as n becomes large.

Remark 2.31 (a) This result can also be written as $F_n = (\alpha^n - \beta^n)/(\alpha - \beta)$ which is symmetric in α and β.

(b) The number α is known as the **golden ratio** or **golden section**. ∎

Proof As F_n is defined in terms of F_{n-2} and F_{n-1}, then we need to know equation (2.11) holds for *two* consecutive Fibonacci numbers in order to be able to deduce anything about the next Fibonacci number. We note

when $n = 0$: LHS of equation (2.11) $= F_0 = 0 = \dfrac{1-1}{\sqrt{5}} = \dfrac{\alpha^0 - \beta^0}{\sqrt{5}} = \text{RHS}$, and

when $n = 1$: LHS of equation (2.11) $= F_1 = 1 = \dfrac{\sqrt{5}}{\sqrt{5}} = \dfrac{\alpha - \beta}{\sqrt{5}} = \text{RHS}$.

---

[15] After the French mathematician Jacques Philippe Marie Binet (1786–1856). Binet made significant contributions to number theory and the early theory of matrices and determinants. (2.11) is commonly attributed to him despite being known to Euler and De Moivre a century earlier.

#355
#365 Suppose now that equation (2.11) holds in the $(n-1)$th and nth cases, with the aim of prov-
#366 ing (2.11) holds for nth (nothing to show!) and $(n+1)$th cases. Based on our assumptions
#367 we may write

$$F_{n+1} = F_n + F_{n-1} = \frac{\alpha^n - \beta^n}{\sqrt{5}} + \frac{\alpha^{n-1} - \beta^{n-1}}{\sqrt{5}} = \frac{\alpha^{n-1}(\alpha+1)}{\sqrt{5}} - \frac{\beta^{n-1}(\beta+1)}{\sqrt{5}}.$$

At this point we note that α and β are the two roots of the quadratic $1 + x = x^2$, giving

$$F_{n+1} = \frac{\alpha^{n-1}\alpha^2}{\sqrt{5}} - \frac{\beta^{n-1}\beta^2}{\sqrt{5}} = \frac{\alpha^{n+1} - \beta^{n+1}}{\sqrt{5}}$$

which is the desired form. So based on our assumption (2.11) holds in the $(n+1)$th case. We now have two new consecutive numbers, n and $n+1$, for which (2.11) is true and so (2.11) is true for all natural numbers n by induction.

That F_{n+1}/F_n approximates to α for large n follows from the fact that $\alpha > 1 > |\beta|$. Hence

$$\frac{F_{n+1}}{F_n} = \frac{\alpha^{n+1} - \beta^{n+1}}{\alpha^n - \beta^n} = \frac{\alpha - (\beta^{n+1}/\alpha^n)}{1 - (\beta^n/\alpha^n)} \approx \alpha \text{ for large } n. \qquad \square$$

In the previous proof, we didn't particularly use induction as laid out formally in Theorems 2.3 and 2.5. We used the truth of (2.11) in two consecutive cases to keep the induction progressing. If we wished to be more formal in our approach we could apply induction, as in Theorem 2.3, to the statements

$$P(n): (2.11) \text{ holds true in the } n\text{th and } (n+1)\text{th cases.}$$

The next proof involves two variables, something we have already tackled with induction (see Remark 2.10). In the following example we proceed by treating at each stage m as arbitrary and inducting through n, but noting, as in Proposition 2.30, that we need two consecutive cases to be true to be able to proceed inductively.

Proposition 2.32 *For $m, n \geqslant 0$*

$$F_{n+m+1} = F_n F_m + F_{n+1} F_{m+1}. \qquad (2.12)$$

Proof For $n \geqslant 0$, we shall take $P(n)$ to be the statement that

$$(2.12) \text{ holds true for all } m \geqslant 0 \text{ in the } n\text{th and } (n+1)\text{th cases.}$$

#346 So we are using induction to progress through n and dealing with m simultaneously at each
#351 stage. To verify $P(0)$, we note that

$$F_{m+1} = F_0 F_m + F_1 F_{m+1} \text{ for all } m \geqslant 0, \qquad F_{m+2} = F_1 F_m + F_2 F_{m+1} \text{ for all } m \geqslant 0,$$

as $F_0 = 0$, $F_1 = F_2 = 1$. For the inductive step we assume $P(n)$, i.e. that for all $m \geqslant 0$,

$$F_{n+m+1} = F_n F_m + F_{n+1} F_{m+1} \qquad \text{and} \qquad F_{n+m+2} = F_{n+1} F_m + F_{n+2} F_{m+1}.$$

To prove $P(n+1)$ it remains to show that for all $m \geqslant 0$,

$$F_{n+m+3} = F_{n+2} F_m + F_{n+3} F_{m+1}. \qquad (2.13)$$

From our $P(n)$ assumptions and the definition of the Fibonacci numbers,

$$
\begin{aligned}
\text{LHS of (2.13)} &= F_{n+m+2} + F_{n+m+1} && \text{[by definition]}\\
&= F_n F_m + F_{n+1}F_{m+1} + F_{n+1}F_m + F_{n+2}F_{m+1} && \text{[by hypothesis]}\\
&= (F_n + F_{n+1})F_m + (F_{n+1} + F_{n+2})F_{m+1} && \text{[rearranging]}\\
&= F_{n+2}F_m + F_{n+3}F_{m+1} && \text{[by definition]}\\
&= \text{RHS of (2.13).} && \square
\end{aligned}
$$

n	M_n	I_n	F_n
1	0	1	1
2	1	0	1
3	1	1	2
4	2	1	3
5	3	2	5
6	5	3	8
7	8	5	13
8	13	8	21
9	21	13	34
10	34	21	55
11	55	34	89
12	89	55	144

Fibonacci. Leonardo of Pisa (c.1170–1250), more commonly known by his nickname *Fibonacci*, was the most talented European medieval mathematician. Besides his own mathematical contributions, his 1202 text, *Liber Abaci* ('book of calculation'), was influential in introducing Arabic numerals (0–9) and place value throughout Europe where the cumbersome Roman numerals were still being used. The Fibonacci numbers appeared in an exercise of *Liber Abaci*, but the sequence was in fact known to Indian mathematicians up to a millennium earlier. The name *Fibonacci numbers* was coined by the French mathematician Édouard Lucas (1842–1891), who also introduced the related *Lucas numbers* (see #366). Fibonacci's exercise that led to the sequence is as follows: how many pairs of rabbits will be produced in a year from a single pair if it is supposed that every mature pair produces a new pair each month and pairs become productive in their second month? The numbers in month n of immature pairs (I_n) and mature pairs (M_n) are given in the table.

Exercises

#344 A Write down two recurrences relating the numbers M_n and I_n as described previously and explain why $F_n = M_n + I_n$.

#345 a Show that F_n is even if and only if n is a multiple of 3.

#346 A† Let $n \geqslant 1$. Use Proposition 2.32 to show that $F_{2n-1} = (F_n)^2 + (F_{n-1})^2$ and deduce $F_{2n} = (F_{n+1})^2 - (F_{n-1})^2$.

#347 a Show, for $n \geqslant m \geqslant 0$, that $\sum_{k=m}^{n} F_k = F_{n+2} - F_{m+1}$.

#348 b Prove by induction the following identities involving the Fibonacci numbers:

(i) $F_1 + F_3 + F_5 + \cdots + F_{2n+1} = F_{2n+2}$.
(ii) $F_2 + F_4 + F_6 + \cdots + F_{2n} = F_{2n+1} - 1$.
(iii) $(F_1)^2 + (F_2)^2 + \cdots + (F_n)^2 = F_n F_{n+1}$.

#349 B Prove **Cassini's identity** [16] which states, for $n \geqslant 1$, that $F_{n+1}F_{n-1} - (F_n)^2 = (-1)^n$.

#350 A The Fibonacci numbers F_n may be extended to negative values for n by requiring that (2.10) holds for all integers n. Show that $F_{-n} = (-1)^{n+1}F_n$ for $n \geqslant 1$.

#351 B Proposition 2.32 holds for all integers m, n in the context of #350. Deduce **d'Ocagne's identity** [17], which generalizes Cassini's Identity, and states

$$F_{m-1}F_{n+1} - F_m F_n = (-1)^n F_{n-m}.$$

#352 b Prove, for $n \geqslant 0$, that $\sum_{k=0}^{n} kF_k = nF_{n+2} - F_{n+3} + 2$.

#353 D† Find an expression for $\sum_{k=1}^{n} kF_{2k-1}$ where $n \geqslant 1$.

#354 D Use Proposition 2.32 to show that $F_{(m+1)k} = F_{mk+1}F_k + F_{k-1}F_{mk}$. Show that if k divides n then F_k divides F_n. Deduce that if F_n is prime [18] then n is a prime or $n = 4$.

#355 B† Use (2.11) to prove **Cesàro's identity** [19] that $\sum_{k=0}^{n} \binom{n}{k} F_k = F_{2n}$.

#356 c Show α and β (as in Proposition 2.30) are roots of the equation $F_{m-1} + F_m x = x^m$ for any $m \geqslant 1$. Hence generalize the result of #355 to show that

$$\sum_{k=0}^{n} \binom{n}{k} (F_m)^k (F_{m-1})^{n-k} F_k = F_{mn}.$$

#357 D† Let $n \geqslant 1$. Show that there are F_{n+2} subsets of $\{1, 2, 3, \ldots n\}$ which contain no consecutive elements. For example, when $n = 3$ the $F_5 = 5$ such subsets of $\{1, 2, 3\}$ are $\varnothing, \{1\}, \{2\}, \{3\}, \{1, 3\}$.

#358 D† A **block fountain** is an arrangement of circles into rows such that each row is a continuous block and such that every circle (except in the first row) touches exactly two circles in the row beneath it. Such a block fountain, with 7 circles in the first row, is pictured in Figure 2.5. Let b_k denote the number of block fountains with k circles in the first row. Determine b_k for $k \leqslant 4$. Explain why

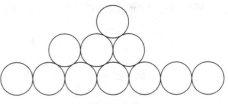

Figure 2.5

$$b_k = \sum_{i=1}^{k-1} (k-i)b_i + 1 \quad \text{for } k \geqslant 1,$$

and deduce that $b_k = F_{2k-1}$.

---

[16] After the Italian mathematician and astronomer Giovanni Cassini (1625–1712), now most remembered for identifying Cassini's Division in the rings of Saturn and four of its moons.

[17] After the French mathematician and engineer Maurice d'Ocagne (1862–1938).

[18] It is currently an open problem as to whether or not infinitely many of the Fibonacci numbers are prime.

[19] After the Italian mathematician Ernesto Cesàro (1859–1906). His work was mainly in differential geometry, though he is now more remembered for his ideas on divergent series. *Cesàro summability* generalizes traditional notions of series convergence; for example, the divergent series $1 - 1 + 1 - 1 + \cdots$ has Cesàro sum $1/2$.

#359 D Define the sequence c_k as follows. Let c_{2k-1} denote the number of block fountains with k circles in the bottom row (so $c_{2k-1} = b_k$) and let c_{2k} denote the number of block fountains with k or fewer circles in the bottom row (so $c_{2k} = b_1 + \cdots + b_k$). Show that $c_{2k} + c_{2k+1} = c_{2k+2}$. By considering whether the leftmost position on the second row is occupied by a circle or not, show also that $c_{2k-1} + c_{2k} = c_{2k+1}$. Deduce that $c_k = F_k$ for all k.

#360 D†(i) Show that there are F_{n-1} compositions $n = n_1 + \cdots + n_k$ of the number n where $n_i \geqslant 2$ for each i.
(ii) Show there are F_{n+1} compositions $n = n_1 + \cdots + n_k$ of the number n with $1 \leqslant n_i \leqslant 2$ for each i.
(iii) Show that there are F_n compositions $n = n_1 + \cdots + n_k$ of the number n where n_i is odd for each i.

#361 B† Let $n \geqslant 0$. Use Lemma 2.17 to prove by induction that

$$F_{n+1} = \binom{n}{0} + \binom{n-1}{1} + \binom{n-2}{2} + \cdots.$$

Re-interpret this result in terms of subsets of $\{1, 2, \ldots, n-1\}$ in light of #357.

#362 c† Let $i, j, k \geqslant 1$. Show that $F_{i+j+k} = F_{i+1}F_{j+1}F_{k+1} + F_i F_j F_k - F_{i-1}F_{j-1}F_{k-1}$.

#363 D Show that the numbers $m = F_{2k-1}F_{2k}$ and $n = F_{2k}F_{2k+1}$ solve the equation $\binom{n}{m-1} = \binom{n-1}{m}$. Deduce that there are infinitely many numbers that appear at least six times in Pascal's triangle.

#364 B† If the nth Fibonacci number F_n has r decimal digits, show that $n \leqslant 5r + 1$.

#365 B Use Proposition 2.30 to show that

$$\sum_{k=0}^{\infty} F_k x^k = \frac{x}{1 - x - x^2}.$$

This is the generating function of the Fibonacci numbers. For what values of x does the infinite series above converge?

#366 b The **Lucas numbers** L_n are defined by $L_0 = 2$ and $L_1 = 1$ and

$$L_n = L_{n-1} + L_{n-2} \text{ for } n \geqslant 2. \tag{2.14}$$

Prove that $L_n = 2F_{n-1} + F_n$ for $n \geqslant 1$. More generally, show that if a sequence of numbers G_n is defined by $G_n = G_{n-1} + G_{n-2}$ for $n \geqslant 2$, and by $G_0 = a$ and $G_1 = b$, show that $G_n = aF_{n-1} + bF_n$ for $n \geqslant 1$.

#367 A Show that $L_n = \alpha^n + \beta^n$ for $n \geqslant 0$, with α, β as defined in Proposition 2.30.

#368 a The Lucas numbers can be extended to negative integers by (2.14). Show for any n that $L_{-n} = (-1)^n L_n$.

#369 c The sequence G_n is determined by $G_n = G_{n-1} + G_{n-2}$ for all integers n and by $G_0 = a$ and $G_1 = b$.

(i) Suppose $G_{-n} = (-1)^n G_n$ for any n. Show that $G_n = bL_n$ for all n.
(ii) Suppose $G_{-n} = (-1)^{n+1}G_n$ for any n. Show that $G_n = bF_n$ for all n.

#370D† Show that the only natural numbers which are both Fibonacci numbers and Lucas numbers are 1, 2 and 3.

#371a Show for an integer n that $(1 + \sqrt{5})^n = 2^{n-1}L_n + 2^{n-1}F_n\sqrt{5}$.

#372A Show that $F_{2n} = F_n L_n$ for $n \geqslant 1$. Deduce the identity $F_{2^n} = L_2 L_4 L_8 \cdots L_{2^{n-1}}$.

#373a Prove, for any integer n, that $(L_n)^2 - 5(F_n)^2 = 4(-1)^n$.

#374D† Show, for $0 \leqslant k \leqslant n$ with k even, that $F_{n-k} + F_{n+k} = L_k F_n$. Can you find, and prove, a similar expression for $F_{n-k} + F_{n+k}$ when k is odd?

#375C Show that $F_{n+k} + (-1)^k F_{n-k} = L_k F_n$ and that $L_{n+k} + (-1)^k L_{n-k} = L_k L_n$.

#376D† Show that if L_n is prime, then $n = 0$, is prime or is a power of 2.

#377D† (**Zeckendorf's theorem** [20]) Show that every integer $n \geqslant 1$ can be written uniquely as a sum of non-consecutive Fibonacci numbers, F_k, where $k \geqslant 2$; for example

$$26 = 5 + 21 = F_5 + F_8, \qquad 107 = 5 + 13 + 89 = F_5 + F_7 + F_{11}.$$

Decompose 900 into such a sum.

#378C Fibonacci multiplication can be defined as if the Fibonacci numbers formed a base like binary or decimal. If we represent two positive integers m, n as $m = \sum_i F_{c_i}$ and $n = \sum_j F_{d_j}$ using Zeckendorf's theorem, then we can define their Fibonacci product as

$$m \circ n = \sum_{i,j} F_{c_i + d_j}. \tag{2.15}$$

Show that $1 \circ 3 = 2 \circ 2$. Determine $5 \circ 7$ and $17 \circ 23$.

#379C Show that $\circ$ is commutative but not distributive; that is, in general,

$$m \circ n = n \circ m, \qquad \text{but} \qquad (l+m) \circ n \neq l \circ m + l \circ n.$$

#380D† Let $\beta = (1 - \sqrt{5})/2$. Given a Zeckendorf representation $n = \sum_t F_{e_t}$ of a positive integer n, we define

$$n_\beta = \sum_t \beta^{e_t}.$$

(i) Show that $n_\beta = a_n + n\beta$ for some integer a_n.
(ii) Show that the interval $-\beta^2 < x < -\beta$ has unit length and that n_β lies in this interval.
(iii) Conversely if $-\beta^2 < a + n\beta < -\beta$ for integers a, n and $n \geqslant 1$, show $n_\beta = a + n\beta$.

#381D Let m and n be positive integers. Show that $(m \circ n)_\beta = m_\beta n_\beta$. Deduce that $\circ$ is associative. [21]

---

[20] After the Belgian medical doctor and mathematician Édouard Zeckendorf (1901–1983).
[21] Associativity was first shown by Donald Knuth in [21], whilst the alternative approach of #380 is due to Pierre Arnoux [1].

#382 D† Associativity means that powers using $\circ$ are well defined. What are $2^{\circ n}$, $3^{\circ n}$ and $5^{\circ n}$? Show that

$$4^{\circ n} = \begin{cases} 5^{n/2} F_{3n} & n \text{ even} \\ 5^{(n-1)/2} L_{3n} & n \text{ odd} \end{cases} \qquad \text{and} \qquad 6^{\circ n} = 2^n F_{4n}.$$

#383 D† Determine $\sum_{n=1}^{\infty} \frac{1}{F_n F_{n+2}}$.

#384 E† Show that $\sum_{n=0}^{\infty} \frac{1}{1+F_{2n+1}} = \sqrt{5}/2$.

#385 E† Show that $\sum_{n=0}^{\infty} \frac{1}{F_{2^n}} = (7 - \sqrt{5})/2$.

2.5 Recurrence Relations*

We begin with the following example.

Example 2.33 Find an expression for the numbers x_n defined recursively by the relation

$$x_{n+2} - 2x_{n+1} + x_n = 2 \text{ for } n \geqslant 0, \qquad \text{with} \qquad x_0 = 1, x_1 = 1. \qquad (2.16)$$

#386
#388
#389
#390

Solution We see that any x_n can be determined by applying this relation sufficiently many times from our initial values of $x_0 = 1$ and $x_1 = 1$. So for example to find x_7 we'd calculate

$$\begin{array}{ll} x_2 = 2x_1 - x_0 + 1 = 2 - 1 + 2 = 3; & x_5 = 2x_4 - x_3 + 1 = 26 - 7 + 2 = 21; \\ x_3 = 2x_2 - x_1 + 1 = 6 - 1 + 2 = 7; & x_6 = 2x_5 - x_4 + 1 = 42 - 13 + 2 = 31; \\ x_4 = 2x_3 - x_2 + 1 = 14 - 3 + 2 = 13; & x_7 = 2x_6 - x_5 + 1 = 62 - 21 + 2 = 43. \end{array}$$

If this was the first time we had seen such a problem, then we might try pattern spotting or qualitatively analyzing the sequence's behaviour, in order to make a guess at a general formula for x_n. Simply looking at the sequence x_n above, no terribly obvious pattern emerges. However, we can see that the x_n are growing, roughly at the same speed as n^2 grows. We might note further that the differences between the numbers $0, 2, 4, 6, 8, 10, 12, \ldots$ are going up linearly. Even if we didn't know how to sum an arithmetic progression, it would seem reasonable to *try* a solution of the form

$$x_n = an^2 + bn + c, \qquad (2.17)$$

where a, b, c are constants, as yet undetermined. *If* a solution of the form (2.17) exists, we can find a, b, c using the first three cases, so that

$$c = x_0 = 1; \qquad a + b + c = x_1 = 1; \qquad 4a + 2b + c = x_2 = 3,$$

giving $a = 1, b = -1, c = 1$. So the only expression of the form (2.17) which works for $n = 0, 1, 2$ is

$$x_n = n^2 - n + 1. \qquad (2.18)$$

If we put $n = 3, 4, 5, 6, 7$ into (2.18) then we get the correct values of x_n calculated at the start of this solution. This is, of course, not a proof, but we could prove this formula to be correct for all values of $n \geqslant 0$ using induction as follows. We have already checked that the formula (2.17) is correct for $n = 0$ and $n = 1$. As our inductive hypothesis let's suppose it was also true for $n = k$ and $n = k + 1$. Then

$$x_{k+2} = 2x_{k+1} - x_k + 2$$
$$= 2\left\{(k+1)^2 - (k+1) + 1\right\} - \left\{k^2 - k + 1\right\} + 2$$
$$= k^2 + 3k + 3 = (k+2)^2 - (k+2) + 1$$

which is the correct formula with $n = k + 2$.

Alternatively, having noted the differences between consecutive x_n go up as $0, 2, 4, 6, 8, \ldots$ we can write, using statement B from the start of this chapter,

$$x_n = 1 + 0 + 2 + 4 + \cdots + (2n - 2) = 1 + \sum_{k=0}^{n-1} 2k = 1 + 2\frac{1}{2}(n-1)n = n^2 - n + 1.$$

To make this proof water-tight we need to check that the pattern $0, 2, 4, 6, 8, \ldots$ seen in the differences carries on forever, and that this wasn't just a fluke. But this follows if we note

$$x_{n+2} - x_{n+1} = x_{n+1} - x_n + 2$$

and so the difference between consecutive terms is increasing by 2 each time. $\square$

As a second (somewhat contrasting) example, recall the result below from the previous section.

Example 2.34 (Proposition 2.30) The recurrence relation

$$x_{n+2} - x_{n+1} - x_n = 0, \qquad x_0 = 0, x_1 = 1,$$

has solution $x_n = F_n = (\alpha^n - \beta^n)/\sqrt{5}$ where $\alpha = (1 + \sqrt{5})/2$ and $\beta = (1 - \sqrt{5})/2$.

It became apparent in the proof of Proposition 2.30 that α and β are crucially the two roots of the quadratic $x^2 - x - 1 = 0$. Because α and β have this property, it follows that $x_n = \alpha^n$ and $x_n = \beta^n$ separately solve the relation $x_{n+2} - x_{n+1} - x_n = 0$. By taking an appropriate linear combination of these two solutions one can meet any initial conditions. Indeed it was shown in #366 that $x_n = aF_{n-1} + bF_n$ is the solution to the recurrence relation

$$x_{n+2} - x_{n+1} - x_n = 0, \qquad x_0 = a, x_1 = b.$$

How does (2.16) compare with the above? One complication is that the relation is not *homogeneous* (there is a non-zero function, in this case 2, on the RHS). Secondly, the quadratic equation associated with (2.16) is $x^2 - 2x + 1 = 0$ which has repeated roots $1, 1$. From this we can note $1^n = 1$ is a solution of the associated homogeneous relation $x_{n+2} - 2x_{n+1} + x_n = 0$ but it remains unclear what a second solution might be. We now address these problems generally with the following theorem.

Theorem 2.35 (*Homogeneous Linear Constant Coefficient Recurrences*) *Let a, b, c be real (or complex) numbers with $a \neq 0$. Consider the recurrence relation below in x_n,*

$$ax_{n+2} + bx_{n+1} + cx_n = 0 \qquad \text{for } n \geqslant 0. \tag{2.19}$$

*Let α and β be the roots of the **auxiliary equation** $ax^2 + bx + c = 0$.*

(a) *If $\alpha \neq \beta$, the general solution of (2.19) has the form $x_n = A\alpha^n + B\beta^n$ for $n \geqslant 0$.*
(b) *If $\alpha = \beta \neq 0$, the general solution of (2.19) has the form $x_n = (An + B)\alpha^n$ for $n \geqslant 0$.*

In each case, the values of A and B are uniquely determined by the values of x_0 and x_1.

#392
#393
#395
#407
#408

Proof First we note that the sequence x_n defined in (2.19) is uniquely determined by the initial values x_0 and x_1. Knowing these values (2.19) gives us x_2, knowing x_1 and x_2 gives us x_3, etc. So if we can find a solution to (2.19) for certain initial values then we have *the* unique solution; if we can find a solution for arbitrary initial values then we have the general solution.

Note that if $\alpha \neq \beta$ then putting $x_n = A\alpha^n + B\beta^n$ into the LHS of (2.19) gives

$$ax_{n+2} + bx_{n+1} + cx_n = a(A\alpha^{n+2} + B\beta^{n+2}) + b(A\alpha^{n+1} + B\beta^{n+1}) + c(A\alpha^n + B\beta^n)$$
$$= A\alpha^n(a\alpha^2 + b\alpha + c) + B\beta^n(a\beta^2 + b\beta + c) = 0$$

as α and β are both roots of the auxiliary equation.

Similarly if $\alpha = \beta$ then putting $x_n = (An + B)\alpha^n$ into the LHS of (2.19) gives

$$ax_{n+2} + bx_{n+1} + cx_n = a(A(n+2) + B)\alpha^{n+2} + b(A(n+1) + B)\alpha^{n+1} + c(An + B)\alpha^n$$
$$= A\alpha^n n((a\alpha^2 + b\alpha + c) + (2a\alpha + b)\alpha) + B\alpha^n(a\alpha^2 + b\alpha + c) = 0$$

because α is a root of the auxiliary equation and also of the derivative of the quadratic in the auxiliary equation, being a repeated root. (Or, if you prefer, you can show that $2a\alpha + b = 0$ by noting that $ax^2 + bx + c = a(x - \alpha)^2$, comparing coefficients and eliminating c.)

So in either case we have a set of solutions. Further, the initial equations $A + B = x_0$ and $A\alpha + B\beta = x_1$ are uniquely solvable for A and B when $\alpha \neq \beta$, whatever the values of x_0 and x_1. Similarly when $\alpha = \beta \neq 0$ then the initial equations $B = x_0$ and $(A + B)\alpha = x_1$, have a unique solution in A and B whatever the values of x_0 and x_1. So in each case our solutions encompassed the general solution. □

Remark 2.36 When $\alpha = \beta = 0$ then (2.19) reads $x_{n+2} = 0$ and has solution x_n given by $x_0, x_1, 0, 0, 0, 0, \dots$. ∎

Let us return now to solving (2.16). The auxiliary equation $x^2 - 2x + 1 = 0$ has repeated roots $1, 1$ and so we know from Theorem 2.35 that $x_n = (An + B)1^n = An + B$ is the general solution of the homogeneous relation $x_{n+2} - 2x_{n+1} + x_n = 0$ but what of the inhomogeneous relation

$$x_{n+2} - 2x_{n+1} + x_n = 2? \tag{2.20}$$

Well, supposing we had *a* solution X_n to (2.20), then $X_n + An + B$ would also be a solution to (2.20); in fact we shall see below that this is the general solution to (2.20). Assuming that fact for now, we still have the problem of finding any solution X_n to begin, a so-called **particular solution**. We shall do this with some educated guesswork.

We are looking to find *any* solution X_n to the recurrence relation (2.20). A sensible starting point would be to *try* a function of a similar form to that appearing on the RHS. So we might try setting $X_n = B$ (a constant) hoping for a solution; however, thinking further, we see that this would yield 0, as constants are part of the general solution to the homogeneous relation. More generally, so is $An + B$, so instead we could try $X_n = Cn^2$. If we substitute this into the LHS of (2.20) we obtain

$$\text{LHS} = C(n+2)^2 - 2C(n+1)^2 + Cn^2 = 2C.$$

As the RHS of (2.16) equals 2 then $C = 1$ and we have a particular solution n^2. Given Theorem 2.37, the general solution to (2.20) is $x_n = n^2 + An + B$ and we can specify A and B by making use of the initial conditions $x_0 = 1$ and $x_1 = 1$.

Theorem 2.37 *(Inhomogeneous Recurrences) Let a, b, c be real (or complex) numbers with $a \neq 0$, and let f be a function defined on the natural numbers. Let X_n be a particular solution of the inhomogeneous recurrence relation*

$$ax_{n+2} + bx_{n+1} + cx_n = f(n) \qquad \text{for } n \geqslant 0. \tag{2.21}$$

Then x_n solves (2.21) if and only if $x_n = X_n + y_n$ where y_n solves the homogeneous recurrence relation (2.19).

Proof We know $aX_{n+2} + bX_{n+1} + cX_n = f(n)$ for $n \geqslant 0$ and hence

$$ax_{n+2} + bx_{n+1} + cx_n = f(n) \iff ax_{n+2} + bx_{n+1} + cx_n = aX_{n+2} + bX_{n+1} + cX_n$$
$$\iff a(x_{n+2} - X_{n+2}) + b(x_{n+1} - X_{n+1}) + c(x_n - X_n) = 0$$
$$\iff ay_{n+2} + by_{n+1} + cy_n = 0$$

where $y_n = x_n - X_n$. $\qquad\qquad\qquad\square$

Remark 2.38 You might have noticed that in the above proof we made no use of the fact that a, b, c are constant. The important point is that recurrence relation is linear in the x_n. Theorem 2.37 more generally holds true when the coefficients a, b, c are functions of n. $\qquad\blacksquare$

We end with two examples of inhomogeneous recurrence relations.

Example 2.39 Find the solution of the following recurrence relation given initial values of $x_0 = x_1 = 3$,

$$x_{n+2} - 5x_{n+1} + 6x_n = 2^n + 4n. \tag{2.22}$$

Solution The auxiliary equation $x^2 - 5x + 6 = 0$ has roots 2 and 3. From Theorem 2.35 we know the general solution of the homogeneous equation

$$x_{n+2} - 5x_{n+1} + 6x_n = 0 \tag{2.23}$$

to be $x_n = A2^n + B3^n$ where A and B are constants. To find a particular solution of the recurrence relation (2.22) we will try various educated guesses $x_n = X_n$, looking for a particular solution which is similar in nature to $2^n + 4n$. We can deal with the $4n$ on the RHS by contributing a term $an + b$ to X_n – what the values of a and b should be will become evident later. But trying to deal with the 2^n on the RHS of (2.22) with contributions to X_n of the form $c2^n$ will not help, as we already know this to be part of the solution to the homogeneous equation (2.23), and as such would just yield 0 on the RHS; rather we need to try instead a multiple of $n2^n$ to deal with the 2^n. So let's try a particular solution of the

#400
#402

form $X_n = an + b + cn2^n$, where a, b, c are constants, as yet undetermined. Putting $x_n = X_n$ into the LHS of (2.22) we get

$$\left\{ a(n+2) + b + c(n+2)2^{n+2} \right\} - 5\left\{ a(n+1) + b + c(n+1)2^{n+1} \right\} + 6\left\{ an + b + cn2^n \right\}$$

$$= 2an + (2b - 3a) - 2c2^n.$$

We need this expression to equal $2^n + 4n$ for all $n \geqslant 0$ and so we see $a = 2, b = 3, c = -1/2$. Hence a particular solution is $x_n = 2n + 3 - n2^{n-1}$, and the general solution of (2.22) is

$$x_n = A2^n + B3^n + 2n + 3 - n2^{n-1}.$$

Recalling the initial conditions $x_0 = x_1 = 3$ we see $A + B + 3 = 3$ and $2A + 3B + 2 + 3 - 1 = 3$, so that $A = 1$ and $B = -1$. Finally then the unique solution of (2.22) is

$$x_n = 2^n - 3^n + 2n + 3 - n2^{n-1}. \qquad \square$$

Example 2.40 Find the solution of the following recurrence relation given initial values of $x_0 = x_1 = x_2 = 0$,

$$x_{n+3} + x_{n+2} - 2x_n = 1. \tag{2.24}$$

Solution Note that this recurrence relation is third order, but the theory we have previously outlined applies just as well. The auxiliary equation here is $x^3 + x^2 - 2 = 0$, which factorizes as

$$x^3 + x^2 - 2 = (x - 1)(x^2 + 2x + 2) = 0$$

and has roots $x = 1, -1 + i, -1 - i$. So the general solution of the homogeneous recurrence relation is

$$x_n = A + B(-1+i)^n + C(-1-i)^n,$$

where A, B, C are constants, potentially complex. We now need to find a particular solution of the inhomogeneous equation. Because constant sequences are solutions of the homogeneous equation, there is no point trying these as particular solutions; instead we try one of the form $x_n = kn$. Putting this into the recurrence relation we obtain

$$k(n+3) + k(n+2) - 2kn = 1 \quad \text{which simplifies to} \quad 5k = 1$$

and so $k = 1/5$. The general solution of the inhomogeneous recurrence relation (2.24) then has the form

$$x_n = \frac{n}{5} + A + B(-1+i)^n + C(-1-i)^n.$$

At first glance this solution does not necessarily look like it is a real sequence, and indeed B and C need to be complex constants for this to be the case. However, from De Moivre's theorem we have

$$(-1+i)^n = \left(\sqrt{2} \operatorname{cis} \frac{3\pi}{4} \right)^n = 2^{n/2} \operatorname{cis} \frac{3n\pi}{4};$$

$$(-1-i)^n = \left(\sqrt{2} \operatorname{cis} \frac{5\pi}{4} \right)^n = 2^{n/2} \operatorname{cis} \frac{5n\pi}{4},$$

and we will use these to rearrange our solution in terms of overtly real sequences. To calculate A, B and C, we use the initial conditions ($n = 0, 1, 2$). We see that

$$A + B + C = 0;$$
$$A + B(-1 + i) + C(-1 - i) = -1/5;$$
$$A + B(-2i) + C(2i) = -2/5,$$

and solving these gives $A = -4/25$, $B = (4 - 3i)/50$ and $C = (4 + 3i)/50$. Hence the unique solution is

$$x_n = \frac{n}{5} - \frac{8}{50} + \left(\frac{4 - 3i}{50}\right)(-1 + i)^n + \left(\frac{4 + 3i}{50}\right)(-1 - i)^n.$$

The last two terms are conjugates of one another and so, recalling that $z + \bar{z} = 2 \operatorname{Re} z$, we have

$$x_n = \frac{1}{50}(10n - 8 + 2 \operatorname{Re}\left[(4 - 3i)(-1 + i)^n\right])$$
$$= \frac{1}{50}\left(10n - 8 + 2 \times 2^{n/2} \operatorname{Re}\left[(4 - 3i)\left(\operatorname{cis}\frac{3n\pi}{4}\right)\right]\right)$$
$$= \frac{1}{25}\left(5n - 4 + 2^{n/2}\left(4\cos\frac{3n\pi}{4} + 3\sin\frac{3n\pi}{4}\right)\right). \qquad \square$$

Exercises

#386a The sequence x_n is defined recursively by $x_{n+2} = x_{n+1} + 2x_n$ and by $x_0 = 1$, and $x_1 = -1$. Calculate x_n for $n \leqslant 6$, and conjecture the value of x_n for general n. Use induction to prove that your conjecture correct.

#387c The sequence x_n is defined recursively by $x_n = 2x_{n-1} - x_{n-2}$ and by $x_0 = a$ and $x_1 = b$. Calculate x_n for $n \leqslant 6$, and conjecture the value of x_n for general n. Use induction to prove your conjecture correct.

#388B† The sequence x_n is defined recursively by $x_n = (6/5)x_{n-1} - x_{n-2}$ with $x_0 = 0$, $x_1 = 4/5$. With the aid of a calculator list the values of x_i for $0 \leqslant i \leqslant 7$. Prove, by induction, that

$$x_n = \operatorname{Im}\{((3 + 4i)/5)^n\} \qquad \text{for} \quad n \geqslant 0.$$

Deduce that $|x_n| \leqslant 1$ for all n. Show also that x_n cannot have the same sign for more than three consecutive n.

#389B The sequence of numbers x_n is defined recursively by $x_{n+3} = 3x_{n+1} + 2x_n$ and by $x_0 = 1$, $x_1 = 3$, $x_2 = 5$. Show by induction that

$$2^n < x_n < 2^{n+1} \text{ for } n \geqslant 1 \qquad \text{and that} \qquad x_{n+1} = 2x_n + (-1)^n.$$

#390B† Solve the recurrence relation $x_{n+2} = 3x_{n+1} - 2x_n$ with $x_0 = 2$ and $x_1 = 4$.

#391c Let a, b be real numbers and $l > k > 0$ be integers. Show that there is a unique solution to the relation $x_{n+2} = 3x_{n+1} - 2x_n$ with $x_k = a$ and $x_l = b$.

#392b Solve the following recurrence relations:

(i) $x_{n+2} + 5x_{n+1} + 6x_n = 0,$ $x_0 = x_1 = 1.$
(ii) $x_{n+2} + 2x_{n+1} + 2x_n = 0,$ $x_0 = x_1 = 1.$

Present your answer to part (ii) so that it is clear the solution is a real sequence.

#393B† Solve the recurrence relation $3x_{n+2} = 2x_{n+1} + x_n$, with $x_0 = 2$ and $x_1 = 1$. Show that $x_n \approx 5/4$ for large n.

#394C† Let a be a real or complex number. The non-zero sequence x_n satisfies the identity

$$x_{n+1} + x_{n-1} = ax_n \qquad \text{for } n \geqslant 1. \tag{2.25}$$

(i) If $x_{n+4} = -x_n$ for all n, find the possible values of a. For such a what is the period of x_n?
(ii) For these values of a, for what values of λ is $x_n = \lambda^n$ a solution of (2.25)?
(iii) What values of a would lead to x_n having period 3?

#395B† In #259 you were asked to show that $x_n = 3 \times 2^{2n} + 2 \times 3^{2n}$ is divisible by 5 for all $n \geqslant 0$. Find a recurrence relation of the form (2.19) and values of x_0, x_1 such that x_n is the unique solution. Deduce directly from this recurrence relation that x_n is divisible by 5 for $n \geqslant 0$.

#396C† Find the solution of the recurrence relation

$$x_{n+2} - (2a+\varepsilon)x_{n+1} + (a+\varepsilon)ax_n = 0, \qquad x_0 = A, x_1 = B.$$

Show that as ε becomes close to 0, this solution approximates to the solution of

$$x_{n+2} - 2ax_{n+1} + a^2 x_n = 0, \qquad x_0 = A, x_1 = B.$$

#397c† Find all solutions to the recurrence relation $x_{n+3} = 2x_{n+2} - x_{n+1} + 2x_n$ with $x_0 = 2$ and $x_1 = 4$.

#398d† Find the general solution of $x_{n+4} + 6x_{n+3} + 18x_{n+2} + 30x_{n+1} + 25x_n = 0$.

#399b Solve the recurrence relation $x_{n+2} = 5 - x_n - 3x_{n+1}$ with $x_0 = 2$ and $x_1 = 4$.

#400B† Obtain particular solutions to the recurrence relation $x_{n+2} + 2x_{n+1} - 3x_n = f(n)$ in each of the following cases: (i) $f(n) = 2^n$; (ii) $f(n) = 5$; (iii) $f(n) = n(-3)^n$.

#401c Solve the following recurrence relations:

(i) $x_{n+2} - 3x_{n+1} + 2x_n = n.$
(ii) $x_{n+2} + 5x_{n+1} + 6x_n = 72n^2 + 5,$ $x_0 = x_1 = 1.$

#402B† Find the general solution of $x_{n+1} = x_n + \sin n$.

#403c Show $x_n = n!$ is a solution of the recurrence relation $x_{n+2} = (n+2)(n+1)x_n$. By making the substitution $x_n = n! u_n$ find a second independent solution. Now find the unique solution given that $x_0 = 1$, $x_1 = 3$.

#404 C(i) Suppose now that $ax^2 + bx + c = 0$ has distinct solutions α, β. The solution to (2.19) is of the form $x_n = A\alpha^n + B\beta^n$. Find A and B in terms of x_0 and x_1. Show that as α and β approximate to 0 that $A\alpha^n + B\beta^n$ becomes closer to the nth term of $x_0, x_1, 0, 0, 0, 0, \ldots$
(ii) Suppose now that $ax^2 + bx + c = 0$ has the repeated non-zero solution α, α. The solution to (2.19) is of the form $x_n = (An + B)\alpha^n$. Find A and B in terms of x_0 and x_1. Show that as α becomes smaller then $(An + B)\alpha^n$ becomes closer to the nth term of $x_0, x_1, 0, 0, 0, 0, \ldots$

#405 D† The sequence x_n satisfies the recurrence relation

$$31x_n = 45x_{n-1} - 17x_{n-2} + x_{n-3} \qquad \text{for } n \geqslant 3; \qquad x_0 = x_1 = x_2 = 1.$$

Show that the roots of the auxiliary equation are reals between 0 and 1. Without working out an explicit formula for x_n, show that

$$x_0 + x_1 + x_2 + \cdots = 10.$$

#406 D With x_n as defined in #405, evaluate

$$\sum_{n=0}^{\infty} x_{2n}, \qquad \sum_{n=0}^{\infty} x_{3n}, \qquad \sum_{n=0}^{\infty} x_{4n}.$$

#407 B† Consider the following simple weather model. [22] On a given day, the weather is in one of two states, sunny or rainy. If it is sunny today then there is a 90% chance of sun tomorrow, 10% chance of rain. If it is rainy today then there is a 50% chance of sun tomorrow, 50% chance of it being rainy.

(i) Using the fact that the probability-of-sun-tomorrow equals

(probability-of-sun-tomorrow-and-rain-today)

+ (probability-of-sun-tomorrow-and-sun-today),

show that, long term, there is a 5/6 chance of sun and 1/6 chance of rain on any day.
(ii) Given that it is sunny on day 0, let S_n denote the probability that it is sunny on day n. Show $S_{n+1} = 0.4S_n + 0.5$ and hence determine S_n. Verify that S_n approximates to the value you found in (i) for large n.
(iii) Find an expression for R_n, the probability that it is sunny on day n given that it is rainy on day 0.

#408 B† (**Gambler's ruin**) A casino runs the following game. A player is given 9 tokens initially. The game played then involves consecutively rolling a die and with each roll one token is added or subtracted from the player's pile. If $1, 2, 3$ or 4 is rolled a token is removed; if 5 or 6 is rolled then the player receives a token. The player loses the game if his/her stock of tokens reaches 0 and wins if the stock reaches 10. This question investigates the probability of the player winning.

(i) For $0 \leqslant n \leqslant 10$, let p_n denote the probability the player wins with a stock of n tokens. What are p_0 and p_{10}?

---

[22] This is an example of a Markov chain, as are the gambler's ruin problem (#408), and random walks (#410). See p.278 for further details.

(ii) Explain why

$$p_n = \frac{1}{3}p_{n+1} + \frac{2}{3}p_{n-1} \qquad \text{for } 1 \leqslant n \leqslant 9.$$

(iii) What is the general solution to the recurrence relation in (ii)? What is the unique solution which gives the correct values of p_0 and p_{10}?

(iv) What is the probability of the player winning?

#409 C Generalize #408 as follows. Suppose that at each turn the player has probability p of winning and probability $q = 1 - p$ of losing. Say, further, that the player starts with K tokens, loses the game by being reduced to zero tokens and wins if the stock of token reaches N tokens (where $N > K$). What is the probability of the player winning? (You will need to treat the cases $p \neq q$ and $p = q$ separately.)

#410 B† A point P follows a **random walk** on the integers as follows. At time $t = 0$, the point P is at 0 and moves according to the following rule: if P is at n at time t there is a half chance that P moves to $n + 1$ at time $t + 1$ and a half chance that P moves to $n - 1$ at time $t + 1$. Let X_t denote the position of P at time t.

(i) For $t \geqslant 0$ and k an integer, what is the probability that $X_t = k$?

(ii) What is the mean value of X_t?

(iii) Show that the mean value of $(X_t)^2$ equals t.

#411 C A fair coin is tossed repeatedly. Let E denote the expected wait for two successive heads. By considering the cases where the first toss is T, the first two tosses are HH and the first two tosses are HT, explain why

$$E = \frac{1}{2}(E + 1) + \frac{1}{4} \times 2 + \frac{1}{4}(E + 2).$$

Hence determine E.

#412 D† A fair coin is tossed repeatedly. Let A_n denote the probability that the first time two successive heads occur is on the $(n-1)$th and nth tosses. Explain why

$$A_n = \frac{1}{2}A_{n-1} + \frac{1}{4}A_{n-2} \qquad \text{for } n \geqslant 3,$$

and deduce that $A_n = F_{n-1}/2^n$. Explain this result in light of #357. Also determine E (as defined in #411) by evaluating an infinite sum.

#413 C A fair coin is tossed repeatedly. Let F denote the expected wait for heads-then-tails on two successive tosses. Let G denote the expected wait for HT given that the very first toss is H. Explain why

$$F = \frac{1}{2}(F + 1) + \frac{1}{2}G; \qquad G = \frac{1}{2} \times 2 + \frac{1}{2}(G + 1).$$

Deduce that $F = 4$.

#414 C† A fair coin is tossed repeatedly. What is the probability f_n that the first time heads-then-tails occurs is on the $(n-1)$th and nth tosses? Rederive the value of F (from #413) by evaluating an infinite sum.

#415C A fair coin is tossed repeatedly. What is the probability that the sequence heads-heads appears before the sequence heads-tails has appeared?

#416D† A fair coin is tossed repeatedly. Person A waits until the first time three consecutive tosses land HTH. Person B waits until they first get HTT consecutively. What is the average wait for each player?

#417D† In the situation of #416, let a_n (resp. b_n) denote the probability that Person A's (resp. B's) sequence concludes first at the nth toss. Show that

$$a_n = a_{n-1} - \frac{1}{4}a_{n-2} + \frac{1}{8}a_{n-3}; \qquad b_n = b_{n-1} - \frac{1}{8}b_{n-3}, \qquad \text{for } n \geqslant 4.$$

#418C With notation as in #417, show that $b_n = (F_n - 1)/2^n$.

#419D In the situation of #416, what is the probability that Person A's sequence occurs before Person B's?

#420B† Let A_n denote the number of ways to cover a $2 \times n$ grid using 2×1 rectangular tiles. For example, $A_2 = 2$ and $A_3 = 3$ can be seen from

1	1
2	2

1	2
1	2

1	2	3
1	2	3

1	1	3
2	2	3

1	2	2
1	3	3

Explain why $A_{n+2} = A_{n+1} + A_n$ for $n \geqslant 1$, and deduce that $A_n = F_{n+1}$.

#421C Let B_n denote the number of ways to cover a $3 \times n$ grid using 3×1 rectangular tiles. Show that

$$B_n = B_{n-1} + B_{n-3} \qquad \text{for } n \geqslant 4$$

and so determine B_{10}.

#422D Let B_n be as defined in #421. Show that

$$B_n = \binom{n}{0} + \binom{n-2}{1} + \binom{n-4}{2} + \binom{n-6}{3} + \cdots \qquad \text{for } n \geqslant 1.$$

#423D† Let C_n denote the number of ways to cover a $3 \times n$ grid using 2×1 rectangular tiles.

(i) What is C_n when n is odd?
(ii) Let D_n denote the number of ways of covering the grid below using 2×1 rectangular tiles:

Explain why $C_n = 2D_{n-1} + C_{n-2}$ for $n \geqslant 3$.
(iii) Find a second recurrence relation between the sequences C_n and D_n and deduce that $C_{n+1} = 4C_{n-1} - C_{n-3}$ for $n \geqslant 4$. Hence find a formula for C_{2n}.

2.6 Further Exercises*

Exercises on Inequalities

#424 B Rederive the result of #8 using the AM–GM inequality.

#425 B A rod of length L is cut into 12 pieces so as to make a frame in the shape of a cuboid. Show that the greatest possible volume of the cuboid is $(L/12)^3$.

#426 C† A cuboid has surface area A. Show that its volume does not exceed $(A/6)^{3/2}$.

#427 C An open box (i.e. a cuboid lacking a top) has volume V. Show that its surface area is at least $3(2V)^{2/3}$.

#428 B Heron's formula [23] for the area A of a triangle with sides a, b and c, states that

$$A^2 = s(s-a)(s-b)(s-c) \qquad \text{where} \qquad s = \frac{1}{2}(a+b+c).$$

Deduce that a triangle of a given perimeter has greatest area when it is equilateral.

#429 C† Let $x_1, \ldots, x_n$ be positive reals with $x_1 x_2 \times \cdots \times x_n = 1$. Show that

$$(1+x_1)(1+x_2) \times \cdots \times (1+x_n) \geqslant 2^n.$$

#430 D† Let $0 < a_1 \leqslant a_2 \leqslant \cdots \leqslant a_n$ and $b_1, \ldots, b_n$ be positive. Prove the **rearrangement inequality**, which states

$$a_1 c_1 + \cdots + a_n c_n \quad \leqslant \quad a_1 b_1 + \cdots + a_n b_n \quad \leqslant \quad a_1 d_1 + \cdots + a_n d_n$$

where $c_1, \ldots, c_n$ is the list of $b_1, \ldots, b_n$ in decreasing order and $d_1, \ldots, d_n$ is the list of $b_1, \ldots, b_n$ in increasing order. Deduce **Chebyshev's inequality** [24], which states whenever $0 < x_1 \leqslant x_2 \leqslant \cdots \leqslant x_n$ and $0 < y_1 \leqslant y_2 \leqslant \cdots \leqslant y_n$ then

$$\left(\frac{x_1 + \cdots + x_n}{n} \right) \left(\frac{y_1 + \cdots + y_n}{n} \right) \quad \leqslant \quad \frac{x_1 y_1 + \cdots + x_n y_n}{n}.$$

#431 C Let $x, y, z > 0$. Use Chebyshev's inequality to show that

$$\frac{x+y+z}{3} \quad \leqslant \quad \frac{x^3 + y^3 + z^3}{x^2 + y^2 + z^2}.$$

#432 D† (See also #1551.) Let $0 < a < b$. Define sequences by setting $a_1 = a$, $b_1 = b$ and then

$$a_{n+1} = \sqrt{a_n b_n}; \qquad b_{n+1} = \frac{a_n + b_n}{2}, \qquad \text{for } n \geqslant 1.$$

(i) Show that $a_1 < a_2 < \cdots < a_n < b_n < \cdots < b_2 < b_1$.
(ii) For $n \geqslant 1$, show that $b_n - a_n \leqslant (b-a)/2^{n-1}$.
   It follows that the sequences a_n and b_n converge to the same number, which is called the **arithmetic-geometric mean** of a and b which we shall denote agm(a,b).
(iii) Show that agm$(a,b) = a \times$ agm$(1, b/a)$.
(iv) Determine agm$(1,2)$ to 5 decimal places.

---

[23] Discovered by Heron of Alexandria (c. 10–c. 75).
[24] See p.278 for a short biography of Chebyshev. This inequality is not be confused with another famous inequality of Chebyshev in probability which states that the probability of a random variable being outside k standard deviations of its mean is at most k^{-2}.

#433D†(i) In a like manner we define the **arithmetic-harmonic mean** ahm(a,b) by setting $a_1 = a, b_1 = b$ and

$$a_{n+1} = \frac{2}{1/a_n + 1/b_n}, \qquad b_{n+1} = \frac{a_n + b_n}{2}, \qquad \text{for } n \geqslant 1.$$

Show that the sequences a_n and b_n converge to the same value, and that this common value ahm(a,b) equals the geometric mean $\sqrt{ab}$.

(ii) Likewise we define the **geometric-harmonic mean** ghm(a,b) by setting $a_1 = a, b_1 = b$ and

$$a_{n+1} = \frac{2}{1/a_n + 1/b_n}, \qquad b_{n+1} = \sqrt{a_n b_n}, \qquad \text{for } n \geqslant 1.$$

Show that the sequences a_n and b_n converge to the same value and that

$$\text{ghm}(a,b) = \frac{1}{\text{agm}(1/a, 1/b)}.$$

Exercises on Catalan Numbers

#434D† (MAT 2010 #7) In a game of *cat and mouse*, a cat starts at position 0, a mouse starts at position m and the mouse's hole is at position h. Here m and h are integers with $0 < m < h$. By way of example, a starting position is shown (Figure 2.6) where $m = 7$ and $h = 12$.

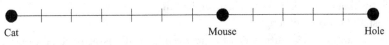

Cat   Mouse   Hole

Figure 2.6

With each turn of the game, one of the mouse or cat (but not both) advances one position towards the hole *on the condition that the cat is always strictly behind the mouse and never catches it*. The game ends when the mouse reaches the safety of its hole at position h. This question is about calculating the number, $g(h,m)$, of different sequences of moves that make a game of cat and mouse.

Let C denote a move of the cat and M denote a move of the mouse. Then, for example, $g(3,1) = 2$ as MM and MCM are the only possible games. Also, $CMCCM$ is *not* a valid game when $h = 4$ and $m = 2$, as the mouse would be caught on the fourth turn.

(i) Write down the five valid games when $h = 4$ and $m = 2$.

(ii) Explain why $g(h, h-1) = h - 1$ for $h \geqslant 2$.

(iii) Explain why $g(h,2) = g(h,1)$ for $h \geqslant 3$.

(iv) By considering the possible first moves of a game, explain why

$$g(h,m) = g(h, m+1) + g(h-1, m-1)$$

when $1 < m < h - 1$.

(v) The table shows certain values of $g(h,m)$ filled in. Complete the remainder of the table and verify that $g(6,1) = 42$.

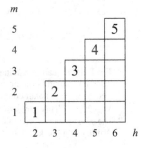

#435 D† With the notation of #434, show that

$$g(h,m) = \binom{2h-m-2}{h-m} - \binom{2h-m-2}{h-m-2} = \frac{(2h-m-1)!\,m}{(h-m)!\,h!}. \qquad (2.26)$$

Definition 2.41 For $n \geqslant 1$, the nth **Catalan** [25] **number** C_n equals $g(n+1,1)$ from #434 and by convention we set $C_0 = 1$. From (2.26) we have $C_n = \frac{1}{n+1}\binom{2n}{n}$. The triangle in the grid in #434 is a version of **Catalan's triangle**. There are generalizations of the Catalan numbers, usually denoted $C_{n,k}$ where $n \geqslant k \geqslant 0$, and in our notation $C_{n,k} = g(n+1, n-k+1)$.

#436 B Show, for $n \geqslant 1$, that $(n+2)C_{n+1} = (4n+2)C_n$.

#437 E† (Adapted MAT 2008 Question 7) *Oxwords* [26] are sequences of letters a and b that are constructed according to the following rules:

  I. The sequence consisting of no letters is an Oxword.
  II. If W is an Oxword, then the sequence beginning with a, followed by W and then b, written aWb, is an Oxword.
  III. If U and V are Oxwords, then the sequence U followed by V, written UV, is an Oxword.

All Oxwords are constructed using these rules. The *length* of an Oxword is the number of letters that occur in it. For example $aabb$ and $abab$ are Oxwords of length 4.

  (i) Show that every Oxword has an even length.
  (ii) List all Oxwords of length 6.
  (iii) Is the number of occurrences of a in an Oxword necessarily equal to the number of occurrences of b?
  (iv) Show that the Oxwords of length $2n$ are precisely those words W which use n as and n bs such that, for any i, the first i letters of W contain at least as many as as bs.
  (v) Show that every Oxword (of positive length) can be written *uniquely* in the form $aXbY$ where X, Y are Oxwords.
  (vi) For $n \geqslant 0$, let C_n be the number of Oxwords of length $2n$. Show that

$$C_{n+1} = \sum_{i=0}^{n} C_i C_{n-i}. \qquad (2.27)$$

Remark 2.42 The C_n of #437(vi) are the Catalan numbers. There C_n is defined as the number of Oxwords of length $2n$ and we saw in part (iv) that such an Oxword includes n as and n bs and any prefix (first i letters of the Oxword) has as many as as bs. If we replace each a with a left bracket and each b with a right bracket then we see that the Oxwords represent the different valid ways of bracketing an expression. For example,

$$aabb \text{ corresponds to } (()), \qquad abab \text{ corresponds to } ()().$$

---

[25] The Belgian mathematician Eugène Catalan (1814–1894), introduced the Catalan numbers in an 1838 paper when investigating the number of ways of dissecting a polygon into triangles (#438). His name is also associated with the Catalan conjecture (1843) that the only positive integer solution of the equation $x^m - y^n = 1$ is $3^2 - 2^3 = 1$. This conjecture was not proved true until 2002 by Preda Mihăilescu.

[26] The so-called Oxwords of this question are usually referred to as *Dyck Words*, after the German mathematician Walther Von Dyck (1856–1934).

These are the only two different ways to validly use two pairs of brackets.

To understand why these numbers are the same as the numbers from #434, note that every game of cat and mouse, where $m = 1$ begins with a move by the mouse. That the cat cannot catch up with the mouse means that the number of Cs never exceeds the number of Ms at any point. For example, the five games where $h = 4$ and $m = 1$ are

$$MMM, \quad MCMM, \quad MMCM, \quad MMCCM, \quad MCMCM.$$

If we replace each M with a left bracket (and each C with a right bracket) and then include at the end of the expression sufficient right brackets to make the bracketing valid, then these five games correspond to

$$((()), \quad ()(), \quad (()), \quad ()), \quad ()(). \quad \blacksquare$$

#438 D† In Figure 2.7 are shown three of the ways in which a hexagon can be decomposed into non-overlapping triangles whose vertices are vertices of the hexagon. Show that there are in all 14 such decompositions of a hexagon. Use (2.27) to show that there are C_n such decompositions of a polygon with $n + 2$ sides.

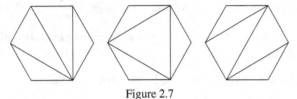

Figure 2.7

Exercises on Continued Fractions and Pell's Equation [27]

Definition 2.43 Let $q_0, q_1, q_2, \ldots, q_n$ be real numbers with (a) $q_0, \ldots, q_{n-1}$ integers, (b) $q_0 \geqslant 0$ and $q_i > 0$ for $1 \leqslant i \leqslant n$. We will then write $[q_0, q_1, \ldots, q_n]$ for the **continued fraction** [28]

$$q_0 + \cfrac{1}{q_1 + \cfrac{1}{q_2 + \cfrac{1}{\ddots + \cfrac{1}{q_n}}}}.$$

So, for example,

$$[q_0] = q_0, \quad [q_0, q_1] = q_0 + \frac{1}{q_1}, \quad [q_0, q_1, q_2] = q_0 + \frac{1}{q_1 + \frac{1}{q_2}}.$$

Some older text books might write $[q_0, q_1, \ldots, q_n]$ as

$$q_0 + \frac{1}{q_1 +} \frac{1}{q_2 +} \cdots \frac{1}{q_{n-1} +} \frac{1}{q_n}.$$

---

[27] For a full treatment on continued fractions or Pell's equation see Burton [6, chapter 15], Rose [27, chapter 7] or Baker [2, chapters 6–8].

[28] Such continued fractions are sometimes referred to as *simple* continued fractions as their numerators are all 1. It is possible to define continued fractions more generally with other numerators.

#439 B Show, for $n \geqslant 1$, that $[1,1,1,\ldots,1] = F_{n+1}/F_n$ where there are n 1s in the continued fraction.

#440 D† (See also #253) (i) Show that $\sqrt{2} = \left[1, 1+\sqrt{2}\right]$ and deduce that

$$\sqrt{2} = \left[1, 2, 2, \ldots, 2, 1+\sqrt{2}\right].$$

(ii) The sequences a_n and b_n are defined recursively by $a_0 = 1 = b_0$ and

$$a_{n+1} = 2b_n + a_n, \qquad b_{n+1} = b_n + a_n, \qquad \text{for } n \geqslant 0.$$

Show that $[1,2,2,\ldots,2] = a_n/b_n$ where there are n 2s in the continued fraction.
(iii) Show that $(a_n)^2 - 2(b_n)^2 = (-1)^{n+1}$.
(iv) Determine explicit formulae for a_n and b_n.

#441 B† Show that if $\left[q_0, q_1, \ldots, q_m\right] = \left[\tilde{q}_0, \tilde{q}_1, \ldots, \tilde{q}_n\right]$, where $q_m > 1$ and $\tilde{q}_n > 1$, then $m = n$ and $q_i = \tilde{q}_i$ for each i.

#442 B† Let $\alpha > 0$. If α is not an integer we may define

$$\alpha_1 = \frac{1}{\alpha - \lfloor \alpha \rfloor} \qquad \text{so that} \qquad \alpha = \lfloor \alpha \rfloor + \frac{1}{\alpha_1} = [\lfloor \alpha \rfloor, \alpha_1],$$

where $\lfloor \alpha \rfloor$ is the integer part of α. If α_1 is not an integer we may similarly define α_2 so that $\alpha_1 = [\lfloor \alpha_1 \rfloor, \alpha_2]$.

 (i) Assuming none of $\alpha, \alpha_1, \ldots, \alpha_{n-1}$ to be integers, show that

$$\alpha = [\lfloor \alpha \rfloor, \lfloor \alpha_1 \rfloor, \lfloor \alpha_2 \rfloor, \ldots, \lfloor \alpha_{n-1} \rfloor, \alpha_n].$$

(ii) Show that if α is irrational then we may recursively define α_n for all $n \geqslant 1$.

In the exercises that follow the notation of #442 will be used without further explanation.

#443 B Let $\alpha = 1 + \sqrt{3}$. Show that $\lfloor \alpha_{2n} \rfloor = 2$ and $\lfloor \alpha_{2n+1} \rfloor = 1$ for $n \geqslant 0$.

#444 B Determine α_i and $\lfloor \alpha_i \rfloor$ when (i) $\alpha = 2/7$, (ii) $\alpha = 13/17$, (iii) $\alpha = 217/99$.

#445 C Show, for $\alpha = \sqrt{5}$ and $i \geqslant 1$, that $\alpha_i = \sqrt{5} + 2$.

#446 B† Determine α_i and $\lfloor \alpha_i \rfloor$ when (i) $\alpha = \sqrt{7}$, (ii) $\alpha = \sqrt{13}$.

#447 D† Let α be positive and irrational. We define the sequences a_n and b_n by

$$a_{n+2} = \lfloor \alpha_{n+2} \rfloor a_{n+1} + a_n, \qquad a_0 = \lfloor \alpha \rfloor, \quad a_1 = \lfloor \alpha \rfloor \lfloor \alpha_1 \rfloor + 1;$$
$$b_{n+2} = \lfloor \alpha_{n+2} \rfloor b_{n+1} + b_n, \qquad b_0 = 1, \quad b_1 = \lfloor \alpha_1 \rfloor.$$

Show the following: (i) for $n \geqslant 1$

$$[\lfloor \alpha \rfloor, \lfloor \alpha_1 \rfloor, \ldots, \lfloor \alpha_n \rfloor, x] \quad = \quad \frac{a_n x + a_{n-1}}{b_n x + b_{n-1}} \qquad \text{for any positive } x$$

and deduce that $[\lfloor \alpha \rfloor, \lfloor \alpha_1 \rfloor, \ldots, \lfloor \alpha_n \rfloor] = a_n/b_n$. The rational a_n/b_n is the nth **convergent** of α.
(ii) $a_n b_{n+1} - a_{n+1} b_n = (-1)^{n+1}$.
(iii) $a_0/b_0 < a_2/b_2 < a_4/b_4 < \cdots < \alpha < \cdots < a_3/b_3 < a_1/b_1$.
(iv) $|\alpha - a_n/b_n| < 1/(b_n b_{n+1})$.

#448D† Let $\alpha > 0$. Show that α is rational if and only if $\alpha = [q_0, q_1, \ldots, q_n]$ for some integers $q_0, q_1, \ldots, q_n$.

#449B Find $\lfloor \alpha \rfloor, \lfloor \alpha_1 \rfloor, \ldots, \lfloor \alpha_4 \rfloor$ when $\alpha = \pi$. Deduce that

$$\left| \pi - \frac{22}{7} \right| < \frac{1}{2 \times 10^2}; \qquad \left| \pi - \frac{355}{113} \right| < \frac{1}{2 \times 10^6}.$$

#450C† Let a, b be positive integers. What real number α has $q_{2n} = a$ and $q_{2n+1} = b$ for $n \geqslant 0$?

#451C† Let $\alpha > 0$ be irrational and such that the continued fraction $[q_0, q_1, q_2, \ldots]$ of α eventually becomes periodic. Show that α is the root of a quadratic equation with integer coefficients. The converse is in fact also true (Baker [2, p.49]).

#452B† Find all positive integers x, y satisfying $x^2 - y^2 = a$ where a equals (i) 105, (ii) 222, (iii) 272.

#453B Let x, y and non-zero k be integers such that $x^2 - k^2 y^2 = 1$. Show $(x, y) = (\pm 1, 0)$.

Definition 2.44 The equation $x^2 - dy^2 = 1$ is known as **Pell's equation** [29], where d is a positive integer which is not a perfect square. The trivial solutions $(x, y) = (\pm 1, 0)$ are always solutions. When d is not a perfect square there are infinitely many pairs of integer solutions (#461).

#454B Let x, y be integer solutions to $x^2 - 2y^2 = 1$. Show that x is odd and that y is even.

#455D†(i) Let $\alpha = \sqrt{3}$ and a_n and b_n be as in #447. Show $a_0 = 1, a_1 = 2, b_0 = 1, b_1 = 1$ and that

$$a_{n+4} - 4a_{n+2} + a_n = 0, \qquad b_{n+4} - 4b_{n+2} + b_n = 0, \qquad \text{for } n \geqslant 0.$$

(ii) Find expressions for a_n and b_n and show that $a_n^2 - 3b_n^2$ equals 1 when n is odd and -2 when n is even. Hence find integer solutions x, y to $x^2 - 3y^2 = 1$, where $x > 1000$.

(iii) Show that $\left(2 + \sqrt{3}\right)^n = a_{2n-1} + b_{2n-1}\sqrt{3}$ for $n \geqslant 1$.

#456D† Let $\sqrt{7} = [q_0, q_1, q_2, \ldots]$ denote the continued fraction of $\sqrt{7}$ found in #446 and define a_n and b_n as in #447. Determine a recurrence relation involving a_{4n+8}, a_{4n+4} and a_{4n}. Hence find expressions for a_{4n} and b_{4n}. Show that $(a_{4n})^2 - 7(b_{4n})^2 = -3$ for each n.

#457D† Using the calculations from #456, show that the sequence $(a_n)^2 - 7(b_n)^2$ equals $-3, 2, -3, 1, -3, 2, -3, 1, \ldots$.

#458B Using the calculations from #456, show that $\left(8 + 3\sqrt{7}\right)^n = a_{4n-1} + b_{4n-1}\sqrt{7}$.

---

[29] Several Western mathematicians worked on the equation in the seventeenth century including Fermat and William Brouncker (1620–1684). Describing the equation as *Pell's equation* is a misnomer because Euler misattributed the equation to John Pell (1611–1685), who had revised an earlier translation of Brounker's solution. In any case the equation had been studied by Indian and Greek mathematicians centuries earlier, with the first general method of solution being due to Bhaksara II (1114–1185) in 1150. He found a method leading to a non-trivial solution whenever d is not a perfect square, something rigorously verified by Lagrange in 1768.

#459 D† Let d be a positive integer which is not a square. Let $\mathbb{Z}[\sqrt{d}]$ denote the set of all real numbers of the form $x + y\sqrt{d}$, where x and y are integers.

  (i) Show that if ζ and η are in $\mathbb{Z}[\sqrt{d}]$ then $\zeta + \eta$, $\zeta - \eta$ and $\zeta\eta$ are each in $\mathbb{Z}[\sqrt{d}]$.

  (ii) Let $\zeta = x + y\sqrt{d}$ be in $\mathbb{Z}[\sqrt{d}]$. Show that $1/\zeta$ is in $\mathbb{Z}[\sqrt{d}]$ if and only if $x^2 - dy^2 = \pm 1$.

 (iii) Show that if $x^2 - dy^2 = 1$ and $x + y\sqrt{d} \geqslant 1$ then $x \geqslant 1$ and $y \geqslant 0$.

 (iv) We define the **norm** by $N(\zeta) = N(x + y\sqrt{d}) = x^2 - dy^2$. Show $N(\zeta\eta) = N(\zeta)N(\eta)$.

  (v) Deduce that if there is a non-trivial integer solution of $x^2 - dy^2 = 1$ then there are infinitely many integer solutions.

#460 D† Let d be a positive integer which is not a square. Assuming there to be a non-trivial solution of $x^2 - dy^2 = 1$, show that there is a solution (X, Y) with $X, Y > 0$ and $X + Y\sqrt{d}$ minimal. This solution is called the **fundamental solution**. Show that if (x, y) is a second non-trivial solution then, for some integer k, we have

$$\pm \left(X + Y\sqrt{d} \right)^k = x + y\sqrt{d}.$$

#461 E†(i) (**Dirichlet's [30] approximation theorem**) Let α be a real number and $Q > 1$ an integer. Show that there are integers p and q with $0 < q < Q$ such that $|q\alpha - p| \leqslant 1/Q$.

(ii) Let d be a natural number which is not a square. Use Dirichlet's theorem to deduce that there are infinitely many $\zeta = p - q\sqrt{d}$ in $\mathbb{Z}[\sqrt{d}]$ such that $|N(\zeta)| \leqslant 3\sqrt{d}$.

(iii) Deduce that there are non-trivial integer solutions to $x^2 - dy^2 = 1$.

So when d is not a perfect square Pell's equation has a non-trivial solution (#461), in fact has infinitely many solutions which are all generated by the fundamental solution (#459). There is no closed form for the fundamental solution (X, Y) and it can be surprisingly large even when d is relatively small (#463). However, there are algorithms that determine the fundamental solution and we quote, without proof, such a method here involving continued fractions. We first recall from #451 that the continued fraction of $\sqrt{d}$ eventually becomes periodic.

Theorem 2.45 *(Burton [6, Theorem 15.13]) Let p denote the period of the continued fraction for $\sqrt{d}$ and let a_n/b_n denote its convergents.*

(a) If p is even then the sequence $\left(a_{np-1}, b_{np-1} \right)$, where $n \geqslant 1$, lists the positive solutions of $x^2 - dy^2 = 1$, with $\left(a_{p-1}, b_{p-1} \right)$ being the fundamental solution.

(b) If p is odd then the sequence $\left(a_{2np-1}, b_{2np-1} \right)$, where $n \geqslant 1$, lists the positive solutions of $x^2 - dy^2 = 1$, with $\left(a_{2p-1}, b_{2p-1} \right)$ being the fundamental solution.

#462 B Use Theorem 2.45 to determine the fundamental solution when d equals (i) 10, (ii) 20, (iii) 41, (iv) 55.

#463 E Show that the continued fraction for $\sqrt{61}$ has period 11 and that the fundamental solution of $x^2 - 61y^2 = 1$ is $X = 1766319049$, $Y = 226153980$.

---

[30] After Gustav Lejeune Dirichlet (1805–1859), a German mathematician who made notable contributions to number theory. This includes his theorem on the infinitude of primes amongst arithmetic progressions (see p.79). His proof used seminal methods involving analysis and thus effectively invented the field of *analytic number theory*.

#464D† Determine all integer solutions of the Pell-like equation $x^2 - 2y^2 = -1$.

#465D† Are there integer solutions to (i) $x^2 - 3y^2 = 2$ or (ii) $x^2 - 3y^2 = 7$?

#466D† By making an appropriate change of variable, show there are infinitely many integer solutions x, y of $6xy + 1 = x^2 + y^2$. Are there integer solutions of $6xy = x^2 + y^2 + 1$?

Exercises on Special Polynomials

#467D† (See also #246) The **Bernoulli numbers** B_n, for $n \geq 0$, are defined as $B_0 = 1$ and by the recursive formula

$$\sum_{k=0}^{n-1} \binom{n}{k} B_k = 0, \qquad \text{for } n \geq 2.$$

The **Bernoulli polynomials** $B_n(x)$, for $n \geq 0$, are then defined as $B_n(x) = \sum_{k=0}^{n} \binom{n}{k} B_k x^{n-k}$.

(i) Determine B_n and $B_n(x)$ for $0 \leq n \leq 4$.
(ii) Show that $B_m(k+1) - B_m(k) = mk^{m-1}$ and deduce that

$$\sum_{k=1}^{n} k^m = \frac{B_{m+1}(n+1) - B_{m+1}}{m+1}.$$

#468E† With the Bernoulli polynomials $B_n(x)$ defined as in #467, show that

(i) $B_n'(x) = nB_{n-1}(x)$.
(ii) $B_n(x+h) = \sum_{k=0}^{n} \binom{n}{k} B_k(x) h^{n-k}$.
(iii) $B_n(1-x) = (-1)^n B_n(x)$.

Deduce that the odd Bernoulli numbers are zero except for B_1.

#469D(i) The **Hermite polynomials**, [31] $H_n(x)$ for $n \geq 0$, are defined recursively by

$$H_{n+1}(x) = 2xH_n(x) - 2nH_{n-1}(x) \qquad \text{for } n \geq 1,$$

with $H_0(x) = 1$ and $H_1(x) = 2x$. Calculate $H_n(x)$ for $n = 2, 3, 4, 5$.
(ii) Show by induction that $H_{2k}(0) = (-1)^k (2k)!/k!$ and $H_{2k+1}(0) = 0$.
(iii) Show by induction that $H_n'(x) = 2nH_{n-1}(x)$.
(iv) Deduce that $y(x) = H_n(x)$ is a solution of the differential equation $y'' - 2xy' + 2ny = 0$.
(v) Use Leibniz's rule for differentiating a product (2.9) to show that the polynomials

$$K_n(x) = (-1)^n e^{x^2} \frac{d^n}{dx^n} (e^{-x^2})$$

satisfy the same recursion as $H_n(x)$ with the same initial conditions and deduce that $H_n(x) = K_n(x)$ for $n \geq 0$.

---

[31] The French mathematician, Charles Hermite (1822–1901), made various important conributions, including showing in 1873 that the number e is transcendental (i.e. not the root of any polynomial with integer coefficients). Most mathematics undergraduates are likely to happen on his name when studying quantum mechanics, where the Hermite polynomials are associated with modelling the quantum harmonic oscillator and observables (such as position or momentum) are represented by *Hermitian* operators or matrices. (A square complex matrix is Hermitian if it equals its conjugate transpose.)

The Bernoullis A Swiss family of mathematicians in the seventeenth and eighteenth centuries who are undoubtedly the first family of mathematics, and in all eight members of the family would become mathematicians. Jacob (1654–1705) was the first to investigate the number e through a study of compound interest, made important contributions to probability, including the *law of large numbers* and introduced the Bernoulli numbers. His brother Johann (1667–1748) would become a tutor of Leonhard Euler and it was actually Johann who proved what is now known as *L'Hôpital's theorem*. It was also Johann who posed and solved the *brachistochrone* problem: find the curve between two points, such that a bead sliding down the curve under gravity takes the least time possible. Knowing the solution, he posed the problem to the mathematical community by way of a challenge. The curve that solves the problem is an upside-down *cycloid* (see #1541). Solutions were received from Leibniz, L'Hôpital and his brother Jacob, and one further solution was anonymously published by the Royal Society which was due to Newton. Johann's son Daniel (1700–1782) was an accomplished applied mathematician, particularly in the study of hydrodynamics (*Bernoulli's equation*) and he also showed that a vibrating string's motion was an infinite sum of its harmonic modes. The years 1727–1733 were a particularly productive time when he and Euler were both in St. Petersburg.

#470 B(i) The **Legendre polynomials** [32] $P_n(x)$ are defined by [33]

$$(n+1)P_{n+1}(x) = (2n+1)xP_n(x) - nP_{n-1}(x) \qquad \text{for } n \geqslant 1$$

and $P_0(x) = 1$, $P_1(x) = x$. Determine $P_n(x)$ for $2 \leqslant n \leqslant 5$.
(ii) Show that $P_n(x)$ is a polynomial of degree n and that $P_n(x)$ is an odd/even function when n is odd/even.
(iii) Show that $P_n(1) = 1$ and $P'_n(1) = n(n+1)/2$.

#471 D† Prove **Rodrigues' formula** [34]

$$P_n(x) = \frac{1}{2^n n!} \frac{d^n}{dx^n} \left[\left(x^2 - 1 \right)^n \right].$$

Deduce that

$$P_n(x) = \frac{1}{2^n} \sum_{k=0}^{n} \binom{n}{k}^2 (x+1)^{n-k}(x-1)^k.$$

[32] Named after Adrien-Marie Legendre (1752–1833), Legendre's polynomials were originally studied as solutions of his differential equation (6.2). Legendre introduced his polynomials in 1784 while investigating planetary orbits. For an undergraduate they are most likely to arise when investigating Laplace's equation in spherical co-ordinates, or as a least squares approximation on account of their orthogonality (#1525).
[33] This recurrence relation is due to the French mathematician Pierre Bonnet (1819–1892).
[34] After the French mathematician Benjamin Olinde Rodrigues (1795–1851).

#472D† Let C_n denote the coefficient of t^n in the expansion of $(1+t+t^2)^n$. These are the **central trinomial coefficients** (see #341). Show that

$$C_n = (i\sqrt{3})^n P_n \left(\frac{1}{i\sqrt{3}} \right)$$

and deduce the recurrence relation $nC_n = (2n-1)C_{n-1} + 3(n-1)C_{n-2}$ for $n \geqslant 2$.

#473D (See also #1529.) The **Chebyshev polynomials** [35] $T_n(x)$ and $U_n(x)$ are defined by the same recurrence relation

$$T_{n+1}(x) = 2xT_n(x) - T_{n-1}(x), \qquad U_{n+1}(x) = 2xU_n(x) - U_{n-1}(x), \qquad \text{for } n \geqslant 1,$$

with $T_0(x) = 1$, $T_1(x) = x$, $U_0(x) = 1$, $U_1(x) = 2x$.

(i) Determine $T_n(x)$ and $U_n(x)$ for $2 \leqslant n \leqslant 5$.
(ii) Show that $T_n(x)$ and $U_n(x)$ are polynomials of degree n and that T_n and U_n are odd/even when n is odd/even.
(iii) Show that

$$T_n(x) = U_n(x) - xU_{n-1}(x), \qquad U_n(x) = \frac{T_{n+2}(x) - xT_{n+1}(x)}{x^2 - 1}.$$

#474D†(i) Show that

$$T_n(\cos\theta) = \cos(n\theta), \qquad U_n(\cos\theta) = \frac{\sin((n+1)\theta)}{\sin\theta}.$$

Deduce that all the roots of $T_n(x)$ and $U_n(x)$ are real and lie in the range $-1 < x < 1$.
(ii) Show that $T_n(T_m(x)) = T_{nm}(x)$.
(iii) Show for $m \geqslant n$ that $2T_m(x)T_n(x) = T_{m+n}(x) + T_{m-n}(x)$.

#475C Prove the following Pell-like identities for the Chebyshev polynomials.

(i) $T_n(x)^2 - (x^2 - 1)U_{n-1}(x)^2 = 1$.
(ii) $T_n(x) + U_{n-1}(x)\sqrt{x^2 - 1} = (x + \sqrt{x^2 - 1})^n$.

#476C The **Laguerre polynomials** [36] $L_n(x)$ are defined by

$$L_n(x) = \frac{e^x}{n!} \frac{d^n}{dx^n} \left(x^n e^{-x} \right) \qquad \text{for } n \geqslant 0. \tag{2.28}$$

Show that

$$L_n(x) = \sum_{k=0}^{n} \binom{n}{k} \frac{(-x)^k}{k!}.$$

Remark 2.46 See p.415 for more exercises on *orthogonal polynomials* such as the ones above. ∎

---

[35] See p.278 for a short biography of Chebyshev.
[36] After the French mathematician Edmond Laguerre (1834–1886) who discovered the polynomials while investigating the integral of e^{-x}/x.

Exercises on Random Walks

#477 D† A triangle has vertices A, B, C. A point P follows a random walk on this triangle as follows. P is at vertex A at time $t = 0$; if P is at a particular vertex at time t then, at time $t + 1$, P moves with equal probability to one of the other two vertices. Find expressions a_t, b_t, c_t for the probabilities of P being at A, B, C at time t.

#478 C† (See also #410.) A point P follows a random walk on the integers as follows. At time $t = 0$, the point P is at 0 and moves according to the following rule: if P is at n at time t, then at time $t + 1$ there is probability p of P being at $n + 1$ and a probability $q = 1 - p$ of P being at $n - 1$. Let X_t denote the position of P at time t.

(i) What is the probability that $X_t = k$?
(ii) What is the mean value of X_t?
(iii) What is the mean value of $(X_t)^2$?

#479 D (See also #341.) A point P follows a random walk on the integers as follows. At time $t = 0$, the point P is at 0 and moves according to the following rule: if P is at n at time t, then at time $t + 1$ there is a one-third chance each of P being at $n + 1$, P being at n and P being at $n - 1$. Let X_t denote the position of P at time t.

(i) Write out X_t's distribution explicitly for $t = 2, 3, 4$.
(ii) Show that the probability that $X_t = k$ is the coefficient of x^k in $3^{-t}(x^{-1} + 1 + x)^t$.
(iii) Determine the mean value of X_t and of $(X_t)^2$.

#480 D Repeat #479, but in the situation where there is probability p of moving to $n + 1$, probability q of remaining at n and probability r of moving to $n - 1$ (where $p + q + r = 1$).

#481 C† A point P follows a random walk on the natural numbers as follows. At time $t = 0$, the point P is at $a > 0$ and moves according to the following rule: if P is at $n > 1$ at time t, then at time $t + 1$ there is probability p of P being at $n + 1$ and a probability $q = 1 - p$ of P being at $n - 1$; the state 0 is an *absorbing state* so that if moves there at any point it remains there. What is the probability P_a that the particle is eventually absorbed at 0? [You may assume, when $p > q$, that P_a approximates to 0 as a becomes large.]

#482 D† A point P follows a random walk on an integer grid as follows. At time $t = 0$, the point P is at $(0,0)$ and moves at $t = 1$ (with equal probabilities) to one of the four adjacent grid points. At each subsequent time, P has a half chance of taking a left or right turn (relative to its previous move) and moving a unit in that direction. Let (X_t, Y_t) denote the position of P at time t.

(i) Show the probability that $(X_t, Y_t) = (0,0)$ is zero unless t is a multiple of 4.
(ii) Determine the probability that $(X_t, Y_t) = (0,0)$.
(iii) What is the probability that $X_t = k$?
(iv) What is the mean value of X_t?
(v) What is the mean value of $(X_t)^2 + (Y_t)^2$?

#483E† Repeat #482(ii)–(v) in the following situation. A point P traces a random walk on an integer grid as follows. At time $t = 0$, the point P is at $(0,0)$ and moves according to the following rule: if P is at (x,y) at time t, then at time $t+1$ there is a one-quarter chance each of P being at

$$(x+1,y), \qquad (x-1,y), \qquad (x,y+1), \qquad (x,y-1).$$

That is, there is an equal chance of P moving right, left, up, down one unit with each move.

#484E† An equilateral triangle of unit length tesselates the complex plane as shown in Figure 2.8. A point P starts at $(0,0)$ and at each time interval moves with equal probability from its current node to one of the six adjacent nodes.

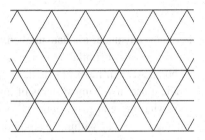

(i) Show that every node can uniquely be written as $a + b\omega$, where $\omega = \operatorname{cis}(2\pi/3)$ and a,b are integers.

(ii) Let Z_t denote the position of P at time t. Show that the mean value of $|Z_t|^2$ equals t.

Figure 2.8

Exercises on Complete Sequences

Definition 2.47 An increasing sequence $a_1, a_2, a_3, \ldots$ of positive integers is said to be **complete** if every positive integer can be written as some sum of the a_i, using each a_i at most once. #248 (with $b = 2$) shows that the sequence $1, 2, 4, 8, \ldots$ is complete. #264 showed that the prime numbers, together with 1, make a complete sequence.

#485D† Show that an increasing sequence $a_1, a_2, a_3, \ldots$ of positive integers is complete if and only if $a_1 = 1$ and

$$1 + a_1 + a_2 + \cdots + a_k \geqslant a_{k+1} \qquad \text{for each } k \geqslant 1.$$

#486D If the increasing sequence $a_1, a_2, a_3, \ldots$ is complete, show that $a_n \leqslant 2^{n-1}$ for each $n \geqslant 1$.

#487E†(i) Zeckendorf's theorem shows that the sequence of Fibonacci numbers $F_1, F_2, F_3, F_4, \ldots$ is complete. Show that, if any Fibonacci number is removed from the sequence, then the sequence is still complete.

(ii) The sequence $b_1, b_2, b_3, \ldots$ has the property that, even when any one element is removed, the sequence remains complete. Show that $b_n \leqslant F_n$ for each $n \geqslant 1$.

3

Vectors and Matrices

In almost any undergraduate mathematics degree there will be one or more courses entitled *linear algebra* or similar. In fact, it is an area most science degrees will cover to some extent or other. At their most practical, linear algebra courses tend to discuss matrices and their efficient use in solving systems of linear equations and the diagonalization of matrices. A somewhat more abstract approach would more likely be taken in a mathematics degree, placing the theory within the study of linear maps between vector spaces over a field.

It is certainly not the intention of these next two chapters to take an abstract approach to introducing matrices, but (with one eye on the fuller treatment they receive at university) more time will be spent here attempting to motivate the relevant linear algebra than can usually be afforded it at the A-level stage.

The unasterisked sections of this chapter cover the fundamentals of vectors and matrices, going over the important definitions and calculations that are typically found in Further Mathematics. The subsequent chapter covers new material, that could be ignored by students just looking to catch up on missed Further Mathematics, but it is still very much concrete, rather than abstract, in design.

Remark 3.1 See the Glossary if you are uncertain about any of the notation relating to sets used in this chapter. ∎

3.1 The Algebra of Vectors

Definition 3.2 By a **vector** we will mean a list of n real numbers $x_1, x_2, x_3, \ldots, x_n$ where n is a positive integer. Commonly this list will be treated as a **row vector** and written as

$$(x_1, x_2, \ldots, x_n).$$

Sometimes (for reasons that will become apparent) the numbers may be arranged as a **column vector**

$$\begin{pmatrix} x_1 \\ x_2 \\ \vdots \\ x_n \end{pmatrix}.$$

Often we will denote a vector by a single letter in bold, say **x**, and refer to x_i as the i**th co-ordinate** of **x**.

Definition 3.3 For a given n, we denote the set of all row vectors with n co-ordinates as $\mathbb{R}^n$, and often refer to $\mathbb{R}^n$ as n-**dimensional co-ordinate space** or simply as n-**dimensional space**.

See Figure 3.1a: If $n = 2$ then we commonly use x and y as co-ordinates and refer to $\mathbb{R}^2 = \{(x,y): x,y \in \mathbb{R}\}$ as the xy-**plane**.

See Figure 3.1b: If $n = 3$ then we commonly use x, y and z as co-ordinates and refer to $\mathbb{R}^3 = \{(x,y,z): x,y,z \in \mathbb{R}\}$ as xyz-**space**.

We will use the notation $\mathbb{R}_n$ to denote [1] the set of all column vectors with n co-ordinates.

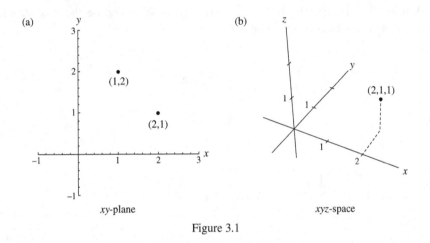

(a) xy-plane   (b) xyz-space

Figure 3.1

- Note that the order of the co-ordinates matters; so, for example, $(2,1)$ and $(1,2)$ are different vectors in $\mathbb{R}^2$.

What geometry there is in the next two chapters relates mainly to the plane or three-dimensional space. But it can be useful to develop the theory more generally, especially when it comes to solving systems of linear equations.

Definition 3.4 There is a special vector $(0,0,\dots,0)$ in $\mathbb{R}^n$ which we denote as **0** and refer to as the **zero vector**.

A vector can be thought of in two different ways. For example, in the case of vectors in $\mathbb{R}^2$, the xy-plane:

- From one point of view a vector can represent the point in $\mathbb{R}^2$ which has co-ordinates x and y. We call this vector the **position vector** of that point (Figure 3.2a). In practice

---

[1] Note that this notation $\mathbb{R}_n$ for the space of $n \times 1$ column vectors is non-standard, but convenient for the purposes of this text. The notation $(\mathbb{R}^n)'$ or $(\mathbb{R}^n)^*$ is sometimes used instead as $\mathbb{R}_n$ can be naturally considered as the dual space of $\mathbb{R}^n$. However, we shall not be considering dual spaces at all in this text.

though, rather than referring to the 'point with position vector **x**', we will simply say 'the point **x**' when the meaning is clear. The point **0** is referred to as the **origin**.

- From a second point of view a vector is a 'movement' or translation. For example, to get from the point $(3,4)$ to the point $(4,5)$ we need to move 'one to the right and one up'; this is the same movement as is required to move from $(-3,2)$ to $(-2,3)$ and from $(1,-2)$ to $(2,-1)$. Thinking about vectors from this second point of view, all three of these movements are the same vector because the same translation 'one right, one up' achieves each of them, even though the 'start' and 'finish' are different in each case (Figure 3.2b). We would write this vector as $(1,1)$. Vectors from this second point of view are sometimes called **translation vectors**. From this point of view **0** represents 'no movement'.

Definition 3.5 The points $(0,0,\ldots,0,x_i,0,\ldots,0)$ in $\mathbb{R}^n$ comprise the x_i-**axis**, with the origin lying at the intersection of all the axes.

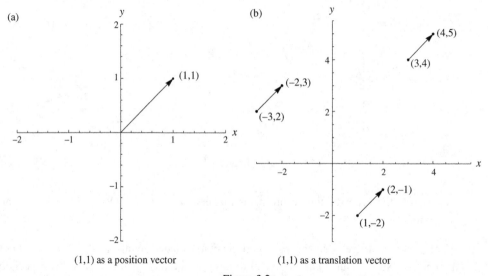

Figure 3.2

Similarly in three (and likewise higher) dimensions, the triple (x,y,z) can be thought of as the point in $\mathbb{R}^3$, which is x units along the x-axis from the origin, y units parallel to the y-axis and z units parallel to the z-axis, or it can represent the translation which would take the origin to that point.

Definition 3.6 Given two vectors $\mathbf{u} = (u_1,u_2,\ldots,u_n)$ and $\mathbf{v} = (v_1,v_2,\ldots,v_n)$ in $\mathbb{R}^n$, we can **add** and **subtract** them, much as you would expect, by separately adding the corresponding co-ordinates. That is

$$\mathbf{u}+\mathbf{v} = (u_1+v_1,u_2+v_2,\ldots,u_n+v_n); \qquad \mathbf{u}-\mathbf{v} = (u_1-v_1,u_2-v_2,\ldots,u_n-v_n).$$

Note that $\mathbf{v} - \mathbf{u}$ is the vector that translates the point (with position vector) $\mathbf{u}$ to the point (with position vector) $\mathbf{v}$.

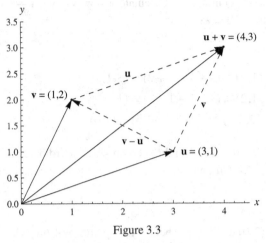

Figure 3.3

Figure 3.3 shows $\mathbf{u} + \mathbf{v}$ and $\mathbf{v} - \mathbf{u}$ for particular choices of $\mathbf{u}, \mathbf{v}$ in the xy-plane.

The sum of two vectors is perhaps easiest to interpret when we consider the vectors as translations. The translation $\mathbf{u} + \mathbf{v}$ is the overall effect of doing the translation $\mathbf{u}$ first and then the translation $\mathbf{v}$, or it can be achieved by doing the translations in the other order – that is, vector addition is *commutative*: $\mathbf{u} + \mathbf{v} = \mathbf{v} + \mathbf{u}$.

Note that two vectors may be added if and only if they have the same number of co-ordinates. No immediate sense can be made of adding a vector in $\mathbb{R}^2$ to one from $\mathbb{R}^3$, for example.

Definition 3.7 Given a vector $\mathbf{v} = (v_1, v_2, \ldots, v_n)$ and a real number k, then the **scalar multiple** $k\mathbf{v}$ is defined as

$$k\mathbf{v} = (kv_1, kv_2, \ldots, kv_n).$$

- When k is a positive integer, then we can think of $k\mathbf{v}$ as the translation achieved when we translate k times by the vector $\mathbf{v}$.
- Note that the points $k\mathbf{v}$, as k varies through the real numbers, make up the line which passes through the origin and the point $\mathbf{v}$. The points $k\mathbf{v}$, where $k \geqslant 0$, lie on one half-line from the origin, the half which includes the point $\mathbf{v}$. And the points $k\mathbf{v}$, where $k < 0$, comprise the remaining half-line.
- We write $-\mathbf{v}$ for $(-1)\mathbf{v} = (-v_1, -v_2, \ldots, -v_n)$. Translating by $-\mathbf{v}$ is the inverse operation of translating by $\mathbf{v}$.

Definition 3.8 The n vectors

$$(1, 0, \ldots, 0), \qquad (0, 1, 0, \ldots, 0), \qquad \cdots \qquad (0, \ldots, 0, 1, 0), \qquad (0, \ldots, 0, 1)$$

in $\mathbb{R}^n$ are known as the **standard** (or **canonical**) basis for $\mathbb{R}^n$. We will denote these as $\mathbf{e}_1, \mathbf{e}_2, \ldots, \mathbf{e}_n$.

- The vectors $(1, 0)$ and $(0, 1)$ form the standard basis for $\mathbb{R}^2$. These are denoted as $\mathbf{i}$ and $\mathbf{j}$ respectively.
- Note that any vector $\mathbf{v} = (x, y)$ can be written uniquely as a linear combination of $\mathbf{i}$ and $\mathbf{j}$: that is $(x, y) = x\mathbf{i} + y\mathbf{j}$ and this is the only way to write (x, y) as a sum of scalar multiples of $\mathbf{i}$ and $\mathbf{j}$.

- The vectors $(1,0,0)$, $(0,1,0)$, $(0,0,1)$ form the standard basis for $\mathbb{R}^3$, being respectively denoted **i, j, k.**

Example 3.9 Let $\mathbf{v} = (2,-1)$ and $\mathbf{w} = (1,2)$ in $\mathbb{R}^2$. (a) Calculate $\mathbf{v} + \mathbf{w}$, $\mathbf{v} - \mathbf{w}$, $2\mathbf{v} + 3\mathbf{w}$. (b) Find the real numbers λ, μ such that $(3,7) = \lambda\mathbf{v} + \mu\mathbf{w}$.

Solution (a)

$$\mathbf{v} + \mathbf{w} = (2,-1) + (1,2) = (2+1,-1+2) = (3,1);$$
$$\mathbf{v} - \mathbf{w} = (2,-1) - (1,2) = (2-1,-1-2) = (1,-3);$$
$$2\mathbf{v} + 3\mathbf{w} = 2(2,-1) + 3(1,2) = (4,-2) + (3,6) = (7,4).$$

(b) If $(3,7) = \lambda(2,-1) + \mu(1,2) = (2\lambda + \mu, -\lambda + 2\mu)$ then, comparing co-ordinates, we have the simultaneous equations $3 = 2\lambda + \mu$ and $7 = -\lambda + 2\mu$. Solving these equations gives $\lambda = -0.2$ and $\mu = 3.4$. □

(margin) #488 #489 #491 #492 #493

Example 3.10 Let $\mathbf{v} = (1,1)$ and $\mathbf{w} = (2,0)$ in $\mathbb{R}^2$. (a) Label the nine points $a\mathbf{v} + b\mathbf{w}$, where a and b are one of $0,1,2$. (b) Sketch the points $a\mathbf{v} + b\mathbf{w}$, where $a + b = 1$. (c) Shade the region of points $a\mathbf{v} + b\mathbf{w}$, where $1 \leqslant a,b \leqslant 2$.

Solution (See Figure 3.4.) (a) We can calculate the nine points as

$(0,0)$, $(2,0)$, $(4,0)$, $(1,1)$, $(3,1)$, $(5,1)$, $(2,2)$, $(4,2)$, $(6,2)$.

Notice that the points $a\mathbf{v} + b\mathbf{w}$, where a and b are integers, make a grid of parallelograms in $\mathbb{R}^2$. (b) Points $a\mathbf{v} + b\mathbf{w}$ with $a + b = 1$ have the form

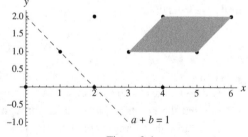

Figure 3.4

$$a\mathbf{v} + (1-a)\mathbf{w} = a(1,1) + (1-a)(2,0) = (2-a,a)$$

which is a general point of the line $x + y = 2$.
(c) The four edges of the shaded parallelogram lie on the lines $b = 1$ (left edge), $b = 2$ (right), $a = 1$ (bottom), $a = 2$ (top). □

Remark 3.11 Vector addition and scalar multiplication satisfy the following properties, which are easily verified. These properties prove that $\mathbb{R}^n$ is a **real vector space**. Given vectors $\mathbf{u}, \mathbf{v}, \mathbf{w}$ in $\mathbb{R}^n$ and real numbers λ, μ, then

$$\mathbf{u} + \mathbf{0} = \mathbf{u}; \qquad \mathbf{u} + \mathbf{v} = \mathbf{v} + \mathbf{u}; \qquad 0\mathbf{u} = \mathbf{0};$$
$$\mathbf{u} + (-\mathbf{u}) = \mathbf{0}; \qquad (\mathbf{u} + \mathbf{v}) + \mathbf{w} = \mathbf{u} + (\mathbf{v} + \mathbf{w}); \qquad 1\mathbf{u} = \mathbf{u};$$
$$(\lambda + \mu)\mathbf{u} = \lambda\mathbf{u} + \mu\mathbf{u}; \qquad \lambda(\mathbf{u} + \mathbf{v}) = \lambda\mathbf{u} + \lambda\mathbf{v}; \qquad \lambda(\mu\mathbf{u}) = (\lambda\mu)\mathbf{u}. \qquad ■$$

Exercises

#488A Let $\mathbf{v} = (1,2)$ and $\mathbf{w} = (2,0)$.

  (i) Plot the points $\mathbf{0}, \mathbf{v}, \mathbf{w}, \mathbf{v} + \mathbf{w}, \frac{1}{2}\mathbf{w}, \mathbf{v} + 2\mathbf{w}, \frac{1}{2}\mathbf{v} + \mathbf{w}$ in the xy-plane.
 (ii) Plot the points $x\mathbf{v} + y\mathbf{w}$, where $x = \frac{1}{2}$ and y is a real number.
(iii) Plot the points $x\mathbf{v} + y\mathbf{w}$, where $x = y$.

#489b Let $\mathbf{v} = (1,0)$ and $\mathbf{w} = (1,1)$ in $\mathbb{R}^2$.

  (i) Sketch the region of points $x\mathbf{v} + y\mathbf{w}$, where $0 \leqslant x, y \leqslant 1$.
 (ii) Sketch the points $x\mathbf{v} + y\mathbf{w}$, where $x + y = 1$.
(iii) Sketch the points $x\mathbf{v} + y\mathbf{w}$, where $x^2 + y^2 = 1$.

#490c Let $\mathbf{v} = (3,4)$ and $\mathbf{w} = (2,-3)$ in $\mathbb{R}^2$.

  (i) Given any vector $\mathbf{x} = (x,y)$ in $\mathbb{R}^2$, determine real numbers α, β such that $\mathbf{x} = \alpha\mathbf{v} + \beta\mathbf{w}$.
 (ii) Describe the region $\{\alpha\mathbf{v} + \beta\mathbf{w} : \alpha > 0, \beta < 0\}$ in $\mathbb{R}^2$.
(iii) Find two distinct non-zero vectors $\mathbf{s}, \mathbf{t}$ in $\mathbb{R}^2$ such that $\{\alpha\mathbf{s} + \beta\mathbf{t} : \alpha, \beta \in \mathbb{R}\} \neq \mathbb{R}^2$.

#491B† Let $\mathbf{v} = (1,0,2)$, $\mathbf{w} = (0,3,1)$ in $\mathbb{R}^3$. Show that $(x,y,z) = \alpha\mathbf{v} + \beta\mathbf{w}$ for some α, β if and only if $6x + y = 3z$.

#492B† Let $\mathbf{u} = (1,1,1)$, $\mathbf{v} = (2,-1,3)$, $\mathbf{w} = (1,-5,3)$ in $\mathbb{R}^3$.

  (i) Show that $(x,y,z) = \alpha\mathbf{u} + \beta\mathbf{v} + \gamma\mathbf{w}$ for some α, β, γ if and only if $4x = y + 3z$.
 (ii) Show that every vector $\alpha\mathbf{u} + \beta\mathbf{v} + \gamma\mathbf{w}$ can be written in the form $\lambda\mathbf{u} + \mu\mathbf{v}$ for some λ, μ and determine expressions for λ and μ in terms of α, β, γ.

#493B Let $\mathbf{v} = (1,-1,3)$, $\mathbf{w} = (1,0,-1)$ in $\mathbb{R}^3$. Under what conditions on x, y, z do there exist real numbers α and β such that $(x,y,z) = \alpha\mathbf{v} + \beta\mathbf{w}$?

#494a Say $\mathbf{v}, \mathbf{w}$ are vectors in $\mathbb{R}^2$ such that $3\mathbf{v} + 4\mathbf{w} = (2,1)$ and $2\mathbf{v} - 3\mathbf{w} = (1,-1)$. Find $\mathbf{v}$ and $\mathbf{w}$.

#495C† Show that the only non-empty subsets of $\mathbb{R}^2$ which are closed[2] under addition and scalar multiplication are

$$\{\mathbf{0}\}, \qquad \text{lines through the origin,} \qquad \mathbb{R}^2.$$

What subsets of $\mathbb{R}^3$ are similarly closed?

3.2 The Geometry of Vectors. The Scalar Product

As vectors represent points and translations, they have important geometric properties as well as algebraic ones.

---

[2] We say that S is closed under addition and scalar multiplication if whenever $\mathbf{x}, \mathbf{y}$ are in S and λ is a real number then $\mathbf{x} + \mathbf{y}$ and $\lambda\mathbf{x}$ are also in S.

Definition 3.12 The **length** (or **magnitude**) of a vector $\mathbf{v} = (v_1, v_2, \ldots, v_n)$, which is written $|\mathbf{v}|$, is defined by

$$|\mathbf{v}| = \sqrt{(v_1)^2 + (v_2)^2 + \cdots + (v_n)^2}. \tag{3.1}$$

We say a vector $\mathbf{v}$ is a **unit vector** if it has length 1.

This formula is exactly what you'd expect it to be from Pythagoras' theorem; we see this is the distance of the point $\mathbf{v}$ from the origin, or equivalently the distance a point moves when it is translated by $\mathbf{v}$. So if $\mathbf{p}$ and $\mathbf{q}$ are points in $\mathbb{R}^n$, then the vector that will translate $\mathbf{p}$ to $\mathbf{q}$ is $\mathbf{q} - \mathbf{p}$, and hence we define:

Definition 3.13 The **distance** between two points $\mathbf{p}, \mathbf{q}$ in $\mathbb{R}^n$ is $|\mathbf{q} - \mathbf{p}|$ (or equally $|\mathbf{p} - \mathbf{q}|$). In terms of their co-ordinates p_i and q_i we have

$$\text{distance between } \mathbf{p} \text{ and } \mathbf{q} = \sqrt{\sum_{i=1}^{n} (p_i - q_i)^2}.$$

- Note that $|\mathbf{v}| \geqslant 0$ and that $|\mathbf{v}| = 0$ if and only if $\mathbf{v} = \mathbf{0}$.
- Also $|\lambda \mathbf{v}| = |\lambda| \, |\mathbf{v}|$ for any real number λ (see #500).

Proposition 3.14 *(Triangle Inequality) Let $\mathbf{u}, \mathbf{v}$ be vectors in $\mathbb{R}^n$. Then*

$$|\mathbf{u} + \mathbf{v}| \leqslant |\mathbf{u}| + |\mathbf{v}|. \tag{3.2}$$

If $\mathbf{v} \neq \mathbf{0}$ then there is equality in (3.2) if and only if $\mathbf{u} = \lambda \mathbf{v}$ for some $\lambda \geqslant 0$.

Remark 3.15 Geometrically, this is intuitively clear. If we review the diagram beneath Definition 3.6 and look at the triangle with vertices $\mathbf{0}, \mathbf{u}, \mathbf{u} + \mathbf{v}$, then we see that its sides have lengths $|\mathbf{u}|, |\mathbf{v}|$ and $|\mathbf{u} + \mathbf{v}|$. So $|\mathbf{u} + \mathbf{v}|$ is the distance along the line from $\mathbf{0}$ to $\mathbf{u} + \mathbf{v}$, whereas $|\mathbf{u}| + |\mathbf{v}|$ is the combined distance from $\mathbf{0}$ to $\mathbf{u}$ to $\mathbf{u} + \mathbf{v}$. This cannot be shorter and will be equal only if we passed through $\mathbf{u}$ on the way to $\mathbf{u} + \mathbf{v}$. Note that this proposition was already proven in two dimensions using complex numbers (Proposition 1.16).　　∎

Proof Let $\mathbf{u} = (u_1, u_2, \ldots, u_n)$, $\mathbf{v} = (v_1, v_2, \ldots, v_n)$. The inequality (3.2) is trivial if $\mathbf{v} = \mathbf{0}$, so suppose $\mathbf{v} \neq \mathbf{0}$. Note that for any real number t,

$$0 \leqslant |\mathbf{u} + t\mathbf{v}|^2 = \sum_{i=1}^{n} (u_i + tv_i)^2 = |\mathbf{u}|^2 + 2t \sum_{i=1}^{n} u_i v_i + t^2 |\mathbf{v}|^2.$$

As $|\mathbf{v}| \neq 0$, the RHS of the above inequality is a quadratic in t which is always non-negative; so it has non-positive discriminant ($b^2 \leqslant 4ac$). Hence

$$4 \left(\sum_{i=1}^{n} u_i v_i \right)^2 \leqslant 4 |\mathbf{u}|^2 |\mathbf{v}|^2, \qquad \text{giving} \qquad \left| \sum_{i=1}^{n} u_i v_i \right| \leqslant |\mathbf{u}| |\mathbf{v}|. \tag{3.3}$$

Finally

$$|\mathbf{u}+\mathbf{v}|^2 = |\mathbf{u}|^2 + 2\sum_{i=1}^{n} u_i v_i + |\mathbf{v}|^2 \leqslant |\mathbf{u}|^2 + 2\left|\sum_{i=1}^{n} u_i v_i\right| + |\mathbf{v}|^2$$
$$\leqslant |\mathbf{u}|^2 + 2|\mathbf{u}||\mathbf{v}| + |\mathbf{v}|^2 = (|\mathbf{u}| + |\mathbf{v}|)^2 \tag{3.4}$$

to give (3.2). We have equality in $b^2 \leqslant 4ac$ if and only if the quadratic $|\mathbf{u}+t\mathbf{v}|^2 = 0$ has a repeated real solution, say $t = t_0$. So $\mathbf{u} + t_0\mathbf{v} = \mathbf{0}$ and we see that $\mathbf{u}$ and $\mathbf{v}$ are multiples of one another. This is required for equality to occur in (3.3). With $\mathbf{u} = -t_0\mathbf{v}$, then equality in (3.4) requires that

$$\sum_{i=1}^{n} u_i v_i = \left|\sum_{i=1}^{n} u_i v_i\right|, \qquad \text{which means} \qquad -t_0 |\mathbf{v}|^2 = |t_0| |\mathbf{v}|^2.$$

This occurs when $-t_0 \geqslant 0$, as $\mathbf{v} \neq \mathbf{0}$. $\square$

Definition 3.16 Given two vectors $\mathbf{u} = (u_1, u_2, \ldots, u_n)$, $\mathbf{v} = (v_1, v_2, \ldots, v_n)$ in $\mathbb{R}^n$, the **scalar product** $\mathbf{u} \cdot \mathbf{v}$, also known as the **dot product** or **Euclidean inner product**, is defined as the real number

$$\mathbf{u} \cdot \mathbf{v} = u_1 v_1 + u_2 v_2 + \cdots + u_n v_n.$$

We then read $\mathbf{u} \cdot \mathbf{v}$ as 'u dot v'; we also often use 'dot' as a verb in this regard.

The following properties of the scalar product are easy to verify and left as #499. Note (e) was proved in (3.3).

#501
#502
#505

Proposition 3.17 *Let* $\mathbf{u}, \mathbf{v}, \mathbf{w}$ *be vectors in* $\mathbb{R}^n$ *and let* λ *be a real number. Then*

(a) $\mathbf{u} \cdot \mathbf{v} = \mathbf{v} \cdot \mathbf{u}$.   (b) $(\lambda\mathbf{u}) \cdot \mathbf{v} = \lambda(\mathbf{u} \cdot \mathbf{v})$.   (c) $(\mathbf{u}+\mathbf{v}) \cdot \mathbf{w} = \mathbf{u} \cdot \mathbf{w} + \mathbf{v} \cdot \mathbf{w}$.

(d) $\mathbf{u} \cdot \mathbf{u} = |\mathbf{u}|^2 \geqslant 0$ *and* $\mathbf{u} \cdot \mathbf{u} = 0$ *if and only if* $\mathbf{u} = \mathbf{0}$.

(e) *Cauchy–Schwarz inequality* [3]

$$|\mathbf{u} \cdot \mathbf{v}| \leqslant |\mathbf{u}||\mathbf{v}|, \tag{3.5}$$

with equality when one of $\mathbf{u}$ *and* $\mathbf{v}$ *is a multiple of the other.*

We see that the length of a vector $\mathbf{u}$ can be written in terms of the scalar product, namely as

$$|\mathbf{u}| = \sqrt{\mathbf{u} \cdot \mathbf{u}}.$$

We can also define the *angle* between two vectors in terms of their scalar product.

Definition 3.18 Given two non-zero vectors $\mathbf{u}, \mathbf{v}$ in $\mathbb{R}^n$ the **angle** between them is given by the expression

$$\cos^{-1}\left(\frac{\mathbf{u} \cdot \mathbf{v}}{|\mathbf{u}||\mathbf{v}|}\right). \tag{3.6}$$

---

[3] Named about Cauchy (see p.60 for some biography) and Hermann Schwarz (1843–1921). The inequality as stated here in $\mathbb{R}^n$ is due to Cauchy (1821), though the inequality generalizes widely and Schwarz proved a version for it involving integrals in 1888 (see #1276).

- The above formula makes sense as $|\mathbf{u} \cdot \mathbf{v}|/(|\mathbf{u}||\mathbf{v}|) \leqslant 1$ by the Cauchy–Schwarz inequality. If we take the principal values of $\cos^{-1}$ to be in the range $0 \leqslant \theta \leqslant \pi$ the formula measures the smaller angle between the vectors.
- Given two vectors $\mathbf{u}$ and $\mathbf{v}$ with angle θ between them, an equivalent definition of the scalar product $\mathbf{u} \cdot \mathbf{v}$ is then

$$\mathbf{u} \cdot \mathbf{v} = |\mathbf{u}|\,|\mathbf{v}|\cos\theta. \tag{3.7}$$

- **(Perpendicularity criterion)** Note $\mathbf{u}$ and $\mathbf{v}$ are perpendicular if and only if $\mathbf{u} \cdot \mathbf{v} = 0$.
- The obvious concern, that Definition 3.18 ties in with our usual notion of angle, is left to #503 and #504.

Example 3.19 Let $\mathbf{u} = (1, 2, -1)$ and $\mathbf{v} = (0, 2, 3)$ in $\mathbb{R}^3$. Find the lengths of $\mathbf{u}$ and $\mathbf{v}$ and the angle θ between them.

#496
#497
#503
#508
#509
#510

Solution We have

$$|\mathbf{u}|^2 = \mathbf{u} \cdot \mathbf{u} = 1^2 + 2^2 + (-1)^2 = 6, \quad \text{giving} \quad |\mathbf{u}| = \sqrt{6}.$$
$$|\mathbf{v}|^2 = \mathbf{v} \cdot \mathbf{v} = 0^2 + 2^2 + 3^2 = 13, \quad \text{giving} \quad |\mathbf{v}| = \sqrt{13}.$$
$$\mathbf{u} \cdot \mathbf{v} = 1 \times 0 + 2 \times 2 + (-1) \times 3 = 1,$$

giving $\theta = \cos^{-1}\left(\frac{1}{\sqrt{6}\sqrt{13}}\right) = \cos^{-1}\frac{1}{\sqrt{78}} \approx 1.457$ radians. $\qquad\square$

Example 3.20 Let $\mathbf{u}$ and $\mathbf{v}$ be vectors in $\mathbb{R}^n$ with $\mathbf{v} \neq \mathbf{0}$. Show there is a unique real number λ such that $\mathbf{u} - \lambda\mathbf{v}$ is perpendicular to $\mathbf{v}$. Then

$$\mathbf{u} = \lambda\mathbf{v} + (\mathbf{u} - \lambda\mathbf{v}),$$

with the vector $\lambda\mathbf{v}$ being called the **component of u in the direction of v** and $\mathbf{u} - \lambda\mathbf{v}$ being called the **component of u perpendicular to the direction of v**.

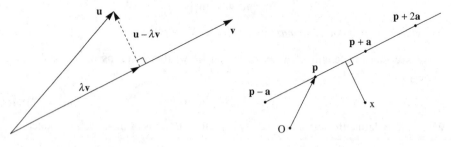

Components of a vector $\mathbf{u}$ relative to $\mathbf{v}$ A line described parametrically

Figure 3.5   Figure 3.6

Solution We have $\mathbf{u} - \lambda\mathbf{v}$ is perpendicular to $\mathbf{v}$ if and only if

$$(\mathbf{u} - \lambda\mathbf{v}) \cdot \mathbf{v} = 0 \quad \Longleftrightarrow \quad \mathbf{u} \cdot \mathbf{v} = \lambda|\mathbf{v}|^2 \quad \Longleftrightarrow \quad \lambda = (\mathbf{u} \cdot \mathbf{v})/|\mathbf{v}|^2.$$

Note that $|\lambda\mathbf{v}| = |\mathbf{u}|\cos\theta$, where θ is the angle between $\mathbf{u}$ and $\mathbf{v}$, as one would expect from Figure 3.5. $\qquad\square$

Example 3.21 (Parametric Form of a Line) Let $\mathbf{p}, \mathbf{a}, \mathbf{x}$ be vectors in $\mathbb{R}^n$ with $\mathbf{a} \neq \mathbf{0}$. Explain why the points

$$\mathbf{r}(\lambda) = \mathbf{p} + \lambda \mathbf{a},$$

where λ varies over all real numbers, comprise a line through $\mathbf{p}$, parallel to $\mathbf{a}$. Show that $|\mathbf{x} - \mathbf{r}(\lambda)|$ is minimal when $\mathbf{x} - \mathbf{r}(\lambda)$ is perpendicular to $\mathbf{a}$.

Proof As commented following Definition 3.7, the points $\lambda \mathbf{a}$ comprise the line which passes through the origin and the point $\mathbf{a}$. So the points $\mathbf{p} + \lambda \mathbf{a}$ comprise the translation of that line by the vector $\mathbf{p}$; that is, they comprise the line through the point $\mathbf{p}$ parallel to $\mathbf{a}$ (see Figure 3.6). We also have

$$\begin{aligned}
|\mathbf{x} - \mathbf{r}(\lambda)|^2 &= (\mathbf{x} - \mathbf{r}(\lambda)) \cdot (\mathbf{x} - \mathbf{r}(\lambda)) \\
&= \mathbf{x} \cdot \mathbf{x} - 2\mathbf{r}(\lambda) \cdot \mathbf{x} + \mathbf{r}(\lambda) \cdot \mathbf{r}(\lambda) \\
&= \mathbf{x} \cdot \mathbf{x} - 2(\mathbf{p} + \lambda \mathbf{a}) \cdot \mathbf{x} + (\mathbf{p} + \lambda \mathbf{a}) \cdot (\mathbf{p} + \lambda \mathbf{a}) \\
&= (\mathbf{x} \cdot \mathbf{x} - 2\mathbf{p} \cdot \mathbf{x} + \mathbf{p} \cdot \mathbf{p}) + \lambda(-2\mathbf{a} \cdot \mathbf{x} + 2\mathbf{a} \cdot \mathbf{p}) + \lambda^2(\mathbf{a} \cdot \mathbf{a}).
\end{aligned}$$

At the minimum value of $|\mathbf{x} - \mathbf{r}(\lambda)|$ we have

$$0 = \frac{d}{d\lambda}\left(|\mathbf{x} - \mathbf{r}(\lambda)|^2\right) = 2(\mathbf{a} \cdot \mathbf{p} - \mathbf{a} \cdot \mathbf{x}) + 2\lambda(\mathbf{a} \cdot \mathbf{a}),$$

which occurs when $0 = (\mathbf{p} + \lambda \mathbf{a} - \mathbf{x}) \cdot \mathbf{a} = (\mathbf{r}(\lambda) - \mathbf{x}) \cdot \mathbf{a}$ as required. $\qquad\square$

#511
#512
#513
#516
#517

Definition 3.22 Let $\mathbf{p}, \mathbf{a}$ be vectors in $\mathbb{R}^n$ with $\mathbf{a} \neq \mathbf{0}$. Then the equation $\mathbf{r}(\lambda) = \mathbf{p} + \lambda \mathbf{a}$, where λ is a real number, is the equation of the **line** through $\mathbf{p}$, parallel to $\mathbf{a}$. It is said to be in **parametric form**, the parameter here being λ. The parameter acts as a co-ordinate on the line, uniquely associating with each point on the line a value of λ.

A plane can similarly be described in parametric form. Whereas just one non-zero vector $\mathbf{a}$ was needed to travel along a line $\mathbf{r}(\lambda) = \mathbf{p} + \lambda \mathbf{a}$, we will need two non-zero vectors to move around a plane. However, we need to be a little careful: if we simply considered those points $\mathbf{r}(\lambda, \mu) = \mathbf{p} + \lambda \mathbf{a} + \mu \mathbf{b}$, for non-zero vectors $\mathbf{a}, \mathbf{b}$ and parameters λ, μ, we wouldn't always get a plane. In the case when $\mathbf{a}$ and $\mathbf{b}$ were scalar multiples of one another, so that they had the same or opposite directions, then the points $\mathbf{r}(\lambda, \mu)$ would just comprise the line through $\mathbf{p}$ parallel to $\mathbf{a}$ (or equivalently $\mathbf{b}$). So we make the definitions:

Definition 3.23 We say that two vectors in $\mathbb{R}^n$ are **linearly independent**, or just simply **independent**, if neither is a scalar multiple of the other. In particular, this means that both vectors are non-zero. Two vectors which aren't independent are said to be **(linearly) dependent.**

Definition 3.24 (Parametric Form of a Plane) Let $\mathbf{p}, \mathbf{a}, \mathbf{b}$ be vectors in $\mathbb{R}^n$ with $\mathbf{a}, \mathbf{b}$ independent. Then

$$\mathbf{r}(\lambda, \mu) = \mathbf{p} + \lambda \mathbf{a} + \mu \mathbf{b}, \tag{3.8}$$

where λ, μ are real numbers, is the equation of the **plane** through $\mathbf{p}$ parallel to the vectors $\mathbf{a}, \mathbf{b}$. The parameters λ, μ act as co-ordinates in the plane, associating to each point of the plane a unique ordered pair (λ, μ) (see #523).

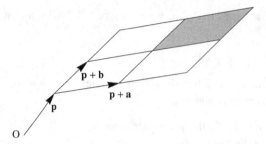

Note that **p** effectively becomes an origin in the plane having co-ordinates $\lambda = \mu = 0$, that **p**+**a** has co-ordinates $(\lambda,\mu) = (1,0)$ and that **p**+**b** = **r**(0,1).

The shaded area in Figure 3.7 comprises those points $\mathbf{r}(\lambda,\mu)$ where it's the case that $1 \leqslant \lambda,\mu \leqslant 2$.

Figure 3.7

Example 3.25 Given three points **a**, **b**, **c** in $\mathbb{R}^n$ which don't lie in a line, then **b** − **a**, **c** − **a** are independent (#522) and we can parametrize the plane Π which contains the points **a**, **b**, **c** as

$$\mathbf{r}(\lambda,\mu) = \mathbf{a} + \lambda(\mathbf{b}-\mathbf{a}) + \mu(\mathbf{c}-\mathbf{a}),$$

noting that $\mathbf{a} = \mathbf{r}(0,0)$, $\mathbf{b} = \mathbf{r}(1,0)$, $\mathbf{c} = \mathbf{r}(0,1)$.

A parametric description of a plane in $\mathbb{R}^n$ may be the most natural starting point but, especially in $\mathbb{R}^3$, planes can be easily described by equations in the Cartesian co-ordinates x, y, z.

Proposition 3.26 (*Cartesian Equation of a Plane in $\mathbb{R}^3$*) *A region Π of $\mathbb{R}^3$ is a plane if and only if it has the equation*

$$\mathbf{r} \cdot \mathbf{n} = c,$$

where $\mathbf{r} = (x,y,z)$, $\mathbf{n} = (n_1,n_2,n_3) \neq \mathbf{0}$ *and c is a real number. In terms of the co-ordinates* x, y, z *this equation reads*

$$n_1 x + n_2 y + n_3 z = c. \tag{3.9}$$

The vector **n** *is normal (i.e. perpendicular) to the plane.*

Proof The proof is left to #527 and #528. □

#522
#523
#524
#525
#529
#533
#535

Example 3.27 Write the equation $\mathbf{r} \cdot (1,1,1) = 1$ in some parametric form. Given your choice, what parameters are associated with the point $(1/3, 1/3, 1/3)$?

Solution The point $(1,0,0)$ lies in the plane. Further, $(1,-1,0)$ and $(0,1,-1)$ are independent vectors perpendicular to the normal $(1,1,1)$. So

$$\mathbf{r}(\lambda,\mu) = (1,0,0) + \lambda(1,-1,0) + \mu(0,1,-1)$$

is a parametric form for the plane. By focusing on the first and third co-ordinates we see that $\mathbf{r}(\lambda,\mu) = (1/3, 1/3, 1/3)$ when $\lambda = -2/3$ and $\mu = -1/3$. □

Example 3.28 Find all the planes that include the point $(\sqrt{2}, 1, \sqrt{2})$, make an angle of $\pi/3$ with the line $x = y = -z$, and an angle of $\pi/4$ with the plane $x = z$.

Solution Suppose that such a plane has equation $\mathbf{r} \cdot \mathbf{n} = c$ where $\mathbf{n}$ is a unit vector. As the point $\left(\sqrt{2}, 1, \sqrt{2}\right)$ lies in the plane then we have

$$\sqrt{2}n_1 + n_2 + \sqrt{2}n_3 = c.$$

As the plane makes an angle of $\pi/3$ with a vector parallel to the line $x = y = -z$, for example $(1, 1, -1)$, then the plane's normal $\mathbf{n}$ makes an angle of $\pi/2 - \pi/3 = \pi/6$ with the line. So we have

$$n_1 + n_2 - n_3 = (1, 1, -1) \cdot \mathbf{n} = \sqrt{3} \times 1 \times \cos(\pi/6) = 3/2.$$

Finally the angle between two planes is the same as the angle between the planes' normals $\mathbf{n}$ and $(1, 0, -1)$. So we have

$$n_1 - n_3 = (1, 0, -1) \cdot \mathbf{n} = \sqrt{2} \times 1 \times \cos(\pi/4) = 1.$$

We can use the three equations above to determine n_1, n_2, n_3 in terms of c, namely

$$n_1 = \frac{c}{\sqrt{8}} - \frac{1}{\sqrt{32}} + \frac{1}{2}; \qquad n_2 = \frac{1}{2}; \qquad n_3 = \frac{c}{\sqrt{8}} - \frac{1}{\sqrt{32}} - \frac{1}{2}.$$

As $(n_1)^2 + (n_2)^2 + (n_3)^2 = 1$, because $\mathbf{n}$ is a unit vector, we have

$$\left(\frac{c}{\sqrt{8}} - \frac{1}{\sqrt{32}} + \frac{1}{2}\right)^2 + \left(\frac{1}{2}\right)^2 + \left(\frac{c}{\sqrt{8}} - \frac{1}{\sqrt{32}} - \frac{1}{2}\right)^2 = 1,$$

which simplifies to $4c^2 - 4c - 3 = 0$, and has roots $c = 3/2$ and $c = -1/2$. Substituting these values of c into our expressions for n_1, n_2, n_3 we see there are two planes with the desired properties, namely

$$\left(\frac{1}{\sqrt{2}} + 1\right)x + y + \left(\frac{1}{\sqrt{2}} - 1\right)z = 3; \qquad \left(\frac{1}{\sqrt{2}} - 1\right)x - y + \left(\frac{1}{\sqrt{2}} + 1\right)z = 1. \qquad \square$$

Exercises

Exercises on Vectors

#496a Show that the vectors $\mathbf{u} = (1, 2, -3)$ and $\mathbf{v} = (6, 3, 4)$ are perpendicular in $\mathbb{R}^3$. Verify directly Pythagoras' theorem for the right-angled triangles with vertices $\mathbf{0}, \mathbf{u}, \mathbf{v}$ and vertices $\mathbf{0}, \mathbf{u}, \mathbf{u} + \mathbf{v}$.

#497A Find the lengths of the vectors $\mathbf{u} = (1, 0, 1)$ and $\mathbf{v} = (3, 2, 1)$ in $\mathbb{R}^3$ and the angle between them.

#498A Let L_1 and L_2 be non-vertical two lines in $\mathbb{R}^2$ respectively parallel to vectors $\mathbf{a}_1$ and $\mathbf{a}_2$. Show that the condition $\mathbf{a}_1 \cdot \mathbf{a}_2 = 0$ is equivalent to the products of the lines' gradients equalling -1.

#499C Verify properties (a)–(d) of the scalar product given in Proposition 3.17.

#500C(i) Let $\mathbf{v}, \mathbf{w}$ be vectors in $\mathbb{R}^n$. Verify that $|\mathbf{v}| \geq 0$ and that $|\mathbf{v}| = 0$ if and only if $\mathbf{v} = \mathbf{0}$.
(ii) Show that $|\lambda \mathbf{v}| = |\lambda| \, |\mathbf{v}|$ for any real number λ.

(iii) If $\mathbf{v} \neq \mathbf{0}$, show that $\frac{1}{|\mathbf{v}|}\mathbf{v}$ is a unit vector.

(iv) If $\mathbf{v} \neq \mathbf{0} \neq \mathbf{w}$ and λ, μ are positive numbers, show that the angle between $\mathbf{v}$ and $\mathbf{w}$ is the same as the angle between $\lambda\mathbf{v}$ and $\mu\mathbf{w}$. How does the angle between $\mathbf{v}$ and $-\mathbf{w}$ relate to the angle between $\mathbf{v}$ and $\mathbf{w}$?

#501A† Let $\mathbf{v}, \mathbf{w}$ be vectors in $\mathbb{R}^n$. Show that if $\mathbf{v} \cdot \mathbf{x} = \mathbf{w} \cdot \mathbf{x}$ for all $\mathbf{x}$ in $\mathbb{R}^n$ then $\mathbf{v} = \mathbf{w}$.

#502b† Let $d(\mathbf{p},\mathbf{q}) = |\mathbf{q} - \mathbf{p}|$ denote the distance between two points $\mathbf{p}$ and $\mathbf{q}$ in $\mathbb{R}^n$. Show

(i) $d(\mathbf{p},\mathbf{q}) \geqslant 0$ and $d(\mathbf{p},\mathbf{q}) = 0$ if and only if $\mathbf{p} = \mathbf{q}$.

(ii) $d(\mathbf{p},\mathbf{q}) = d(\mathbf{q},\mathbf{p})$.

(iii) $d(\mathbf{p},\mathbf{q}) + d(\mathbf{q},\mathbf{r}) \geqslant d(\mathbf{p},\mathbf{r})$ for points $\mathbf{p}$, $\mathbf{q}$ and $\mathbf{r}$ in $\mathbb{R}^n$. When is there equality in (iii)?

Properties (i), (ii), (iii) show that d is a **metric** (see p.143) on $\mathbb{R}^n$.

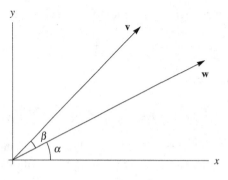

Figure 3.8

#503A Consider Figure 3.8. Write down the xy-co-ordinates of $\mathbf{v}$ and $\mathbf{w}$ in terms of $|\mathbf{v}|$, $|\mathbf{w}|$, α, β. Verify that the formulas in Definitions 3.16 and 3.18 do indeed calculate the angle between $\mathbf{v}$ and $\mathbf{w}$ as β.

#504c We may identify the complex plane $\mathbb{C}$ with $\mathbb{R}^2$ by identifying $x + yi$ with (x, y). If we identify non-zero v and w with vectors $\mathbf{v}$ and $\mathbf{w}$ in this way, verify (as in Definition 1.34) that

$$\frac{\mathbf{v} \cdot \mathbf{w}}{|\mathbf{v}||\mathbf{w}|} = \cos \arg\left(\frac{w}{v}\right).$$

#505B† (See also #1397) Deduce from (3.2) that $|\mathbf{v}_1 + \cdots + \mathbf{v}_n| \leqslant |\mathbf{v}_1| + \cdots + |\mathbf{v}_n|$ for n vectors $\mathbf{v}_1, \ldots, \mathbf{v}_n$ in $\mathbb{R}^m$. By choosing appropriate $\mathbf{v}_k$ in $\mathbb{R}^2$, show that

$$\sum_{k=1}^{n} \sqrt{k^2 + 1} \geqslant \frac{1}{2}n\sqrt{n^2 + 2n + 5}.$$

#506b† (i) What is the maximum value of $2x + 3y - 4z$ given that $x^2 + y^2 + z^2 = 1$?

(ii) What is the maximum value of $3x - 2y - 5z$ given that $2x^2 + 3y^2 + z^2 = 2$?

(iii) What is the maximum value of $3x - 7y$ given that $x^2 + y^2 + 2x - 4y = 2$?

#507b† Show that any unit vector in three dimensions can be written in the form $\mathbf{r}(\theta, \phi) = (\sin\theta \cos\phi, \sin\theta \sin\phi, \cos\theta)$, where $0 \leqslant \phi < 2\pi$ and $0 \leqslant \theta \leqslant \pi$. Show that each unit vector is associated[4] with unique θ and ϕ except for $(0, 0, \pm 1)$.

#508a Find all vectors in $\mathbb{R}^3$ which are perpendicular to $\mathbf{u} = (1, -1, 2)$ and $\mathbf{v} = (3, -1, 2)$.

---

[4] The parametrization $(r\cos\phi \sin\theta, r\sin\phi \sin\theta, r\cos\theta)$ of three-dimensional space (where $0 < \phi < 2\pi, 0 < \theta < \pi$ and $r > 0$) uses *spherical polar co-ordinates*. Fixing r constrains the point to a sphere of radius r, centred at the origin, with meridians being curves of constant ϕ and lines of latitude being curves of constant θ. Unlike ϕ and θ, longitude is usually in the range $-\pi$ to π (or $180°$W to $180°$E) and latitude in the range $\pi/2$ to $-\pi/2$ (or $90°$N to $90°$S).

#509a Let $\mathbf{a} = (a_1, a_2, a_3)$ and $\mathbf{b} = (b_1, b_2, b_3)$ be two vectors in $\mathbb{R}^3$. Show directly that the vector

$$\mathbf{n} = (a_2 b_3 - a_3 b_2, a_3 b_1 - a_1 b_3, a_1 b_2 - a_2 b_1) \qquad (3.10)$$

is perpendicular to both $\mathbf{a}$ and $\mathbf{b}$. Show further that $\mathbf{n} = \mathbf{0}$ if $\mathbf{a} = \lambda \mathbf{b}$ for some λ.

#510B Let $\mathbf{a}$ and $\mathbf{b} \neq \mathbf{0}$ be two vectors in $\mathbb{R}^3$. If the vector $\mathbf{n} = \mathbf{0}$, where $\mathbf{n}$ is as given in (3.10), show that $\mathbf{a} = \lambda \mathbf{b}$ for some real number λ.

Exercises on Lines

#511B† Show that (x, y, z) lies on the line parametrized by $\mathbf{r}(\lambda) = (1, 2, -3) + \lambda(2, 1, 4)$ if and only if $(x - 1)/2 = y - 2 = (z + 3)/4$.

#512b† Give, in parametric form, the equation of the line $(x - 2)/3 = (y + 1)/2$, $z = 2$.

#513a† Let $\mathbf{a}, \mathbf{b}$ be distinct points in $\mathbb{R}^n$. Find the equation of the line through $\mathbf{a}$ and $\mathbf{b}$ in parametric form in such a way that $\mathbf{a}$ is assigned parameter $\lambda = 0$ and $\mathbf{b}$ is assigned parameter $\lambda = 1$. Find, in parametric form, the equation of the line which assigns parameter $\lambda = 2$ to $\mathbf{a}$ and parameter $\lambda = -1$ to $\mathbf{b}$.

#514c Let $\mathbf{p}, \mathbf{a}$ be vectors in $\mathbb{R}^n$ with $\mathbf{a} \neq \mathbf{0}$. Parametrize the line through $\mathbf{p}$, parallel to $\mathbf{a}$, so as to assign parameter $\lambda = 2$ to the point $\mathbf{p} + 3\mathbf{a}$ and parameter $\lambda = -3$ to the point $\mathbf{p} + 7\mathbf{a}$. What parameter is assigned to the point $\mathbf{p} + t\mathbf{a}$?

#515B Find s such that the points $(3 - s^2, s), (2s - s^2, 1 + s)$ and $(2s - 9, 2s - 2)$ are collinear in $\mathbb{R}^2$ and distinct.

#516A† Given the points $P = (1, -2, 3)$ and $Q = (-3, 0, 2)$ in $\mathbb{R}^3$, find the point R which divides the line segment PQ in the ratio $2 : 3$. That is, find R between P and Q such that $|PR| = \frac{2}{3}|QR|$. Show that there is precisely one other point S on the line PQ such that $|PS| = \frac{2}{3}|QS|$.

#517b† Find the point on the line $\mathbf{r}(\lambda) = (-3, 1, 1) + \lambda(0, -2, 4)$ which is closest to the point $(1, 1, 6)$.

#518B† Show that the shortest distance from the point (x_0, y_0) in $\mathbb{R}^2$ to the line with equation $ax + by + c = 0$ equals

$$\frac{|ax_0 + by_0 + c|}{\sqrt{a^2 + b^2}}.$$

#519c Show that the points on the line $\mathbf{r}(\lambda) = (1, 2, -1) + \lambda(1, -3, 1)$, which lie within distance 5 of the origin, form a line segment and determine its length.

#520D† Find the shortest distance between the two lines $\mathbf{r}(\lambda) = (1, 3, 0) + \lambda(2, 3, 2)$ and $\mathbf{s}(\mu) = (2, 1, 0) + \mu(0, 2, 1)$.

#521D Let $\mathbf{p}, \mathbf{q}, \mathbf{a}, \mathbf{b}$ be vectors in $\mathbb{R}^n$ with $\mathbf{a}, \mathbf{b}$ non-zero. Show that the parametric equations $\mathbf{r}(\lambda) = \mathbf{p} + \lambda\mathbf{a}$ and $\mathbf{s}(\mu) = \mathbf{q} + \mu\mathbf{b}$ define the same line if and only if $\mathbf{q} - \mathbf{p} = k\mathbf{a}$ for some k and $\mathbf{b} = l\mathbf{a}$ for some l. If these conditions hold, and $\mathbf{r}(\lambda) = \mathbf{s}(\mu)$ represent the same point, how are the parameters λ and μ related?

Exercises on Planes

#522A Show that points $\mathbf{a}, \mathbf{b}, \mathbf{c}$ are collinear in $\mathbb{R}^n$ if and only if the vectors $\mathbf{b} - \mathbf{a}$ and $\mathbf{c} - \mathbf{a}$ are linearly dependent.

#523A Let $\mathbf{p}, \mathbf{a}, \mathbf{b}$ be in $\mathbb{R}^n$ with $\mathbf{a}, \mathbf{b}$ independent and $\mathbf{p} + \lambda_1 \mathbf{a} + \mu_1 \mathbf{b} = \mathbf{p} + \lambda_2 \mathbf{a} + \mu_2 \mathbf{b}$ for some real numbers $\lambda_1, \mu_1, \lambda_2, \mu_2$. Show that $\lambda_1 = \lambda_2$ and $\mu_1 = \mu_2$.

#524B† Let $\mathbf{a}, \mathbf{b}$ be independent vectors in $\mathbb{R}^2$. Then every point in $\mathbb{R}^2$ has associated parameters (λ, μ), as it can be uniquely written as $\mathbf{r}(\lambda, \mu) = \lambda \mathbf{a} + \mu \mathbf{b}$ for some λ, μ.

(i) Show that the equation $\lambda + \mu = 1$ describes a line.
(ii) Show that the four points with parameters $(0,0), (1,0), (1,1), (0,1)$ form the vertices of a parallelogram.
(iii) Show that the equation $\lambda^2 + \mu^2 = 1$ describes a circle if and only if $\mathbf{a} \cdot \mathbf{b} = 0$ and $|\mathbf{a}| = |\mathbf{b}|$.

#525b Show the three vectors $\mathbf{u} = (1, -1, 0)/\sqrt{2}$, $\mathbf{v} = (1, 1, 1)/\sqrt{3}$, $\mathbf{w} = (1, 1, -2)/\sqrt{6}$ in $\mathbb{R}^3$ are each of unit length and are mutually perpendicular. Show that $(x, y, z) = \lambda \mathbf{u} + \mu \mathbf{v}$, for real numbers λ, μ if and only if $x + y = 2z$. Show further, that if $\mathbf{r} = \lambda \mathbf{u} + \mu \mathbf{v}$, then $\lambda = \mathbf{r} \cdot \mathbf{u}$ and $\mu = \mathbf{r} \cdot \mathbf{v}$.

#526c Let $\mathbf{a}, \mathbf{b}$ be independent vectors in $\mathbb{R}^n$. Then $\mathbf{r} = \lambda \mathbf{a} + \mu \mathbf{b}$ is the equation of a plane in parametric form. Show that $\lambda = \mathbf{r} \cdot \mathbf{a}$ and $\mu = \mathbf{r} \cdot \mathbf{b}$ hold true for all points of the plane if and only if $|\mathbf{a}| = |\mathbf{b}| = 1$ and $\mathbf{a} \cdot \mathbf{b} = 0$.

#527D† By eliminating λ and μ from the three corresponding scalar equations, show that (3.8) is equivalent to a single equation of the form $\mathbf{r} \cdot \mathbf{N} = c$, where $\mathbf{N}$ is some non-zero multiple of the vector $\mathbf{n}$ in (3.10).

#528A Let $\mathbf{p}$ and $\mathbf{q}$ be two points which lie in the plane $\mathbf{r} \cdot \mathbf{n} = c$. Show that $\mathbf{p} - \mathbf{q}$ is perpendicular to $\mathbf{n}$.

#529b Determine, in the form $\mathbf{r} \cdot \mathbf{n} = c$, (i) the plane containing the points $(1,0,0), (1,1,0), (0,1,1)$; (ii) the plane containing the point $(2,1,0)$ and the line $x = y = z$; (iii) the plane which is perpendicular to the line PQ and which passes through the midpoint of PQ where $P = (1, 0, -3)$ and $Q = (0, 2, 1)$.

#530B Show that the distance of a point $\mathbf{p}$ from the plane with equation $\mathbf{r} \cdot \mathbf{n} = c$ equals $|\mathbf{p} \cdot \mathbf{n} - c|/|\mathbf{n}|$.

#531C† Determine, in the form $\mathbf{r} \cdot \mathbf{n} = c$, the two planes containing the points $(1,0,1)$, $(0,1,1)$ and which are tangential to the unit sphere, centre $\mathbf{0}$.

#532c A sphere with unit radius is tangential to the both the plane $x + 2y + 2z = 5$ and also $2x + y - 2z = 1$. Show that the sphere's centre must lie on one of four parallel lines and find the equations of the four lines.

#533a Show that the planes $x + 2y + 2z = 5$ and $2x + y - 2z = 1$ meet in the line with equation $(x - 1)/2 = (1 - y)/2 = z - 1$.

Metrics, Norms and Inner Products A metric is a distance function. Because many important ideas of analysis and geometry rely on a notion of distance, for example, convergence and continuity, metrics are useful in introducing such notions, and analytical and geometric methods, to address a wide range of problems. We have already met the Euclidean metric in $\mathbb{R}^n$; given two points $\mathbf{p}$ and $\mathbf{q}$ we defined the distance between them to be

$$d(\mathbf{p},\mathbf{q}) = \sqrt{\sum_{i=1}^{n} (p_i - q_i)^2}.$$

There are other metrics which we might naturally use on $\mathbb{R}^n$. Two such are

$$D(\mathbf{p},\mathbf{q}) = \sum_{i=1}^{n} |p_i - q_i|, \qquad \delta(\mathbf{p},\mathbf{q}) = \max_{1 \leqslant i \leqslant n} |p_i - q_i|.$$

But metrics can be introduced much more variously. It's less obvious what the distance between two functions might be. If V is the set of polynomials $p(x)$ defined on the interval $-1 \leqslant x \leqslant 1$ then we can define

$$d_1(p,q) = \int_{-1}^{1} |p(x) - q(x)| \, \mathrm{d}x, \qquad d_2(p,q) = \sqrt{\int_{-1}^{1} (p(x) - q(x))^2 \, \mathrm{d}x},$$

$$d_\infty(p,q) = \max_{-1 \leqslant x \leqslant 1} |p(x) - q(x)|,$$

as three possible metrics. The distance function d_D defined on the hyperbolic plane is also a metric and distance on a sphere can similarly be defined as the shortest distance along a great circle between two points (see p.332 for exercises on spherical geometry). The required properties of a **metric** d on a non-empty set M are that

- (M1) $d(m_1,m_2) \geqslant 0$ and $d(m_1,m_2) = 0$ if and only if $m_1 = m_2$.
- (M2) (Symmetry) $d(m_1,m_2) = d(m_2,m_1)$.
- (M3) (Triangle inequality) $d(m_1,m_2) + d(m_2,m_3) \geqslant d(m_1,m_3)$.

When the set M has more algebraic structure – for example, addition and scalar multiplication – then we might wish our distance function to reflect this. A **norm** $||\ \ ||$ on such a set M is a real-valued function satisfying

- (N1) $||v|| \geqslant 0$ and $||v|| = 0$ if and only if $v = 0$.
- (N2) $||\alpha v|| = |\alpha|\,||v||$ when α is a scalar.
- (N3) $||v + w|| \leqslant ||v|| + ||w||$.

Given a norm $||\ \ ||$ then $d(v,w) = ||v - w||$ defines a metric (#962). Alternatively, we might define an **inner product** on M which will also allow us to define angles as well. An inner product is a generalization of the dot product and must satisfy for scalars α_1, α_2 and u, v, w in M

- (IP1) $(\alpha_1 v_1 + \alpha_2 v_2) \cdot w = \alpha_1 v_1 \cdot w + \alpha_2 v_2 \cdot w.$

- (IP2) $v \cdot w = w \cdot v$.
- (IP3) $v \cdot v \geqslant 0$ and $v \cdot v = 0$ if and only if $v = 0$.

(More commonly the notation $\langle v, w \rangle$ is used instead of $v \cdot w$, when the inner product in question is not the dot product.) Given an inner product we can define a norm by $||v|| = \sqrt{v \cdot v}$ (#963). From these axioms we can prove the Cauchy–Schwarz inequality and so define the angle between v and w as $\cos^{-1}\left(\frac{v \cdot w}{||v||\,||w||}\right)$. (See #961 et seq.)

#534 B† Find a parametric form $\mathbf{r}(\lambda, \mu)$ for the plane $3x + 2y + z = 6$. With your choice of parametric form, what are λ, μ for the points $(2,0,0)$, $(0,3,0)$ and $(0,0,6)$?

#535 B Let $(x,y,z) = (2s+t-3, 3s-t+4, s+t-2)$. Show that, as s and t vary over the real numbers, the points (x,y,z) range over a plane whose equation, in the form $ax+by+cz = d$, you should determine.

#536 B† Rewrite, as $\mathbf{r} \cdot \mathbf{n} = c$, the plane $\mathbf{r}(\lambda, \mu) = (0,2,-1) + \lambda(1,0,3) + \mu(2,3,-1)$.

#537 b† Rewrite the equation of the plane $3x + 2y + z = 1$ in parametric form, in such a way as to assign the points $(1,-1,0)$, $(0,0,1)$, $(0,1,-1)$ the parameters $(\lambda, \mu) = (0,0)$, $(1,0)$ and $(0,1)$ respectively.

#538 C Rewrite the equation of the plane $3x + 2y + z = 1$ in parametric form, in such a way as to assign the points $(0,0,1)$, $(1,0,-2)$, $(1,1,-4)$ the parameters $(\lambda', \mu') = (1,-1)$, $(1,1)$ and $(2,1)$ respectively. How are the parameters λ', μ' of a given point in the plane related to those λ, μ (for the same point) from #537?

#539 D† Suppose that $\mathbf{r}_1(\lambda_1, \mu_1)$ and $\mathbf{r}_2(\lambda_2, \mu_2)$ are parametrizations which describe the same plane Π in parametric form in two different ways. Show that there exist a,b,c,d,e,f with $ad - bc \neq 0$ such that

$$\mathbf{r}_1(\lambda_1, \mu_1) = \mathbf{r}_2(\lambda_2, \mu_2) \iff \lambda_1 = a\lambda_2 + b\mu_2 + e \quad \text{and} \quad \mu_1 = c\lambda_2 + d\mu_2 + f.$$

Some Geometric Theory

#540 B† Let $\mathbf{n}$ be a unit vector. Let P denote orthogonal projection [5] on to the plane $\mathbf{r} \cdot \mathbf{n} = 0$ and Q denote reflection in that plane. Show for any $\mathbf{v}$ that

(i) $P(\mathbf{v}) = \mathbf{v} - (\mathbf{v} \cdot \mathbf{n})\mathbf{n}$. (ii) $P^2(\mathbf{v}) = P(\mathbf{v})$. (iii) $Q(\mathbf{v}) = \mathbf{v} - 2(\mathbf{v} \cdot \mathbf{n})\mathbf{n}$. (iv) $Q^2(\mathbf{v}) = \mathbf{v}$.

#541 B† Let $\mathbf{n}$ be a unit vector. Show that reflection in the plane $\mathbf{r} \cdot \mathbf{n} = c$ is given by

$$Q_c(\mathbf{v}) = \mathbf{v} - 2(\mathbf{v} \cdot \mathbf{n})\mathbf{n} + 2c\mathbf{n}.$$

Let Q_c denote reflection in the plane $2x + y + 2z = c$. Show that

$$Q_c(x,y,z) = \frac{1}{9}(x - 4y - 8z + 12c, -4x + 7y - 4z + 6c, -8x - 4y + z + 12c).$$

---

[5] The *orthogonal projection* of a point A to a plane Π is the unique point B in Π such that AB is normal to Π.

#542C Let **n** be a unit vector in $\mathbb{R}^3$ and k,c be real numbers. Show that every vector **v** in $\mathbb{R}^3$ can be uniquely written $\mathbf{v} = \mathbf{p} + z\mathbf{n}$ for some real number z and some **p** satisfying $\mathbf{p} \cdot \mathbf{n} = c$. The **stretch** S_k with invariant plane $\mathbf{r} \cdot \mathbf{n} = c$ and stretch factor k is then defined by $S_k(\mathbf{v}) = \mathbf{p} + kz\mathbf{n}$.

 (i) Describe the maps S_1, S_0 and S_{-1}.
 (ii) Show that $S_{kl} = S_k S_l$ for any real numbers k, l.
 (iii) Show that $S_k(\mathbf{v}) = \mathbf{v} + (k-1)[(\mathbf{v} \cdot \mathbf{n}) - c]\mathbf{n}$.

#543D Under what circumstances do two stretches of $\mathbb{R}^3$, with different invariant planes, commute?

#544b (See also Theorem 1.44.) Given **a** in $\mathbb{R}^2$ and $0 < \lambda < 1$, define $\mathbf{b} = \mathbf{a}/(1 - \lambda^2)$ and prove that

$$\frac{|\mathbf{r} - \mathbf{a}|^2 - \lambda^2 |\mathbf{r}|^2}{1 - \lambda^2} = |\mathbf{r} - \mathbf{b}|^2 - \lambda^2 |\mathbf{b}|^2.$$

Deduce that if O and A are fixed points in a plane, then the locus of all points X, such that $|AX| = \lambda |OX|$, is a circle.

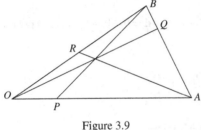

Figure 3.9

#545D† Let OAB be a triangle in the plane and let P, Q and R be points on the triangle respectively on OA, AB and BO, as shown in Figure 3.9. Prove Ceva's theorem – named after the Italian mathematician Giovanni Ceva (1647–1734) – which states that AR, BP and OQ are concurrent if and only if

$$|OP| \times |AQ| \times |BR| = |PA| \times |QB| \times |RO|.$$

#546B† Show that the three lines connecting the midpoints of opposite edges of a tetrahedron meet in a point. Show also that the four lines connecting the vertices to the centroids of the opposite faces meet in the same point.

#547B†(i) Show, for any four vectors **a, b, c, d**, that

$$(\mathbf{a} - \mathbf{d}) \cdot (\mathbf{b} - \mathbf{c}) - (\mathbf{a} - \mathbf{c}) \cdot (\mathbf{b} - \mathbf{d}) + (\mathbf{a} - \mathbf{b}) \cdot (\mathbf{c} - \mathbf{d}) = 0.$$

(ii) Let ABC be a triangle: an altitude of the triangle is a line through one vertex perpendicular to the opposite side. Use the identity above to prove that the three altitudes of ABC are concurrent, at a point called the **orthocentre.**
(iii) Let XYZ be another triangle: by considering the triangle whose vertices are the midpoints of XYZ, prove that the three perpendicular bisectors of the sides of XYZ meet at a point, called the **circumcentre.** Show that there is a circle centred on the circumcentre passing through each of X, Y, Z.

#548D† Let ABC be a triangle. Let $\mathbf{a}, \mathbf{b}, \mathbf{c}$ be the position vectors of A, B, C from an origin O.

 (i) Write down the position vector of the centroid of the triangle.
 (ii) Show that if we take the orthocentre as the origin for the plane, then $\mathbf{a} \cdot \mathbf{b} = \mathbf{b} \cdot \mathbf{c} = \mathbf{c} \cdot \mathbf{a}$.
 (iii) With the orthocentre still as origin, show that the circumcentre has position vector $(\mathbf{a} + \mathbf{b} + \mathbf{c})/2$. Deduce that the centroid, circumcentre and orthocentre are collinear. The line on which they lie is called the **Euler line**.

#549c Let $\mathbf{u} = (1, 0, 2, -3)$, $\mathbf{v} = (1, 3, -1, 0)$. Find two independent vectors $\mathbf{a}, \mathbf{b}$ in $\mathbb{R}^4$ which are perpendicular to both $\mathbf{u}$ and $\mathbf{v}$. Show that $\mathbf{w}$ in $\mathbb{R}^4$ is perpendicular to both $\mathbf{u}, \mathbf{v}$ if and only if $\mathbf{w} = \lambda \mathbf{a} + \mu \mathbf{b}$ for some λ, μ.

#550c† The plane Π in $\mathbb{R}^4$ has equation $\mathbf{r} = (0, 1, -2, 3) + \lambda(1, 2, 0, 1) + \mu(2, -1, 0, 3)$. Find two vectors $\mathbf{n}_1, \mathbf{n}_2$ in $\mathbb{R}^4$ and real scalars c_1, c_2 such that $\mathbf{r}$ lies in Π if and only if $\mathbf{r} \cdot \mathbf{n}_1 = c_1$ and $\mathbf{r} \cdot \mathbf{n}_2 = c_2$.

#551D† A tetrahedron $ABCD$ has vertices with respective position vectors $\mathbf{a}, \mathbf{b}, \mathbf{c}, \mathbf{d}$ from an origin O inside the tetrahedron. The lines AO, BO, CO, DO meet the opposite faces in E, F, G, H.

 (i) Show a point lies in the plane BCD if and only if it has position vector $\lambda \mathbf{b} + \mu \mathbf{c} + \nu \mathbf{d}$ where $\lambda + \mu + \nu = 1$.
 (ii) There are non-zero $\alpha, \beta, \gamma, \delta$ such that $\alpha \mathbf{a} + \beta \mathbf{b} + \gamma \mathbf{c} + \delta \mathbf{d} = \mathbf{0}$. Show that E has position vector $-\alpha \mathbf{a}/(\beta + \gamma + \delta)$.
 (iii) Deduce that

$$\frac{|AO|}{|AE|} + \frac{|BO|}{|BF|} + \frac{|CO|}{|CG|} + \frac{|DO|}{|DH|} = 3.$$

3.3 The Algebra of Matrices

At its simplest, a *matrix* is just a two-dimensional array of numbers; for example,

$$\begin{pmatrix} 1 & 2 & -3 \\ \sqrt{2} & \pi & 0 \end{pmatrix}, \qquad \begin{pmatrix} 1 \\ -1.2 \\ -1 \end{pmatrix}, \qquad \begin{pmatrix} 0 & 0 \\ 0 & 0 \end{pmatrix} \qquad (3.11)$$

are all matrices. The examples above are respectively a 2×3 matrix, a 3×1 matrix and a 2×2 matrix (read '2 by 3' etc.); the first figure refers to the number of horizontal *rows* and the second to the number of vertical *columns* in the matrix. Row vectors in $\mathbb{R}^n$ are $1 \times n$ matrices and columns vectors in $\mathbb{R}_n$ are $n \times 1$ matrices.

Definition 3.29 Let m, n be positive integers. An $m \times n$ **matrix** is an array of real numbers arranged into m **rows** and n **columns**.

Example 3.30 Consider the first matrix above. Its second row is $\begin{pmatrix} \sqrt{2} & \pi & 0 \end{pmatrix}$ and its third column is $\begin{pmatrix} -3 \\ 0 \end{pmatrix}$.

Definition 3.31 The numbers in a matrix are its **entries**. Given an $m \times n$ matrix A, we will write $[A]_{ij}$ for the entry in the ith row and jth column. Note that i can vary between 1 and

m, and that j can vary between 1 and n. So

$$\text{ith row} = ([A]_{i1}, \ldots, [A]_{in}) \quad \text{and} \quad \text{jth column} = \begin{pmatrix} [A]_{1j} \\ \vdots \\ [A]_{mj} \end{pmatrix}.$$

It is also common to write 'let $A = (a_{ij})$ be a matrix' as a shorthand for '$[A]_{ij} = a_{ij}$'.

Notation 3.32 We shall denote the set of real $m \times n$ matrices as M_{mn}, or $M_{mn}(\mathbb{R})$ if we wish to stress that the entries are real numbers. Note that $M_{1n} = \mathbb{R}^n$ and that $M_{n1} = \mathbb{R}_n$. From time to time we shall also consider matrices whose entries are complex numbers; we shall be sure to refer to such matrices as **complex matrices** to avoid any ambiguity. We denote the set of complex $m \times n$ matrices as $M_{mn}(\mathbb{C})$.

Example 3.33 If we write A for the first matrix in (3.11) then we have $[A]_{23} = 0$ and $[A]_{12} = 2$.

There are three important operations that can be performed with matrices: *matrix addition, scalar multiplication* and *matrix multiplication*. As with vectors, not all pairs of matrices can be meaningfully added or multiplied.

Definition 3.34 Addition Let A be an $m \times n$ matrix (recall: m rows and n columns) and B be a $p \times q$ matrix. As with vectors, matrices are added by adding their corresponding entries. So, as with vectors, to add two matrices they have to be the same size – that is, to add A and B, we must have $m = p$ and $n = q$ and we then have

$$[A + B]_{ij} = [A]_{ij} + [B]_{ij} \quad \text{for } 1 \leqslant i \leqslant m \text{ and } 1 \leqslant j \leqslant n.$$

Example 3.35 Let

$$A = \begin{pmatrix} 1 & 2 \\ -1 & 0 \end{pmatrix}, \quad B = \begin{pmatrix} 1 & 2 & 3 \\ 3 & 2 & 1 \end{pmatrix}, \quad C = \begin{pmatrix} 1 & -1 \\ 1 & -1 \end{pmatrix}. \tag{3.12}$$

$$\underbrace{\phantom{A = \begin{pmatrix} 1 & 2 \\ -1 & 0 \end{pmatrix}}}_{2 \times 2} \quad \underbrace{\phantom{B = \begin{pmatrix} 1 & 2 & 3 \\ 3 & 2 & 1 \end{pmatrix}}}_{2 \times 3} \quad \underbrace{\phantom{C = \begin{pmatrix} 1 & -1 \\ 1 & -1 \end{pmatrix}}}_{2 \times 2}$$

Of the possible sums involving these matrices, only $A + C$ and $C + A$ make sense as B is a different size. Note that

$$A + C = \begin{pmatrix} 2 & 1 \\ 0 & -1 \end{pmatrix} = C + A.$$

Remark 3.36 In general, matrix addition is **commutative** as for matrices M and N of the same size we have

$$[M + N]_{ij} = [M]_{ij} + [N]_{ij} = [N]_{ij} + [M]_{ij} = [N + M]_{ij}$$

as addition of real numbers is commutative.   ■

Definition 3.37 The $m \times n$ **zero matrix** is the matrix with m rows and n columns whose every entry is 0. This matrix is simply denoted as 0 unless we need to specify its size, in which case it is written 0_{mn}. For example,

$$0_{23} = \begin{pmatrix} 0 & 0 & 0 \\ 0 & 0 & 0 \end{pmatrix}.$$

A simple check shows that $A + 0_{mn} = A = 0_{mn} + A$ for any $m \times n$ matrix A.

Definition 3.38 Scalar Multiplication Let A be an $m \times n$ matrix and k be a real number (a scalar). Then the matrix kA is defined by

$$[kA]_{ij} = k[A]_{ij} \qquad \text{for } 1 \leqslant i \leqslant m \text{ and } 1 \leqslant j \leqslant n.$$

So, to obtain the matrix kA, we multiply each of the entries of A by k.

Example 3.39 Show that $2(A + B) = 2A + 2B$ for the following matrices:

$$A = \begin{pmatrix} 1 & 2 \\ 3 & 4 \end{pmatrix}; \qquad B = \begin{pmatrix} 0 & -2 \\ 5 & 1 \end{pmatrix}.$$

Solution Here we are checking the **distributive law** in a specific example. We note that

$$A + B = \begin{pmatrix} 1 & 0 \\ 8 & 5 \end{pmatrix}, \quad \text{and so} \quad 2(A+B) = \begin{pmatrix} 2 & 0 \\ 16 & 10 \end{pmatrix};$$

$$2A = \begin{pmatrix} 2 & 4 \\ 6 & 8 \end{pmatrix}, \quad \text{and} \quad 2B = \begin{pmatrix} 0 & -4 \\ 10 & 2 \end{pmatrix}, \quad \text{so} \quad 2A + 2B = \begin{pmatrix} 2 & 0 \\ 16 & 10 \end{pmatrix}. \quad \square$$

Remark 3.40 More generally, the following identities hold. Let A, B, C be $m \times n$ matrices and λ, μ be real numbers.

$$\begin{array}{lll} A + 0_{mn} = A; & A + B = B + A; & 0A = 0_{mn}; \\ A + (-A) = 0_{mn}; & (A+B) + C = A + (B+C); & 1A = A; \\ (\lambda + \mu)A = \lambda A + \mu A; & \lambda(A+B) = \lambda A + \lambda B; & \lambda(\mu A) = (\lambda \mu)A. \end{array}$$

These are readily verified and show that M_{mn} is a real vector space (see also Remark 3.11). ■

Based on how we added matrices then you might think that we multiply matrices in a similar fashion, namely multiplying corresponding entries, but we do not. At first glance the rule for multiplying matrices is going to seem rather odd but, in due course (§3.7), we will see why matrix multiplication is done as follows and that this is natural in the context of matrices representing linear maps.

Definition 3.41 Matrix Multiplication We can multiply an $m \times n$ matrix A with an $p \times q$ matrix B if $n = p$. That is, A must have as many columns as B has rows. If this is the case then the product AB is the $m \times q$ matrix with entries

$$[AB]_{ij} = \sum_{k=1}^{n} [A]_{ik}[B]_{kj} \qquad \text{for } 1 \leqslant i \leqslant m \text{ and } 1 \leqslant j \leqslant q. \tag{3.13}$$

It may help to write the rows of A as $\mathbf{r}_1,\dots,\mathbf{r}_m$ and the columns of B as $\mathbf{c}_1,\dots,\mathbf{c}_q$. Rule (3.13) then states that

$$\text{the } (i,j)\text{th entry of } AB = \mathbf{r}_i \cdot \mathbf{c}_j \qquad \text{for } 1 \leqslant i \leqslant m \text{ and } 1 \leqslant j \leqslant q. \tag{3.14}$$

We dot (i.e. take the scalar product of) the rows of A with the columns of B; specifically to find the (i,j)th entry of AB we dot the ith row of A with the jth column of B.

Remark 3.42 We shall give full details in §3.7 as to why it makes sense (and, in fact, is quite natural) to multiply matrices as in (3.13). For now, it is worth noting the following. Let A be an $m \times n$ matrix and B be $n \times p$ so that AB is $m \times p$. There is a map μ_A from $\mathbb{R}_n$ to $\mathbb{R}_m$ associated with A, as given an $n \times 1$ column vector $\mathbf{v}$ in $\mathbb{R}_n$ then $A\mathbf{v}$ is a $m \times 1$ column vector in $\mathbb{R}_m$. So we have associated maps

$$\mu_A \text{ from } \mathbb{R}_n \text{ to } \mathbb{R}_m, \qquad \mu_B \text{ from } \mathbb{R}_p \text{ to } \mathbb{R}_n, \qquad \mu_{AB} \text{ from } \mathbb{R}_p \text{ to } \mathbb{R}_m.$$

Multiplying matrices as we have, it turns out that

$$\mu_{AB} = \mu_A \circ \mu_B.$$

This is equivalent to $(AB)\mathbf{v} = A(B\mathbf{v})$, which follows from the associativity of matrix multiplication (Proposition 3.46(c)). So matrix multiplication is best thought of as composition: performing μ_{AB} is equal to the performing μ_B then μ_A. ∎

Example 3.43 Calculate the possible products of the pairs of matrices in (3.12).

Solution Recall that a matrix product MN makes sense if M has the same number of columns as N has rows. A,B,C are respectively 2×2, 2×3, 2×2 matrices. So the products we can form are AA, AB, AC, CA, CB, CC. Let's slowly go through the product AC.

$$\begin{pmatrix} \boxed{\begin{array}{cc} 1 & 2 \end{array}} \\ -1 & 0 \end{pmatrix} \begin{pmatrix} \boxed{\begin{array}{c} 1 \\ 1 \end{array}} & -1 \\ & -1 \end{pmatrix} = \begin{pmatrix} 1 \times 1 + 2 \times 1 & ?? \\ ?? & ?? \end{pmatrix} = \begin{pmatrix} 3 & ?? \\ ?? & ?? \end{pmatrix}.$$

This is how we calculate the $(1,1)$th entry of AC. We take the first row of A and the first column of C and dot them together. We complete the remainder of the product as follows:

$$\begin{pmatrix} \boxed{\begin{array}{cc} 1 & 2 \end{array}} \\ -1 & 0 \end{pmatrix} \begin{pmatrix} 1 & \boxed{-1} \\ 1 & \boxed{-1} \end{pmatrix} = \begin{pmatrix} 2 & \boxed{1 \times (-1) + 2 \times (-1)} \\ ?? & ?? \end{pmatrix} = \begin{pmatrix} 3 & \boxed{-3} \\ ?? & ?? \end{pmatrix};$$

$$\begin{pmatrix} 1 & 2 \\ \boxed{\begin{array}{cc} -1 & 0 \end{array}} \end{pmatrix} \begin{pmatrix} \boxed{\begin{array}{c} 1 \\ 1 \end{array}} & -1 \\ & -1 \end{pmatrix} = \begin{pmatrix} 2 & 4 \\ \boxed{(-1) \times 1 + 0 \times 1} & ?? \end{pmatrix} = \begin{pmatrix} 3 & -3 \\ \boxed{-1} & ?? \end{pmatrix};$$

$$\begin{pmatrix} 1 & 2 \\ \boxed{\begin{array}{cc} -1 & 0 \end{array}} \end{pmatrix} \begin{pmatrix} 1 & \boxed{-1} \\ 1 & \boxed{-1} \end{pmatrix} = \begin{pmatrix} 2 & 4 \\ 0 & \boxed{(-1) \times (-1) + 0 \times (-1)} \end{pmatrix} = \begin{pmatrix} 3 & -3 \\ -1 & \boxed{1} \end{pmatrix}.$$

So finally

$$\begin{pmatrix} 1 & 2 \\ -1 & 0 \end{pmatrix} \begin{pmatrix} 1 & -1 \\ 1 & -1 \end{pmatrix} = \begin{pmatrix} 3 & -3 \\ -1 & 1 \end{pmatrix}.$$

We complete the remaining examples more quickly but still leaving a middle stage in the calculation to help see the process.

$$AA = \begin{pmatrix} 1 & 2 \\ -1 & 0 \end{pmatrix} \begin{pmatrix} 1 & 2 \\ -1 & 0 \end{pmatrix} = \begin{pmatrix} 1-2 & 2+0 \\ -1+0 & -2+0 \end{pmatrix} = \begin{pmatrix} -1 & 2 \\ -1 & -2 \end{pmatrix};$$

$$AB = \begin{pmatrix} 1 & 2 \\ -1 & 0 \end{pmatrix} \begin{pmatrix} 1 & 2 & 3 \\ 3 & 2 & 1 \end{pmatrix} = \begin{pmatrix} 1+6 & 2+4 & 3+2 \\ -1+0 & -2+0 & -3+0 \end{pmatrix}$$

$$= \begin{pmatrix} 7 & 6 & 5 \\ -1 & -2 & -3 \end{pmatrix};$$

$$CA = \begin{pmatrix} 1 & -1 \\ 1 & -1 \end{pmatrix} \begin{pmatrix} 1 & 2 \\ -1 & 0 \end{pmatrix} = \begin{pmatrix} 1+1 & 2-0 \\ 1+1 & 2-0 \end{pmatrix} = \begin{pmatrix} 2 & 2 \\ 2 & 2 \end{pmatrix};$$

$$CB = \begin{pmatrix} 1 & -1 \\ 1 & -1 \end{pmatrix} \begin{pmatrix} 1 & 2 & 3 \\ 3 & 2 & 1 \end{pmatrix} = \begin{pmatrix} 1-3 & 2-2 & 3-1 \\ 1-3 & 2-2 & 3-1 \end{pmatrix} = \begin{pmatrix} -2 & 0 & 2 \\ -2 & 0 & 2 \end{pmatrix};$$

$$CC = \begin{pmatrix} 1 & -1 \\ 1 & -1 \end{pmatrix} \begin{pmatrix} 1 & -1 \\ 1 & -1 \end{pmatrix} = \begin{pmatrix} 1-1 & -1+1 \\ 1-1 & -1+1 \end{pmatrix} = \begin{pmatrix} 0 & 0 \\ 0 & 0 \end{pmatrix}. \qquad \square$$

Definition 3.44 The $n \times n$ **identity matrix** I_n is the $n \times n$ matrix with entries

$$[I_n]_{ij} = \begin{cases} 1 & \text{if} \quad i=j, \\ 0 & \text{if} \quad i \neq j. \end{cases}$$

For example,

$$I_2 = \begin{pmatrix} 1 & 0 \\ 0 & 1 \end{pmatrix}, \qquad I_3 = \begin{pmatrix} 1 & 0 & 0 \\ 0 & 1 & 0 \\ 0 & 0 & 1 \end{pmatrix}.$$

The identity matrix will be simply denoted as I unless we need to specify its size. The (i,j)th entry of I is denoted as δ_{ij} which is referred to as the **Kronecker delta.** [6]

Remark 3.45 (Sifting Property of the Kronecker Delta) Let $x_1,\ldots,x_n$ be n real numbers, and $1 \leqslant k \leqslant n$. Then

$$\sum_{i=1}^{n} x_i \delta_{ik} = x_k.$$

This is because $\delta_{ik} = 0$ when $i \neq k$ and $\delta_{kk} = 1$. Thus the above sum sifts out (i.e. selects) the kth element x_k. ∎

There are certain important points to highlight from Example 3.43, some of which make matrix algebra crucially different from the algebra of real numbers.

Proposition 3.46 (*Properties of Matrix Multiplication*) (*a*) *For an* $m \times n$ *matrix* A *and positive integers* l,p,

$$A0_{np} = 0_{mp}; \qquad 0_{lm}A = 0_{ln}; \qquad AI_n = A; \qquad I_mA = A.$$

---

[6] After the German mathematician Leopold Kronecker (1823–1891).

*(b) In general, matrix multiplication is **not commutative**; $AB \neq BA$ in general, even if both products meaningfully exist and have the same size.*

*(c) Matrix multiplication is **associative**; for matrices A, B, C, which are respectively of sizes $m \times n$, $n \times p$ and $p \times q$, we have*

$$A(BC) = (AB)C.$$

*(d) The **distributive** laws hold for matrix multiplication; whenever the following products and sums make sense,*

$$A(B+C) = AB+AC, \quad \text{and} \quad (A+B)C = AC+BC.$$

*(e) In Example 3.43 we saw $CC = 0$ even though $C \neq 0$ – so one **cannot** conclude from $MN = 0$ that either matrix M or N is zero.*

#560
#563
#564
#565
#569
#572

Proof (a) To find an entry of the product $A0_{np}$ we dot a row of A with a zero column of 0_{np} and likewise in the product $0_{lm}A$ we are dotting with zero rows. Also, by the sifting property,

$$[AI_n]_{ij} = \sum_{k=1}^{n} [A]_{ik}\delta_{kj} = [A]_{ij}; \quad [I_nA]_{ij} = \sum_{k=1}^{n} \delta_{ik}[A]_{kj} = [A]_{ij}, \quad (1 \leqslant i \leqslant m, 1 \leqslant j \leqslant n).$$

(b) In Example 3.43, we saw that $AC \neq CA$. More generally, if A is $m \times n$ and B is $n \times p$ then the product AB exists but BA doesn't even make sense as a matrix product unless $m = p$.

(c) Given i, j in the ranges $1 \leqslant i \leqslant m, 1 \leqslant j \leqslant q$, applying (3.13) repeatedly we see

$$[(AB)C]_{ij} = \sum_{r=1}^{p} [AB]_{ir}[C]_{rj} = \sum_{r=1}^{p} \left(\sum_{s=1}^{n} [A]_{is}[B]_{sr} \right) [C]_{rj} = \sum_{r=1}^{p}\sum_{s=1}^{n} [A]_{is}[B]_{sr}[C]_{rj}$$

$$= \sum_{s=1}^{n}\sum_{r=1}^{p} [A]_{is}[B]_{sr}[C]_{rj} = \sum_{s=1}^{n} [A]_{is} \left(\sum_{r=1}^{p} [B]_{sr}[C]_{rj} \right)$$

$$= \sum_{s=1}^{n} [A]_{is}[BC]_{sj} = [A(BC)]_{ij}.$$

(d) This is left to #559. $\qquad\square$

Definition 3.47 A **square** matrix is a matrix with an equal number of rows and columns. The **diagonal** of an $n \times n$ matrix A comprises the entries $[A]_{11}, [A]_{22}, \ldots, [A]_{nn}$ – that is, the n entries running diagonally from the top left to the bottom right. A **diagonal** matrix is a square matrix whose non-diagonal entries are all zero.

Notation 3.48 We shall write $\mathrm{diag}(c_1, c_2, \ldots, c_n)$ for the $n \times n$ diagonal matrix whose (i, i)th entry is c_i. More generally if A is a $k \times k$ matrix and B is an $n \times n$ matrix we shall write $\mathrm{diag}(A, B)$ for the $(k+n) \times (k+n)$ matrix

$$\mathrm{diag}(A, B) = \begin{pmatrix} A & 0_{kn} \\ 0_{nk} & B \end{pmatrix},$$

and extend the notation more generally to $\mathrm{diag}(A_1, A_2, \ldots, A_r)$ when $A_1, \ldots, A_r$ are square matrices.

Example 3.49 With A, B, C as in Example 3.35, $\mathrm{diag}(A, B)$ does not make sense as B is not square and

$$\mathrm{diag}(2,1,0) = \begin{pmatrix} 2 & 0 & 0 \\ 0 & 1 & 0 \\ 0 & 0 & 0 \end{pmatrix}, \qquad \mathrm{diag}(3,A) = \begin{pmatrix} 3 & 0 & 0 \\ 0 & 1 & 2 \\ 0 & -1 & 0 \end{pmatrix},$$

$$\mathrm{diag}(A,C) = \begin{pmatrix} 1 & 2 & 0 & 0 \\ -1 & 0 & 0 & 0 \\ 0 & 0 & 1 & -1 \\ 0 & 0 & 1 & -1 \end{pmatrix}.$$

Because matrix multiplication is not commutative, we need to be clearer than usual in what we might mean by a phrase like 'multiply by the matrix A'; typically we need to give some context as to whether we have multiplied on the left or on the right.

Definition 3.50 Let A and M be matrices.

(a) To **premultiply** M by A is to form the product AM – i.e. premultiplication is multiplication on the left.

(b) To **postmultiply** M by A is to form the product MA – i.e. postmultiplication is multiplication on the right.

Notation 3.51 We write A^2 for the product AA and similarly, for n a positive integer, we write A^n for the product

$$\underbrace{AA \cdots A}_{n \text{ times}}.$$

Note that A must be a square matrix for this to make sense. We also define $A^0 = I$. Note also that $A^m A^n = A^{m+n}$ for natural numbers m, n (#571). Further, given a real polynomial $p(x) = a_k x^k + a_{k-1} x^{k-1} + \cdots + a_1 x + a_0$, then we define

$$p(A) = a_k A^k + a_{k-1} A^{k-1} + \cdots + a_1 A + a_0 I.$$

Example 3.52 Let

$$A = \begin{pmatrix} \cos\alpha & \sin\alpha \\ \sin\alpha & -\cos\alpha \end{pmatrix} \qquad \text{and} \qquad B = \begin{pmatrix} 0 & 1 \\ 0 & 0 \end{pmatrix}. \tag{3.15}$$

Then $A^2 = I_2$ for any choice of α. Also there is no matrix C (with real or complex entries) such that $C^2 = B$. This shows that the idea of a square root is a much more complicated issue for matrices than for real or complex numbers. A square matrix may have none or many, even infinitely many, different square roots.

Solution We note for any α that

$$A^2 = \begin{pmatrix} \cos^2\alpha + \sin^2\alpha & \cos\alpha\sin\alpha - \sin\alpha\cos\alpha \\ \sin\alpha\cos\alpha - \cos\alpha\sin\alpha & \sin^2\alpha + (-\cos\alpha)^2 \end{pmatrix} = \begin{pmatrix} 1 & 0 \\ 0 & 1 \end{pmatrix} = I_2.$$

To show B has no square roots, say a, b, c, d are real (or complex) numbers such that

$$B = \begin{pmatrix} 0 & 1 \\ 0 & 0 \end{pmatrix} = \begin{pmatrix} a & b \\ c & d \end{pmatrix}^2 = \begin{pmatrix} a^2 + bc & b(a+d) \\ c(a+d) & bc + d^2 \end{pmatrix}.$$

Looking at the $(2,1)$ entry, we see $c = 0$ or $a + d = 0$. But $a + d = 0$ contradicts $b(a+d) = 1$ from the $(1,2)$ entry and so $c = 0$. From the $(1,1)$ entry we see $a = 0$ and from the $(2,2)$ entry we see $d = 0$, but these lead to the same contradiction. □

Definition 3.53 Given an $m \times n$ matrix A, then its **transpose** A^T is the $n \times m$ matrix with $\left[A^T\right]_{ij} = [A]_{ji}$.

Proposition 3.54 *(Properties of Transpose)*

(a) *(**Addition and Scalar Multiplication Rules**) Let A, B be $m \times n$ matrices and λ a real number. Then*

$$(A + B)^T = A^T + B^T; \qquad (\lambda A)^T = \lambda A^T.$$

(b) *(**Product Rule**) Let A be an $m \times n$ matrix and B be an $n \times p$ matrix. Then $(AB)^T = B^T A^T$.*

(c) *(**Involution Rule**) Let A be an $m \times n$ matrix. Then $(A^T)^T = A$.*

Proof These are left to #586. □

Definition 3.55 A square matrix A is said to be

- **symmetric** if $A^T = A$.
- **skew-symmetric** (or **antisymmetric**) if $A^T = -A$.
- **upper triangular** if $[A]_{ij} = 0$ when $i > j$. Entries below the diagonal are zero.
- **strictly upper triangular** if $[A]_{ij} = 0$ when $i \geqslant j$. Entries on or below the diagonal are zero.
- **lower triangular** if $[A]_{ij} = 0$ when $i < j$. Entries above the diagonal are zero.
- **strictly lower triangular** if $[A]_{ij} = 0$ when $i \leqslant j$. Entries on or above the diagonal are zero.
- **triangular** if it is either upper or lower triangular.

Example 3.56 Let

$$A = \begin{pmatrix} 1 & 2 \\ 0 & 3 \end{pmatrix}, \quad B = \begin{pmatrix} 1 & 0 \\ 2 & -1 \\ 1 & -1 \end{pmatrix}, \quad C = \begin{pmatrix} 0 & 1 \\ -1 & 0 \end{pmatrix}, \quad D = \begin{pmatrix} 1 & 0 & 0 \\ 0 & 2 & 0 \\ 0 & 0 & 3 \end{pmatrix}.$$

Then

$$A^T = \begin{pmatrix} 1 & 0 \\ 2 & 3 \end{pmatrix}, \qquad B^T = \begin{pmatrix} 1 & 2 & 1 \\ 0 & -1 & -1 \end{pmatrix},$$

$$C^T = \begin{pmatrix} 0 & -1 \\ 1 & 0 \end{pmatrix}, \qquad D^T = \begin{pmatrix} 1 & 0 & 0 \\ 0 & 2 & 0 \\ 0 & 0 & 3 \end{pmatrix}.$$

Note that A is upper triangular and so A^T is lower triangular. Also C and C^T are skew-symmetric. And D is diagonal and so also symmetric, upper triangular and lower triangular.

We conclude this section with the following theorem. On a first reading you might wish to omit the proof, as it is somewhat technical, and instead just remember the statement of the theorem. The proof does, though, demonstrate the power of the sigma-notation for matrix multiplication introduced in (3.13) and that of the Kronecker delta. In this proof we will make use of the *standard basis for matrices*.

Notation 3.57 For I, J in the range $1 \leqslant I \leqslant m, 1 \leqslant J \leqslant n$, we denote by E_{IJ} the $m \times n$ matrix with entry 1 in the Ith row and Jth column and 0s elsewhere. Then the (i,j)th entry of E_{IJ} is

$$[E_{IJ}]_{ij} = \delta_{Ii}\delta_{Jj}$$

as $\delta_{Ii}\delta_{Jj} = 0$ unless $i = I$ and $j = J$ in which case it is 1. These matrices form the **standard basis** for M_{mn}.

Theorem 3.58 *Let A be an $n \times n$ matrix such that $AM = MA$ for all $n \times n$ matrices M. i.e. A commutes with all $n \times n$ matrices. Then $A = \lambda I_n$ for some real number λ.*

Proof As A commutes with every $n \times n$ matrix, then in particular it commutes with each of the n^2 basis matrices E_{IJ}. So $[AE_{IJ}]_{ij} = [E_{IJ}A]_{ij}$ for every I, J, i, j. Using the sifting property

$$[AE_{IJ}]_{ij} = \sum_{k=1}^{n} [A]_{ik}[E_{IJ}]_{kj} = \sum_{k=1}^{n} [A]_{ik}\delta_{Ik}\delta_{Jj} = [A]_{Ii}\delta_{Jj};$$

$$[E_{IJ}A]_{ij} = \sum_{k=1}^{n} [E_{IJ}]_{ik}[A]_{kj} = \sum_{k=1}^{n} \delta_{Ii}\delta_{Jk}[A]_{kj} = \delta_{Ii}[A]_{Jj}.$$

Hence for all I, J, i, j,

$$[A]_{Ii}\delta_{Jj} = \delta_{Ii}[A]_{Jj}. \tag{3.16}$$

Let $i \neq j$. If we set $I = J = i$, then (3.16) becomes $0 = [A]_{ij}$ showing that the non-diagonal entries of A are zero. If we set $I = i$ and $J = j$, then (3.16) becomes $[A]_{ii} = [A]_{jj}$, which shows that all the diagonal entries of A are equal – call this shared value λ and we have shown $A = \lambda I_n$. This shows that any such M is necessarily of the form λI_n, and conversely such matrices do indeed commute with every other $n \times n$ matrix. □

#566
#567
#593

Exercises

#552a For the following matrices, where it makes sense to do so, calculate their pairwise sums and products.

$$A = \begin{pmatrix} 1 & 2 \\ 0 & 3 \end{pmatrix}, \quad B = \begin{pmatrix} 0 & 1 \\ 3 & 2 \\ 1 & 0 \end{pmatrix}, \quad C = \begin{pmatrix} -1 & 2 & 3 \\ -2 & 1 & 0 \end{pmatrix}, \quad D = \begin{pmatrix} -1 & 12 \\ 6 & 0 \end{pmatrix}.$$

#553A Show that $AB = 0$, but that $BA \neq 0$, where

$$A = \begin{pmatrix} 4 & 2 \\ 2 & 1 \end{pmatrix}, \quad B = \begin{pmatrix} -2 & -1 \\ 4 & 2 \end{pmatrix}.$$

#554c Calculate the products AB, BA, CA, BC, C^2 where

$$A = \begin{pmatrix} 1 & 2 & 0 \\ 0 & 1 & 1 \end{pmatrix}, \qquad B = \begin{pmatrix} -1 & 0 \\ 0 & 3 \\ -1 & 0 \end{pmatrix}, \qquad C = \begin{pmatrix} 2 & 1 \\ 0 & -1 \end{pmatrix}.$$

#555a How many arithmetic operations (additions, subtractions, multiplications, divisions) are involved in multiplying an $m \times n$ matrix by an $n \times p$ matrix?

#556A Write out the 3×3 matrices A and B where $[A]_{ij} = i + j^2$ and $[B]_{ij} = i - j$. Calculate AB and BA. Determine $[AB]_{ij}$ and $[BA]_{ij}$ as polynomials in i and j.

#557C Let A be the $n \times n$ matrix with entries $[A]_{ij} = (-1)^{i+j}$. Use the summation notation from (3.13) to show that $A^2 = nA$. What is A^k for $k \geqslant 1$?

#558a Verify directly the distributive law $A(B + C) = AB + AC$, where

$$A = \begin{pmatrix} 1 & 3 & 0 \\ 2 & 1 & 1 \end{pmatrix}, \qquad B = \begin{pmatrix} 1 & 0 \\ 2 & 1 \\ -1 & -1 \end{pmatrix}, \qquad C = \begin{pmatrix} 2 & 1 \\ -1 & 1 \\ 0 & 1 \end{pmatrix}.$$

#559C Prove the distributive laws given in Proposition 3.46(d).

#560B† Show no square matrix A exists such that $A^m = I$ and $A^n = 0$ for some $m, n \geqslant 1$.

#561B (See also #170.) Define the complex 2×2 matrices $\mathcal{I}, \mathcal{J}, \mathcal{K}$ by

$$\mathcal{I} = \begin{pmatrix} 0 & 1 \\ -1 & 0 \end{pmatrix}, \qquad \mathcal{J} = \begin{pmatrix} 0 & i \\ i & 0 \end{pmatrix}, \qquad \mathcal{K} = \begin{pmatrix} i & 0 \\ 0 & -i \end{pmatrix}.$$

Show that

$$\mathcal{I}^2 = \mathcal{J}^2 = \mathcal{K}^2 = -I_2; \quad \mathcal{I}\mathcal{J} = -\mathcal{J}\mathcal{I} = \mathcal{K}; \quad \mathcal{J}\mathcal{K} = -\mathcal{K}\mathcal{J} = \mathcal{I}; \quad \mathcal{K}\mathcal{I} = -\mathcal{I}\mathcal{K} = \mathcal{J}.$$

#562B We can identify the complex number $z = x + yi$ with the 2×2 real matrix Z by

$$x + yi = z \quad \leftrightarrow \quad Z = \begin{pmatrix} x & y \\ -y & x \end{pmatrix}.$$

Let z, w be two complex numbers such that $z \leftrightarrow Z$ and $w \leftrightarrow W$. Show that $z + w \leftrightarrow Z + W$ and $zw \leftrightarrow ZW$.

#563A† Let A be an $m \times n$ matrix. Show that
(i) $e_i A$ is the ith row of A.   (ii) Ae_j^T is the jth column of A.   (iii) $e_i Ae_j^T = [A]_{ij}$.

#564B† Given $v_1, \ldots, v_n$ in $\mathbb{R}_m$, show there is a unique $m \times n$ matrix A such that $Ae_j^T = v_j$ for $1 \leqslant j \leqslant n$.

#565B Let A be an $m \times n$ matrix and B be an $n \times p$ matrix.

(i) Let the columns of B be $c_1, c_2, \ldots, c_p$. Show that the columns of AB are $Ac_1, Ac_2, \ldots, Ac_p$.
(ii) Let the rows of A be $r_1, r_2, \ldots, r_m$. Show that the rows of AB are $r_1 B, r_2 B, \ldots, r_m B$.

#566b† Simplify the following sums as single entries in a matrix product. In each case, make clear what is known about the number of rows or columns of the matrices involved in the products.

$$\sum_{i=1}^{n}[A]_{ij}[B]_{ik}, \qquad \sum_{i=1}^{m}\sum_{j=1}^{n}[A]_{ji}[B]_{ik}[C]_{jk},$$

$$\sum_{i=1}^{m}\sum_{j=1}^{n}\sum_{k=1}^{m}[A]_{i1}[B]_{ij}[C]_{kj}[D]_{k1}, \qquad \sum_{i=1}^{n}\sum_{j=1}^{n}\sum_{k=1}^{n}[A]_{ki}[B]_{jn}[C]_{jk}[D]_{in}.$$

#567b† Let δ_{ij} denote the Kronecker delta. Simplify the following sums.

$$\sum_{j=1}^{n}\delta_{ij}, \qquad \sum_{i=1}^{n}\sum_{j=1}^{n}\delta_{ij}, \qquad \sum_{i=1}^{n}\sum_{j=1}^{n}\delta_{kj}\delta_{ki}, \qquad \sum_{k=1}^{n}\delta_{kj}\delta_{ki}.$$

#568C Suppose that two matrices A, B commute – that is, $AB = BA$. Show that A and B are both square and of the same size. Show further that A^m commutes with B^n for any natural numbers m, n.

#569a Let A and B be $n \times n$ real matrices. Show that A and B commute if and only if

$$(A + B)^2 = A^2 + 2AB + B^2. \tag{3.17}$$

Show further that if A and B commute then

$$(A + B)^3 = A^3 + 3A^2B + 3AB^2 + B^3. \tag{3.18}$$

#570D† Give examples of matrices A and B which do not commute but which satisfy (3.18).

#571C Let A be a square matrix, m, n natural numbers and $p(x), q(x)$ be real polynomials.

(i) Show that $A^m A^n = A^{m+n} = A^n A^m$.
(ii) Show that $p(A)q(A) = (pq)(A) = q(A)p(A)$.

#572A A square matrix M is a **projection** if $M^2 = M$ and is an **involution** if $M^2 = I$. If P is a projection show that $Q = 2P - I$ is an involution. Conversely, if Q is an involution show that $P = (I + Q)/2$ is a projection.

#573C† Let P_1 and P_2 be $n \times n$ projections. Show that $P_1 + P_2$ is a projection matrix if and only if $P_1 P_2 = P_2 P_1 = 0$.

#574C† Let A be as in Example 3.52 and $p(x)$ a polynomial. Show that $p(A) = 0$ if and only if $p(1) = 0 = p(-1)$.

#575B Let A be as in Example 3.52. Show that there is no 2×2 matrix M with real entries such that $M^2 = A$. Are there matrices with complex entries whose square is A?

#576b Find all the 2×2 matrices which commute with A, where

$$(i) \quad A = \begin{pmatrix} 1 & 0 \\ 0 & 2 \end{pmatrix}; \qquad (ii) \quad A = \begin{pmatrix} 1 & 1 \\ 0 & 1 \end{pmatrix}.$$

#577c Find all the 3×3 matrices which commute with A where (i) $A = \mathrm{diag}(1,2,2)$; (ii) $A = \mathrm{diag}(1,2,3)$.

#578C† Let $A_1,\ldots,A_n,B_1,\ldots,B_n$ be square matrices such that A_i has the same size as B_i for each i. Show that

$$\mathrm{diag}(A_1,\ldots,A_n)\mathrm{diag}(B_1,\ldots,B_n) = \mathrm{diag}(A_1B_1,\ldots,A_nB_n).$$

#579C Let A be a diagonal $n \times n$ matrix with distinct diagonal entries. Show that if $AB = BA$ then B is diagonal.

#580c Let $A,B,\ldots,G,H$ be eight 2×2 matrices and let M,N be the 4×4 matrices given by

$$M = \begin{pmatrix} A & B \\ C & D \end{pmatrix}, \qquad N = \begin{pmatrix} E & F \\ G & H \end{pmatrix}.$$

Show that

$$MN = \begin{pmatrix} AE+BG & AF+BH \\ CE+DG & CF+DH \end{pmatrix}.$$

#581d Generalize the result of #580 to $n \times n$ matrices M and N with A,E being $k \times k$, B,F being $k \times (n-k)$, C,G being $(n-k) \times k$ and D,H being $(n-k) \times (n-k)$.

#582A† Show that if A and B are diagonal $n \times n$ matrices then so also are $A+B$, cA (where c is a real number) and AB. Show further that $AB = BA$.

#583c(i) Show that if $A+B$ and $A-B$ are diagonal $n \times n$ matrices then so are A and B.
(ii) Find two 2×2 matrices such that $A+B$ and AB are diagonal though neither A nor B is diagonal.

#584A Let A,B be $n \times n$ upper triangular matrices. Show that $A+B$ and AB are also upper triangular.

#585b† Show that if A is $n \times n$ and strictly upper triangular then $A^n = 0$.

#586A Verify Proposition 3.54(a), (b) and (c).

#587a Let $\mathbf{v}$ and $\mathbf{w}$ be in $\mathbb{R}_n$. Show that $\mathbf{v} \cdot \mathbf{w} = \mathbf{v}^T\mathbf{w}$.

#588B† An $n \times n$ matrix A is said to be **orthogonal** if $A^TA = I_n$.

(i) Show that $A\mathbf{v} \cdot A\mathbf{w} = \mathbf{v} \cdot \mathbf{w}$, for all $\mathbf{v}$ and $\mathbf{w}$ in $\mathbb{R}_n$, if and only if A is orthogonal.
(ii) Show that A is orthogonal if and only if its columns are mutually perpendicular unit vectors.
(iii) Show that $|A\mathbf{x}| = |\mathbf{x}|$ for all $\mathbf{x}$ in $\mathbb{R}_n$ if and only if A is orthogonal.

#589A Let A be a square matrix. (i) Show that A is symmetric if and only if A^T is symmetric.
(ii) Show that $A+A^T$ is symmetric and that $A-A^T$ is skew-symmetric.
(iii) Show that A can be written, in a unique way, as the sum of a symmetric matrix and a skew-symmetric matrix.
(iv) Let B be an $m \times n$ matrix. Show that B^TB is a symmetric $n \times n$ matrix.

#590a Let A, B be symmetric $n \times n$ matrices. Show that $A + B$ is symmetric. Show that AB is symmetric if and only if A and B commute. Find 2×2 matrices C and D, neither symmetric, such that CD is symmetric.

#591D† Let A be a symmetric $n \times n$ matrix.

(i) Let $\mathbf{v}$ be an $n \times 1$ column vector such that $A^2\mathbf{v} = \mathbf{0}$. Show that $A\mathbf{v} = \mathbf{0}$.
(ii) Suppose now that $A^k\mathbf{v} = \mathbf{0}$ for some $k \geqslant 2$. Show that $A\mathbf{v} = \mathbf{0}$.
(iii) Deduce that if $A^k = 0$ for some k then $A = 0$.
(iv) Give an example to show that (iii) is not generally true for square matrices.

#592B Show that no 2×2 matrix A satisfies $A^T A = \mathrm{diag}(1, -1)$. Does any $m \times 2$ matrix A satisfy this equation?

#593B† The **trace** of an $n \times n$ matrix A equals the sum of its diagonal elements. That is $\mathrm{trace}\, A = \sum_{k=1}^{n} [A]_{kk}$.

(i) Let A, B be $n \times n$ matrices and c be a real number. Show that

$$\mathrm{trace}(A + B) = \mathrm{trace}\, A + \mathrm{trace}\, B; \qquad \mathrm{trace}(cA) = c\,\mathrm{trace}\, A.$$

(ii) (**Product Rule**) Let C be an $m \times n$ matrix and D an $n \times m$ matrix. Show that $\mathrm{trace}(CD) = \mathrm{trace}(DC)$.

#594C† Let A be an $m \times n$ matrix such that $\mathrm{trace}(AB) = 0$ for all $n \times m$ matrices B. Show that $A = 0$.

#595b Let n be a natural number. Find formulae, which you should verify by induction, for A^n and B^n where

$$A = \begin{pmatrix} 1 & 1 \\ 0 & 1 \end{pmatrix} \qquad \text{and} \qquad B = \begin{pmatrix} 1 & 2 \\ 0 & 3 \end{pmatrix}.$$

#596B Show that

$$\begin{pmatrix} F_{n+1} \\ F_n \end{pmatrix} = A \begin{pmatrix} F_n \\ F_{n-1} \end{pmatrix}, \qquad \text{where} \quad A = \begin{pmatrix} 1 & 1 \\ 1 & 0 \end{pmatrix},$$

and F_n denotes the nth Fibonacci number. Hence determine A^n and rederive Proposition 2.32.

#597b† Let n be a positive integer. Show that

$$A^n = 3^{n-1} \begin{pmatrix} 2n + 3 & -n \\ 4n & 3 - 2n \end{pmatrix} \qquad \text{where} \quad A = \begin{pmatrix} 5 & -1 \\ 4 & 1 \end{pmatrix}.$$

Can you find a matrix B such that $B^2 = A$?

Arthur Cayley (Bettmann/Getty Images)

Modern Algebra and Abstraction
Matrix algebra is clearly very different to that of everyday numbers like the integers, real numbers or even the complex numbers: matrix multiplication is not commutative, not all non-zero matrices are invertible, the product of two matrices can be zero without either being zero and square matrices can have no, finitely many or infinitely many square roots. During the nineteenth century there became an increasing awareness of various 'non-traditional' algebraic structures and an appreciation that theorems would be most general when applying to all those examples that had certain properties in common – a move to *abstract* thinking.

Many of these examples came from attempts to address concrete problems. Hamilton's quaternions are described on p.53, and were employed to describe rotations in three dimensions. Around the same time, George Boole (1815–1864) was interested in the algebra associated with sets and subsets. Given a set X, addition and multiplication operations can be defined on its subsets by $A + B = (A \backslash B) \cup (B \backslash A)$ and $A \times B = A \cap B$, where $\backslash$ denotes complement. This then leads to unusual algebraic properties, where $A + A = 0$ and $A^2 = A$ for each subset A. As described on p.81, incorrect 'proofs' of Fermat's Last Theorem led to an interest in certain algebras amongst the complex numbers, many of which had unfortunate factorization properties.

Abstract algebra then began by bringing together many such observations and recognizing common properties amongst them. Abstraction would in due course become widespread throughout mathematics. Thinking abstractly, mathematicians began looking at the underlying rules, logic and structures that drove seemingly disparate results. The power of abstraction, then, is its generality: beginning with the rules of an abstract structure, one can begin to demonstrate results that apply to all examples with that structure. Whilst the nature of a specific structure is likely grounded in some important concrete examples, proofs now emerge independent of any particular examples. Still more, with luck, these proofs will be that much more apparent as they focus on the structure's rules only and there is no distraction from superficial clutter. Most undergraduate degrees in mathematics will introduce students to *groups*, *rings* and *fields*. These concepts had all been implicitly or explicitly used by certain mathematicians through the nineteenth century. Groups had been apparent in the work of Gauss, Cauchy, Abel et al. in the early nineteenth century, and had been important in Galois' proof that a quintic is generally insolvable by radicals (p.33). A group is a set G with an operation $*$ and an identity element e such that

- for any g and h in G, then $g * h$ is in G;
- $*$ is associative – that is, $g * (h * k) = (g * h) * k$ for any g, h, k in G;
- $e * g = g = g * e$ for any g;
- for each g, there is an inverse g^{-1} with $g * g^{-1} = e = g^{-1} * g$;

and the group is called *commutative* (or *Abelian*) if $g * h = h * g$ for all g, h. Examples are plentiful within mathematics, the integers under addition and the non-zero reals under multiplication being commutative groups, and the invertible 2×2 matrices under multiplication being a non-commutative example. The axioms for what it is to be, abstractly, a group were first written down by Arthur Cayley (1821–1895) (pictured) in 1849, but their importance wasn't acknowledged at the time; it wasn't until the 1870s that an appreciation of the merit of abstract structures was showing. The father of abstract algebra was arguably the German mathematician Richard Dedekind (1831–1916); certainly he was one of the first to fully appreciate the power of abstract structures. Whilst working on the nascent algebraic number theory, Dedekind came up with the abstract idea of a ring – examples are the integers, polynomials, matrices – where we have addition, substraction, multiplication but not always division – and that of a field. Examples of fields include the rational numbers, real numbers and complex numbers; there are also fields with just finitely many elements. The theory of linear algebra we have met, and will continue to meet, including matrices and simultaneous equations, generally applies when the scalars are taken from a field. To be precise a **field** is a set F together with *two* operations $+$ and $\times$ and two distinct elements 0 and 1 such that

- F and $+$ is a commutative group with (additive) identity 0;
- $F \backslash \{0\}$ and $\times$ is a commutative group with (multiplicative) identity 1;
- $a \times (b + c) = (a \times b) + (a \times c)$ for all a, b, c in F.

#598 B A **Jordan block** is a square matrix $J(\lambda, r)$ such as the $r \times r$ matrix shown below and where λ is a complex number. Determine $J(\lambda, r)^n$ for positive integers n.

$$J(\lambda, r) = \begin{pmatrix} \lambda & 1 & 0 & \cdots & 0 \\ 0 & \lambda & 1 & \ddots & \vdots \\ \vdots & 0 & \lambda & \ddots & 0 \\ \vdots & \vdots & \ddots & \ddots & 1 \\ 0 & 0 & \cdots & 0 & \lambda \end{pmatrix}.$$

#599 a Let A be as in #595. Find all 2×2 matrices M such that $M^2 = A$.

#600 C Show no real matrix B satisfies $B^2 = \text{diag}(1, -1)$. How many such complex matrices B are there?

#601 c Let $a > d > 0$. Show that the matrix $\text{diag}(a, d)$ has precisely four square roots.

#602 A Let α, β be real numbers with $-\pi/2 < \alpha < \pi/2$ and $0 \leqslant \beta < 2\pi$. Show that $(A_{\alpha,\beta})^2 = I_2$, where

$$A_{\alpha,\beta} = \begin{pmatrix} \sec\alpha\cos\beta & \sec\alpha\sin\beta + \tan\alpha \\ \sec\alpha\sin\beta - \tan\alpha & -\sec\alpha\cos\beta \end{pmatrix}.$$

#603 D† Show that the square roots of I_2 are $A_{\alpha,\beta}$ and $\pm I_2$.

#604 d† Find all the square roots of 0_{22}.

#605 E† Let $D = \text{diag}(\lambda_1, \lambda_2, \ldots, \lambda_n)$, where the λ_i are distinct. Show that D has 2^n square roots if the λ_i are non-negative and no square roots if any of the λ_i are negative.

#606 B Consider the weather model described in #407. Let p_n denote the probability of sun on day n and $q_n = 1 - p_n$ the probability of rain. Show that[7]

$$\left(\begin{matrix} p_{n+1} & q_{n+1} \end{matrix} \right) = \left(\begin{matrix} p_n & q_n \end{matrix} \right) \begin{pmatrix} 0.9 & 0.1 \\ 0.5 & 0.5 \end{pmatrix}. \tag{3.19}$$

With S_n and R_n defined as in #407, deduce that

$$\begin{pmatrix} 0.9 & 0.1 \\ 0.5 & 0.5 \end{pmatrix}^n = \begin{pmatrix} S_n & 1 - S_n \\ R_n & 1 - R_n \end{pmatrix}.$$

#607 D† (Adapted from MAT 2011 #2) Let

$$X = \begin{pmatrix} 0 & 1 & -2 \\ 1 & 1 & 3 \\ 0 & 0 & -1 \end{pmatrix} \quad \text{and} \quad Y = \begin{pmatrix} 0 & 0 & 1 \\ 1 & 0 & 2 \\ 0 & 1 & 0 \end{pmatrix}.$$

(i) Show that $X^3 = 2X + I$. Show further, for each positive integer n there are integers a_n, b_n, c_n such that

$$X^n = a_n I + b_n X + c_n X^2 \tag{3.20}$$

and that a_n, b_n, c_n are uniquely determined by (3.20). Find a_k, b_k, c_k for $1 \leqslant k \leqslant 5$.

(ii) Show that

$$a_{n+1} = c_n, \qquad b_{n+1} = a_n + 2c_n, \qquad c_{n+1} = b_n.$$

(iii) Let $d_n = a_n - b_n + c_n$. Show that $d_{n+1} = -d_n$ and deduce that $d_n = (-1)^n$.

(iv) Show that $a_n + c_n = F_{n-1}$, the $(n-1)$th Fibonacci number, and hence determine expressions for a_n, b_n, c_n, X^n.

(v) Determine Y^n.

---

[7] The 2×2 matrix in (3.19) is called a **transition matrix** as it relates the chances of being in a particular state one day to those for the following day. A square matrix is a transition matrix if its entries are non-negative and the sum of each row is 1.

3.4 Simultaneous Equations Inverses

By a **linear system**, or **linear system of equations**, we will mean a set of m simultaneous equations in n variables $x_1, x_2, \ldots, x_n$ which are of the form

$$
\begin{aligned}
a_{11}x_1 &+ a_{12}x_2 + \cdots + a_{1n}x_n &= b_1; \\
a_{21}x_1 &+ a_{22}x_2 + \cdots + a_{2n}x_n &= b_2; \\
&\;\;\vdots \\
a_{m1}x_1 &+ a_{m2}x_2 + \cdots + a_{mn}x_n &= b_m,
\end{aligned}
$$

where a_{ij} and b_i are real constants. Linear systems fall naturally within the theory of matrices, as we can succinctly rewrite the above m *scalar* equations in n variables as a single *vector* equation in $\mathbf{x}$, namely

$$A\mathbf{x} = \mathbf{b} \tag{3.21}$$

where

$$
A = \begin{pmatrix} a_{11} & a_{12} & \cdots & a_{1n} \\ a_{21} & a_{22} & \cdots & a_{2n} \\ \vdots & \vdots & & \vdots \\ a_{m1} & a_{m2} & \cdots & a_{mn} \end{pmatrix}, \quad
\mathbf{x} = \begin{pmatrix} x_1 \\ x_2 \\ \vdots \\ x_m \end{pmatrix}, \quad
\mathbf{b} = \begin{pmatrix} b_1 \\ b_2 \\ \vdots \\ b_m \end{pmatrix}.
$$

Any vector $\mathbf{x}$ which satisfies (3.21) is said to be a **solution**; if the linear system has one or more solutions then it is said to be **consistent**. The **general solution** to the system is any description of *all* the solutions of the system. We shall investigate the general theory of linear systems in the next section on elementary row operations and shall see that such systems can have no, one or infinitely many solutions, but we will first look at what can happen in the cases when $n = 2, 3$ and $m = 1, 2, 3$. Remember that m is the number of equations and n is the number of variables.

- *Case $n = 2$:*

When dealing with two variables, which we'll call x and y, the equation $ax + by = c$ represents a line in the xy-plane (unless $a = b = 0$). When $m = 1, n = 2$ this 'system' has infinitely many solutions (x, y) which represent the points on the line. In the case when we have a second equation ($m = 2, n = 2$) various different cases can arise.

Example 3.59 Consider the following two simultaneous linear equations in two variables x and y:

$$2x + 3y = 1; \tag{3.22}$$
$$3x + 2y = 2. \tag{3.23}$$

Solution To solve these we might argue:

$$\text{eliminate } x \text{ by using } 3 \times (3.22) - 2 \times (3.23) : (6x + 9y) - (6x + 4y) = 3 - 4;$$

$$\text{simplify} : 5y = -1;$$

$$\text{so} : y = -1/5;$$

$$\text{substitute back in } (3.22) : 2x - 3/5 = 1;$$

$$\text{solving} : x = 4/5.$$

(a)

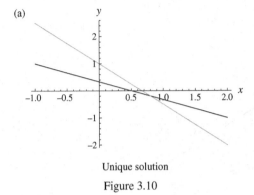

Unique solution

Figure 3.10

So we see we have a unique solution

$$x = 4/5, \quad y = -1/5.$$

Alternatively, we might resolve the problem graphically by drawing, in the xy-plane, the lines which the two equations represent.

The solution $(x, y) = (4/5, -1/5)$ then gives the co-ordinates of the lines' unique point of intersection (Figure 3.10). □

When we are in the more general $m = n = 2$ situation and have two simultaneous linear equations in two variables

$$ax + by = e; \qquad cx + dy = f, \tag{3.24}$$

then we can go through the same process as in Example 3.59 and *typically* find a unique solution (x, y) given by

$$x = \frac{de - bf}{ad - bc}; \qquad y = \frac{af - ce}{ad - bc}. \tag{3.25}$$

#608
#611
#612

(The full details are left to #611.) However, if $ad - bc = 0$ then this solution is meaningless. It's probably easiest to appreciate geometrically why this is (Figure 3.11): the equations in (3.24) represent lines in the xy-plane with gradients $-a/b$ and $-c/d$ respectively, and hence the two lines are parallel if $ad - bc = 0$. (Notice that this is still the correct condition when $b = d = 0$ and the lines are parallel and vertical.) If the lines are parallel then there cannot be a unique solution.

- If the two equations are multiples of one another, that is $a : c = b : d = e : f$, then we essentially have just one equation, with the second equation providing no further information about x and y. The equations represent the same line and there are infinitely many solutions (x, y) representing the points on this common line (Figure 3.11c).
- On the other hand, if $a : c = b : d \neq e : f$, then the second equation contains information about x and y which contradicts information from the first and there is no simultaneous solution (x, y) to the two equations. In this case, geometrically, the two equations represent two parallel and distinct lines in the xy-plane which never meet (Figure 3.11b).

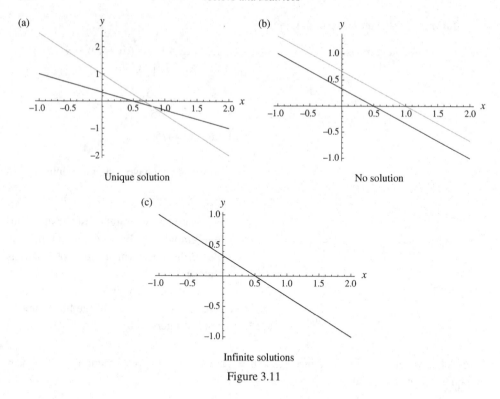

(a) Unique solution

(b) No solution

(c) Infinite solutions

Figure 3.11

Consider the $m = 3, n = 2$ case now, where we have three equations in two variables. Typically there will be no solution (x, y) which simultaneously satisfies the three equations; this is not surprising as the first two equations will typically determine a unique solution which then won't, by and large, happen to satisfy the third equation. From a geometric point of view the three equations represent lines in the xy-plane which normally won't be concurrent. If they are concurrent in a point, there will be a unique solution; if the three equations all represent the same line (each equation being a multiple of the others) there will be infinitely many solutions.

We return now to the $m = 2, n = 2$ case of two equations in two variables. We found that, typically, the two equations (3.24) had a unique solution (3.25); at least this is the case when $ad - bc \neq 0$. We can represent the two scalar equations in (3.24) and (3.25) by a single vector equation in each case:

$$\begin{pmatrix} a & b \\ c & d \end{pmatrix} \begin{pmatrix} x \\ y \end{pmatrix} = \begin{pmatrix} e \\ f \end{pmatrix}; \tag{3.26}$$

$$\begin{pmatrix} x \\ y \end{pmatrix} = \frac{1}{ad - bc} \begin{pmatrix} d & -b \\ -c & a \end{pmatrix} \begin{pmatrix} e \\ f \end{pmatrix}. \tag{3.27}$$

Equation (3.26) is just a rewriting of the linear system (3.24) as we more generally did in equation (3.21). Equation (3.27) is a similar rewriting of the unique solution found in (3.25) and something we *typically* can do in the $m = n = 2$ case. It also introduces us to the notion of the *inverse* of a matrix. Note that

$$\begin{pmatrix} d & -b \\ -c & a \end{pmatrix} \begin{pmatrix} a & b \\ c & d \end{pmatrix} = (ad - bc)I_2 = \begin{pmatrix} a & b \\ c & d \end{pmatrix} \begin{pmatrix} d & -b \\ -c & a \end{pmatrix}. \tag{3.28}$$

So if $ad - bc \neq 0$ and we set

$$A = \begin{pmatrix} a & b \\ c & d \end{pmatrix} \quad \text{and} \quad B = \frac{1}{ad - bc} \begin{pmatrix} d & -b \\ -c & a \end{pmatrix},$$

then $BA = I_2$ and $AB = I_2$.

Definition 3.60 Let A be a square matrix. We say that B is an **inverse** of A if $BA = AB = I$. We refer to a matrix with an inverse as **invertible** and otherwise the matrix is said to be **singular**.

Proposition 3.61 *(**Properties of Inverses**)* [8]

(a) *(**Uniqueness**) If a square matrix A has an inverse, then it is unique. We write A^{-1} for this inverse.*
(b) *(**Product Rule**) If A, B are invertible $n \times n$ matrices then AB is invertible with the inverse given by $(AB)^{-1} = B^{-1}A^{-1}$.*
(c) *(**Transpose Rule**) A square matrix A is invertible if and only if A^T is invertible. In this case $(A^T)^{-1} = (A^{-1})^T$.*
(d) *(**Involution Rule**) If A is invertible then so is A^{-1} with $(A^{-1})^{-1} = A$.*

Proof (a) Suppose B and C were two inverses for an $n \times n$ matrix A then

$$C = I_n C = (BA)C = B(AC) = BI_n = B$$

as matrix multiplication is associative. Parts (b) and (c) are left as #614 and #615. To verify (d) note that

$$\left(A^{-1}\right) A = A(A^{-1}) = I$$

and so $\left(A^{-1}\right)^{-1} = A$ by uniqueness.   □

Definition 3.62 If A is $m \times n$ and $BA = I_n$ then B is said to be a **left inverse**; if C satisfies $AC = I_m$ then C is said to be a **right inverse**.

- If A is $m \times n$ where $m \neq n$ then A cannot have both left and right inverses (#709).
- If A, B are $n \times n$ matrices with $BA = I_n$ then, in fact, $AB = I_n$ (Proposition 3.95).

Inverses, in the 2×2 case, are a rather simple matter to deal with.

---

[8] The invertible $n \times n$ matrices form a group, denoted $GL(n, \mathbb{R})$, under matrix multiplication; this is a consequence of (b), (d), the presence of the identity matrix and associativity of matrix multiplication.

Proposition 3.63 *The matrix* $A = \begin{pmatrix} a & b \\ c & d \end{pmatrix}$ *has an inverse if and only if* $ad - bc \neq 0$. *If* $ad - bc \neq 0$ *then*

$$A^{-1} = \frac{1}{ad - bc} \begin{pmatrix} d & -b \\ -c & a \end{pmatrix}.$$

Remark 3.64 The scalar quantity $ad - bc$ is called the *determinant* of A, written $\det A$. We shall see more generally that a square matrix is invertible if and only if its determinant is non-zero (Corollary 3.159). ∎

Proof We have already seen in (3.28) that if $ad - bc \neq 0$ then $AA^{-1} = I_2 = A^{-1}A$. If, however, $ad - bc = 0$ then

$$B = \begin{pmatrix} d & -b \\ -c & a \end{pmatrix}$$

satisfies $BA = 0$. If an inverse C for A existed then $0 = 0C = (BA)C = B(AC) = BI_2 = B$ by associativity. So each of a, b, c and d would be zero and consequently $A = 0$ which contradicts $AC = I_2$. □

If we return to the simultaneous equations represented by (3.26) and premultiply both sides by A^{-1} we see

$$A \begin{pmatrix} x \\ y \end{pmatrix} = \begin{pmatrix} e \\ f \end{pmatrix} \implies \begin{pmatrix} x \\ y \end{pmatrix} = A^{-1} \begin{pmatrix} e \\ f \end{pmatrix}$$

and find afresh our unique solution, as we did in (3.27). You may also note that the condition $ad - bc \neq 0$ is equivalent to A's columns being linearly independent (#624). When A's columns are dependent, varying x and y will only move

$$A \begin{pmatrix} x \\ y \end{pmatrix} = x \begin{pmatrix} a \\ b \end{pmatrix} + y \begin{pmatrix} c \\ d \end{pmatrix}, \tag{3.29}$$

back and forth along a line; however, when the columns are independent and A is invertible, we have shown the above vector ranges over all of $\mathbb{R}_2$ and indeed equals $(e, f)^T$ for unique choices of x and y.

Note that if A is singular then the above method cannot apply to solving $A\mathbf{x} = \mathbf{b}$; in any case, with more than two variables, even if A is invertible, the above is not an efficient approach. In the next section we will meet a more general and efficient method involving elementary row operations applicable to all linear systems.

- *Case* $n = 3$:

Suppose now that our linear system involves three variables x, y, z and we have m equations where $1 \leqslant m \leqslant 3$. Recall from Proposition 3.26 that a linear equation $ax + by + cz = j$ represents a plane in three-dimensional xyz-space; there will be infinitely many solutions to this single equation. For example, if we consider the equation

$$2x - y + 3z = 2, \tag{3.30}$$

then any one of x,y,z is determined by knowing the values of the other two variables which may be freely chosen. So all the solutions to (3.30) can be written in the form $(x,y,z) = (s, 2s + 3t - 2, t)$, where s,t are parameters which may independently take any real value. This, of course, is not the only way to write the solutions parametrically; for example, we could have assigned parameters to x and y and solved for z using (3.30).

When we have two equations in three variables ($m = 2, n = 3$), such as $ax + by + cz = j$ and $dx + ey + fz = k$, then *typically* the planes will meet in a line and we will have a one-parameter family of solutions. For example, if we consider the equations

$$2x - y + 3z = 2; \qquad 2x - y + 2z = 1, \tag{3.31}$$

then we see (by subtracting the second equation from the first) that $z = 1$ and aside from this we only know that $2x - y = -1$. So the general solution is a one-parameter family $(x,y,z) = (s, 2s + 1, 1)$ where s can take any real value.

So the typical $m = 2, n = 3$ situation is to have a one-parameter line of solutions. However, in the special case that $a : d = b : e = c : f$ then the planes are parallel; either the planes are parallel and distinct, when $a : d = b : e = c : f \neq j : k$, with there being no solutions as the second equation gives contradictory information to that provided by the first, or the two equations represent the same plane, $a : d = b : e = c : f = j : k$, with there being a two-parameter set of solutions and the second equation providing no more information about x, y, z than the first equation did.

What happens when we have three equations in three variables ($m = 3, n = 3$)? The answer is that *typically* there will be a unique solution.

Example 3.65 Determine the solutions (if any) to the following equations:

$$3x + y - 2z = -2; \qquad x + y + z = 2; \qquad 2x + 4y + z = 0.$$

Solution We can substitute $z = 2 - x - y$ from the second equation into the first and third to find

$$3x + y - 2(2 - x - y) = 5x + 3y - 4 = -2 \implies 5x + 3y = 2;$$
$$2x + 4y + (2 - x - y) = x + 3y + 2 = 0 \implies x + 3y = -2.$$

Subtracting the second of these equations from the first gives $4x = 4$ and so we see

$$x = 1, \qquad y = (-2 - x)/3 = -1, \qquad z = 2 - x - y = 2. \tag{3.32}$$

Thus there is a unique solution $(x,y,z) = (1, -1, 2)$. We can verify easily that this is indeed a solution (just to check that the system contains no contradictory information elsewhere that we haven't used). □

In fact, three linear equations

$$ax + by + cz = j; \qquad dx + ey + fz = k; \qquad gx + hy + iz = l,$$

#609
#610
will have a unique solution if and only if $aei + bfg + cdh - afh - bdi - ceg \neq 0$ (see #632).

Geometrically, three linear equations represent three planes in three-dimensional xyz-space. Typically the first two planes will intersect in a line; the third plane will intersect this line in a point *unless* the third plane is parallel to the line in which case it either contains it (infinite solutions) or does not intersect the line (no solutions). So the possible arrangements of three planes which *don't* lead to a unique solution are (a) the three planes are all the same; (b) three planes meet in a line; (c) two planes are parallel and distinct with the third plane meeting both; (d) the three planes pairwise meet in distinct parallel lines; (e) three planes all parallel and not all the same. In cases (a) and (b) there are infinitely many solutions; cases (c), (d), (e) represent inconsistent systems.

Exercises

#608 a Find, if any, all the solutions of the following linear systems:

(i) $2x + 7y = 3$; $x - 2y = 1$.
(ii) $2x - 3y = 1$; $4x - 6y = 2$.
(iii) $x + y = 3$; $3x + 3y = 4$.
(iv) $x + y = 3$; $4x - y = 2$; $5x - 2y = 1$.

#609 b Find, if any, all the solutions of the following linear systems.

(i) $3x + 2y + z = 7$; $x - 2y + 3z = 5$; $2x + 3y = 4$.
(ii) $x - 2y = 3z$; $4y + 6z = 2x + 1$.
(iii) $x + 2y - z = 1$; $x + y + 2z = 1; x - y + 8z = 0$.
(iv) $x + 2y - z = x + y + 2z$ $= x - y + 8z = 1$.
(v) $x + y + z = 1$; $2x - 6y + z = 2$; $3x + z = 3$; $y + 2z = 0$.

#610 B† The following systems of equations each represent three planes in xyz-space. In each case there is not a unique solution. Classify, geometrically, each system as one of (a)–(e) as described in the final paragraph in the solution to Example 3.65.

(i) $x - 2y = z + 1$; $2x + y = 3$; $x = 2y + z$.
(ii) $2x + y = 3z + 2$; $x + 2z = 3y + 1$; $y + z + 2 = 2x$.
(iii) $2x + 3y = 3z - 2$; $x + y + 1 = z$; $x + 2y = 2z$.
(iv) $x + y + z = 4$; $x + 2y = 3$; $x + 2z = 1$.

#611 A(i) Assuming $ad - bc \neq 0$, show directly that the equations in (3.24) have the unique solution given in (3.25).
(ii) Assuming $ad - bc = 0$, show directly that the equations in (3.24) have no solution or infinitely many solutions.

#612 B†(i) Let a, b be real numbers. Determine x and y in terms of a and b, given the equations $2x + 3y = a$ and $3x - 2y = b$.
(ii) Find 2×2 matrices M and N which satisfy

$$2M + 3N = \begin{pmatrix} 1 & -2 \\ 1 & 0 \end{pmatrix}; \qquad 3M - 2N = \begin{pmatrix} 0 & 1 \\ 2 & -2 \end{pmatrix}.$$

(iii) Find polynomials $p(x)$ and $q(x)$ which satisfy $2p(x) + 3q(x) = 12 + x + 3x^3$ and also $3p(x) - 2q(x) = x + x^2 - 2x^3$.

#613 C† Let a,b,c,d,e,f be integers such that $ad - bc \neq 0$. Show that (3.24) has integer solutions x,y if $ad - bc = \pm 1$. Show that, if the solutions x,y are integers for all integer choices of e,f then $ad - bc = \pm 1$.

#614 A Let A and B be $n \times n$ invertible matrices. Show that AB is invertible and that $(AB)^{-1} = B^{-1}A^{-1}$.

#615 A† Show that a square matrix A is invertible if and only if A^T is invertible, in which case $(A^T)^{-1} = (A^{-1})^T$.

#616 B† For an invertible square matrix A and positive integer m, we define $A^{-m} = (A^{-1})^m$. For such A show that $A^r A^s = A^{r+s}$ for all integers r,s and in particular that $(A^r)^{-1} = A^{-r}$.

#617 b Verify #597 is valid for $n < 0$ in the context of $A^n = (A^{-1})^{-n}$.

#618 A Let A,P be $n \times n$ matrices with P invertible. Show that $P^{-1}A^k P = (P^{-1}AP)^k$ for $k \geqslant 1$.

#619 B† Show that $P^{-1}AP$ is diagonal where

$$A = \begin{pmatrix} 2 & 1 \\ 1 & 2 \end{pmatrix} \quad \text{and} \quad P = \begin{pmatrix} 1 & 1 \\ -1 & 1 \end{pmatrix}.$$

Use #618 to calculate a general formula for A^n where $n \geqslant 1$. Find a matrix B such that $B^3 = A$.

#620 B In the notation of #562, show that Z is invertible if and only if $z \neq 0$, in which case $z^{-1} \leftrightarrow Z^{-1}$.

#621 b Let a,b,c,d be reals, not all zero. Show the complex matrix $aI_2 + b\mathcal{I} + c\mathcal{J} + d\mathcal{K}$ is invertible where $\mathcal{I}, \mathcal{J}, \mathcal{K}$ are as defined in #561.

#622 A† Let A be a square matrix with a zero row or a zero column. Show that A is singular.

#623 B†

(i) Given the square matrix A is invertible, show that the only solution of $A\mathbf{x} = \mathbf{0}$ is $\mathbf{x} = \mathbf{0}$.
(ii) Suppose that the square matrix B has columns $\mathbf{c}_1, \ldots, \mathbf{c}_n$ and that $k_1\mathbf{c}_1 + \cdots + k_n\mathbf{c}_n = \mathbf{0}$ where $k_1, \ldots, k_n$ are real numbers, not all zero. Show that B is singular.
(iii) Prove the corresponding result to (ii) for the rows of B.

#624 C Show a 2×2 matrix A is invertible if and only if its columns are independent. Prove the same for its rows.

#625 c Find a 2×3 matrix A and a 3×2 matrix B such that AB is invertible and BA isn't invertible.

#626 c† Show that B is a left inverse for a matrix A if and only if B^T is a right inverse for A^T.

Definition 3.66 Two $n \times n$ matrices A and B are **similar** if there is an invertible $n \times n$ matrix P such that $A = P^{-1}BP$.

#627 B [9] Let A, B, C be $n \times n$ matrices. Show

  (i) A is similar to A;

 (ii) if A is similar to B then B is similar to A;

(iii) if A is similar to B and B is similar to C then A is similar to C.

#628 B†(i) Show that if A and B are similar matrices then so are A^2 and B^2 (with the same choice of P).

(ii) Show that if A and B are similar matrices then trace$(A) = $ trace(B).

(iii) Show that if A and B are similar matrices then so are A^T and B^T (with generally a different choice of P).

#629 C(i) Show that a square matrix and its transpose have the same number of square roots.

(ii) Show that similar matrices have the same number of square roots.

#630 B† Let A be a square matrix such that $A^n = 0$. Show that $I - A$ is invertible.

#631 c† Let A be a square matrix such that $A^2 = A$. Show that $I + A$ is invertible.

#632 C Verify directly that $AB = BA = (aei + bfg + cdh - afh - bdi - ceg)I_3$ where

$$A = \begin{pmatrix} a & b & c \\ d & e & f \\ g & h & i \end{pmatrix}, \qquad B = \begin{pmatrix} ei - fh & ch - bi & bf - ce \\ fg - di & ai - cg & cd - af \\ dh - eg & bg - ah & ae - bd \end{pmatrix}.$$

Deduce that A is invertible if and only if $aei + bfg + cdh - afh - bdi - ceg \neq 0$.

Remark 3.67 The expression $aei + bfg + cdh - afh - bdi - ceg$ in #632 is the determinant of the 3×3 matrix A. Determinants of 2×2 matrices were already discussed in Remark 3.64, and more generally determinants will be covered in §3.8. The matrix B is called the *adjugate* of A. ∎

#633 A Verify, for 2×2 and 3×3 matrices, that $\det A = \det A^T$. (This result generalizes; see Corollary 3.161.)

#634 A Show, for 2×2 matrices A and B, that $\det AB = \det A \det B$. Deduce that A in (3.15) has no square roots.

#635 D† [10] Let A be a 2×2 matrix. Show that $A^2 - (\text{trace}A)A + (\det A)I_2 = 0$. Suppose now that $A^n = 0$ for some $n \geqslant 2$. Prove that $\det A = 0$ and deduce that $A^2 = 0$.

#636 D† Let A be the matrix from #619 and let $x_n = [A^n]_{11}$. Explain why

$$x_{n+2} - 4x_{n+1} + 3x_n = 0, \qquad x_0 = 1, x_1 = 2.$$

Solve this recurrence relation. Hence find an expression for A^n.

#637 B Find two different right inverses for $\begin{pmatrix} 1 & 2 & 0 \\ 0 & 1 & 1 \end{pmatrix}$.

---

[9] This exercise shows that similarity is an *equivalence relation*.
[10] This exercise verifies the Cayley–Hamilton theorem for 2×2 matrices. See also #998.

#638 B† Consider the matrices

$$A = \begin{pmatrix} 1 & 1 \\ 3 & 4 \\ 2 & 3 \end{pmatrix}, \qquad \mathbf{x} = \begin{pmatrix} x \\ y \end{pmatrix}, \qquad \mathbf{b} = \begin{pmatrix} b_1 \\ b_2 \\ b_3 \end{pmatrix}.$$

(i) Show that if $A\mathbf{x} = \mathbf{0}$ then $\mathbf{x} = \mathbf{0}$.

(ii) Show that B_1 and B_2 are both left inverses of A, where

$$B_1 = \begin{pmatrix} 8 & -5 & 4 \\ -6 & 4 & -3 \end{pmatrix}, \qquad B_2 = \begin{pmatrix} -4 & 7 & -8 \\ 3 & -5 & 6 \end{pmatrix}.$$

(iii) Show that the system $A\mathbf{x} = \mathbf{b}$ is consistent if and only if $b_1 + b_3 = b_2$, in which case there is a unique solution $\mathbf{x}$.

(iv) Let $\mathbf{b} = (2,7,5)^T$. Find the unique solution $\mathbf{x}$ and show that $\mathbf{x} = B_1 \mathbf{b} = B_2 \mathbf{b}$.

(v) As x and y vary, over what subset of $\mathbb{R}_3$ does the point $A\mathbf{x}$ range?

#639 c† Let A denote the matrix in #638. Determine a 1×3 vector $\mathbf{x}$ such that $\mathbf{x}A = (0,0)$. Hence find a left inverse B_3 of A which is not a linear combination of the matrices B_1 and B_2.

#640 C† Let A, B_1, B_2 be as in #638 and B_3 the matrix found in #639. Find all the left inverses of A and show that they are all of the form $c_1 B_1 + c_2 B_2 + c_3 B_3$ where c_1, c_2, c_3 satisfy a condition which you should determine.

#641 B Show that there is no right inverse for the matrix A in #638.

#642 b Find all 2×2 matrices X which solve the equation $AX = B$ in each of the following cases:

$$A = \begin{pmatrix} 1 & 2 \\ 3 & 4 \end{pmatrix}, \quad B = \begin{pmatrix} 3 & 2 \\ 1 & 1 \end{pmatrix}; \quad A = \begin{pmatrix} 1 & 1 \\ 2 & 2 \end{pmatrix}, \quad B = \begin{pmatrix} 3 & 2 \\ 6 & 4 \end{pmatrix};$$

$$A = \begin{pmatrix} 1 & 2 \\ 2 & 4 \end{pmatrix}, \quad B = \begin{pmatrix} 1 & 0 \\ 0 & 1 \end{pmatrix}.$$

#643 C Let

$$M = \begin{pmatrix} 1 & 0 \\ 0 & -1 \end{pmatrix} \qquad \text{and} \qquad N = \begin{pmatrix} 1 & 1 \\ 0 & 0 \end{pmatrix}.$$

Show that there aren't 2×1 matrices $\mathbf{v}, \mathbf{w}$ such that $\mathbf{v}\mathbf{w}^T = M$ and find all such $\mathbf{v}, \mathbf{w}$ satisfying $\mathbf{v}\mathbf{w}^T = N$.

#644 C (i) Show that if $A = \mathbf{v}\mathbf{w}^T$ where $\mathbf{v}, \mathbf{w}$ are 2×1 matrices then A is singular. (ii) Show, conversely, that if A is a non-invertible 2×2 matrix then $A = \mathbf{v}\mathbf{w}^T$ for some 2×1 matrices $\mathbf{v}, \mathbf{w}$.

#645 C† Show that the 3×3 matrix A on the right is singular. Show further that A cannot be written $A = \mathbf{v}\mathbf{w}^T$ for some 3×1 matrices $\mathbf{v}, \mathbf{w}$.

$$A = \begin{pmatrix} 1 & 0 & 3 \\ 1 & 2 & 1 \\ 1 & 1 & 2 \end{pmatrix}$$

3.5 Elementary Row Operations

In Example 3.65, we showed that the only solution to the three equations

$$3x+y-2z=-2; \qquad x+y+z=2; \qquad 2x+4y+z=0, \tag{3.33}$$

is $(x,y,z)=(1,-1,2)$. Whilst we solved it rigorously, our approach was a little ad hoc; at least, it's not hard to appreciate that if we were presented with 1969 equations in 2017 variables then we would need a much more systematic approach to treat them – or more likely we would need to be more methodical while programming our computers to determine any solutions for us. We introduce such a process, called *row-reduction*, here.

Our first introduction is an improvement in notation. We will now write the three equations in (3.33) as

$$\begin{pmatrix} 3 & 1 & -2 & \bigm| & -2 \\ 1 & 1 & 1 & \bigm| & 2 \\ 2 & 4 & 1 & \bigm| & 0 \end{pmatrix}. \tag{3.34}$$

The first row $\begin{pmatrix} 3 & 1 & -2 & | & -2 \end{pmatrix}$ represents the first equation $3x+y-2z=-2$ and similarly the second and third rows represent the second and third equations. More generally a linear system $A\mathbf{x}=\mathbf{b}$, of m equations in n variables as in (3.21), will now be written as $(A|\mathbf{b})$, where A is an $m \times n$ matrix and $\mathbf{b}$ is an $m \times 1$ column vector. All that has been lost in this representation are the names of the variables, but these names are unchanging and unimportant in the actual handling of the equations. The advantages, we shall see, are that we will be able to progress systematically towards any solution and at each stage we shall retain all the information that the system contains – any redundancies (superfluous, unnecessary equations) or contradictions will naturally appear as part of the calculation.

This process is called *row-reduction*. It relies on three types of operation, called *elementary row operations* or *EROs*, which importantly do not affect the set of solutions of a linear system as we apply them.

Definition 3.68 Given a linear system of equations, an **elementary row operation** or **ERO** is an operation of one of the following three kinds:

(a) The ordering of two equations (or rows) may be swapped – for example, one might reorder the writing of the equations so that the first equation now appears third and vice versa.

(b) An equation may be multiplied by a non-zero scalar – for example, one might replace $2x-y+z=3$ by $x-\frac{1}{2}y+\frac{1}{2}z=\frac{3}{2}$ from multiplying both sides of the equation by $\frac{1}{2}$.

(c) A multiple of one equation might be added to a different equation – for example, one might replace the second equation by the second equation plus twice the third equation.

Notation 3.69 (a) Let S_{ij} denote the ERO which swaps rows i and j (or equivalently the ith and jth equations).

(b) Let $M_i(\lambda)$ denote the ERO which multiplies row i by $\lambda \neq 0$ (or equivalently both sides of the ith equation).

(c) For $i \neq j$, let $A_{ij}(\lambda)$ denote the ERO which adds λ times row i to row j (or does the same to the equations).

All these operations may well seem uncontroversial (their validity will be shown in Corollary 3.77) but it is probably not yet clear that these three simple operations are powerful enough to *reduce* any linear system to a point where any solutions can just be read off (Theorem 3.89). Before treating the general case, we will see how the three equations (3.33) can be solved using EROs to get an idea of the process.

Example 3.70 Find all solutions of the linear system (3.34).

Solution If we use S_{12} to swap the first two rows the system becomes

$$\left(\begin{array}{ccc|c} 3 & 1 & -2 & -2 \\ 1 & 1 & 1 & 2 \\ 2 & 4 & 1 & 0 \end{array}\right) \xrightarrow{S_{12}} \left(\begin{array}{ccc|c} 1 & 1 & 1 & 2 \\ 3 & 1 & -2 & -2 \\ 2 & 4 & 1 & 0 \end{array}\right).$$

Now subtract three times the first row from the second, i.e. $A_{12}(-3)$ and follow this by subtracting twice the first row from the third, i.e. $A_{13}(-2)$, so that

$$\left(\begin{array}{ccc|c} 1 & 1 & 1 & 2 \\ 3 & 1 & -2 & -2 \\ 2 & 4 & 1 & 0 \end{array}\right) \xrightarrow{A_{12}(-3)} \left(\begin{array}{ccc|c} 1 & 1 & 1 & 2 \\ 0 & -2 & -5 & -8 \\ 2 & 4 & 1 & 0 \end{array}\right) \xrightarrow{A_{13}(-2)} \left(\begin{array}{ccc|c} 1 & 1 & 1 & 2 \\ 0 & -2 & -5 & -8 \\ 0 & 2 & -1 & -4 \end{array}\right).$$
(3.35)

We can now divide the second row by -2, i.e. $M_2(-1/2)$ to find

$$\left(\begin{array}{ccc|c} 1 & 1 & 1 & 2 \\ 0 & -2 & -5 & -8 \\ 0 & 2 & -1 & -4 \end{array}\right) \xrightarrow{M_2(-1/2)} \left(\begin{array}{ccc|c} 1 & 1 & 1 & 2 \\ 0 & 1 & 2\frac{1}{2} & 4 \\ 0 & 2 & -1 & -4 \end{array}\right).$$

We then subtract the second row from the first, i.e. $A_{21}(-1)$, and follow this by subtracting twice the second row from the third, i.e. $A_{23}(-2)$, to obtain

$$\left(\begin{array}{ccc|c} 1 & 1 & 1 & 2 \\ 0 & 1 & 2\frac{1}{2} & 4 \\ 0 & 2 & -1 & -4 \end{array}\right) \xrightarrow{A_{21}(-1)} \left(\begin{array}{ccc|c} 1 & 0 & -1\frac{1}{2} & -2 \\ 0 & 1 & 2\frac{1}{2} & 4 \\ 0 & 2 & -1 & -4 \end{array}\right) \xrightarrow{A_{23}(-2)} \left(\begin{array}{ccc|c} 1 & 0 & -1\frac{1}{2} & -2 \\ 0 & 1 & 2\frac{1}{2} & 4 \\ 0 & 0 & -6 & -12 \end{array}\right).$$
(3.36)

If we divide the third row by -6, i.e. $M_3(-1/6)$, the system becomes

$$\left(\begin{array}{ccc|c} 1 & 0 & -1\frac{1}{2} & -2 \\ 0 & 1 & 2\frac{1}{2} & 4 \\ 0 & 0 & -6 & -12 \end{array}\right) \xrightarrow{M_3(-1/6)} \left(\begin{array}{ccc|c} 1 & 0 & -1\frac{1}{2} & -2 \\ 0 & 1 & 2\frac{1}{2} & 4 \\ 0 & 0 & 1 & 2 \end{array}\right).$$

Finally, we subtract $2\frac{1}{2}$ times the third row from the second, i.e. $A_{32}(-2\frac{1}{2})$, and follow this by adding $1\frac{1}{2}$ times the third row to the first, i.e. $A_{31}(1\frac{1}{2})$.

$$\left(\begin{array}{ccc|c} 1 & 0 & -1\frac{1}{2} & -2 \\ 0 & 1 & 2\frac{1}{2} & 4 \\ 0 & 0 & 1 & 2 \end{array}\right) \xrightarrow{A_{32}(-5/2)} \left(\begin{array}{ccc|c} 1 & 0 & -1\frac{1}{2} & -2 \\ 0 & 1 & 0 & -1 \\ 0 & 0 & 1 & 2 \end{array}\right) \xrightarrow{A_{31}(3/2)} \left(\begin{array}{ccc|c} 1 & 0 & 0 & 1 \\ 0 & 1 & 0 & -1 \\ 0 & 0 & 1 & 2 \end{array}\right).$$

The rows of the final matrix represent the equations $x = 1$, $y = -1$, $z = 2$ as expected from (3.32). $\qquad \square$

Remark 3.71 In case the systematic nature of the previous example isn't apparent, note that the first three operations $S_{12}, A_{12}(-3), A_{13}(-2)$ were chosen so that the first column became e_1^T in (3.35). There were many other ways to achieve this: for example, we could have begun with $M_1(1/3)$ to divide the first row by 3, then used $A_{12}(-1)$ and $A_{13}(-2)$ to clear the rest of the column. Once done, we then produced a similar leading entry of 1 in the second row with $M_2(-1/2)$ and used $A_{21}(-1)$ and $A_{23}(-2)$ to turn the second column into e_2^T in (3.36). The final three EROs were chosen to transform the third column to e_3^T at which point we could simply read off the solutions. ∎

Here are two slightly different examples, the first where we find that there are infinitely many solutions, whilst in the second example we see that there are no solutions.

Example 3.72 Find the general solution of the following systems of equations in variables x_1, x_2, x_3, x_4.

(a) $x_1 - x_2 + x_3 + 3x_4 = 2$; $\qquad 2x_1 - x_2 + x_3 + 2x_4 = 4$; $\qquad 4x_1 - 3x_2 + 3x_3 + 8x_4 = 8$.
(b) $x_1 + x_2 + x_3 + x_4 = 4$; $\qquad 2x_1 + 3x_2 - 2x_3 - 3x_4 = 1$; $\qquad x_1 + 5x_3 + 6x_4 = 1$.

Solution (a) This time we will not spell out at quite so much length which EROs are being used. But we continue in a similar vein to the previous example and proceed by the method outlined in Remark 3.71.

$$\left(\begin{array}{cccc|c} 1 & -1 & 1 & 3 & 2 \\ 2 & -1 & 1 & 2 & 4 \\ 4 & -3 & 3 & 8 & 8 \end{array} \right) \xrightarrow[A_{13}(-4)]{A_{12}(-2)} \left(\begin{array}{cccc|c} 1 & -1 & 1 & 3 & 2 \\ 0 & 1 & -1 & -4 & 0 \\ 0 & 1 & -1 & -4 & 0 \end{array} \right)$$

$$\xrightarrow[A_{23}(-1)]{A_{21}(1)} \left(\begin{array}{cccc|c} 1 & 0 & 0 & -1 & 2 \\ 0 & 1 & -1 & -4 & 0 \\ 0 & 0 & 0 & 0 & 0 \end{array} \right).$$

We have manipulated our system of three equations to two equations equivalent to the original system, namely

$$x_1 - x_4 = 2; \qquad x_2 - x_3 - 4x_4 = 0. \tag{3.37}$$

#646
#647
#648
#649
#656
The presence of the zero row in the last matrix means that there was some redundancy in the system. Note, for example, that the third equation can be deduced from the first two (it's the second equation added to twice the first) and so it provides no new information. As there are now only two equations in four variables, it's impossible for each column to contain a row's leading entry. In this example, the third and fourth columns lack such an entry. To describe all the solutions to a consistent system, we assign parameters to the columns/variables without leading entries. In this case that's x_3 and x_4 and we'll assign parameters by setting $x_3 = s, x_4 = t$, and then use the two equations in (3.37) to read off x_1 and x_2. So

$$x_1 = t + 2, \qquad x_2 = s + 4t, \qquad x_3 = s, \qquad x_4 = t, \tag{3.38}$$

or we could write

$$(x_1, x_2, x_3, x_4) = (t + 2, s + 4t, s, t) = (2, 0, 0, 0) + s(0, 1, 1, 0) + t(1, 4, 0, 1). \tag{3.39}$$

For each choice of s and t we have a solution as in (3.38) and this is one way of representing the general solution. (3.39) makes more apparent that these solutions form a plane in $\mathbb{R}^4$, a plane which passes through $(2,0,0,0)$ is parallel to $(0,1,1,0)$ and $(1,4,0,1)$ with s,t parametrizing the plane.

(b) Applying EROs again in a like manner, we find

$$
\begin{pmatrix} 1 & 1 & 1 & 1 & | & 4 \\ 2 & 3 & -2 & -3 & | & 1 \\ 1 & 0 & 5 & 6 & | & 1 \end{pmatrix}
\xrightarrow{\substack{A_{12}(-2) \\ A_{13}(-1)}}
\begin{pmatrix} 1 & 1 & 1 & 1 & | & 4 \\ 0 & 1 & -4 & -5 & | & -7 \\ 0 & -1 & 4 & 5 & | & -3 \end{pmatrix}
\xrightarrow{A_{23}(1)}
\begin{pmatrix} 1 & 1 & 1 & 1 & | & 4 \\ 0 & 1 & -4 & -5 & | & -7 \\ 0 & 0 & 0 & 0 & | & -10 \end{pmatrix}
$$

$$
\xrightarrow{\substack{M_3(-1/10) \\ A_{21}(-1)}}
\begin{pmatrix} 1 & 0 & -5 & -6 & | & -11 \\ 0 & 1 & -4 & -5 & | & -7 \\ 0 & 0 & 0 & 0 & | & 1 \end{pmatrix}.
$$

Note that any $\mathbf{x} = (x_1,x_2,x_3,x_4)$ which solves the final equation must satisfy

$$0x_1 + 0x_2 + 0x_3 + 0x_4 = 1.$$

There clearly are no such x_i and so there are no solutions to this equation. Any solution to the system has, in particular, to solve the third equation and so this system has no solutions. In fact, this was all apparent once the third row had become $(\,0\ \ 0\ \ 0\ \ 0\ |\ -10\,)$ as the equation it represents is clearly insolvable also. The final two EROs were simply done to put the matrix into what is called *reduced row echelon form* (see Definition 3.78). $\square$

Examples 3.70, 3.72(a) and 3.72(b) are specific examples of the following general cases:

- A *linear system can have no, one or infinitely many solutions.*

We shall prove this in due course (Corollary 3.92). We finish our examples though with a linear system that involves a parameter – so really we have a family of linear systems, one for each value of that parameter. What EROs may be permissible at a given stage may well depend on the value of the parameter and so we may see (as below) that such a family can exhibit all three of the possible scenarios just described.

Example 3.73 Consider the system of equations in x,y,z,

$$x + z = -5; \qquad 2x + \alpha y + 3z = -9; \qquad -x - \alpha y + \alpha z = \alpha^2,$$

where α is a constant. For which values of α has the system one solution, none or infinitely many?

Solution Writing this system in matrix form and applying EROs we can argue as follows:

$$
\begin{pmatrix} 1 & 0 & 1 & | & -5 \\ 2 & \alpha & 3 & | & -9 \\ -1 & -\alpha & \alpha & | & \alpha^2 \end{pmatrix}
\xrightarrow{\substack{A_{12}(-2) \\ A_{13}(1)}}
\begin{pmatrix} 1 & 0 & 1 & | & -5 \\ 0 & \alpha & 1 & | & 1 \\ 0 & -\alpha & \alpha+1 & | & \alpha^2 - 5 \end{pmatrix}
$$

$$
\xrightarrow{A_{23}(1)}
\begin{pmatrix} 1 & 0 & 1 & | & -5 \\ 0 & \alpha & 1 & | & 1 \\ 0 & 0 & \alpha+2 & | & \alpha^2 - 4 \end{pmatrix}. \tag{3.40}
$$

At this point, which EROs are permissible depends on the value of α. We would like to divide the second equation by α and the third by $\alpha+2$. Both these are permissible provided

#650
#651
#655

that $\alpha \neq 0$ and $\alpha \neq -2$. We will have to treat separately those particular cases but, assuming for now that $\alpha \neq 0, -2$, we obtain

$$\left(\begin{array}{ccc|c} 1 & 0 & 1 & -5 \\ 0 & \alpha & 1 & 1 \\ 0 & 0 & \alpha+2 & \alpha^2-4 \end{array}\right) \xrightarrow[M_3(1/(\alpha+2))]{M_2(1/\alpha)} \left(\begin{array}{ccc|c} 1 & 0 & 1 & -5 \\ 0 & 1 & 1/\alpha & 1/\alpha \\ 0 & 0 & 1 & \alpha-2 \end{array}\right)$$

$$\xrightarrow[A_{32}(-1/\alpha)]{A_{31}(-1)} \left(\begin{array}{ccc|c} 1 & 0 & 0 & -\alpha-3 \\ 0 & 1 & 0 & 3/\alpha-1 \\ 0 & 0 & 1 & \alpha-2 \end{array}\right)$$

and we see that the system has a unique solution when $\alpha \neq 0, -2$. Returning though to the last matrix of (3.40) for our two special cases, we would proceed as follows:

$\alpha = 0:$
$$\left(\begin{array}{ccc|c} 1 & 0 & 1 & -5 \\ 0 & 0 & 1 & 1 \\ 0 & 0 & 2 & -4 \end{array}\right) \xrightarrow[M_3(-1/6)]{A_{23}(-2)} \left(\begin{array}{ccc|c} 1 & 0 & 1 & -5 \\ 0 & 0 & 1 & 1 \\ 0 & 0 & 0 & 1 \end{array}\right).$$

$\alpha = -2:$
$$\left(\begin{array}{ccc|c} 1 & 0 & 1 & -5 \\ 0 & -2 & 1 & 1 \\ 0 & 0 & 0 & 0 \end{array}\right) \xrightarrow{M_2(-1/2)} \left(\begin{array}{ccc|c} 1 & 0 & 1 & -5 \\ 0 & 1 & -1/2 & -1/2 \\ 0 & 0 & 0 & 0 \end{array}\right).$$

We see then that the system is inconsistent when $\alpha = 0$ (because of the insolvability of the third equation) while there are infinitely many solutions $x = -5 - t$, $y = (t-1)/2$, $z = t$, when $\alpha = -2$. We assign a parameter, here t, to the variable z as the third column has no leading entry. □

Now looking to treat linear systems more generally, we will first show that the set of solutions of a linear system does not change under the application of EROs. We shall see that applying any ERO to a linear system $(A|\mathbf{b})$ is equivalent to premultiplying by an invertible *elementary* matrix E to obtain $(EA|E\mathbf{b})$, and it is the invertibility of elementary matrices that means the set of solutions remains unchanged when we apply EROs.

Proposition 3.74 *(Elementary Matrices) Let A be an $m \times n$ matrix. Applying any of the EROs S_{IJ}, $M_I(\lambda)$ and $A_{IJ}(\lambda)$ is equivalent to pre-multiplying A by certain matrices, which we also denote as S_{IJ}, $M_I(\lambda)$ and $A_{IJ}(\lambda)$. Specifically these matrices have entries*

$$[S_{IJ}]_{ij} = \begin{cases} 1 & i=j \neq I,J, \\ 1 & i=J, j=I, \\ 1 & i=I, j=J, \\ 0 & \text{otherwise.} \end{cases} \qquad [M_I(\lambda)]_{ij} = \begin{cases} 1 & i=j \neq I, \\ \lambda & i=j=I, \\ 0 & \text{otherwise.} \end{cases}$$

$$[A_{IJ}(\lambda)]_{ij} = \begin{cases} 1 & i=j, \\ \lambda & i=J, j=I, \\ 0 & \text{otherwise} \end{cases}$$

*The above matrices are known as **elementary matrices**.*

Proof The proof is left to #682. □

Example 3.75 When $m = 3$ we see

$$S_{21} = \begin{pmatrix} 0 & 1 & 0 \\ 1 & 0 & 0 \\ 0 & 0 & 1 \end{pmatrix}, \quad M_3(7) = \begin{pmatrix} 1 & 0 & 0 \\ 0 & 1 & 0 \\ 0 & 0 & 7 \end{pmatrix}, \quad A_{31}(-2) = \begin{pmatrix} 1 & 0 & -2 \\ 0 & 1 & 0 \\ 0 & 0 & 1 \end{pmatrix}.$$

Note that these elementary matrices are the results of performing the corresponding EROs $S_{21}, M_3(7), A_{31}(-2)$ on the identity matrix I_3. This is generally true of elementary matrices.

Proposition 3.76 *Elementary matrices are invertible.*

Proof This follows from noting that

$$(S_{ij})^{-1} = S_{ji} = S_{ij}; \qquad (A_{ij}(\lambda))^{-1} = A_{ij}(-\lambda); \qquad (M_i(\lambda))^{-1} = M_i(\lambda^{-1}),$$

whether considered as EROs or their corresponding matrices. □

Corollary 3.77 *(Invariance of Solution Space under EROs) Let $(A|\mathbf{b})$ be a linear system of m equations and E an elementary $m \times m$ matrix. Then $\mathbf{x}$ is a solution of $(A|\mathbf{b})$ if and only if $\mathbf{x}$ is a solution of $(EA|E\mathbf{b})$.*

Proof The important point here is that E is invertible. So if $A\mathbf{x} = \mathbf{b}$ then $EA\mathbf{x} = E\mathbf{b}$ follows by premultiplying by E. But likewise if $EA\mathbf{x} = E\mathbf{b}$ is true then it follows that $A\mathbf{x} = \mathbf{b}$ by premultiplying by E^{-1}. □

So applying an ERO, or any succession of EROs, won't alter the set of solutions of a linear system. The next key result is that, systematically using EROs, it is possible to reduce any system $(A|\mathbf{b})$ to *reduced row echelon form*. Once in this form it is simple to read off the system's solutions.

Definition 3.78 A matrix A is said to be in **reduced row echelon form** (or simply **RRE form**) if

(a) the first non-zero entry of any non-zero row is 1;

(b) in a column [11] that contains such a leading 1, all other entries are zero;

(c) the leading 1 of a non-zero row appears to the right of the leading 1s of the rows above it;

(d) any zero rows appear below the non-zero rows.

#676
#677
#683
#685

Definition 3.79 The process of applying EROs to transform a matrix into RRE form is called **row-reduction**, or just simply **reduction.** It is also commonly referred to as **Gaussian** [12] **elimination**.

---

[11] Some texts refer to such columns as *pivot columns* and to leading 1s as *pivots*.

[12] After the German mathematician Gauss, though the method was known to Newton, and earlier still to Chinese mathematicians, certainly by the second century AD. Wilhelm Jordan (1842–1899) adapted the method to require that leading non-zero entries be 1 and so the method is also sometimes called Gauss–Jordan elimination.

Example 3.80 Of the following matrices

$$
\begin{pmatrix} 0 & 1 & 2 & 0 & -4 \\ 0 & 0 & 0 & 1 & \pi \\ 0 & 0 & 0 & 0 & 0 \end{pmatrix}, \qquad
\begin{pmatrix} 1 & 0 & \sqrt{2} & 0 \\ 0 & 1 & 2 & 0 \\ 0 & 0 & 0 & 1 \end{pmatrix}, \qquad
\begin{pmatrix} 1 & 0 \\ 0 & 1 \\ 0 & 0 \end{pmatrix},
$$

$$
\begin{pmatrix} 1 & 2 \\ 0 & 1 \\ 0 & 0 \end{pmatrix}, \qquad
\begin{pmatrix} 1 & 0 & 0 & \sqrt{3} \\ 0 & 1 & 0 & 0 \\ 0 & 0 & 2 & 1 \end{pmatrix},
$$

the first three are in RRE form. The fourth is not as the second column contains a leading 1 but not all other entries of that column are 0. The fifth matrix is not in RRE form as the leading entry of the third row is not 1.

We have yet to show that any matrix can be uniquely put into RRE form using EROs (Theorem 3.89) but – as we have already seen examples covering the range of possibilities – it seems timely to prove the following result here.

Proposition 3.81 *(Solving Systems in RRE Form) Let $(A|\mathbf{b})$ be a matrix in RRE form which represents a linear system $A\mathbf{x} = \mathbf{b}$ of m equations in n variables. Then*

(a) The system has no solutions if and only if the last non-zero row of $(A|\mathbf{b})$ is

$$
\begin{pmatrix} 0 & 0 & \cdots & 0 & | & 1 \end{pmatrix}.
$$

(b) The system has a unique solution if and only if the non-zero rows of A form the identity matrix I_n. In particular, this case is only possible if $m \geqslant n$.

(c) The system has infinitely many solutions if $(A|\mathbf{b})$ has as many non-zero rows as A, and not every column of A contains a leading 1. The set of solutions can be described with k parameters where k is the number of columns not containing a leading 1.

Proof If $(A|\mathbf{b})$ contains the row $\begin{pmatrix} 0 & 0 & \cdots & 0 & | & 1 \end{pmatrix}$ then the system is certainly inconsistent as no $\mathbf{x}$ satisfies the equation

$$
0x_1 + 0x_2 + \cdots + 0x_n = 1.
$$

As $(A|\mathbf{b})$ is in RRE form, then this is the only way in which $(A|\mathbf{b})$ can have more non-zero rows than A. We will show that whenever $(A|\mathbf{b})$ has as many non-zero rows as A then the system $(A|\mathbf{b})$ is consistent.

Say, then, that both $(A|\mathbf{b})$ and A have r non-zero rows, so there are r leading 1s within these rows and we have $k = n - r$ columns without leading 1s. By reordering the numbering of the variables $x_1, \ldots, x_n$ if necessary, we can assume that the leading 1s appear in the first r columns. So, ignoring any zero rows, and remembering the system is in RRE form, the system now reads as the r equations:

$$
x_1 + a_{1(r+1)}x_{r+1} + \cdots + a_{1n}x_n = b_1; \qquad \cdots \qquad x_r + a_{r(r+1)}x_{r+1} + \cdots + a_{rn}x_n = b_r.
$$

We can see that if we assign $x_{r+1}, \ldots, x_n$ the k parameters $s_{r+1}, \ldots, s_n$, then we can read off from the r equations the values for $x_1, \ldots, x_r$. So for any values of the parameters we have a solution $\mathbf{x}$. Conversely though if $\mathbf{x} = (x_1, \ldots, x_n)$ is a solution, then it appears amongst the solutions we've just found when we assign values $s_{r+1} = x_{r+1}, \ldots, s_n = x_n$ to the parameters. We see that we have an infinite set of solutions associated with $k = n - r$ independent parameters when $n > r$ and a unique solution when $r = n$, in which case the non-zero rows of A are the matrix I_n. □

Remark 3.82 Note we showed in this proof that

- a system $(A|\mathbf{b})$ in RRE form is consistent if and only if $(A|\mathbf{b})$ has as many non-zero rows as A;
- all the solutions of a consistent system can be found by assigning parameters to the variables corresponding to the columns without leading 1s. ∎

Example 3.83 For the following linear systems which have been put in RRE form, we note:

$$\left(\begin{array}{cccc|c} 1 & -2 & 0 & 2 & 3 \\ 0 & 0 & 1 & 1 & -2 \\ 0 & 0 & 0 & 0 & 1 \end{array} \right), \qquad \left(\begin{array}{ccc|c} 1 & 0 & 0 & 2 \\ 0 & 1 & 0 & -1 \\ 0 & 0 & 1 & 3 \\ 0 & 0 & 0 & 0 \end{array} \right),$$

no solutions   unique solution

$$\left(\begin{array}{cccc|c} 1 & 2 & 0 & 0 & 3 \\ 0 & 0 & 1 & 0 & 2 \\ 0 & 0 & 0 & 1 & 1 \end{array} \right), \qquad \left(\begin{array}{cccc|c} 1 & -2 & 0 & 2 & 3 \\ 0 & 0 & 1 & 1 & -2 \\ 0 & 0 & 0 & 0 & 0 \end{array} \right).$$

one parameter family of solutions   two parameter family of solutions
$(3-2s,s,2,1)$ $(3+2s-2t,s,-2-t,t)$

We return now to the issue of determining the invertibility of a square matrix. There is no neat expression for the inverse of an $n \times n$ matrix in general – we have seen that the $n = 2$ case is easy enough (Proposition 3.63) though the $n = 3$ case is already messy (#632) – but the following method shows how to determine efficiently, using EROs, whether an $n \times n$ matrix is invertible and, in such a case, how to find the inverse. For the moment, we will assume the as-yet-unproven fact that every matrix can be uniquely put into RRE form.

Algorithm 3.84 *(Determining Invertibility) Let A be an $n \times n$ matrix. Place A side-by-side with I_n as an $n \times 2n$ matrix $(A \mid I_n)$. By assumption there are EROs that will reduce A to a matrix R in RRE form. We will simultaneously apply these EROs to both sides of $(A \mid I_n)$ until we arrive at $(R \mid P)$.*

- *If $R = I_n$ then A is invertible and $P = A^{-1}$.*
- *If $R \neq I_n$ then A is singular.*

Proof Denote the elementary matrices representing the EROs that reduce A as $E_1, E_2, \ldots, E_k$, so that $(A \mid I_n)$ becomes

$$(E_k E_{k-1} \cdots E_1 A \mid E_k E_{k-1} \cdots E_1) = (R \mid P) \qquad (3.41)$$

and we see that $R = PA$ and $E_k E_{k-1} \cdots E_1 = P$. If $R = I_n$ then

$$(E_k E_{k-1} \cdots E_1)A = I_n \quad \Longrightarrow \quad A^{-1} = E_k E_{k-1} \cdots E_1 = P$$

as elementary matrices are (left and right) invertible. If $R \neq I_n$ then, as R is in RRE form and square, R must have at least one zero row. It follows that $\mathbf{e}_n(PA) = \mathbf{0}$. As P is invertible, if A were also invertible, we could postmultiply by $A^{-1}P^{-1}$ to conclude $\mathbf{e}_n = \mathbf{0}$, a contradiction. Hence A is singular; indeed we can see from this proof that as soon as a zero row appears when reducing A then we know that A is singular.   □

Example 3.85 Determine whether the following matrices are invertible, finding any inverses that exist.

$$A = \begin{pmatrix} 1 & 2 & 1 \\ 2 & 1 & 0 \\ 1 & 3 & 1 \end{pmatrix}, \qquad B = \begin{pmatrix} 1 & 3 & -1 & 0 \\ 0 & 2 & 1 & 1 \\ 3 & 1 & 2 & 1 \\ 0 & 1 & 5 & 3 \end{pmatrix}.$$

Solution Quickly applying a sequence of EROs leads to

$$(A|I_3) \overset{\substack{A_{12}(-2) \\ A_{13}(-1)}}{\longrightarrow} \left(\begin{array}{ccc|ccc} 1 & 2 & 1 & 1 & 0 & 0 \\ 0 & -3 & -2 & -2 & 1 & 0 \\ 0 & 1 & 0 & -1 & 0 & 1 \end{array} \right) \overset{\substack{A_{31}(-2) \\ A_{32}(3) \\ S_{23}}}{\longrightarrow} \left(\begin{array}{ccc|ccc} 1 & 0 & 1 & 3 & 0 & -2 \\ 0 & 1 & 0 & -1 & 0 & 1 \\ 0 & 0 & -2 & -5 & 1 & 3 \end{array} \right)$$

$$\overset{\substack{M_3(-1/2) \\ A_{31}(-1)}}{\longrightarrow} \left(\begin{array}{c|ccc} & 1/2 & 1/2 & -1/2 \\ I_3 & -1 & 0 & 1 \\ & 5/2 & -1/2 & -3/2 \end{array} \right).$$

#664
#665
#666
#667
Hence

$$\begin{pmatrix} 1 & 2 & 1 \\ 2 & 1 & 0 \\ 1 & 3 & 1 \end{pmatrix}^{-1} = \begin{pmatrix} 1/2 & 1/2 & -1/2 \\ -1 & 0 & 1 \\ 5/2 & -1/2 & -3/2 \end{pmatrix}.$$

For B we note

$$(B|I_4) \overset{\substack{A_{13}(-3) \\ S_{24}}}{\longrightarrow} \left(\begin{array}{cccc|cccc} 1 & 3 & -1 & 0 & 1 & 0 & 0 & 0 \\ 0 & 1 & 5 & 3 & 0 & 0 & 0 & 1 \\ 0 & -8 & 5 & 1 & -3 & 0 & 1 & 0 \\ 0 & 2 & 1 & 1 & 0 & 1 & 0 & 0 \end{array} \right)$$

$$\overset{\substack{A_{23}(8) \\ A_{24}(-2) \\ A_{34}(1/5)}}{\longrightarrow} \left(\begin{array}{cccc|cccc} 1 & 3 & -1 & 0 & 1 & 0 & 0 & 0 \\ 0 & 1 & 5 & 3 & 0 & 0 & 0 & 1 \\ 0 & 0 & 45 & 25 & -3 & 0 & 1 & 8 \\ 0 & 0 & 0 & 0 & -3/5 & 1 & 1/5 & -2/5 \end{array} \right).$$

The left matrix is not yet in RRE form, but the presence of a zero row is sufficient to show that B is singular.   □

The remainder of this section contains some rather dense material, the main points of note being Theorem 3.89:

- *Any matrix can be put into a unique RRE form by means of EROs.*

We have already seen the main techniques associated with EROs, namely finding the general solution of a linear system and testing for the invertibility of a matrix. A further consequence of Theorem 3.89 is that *row rank* is well defined; the row rank of a matrix is the number of non-zero rows in the matrix's RRE form. *At this stage a first-time reader may wish to skip the remainder of this section and move straight on to the exercises and then also skip the next section on rank and nullity.*

Definition 3.86 (a) A **linear combination** of vectors $v_1, \ldots, v_m$ in $\mathbb{R}^n$ is any vector of the form

$$c_1 v_1 + \cdots + c_m v_m$$

where $c_1, \ldots, c_m$ are real numbers.

(b) $v_1, \ldots, v_m$ are said to be **linearly independent** (or simply **independent**) if the equation

$$c_1 v_1 + \cdots + c_m v_m = 0$$

implies $c_1 = \cdots = c_m = 0$. This is equivalent to saying that no v_i can be written as a linear combination of the other vectors (#696). Otherwise the vectors are said to be **linearly dependent.**

(c) Given an $m \times n$ matrix A with rows $v_1, \ldots, v_m$, then the **rowspace** of A, written Row(A), is the set of all linear combinations of the rows of A; that is,

$$\text{Row}(A) = \{c_1 v_1 + \cdots + c_m v_m : c_1, \ldots, c_m \in \mathbb{R}\}.$$

Example 3.87 (a) The vectors $v_1 = (1, 2, -1, 0)$, $v_2 = (2, 1, 0, 3)$, $v_3 = (0, 1, 1, 1)$ in $\mathbb{R}^4$ are linearly independent. To show this, we need to show that the vector equation

$$c_1 (1, 2, -1, 0) + c_2 (2, 1, 0, 3) + c_3 (0, 1, 1, 1) = 0$$

only has solution $(c_1, c_2, c_3) = 0$. One way is to note that this single vector equation is equivalent to a linear system of four equations in c_1, c_2, c_3, namely the first system below, and to use EROs. The details are left to #700(i).

$$\begin{pmatrix} 1 & 2 & 0 & 0 \\ 2 & 1 & 1 & 0 \\ -1 & 0 & 1 & 0 \\ 0 & 3 & 1 & 0 \end{pmatrix}, \quad \begin{pmatrix} 1 & 2 & 0 & x_1 \\ 2 & 1 & 1 & x_2 \\ -1 & 0 & 1 & x_3 \\ 0 & 3 & 1 & x_4 \end{pmatrix}. \tag{3.42}$$

(b) A vector $x = (x_1, x_2, x_3, x_4)$ is a linear combination of v_1, v_2, v_3 if and only if the condition $8x_1 + 6x_3 = x_2 + 5x_4$ holds. One way to see this is to note that x is a linear combination of the three vectors if and only if the second linear system in (3.42) is consistent. The details are left to #700(ii).

Proposition 3.88 *Let A be an $m \times n$ matrix and let B be a $k \times m$ matrix. Suppose also that a matrix R in RRE form can be obtained by EROs from A.*

(a) *The non-zero rows of R are independent.*

(b) *The rows of R are linear combinations of the rows of A.*

(c) *Row(BA) is contained in Row(A).*

(d) *If $k = m$ and B is invertible then Row(BA) = Row(A).*

(e) *Row(R) = Row(A).*

Proof (a) Denote the non-zero rows of R as $\mathbf{r}_1, \ldots, \mathbf{r}_r$ and suppose that $c_1\mathbf{r}_1 + \cdots + c_r\mathbf{r}_r = \mathbf{0}$. Say the leading 1 of $\mathbf{r}_1$ appears in the jth column. Then

$$c_1 + c_2[R]_{2j} + c_3[R]_{3j} + \cdots + c_r[R]_{rj} = 0.$$

But as R is in RRE form each of $[R]_{2j}, [R]_{3j}, \ldots, [R]_{rj}$ is zero, being entries under a leading 1. It follows that $c_1 = 0$. By focusing on the column which contains the leading 1 of $\mathbf{r}_2$ we can likewise show that $c_2 = 0$ and so on. As $c_i = 0$ for each i then the non-zero rows $\mathbf{r}_i$ are independent.

We shall prove (c) first and then (b) follows from it. Recall that

$$[BA]_{ij} = \sum_{s=1}^{m} [B]_{is}[A]_{sj} \qquad (1 \leqslant i \leqslant k, \, 1 \leqslant j \leqslant n).$$

Thus the ith row of BA is the row vector

$$\left(\sum_{s=1}^{m} [B]_{is}[A]_{s1}, \ldots, \sum_{s=1}^{m} [B]_{is}[A]_{sn} \right) = \sum_{s=1}^{m} [B]_{is} \underbrace{([A]_{s1}, [A]_{s2}, \cdots, [A]_{sn})}_{s\text{th row of } A}, \tag{3.43}$$

which is a linear combination of the rows of A. So every row of BA is in Row(A). A vector in the rowspace Row(BA) is a linear combination of BA's rows which, in turn, are linear combinations of A's rows. Hence, by #698, Row(BA) is contained in Row(A). Because $R = E_k \cdots E_1 A$ for some elementary matrices $E_1, E_2, \ldots, E_k$ then (b) follows from (c) with $B = E_k \cdots E_1$.

Now (d) also follows from (c). We know Row(BA) is contained in Row(A) and likewise Row$(A) = $ Row$(B^{-1}(BA))$ is contained in Row(BA). In fact, for (d), we only need B to have a left inverse and the hypothesis $k = m$ is unnecessary. Finally (e) follows from (d) by taking $B = E_k \cdots E_1$ which is invertible as elementary matrices are invertible. $\square$

Theorem 3.89 *(Existence and Uniqueness of RRE Form)*

(a) Every $m \times n$ matrix A can be reduced by EROs to a matrix in RRE form.

(b) The reduced row echelon form of an $m \times n$ matrix A is unique; we will denote this as RRE(A).

Proof (a) *Existence*: Note that a $1 \times n$ matrix is either zero or can be put into RRE form by dividing by its leading entry. Suppose, as our inductive hypothesis, that any matrix with fewer than m rows can be transformed with EROs into RRE form. Let A be an $m \times n$ matrix. If A is the zero matrix, then it is already in RRE form. Otherwise there is a first column $\mathbf{c}_j$ which contains a non-zero element α. With an ERO we can swap the row containing α with the first row and then divide the first row by $\alpha \neq 0$ so that the $(1, j)$th entry now equals 1. Our matrix now takes the form

$$\begin{pmatrix} 0 & \cdots & 0 & 1 & \tilde{a}_{1(j+1)} & \cdots & \tilde{a}_{1n} \\ 0 & \cdots & 0 & \tilde{a}_{2j} & \vdots & \vdots & \vdots \\ \vdots & \cdots & \vdots & \vdots & \vdots & \vdots & \vdots \\ 0 & \cdots & 0 & \tilde{a}_{mj} & \tilde{a}_{m(j+1)} & \cdots & \tilde{a}_{mn} \end{pmatrix},$$

for new entries $\tilde{a}_{1(j+1)}, \ldots, \tilde{a}_{mn}$. Applying $A_{12}(-\tilde{a}_{2j}), A_{13}(-\tilde{a}_{3j}), \ldots, A_{1m}(-\tilde{a}_{mj})$ consecutively leaves column $\mathbf{c}_j = \mathbf{e}_1^T$ so that our matrix has become

$$
\begin{pmatrix}
0 & \cdots & 0 & 1 & \tilde{a}_{1(j+1)} & \cdots & \tilde{a}_{1n} \\
0 & \cdots & 0 & 0 & & & \\
\vdots & \cdots & \vdots & \vdots & & B & \\
0 & \cdots & 0 & 0 & & &
\end{pmatrix}.
$$

By induction, the $(m-1) \times (n-j)$ matrix B can be put into some RRE form by means of EROs. Applying these same EROs to the bottom $m-1$ rows of the above matrix we would have reduced A to

$$
\begin{pmatrix}
0 & \cdots & 0 & 1 & \tilde{a}_{1(j+1)} & \cdots & \tilde{a}_{1n} \\
0 & \cdots & 0 & 0 & & & \\
\vdots & \cdots & \vdots & \vdots & & RRE(B) & \\
0 & \cdots & 0 & 0 & & &
\end{pmatrix}.
$$

To get the above matrix into RRE form we need to make zero any of $\tilde{a}_{1(j+1)}, \ldots, \tilde{a}_{1n}$ which are above a leading 1 in $RRE(B)$; if $\tilde{a}_{1k}$ is the first such entry to lie above a leading 1 in row l then $A_{1l}(-\tilde{a}_{1k})$ will make the required edit and in due course we will have transformed A into RRE form. Then (a) follows by induction.

(b) *Uniqueness*: The proof below follows by fixing the number of rows m and arguing by induction on the number of columns n. The only $m \times 1$ matrices which are in RRE form are $\mathbf{0}$ and $\mathbf{e}_1^T$. The zero $m \times 1$ matrix will reduce to the former and non-zero $m \times 1$ matrices to the latter. In particular, the RRE form of an $m \times 1$ matrix is unique.

Suppose, as our inductive hypothesis, any $m \times (n-1)$ matrix M has a unique reduced row echelon form $RRE(M)$. Let A be an $m \times n$ matrix and let $\tilde{A}$ denote the $m \times (n-1)$ matrix comprising the first $n-1$ columns of A. Given *any* EROs which reduce A to RRE form, these EROs also reduce $\tilde{A}$ to $RRE(\tilde{A})$ which is unique by hypothesis. Say $RRE(\tilde{A})$ has r non-zero rows.

There are two cases to consider: (i) any RRE form of A has one more non-zero row than $RRE(\tilde{A})$; (ii) any RRE form of A has the same number of non-zero rows as $RRE(\tilde{A})$. These can be the only cases as the first $n-1$ columns of an RRE form of A are those of $RRE(\tilde{A})$ and both matrices are in RRE form; note further that an extra non-zero row in any RRE form of A, if it exists, must equal $\mathbf{e}_n$. Case (i) occurs if $\mathbf{e}_n$ is in the rowspace of A and case (ii) if not. In particular, it is impossible that different sets of EROs might reduce a given A to both cases (i) and (ii).

So $RRE(A)$ has one of the following two forms:

$$
\text{(i)} \quad
\begin{pmatrix}
\boxed{\begin{matrix} \text{non-zero} \\ RRE(\tilde{A}) \\ \text{rows} \end{matrix}} & \begin{matrix} 0 \\ \vdots \\ 0 \end{matrix} \\
0 \ \cdots \ 0 & 1 \\
\multicolumn{2}{c}{m-r-1 \text{ zero rows}}
\end{pmatrix},
\quad
\text{(ii)} \quad
\begin{pmatrix}
\boxed{\begin{matrix} \text{non-zero} \\ RRE(\tilde{A}) \\ \text{rows} \end{matrix}} & \begin{matrix} * \\ \vdots \\ * \end{matrix} \\
\multicolumn{2}{c}{m-r \text{ zero rows}}
\end{pmatrix}
=
\begin{pmatrix}
\mathbf{r}_1(R) \\
\vdots \\
\mathbf{r}_r(R) \\
m-r \text{ zero rows}
\end{pmatrix}.
$$

In case (i) the last column of any RRE form of A is $\mathbf{e}_{r+1}^T$ and so we see that RRE(A) is uniquely determined, as we also know the first $n-1$ columns to be RRE($\tilde{A}$) by our inductive hypothesis. In case (ii), then any RRE form of A and RRE($\tilde{A}$) both have r non-zero rows. Let R_1 and R_2 be RRE forms of A. By hypothesis, their first $n-1$ columns agree and equal RRE($\tilde{A}$). By Proposition 3.88(e),

$$\text{Row}(R_1) = \text{Row}(A) = \text{Row}(R_2).$$

In particular, this means that the rows $\mathbf{r}_k(R_1)$ of R_1 are linear combinations of the rows $\mathbf{r}_k(R_2)$ of R_2. So, for any $1 \leqslant i \leqslant r$, there exist real numbers $\lambda_1,\ldots,\lambda_r$ such that

$$\mathbf{r}_i(R_1) = \sum_{k=1}^{r} \lambda_k \mathbf{r}_k(R_2) \qquad \text{and hence} \qquad \mathbf{r}_i(\text{RRE}(\tilde{A})) = \sum_{k=1}^{r} \lambda_k \mathbf{r}_k(\text{RRE}(\tilde{A}))$$

by focusing on the first $n-1$ columns. RRE($\tilde{A}$) is in RRE form and so its non-zero rows are independent; it follows that $\lambda_i = 1$ and $\lambda_j = 0$ for $j \neq i$. In particular $\mathbf{r}_i(R_1) = \mathbf{r}_i(R_2)$ for each i and hence $R_1 = R_2$ as required. $\qquad\square$

Corollary 3.90 *(Test for Independence) Let A be an $m \times n$ matrix. Then RRE(A) contains a zero row if and only if the rows of A are dependent.*

Proof We have that RRE(A) $= BA$ where B is a product of elementary matrices and so invertible. Say the ith row of BA is $\mathbf{0}$. By (3.43)

$$\mathbf{0} = \sum_{s=1}^{m} [B]_{is}(\text{sth row of A}).$$

Now $[B]_{is}$ are the entries of the ith row of B which, as B is invertible, cannot all be zero (#622). The above then shows the rows of A are linearly dependent.

Conversely suppose that the rows of A are linearly dependent. Let $\mathbf{r}_1, \mathbf{r}_2,\ldots, \mathbf{r}_m$ denote the rows of A and, without any loss of generality, assume that $\mathbf{r}_m = c_1\mathbf{r}_1 + \cdots + c_{m-1}\mathbf{r}_{m-1}$ for real numbers $c_1,\ldots,c_{m-1}$. By performing the EROs $A_{1m}(-c_1),\ldots,A_{(m-1)m}(-c_{m-1})$ we arrive at a matrix whose mth row is zero. We can continue to perform EROs on the top $m-1$ rows, leaving the bottom row untouched, until we arrive at a matrix in RRE form, whose bottom row is zero and which, by uniqueness, is RRE(A). $\qquad\square$

- The above is a useful test for the independence of a set of vectors. By setting the vectors as the rows of a matrix, and then reducing that matrix, we see that the vectors are dependent if and only if a zero row arises.

We may now define:

Definition 3.91 The **row rank,** or simply **rank,** of a matrix A is the number of non-zero rows in RRE(A). We write this as rank(A). The uniqueness of RRE(A) means row rank is well defined.

Corollary 3.92 *Let $(A|\mathbf{b})$ be the matrix representing the linear system $A\mathbf{x} = \mathbf{b}$. Then the system is consistent (i.e. has at least one solution) if and only if* rank($A|\mathbf{b}$) $=$ rank(A).

Proof Note this result was already demonstrated for systems in RRE form during the proof of Proposition 3.81. Say that $RRE(A) = PA$ where P is a product of elementary matrices that reduce A.

Now if E is an elementary matrix then $RRE(EA) = RRE(A)$ by the uniqueness of RRE form and so $rank(EA) = rank(A)$. We then have

$$\begin{aligned} rank(A|\mathbf{b}) = rank(A) &\iff rank(PA|P\mathbf{b}) = rank(PA) \\ &\iff \text{the system } PA\mathbf{x} = P\mathbf{b} \text{ is consistent} \\ &\iff \text{the system } A\mathbf{x} = \mathbf{b} \text{ is consistent} \end{aligned}$$

as the set of solutions to $A\mathbf{x} = \mathbf{b}$ is unaffected by EROs. $\qquad\square$

Proposition 3.93 *Let A be an $m \times n$ matrix and $\mathbf{b}$ in $\mathbb{R}_m$.*

(a) The system $A\mathbf{x} = \mathbf{b}$ has no solutions if and only if $\begin{pmatrix} 0 & 0 & \cdots & 0 & | & 1 \end{pmatrix}$ is in $row(A|\mathbf{b})$.
If the system $A\mathbf{x} = \mathbf{b}$ is consistent then:
(b) There is a unique solution if and only if $rank(A) = n$. It follows that $m \geqslant n$.
(c) There are infinitely many solutions if $rank(A) < n$. The set of solutions is an $n - rank(A)$ parameter family.

Proof As we know that $(A|\mathbf{b})$ can be put into RRE form, and that EROs affect neither the rowspace nor the set of solutions, the above is just a rephrasing of Proposition 3.81. $\qquad\square$

Remark 3.94 One might rightly guess that there is the equivalent notion of column rank. Namely the number of non-zero columns remaining when a matrix is similarly reduced using ECOs (elementary column operations, see #688). It is the case, in fact, that column rank and row rank are equal (Theorem 3.126). So we may refer to the rank of a matrix without ambiguity. $\qquad\blacksquare$

We conclude this section with the following results regarding the invertibility of matrices.

Proposition 3.95 *(**Criteria for Invertibility**) Let A be an $n \times n$ matrix. The following statements are equivalent:*

(a) A is invertible.
(b) A has a left inverse.
(c) $Row(A) = \mathbb{R}^n$.
(d) The only solution $\mathbf{x}$ in $\mathbb{R}_n$ to the system $A\mathbf{x} = \mathbf{0}$ is $\mathbf{x} = \mathbf{0}$.
(e) The columns of A are linearly independent.
(f) A has a right inverse.
(g) The rows of A are linearly independent.
(h) The row rank of A is n.
(i) $RRE(A) = I_n$.

#709
#712
#713 **Proof** It is left to #709 to show that (b), (c), (d), (e) are equivalent in general for $m \times n$ matrices, and that (f), (g), (h) are likewise equivalent. We shall write $(*)$ for the truth of any of (b), (c), (d), (e) and $(+)$ for the truth of (f), (g), (h) so that we effectively have four statements (a), $(*)$, $(+)$, (i) that we need to show are equivalent.

(a) $\implies$ ($+$) As A is invertible then its inverse is also a right inverse, thus proving (f).

($+$) $\implies$ (i) As the row rank of A is n, there are n leading 1s in RRE(A) and hence RRE(A) = I_n.

(i) $\implies$ ($*$) Suppose that RRE(A) = I_n. This means that there exist elementary matrices, say with overall effect E, such that $EA = I_n$, and hence (b) follows.

($*$) $\implies$ (a) Suppose that Row(A) = $\mathbb{R}^n$. As EROs do not affect rowspace, then Row(RRE(A)) = $\mathbb{R}^n$ and hence RRE(A) = I_n. (If the ith column of RRE(A) did not include a leading 1, this would mean that $\mathbf{e}_i$ was not in the rowspace.) Hence there are elementary matrices $E_1, E_2, \ldots, E_k$ such that $E_k E_{k-1} \cdots E_1 A = I_n$. As elementary matrices are invertible then so is A with $A^{-1} = E_k E_{k-1} \cdots E_1$. $\qquad\square$

Corollary 3.96 *Every invertible matrix can be written as a product of elementary matrices.*

Proof Let A be invertible. By Proposition 3.95(i) there are elementary matrices $E_1, \ldots, E_k$ such that $E_k \cdots E_1 A = I$. As elementary matrices have elementary inverses then we have $A = E_1^{-1} \cdots E_k^{-1}$ is a product of elementary matrices. $\qquad\square$

Corollary 3.97 *Let A, B be square matrices of the same size. If AB is invertible then A and B are invertible.*

Proof As AB is invertible there exists M such that $ABM = I$ and $MAB = I$. Then BM is a right inverse of A and MA is a left inverse of B. By Proposition 3.95(f) and (b) A and B are invertible. $\qquad\square$

Exercises

Exercises on Linear Systems

#646A Say $\mathbf{x}_0$ solves the linear system $A\mathbf{x} = \mathbf{b}$. Show every solution of $A\mathbf{x} = \mathbf{b}$ is of the form $\mathbf{x}_0 + \mathbf{y}$, where $A\mathbf{y} = \mathbf{0}$.

#647a Use row-reduction to find all solutions of the following linear systems:

(i) $\begin{aligned} x_1 + 2x_2 - 3x_3 &= 4; \\ 2x_1 + 5x_2 - 4x_3 &= 13; \end{aligned}$ $\qquad \begin{aligned} x_1 + 3x_2 + x_3 &= 11; \\ 2x_1 + 6x_2 + 2x_3 &= 22. \end{aligned}$

(ii) $2x_1 + x_2 - 2x_3 + 3x_4 = 1;$ $\quad 3x_1 + 2x_2 - x_3 + 2x_4 = 4;$ $\quad 3x_1 + 3x_2 + 3x_3 - 3x_4 = 5.$

#648A Find all solutions to the following linear systems:

(i) $\left(\begin{array}{cc|c} 1 & 2 & 2 \\ 2 & 4 & 2 \end{array} \right)$

(ii) $\left(\begin{array}{cccc|c} 1 & 0 & 2 & 1 & 3 \\ 2 & 1 & 2 & 4 & -1 \\ 0 & 2 & 3 & -1 & 2 \end{array} \right)$

(iii) $\left(\begin{array}{cccc|c} 2 & 2 & 1 & -1 & 4 \\ 2 & 1 & 3 & 2 & 1 \\ 2 & 3 & -1 & -4 & 7 \end{array} \right)$

#649 A Solve the following linear system, expressing the solutions in terms of parameters.

$$x_1 + 2x_2 + 4x_3 - 3x_4 + x_5 = 2; \quad 2x_1 + 4x_2 - 2x_3 - 3x_4 + x_5 = 5;$$
$$3x_1 + 6x_2 + x_3 - 2x_4 - x_5 = 5; \quad 2x_3 - x_4 + 3x_5 = 5.$$

#650 b† For what values of α do the simultaneous equations

$$x + 2y + \alpha^2 z = 0; \qquad x + \alpha y + z = 0; \qquad x + \alpha y + \alpha^2 z = 0$$

have a solution other than $x = y = z = 0$? Find, for each such α, the general solution of the above equations.

#651 B† For which values of a is the system below consistent? For each such a, find the general solution.

$$x - y + z - t = a^2; \qquad 2x + y + 5z + 4t = a; \qquad x + 2z + t = 4.$$

#652 c Find all values of a for which the following system has (i) one, (ii) no, (iii) infinitely many solutions.

$$x_1 + x_2 + x_3 = a; \qquad ax_1 + x_2 + 2x_3 = 2; \qquad x_1 + ax_2 + x_3 = 4.$$

#653 c For what values of α and β is the following system consistent? For such values, find the general solution.

$$x + z = 1; \qquad 2x + \alpha y + 4z = 1; \qquad -x - \alpha y + \alpha z = \beta.$$

#654 C Let a, b, c be real numbers. Show that

$$(x_1, x_2, x_3, x_4) = (1 + a - 3b + 4c, -5a + 5b - 6c, a + 2b - 3c, 10a - 5b + 5c)$$

is a solution of the three equations

$$x_1 + 2x_2 - x_3 + x_4 = 1; \qquad 2x_1 - x_2 + 3x_3 - x_4 = 2; \qquad x_1 - 8x_2 + 9x_3 - 5x_4 = 1.$$

Show that the solutions of the linear system can, in fact, be described using just two parameters instead of three.

#655 B† Find conditions which the real numbers α, β, γ must satisfy for the system $(A|\mathbf{b})$ to be consistent, where

$$A = \begin{pmatrix} 1 & 4 & 1 \\ 2 & 3 & 1 \\ 3 & 2 & 1 \\ 4 & 1 & 1 \end{pmatrix}, \qquad \mathbf{b} = \begin{pmatrix} 1 \\ \alpha \\ \beta \\ \gamma \end{pmatrix}, \qquad B = \begin{pmatrix} 1 & 1 & 1 \\ 2 & 3 & 4 \\ 3 & 5 & 7 \\ 4 & 7 & 10 \end{pmatrix},$$

and find the general solution when these conditions are met. Show that the equation $AX = B$ in the 3×3 matrix X is consistent. Find a singular 3×3 matrix X which solves the system $AX = B$.

#656 B† (i) Let A be an $m \times n$ matrix where $m < n$. Show that the system $(A \,|\, \mathbf{0})$ has infinitely many solutions.
(ii) Let A be an $m \times n$ matrix such that the system $A\mathbf{x} = \mathbf{b}$ is consistent for all choices of $\mathbf{b}$. Show that $m \leqslant n$.

#657c† Let

$$A_1 = \begin{pmatrix} 10 & 3 \\ 7 & 2 \end{pmatrix}, \quad B_1 = \begin{pmatrix} 2 & -1 \\ 0 & -3 \end{pmatrix}; \quad A_2 = \begin{pmatrix} 1 & 2 \\ 2 & 4 \end{pmatrix}, \quad B_2 = \begin{pmatrix} 5 & 3 \\ 10 & 6 \end{pmatrix}.$$

(i) Find the inverse of A_1 and hence solve the matrix equation $A_1 X = B_1$.
(ii) Rederive your answer to (i) by treating $A_1 X = B_1$ as four equations in the entries of X.
(iii) Show that A_2 is singular. Find all the matrices X which solve $A_2 X = B_2$.
(iv) Find a 2×2 matrix B_3 such that the equation $A_2 X = B_3$ has no solution X.

#658D† For what values of a and b is the matrix equation $AX = B$ consistent, where

$$A = \begin{pmatrix} 1 & 1 & a \\ 2 & 1 & -1 \\ 1 & 1 & 0 \end{pmatrix} \quad \text{and} \quad B = \begin{pmatrix} 1 & 0 & b \\ -1 & b & -1 \\ b & 0 & 1 \end{pmatrix} ?$$

In those cases where the system is consistent, find all solutions X.

#659D†(i) Show that $AD - BC = 0$, where

$$A = \begin{pmatrix} 8 & -5 \\ 20 & -13 \end{pmatrix}, \quad B = \begin{pmatrix} 1 & 2 \\ 3 & 4 \end{pmatrix}, \quad C = \begin{pmatrix} 4 & 3 \\ 2 & 1 \end{pmatrix}, \quad D = \begin{pmatrix} 1 & 0 \\ 0 & -1 \end{pmatrix}.$$

(ii) Show that the 4×4 matrix $M = \begin{pmatrix} A & B \\ C & D \end{pmatrix}$ is invertible, and find its inverse.

(iii) Deduce that, for any pair of vectors $\mathbf{e}, \mathbf{f}$ in $\mathbb{R}_2$, there are unique solutions $\mathbf{x}, \mathbf{y}$ in $\mathbb{R}_2$ such that

$$A\mathbf{x} + B\mathbf{y} = \mathbf{e}, \quad C\mathbf{x} + D\mathbf{y} = \mathbf{f}.$$

(iv) Why is this conclusion different from (3.24) where there is a unique solution (x, y) if and only if $ad - bc \neq 0$?

#660D (See also #938.) Let M be a fixed $n \times n$ matrix. Show that the matrix equation $MX = XM$ is equivalent to n^2 linear equations in the entries of X. Hence find all the matrices which commute with the matrix M to the right.
$$M = \begin{pmatrix} 1 & 2 & 1 \\ 0 & 2 & 2 \\ 0 & 0 & -1 \end{pmatrix}.$$

#661D† (See also #1000.) Let

$$A_1 = \begin{pmatrix} 2 & 1 \\ 1 & 2 \end{pmatrix}, \quad B_2 = \begin{pmatrix} 1 & 2 \\ 0 & 1 \end{pmatrix}; \quad A_2 = \begin{pmatrix} 2 & 1 \\ 1 & 2 \end{pmatrix}, \quad B_2 = \begin{pmatrix} 1 & 2 \\ 0 & -1 \end{pmatrix}.$$

Show that for any 2×2 matrix C_1, there is a unique 2×2 matrix X solving $A_1 X + X B_1 = C_1$. This is known as **Sylvester's equation.** [13] Give an example of a matrix C_2 such that the equation $A_2 X + X B_2 = C_2$ has no solution X.

---

[13] After James Joseph Sylvester (1814–1897). Sylvester made important contrubutions to the theory of matrices, and the word "matrix" first appears in an 1850 paper of his. He also in 1851 defined the discriminant of a cubic and other polynomials.

#662D (Bézout's [14] Lemma [15] over ℂ) (i) Let n be a positive integer and $p(x), q(x)$ be non-zero complex polynomials of degree n. Assuming the fundamental theorem of algebra, show that if there are non-zero complex polynomials $a(x), b(x)$ of degree less than n such that

$$a(x)p(x) + b(x)q(x) = 0,$$

then there exists a complex number α such that $p(\alpha) = q(\alpha) = 0$.

(ii) Let $P(x)$ and $Q(x)$ be non-zero complex polynomials which have no complex root in common. By treating $A(x)P(x) + B(x)Q(x) = 0$ as a linear system in the coefficients of $A(x)$ and $B(x)$, show that there exist complex polynomials $A(x), B(x)$ such that

$$A(x)P(x) + B(x)Q(x) = 1.$$

(iii) Let $P_1(x), P_2(x), P_3(x)$ be non-zero complex polynomials with no root common to all three polynomials. Show that there exist complex polynomials $A_1(x), A_2(x), A_3(x)$ such that

$$A_1(x)P_1(x) + A_2(x)P_2(x) + A_3(x)P_3(x) = 1.$$

#663D† (Bézout's Lemma over ℝ) (i) Show #662(ii) is false if we replace 'complex' with 'real' at each stage.

(ii) Let $P(x)$ and $Q(x)$ be non-zero real polynomials which have no real or complex root in common. Show there exist real polynomials $A(x), B(x)$ such that $A(x)P(x) + B(x)Q(x) = 1$.

Exercises on Inverses

#664A Determine whether the following matrices are invertible and find those inverses that do exist.

$$\begin{pmatrix} 2 & 3 \\ 3 & 4 \end{pmatrix}, \quad \begin{pmatrix} 2 & 1 & 3 \\ 1 & 0 & 2 \\ 4 & 5 & 4 \end{pmatrix}, \quad \begin{pmatrix} -1 & 1 & 1 \\ 2 & 0 & 1 \\ 1 & 3 & 5 \end{pmatrix}, \quad \begin{pmatrix} 1 & 1 & 2 \\ 1 & 2 & 1 \\ 2 & 1 & 1 \end{pmatrix}.$$

#665b Determine whether the following matrices are invertible and find those inverses that do exist.

$$\begin{pmatrix} 0 & 2 & 1 & 0 \\ 3 & 0 & 1 & -1 \\ 1 & 2 & 2 & 0 \\ 2 & 1 & 2 & 0 \end{pmatrix}, \quad \begin{pmatrix} 1 & 3 & -1 & 1 \\ 8 & -3 & -2 & 2 \\ 1 & 2 & 1 & 1 \\ 0 & 3 & 2 & 1 \end{pmatrix}, \quad \begin{pmatrix} 1 & 2 & 3 & 0 \\ 3 & 1 & 0 & 2 \\ 1 & 0 & 2 & 3 \\ 3 & 2 & 1 & 0 \end{pmatrix}.$$

---

[14] After the French mathematician Étienne Bézout (1730–1783). He is also known for *Bézout's theorem* in geometry: given two curves in $ℂ^2$ defined by polynomials $f(x, y) = 0$ and $g(x, y) = 0$ of degrees m and n, then the curves will have at most mn intersections. With care this can be improved to *exactly* mn intersections by counting the *multiplicities* of intersections and including so-called *points at infinity*.

[15] #662 and #663 concern Bézout's lemma for complex and real polynomials. The result holds more generally for coprime polynomials over other fields. That is, if the only polynomials to divide $P(x)$ and $Q(x)$ are the constant polynomials, then there are polynomials $A(x)$ and $B(x)$ such that $A(x)P(x) + B(x)Q(x) = 1$. This more general result requires a greater understanding of the algebra of polynomials than we have developed here.

#666 A† Calculate A^{-1} and express your answer as a product of elementary matrices

$$A = \begin{pmatrix} 1 & -3 & 0 \\ 1 & -2 & 4 \\ 2 & -5 & 4 \end{pmatrix},$$

#667 b† Solve $BX = C$.

$$B = \begin{pmatrix} 1 & 0 & 3 \\ 1 & 2 & 1 \\ 1 & 1 & 2 \end{pmatrix},$$

$$C = \begin{pmatrix} 6 & 4 & 5 \\ 0 & 2 & -1 \\ 3 & 3 & 2 \end{pmatrix}.$$

#668 C† Let A and B denote the matrices below. Find a left inverse C for A and show that if the 2×2 matrix X satisfies $AX = B$ then $X = CB$. Show that there are in fact no solutions to the equation $AX = B$. How do you resolve this seeming contradiction?

$$A = \begin{pmatrix} 2 & 1 \\ 1 & 2 \\ 3 & 1 \end{pmatrix}, \qquad B = \begin{pmatrix} 1 & 2 \\ 3 & -1 \\ 2 & 0 \end{pmatrix}.$$

#669 b For what values of x are the following matrices singular?

$$\begin{pmatrix} 1 & x & x^2 \\ x & x^2 & 1 \\ x^2 & 1 & x \end{pmatrix}, \qquad \begin{pmatrix} 1-x & 1 & 1 \\ 1 & 1-x & 1 \\ 1 & 1 & 1-x \end{pmatrix}, \qquad \begin{pmatrix} 1 & 3 & -1 & 0 \\ 0 & 1 & x & -2 \\ 1 & 2 & 2 & 1 \\ 1 & 0 & 1 & 2 \end{pmatrix}.$$

#670 C† Let A be the matrix in #638. Reduce $(A \,|\, I_3)$ and hence find a left inverse of A, i.e. a 2×3 matrix B such that $BA = I_2$. Without further calculation, write down all the left inverses of A.

#671 c Use the method of #670 to find all the left inverses of the matrix A below.

$$A = \begin{pmatrix} 1 & -2 & 4 \\ 3 & 2 & 0 \\ 1 & 0 & 1 \\ 0 & 2 & -3 \\ 1 & 1 & 1 \end{pmatrix}$$

#672 C† Let $A_1, \ldots, A_k$ be square matrices. Show that $\text{diag}(A_1, \ldots, A_k)$ is invertible if and only if each A_i is invertible.

#673 D† Show that a triangular matrix is invertible if and only if all its diagonal entries are non-zero. In this case show that the inverse is also triangular.

#674D† Determine the invertibility of the $n \times n$ matrix below.

$$\begin{pmatrix} 1 & 1 & 0 & 0 & \cdots & 0 \\ 1 & 1 & 1 & 0 & \ddots & \vdots \\ 0 & 1 & 1 & 1 & \ddots & 0 \\ \vdots & \ddots & \ddots & \ddots & \ddots & 0 \\ 0 & 0 & \ddots & 1 & 1 & 1 \\ 0 & 0 & \cdots & 0 & 1 & 1 \end{pmatrix}$$

#675D Show that the complex $n \times n$ matrix F, with entries given below, is invertible with inverse as given.

$$[F]_{ij} = n^{-1/2}\mathrm{cis}(2(i-1)(j-1)\pi/n), \qquad [F^{-1}]_{ij} = n^{-1/2}\mathrm{cis}(-2(i-1)(j-1)\pi/n).$$

Exercises on EROs

#676a Put the following matrices into RRE form.

$$\begin{pmatrix} 0 & 2 & 3 & -1 \\ 3 & 1 & 0 & 2 \end{pmatrix}, \qquad \begin{pmatrix} 1 & 0 & 3 \\ 2 & -1 & 0 \\ -1 & -1 & 3 \\ 3 & 1 & 1 \end{pmatrix},$$

$$\begin{pmatrix} 3 & -1 & 2 & -3 \\ -1 & 0 & -1 & 2 \\ 0 & 2 & 2 & 2 \end{pmatrix}, \qquad \begin{pmatrix} 0 & 2 & -3 & 2 & 7 \\ 0 & 4 & -6 & 1 & 2 \\ 0 & 0 & 0 & 3 & 12 \end{pmatrix}.$$

#677B(i) List the 2×2 matrices that are in RRE form.

(ii) List the 3×2 matrices that are in RRE form.
(iii) List the 2×3 matrices that are in RRE form.
(iv) For each such matrix in (i), explain which 2×2 matrices reduce to that matrix.

#678D† Put the following $n \times n$ matrices into RRE form.

$$\begin{pmatrix} 1 & 2 & \cdots & n \\ 2 & 3 & \cdots & n+1 \\ \vdots & \vdots & \cdots & \vdots \\ n & n+1 & \cdots & 2n-1 \end{pmatrix}, \qquad \begin{pmatrix} 1 & 4 & \cdots & n^2 \\ 4 & 9 & \cdots & (n+1)^2 \\ \vdots & \vdots & \cdots & \vdots \\ n^2 & (n+1)^2 & \cdots & (2n-1)^2 \end{pmatrix}.$$

#679D† Let $m \leqslant n$. Find an upper bound $P(m,n)$ for the number of arithmetic operations (additions, subtractions, multiplications, divisions) needed to row-reduce an $m \times n$ matrix and such that $P(n,n)/n^3$ approximates to some constant for large n. Hence determine an upper bound for the number of arithmetic operations needed to invert an $n \times n$ matrix using Algorithm 3.84.

#680c Find all the matrices that reduce to the following matrices:

$$\begin{pmatrix} 1 & 2 \\ 0 & 0 \\ 0 & 0 \end{pmatrix}, \qquad \begin{pmatrix} 1 & 0 & 0 & 0 \\ 0 & 1 & 0 & 0 \\ 0 & 0 & 0 & 1 \end{pmatrix}, \qquad \begin{pmatrix} 1 & 0 & 3 & 0 \\ 0 & 0 & 0 & 1 \end{pmatrix}.$$

#681c† Let v_1 and v_2 be two independent vectors in $\mathbb{R}^3$ and let A be the 2×3 matrix with rows v_1, v_2. Under what circumstances (geometrically) are the leading 1s of RRE(A) in the (i) first and second columns, (ii) first and third columns, (iii) second and third columns?

#682C† Prove Proposition 3.74, showing that premultiplication by each of the three matrices does indeed have the same effect as applying the corresponding ERO.

#683A† Show that $A_{ij}(\lambda)$ commutes with $A_{ik}(\mu)$, but that $A_{ij}(\lambda)$ and $A_{kl}(\mu)$ in general don't commute.

#684b† Under what conditions on i,j,k,l do S_{ij} and S_{kl} commute?

#685B Let i,j,k,l be distinct and λ, μ be non-zero real scalars. Evaluate the following products:

$$M_i(\lambda)M_i(\mu), \qquad A_{ij}(\lambda)A_{ij}(\mu), \qquad M_i(\mu^{-1})A_{ij}(\lambda)M_i(\mu), \qquad S_{ij}M_k(\lambda)S_{ij},$$
$$S_{ij}M_j(\lambda)S_{ij}, \qquad S_{ij}A_{kl}(\lambda)S_{ij}, \qquad S_{ij}A_{ik}(\lambda)S_{ij}, \qquad S_{ij}A_{ij}(\lambda)S_{ij}.$$

#686D† A matrix is said to be in **row echelon form** if the leading entry of a non-zero row, which need not be 1, appears to the right of the leading entries of the rows above it and any zero rows appear below the non-zero rows. Show that a matrix may be put into row echelon form by a sequence of EROs of the form S_{ij} followed by a sequence of EROs of the form $A_{ij}(\lambda)$ where $i < j$.

#687C Put the following matrices into row echelon form in the manner described in #686.

$$\begin{pmatrix} 0 & 0 & 2 & 1 \\ 1 & 2 & 0 & 1 \\ 2 & 4 & 2 & 3 \end{pmatrix}, \qquad \begin{pmatrix} 0 & 0 & 0 & 4 \\ 0 & 0 & 3 & 0 \\ 0 & 2 & 0 & 0 \\ 1 & 0 & 0 & 0 \end{pmatrix}, \qquad \begin{pmatrix} 0 & 2 & 4 & 6 & 2 \\ 0 & 1 & 2 & 3 & 1 \\ 1 & 1 & 1 & 1 & 1 \\ 0 & 2 & 4 & 6 & 3 \end{pmatrix}.$$

#688A Describe precisely the effect on a matrix of *postmultiplying* it by the elementary matrices $S_{IJ}, M_I(\lambda), A_{IJ}(\lambda)$. These then are the **elementary column operations** (ECOs), and a matrix can be similarly **column-reduced**.

#689c Column-reduce the following matrices, verifying in each case that the row rank and column rank are equal.

$$\begin{pmatrix} 1 & 0 & 2 & 2 \\ -1 & 2 & 3 & 1 \\ 1 & -1 & 3 & 2 \end{pmatrix}, \qquad \begin{pmatrix} 1 & 2 & 1 \\ 0 & -1 & 2 \\ 2 & 0 & 3 \\ 1 & 1 & 1 \end{pmatrix}, \qquad \begin{pmatrix} 2 & 0 & -3 & 1 \\ 1 & 2 & 3 & 1 \\ 3 & 2 & 0 & 2 \end{pmatrix}.$$

#690c† Let

$$A = \begin{pmatrix} 1 & 3 \\ 2 & 6 \end{pmatrix}, \qquad B = \begin{pmatrix} 1 & 7 \\ 2 & 3 \end{pmatrix}, \qquad C = \begin{pmatrix} 1 & 2 & 3 \\ 2 & 3 & 4 \\ 3 & 4 & 5 \end{pmatrix}, \qquad D = \begin{pmatrix} 1 & 0 & 2 \\ 1 & 2 & 3 \\ 1 & 6 & 2 \end{pmatrix}.$$

Use elementary column operations on the matrices $\left(\dfrac{A}{B} \right)$ and $\left(\dfrac{C}{D} \right)$ to reduce the lower matrices (B and D) to the identity. Explain why the top matrices become AB^{-1} and CD^{-1} respectively.

#691B† Let A be an $m \times n$ matrix with $r = \text{rank}(A)$. Show there is an invertible $m \times m$ matrix P and an invertible $n \times n$ matrix Q such that

$$PAQ = \begin{pmatrix} I_r & 0_{r(n-r)} \\ 0_{(m-r)r} & 0_{(m-r)(n-r)} \end{pmatrix}.$$

#692B† Find invertible matrices P and Q such that

$$PAQ = \begin{pmatrix} 1 & 0 & 0 & 0 \\ 0 & 1 & 0 & 0 \\ 0 & 0 & 0 & 0 \end{pmatrix} \qquad \text{where} \qquad A = \begin{pmatrix} 1 & 4 & -2 & 1 \\ -1 & -6 & 8 & -5 \\ 2 & 4 & 8 & -6 \end{pmatrix}.$$

#693c Find matrices P and Q, as described in #691, for each of the three matrices in #689.

#694D† Let A be an $m \times n$ matrix. Show that Row(A) is unaffected by ECOs if and only if rank(A) = n.

Exercises on Linear Independence

#695B† Let $v_1, v_2, \ldots, v_m$ be vectors in $\mathbb{R}^n$ where $m > n$. Show that $v_1, v_2, \ldots, v_m$ are linearly dependent.

#696A Show that vectors $v_1, \ldots, v_m$ in $\mathbb{R}^n$ are linearly independent if and only if none of the v_i can be written as a linear combination of the others.

#697A Let $v_1, \ldots, v_m$ be independent vectors in $\mathbb{R}^n$. Show that if

$$c_1 v_1 + \cdots + c_m v_m = d_1 v_1 + \cdots + d_m v_m$$

for real numbers c_i, d_i then $c_i = d_i$ for each i.

#698A Let $v_1, \ldots, v_k$ be vectors in $\mathbb{R}^n$. Let $w_1, \ldots, w_l$ be linear combinations of $v_1, \ldots, v_k$ and let $x_1, \ldots, x_m$ be linear combinations of $w_1, \ldots, w_l$. Show that $x_1, \ldots, x_m$ are linear combinations of $v_1, \ldots, v_k$.

#699c Let $v_1, \ldots, v_k$ be vectors in $\mathbb{R}^n$. Let $w_1, \ldots, w_l$ be linear combinations of $v_1, \ldots, v_k$ and let $x_1, \ldots, x_m$ be linear combinations of $v_1, \ldots, v_k$. Does it follow that $x_1, \ldots, x_m$ are linear combinations of $w_1, \ldots, w_l$?

#700 B(i) Show that the vectors $v_1 = (1,2,-1,0)$, $v_2 = (2,1,0,3)$, $v_3 = (0,1,1,1)$ in $\mathbb{R}^4$ are linearly independent by reducing the first system in (3.42).

(ii) By reducing the second system in (3.42), show that $x = (x_1,x_2,x_3,x_4)$ is a linear combination of v_1, v_2, v_3 if and only if $8x_1 + 6x_3 = x_2 + 5x_4$.

(iii) Show that v_1,v_2,v_3 from part (i) are linearly independent using the method of Corollary 3.90.

#701 A Let $a_0,a_1,\ldots,a_n$ be reals such that $a_0 + a_1 x + \cdots + a_n x^n = 0$ for all real x. Carefully show each $a_i = 0$.

#702 c† Under what conditions on the parameters a and b, are the following vectors dependent?

$$(1,2,a,-3,b), \qquad (1,3,2b,a,0), \qquad (2,0,3,b,2), \qquad (0,5,a-b,a-4,-1).$$

#703 B† Consider the system $(A \mid 0)$, where A is an $m \times n$ matrix in RRE form of rank $m < n$. In the proof of Proposition 3.81, it was shown that every solution could be written $x = s_1 v_1 + \cdots + s_{n-m} v_{n-m}$ for some v_i in $\mathbb{R}^n$ and parameters $s_1,s_2,\ldots,s_{n-m}$. Show that the vectors $v_1,v_2,\ldots,v_{n-m}$ are linearly independent.

#704 D† Let A be an $m \times n$ matrix with linearly independent columns (so $m \geqslant n$). Show that $A^T A$ is invertible. Is the converse true?

#705 C† Let A be an $n \times n$ matrix. Show that $I,A,A^2,\ldots,A^{n^2}$ is a linearly dependent subset of M_{nn}. Deduce that there exists a unique monic [16] least degree polynomial $m(x)$, called the **minimal polynomial**, such that $m(A) = 0$.

#706 c† Show that similar matrices have equal minimal polynomials.

#707 D Find the minimal polynomial for each of the following matrices:

(i) $A = \begin{pmatrix} 1 & 2 \\ 3 & 4 \end{pmatrix}$.   (ii) $B = \begin{pmatrix} 1 & 1 & 1 \\ 1 & 1 & 1 \\ 1 & 1 & 1 \end{pmatrix}$.   (iii) $C = \begin{pmatrix} 1 & 1 & 1 \\ 0 & 2 & 1 \\ 0 & 0 & 3 \end{pmatrix}$.

#708 D† Let A be an $n \times n$ upper triangular matrix with diagonal entries $\lambda_1,\lambda_2,\ldots\lambda_n$. Show that

$$(A - \lambda_1 I)(A - \lambda_2 I)\cdots(A - \lambda_n I) = 0.$$

Exercises on Rowspace and Rank

#709 B† Let A be an $m \times n$ matrix. Show that the following are equivalent:

(i) A has a left inverse.
(ii) The row space of A is $\mathbb{R}^n$.
(iii) The only solution x in $\mathbb{R}_n$ of $Ax = 0$ is $x = 0$.

---

[16] A polynomial is said to be *monic* if its leading coefficient is 1.

(iv) The columns of A are independent.

In particular, if any of (i)–(iv) apply, then $m \geqslant n$.

Show also that the following are equivalent:

(iv) The columns of A are independent.

(v) A has a right inverse.

(vi) The rows of A are independent.

(vii) The row rank of A is m.

In particular, if any of (v)–(vii) apply, then $m \leqslant n$.

#710B† Let $\mathbf{v}_1, \ldots, \mathbf{v}_k$ be k independent vectors in $\mathbb{R}^n$. Show that there exist a further $n - k$ vectors $\mathbf{v}_{k+1}, \ldots, \mathbf{v}_n$ such that $\mathbf{v}_1, \ldots, \mathbf{v}_n$ are independent.

#711D† Let A be an $m \times n$ matrix with $r = \mathrm{rank}(A)$. Show that there are r rows of A which are linearly independent. If $r < m$, show that any $r + 1$ rows of A are linearly dependent.

#712B† Let $\mathbf{v}$ and $\mathbf{v}_1, \ldots, \mathbf{v}_k$ be vectors in $\mathbb{R}^n$. Let A be the matrix with rows $\mathbf{v}_1, \ldots, \mathbf{v}_k$ and B be the matrix with rows $\mathbf{v}_1, \ldots, \mathbf{v}_k, \mathbf{v}$. Show that $\mathbf{v}$ is a linear combination of $\mathbf{v}_1, \ldots, \mathbf{v}_k$ if and only if $\mathrm{rank}(A) = \mathrm{rank}(B)$.

#713B† Let A and B be $m \times n$ matrices. Show that $\mathrm{RRE}(A) = \mathrm{RRE}(B)$ if and only if $A = PB$ for some invertible $m \times m$ matrix P.

#714D† Let A and B be $m \times n$ matrices. Show that $\mathrm{Row}(A) = \mathrm{Row}(B)$ if and only if $\mathrm{RRE}(A) = \mathrm{RRE}(B)$.

#715C† Let A and B be $m \times n$ matrices. Show that if $\mathrm{Row}(A + B) = \mathrm{Row}(A)$ then $\mathrm{Row}(B)$ is contained in $\mathrm{Row}(A)$. Is the converse true?

#716C† Let A be an $m_1 \times n$ matrix, B be an $m_2 \times n$ matrix with $m_1 < m_2$ and further assume $\mathrm{Row}(A) = \mathrm{Row}(B)$. Show that

$$\mathrm{RRE}(B) = \begin{pmatrix} \mathrm{RRE}(A) \\ 0_{(m_2 - m_1)n} \end{pmatrix}.$$

#717B† Let A and B be square matrices. Show that $\mathrm{rank}\,(\mathrm{diag}(A, B)) = \mathrm{rank}(A) + \mathrm{rank}(B)$.

#718D† Let A and B respectively be $m_1 \times n$ and $m_2 \times n$ matrices. Show that

$$\max\{\mathrm{rank}(A), \mathrm{rank}(B)\} \leqslant \mathrm{rank} \begin{pmatrix} A \\ B \end{pmatrix} \leqslant \mathrm{rank}(A) + \mathrm{rank}(B).$$

#719D Let A be an $m \times n$ matrix and B an $n \times m$ matrix where $m < n$. Show that BA cannot be invertible. Show that AB may be invertible in which case the ranks of A and B are both m.

#720c Let A be an $m \times n$ matrix and B an $n \times m$ matrix where $m < n$. If the ranks of A and B are both m, need it be the case that AB is invertible?

3.6 Dimension. Rank and Nullity*

In this section we shall make rigorous the notion of *dimension*. This is an intuitively clear notion in linear algebra: $\mathbb{R}$ is one-dimensional, the plane $x + y = z$ in $\mathbb{R}^3$ is two-dimensional,

$\mathbb{R}^n$ is n-dimensional, etc. From another perspective we shall see that dimension equates with the number of co-ordinates needed to parametrize a space. We first introduce some notation which will be of considerable use in this section.

Notation 3.98 Let $c_1,\ldots,c_n$ be n column vectors in $\mathbb{R}_m$. We will write $(c_1|\cdots|c_n)$ for the $m \times n$ matrix with columns $c_1,\ldots,c_n$. Let $r_1,\ldots,r_m$ be m row vectors in $\mathbb{R}^n$. We will write $(r_1/\cdots/r_m)$ for the $m \times n$ matrix with rows $r_1,\ldots,r_m$.

Recall that in the previous section we made two definitions (Definition 3.86(a) and (c)).

- A vector $\mathbf{x}$ is a *linear combination* of vectors $\mathbf{v}_1,\ldots,\mathbf{v}_k$ if $\mathbf{x} = c_1\mathbf{v}_1 + \cdots + c_k\mathbf{v}_k$ for some real numbers $c_1,\ldots,c_k$.
- The *rowspace* of a matrix is the set of linear combinations of its rows.

An important concept of linear algebra is that of a *span*.

Definition 3.99 The **span** of vectors $\mathbf{v}_1,\ldots,\mathbf{v}_k$ in $\mathbb{R}^n$ is the set of linear combinations of $\mathbf{v}_1,\ldots,\mathbf{v}_k$. Given any finite subset S of $\mathbb{R}^n$, we write $\langle S \rangle$ for its span. So

$$\langle \mathbf{v}_1,\ldots,\mathbf{v}_k \rangle = \{c_1\mathbf{v}_1 + \cdots + c_k\mathbf{v}_k : c_1,\ldots,c_k \in \mathbb{R}\}.$$

Remark 3.100 Though not of much relevance here, if $k = 0$ the 'span of no vectors' is defined to be the set containing just the zero vector $\mathbf{0}$. If S is infinite then we define $\langle S \rangle$ to be the set of finite linear combinations from S. ∎

Example 3.101 (a) The span of the zero vector $\mathbf{0}$ in $\mathbb{R}^3$ consists of $\mathbf{0}$ alone.
(b) If $\mathbf{v}$ is a non-zero vector in $\mathbb{R}^3$, then $\langle \mathbf{v} \rangle$ is the set of scalar multiples of $\mathbf{v}$, i.e. it is the line through $\mathbf{0}$ and $\mathbf{v}$.
(c) If $\mathbf{v},\mathbf{w}$ are independent vectors in $\mathbb{R}^3$, then $\langle \mathbf{v},\mathbf{w} \rangle$ is the plane containing the points $\mathbf{0}$, $\mathbf{v}$, $\mathbf{w}$. If $\mathbf{v},\mathbf{w}$ are dependent then $\langle \mathbf{v},\mathbf{w} \rangle$ is a line, unless $\mathbf{v} = \mathbf{w} = \mathbf{0}$.
(d) If $\mathbf{v},\mathbf{w},\mathbf{x}$ are independent vectors in $\mathbb{R}^3$ then $\langle \mathbf{v},\mathbf{w},\mathbf{x} \rangle = \mathbb{R}^3$. If $\mathbf{v},\mathbf{w},\mathbf{x}$ are dependent vectors then $\langle \mathbf{v},\mathbf{w},\mathbf{x} \rangle$ is a plane, line or the origin depending on whether the matrix $(\mathbf{v}/\mathbf{w}/\mathbf{x})$ has rank $2, 1$ or 0 respectively.
(e) The rowspace of a matrix is the span of its rows.

You may wish to compare the following with the method of Example 3.87(b) and #700.

Example 3.102 Under what circumstances is $\mathbf{x} = (x_1,x_2,x_3,x_4)$ in the span of

$$\mathbf{v}_1 = (1,1,1,-1), \qquad \mathbf{v}_2 = (2,0,1,1), \qquad \mathbf{v}_3 = (1,-3,-1,5)?$$

Solution From #712, $\mathbf{x}$ is in $\langle \mathbf{v}_1,\mathbf{v}_2,\mathbf{v}_3 \rangle$ if and only if the matrix $(\mathbf{v}_1/\mathbf{v}_2/\mathbf{v}_3/\mathbf{x})$ has the same rank as $(\mathbf{v}_1/\mathbf{v}_2/\mathbf{v}_3)$. If we reduce $(\mathbf{v}_1/\mathbf{v}_2/\mathbf{v}_3/\mathbf{x})$, we find

$$\begin{pmatrix} 1 & 1 & 1 & -1 \\ 2 & 0 & 1 & 1 \\ 1 & -3 & -1 & 5 \\ x_1 & x_2 & x_3 & x_4 \end{pmatrix} \xrightarrow[\;\substack{A_{12}(-2) \\ A_{13}(-1) \\ A_{14}(-x_1)}\;]{} \begin{pmatrix} 1 & 1 & 1 & -1 \\ 0 & -2 & -1 & 3 \\ 0 & -4 & -2 & 6 \\ 0 & x_2-x_1 & x_3-x_1 & x_4+x_1 \end{pmatrix}$$

$$A_{23}(-2) \quad \begin{pmatrix} 1 & 1 & 1 & -1 \\ 0 & -2 & -1 & 3 \\ 0 & 0 & 0 & 0 \\ 0 & x_2-x_1 & x_3-x_1 & x_4+x_1 \end{pmatrix}$$

$$\begin{matrix} S_{34} \\ M_2(-1/2) \end{matrix} \quad \begin{pmatrix} 1 & 1 & 1 & -1 \\ 0 & 1 & \frac{1}{2} & -1\frac{1}{2} \\ 0 & x_2-x_1 & x_3-x_1 & x_4+x_1 \\ 0 & 0 & 0 & 0 \end{pmatrix}$$

$$\begin{matrix} A_{21}(-1) \\ A_{23}(x_1-x_2) \end{matrix} \quad \begin{pmatrix} 1 & 0 & \frac{1}{2} & \frac{1}{2} \\ 0 & 1 & \frac{1}{2} & -1\frac{1}{2} \\ 0 & 0 & x_3-\frac{1}{2}x_1-\frac{1}{2}x_2 & x_4-\frac{1}{2}x_1+\frac{3}{2}x_2 \\ 0 & 0 & 0 & 0 \end{pmatrix}.$$

Here the zero row appears because the vectors v_1, v_2, v_3 are themselves dependent (note $2v_2 = 3v_1 + v_3$) and so the matrix $(v_1/v_2/v_3)$ has rank two. So x is dependent on v_1, v_2, v_3 if and only if the final matrix above also has rank two, i.e. if and only if the third row is also a zero row. So we need

$$x_3 - \frac{1}{2}x_1 - \frac{1}{2}x_2 = 0 = x_4 - \frac{1}{2}x_1 + \frac{3}{2}x_2,$$

or, rearranged, two conditions for x to be in the span of v_1, v_2, v_3 are $x_1 + x_2 = 2x_3$ and $x_1 = 3x_2 + 2x_4$. □

It is a consequence of #698 that spans are closed under addition and scalar multiplication – that is, for S a finite subset of $\mathbb{R}^n$, v, w in $\langle S \rangle$ and reals λ, μ, it follows that $\lambda v + \mu w$ is in $\langle S \rangle$. Put another way, spans are examples of *subspaces*.

Definition 3.103 A subset V of $\mathbb{R}^n$ is said to be a **subspace** when (a) 0 is in V, (b) if v, w are in V and λ, μ are reals then $\lambda v + \mu w$ is in V.

#721
#726
#727
#732
#737

- Note consequently that any linear combination of vectors from a subspace is also in that subspace.

Example 3.104 (See also #495.)

(a) The subspaces of $\mathbb{R}^2$ are the origin, lines through the origin and $\mathbb{R}^2$ itself.
(b) The subspaces of $\mathbb{R}^3$ are the origin, lines through the origin, planes containing the origin and $\mathbb{R}^3$ itself.

Example 3.105 The set of integers $\mathbb{Z}$ is not a subspace of $\mathbb{R}$, as it is not closed under scalar multiplication. The set of non-negative numbers is not a subspace either – it also is not closed under scalar multiplication as $(-1)1 < 0$, treating 1 here as a vector and -1 a scalar. In fact, the only subspaces of $\mathbb{R}$ are $\{0\}$ and $\mathbb{R}$ itself.

Example 3.106 The **hyperplane** $a \cdot x = c$ in $\mathbb{R}^n$, where $a \neq 0$, is a subspace if and only if $c = 0$ (see #726).

Example 3.107 We similarly say a subset V of M_{mn}, the set of $m \times n$ real matrices, is a subspace of M_{mn} if it contains the zero matrix and is closed under addition and scalar multiplication.

(a) The 2×2 symmetric matrices form a subspace of M_{22}. As any such matrix is of the form

$$\begin{pmatrix} a & b \\ b & c \end{pmatrix} = aE_{11} + b(E_{12} + E_{21}) + cE_{22},$$

then this subspace is the span of $\{E_{11}, E_{12} + E_{21}, E_{22}\}$.

(b) The upper triangular 2×2 matrices form a subspace of M_{22} spanned by E_{11}, E_{12}, E_{22}.

(c) The invertible 2×2 matrices do not form a subspace, as the set does not contain 0_{22}. Nor is it closed under addition, or under scalar multiplication.

An important characteristic of any subspace is its *dimension*. This is a very intuitive notion: two-dimensional subspaces are planes, one-dimensional subspaces are lines, $\mathbb{R}^3$ is three-dimensional and more generally the dimension of $\mathbb{R}^n$ is n. Loosely speaking, we will see that the dimension of a subspace is the number of co-ordinates needed to uniquely identify points within the subspace. We need first to show that the dimension of a subspace is well defined.

The next result is again reasonably intuitive geometrically – a line might be contained in a plane or a plane in $\mathbb{R}^3$ but a line should not be able to contain a plane.

Proposition 3.108 *(Steinitz* [17] *Exchange Lemma) Let* $\mathbf{v}_1, \ldots, \mathbf{v}_k$ *and* $\mathbf{w}_1, \ldots, \mathbf{w}_l$ *be two sets of linearly independent vectors in* $\mathbb{R}^n$ *such that* $\langle \mathbf{w}_1, \ldots, \mathbf{w}_l \rangle$ *is contained in* $\langle \mathbf{v}_1, \ldots, \mathbf{v}_k \rangle$. *Then* $l \leqslant k$.

Proof Let $A = (\mathbf{v}_1/\ldots/\mathbf{v}_k)$ and $B = (\mathbf{w}_1/\ldots/\mathbf{w}_l)$. Note that both matrices have n columns. Say

$$\text{RRE}(A) = (\tilde{\mathbf{v}}_1/\ldots/\tilde{\mathbf{v}}_k), \qquad \text{RRE}(B) = (\tilde{\mathbf{w}}_1/\ldots/\tilde{\mathbf{w}}_l).$$

Note, by the test for independence, that none of the rows $\tilde{\mathbf{v}}_i$ or $\tilde{\mathbf{w}}_i$ is zero. The rest of this proof relies on showing that those columns of $\text{RRE}(B)$ which contain a leading 1 are a subset of the columns of $\text{RRE}(A)$ which contain a leading 1. Then $\text{RRE}(B)$ has no more leading 1s than $\text{RRE}(A)$ does. But $\text{RRE}(B)$ has l leading 1s, $\text{RRE}(A)$ has k leading 1s and hence $l \leqslant k$.

By Proposition 3.88(e), $\text{Row}(\text{RRE}(B)) = \text{Row}(B)$ which we know to be contained in $\text{Row}(A) = \text{Row}(\text{RRE}(A))$. So for any row $\tilde{\mathbf{w}}_i$ there are reals $c_1, \ldots, c_k$, not all zero, such that

$$\tilde{\mathbf{w}}_i = c_1 \tilde{\mathbf{v}}_1 + c_2 \tilde{\mathbf{v}}_2 + \cdots + c_k \tilde{\mathbf{v}}_k. \tag{3.44}$$

---

[17] This result more usually states that if $\mathbf{w}_1, \ldots, \mathbf{w}_l$ are independent and $\mathbf{v}_1, \ldots, \mathbf{v}_k$ have span $\mathbb{R}^n$ then $l \leqslant k$. This follows from Proposition 3.108 as the $\mathbf{v}_i$ can be reduced to a set of $k' \leqslant k$ independent vectors with the same span and then $l \leqslant k' \leqslant k$. An alternative proof appears as #760 which more readily explains the lemma's name. The result is named after the German mathematician Ernst Steinitz (1871–1928), who made important contributions in graph theory (p.274) and field theory (p.160). In an influential 1910 paper he introduced many important aspects of field theory and proved that every field has an algebraic closure (in the same way that $\mathbb{C}$ is a field containing all the roots of polynomials with coefficients from the field $\mathbb{R}$).

If $c_1 \neq 0$ then the first non-zero entry of $\tilde{\mathbf{w}}_i$ is in the same column as the leading 1 of $\tilde{\mathbf{v}}_1$ (as the $\tilde{\mathbf{v}}_i$ are the rows of a matrix in RRE form). If $c_1 = 0 \neq c_2$ then the first non-zero entry of $\tilde{\mathbf{w}}_i$ is in the same column as the leading 1 of $\tilde{\mathbf{v}}_2$. Thus we see that the first non-zero entry of $\tilde{\mathbf{w}}_i$ (which is a leading 1) is in the same column as the leading 1 of some $\tilde{\mathbf{v}}_j$, completing the proof. $\square$

This leads to the following important definition.

Definition 3.109 Let V be a subspace of $\mathbb{R}^n$ such that $V = \langle \mathbf{v}_1, \ldots, \mathbf{v}_k \rangle$ for k independent vectors $\mathbf{v}_1, \ldots, \mathbf{v}_k$. Then we define the **dimension** of V to be k and write $\dim V = k$. Such a set of vectors $\mathbf{v}_1, \ldots, \mathbf{v}_k$ is called a **basis** [18] for V.

Remark 3.110 There are various potential points of ambiguity with this definition that need resolving.

(a) Firstly, note that if $\mathbf{v}_1, \ldots, \mathbf{v}_k$ and $\mathbf{w}_1, \ldots, \mathbf{w}_l$ are two different sets of independent vectors in $\mathbb{R}^n$ such that

$$\langle \mathbf{v}_1, \ldots, \mathbf{v}_k \rangle = \langle \mathbf{w}_1, \ldots, \mathbf{w}_l \rangle,$$

then $k \leqslant l$ as $\langle \mathbf{v}_1, \ldots, \mathbf{v}_k \rangle$ is contained in $\langle \mathbf{w}_1, \ldots, \mathbf{w}_l \rangle$ and similarly $l \leqslant k$. So $k = l$ and there is ultimately no ambiguity here: any two bases for a subspace contain the same number of elements, that being the dimension of the subspace.

(b) It is important to note that bases are not unique. We should be careful to refer to '*a* basis for a subspace' and not '*the* basis for a subspace', though we may find that some subspaces have standard bases to use.

(c) Finally we haven't yet demonstrated that a basis for a subspace need exist at all, and this is what we must focus on now. This will be resolved with Corollary 3.113. However, we can, for now, refer in a well-defined fashion to the dimension of a subspace that is known to have a basis.

(d) If $\mathbf{w}_1, \ldots, \mathbf{w}_l$ are dependent they can be reduced to independent vectors $\tilde{\mathbf{w}}_1, \ldots, \tilde{\mathbf{w}}_k$ (where $k < l$) using EROs. By Proposition 3.88(e), $\langle \mathbf{w}_1, \ldots, \mathbf{w}_l \rangle = \langle \tilde{\mathbf{w}}_1, \ldots, \tilde{\mathbf{w}}_k \rangle$ and so

$$\dim \langle \mathbf{w}_1, \ldots, \mathbf{w}_l \rangle = \dim \langle \tilde{\mathbf{w}}_1, \ldots, \tilde{\mathbf{w}}_k \rangle = k.$$

So we also know now that spans have well-defined dimensions.

(e) Note that, by definition, the rank of a matrix is the dimension of its rowspace. ∎

Example 3.111 Find $\dim V$, and a basis for V, where V is the subspace of $\mathbb{R}^5$ spanned by

$$(1, 2, -2, 3, 1), \ (0, 1, -1, 2, -1), \ (0, 1, -2, -1, -1), \ (0, 2, -1, 7, -2), \ (1, 0, -1, -4, 3).$$

#728
#729

Solution V is spanned by five vectors, but they may not be independent so we cannot simply conclude $\dim V = 5$. If we reduce the above five vectors with EROs, we will find

---

[18] The plural of basis is bases, pronounced 'bay-sees'.

independent vectors which still span V and so form a basis.

$$\begin{pmatrix} 1 & 2 & -2 & 3 & 1 \\ 0 & 1 & -1 & 2 & -1 \\ 0 & 1 & -2 & -1 & -1 \\ 0 & 2 & -1 & 7 & -2 \\ 1 & 0 & -1 & -4 & 3 \end{pmatrix} \xrightarrow[\substack{A_{34}(1) \\ A_{23}(-1) \\ A_{24}(-2)}]{A_{15}(-1)} \begin{pmatrix} 1 & 2 & -2 & 3 & 1 \\ 0 & 1 & -1 & 2 & -1 \\ 0 & 0 & -1 & -3 & 0 \\ 0 & 0 & 1 & 3 & 0 \\ 0 & 0 & 0 & 0 & 0 \end{pmatrix}$$

$$\xrightarrow[\substack{M_3(-1) \\ A_{21}(-2) \\ A_{32}(1)}]{A_{34}(1)} \begin{pmatrix} 1 & 0 & 0 & -1 & 3 \\ 0 & 1 & 0 & 5 & -1 \\ 0 & 0 & 1 & 3 & 0 \\ 0 & 0 & 0 & 0 & 0 \\ 0 & 0 & 0 & 0 & 0 \end{pmatrix}.$$

Hence, $\dim V = 3$ and a basis for V is $(1,0,0,-1,3)$, $(0,1,0,5,-1)$, $(0,0,1,3,0)$. $\qquad\square$

We finally address the matter raised in Remark 3.110(c). As was noted there, if a subspace V can be written as the span of finitely many vectors (whether or not those vectors are independent), then $\dim V$ is well defined. But the question remains as to whether all subspaces are spans.

Proposition 3.112 *Let* $\mathbf{v}_1,\ldots,\mathbf{v}_k,\mathbf{v}_{k+1}$ *be vectors in* $\mathbb{R}^n$.

(a) $\dim\langle\mathbf{v}_1,\ldots,\mathbf{v}_k,\mathbf{v}_{k+1}\rangle = \dim\langle\mathbf{v}_1,\ldots,\mathbf{v}_k\rangle$ *if and only if* $\mathbf{v}_{k+1}$ *is in* $\langle\mathbf{v}_1,\ldots,\mathbf{v}_k\rangle$.
(b) *If* $\mathbf{v}_{k+1}$ *is not in* $\langle\mathbf{v}_1,\ldots,\mathbf{v}_k\rangle$ *then* $\dim\langle\mathbf{v}_1,\ldots,\mathbf{v}_k,\mathbf{v}_{k+1}\rangle = \dim\langle\mathbf{v}_1,\ldots,\mathbf{v}_k\rangle + 1$.
(c) $\dim\langle\mathbf{v}_1,\ldots,\mathbf{v}_k\rangle \leqslant k$.
(d) $\dim\langle\mathbf{v}_1,\ldots,\mathbf{v}_k\rangle = k$ *if and only if* $\mathbf{v}_1,\ldots,\mathbf{v}_k$ *are independent.*
(e) $\dim\mathbb{R}^n = n$.

Proof (a) and (b): If $\mathbf{v}_{k+1}$ is in $\langle\mathbf{v}_1,\ldots,\mathbf{v}_k\rangle$ then $\langle\mathbf{v}_1,\ldots,\mathbf{v}_k,\mathbf{v}_{k+1}\rangle = \langle\mathbf{v}_1,\ldots,\mathbf{v}_k\rangle$ by #698 and so their dimensions agree. Suppose instead that $\mathbf{v}_{k+1}$ is not in $\langle\mathbf{v}_1,\ldots,\mathbf{v}_k\rangle$. Denote as $\tilde{\mathbf{v}}_1,\ldots,\tilde{\mathbf{v}}_l$ (where $l \leqslant k$) the independent vectors to which $\mathbf{v}_1,\ldots,\mathbf{v}_k$ reduce. Then $\mathbf{v}_{k+1}$ is not in $\langle\tilde{\mathbf{v}}_1,\ldots,\tilde{\mathbf{v}}_l\rangle = \langle\mathbf{v}_1,\ldots,\mathbf{v}_k\rangle$ so that $\tilde{\mathbf{v}}_1,\ldots,\tilde{\mathbf{v}}_l,\mathbf{v}_{k+1}$ are independent by #722. Hence by the definition of dimension we have

$$\dim\langle\mathbf{v}_1,\ldots,\mathbf{v}_k,\mathbf{v}_{k+1}\rangle = \dim\langle\tilde{\mathbf{v}}_1,\ldots,\tilde{\mathbf{v}}_l,\mathbf{v}_{k+1}\rangle = l+1 = \dim\langle\mathbf{v}_1,\ldots,\mathbf{v}_k\rangle + 1.$$

(c) From the previous parts, for any vectors $\mathbf{v}_1,\ldots,\mathbf{v}_k$, we have

$$\dim\langle\mathbf{v}_1,\ldots,\mathbf{v}_{k-1},\mathbf{v}_k\rangle \leqslant \dim\langle\mathbf{v}_1,\ldots,\mathbf{v}_{k-1}\rangle + 1 \leqslant \cdots \leqslant \dim\langle\mathbf{v}_1\rangle + (k-1) \leqslant k.$$

(d) If $\mathbf{v}_1,\ldots,\mathbf{v}_k$ are independent, $\dim\langle\mathbf{v}_1,\ldots,\mathbf{v}_k\rangle = k$ by definition. On the other hand, from part (c), we can see that $\dim\langle\mathbf{v}_1,\ldots,\mathbf{v}_k\rangle = k$ can occur only if

$$\dim\langle\mathbf{v}_1,\ldots,\mathbf{v}_{i-1},\mathbf{v}_i\rangle = \dim\langle\mathbf{v}_1,\ldots,\mathbf{v}_{i-1}\rangle + 1$$

at each stage $i = 1,\ldots,k$. That is to say, $\mathbf{v}_i$ is not in $\langle\mathbf{v}_1,\ldots,\mathbf{v}_{i-1}\rangle$ for $i = 1,\ldots,k$. But, by #723, this is the case if and only if $\mathbf{v}_1,\ldots,\mathbf{v}_k$ are independent.
(e) $\mathbf{e}_1,\mathbf{e}_2,\ldots,\mathbf{e}_n$ are independent and $\mathbb{R}^n = \langle\mathbf{e}_1,\mathbf{e}_2,\ldots,\mathbf{e}_n\rangle$. $\qquad\square$

Corollary 3.113 *(Subspaces have bases)* *A subspace* V *of* $\mathbb{R}^n$ *has a basis* $\mathbf{v}_1,\ldots,\mathbf{v}_k$ *where* $0 \leqslant k \leqslant n$.

Proof If V is just the origin $\mathbf{0}$ then we take $k = 0$ and choose 'no vectors' as a basis (see Remark 3.100). Otherwise, there is a non-zero vector $\mathbf{v}_1$ in V. If $V = \langle \mathbf{v}_1 \rangle$ then we are done as $\mathbf{v}_1$ forms a basis. Otherwise we can choose $\mathbf{v}_2$ in V which is not in $\langle \mathbf{v}_1 \rangle$, noting $\langle \mathbf{v}_1, \mathbf{v}_2 \rangle$ is contained in V as V is a subspace. If $V = \langle \mathbf{v}_1, \mathbf{v}_2 \rangle$ then we are done or alternatively we can choose $\mathbf{v}_3$ in V which is not in $\langle \mathbf{v}_1, \mathbf{v}_2 \rangle$. Note that this process produces independent vectors $\mathbf{v}_1, \mathbf{v}_2, \mathbf{v}_3, \ldots$ (#723). If this process terminates at some point then we have $\langle \mathbf{v}_1, \ldots, \mathbf{v}_k \rangle = V$ as required. However, as, at each stage, $\langle \mathbf{v}_1, \ldots, \mathbf{v}_i \rangle$ is a subset of $\mathbb{R}^n = \langle \mathbf{e}_1, \mathbf{e}_2, \ldots, \mathbf{e}_n \rangle$, then it must be that $i \leqslant n$ by the Steinitz exchange lemma. So the process must terminate at some $k \leqslant n$ at which stage a basis $\mathbf{v}_1, \ldots, \mathbf{v}_k$ has been produced. $\qquad \square$

Corollary 3.114 *Let U and V be subspaces of $\mathbb{R}^n$ with U contained in V.*

(a) $\dim U \leqslant \dim V$.
(b) If $U \neq V$ then $\dim U < \dim V$.
(c) If $\dim U = \dim V$ then $U = V$.

Proof (a) Let $\mathbf{u}_1, \ldots, \mathbf{u}_l$ be a basis for U and $\mathbf{v}_1, \ldots, \mathbf{v}_k$ be a basis for V, so that $l = \dim U$ and $k = \dim V$. As $\langle \mathbf{u}_1, \ldots \mathbf{u}_l \rangle = U$ is contained in $\langle \mathbf{v}_1, \ldots \mathbf{v}_k \rangle = V$ then by the Steinitz exchange lemma we have $\dim U = l \leqslant k = \dim V$.
(b) Now if $U \neq V$ then there exists $\mathbf{v}$ in V that is not in U and we have by (a) above and Proposition 3.112(b)

$$\dim U < \dim U + 1 = \dim \langle \mathbf{u}_1, \ldots, \mathbf{u}_l, \mathbf{v} \rangle \leqslant \dim V.$$

(c) This is equivalent to (b) as it is impossible to have $\dim U > \dim V$. $\qquad \square$

Corollary 3.115 *n vectors in $\mathbb{R}^n$ are linearly independent if and only if their span equals $\mathbb{R}^n$.*

Proof If $\mathbf{v}_1, \ldots, \mathbf{v}_n$ in $\mathbb{R}^n$ are independent then $\dim \langle \mathbf{v}_1, \ldots, \mathbf{v}_n \rangle = n$ and $\langle \mathbf{v}_1, \ldots, \mathbf{v}_n \rangle = \mathbb{R}^n$. If $\langle \mathbf{v}_1, \ldots, \mathbf{v}_n \rangle = \mathbb{R}^n$ then $\dim \langle \mathbf{v}_1, \ldots, \mathbf{v}_n \rangle = n$ and so $\mathbf{v}_1, \ldots, \mathbf{v}_n$ are independent by Proposition 3.112(d). $\qquad \square$

Corollary 3.116 *A linearly independent subset of a subspace can be extended to a basis.*

Proof If we have an independent subset $\mathbf{v}_1, \ldots, \mathbf{v}_i$ in a subspace V, then we may proceed as in the proof of Corollary 3.113 to extend it to a basis $\mathbf{v}_1, \ldots, \mathbf{v}_k$ for V. $\qquad \square$

Corollary 3.117 *(See also Corollary 3.128.) Let A be an $n \times p$ matrix and B be an $m \times n$ matrix. Then*

$$\mathrm{rank}(BA) \leqslant \mathrm{rank}(A).$$

#748
#749
#750
#755

Proof By Proposition 3.88(c), $\mathrm{Row}(BA)$ is contained in $\mathrm{Row}(A)$ and so

$$\mathrm{rank}(BA) = \dim \mathrm{Row}(BA) \leqslant \dim \mathrm{Row}(A) = \mathrm{rank}(A). \qquad \square$$

A subspace V of $\mathbb{R}^n$ can be described naturally in two different ways. We could give a basis $\mathbf{v}_1, \ldots, \mathbf{v}_k$ for V, where $k = \dim V$, or we might describe V as the solution space of $n - k$ homogeneous linear equations. If V is defined by homogeneous linear equations then we can find a basis by finding their general solution in terms of parameters. Conversely if a basis (or more generally a spanning set) is given for V we can find homogeneous linear equations determining the span V as done in Example 3.102.

Example 3.118 Determine $\dim W$, and find a basis for W, where W is the subspace of $\mathbb{R}^5$ given by the equations

$$x_1 + 2x_2 = 3x_3 + x_4; \quad x_2 + x_4 = x_1 + x_5; \quad x_1 + 5x_2 = 4x_3 + 2x_5; \quad x_3 + x_5 = x_1 + 2x_2.$$

Solution This question amounts to finding the general solution of the given equations. So, reducing the system,

$$\begin{pmatrix} 1 & 2 & -3 & -1 & 0 \\ -1 & 1 & 0 & 1 & -1 \\ 1 & 5 & -4 & 0 & -2 \\ -1 & -2 & 1 & 0 & 1 \end{pmatrix} \xrightarrow[\begin{subarray}{l} A_{12}(1) \\ A_{13}(-1) \\ A_{14}(1) \end{subarray}]{} \begin{pmatrix} 1 & 2 & -3 & -1 & 0 \\ 0 & 3 & -3 & 0 & -1 \\ 0 & 3 & -1 & 1 & -2 \\ 0 & 0 & -2 & -1 & 1 \end{pmatrix}$$

$$\xrightarrow[\begin{subarray}{l} A_{23}(-1) \\ A_{34}(1) \end{subarray}]{} \begin{pmatrix} 1 & 2 & -3 & -1 & 0 \\ 0 & 3 & -3 & 0 & -1 \\ 0 & 0 & 2 & 1 & -1 \\ 0 & 0 & 0 & 0 & 0 \end{pmatrix} \xrightarrow[\begin{subarray}{l} M_2(1/3) \\ M_3(1/2) \end{subarray}]{} \begin{pmatrix} 1 & 2 & -3 & -1 & 0 \\ 0 & 1 & -1 & 0 & -\frac{1}{3} \\ 0 & 0 & 1 & \frac{1}{2} & -\frac{1}{2} \\ 0 & 0 & 0 & 0 & 0 \end{pmatrix}$$

$$\xrightarrow[A_{21}(-2)]{} \begin{pmatrix} 1 & 0 & -1 & -1 & \frac{2}{3} \\ 0 & 1 & -1 & 0 & -\frac{1}{3} \\ 0 & 0 & 1 & \frac{1}{2} & -\frac{1}{2} \\ 0 & 0 & 0 & 0 & 0 \end{pmatrix} \xrightarrow[\begin{subarray}{l} A_{31}(1) \\ A_{32}(1) \end{subarray}]{} \begin{pmatrix} 1 & 0 & 0 & -\frac{1}{2} & \frac{1}{6} \\ 0 & 1 & 0 & \frac{1}{2} & -\frac{5}{6} \\ 0 & 0 & 1 & \frac{1}{2} & -\frac{1}{2} \\ 0 & 0 & 0 & 0 & 0 \end{pmatrix}.$$

Assigning parameters to $x_4 = s$ and $x_5 = t$ we see that the general solution is

$$\mathbf{x} = \left(\frac{s}{2} - \frac{t}{6}, -\frac{s}{2} + \frac{5t}{6}, -\frac{s}{2} + \frac{t}{2}, s, t \right) = \frac{s}{2}(1, -1, -1, 2, 0) + \frac{t}{6}(-1, 5, 3, 0, 6).$$

So $\dim W = 2$ and $\mathbf{v}_1 = (1, -1, -1, 2, 0)$, $\mathbf{v}_2 = (-1, 5, 3, 0, 6)$ form a basis for W. $\square$

Any subspace V can then be written as a span of independent vectors, a basis – moreover all V's bases contain the same number of elements, namely $\dim V$. As we see now, a basis can be used to assign co-ordinates to a subspace.

Definition 3.119 Let $\mathbf{v}_1, \ldots, \mathbf{v}_k$ be a basis for V and $\mathbf{x}$ any vector in V. Then, by #697,

there exist unique real numbers $X_1, \ldots, X_k$ such that $\mathbf{x} = X_1 \mathbf{v}_1 + \cdots + X_k \mathbf{v}_k$.

The numbers $X_1, \ldots, X_k$ are called the **co-ordinates** of $\mathbf{x}$ with respect to the basis $\mathbf{v}_1, \ldots, \mathbf{v}_k$.

In many ways, bases, linear independence and spans, are best thought of in terms of co-ordinates. Given a basis $\mathbf{v}_1, \ldots, \mathbf{v}_k$ of a subspace V then we can uniquely assign co-ordinates $X_1, \ldots, X_k$ to a given vector $\mathbf{x}$ in V. However, if the vectors had been independent but their span had not been all of V, then there would be certain $\mathbf{x}$ in V which could not be written as a linear combination of the $\mathbf{v}_i$ and so could not be assigned co-ordinates – it is the independence of the vectors that guarantees the uniqueness of co-ordinates, but not their existence. On the other hand, if the span of the $\mathbf{v}_i$ equalled V but the $\mathbf{v}_i$ were not independent, then it would be possible to assign different sets of co-ordinates to the same point. By way of examples,

- $v_1 = (1,0,1)$ and $v_2 = (1,1,0)$ are independent in $V = \mathbb{R}^3$. The vector $X_1v_1 + X_2v_2$ can uniquely be assigned the co-ordinates (X_1,X_2) but v_1 and v_2 span only the plane $x = y + z$ and not all of $\mathbb{R}^3$. We could not, for example, assign the point $(1,0,0)$ any X_1,X_2 co-ordinates with respect to just these two vectors.
- $v_1 = (1,1)$, $v_2 = (1,-1)$, $v_3 = (2,0)$ are such that $\langle v_1, v_2, v_3 \rangle = \mathbb{R}^2$ so that every point of the plane can be written as a linear combination of the vectors and so assigned co-ordinates. However, as

$$0v_1 + 0v_2 + 1v_3 = 1v_1 + 1v_2 + 0v_3,$$

we might, confusingly, assign v_3 either the co-ordinates $(X_1,X_2,X_3) = (0,0,1)$ or $(X_1,X_2,X_3) = (1,1,0)$.

Example 3.120 Let W be the subspace of $\mathbb{R}^5$ from Example 3.118. Verify it's the case that $v = (-1,17,9,6,24)$ is in W and find the co-ordinates of v with respect to the basis previously found.

Solution We found $v_1 = (1,-1,-1,2,0)$, $v_2 = (-1,5,3,0,6)$ form a basis for W. To see that $(-1,17,9,6,24)$ is in W we can verify that $(-1,17,9,6,24) = 3v_1 + 4v_2$ (where finding these co-ordinates is easiest done focusing on x_4 and x_5). So the given vector has co-ordinates $(X_1,X_2) = (3,4)$ with respect to our basis. □

Example 3.121 Verify that $v_1 = (2,1,5)$, $v_2 = (1,-1,1)$, $v_3 = (1,2,3)$ form a basis for $\mathbb{R}^3$ and find the co-ordinates of $(x,y,z) = x\mathbf{i} + y\mathbf{j} + z\mathbf{k}$ with respect to this basis.

Solution We aim to find co-ordinates X_1,X_2,X_3 such that $(x,y,z) = X_1v_1 + X_2v_2 + X_3v_3$ or equivalently to solve the linear system below in the variables X_1,X_2,X_3.

$$(Q\,|\,\mathbf{x}) = \begin{pmatrix} 2 & 1 & 1 & x \\ 1 & -1 & 2 & y \\ 5 & 1 & 3 & z \end{pmatrix}. \tag{3.45}$$

If v_1, v_2, v_3 aren't a basis then this will become apparent during reduction. Reducing this system we obtain

$$\begin{pmatrix} 2 & 1 & 1 & x \\ 1 & -1 & 2 & y \\ 5 & 1 & 3 & z \end{pmatrix} \xrightarrow[\substack{A_{12}(-2) \\ A_{13}(-5)}]{S_{12}} \begin{pmatrix} 1 & -1 & 2 & y \\ 0 & 3 & -3 & x-2y \\ 0 & 6 & -7 & z-5y \end{pmatrix}$$

$$\xrightarrow[M_2(1/3)]{A_{23}(-2)} \begin{pmatrix} 1 & -1 & 2 & y \\ 0 & 1 & -1 & (x-2y)/3 \\ 0 & 0 & -1 & z-2x-y \end{pmatrix} \xrightarrow[M_3(-1)]{A_{32}(-1)} \begin{pmatrix} 1 & -1 & 2 & y \\ 0 & 1 & 0 & (7x+y-3z)/3 \\ 0 & 0 & 1 & 2x+y-z \end{pmatrix}$$

$$\xrightarrow[A_{31}(-2)]{A_{21}(1)} \begin{pmatrix} 1 & 0 & 0 & (3z-5x-2y)/3 \\ 0 & 1 & 0 & (7x+y-3z)/3 \\ 0 & 0 & 1 & 2x+y-z \end{pmatrix}.$$

Hence

$$X_1 = \frac{1}{3}(-5x - 2y + 3z); \qquad X_2 = \frac{1}{3}(7x + y - 3z); \qquad X_3 = 2x + y - z,$$

or phrased as a single matrix equation

$$\begin{pmatrix} X_1 \\ X_2 \\ X_3 \end{pmatrix} = P \begin{pmatrix} x \\ y \\ z \end{pmatrix} \qquad \text{where} \qquad P = \frac{1}{3} \begin{pmatrix} -5 & -2 & 3 \\ 7 & 1 & -3 \\ 6 & 3 & -3 \end{pmatrix}.$$

Note $X_1 \mathbf{v}_1 + X_2 \mathbf{v}_2 + X_3 \mathbf{v}_3 = x\mathbf{i} + y\mathbf{j} + z\mathbf{k}$ are the same vector, representing the same point in $\mathbb{R}^3$. But the X_1-, X_2-, X_3-axes are in a different arrangement to the x-, y-, z-axes and so this same point has different co-ordinates according to the two different bases. The matrix P describes the **change of co-ordinates** from the x, y, z system to the X_1, X_2, X_3 system. The change of co-ordinates back, from X_1, X_2, X_3 to x, y, z, is given by Q as in (3.45). Note that the calculation at the heart of this example amounts to inverting Q to find its inverse P. $\square$

We conclude this section by introducing two important subspaces associated with matrices.

Definition 3.122 Let A be an $m \times n$ matrix. The **null space** (or **kernel**) of A, denoted Null(A), is the set of solutions $\mathbf{x}$ in $\mathbb{R}_n$ to the equation $A\mathbf{x} = \mathbf{0}$. The null space of A is a subspace of $\mathbb{R}_n$ (#761) and its dimension is called the **nullity** of A.

Definition 3.123 The **column space** (or **image**) of a matrix A, denoted Col(A), is the span of its columns and its dimension is called the **column rank** of A.

Example 3.124 Let A denote the first matrix in the solution of Example 3.111. From the reduction done there, we can find the general solution of $A\mathbf{x} = \mathbf{0}$ and so see that Null(A) consists of the vectors $(s - 3t, -5s + t, -3s, s, t)$ where s and t are real parameters assigned to the fourth and fifth columns. In particular, the nullity of A is 2. The span of the columns of A equals Col(A). One can readily check that the first three columns $\mathbf{c}_1, \mathbf{c}_2, \mathbf{c}_3$ of A are independent and that

$$\mathbf{c}_4 = -\mathbf{c}_1 + 5\mathbf{c}_2 + 3\mathbf{c}_3; \qquad \mathbf{c}_5 = 3\mathbf{c}_1 - \mathbf{c}_2,$$

showing that the column space equals $\langle \mathbf{c}_1, \mathbf{c}_2, \mathbf{c}_3 \rangle$ and has dimension 3.

Proposition 3.125 *Let A be a matrix. The system $A\mathbf{x} = \mathbf{b}$ is consistent if and only if $\mathbf{b}$ is in the column space of A. This is equivalent to saying that Col(A) is the image of μ_A.*

Proof Let $A = (\mathbf{c}_1 | \cdots | \mathbf{c}_n)$. To say that $A\mathbf{x} = \mathbf{b}$ is consistent is equivalent to saying that there is at least one solution $\mathbf{x} = (x_1, \ldots, x_n)^T$. In turn, this is equivalent to saying that there are real numbers x_i such that $x_1\mathbf{c}_1 + \cdots + x_n\mathbf{c}_n = \mathbf{b}$. And this is exactly what it means for $\mathbf{b}$ to be in the column space of A. $\square$

Theorem 3.126 *The column rank of a matrix equals its row rank.*

#761
#764
#766
#770
Proof (See #777, #778, #845 for alternative proofs.) We prove this by induction on the number of columns in the matrix. A non-zero $m \times 1$ matrix has column rank 1 and also row rank 1 as the matrix reduces to $\mathbf{e}_1^T$; the column rank and row rank of 0_{m1} are both 0.

So the $n = 1$ case is true. Suppose, as our inductive hypothesis, that column rank and row rank are equal for $m \times n$ matrices. Any $m \times (n + 1)$ matrix $(A|\mathbf{b})$ can be considered as an $m \times n$ matrix A alongside $\mathbf{b}$ in $\mathbb{R}_m$. If the system $(A|\mathbf{b})$ is consistent then

column rank of $(A|\mathbf{b})$ = column rank of A	[by Propositions 3.125 and 3.112(a)]
= row rank of A	[by inductive hypothesis]
= row rank of $(A|\mathbf{b})$	[see Remark 3.82].

On the other hand, if the system $(A|\mathbf{b})$ has no solutions then

column rank of $(A|\mathbf{b})$ = (column rank of A) + 1	[by Propositions 3.125 and 3.112(b)]
= (row rank of A) + 1	[by inductive hypothesis]
= row rank of $(A|\mathbf{b})$	[by Proposition 3.81(a) and Remark 3.82].

So if the system is consistent the row rank and column rank maintain their common value. If inconsistent, then $\mathbf{b}$ adds a further dimension to the column space and $(00\cdots0|1)$ adds an extra dimension to the row space. Either way the column rank and row rank of $(A|\mathbf{b})$ still agree and the proof follows by induction. $\qquad\square$

Corollary 3.127 *Let A be a matrix. Then* $\mathrm{rank}(A) = \mathrm{rank}(A^T)$.

Proof The rows of A^T are the columns of A. Hence

$$\mathrm{rank}(A^T) = \dim \mathrm{Row}(A^T) = \dim \mathrm{Col}(A) = \dim \mathrm{Row}(A) = \mathrm{rank}(A). \qquad\square$$

Corollary 3.128 *(See also Corollary 3.117.) Let A be an $n \times p$ matrix and B be an $m \times n$ matrix. Then*

$$\mathrm{rank}(BA) \leqslant \mathrm{rank}(B).$$

Proof By Corollaries 3.117 and 3.127, we have

$$\mathrm{rank}(BA) = \mathrm{rank}((BA)^T) = \mathrm{rank}(A^T B^T) \leqslant \mathrm{rank}(B^T) = \mathrm{rank}(B). \qquad\square$$

Finally we come to an important theorem relating the dimensions of the null space and column space. Given an $m \times n$ matrix A, recall that its nullity is the dimension of its null space and its rank is the dimension of its column space (or equally of its row space). We shall see that the sum of these dimensions is n.

Intuitively, though very loosely, it may help to consider the associated map μ_A from $\mathbb{R}_n$ to $\mathbb{R}_m$ given by $\mu_A(\mathbf{v}) = A\mathbf{v}$. The null space is the subspace of $\mathbb{R}_n$ mapped to $\mathbf{0}$ by μ_A and the column space of A,

$$\mathrm{Col}(A) = \{\mathbf{b} \in \mathbb{R}_m : A\mathbf{x} = \mathbf{b} \text{ is consistent}\} = \mu_A(\mathbb{R}_n),$$

is the *image* of the map μ_A by Proposition 3.125. Thus, in a rather loose sense, we begin with the n dimensions of the domain $\mathbb{R}_n$, the map μ_A collapses nullity(A) of these dimensions, and we are left with the rank(A) dimensions of the image $\mu_A(\mathbb{R}_n)$ within $\mathbb{R}_m$. A rigorous proof of this result is given below. Note that the proof itself does not make use of row rank and column rank being equal, but we do need that fact to appreciate rank as the dimension of the image.

Theorem 3.129 *(Rank-Nullity Theorem) Let A be an $m \times n$ matrix. Then*

$$\text{rank}(A) + \text{nullity}(A) = n.$$

#763
#774
#776

Proof By definition we have that $\text{rank}(A)$ is the number of non-zero rows in $\text{RRE}(A)$, which is equal to the number of leading 1s. We also know from Remark 3.82 and #703 that the dimension of the solution space $A\mathbf{x} = \mathbf{0}$ equals the number of columns in $\text{RRE}(A)$ without a leading 1. So

$n =$ number of columns of $\text{RRE}(A)$

   $=$ (number of columns of $\text{RRE}(A)$ with a leading 1)

      $+$(number of columns of $\text{RRE}(A)$ without a leading 1)

   $=$ (row rank of A) $+$ (dimension of null space of A)

   $=$ rank$(A) +$ nullity(A).   □

Exercises

Exercises on Spans and Subspaces

#721A Which of the following subsets of $\mathbb{R}^4$ are subspaces? Justify your answers.

$$\{\mathbf{x} \in \mathbb{R}^4 : x_1 + x_2 = x_3 + x_4\}, \qquad \{\mathbf{x} \in \mathbb{R}^4 : x_1 = 1\},$$
$$\{\mathbf{x} \in \mathbb{R}^4 : x_1 = x_3 = 0\}, \qquad \{\mathbf{x} \in \mathbb{R}^4 : x_1 = (x_2)^2\}.$$

#722A† Let $\mathbf{v}_1, \ldots, \mathbf{v}_k, \mathbf{v}$ be vectors in $\mathbb{R}^n$ with $\mathbf{v}_1, \ldots, \mathbf{v}_k$ independent. Show that $\mathbf{v}_1, \ldots, \mathbf{v}_k, \mathbf{v}$ are independent if and only $\mathbf{v}$ is not in $\langle \mathbf{v}_1, \ldots, \mathbf{v}_k \rangle$.

#723A Let $\mathbf{v}_1, \ldots, \mathbf{v}_k$ be k vectors in $\mathbb{R}^n$. Show that $\mathbf{v}_1, \ldots, \mathbf{v}_k$ are independent if and only $\mathbf{v}_i$ is not in $\langle \mathbf{v}_1, \ldots, \mathbf{v}_{i-1} \rangle$ for each i in the range $2 \leqslant i \leqslant k$.

#724B† Let $\mathbf{v}_1, \ldots, \mathbf{v}_k$ be k (not necessarily independent) vectors in $\mathbb{R}^n$ whose span is $\mathbb{R}^n$. Show that $\mathbf{v}_1, \ldots, \mathbf{v}_k$ contains a basis for $\mathbb{R}^n$.

#725c† Let A be an $m \times n$ matrix and B be an $n \times p$ matrix. Show that $\text{Col}(AB)$ is contained in $\text{Col}(A)$ with equality if B has a right inverse.

#726B† Show that the hyperplane $\mathbf{a} \cdot \mathbf{x} = c$ in $\mathbb{R}^n$ is a subspace if and only if $c = 0$. Show that if $c = 0$ then the hyperplane's dimension is $n - 1$.

#727B Let S be a finite subset of $\mathbb{R}^n$. Show that $\langle S \rangle$ is the smallest subspace of $\mathbb{R}^n$ which contains S; that is, if V is a subspace which contains S then V also contains $\langle S \rangle$.

#728b Let V be the subspace of $\mathbb{R}^5$ described in Example 3.111. Use the method of Example 3.102 to find two independent homogeneous linear conditions on $x_1, \ldots, x_5$ which determine when a vector $(x_1, \ldots, x_5)$ is in V.

#729b(i) Under what conditions is (x, y, z) in the span of $(1, 2, -1)$ and $(2, 1, 3)$?
(ii) Under what conditions is (x, y, z, t) in the span of $(1, 1, 3, -1)$, $(1, 0, 2, 1)$ and $(2, 3, -1, 1)$?
(iii) Under what conditions is (x, y, z, t) in the span of $(1, 1, 1, 0)$ and $(1, 0, 1, 1)$?
(iv) Under what conditions is (x, y, z, t) in the span of $(-1, 3, 2, 1)$, $(2, -5, 1, 1)$ and $(0, 1, 5, 3)$?

#730 A† Let $\mathbf{v}_1, \ldots, \mathbf{v}_k$ be k vectors in $\mathbb{R}^n$.

(i) Show, for $\alpha \neq 0$, that $\langle \mathbf{v}_1, \ldots, \mathbf{v}_{k-1}, \alpha \mathbf{v}_k \rangle = \langle \mathbf{v}_1, \ldots, \mathbf{v}_{k-1}, \mathbf{v}_k \rangle$.

(ii) Show, for any real β, that $\langle \mathbf{v}_1, \ldots, \mathbf{v}_{k-1}, \mathbf{v}_k + \beta \mathbf{v}_1 \rangle = \langle \mathbf{v}_1, \ldots, \mathbf{v}_{k-1}, \mathbf{v}_k \rangle$.

#731 B† (**Test for a Spanning Set**) Let $\mathbf{v}_1, \ldots, \mathbf{v}_k$ be vectors in $\mathbb{R}^n$. Show $\langle \mathbf{v}_1, \ldots, \mathbf{v}_k \rangle = \mathbb{R}^n$ if and only if

$$\mathrm{RRE}\,(\mathbf{v}_1/\mathbf{v}_2/\cdots/\mathbf{v}_k) = \left(\begin{array}{c} I_n \\ 0_{(k-n)n} \end{array} \right).$$

#732 B For which of the following sets S_i is it true that $\langle S_i \rangle = \mathbb{R}^4$?

$S_1 = \{(0,1,0,0),(0,0,1,0),(0,1,1,0),(3,0,0,4),(4,0,0,3)\}$.

$S_2 = \{(1,1,1,1),(0,0,1,1),(1,1,0,0)\}$.

$S_3 = \{(-2,-1,1,2),(-1,1,2,-2),(1,2,-2,-1),(2,-2,-1,1)\}$.

$S_4 = \{(0,1,2,3),(1,2,3,4),(2,3,4,5),(3,4,5,6)\}$.

$S_5 = \{(1,1,0,0),(1,0,1,0),(1,0,0,1),(0,1,1,0),(0,1,0,1),(0,0,1,1)\}$.

#733 B† For those sets S_i in #732 with span $\mathbb{R}^4$, find all the subsets of S_i that are a basis.

#734 C† Under what conditions is a 2×2 matrix in the span $V = \langle I_2, A, A^2, A^3, \ldots \rangle$ or a 3×3 matrix in the span $W = \langle I_3, B, B^2, B^3, \ldots \rangle$ where

$$A = \left(\begin{array}{cc} 1 & 3 \\ 2 & 1 \end{array} \right), \qquad B = \left(\begin{array}{ccc} 1 & -1 & 2 \\ 0 & 2 & -3 \\ 0 & 0 & 3 \end{array} \right)?$$

Note that this exercise is equivalent to ascertaining which matrices can be written as polynomials in A or B.

#735 C Let A, B be $n \times n$ matrices. Show that the following are subspaces of M_{nn}.

$$\{X \in M_{nn} : AX = 0\}, \qquad \{X \in M_{nn} : AX = XA\}, \qquad \{X \in M_{nn} : AX + XB = 0\}.$$

#736 C Let A be an $n \times n$ matrix and let $V_A = \{B \in M_{nn} : BA = AB\}$. That is, V_A is the set of $n \times n$ matrices which commute with A.

(i) Show V_A is, in fact, a subalgebra of M_{nn} – that is if X, Y are in V_A then XY is in V_A.

(ii) Let P be an invertible $n \times n$ matrix. Show that $V_{P^{-1}AP} = P^{-1}V_AP$.

(iii) Show that V_A is contained in V_{A^2} but need not be equal to it.

#737 B(i) Show that, if X and Y are subspaces of $\mathbb{R}^n$, then $X + Y$ and $X \cap Y$ are subspaces where

$$X + Y = \{\mathbf{x} + \mathbf{y} : \mathbf{x} \in X, \mathbf{y} \in Y\} \qquad \text{and} \qquad X \cap Y = \{\mathbf{z} : \mathbf{z} \in X \text{ and } \mathbf{z} \in Y\}.$$

(ii) Show $X + Y$ is the smallest subspace containing both X and Y – that is, if W is a subspace of $\mathbb{R}^n$ containing X and Y, then W contains $X + Y$. Show $X \cap Y$ is the largest subspace contained in both X and Y.

#738 C† Let H_1 and H_2 be two distinct hyperplanes in $\mathbb{R}^n$. What are the dimensions of $H_1 \cap H_2$ and $H_1 + H_2$?

#739 b† Show that

$X = \{$set of symmetric $n \times n$ matrices$\}$, $Y = \{$set of skew-symmetric $n \times n$ matrices$\}$,

are subspaces of M_{nn}. Determine $X + Y$ and $X \cap Y$.

#740 B† Find conditions for (x, y, z, t) to be in $X + Y$ and $X \cap Y$ where X and Y are the following subspaces of $\mathbb{R}^4$.

$$X = \langle (1, 2, 1, 0), (2, 3, 0, -1) \rangle, \qquad Y = \langle (1, 1, 1, 1), (1, 1, -1, -1) \rangle.$$

#741 c Show that $X + Y = \mathbb{R}^4$ and find a basis for $X \cap Y$ where X and Y are the following subspaces of $\mathbb{R}^4$:

$$X = \{(x_1, x_2, x_3, x_4) : x_1 + x_2 = x_3 + x_4\}, \qquad Y = \{(x_1, x_2, x_3, x_4) : 2x_1 + 3x_2 = 3x_3 + 2x_4\}.$$

#742 C Let S and T be finite subsets of $\mathbb{R}^n$. Show that $\langle S \cup T \rangle = \langle S \rangle + \langle T \rangle$. Either prove or disprove (with a counter-example) that $\langle S \cap T \rangle = \langle S \rangle \cap \langle T \rangle$.

#743 c† Let X and Y be subsets of $\mathbb{R}^n$ such that $X + Y$ and $X \cap Y$ are both subspaces. Need X and Y be subspaces?

Definition 3.130 Given an $n \times n$ real (or complex) matrix A, we say that a subspace V of $\mathbb{R}_n$ (or $\mathbb{C}_n$) is **invariant**, or A**-invariant**, if whenever $\mathbf{v}$ is in V then $A\mathbf{v}$ is in $\mathbf{v}$.

#744 B† (i) Find all the invariant subspaces of $\mathrm{diag}(1, 2, 3)$ and $\mathrm{diag}(1, 1, 2)$.
(ii) If $B = P^{-1}AP$, how do the B-invariant subspaces relate to the A-invariant subspaces?

#745 D† (i) Let A be an upper triangular $n \times n$ matrix. Show that there are A-invariant subspaces W_i of $\mathbb{R}_n$, for $i = 1, \ldots, n$, such that $\dim W_i = i$ and W_i is contained in W_{i+1} for $1 \leqslant i < n$.
(ii) Let B be an $n \times n$ matrix with B-invariant subspaces V_i of $\mathbb{R}_n$, for $i = 1, \ldots, n$, such that $\dim V_i = i$ and V_i is contained in V_{i+1} for $1 \leqslant i < n$. Show that there is an invertible matrix P such that $P^{-1}BP$ is upper triangular.

#746 D† Let C be an $n \times n$ matrix such that $C^n = 0$. Show that there is an invertible matrix Q such that $Q^{-1}CQ$ is strictly upper triangular. (The converse is also true – see #585.)

Exercises on Bases and Dimension

#747 a What is $\dim M_{mn}$?

#748 A† Show that $\mathrm{rank}(AB) = \mathrm{rank}(A)$ where A is an $m \times n$ matrix and B an invertible $n \times n$ matrix.

#749 A Find a basis for $\langle (1, 1, 2, 1), (2, 4, 4, 1), (0, 2, 0, -1) \rangle$.

#750 a† Find a basis for $\{(x_1, x_2, x_3, x_4) : x_1 - x_2 = x_3 - x_4 = x_2 + x_3\}$.

#751 a Show $\{(1, 1, 1), (1, 1, 0), (1, 0, 0)\}$ is a basis for $\mathbb{R}^3$ and find the co-ordinates of (x, y, z) with respect to it.

#752B Show that $\{(1,0,1,0),(0,1,1,0),(0,0,1,1)\}$ is a basis for the hyperplane H with equation $x_1+x_2+x_4=x_3$ in $\mathbb{R}^4$. Find the co-ordinates X_1,X_2,X_3 for $(2,3,6,1)$ with respect to this basis. Find the equation in X_1,X_2,X_3 for the intersection of H with the hyperplane $x_1+x_3=x_2+x_4$.

#753b Find a basis for $\{(x_1,x_2,x_3,x_4): x_1-x_2=x_2-x_3=x_3-x_4\}$. Find the co-ordinates of $(4,3,2,1)$ with respect to your basis.

#754D† Let A,B,V,W be as in #734, and n a natural number.

(i) Show that $\{I_2,A\}$ is a basis for V and write down the co-ordinates of A^n in terms of this basis.

(ii) Show that $\{I_3,B,B^2\}$ is a basis for W and write down the co-ordinates of B^n in terms of this basis.

#755b† What is the dimension of the subspace of (i) symmetric $n \times n$ matrices? (ii) skew-symmetric $n \times n$ matrices? (iii) $n \times n$ matrices with trace equal to zero?

#756D† Find a basis for $\mathbb{R}^4$ which includes bases for each of X and Y where

$$X = \{(x_1,x_2,x_3,x_4): x_1+2x_2+x_4=0=x_3+2x_4\}$$
$$\text{and} \qquad Y = \langle(1,0,1,0),(0,0,1,-1),(1,0,0,1)\rangle.$$

#757D† (From Oxford 2008 Paper A #1) Let B be the $n \times n$ matrix on the right.

(i) For each integer k in the range $1 \leqslant k \leqslant n$, show that
$$\dim\langle e_k^T, Be_k^T, B^2e_k^T, B^3e_k^T, \ldots, B^n e_k^T\rangle = k.$$

(ii) Suppose that $v_1,v_2,\ldots,v_n$ are vectors in $\mathbb{R}_n$ such that
$$\dim\langle v_k, Bv_k, B^2v_k, B^3v_k, \ldots, B^n v_k\rangle = k \quad \text{for} \quad 1\leqslant k\leqslant n.$$
Show that $v_1,v_2,\ldots,v_n$ is a basis for $\mathbb{R}_n$.

$$B= \begin{pmatrix} 0 & 1 & 0 & \cdots & 0 \\ 0 & 0 & 1 & \ddots & \vdots \\ 0 & 0 & 0 & \ddots & 0 \\ \vdots & \vdots & \vdots & \ddots & 1 \\ 0 & 0 & 0 & \cdots & 0 \end{pmatrix}$$

#758C† If we identify $ax^3 + bx^2 + cx + d$ with (a,b,c,d), this identifies the space V of polynomials in x of degree three or less with $\mathbb{R}^4$. Show that the following are subspaces of V and find a basis for each subspace and its dimension.

$$W_1 = \{p(x) \in V : p(1)=0\}; \qquad W_2 = \{p(x) \in V : p(1)=p(2)=0\};$$
$$W_3 = \{p(x) \in V : p(1)=p'(1)=0\}; \qquad W_4 = \{p(x) \in V : xp'(x)=2p(x)\}.$$

#759c† Show that the polynomials $(x-1)^k$, where $0\leqslant k\leqslant 3$, form a basis for V (as defined in #758). Find the co-ordinates for each of $1,x,x^2,x^3$ with respect to the given basis.

#760D† (**Alternative proof of Steinitz exchange lemma**) Let $v_1, \ldots, v_k$ and $w_1, \ldots, w_l$ be two sets of vectors in $\mathbb{R}^n$ such that the v_i span $\mathbb{R}^n$ and that the w_j are independent.

(i) Show there is an i in the range $1 \leqslant i \leqslant k$ such that $\langle v_1, \ldots, v_{i-1}, w_1, v_{i+1}, \ldots, v_k \rangle = \mathbb{R}^n$.
(ii) Show how this process can be repeated, successively exchanging the next w_r for some v_s to produce a new set whose span is still $\mathbb{R}^n$. Deduce that $l \leqslant k$.

Exercises on Rank and Nullity

#761A Show that the null space of a matrix is a subspace.

#762B† Use Corollary 3.117 to show that a matrix with both left and right inverses is square.

#763b† For each of the following matrices, find a basis for the null space and also for the column space. Verify the rank-nullity theorem in each case.

$$\begin{pmatrix} 1 & 2 & 3 \\ 2 & 3 & 4 \\ 3 & 4 & 5 \end{pmatrix}, \quad \begin{pmatrix} 2 & 3 & 1 & 1 \\ 0 & 1 & 2 & 5 \\ 1 & 3 & 2 & 1 \end{pmatrix}, \quad \begin{pmatrix} 2 & 2 & 0 \\ 3 & 2 & 3 \\ 1 & 1 & 2 \\ 0 & 2 & 2 \end{pmatrix}, \quad \begin{pmatrix} 2 & 2 & 0 \\ 1 & 2 & 1 \\ 1 & 3 & 2 \\ 0 & 3 & 3 \end{pmatrix}.$$

#764b For each of the following matrices, find a basis for the row space and column space. Verify in each case that the row rank and column rank are equal.

$$\begin{pmatrix} 2 & 2 \\ 1 & 3 \\ 3 & 0 \end{pmatrix}, \quad \begin{pmatrix} 1 & 0 & 2 & 6 \\ 0 & 1 & 3 & 2 \\ 1 & 2 & 2 & 2 \end{pmatrix}, \quad \begin{pmatrix} 1 & 0 & 1 \\ 0 & 1 & 2 \\ 2 & 3 & 2 \\ 6 & 2 & 2 \end{pmatrix}, \quad \begin{pmatrix} 1 & 1 & 0 & 2 & 2 \\ 1 & -1 & 2 & 2 & 3 \\ 2 & 0 & 2 & 4 & 5 \end{pmatrix}.$$

#765B Let A and B be $m \times n$ matrices. Show that $\operatorname{rank}(A + B) \leqslant \operatorname{rank}(A) + \operatorname{rank}(B)$. Give two examples, one with equality, one strict.

#766b† (Oxford 1995 Paper 1 #2) (i) Let A denote the matrix on the right. Consider the linear system $Ax = y$ for x in $\mathbb{R}_4$ and y in $\mathbb{R}_3$. Find the conditions y must satisfy for this system to be consistent, and find the general solution x when these conditions hold.

$$\begin{pmatrix} 1 & 1 & 1 & 0 \\ 1 & 2 & 3 & 1 \\ 1 & 0 & -1 & -1 \end{pmatrix}$$

(ii) Describe Col(A) in two ways: as the set of solutions of a system of independent homogeneous linear equations; by giving a basis. Do the same for Null(A).

#767B† Let A be an $m \times n$ matrix. Show that $\operatorname{rank}(A^T A) = \operatorname{rank}(A)$ and deduce that $\operatorname{nullity}(A^T A) = \operatorname{nullity}(A)$.

#768D† Let A be an $m \times n$ matrix of rank r. Show $V = \left\{ X \in M_{np} : AX = 0 \right\}$ is a subspace of M_{np} and find $\dim V$.

#769d† Let A be an $n \times n$ matrix such that $A^n = 0 \neq A^{n-1}$. Show that $\operatorname{rank}(A) = n - 1$.

#770C Given matrices A, B of appropriate sizes, show that $AB = 0$ if and only if $\text{Col}(B)$ is contained in $\text{Null}(A)$.

#771B† Let P be a projection matrix, so that $P^2 = P$. Show that $\text{Col}(P) = \text{Null}(P - I)$.

#772b† Suppose that A is an $n \times n$ matrix which is singular. Show that there are *non-zero* $n \times n$ matrices P and Q such that $AP = 0$ and $QA = 0$.

#773D† Let A and B be $m \times n$ and $n \times p$ matrices. Show that

$$\text{rank}(AB) \geqslant \text{rank}(A) + \text{rank}(B) - n.$$

Give an example where equality holds.

#774b† (From Oxford 2008 Paper A #3) Let A denote the matrix on the right.

(i) Find a basis for $\text{Null}(A)$. Find the rank of A and verify the rank-nullity theorem for A.

(ii) Show that there does not exist a 4×3 matrix B such that $BA = I_4$.

(iii) Find all 4×3 matrices C such that $AC = I_3$.

$$\begin{pmatrix} 1 & 0 & 2 & 1 \\ 2 & 3 & 7 & 0 \\ 1 & 0 & 3 & 2 \end{pmatrix}$$

#775B Let A, B, C be $m \times n$ matrices. We say that A and B are **equivalent** if there is an invertible $m \times m$ matrix P and an invertible $n \times n$ matrix Q such that $PAQ = B$. Show that [19]
(i) A is equivalent to A; (ii) if A is equivalent to B then B is equivalent to A; (iii) if A is equivalent to B and B is equivalent to C then A is equivalent to C.
(iv) Show that A and B are equivalent if and only if they have the same rank.

#776b Let A denote the matrix in #766. Find the row rank r of A, and write A as the product of a $3 \times r$ matrix and an $r \times 4$ matrix.

#777D† Let $P = QR$ where P, Q, R are respectively $k \times l$, $k \times m$, $m \times l$ real matrices.

(i) Show that the row rank of P is at most m.

(ii) Let p be the row rank of P. Show that P may be written as the product of a $k \times p$ matrix and a $p \times l$ matrix.

(iii) Deduce that the row rank and column rank of a matrix are equal.

#778D† Show that the row rank of a matrix is not changed by ECOs and deduce that the column rank of a matrix is not changed by EROs. Deduce from #691 that the row rank and column rank of a matrix are equal.

3.7 Matrices as Maps

§3.5 focused on the use of matrices to solve linear systems of equations. But an equally important aspect of matrices is their use describing *linear maps*. Further, thinking of

---

[19] (i), (ii) and (iii) show that equivalence of matrices is an *equivalence relation*.

matrices as maps best explains why matrix multiplication is defined as it is (and not, for example, by multiplying corresponding entries).

Given a *column* vector $\mathbf{v}$ in $\mathbb{R}_n$ and an $m \times n$ matrix A then $A\mathbf{v}$ is in $\mathbb{R}_m$. So we make the following definition. [20]

Definition 3.131 To any $m \times n$ matrix A we can associate a map μ_A from $\mathbb{R}_n$ to $\mathbb{R}_m$ which sends $\mathbf{v}$ to $A\mathbf{v}$. This map is known as the **associated map** of A or the **map represented by** A.

Example 3.132 Let

$$A = \begin{pmatrix} a & b \\ c & d \end{pmatrix} \qquad \text{so that} \qquad A\begin{pmatrix} x \\ y \end{pmatrix} = \begin{pmatrix} a & b \\ c & d \end{pmatrix}\begin{pmatrix} x \\ y \end{pmatrix} = \begin{pmatrix} ax + by \\ cx + dy \end{pmatrix}.$$

Note that

$$A\mathbf{e}_1^T = \begin{pmatrix} a & b \\ c & d \end{pmatrix}\begin{pmatrix} 1 \\ 0 \end{pmatrix} = \begin{pmatrix} a \\ c \end{pmatrix}; \qquad A\mathbf{e}_2^T = \begin{pmatrix} a & b \\ c & d \end{pmatrix}\begin{pmatrix} 0 \\ 1 \end{pmatrix} = \begin{pmatrix} b \\ d \end{pmatrix}.$$

So $A\mathbf{e}_1^T$ is the first column of A and $A\mathbf{e}_2^T$ is the second column of A.

Note that $(x, y)^T$ maps to $(ax + by, cx + dy)^T$. So there are lots of maps from $\mathbb{R}_2$ to $\mathbb{R}_2$ which cannot be represented by matrices. For example, the map from $(x, y)^T$ to $(x^2, y^2)^T$ cannot; nor can the constant map which sends every vector to $(1, 0)^T$ be so represented.

- Note in particular that the associated map of an $m \times n$ matrix sends the origin in $\mathbb{R}_n$ to the origin in $\mathbb{R}_m$.

Example 3.133 (**Rotations and Reflections of the Plane**) Find the 2×2 matrices whose associated maps are (a) R_θ, which denotes rotation by θ anticlockwise about the origin; (b) S_θ, which is reflection in the line $y = x \tan \theta$.

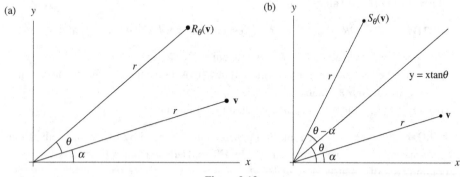

Figure 3.12

<hr />

[20] When considering matrices as maps, we represent vectors as column vectors. We have, till now, largely written vectors as row vectors, one reason being that row vectors are more conveniently presented on the page. We could continue using row vectors if we wished and instead postmultiply with a matrix: so if $\mathbf{v}$ is in $\mathbb{R}^m$, and A is an $m \times n$ matrix, then $\mathbf{v}A$ is in $\mathbb{R}^n$. In this case A represents a map from $\mathbb{R}^m$ to $\mathbb{R}^n$. However, the more common convention is to use column vectors and write matrices on the left, in a comparable way to how we write functions on the left as in the notation $f(x)$.

In § 4.1 we will address the more general issue of matrices representing maps with respect to co-ordinates other than the standard ones.

Solution (a) Given that we are describing a rotation about the origin then polar co-ordinates seem a natural way to help describe the map. Say that $\mathbf{v} = (r\cos\alpha, r\sin\alpha)^T$, as in the Figure 3.12a above; then

$$R_\theta(\mathbf{v}) = \begin{pmatrix} r\cos(\alpha+\theta) \\ r\sin(\alpha+\theta) \end{pmatrix} = \begin{pmatrix} \cos\theta(r\cos\alpha) - \sin\theta(r\sin\alpha) \\ \cos\theta(r\sin\alpha) + \sin\theta(r\cos\alpha) \end{pmatrix} = \begin{pmatrix} \cos\theta & -\sin\theta \\ \sin\theta & \cos\theta \end{pmatrix} \mathbf{v}.$$

(b) From Figure 3.12b we see S_θ maps $\mathbf{v}$ to the point $(r\cos(2\theta-\alpha), r\sin(2\theta-\alpha))^T$. So

$$S_\theta(\mathbf{v}) = \begin{pmatrix} r\cos(2\theta-\alpha) \\ r\sin(2\theta-\alpha) \end{pmatrix} = \begin{pmatrix} \cos2\theta(r\cos\alpha) + \sin2\theta(r\sin\alpha) \\ \sin2\theta(r\cos\alpha) - \cos2\theta(r\sin\alpha) \end{pmatrix}$$

$$= \begin{pmatrix} \cos2\theta & \sin2\theta \\ \sin2\theta & -\cos2\theta \end{pmatrix} \mathbf{v}.$$

Hence, with a slight abuse of notation, we can write

$$R_\theta = \begin{pmatrix} \cos\theta & -\sin\theta \\ \sin\theta & \cos\theta \end{pmatrix}; \qquad S_\theta = \begin{pmatrix} \cos2\theta & \sin2\theta \\ \sin2\theta & -\cos2\theta \end{pmatrix}. \tag{3.46}$$

More precisely we should say that R_θ and S_θ are the associated maps of these two matrices. □

Example 3.134

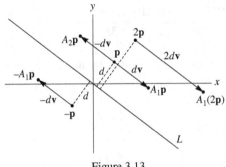

Figure 3.13

A **shear** of $\mathbb{R}_2$ is a map which moves points at a distance d on one side of a given (invariant) line L by the vector $d\mathbf{v}$ where $\mathbf{v}$ is a given vector parallel to L (Figure 3.13).

Note that the shear then fixes the given line (as $d = 0$). Points a distance d away on the other side of the line move by $-d\mathbf{v}$.

For a shear to be the associated map of a matrix the line L must go through the origin. A shear of $\mathbb{R}_3$ is similarly defined with a plane replacing the line, or more generally in $\mathbb{R}_n$ with a hyperplane replacing the line.

Determine the two matrices, A_1 and A_2, of the two shears of $\mathbb{R}_2$ which have invariant line $3x + 4y = 0$ and which move the point $\mathbf{p} = (1, 1)^T$ a distance of 2. Show that both matrices have determinant 1 and are inverses of one another.

Solution As the shear's invariant line is $3x + 4y = 0$ then the shear fixes the point $(4, -3)^T$ which lies on this line. The vector $(4, -3)^T$ is also parallel to the line and has length 5; as the point $(1, 1)^T$ moves a distance of 2 parallel to the line, then it moves by $\pm\frac{2}{5}(4, -3)^T$ and hence to

$$\begin{pmatrix} 1 \\ 1 \end{pmatrix} \pm \frac{2}{5} \begin{pmatrix} 4 \\ -3 \end{pmatrix} = \frac{1}{5} \begin{pmatrix} 13 \\ -1 \end{pmatrix} \quad \text{or} \quad \frac{1}{5} \begin{pmatrix} -3 \\ 11 \end{pmatrix}.$$

By #565, if A_1 represents the first shear then

$$A_1 \begin{pmatrix} 4 & 1 \\ -3 & 1 \end{pmatrix} = \begin{pmatrix} 4 & 13/5 \\ -3 & -1/5 \end{pmatrix}.$$

Recalling the formula for the inverse of a 2×2 matrix, we have

$$A_1 = \begin{pmatrix} 4 & 13/5 \\ -3 & -1/5 \end{pmatrix} \begin{pmatrix} 4 & 1 \\ -3 & 1 \end{pmatrix}^{-1} = \begin{pmatrix} 4 & 13/5 \\ -3 & -1/5 \end{pmatrix} \frac{1}{7} \begin{pmatrix} 1 & -1 \\ 3 & 4 \end{pmatrix}$$

$$= \frac{1}{35} \begin{pmatrix} 59 & 32 \\ -18 & 11 \end{pmatrix}.$$

Similarly we find

$$A_2 = \begin{pmatrix} 4 & -3/5 \\ -3 & 11/5 \end{pmatrix} \begin{pmatrix} 4 & 1 \\ -3 & 1 \end{pmatrix}^{-1} = \begin{pmatrix} 4 & -3/5 \\ -3 & 11/5 \end{pmatrix} \frac{1}{7} \begin{pmatrix} 1 & -1 \\ 3 & 4 \end{pmatrix}$$

$$= \frac{1}{35} \begin{pmatrix} 11 & -32 \\ 18 & 59 \end{pmatrix}.$$

Note that the determinant of A_1 equals

$$\frac{59 \times 11 + 18 \times 32}{35^2} = \frac{(35+24)(35-24)+24^2}{35^2} = \frac{35^2}{35^2} = 1,$$

and A_2's determinant leads to the same calculation. So $A_2 = A_1^{-1}$ by the 2×2 inverse formula. □

Example 3.135 Let

$$A_1 = \begin{pmatrix} 1 & 1 \\ 0 & 2 \end{pmatrix}, \qquad A_2 = \begin{pmatrix} -1 & 1 \\ -2 & 2 \end{pmatrix}, \qquad A_3 = \begin{pmatrix} 1 & -1 \\ 1 & 1 \end{pmatrix},$$

and

$$S_1 = \{(x,y)^T : x+y = 1\}, \qquad S_2 = \{(x,y)^T : x,y > 0\}, \qquad S_3 = \{(x,y)^T : x^2 + y^2 = 1\}.$$

For each of $i,j = 1,2,3$, find the image of S_j under A_i – that is, $A_i(S_j) = \{A_i\mathbf{v} : \mathbf{v} \in S_j\}$.

Solution
Full details are left to #792–#796.

- S_1 is a line; S_2 is a quadrant; S_3 is a circle (Figure 3.14a).

After applying A_1 (see Figure 3.14b)

- S_1 becomes the line $x = 1$;

- S_2 maps to the region $2x > y > 0$, an infinite sector with apex at the origin;
- S_3 maps to the curve $2x^2 - 2xy + y^2 = 2$ which is an ellipse (see #796).

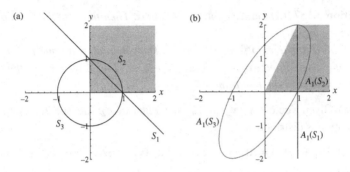

Figure 3.14

- A_2 is singular and maps the whole plane to the line $y = 2x$ (see Figure 3.15).
- The entire line $y = x + k$ maps to the same point $(k, 2k)^T$.
- A_2 is an example of a **projection** (see #572).
- The image of S_1 is all of the line $y = 2x$.
- The image of S_2 is all of the line $y = 2x$.
- S_3 maps to the segment of the line $y = 2x$ where $-\sqrt{2} \leqslant x \leqslant \sqrt{2}$.

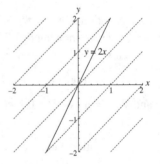

Figure 3.15

- As

$$A_3 = \begin{pmatrix} \frac{1}{\sqrt{2}} & -\frac{1}{\sqrt{2}} \\ \frac{1}{\sqrt{2}} & \frac{1}{\sqrt{2}} \end{pmatrix} \begin{pmatrix} \sqrt{2} & 0 \\ 0 & \sqrt{2} \end{pmatrix},$$

then A_3 is the composition of rotation about the origin anticlockwise by $\pi/4$ with dilation of scale factor $\sqrt{2}$ from the origin (see Figure 3.16).
- S_1 maps to the line $y = 1$.
- S_2 maps to the region $y > x > -y$.
- S_3 maps to the circle with radius $\sqrt{2}$ centred at the origin. □

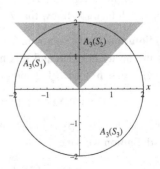

Figure 3.16

As has already been commented, most maps from $\mathbb{R}_n$ to $\mathbb{R}_m$ cannot be represented by matrices. But the ones that can be, namely the *linear maps*, form a large class involving many important examples.

Definition 3.136 A **linear map** (or **linear transformation**) T from $\mathbb{R}_n$ to $\mathbb{R}_m$ is a map which satisfies

#779
#780
#782

$$T(c_1\mathbf{v}_1 + c_2\mathbf{v}_2) = c_1 T\mathbf{v}_1 + c_2 T\mathbf{v}_2$$

for all vectors $\mathbf{v}_1, \mathbf{v}_2$ in $\mathbb{R}_n$ and real numbers c_1, c_2.

#783
#788
#797
#799

Proposition 3.137 *(a) Let A be an $m \times n$ matrix. Then the associated map μ_A is a linear map from $\mathbb{R}_n$ to $\mathbb{R}_m$.*

(b) Let T be a linear map from $\mathbb{R}_n$ to $\mathbb{R}_m$. Then there is a unique $m \times n$ matrix A such that $T = \mu_A$. The columns of A are $T(\mathbf{e}_1^T), T(\mathbf{e}_2^T), \ldots, T(\mathbf{e}_n^T)$. We refer to the matrix A as the **associated matrix** *of the linear map T.*

Proof (a) Given vectors $\mathbf{v}_1, \mathbf{v}_2$ in $\mathbb{R}_n$ and reals c_1, c_2, we see μ_A is linear, as matrix multiplication is distributive:

$$\mu_A(c_1\mathbf{v}_1 + c_2\mathbf{v}_2) = A(c_1\mathbf{v}_1 + c_2\mathbf{v}_2) = c_1 A\mathbf{v}_1 + c_2 A\mathbf{v}_2 = c_1\mu_A(\mathbf{v}_1) + c_2\mu_A(\mathbf{v}_2).$$

(b) Let T be a linear map from $\mathbb{R}_n$ to $\mathbb{R}_m$. Let

$$A = (T(\mathbf{e}_1^T) \,|\, T(\mathbf{e}_2^T) \,|\, \cdots \,|\, T(\mathbf{e}_n^T))$$

as in Notation 3.98. By #563(ii), the jth column of A is also $A\mathbf{e}_j^T$. So $\mu_A(\mathbf{e}_j^T) = A\mathbf{e}_j^T = T(\mathbf{e}_j^T)$ for $1 \leqslant j \leqslant n$, showing that $T = \mu_A$ on the basis $\mathbf{e}_1^T, \ldots, \mathbf{e}_n^T$. Now any vector $\mathbf{x}$ in $\mathbb{R}_n$ can be written $\mathbf{x} = x_1\mathbf{e}_1^T + \cdots + x_n\mathbf{e}_n^T$ for some reals $x_1, \ldots, x_n$. By the linearity of T we have

$$T(\mathbf{x}) = x_1 T(\mathbf{e}_1^T) + \cdots + x_n T(\mathbf{e}_n^T) = x_1 A\mathbf{e}_1^T + \cdots + x_n A\mathbf{e}_n^T = A(x_1\mathbf{e}_1^T + \cdots + x_n\mathbf{e}_n^T) = A\mathbf{x},$$

and so $T = \mu_A$ on $\mathbb{R}_n$. If there were a second $m \times n$ matrix B such that $T = \mu_B$ then again we'd have

$$j\text{th column of } B = B\mathbf{e}_j^T = \mu_B(\mathbf{e}_j^T) = T(\mathbf{e}_j^T) = A\mathbf{e}_j^T = j\text{th column of } A$$

for each $1 \leqslant j \leqslant n$ and so $B = A$. $\qquad\square$

Example 3.138 Let T_k denote the stretch of $\mathbb{R}_3$ with invariant plane $x + y + z = 0$ and stretch factor k. Show that T_k is a linear map and find the matrix A_k of T_k.

Solution Every vector in $\mathbb{R}_3$ can be written in a unique fashion as $\mathbf{p} + \mathbf{n}$ where $\mathbf{p}$ is in the invariant plane and $\mathbf{n}$ is normal to the plane (#781), so that $T_k(\mathbf{p} + \mathbf{n}) = \mathbf{p} + k\mathbf{n}$. Say $\mathbf{v}_i = \mathbf{p}_i + \mathbf{n}_i$ for $i = 1, 2$ and c_1, c_2 are scalars; then

$$c_1\mathbf{v}_1 + c_2\mathbf{v}_2 = c_1(\mathbf{p}_1 + \mathbf{n}_1) + c_2(\mathbf{p}_2 + \mathbf{n}_2) = (c_1\mathbf{p}_1 + c_2\mathbf{p}_2) + (c_1\mathbf{n}_1 + c_2\mathbf{n}_2).$$

Note $c_1\mathbf{p}_1 + c_2\mathbf{p}_2$ is in the invariant plane and $c_1\mathbf{n}_1 + c_2\mathbf{n}_2$ is normal to that plane, so by definition we have

$$T_k(c_1\mathbf{v}_1 + c_2\mathbf{v}_2) = (c_1\mathbf{p}_1 + c_2\mathbf{p}_2) + k(c_1\mathbf{n}_1 + c_2\mathbf{n}_2)$$
$$= c_1(\mathbf{p}_1 + k\mathbf{n}_1) + c_2(\mathbf{p}_2 + k\mathbf{n}_2) = c_1 T_k\mathbf{v}_1 + c_2 T_k\mathbf{v}_2,$$

and thus T_k is linear. Now $(1, -1, 0)^T$ and $(0, 1, -1)^T$ are in the invariant plane and so fixed by T_k, whilst $(1, 1, 1)^T$ is normal to the plane and so mapped to $(k, k, k)^T$. Hence

$$A_k \begin{pmatrix} 1 & 0 & 1 \\ -1 & 1 & 1 \\ 0 & -1 & 1 \end{pmatrix} = \begin{pmatrix} 1 & 0 & k \\ -1 & 1 & k \\ 0 & -1 & k \end{pmatrix},$$

and $T_k = \mu_{A_k}$ where A_k equals

$$\begin{pmatrix} 1 & 0 & k \\ -1 & 1 & k \\ 0 & -1 & k \end{pmatrix} \begin{pmatrix} 1 & 0 & 1 \\ -1 & 1 & 1 \\ 0 & -1 & 1 \end{pmatrix}^{-1} = \begin{pmatrix} 1 & 0 & k \\ -1 & 1 & k \\ 0 & -1 & k \end{pmatrix} \frac{1}{3} \begin{pmatrix} 2 & -1 & -1 \\ 1 & 1 & -2 \\ 1 & 1 & 1 \end{pmatrix}$$

$$= \frac{1}{3} \begin{pmatrix} k+2 & k-1 & k-1 \\ k-1 & k+2 & k-1 \\ k-1 & k-1 & k+2 \end{pmatrix}. \qquad \square$$

Once we begin thinking of matrices in terms of maps, it becomes clear that the product of two matrices is very natural: the map associated with a matrix product is the *composition* of the maps associated with the matrices. That is, the matrix product AB represents the map having the effect of performing B's map first, followed by A's map or more succinctly $\mu_{AB} = \mu_A \circ \mu_B$.

Proposition 3.139 *(Matrix Multiplication Represents Composition) Let A and B respectively be $m \times n$ and $n \times p$ matrices, so that AB is an $m \times p$ matrix. Then*

$$\mu_{AB}(\mathbf{v}) = (\mu_A \circ \mu_B)(\mathbf{v}) = \mu_A(\mu_B(\mathbf{v})) \qquad \text{for } \mathbf{v} \text{ in } \mathbb{R}_p.$$

Proof Ultimately this amounts to matrix multiplication being associative, as we have

$$\mu_{AB}(\mathbf{v}) = (AB)\mathbf{v} = A(B\mathbf{v}) = \mu_A(\mu_B(\mathbf{v})). \square$$

#781
#784
#785
#786
#787

Corollary 3.140 *Let A be a square matrix and n a natural number.*

(a) $\mu_{A^n} = \mu_A \circ \mu_A \circ \cdots \circ \mu_A$ (n times).
(b) μ_A is invertible if and only if A is invertible, and in this case $\mu_{A^{-1}} = (\mu_A)^{-1}$.

Proof (a) follows from Proposition 3.139 by a straightforward induction. (b) is left as #787. $\square$

Example 3.141 In the notation of (3.46):

(a) Show that $R_\theta R_\phi = R_{\theta+\phi}$ and determine $(R_\theta)^n$ where n is an integer.
(b) Show that $S_\theta S_\phi$ is a rotation and determine $(S_\theta)^2$.
(c) Show that $R_\theta S_\phi$ is a reflection and determine $(S_\phi)^{-1} R_\theta S_\phi$.

Solution (a) Writing c_θ for $\cos\theta$ etc. we have

$$R_\theta R_\phi = \begin{pmatrix} c_\theta & -s_\theta \\ s_\theta & c_\theta \end{pmatrix} \begin{pmatrix} c_\phi & -s_\phi \\ s_\phi & c_\phi \end{pmatrix} = \begin{pmatrix} c_\theta c_\phi - s_\theta s_\phi & -c_\theta s_\phi - s_\theta c_\phi \\ c_\theta s_\phi + s_\theta c_\phi & c_\theta c_\phi - s_\theta s_\phi \end{pmatrix}$$

$$= \begin{pmatrix} c_{\theta+\phi} & -s_{\theta+\phi} \\ s_{\theta+\phi} & c_{\theta+\phi} \end{pmatrix} = R_{\theta+\phi},$$

which is as we'd expect given that matrix multiplication is composition. Consequently $(R_\theta)^n = R_{n\theta}$.

(b) Similarly

$$S_\theta S_\phi = \begin{pmatrix} c_{2\theta} & s_{2\theta} \\ s_{2\theta} & -c_{2\theta} \end{pmatrix} \begin{pmatrix} c_{2\phi} & s_{2\phi} \\ s_{2\phi} & -c_{2\phi} \end{pmatrix}$$

$$= \begin{pmatrix} c_{2\theta}c_{2\phi} + s_{2\phi}s_{2\theta} & c_{2\theta}s_{2\phi} - s_{2\theta}c_{2\phi} \\ s_{2\theta}c_{2\phi} - c_{2\theta}s_{2\phi} & s_{2\theta}s_{2\phi} + c_{2\theta}c_{2\phi} \end{pmatrix} = \begin{pmatrix} c_{2\theta-2\phi} & -s_{2\theta-2\phi} \\ s_{2\theta-2\phi} & c_{2\theta-2\phi} \end{pmatrix} = R_{2\theta-2\phi}.$$

So $(S_\theta)^2 = R_{2\theta-2\theta} = R_0 = I_2$ (as expected because reflections are self-inverse).

(c) Now

$$R_\theta S_\phi = \begin{pmatrix} c_\theta & -s_\theta \\ s_\theta & c_\theta \end{pmatrix} \begin{pmatrix} c_{2\phi} & s_{2\phi} \\ s_{2\phi} & -c_{2\phi} \end{pmatrix} = \begin{pmatrix} c_\theta c_{2\phi} - s_\theta s_{2\phi} & c_\theta s_{2\phi} + s_\theta c_{2\phi} \\ s_\theta c_{2\phi} + c_\theta s_{2\phi} & s_\theta s_{2\phi} - c_\theta c_{2\phi} \end{pmatrix}$$

$$= \begin{pmatrix} c_{\theta+2\phi} & s_{\theta+2\phi} \\ s_{\theta+2\phi} & -c_{\theta+2\phi} \end{pmatrix} = S_{\phi+\theta/2}.$$

So by (b) we have $(S_\phi)^{-1} R_\theta S_\phi = S_\phi S_{\phi+\theta/2} = R_{2\phi-(2\phi+\theta)} = R_{-\theta}.$ $\qquad\square$

Example 3.142 Transpose is a linear map from M_{22} to M_{22} as $(\lambda A + \mu B)^T = \lambda A^T + \mu B^T$. We can identify M_{22} with $\mathbb{R}_4$ as follows

$$\begin{pmatrix} a & b \\ c & d \end{pmatrix} \leftrightarrow (a,b,c,d)^T. \qquad \text{so } E_{11} \leftrightarrow e_1^T, \text{ etc.}$$

As $(E_{12})^T = E_{21}, (E_{21})^T = E_{12}$ and E_{11} and E_{22} are symmetric, the associated matrix for transpose is

$$\begin{pmatrix} 1 & 0 & 0 & 0 \\ 0 & 0 & 1 & 0 \\ 0 & 1 & 0 & 0 \\ 0 & 0 & 0 & 1 \end{pmatrix}.$$

Example 3.143 Let

$$A = \frac{1}{5}\begin{pmatrix} 1 & 2 \\ 2 & 4 \end{pmatrix} \qquad \text{and} \qquad \begin{pmatrix} X \\ Y \end{pmatrix} = \frac{1}{\sqrt{5}}\begin{pmatrix} 2x-y \\ x+2y \end{pmatrix} = \frac{1}{\sqrt{5}}\begin{pmatrix} 2 & -1 \\ 1 & 2 \end{pmatrix}\begin{pmatrix} x \\ y \end{pmatrix}.$$

(a) Determine those vectors $\mathbf{v} = (v_1, v_2)^T$ such that $A\mathbf{x} = \mathbf{v}$ is consistent. Find the general solution when consistent.

(b) What are the XY-co-ordinates of the point with xy-co-ordinates $(2,1)$? What are the xy-co-ordinates of the point with XY-co-ordinates $(1,1)$?

(c) What is the equation of the Y-axis in terms of x and y and the equation of the y-axis in terms of X and Y?

(d) Say P is the point such that $(X, Y) = (\alpha, \beta)$. Find the XY-co-ordinates of the point $Q = \mu_A(P)$.

Solution (a) Note that A is singular. If we consider the linear system $(A \mid v)$, it reduces as

$$\left(\begin{array}{cc|c} 1/5 & 2/5 & v_1 \\ 2/5 & 4/5 & v_2 \end{array} \right) \xrightarrow{\text{RRE}} \left(\begin{array}{cc|c} 1 & 2 & 5v_1 \\ 0 & 0 & v_2 - 2v_1 \end{array} \right)$$

and so the system is consistent if and only if $v_2 = 2v_1$.

Equally what we have just shown is that μ_A has the line $y = 2x$ for its image. Now if $v_2 = 2v_1$, so the system is consistent, we can note the general solution $(x, y) = (5v_1 - 2t, t)$, where t is a parameter. These points parametrize the line $x + 2y = 5v_1$ and means μ_A maps all (and only) this line to $(v_1, v_2) = (v_1, 2v_1)$.

(b) If $x = 2$ and $y = 1$ then $X = 3/\sqrt{5}$ and $Y = 4/\sqrt{5}$ as

$$\left(\begin{array}{c} X \\ Y \end{array} \right) = \frac{1}{\sqrt{5}} \left(\begin{array}{cc} 2 & -1 \\ 1 & 2 \end{array} \right) \left(\begin{array}{c} 2 \\ 1 \end{array} \right) = \frac{1}{\sqrt{5}} \left(\begin{array}{c} 3 \\ 4 \end{array} \right).$$

If $X = 1$ and $Y = 1$ then $x = 3/\sqrt{5}$ and $y = 1/\sqrt{5}$ as

$$\left(\begin{array}{c} x \\ y \end{array} \right) = \left[\frac{1}{\sqrt{5}} \left(\begin{array}{cc} 2 & -1 \\ 1 & 2 \end{array} \right) \right]^{-1} \left(\begin{array}{c} 1 \\ 1 \end{array} \right) = \frac{1}{\sqrt{5}} \left(\begin{array}{cc} 2 & 1 \\ -1 & 2 \end{array} \right) \left(\begin{array}{c} 1 \\ 1 \end{array} \right) = \frac{1}{\sqrt{5}} \left(\begin{array}{c} 3 \\ 1 \end{array} \right).$$

(c) Now a point lies on the Y-axis if and only if $X = 0$ or, in terms of the xy-co-ordinates, if $2x = y$. Similarly $(0, y)$ is a general point of the y-axis and its XY-co-ordinates are

$$\left(\begin{array}{c} X \\ Y \end{array} \right) = \frac{1}{\sqrt{5}} \left(\begin{array}{cc} 2 & -1 \\ 1 & 2 \end{array} \right) \left(\begin{array}{c} 0 \\ y \end{array} \right) = \frac{1}{\sqrt{5}} \left(\begin{array}{c} -y \\ 2y \end{array} \right)$$

which is a general point of $Y = -2X$. This then is the equation of the y-axis in terms of XY-co-ordinates.

(d) If the point P has XY-co-ordinates $X_P = \alpha$, $Y_P = \beta$ then P's xy-co-ordinates are

$$\left(\begin{array}{c} x_P \\ y_P \end{array} \right) = \frac{1}{\sqrt{5}} \left(\begin{array}{cc} 2 & 1 \\ -1 & 2 \end{array} \right) \left(\begin{array}{c} \alpha \\ \beta \end{array} \right) = \left(\begin{array}{c} (2\alpha + \beta)/\sqrt{5} \\ (2\beta - \alpha)/\sqrt{5} \end{array} \right).$$

Hence $Q = \mu_A(P)$ has xy-co-ordinates

$$\left(\begin{array}{c} x_Q \\ y_Q \end{array} \right) = \left(\begin{array}{cc} 1/5 & 2/5 \\ 2/5 & 4/5 \end{array} \right) \left(\begin{array}{c} (2\alpha + \beta)/\sqrt{5} \\ (2\beta - \alpha)/\sqrt{5} \end{array} \right) = \frac{1}{5\sqrt{5}} \left(\begin{array}{c} 5\beta \\ 10\beta \end{array} \right) = \frac{1}{\sqrt{5}} \left(\begin{array}{c} \beta \\ 2\beta \end{array} \right).$$

Finally we see $\mu_A(P)$ has XY-co-ordinates

$$\left(\begin{array}{c} X_Q \\ Y_Q \end{array} \right) = \frac{1}{\sqrt{5}} \left(\begin{array}{cc} 2 & -1 \\ 1 & 2 \end{array} \right) \left(\begin{array}{c} \beta/\sqrt{5} \\ 2\beta/\sqrt{5} \end{array} \right) = \frac{1}{5} \left(\begin{array}{cc} 2 & -1 \\ 1 & 2 \end{array} \right) \left(\begin{array}{c} \beta \\ 2\beta \end{array} \right)$$

$$= \frac{1}{5} \left(\begin{array}{c} 0 \\ 5\beta \end{array} \right) = \left(\begin{array}{c} 0 \\ \beta \end{array} \right).$$

So, in terms of the XY-co-ordinates, we see that (α, β) maps to $(0, \beta)$. □

#790
#791
#802
#808
#809
#810

The previous example highlights an important point which will be addressed further in §3.11 and §4.1. When using the standard xy-co-ordinates, the map μ_A in Example 3.143 has the rather complicated matrix A, but if we instead use the XY-co-ordinates then the map has the much simpler associated matrix $\mathrm{diag}(0, 1)$ as the point with co-ordinates $X = \alpha, Y = \beta$ maps to the point with co-ordinates $X = 0, Y = \beta$. Using the (perpendicular) XY-axes we can see that the linear map is in fact orthogonal projection on to the Y-axis (Figure 3.17). However, this is far from apparent initially as the already-in-place xy-co-ordinates have nothing in particular to do with how this linear map transforms the plane. Here a different choice of co-ordinates in the plane much simplifies this linear map, so we might more generally ask the question: how might we choose co-ordinates to most clearly describe a linear map? That question will lead us to the notion of eigenvectors in §3.11.

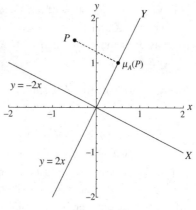

Figure 3.17

Exercises

#779 A (i) Let S, T be linear maps from $\mathbb{R}_n$ to $\mathbb{R}_m$ and from $\mathbb{R}_m$ to $\mathbb{R}_k$. Show the composition $T \circ S$ is linear.

(ii) Let S and T be linear maps from $\mathbb{R}_n$ and $\mathbb{R}_m$ and λ, μ real scalars. Show that $\lambda S + \mu T$ is a linear map.

#780 B† Let T be an invertible linear map from $\mathbb{R}_n$ to $\mathbb{R}_n$. Show that T^{-1} is linear.

#781 A Show, as claimed in Example 3.138, that every $\mathbf{v}$ in $\mathbb{R}_3$ can be uniquely written as $\mathbf{p} + \mathbf{n}$ where $\mathbf{p}$ is in the invariant plane of A_k and $\mathbf{n}$ is normal to that plane. For real r, s, what is $A_r A_s$? Show $(A_r)^{-1} = A_{1/r}$ when $r \neq 0$.

#782 A Show, in the notation of Example 3.134, that $\mathbf{v} = \frac{2}{7}(4, -3)^T$ for A_1.

#783 b† Let A_1 be the matrix in Example 3.134 and let n be a positive integer. What is $(A_1)^n (4, -3)^T$? What is $(A_1)^n (1, 1)^T$? Hence determine the matrix $(A_1)^n$.

#784 A† Let T be a linear map from $\mathbb{R}_2$ to $\mathbb{R}_2$. If $T(2, 1)^T = (0, 3)^T$ and $T(1, 2)^T = (2, 1)^T$, find $T(3, -1)^T$.

#785 a Determine the associated matrix of T where T is the linear map from $\mathbb{R}_3$ to $\mathbb{R}_3$ such that

$$T(1, 2, 3)^T = (1, 2, 3)^T, \qquad T(2, 1, 0)^T = (2, 1, 2)^T, \qquad T(1, 1, 0)^T = (0, 1, 2)^T.$$

#786 B† Show that there is no linear map T from $\mathbb{R}_3$ to $\mathbb{R}_3$ such that

$$T(2, 3, 4)^T = (1, 2, 3)^T, \qquad T(1, 2, 3)^T = (2, 1, 2)^T, \qquad T(0, 1, 2)^T = (0, 1, 2)^T.$$

#787B Let A be a square matrix. Show μ_A is invertible if and only if A is invertible, and then $\mu_{A^{-1}} = (\mu_A)^{-1}$.

#788B† Let $\mathbf{v}_1, \ldots, \mathbf{v}_n, \mathbf{w}_1, \ldots, \mathbf{w}_n$ be vectors in $\mathbb{R}_n$ with $\mathbf{v}_1, \ldots, \mathbf{v}_n$ being independent. Show that there is a unique linear map T from $\mathbb{R}_n$ to $\mathbb{R}_n$ such that $T(\mathbf{v}_i) = \mathbf{w}_i$ for each i.

#789D† Find all the linear maps from $\mathbb{R}_3$ to $\mathbb{R}_3$ such that

$$T(2,3,4)^T = (4,2,3)^T, \qquad T(1,2,3)^T = (2,1,2)^T, \qquad T(0,1,2)^T = (0,0,1)^T.$$

Which of these maps are invertible?

#790A In the notation of (3.46), show $R_\theta(\mathbf{v}) \cdot R_\theta(\mathbf{w}) = \mathbf{v} \cdot \mathbf{w} = S_\theta(\mathbf{v}) \cdot S_\theta(\mathbf{w})$ for any $\mathbf{v}$ and $\mathbf{w}$ in $\mathbb{R}_2$.

#791B† Find the matrix for the transpose map (see Example 3.142) when we identify M_{22} with $\mathbb{R}_4$ by

$$\begin{pmatrix} a & b+c \\ b-c & d \end{pmatrix} \leftrightarrow (a,b,c,d)^T.$$

#792B† Let A be an $n \times n$ matrix.

(i) Say that A is invertible. Show that the image of a line in $\mathbb{R}^n$ under μ_A is still a line.
(ii) Say now that A is singular. Show that the image of a line in $\mathbb{R}^n$ under μ_A is a line or a point.
(iii) Give an example of a singular 2×2 matrix A, and lines L_1, L_2, such that the image of L_1 under μ_A is a line but the image of L_2 is a point.

#793c† Let A be a 2×2 matrix. Show that the image of the quadrant $\{(x,y): x,y > 0\}$ under A is (i) an infinite sector, (ii) a line, (iii) a half-line or (iv) a point. Give conditions on A which determine when each case arises.

#794C† From Example 3.135, verify the claims made about $A_1(S_1), A_1(S_2), A_3(S_1), A_3(S_2)$, and $A_3(S_3)$.

#795C(i) From Example 3.135, verify that the image of the xy-plane under A_2 is indeed the line $y = 2x$.
(ii) Show that the line $y = x + k$, and only this line, maps to the point $(k, 2k)$.
(iii) Verify the claims made about $A_2(S_1), A_2(S_2)$, and $A_2(S_3)$.
(iv) Show further that $(A_2)^2 = A_2$ and that $A_2(x,y)^T = (x,y)^T$ if and only if $y = 2x$.

#796D†(i) From Example 3.135, verify that $A_1(S_3)$ is the curve $2x^2 - 2xy + y^2 = 2$.
(ii) If we set $x = X\cos\alpha - Y\sin\alpha$ and $y = X\sin\alpha + Y\cos\alpha$ into this curve's equation, find a value of α such that the mixed term XY disappears.
(iii) Deduce that the curve $A_1(S_3)$ is an ellipse and find its area.

#797b Find the associated matrix of the shear which has invariant line $2x + 5y = 0$ and maps $(5,0)^T$ to $(0,2)^T$.

#798 C Find the general form of associated matrix of a shear of $\mathbb{R}_2$ which has invariant line $x + 2y = 0$.

#799 B† Find the general form of associated matrix of a shear of $\mathbb{R}_3$ which has invariant plane $z = 0$.

#800 d Find the associated matrix A of the shear which has invariant plane $2x + z = y$ and maps $(1, 1, 1)^T$ to $(0, -1, 1)^T$. Determine A^n for an integer n.

#801 c Find the general form of associated matrix of a shear of $\mathbb{R}_3$ which has invariant plane $x + 2y + 2z = 0$.

#802 B† Let A be the associated matrix of a shear of $\mathbb{R}_2$. Show that $\det A = 1$ and $\mathrm{trace} A = 2$.

#803 D† Let S and T be two shears of the plane.

(i) Assuming they have the same invariant line through the origin, show that $S \circ T$ is a shear.

(ii) Assuming $S \circ T$ is a shear, show that S and T have the same invariant line.

#804 c† Identify the space V of polynomials in x of degree 3 or less with $\mathbb{R}_4$ as in #758. Show that the following are linear maps of V and find their associated matrices.

$$S(p(x)) = p'(x); \qquad T(p(x)) = p(x + 1); \qquad U(p(x)) = (x + 1)p'(x).$$

#805 D† Let X be an $m \times n$ such that $m \geqslant n$ and $X^T X$ is invertible. Let $P = X(X^T X)^{-1} X^T$. Show that

(i) $P^2 = P$;   (ii) $P^T = P$;   (iii) $\mathrm{trace}(P) = n$;   (iv) $X^T(I_m - P) = 0$.

(v) For $\mathbf{y}$ in $\mathbb{R}_m$, show that $P\mathbf{y} = \mathbf{y}$ if and only if $\mathbf{y} = X\mathbf{v}$ for some $\mathbf{v}$.

(vi) Show that P is orthogonal projection on to the column space of X.

#806 c† Find the 4×4 matrix for orthogonal projection on to the plane

$$x_1 + 2x_2 - x_4 = 0 = x_2 + x_3 + 2x_4 \qquad \text{in } \mathbb{R}_4.$$

#807 D† Let P be a projection of $\mathbb{R}_2$, i.e. $P^2 = P$. Show that the associated matrix has one of the following forms:

$$\begin{pmatrix} 0 & 0 \\ 0 & 0 \end{pmatrix}, \quad \begin{pmatrix} \frac{1}{2}(1 + \cosh u \cos v) & \frac{1}{2}(\cosh u \sin v + \sinh u) \\ \frac{1}{2}(\cosh u \sin v - \sinh u) & \frac{1}{2}(1 - \cosh u \cos v) \end{pmatrix}, \quad \begin{pmatrix} 1 & 0 \\ 0 & 1 \end{pmatrix}.$$

#808 b Let P and Q be the projection and reflection as defined in #540. Find the associated matrices of P and Q.

#809 B Let S_θ be as in (3.46), $\mathbf{v}_1 = (\cos\theta, \sin\theta)^T$ and $\mathbf{v}_2 = (-\sin\theta, \cos\theta)^T$. For real X, Y, determine $S_\theta(X\mathbf{v}_1 + Y\mathbf{v}_2)$ as a linear combination of $\mathbf{v}_1$ and $\mathbf{v}_2$. What is the associated matrix of S_θ when using the XY-co-ordinates?

#810 b Let T be the shear in Example 3.134 which has matrix A_1. Let $\mathbf{v}_1 = (4, -3)/5$ and $\mathbf{v}_2 = (3, 4)/5$. For real X, Y, determine $T(X\mathbf{v}_1 + Y\mathbf{v}_2)$ as a linear combination of $\mathbf{v}_1$ and $\mathbf{v}_2$. What is the matrix of T when using XY-co-ordinates?

#811c Let T_k denote the stretch in Example 3.138. Find three mutually perpendicular unit vectors $\mathbf{v}_1, \mathbf{v}_2, \mathbf{v}_3$ such that $T_k(X\mathbf{v}_1 + Y\mathbf{v}_2 + Z\mathbf{v}_3) = X\mathbf{v}_1 + Y\mathbf{v}_2 + kZ\mathbf{v}_3$. What is the associated matrix of T_k when using the *XYZ*-co-ordinates?

#812D† Let M be a fixed $n \times n$ matrix. Show that the following are linear maps of M_{nn} and find their ranks.

$$S(A) = A + A^T, \qquad T(A) = A - A^T, \qquad U(A) = MA.$$

3.8 Determinants

A square matrix has a number associated with it called its *determinant*. There are various different ways of introducing determinants, each of which has its advantages but none of which is wholly ideal, as will become clearer below. The definition we shall use is an inductive one, defining the determinant of an $n \times n$ matrix in terms of $(n-1) \times (n-1)$ determinants. Quite what the determinant of a matrix signifies will be discussed shortly in Remark 3.154.

Notation 3.144 Given a square matrix A and $1 \leqslant I, J \leqslant n$ then we will write A_{IJ} for the $(n-1) \times (n-1)$ matrix formed by removing the Ith row and the Jth column from A.

Example 3.145 Let

$$A = \begin{pmatrix} 1 & -3 & 2 \\ 0 & 7 & 1 \\ -5 & 1 & 3 \end{pmatrix}.$$

Then (a) removing the second row and third column or (b) removing the third row and first column, we get

$$\text{(a)} \quad A_{23} = \begin{pmatrix} 1 & -3 \\ -5 & 1 \end{pmatrix}; \qquad \text{(b)} \quad A_{31} = \begin{pmatrix} -3 & 2 \\ 7 & 1 \end{pmatrix}.$$

Our inductive definition of a determinant is then:

Definition 3.146 The determinant of a 1×1 matrix (a_{11}) is simply a_{11} itself. The **determinant** $\det A$ of an $n \times n$ matrix $A = (a_{ij})$ is then given by

$$\det A = a_{11} \det A_{11} - a_{21} \det A_{21} + a_{31} \det A_{31} - \cdots + (-1)^{n+1} a_{n1} \det A_{n1}.$$

Notation 3.147 The determinant of a square matrix A is denoted as $\det A$ and also sometimes as $|A|$. So we may also write the determinant of the matrix A in Example 3.145 as

$$\begin{vmatrix} 1 & -3 & 2 \\ 0 & 7 & 1 \\ -5 & 1 & 3 \end{vmatrix}.$$

Proposition 3.148 *The determinants of 2×2 and 3×3 matrices are given by the following formulae:*

(a) For 2×2 matrices

$$\det \begin{pmatrix} a_{11} & a_{12} \\ a_{21} & a_{22} \end{pmatrix} = a_{11}a_{22} - a_{12}a_{21}.$$

(b) For 3×3 matrices

$$\det \begin{pmatrix} a_{11} & a_{12} & a_{13} \\ a_{21} & a_{22} & a_{23} \\ a_{31} & a_{32} & a_{33} \end{pmatrix}$$

$$= a_{11}a_{22}a_{33} + a_{12}a_{23}a_{31} + a_{13}a_{21}a_{32} - a_{12}a_{21}a_{33} - a_{13}a_{22}a_{31} - a_{11}a_{23}a_{32}. \quad (3.47)$$

Proof (a) We already met the 2×2 determinant in Proposition 3.63, but applying the above inductive definition, we have $\det A_{11} = \det(a_{22}) = a_{22}$ and $\det A_{21} = \det(a_{12}) = a_{12}$, so that

$$\det \begin{pmatrix} a_{11} & a_{12} \\ a_{21} & a_{22} \end{pmatrix} = a_{11}\det A_{11} - a_{21}\det A_{21} = a_{11}a_{22} - a_{12}a_{21}.$$

(b) For the 3×3 case

$$\det A_{11} = \begin{vmatrix} a_{22} & a_{23} \\ a_{32} & a_{33} \end{vmatrix}, \qquad \det A_{21} = \begin{vmatrix} a_{12} & a_{13} \\ a_{32} & a_{33} \end{vmatrix}, \qquad \det A_{31} = \begin{vmatrix} a_{12} & a_{13} \\ a_{22} & a_{23} \end{vmatrix},$$

so that

$$\det \begin{pmatrix} a_{11} & a_{12} & a_{13} \\ a_{21} & a_{22} & a_{23} \\ a_{31} & a_{32} & a_{33} \end{pmatrix} = a_{11} \begin{vmatrix} a_{22} & a_{23} \\ a_{32} & a_{33} \end{vmatrix} - a_{21} \begin{vmatrix} a_{12} & a_{13} \\ a_{32} & a_{33} \end{vmatrix} + a_{31} \begin{vmatrix} a_{12} & a_{13} \\ a_{22} & a_{23} \end{vmatrix}$$

$$= a_{11}(a_{22}a_{33} - a_{23}a_{32}) - a_{21}(a_{12}a_{33} - a_{13}a_{32}) + a_{31}(a_{12}a_{23} - a_{13}a_{22})$$

using the formula for 2×2 determinants. This rearranges to (3.47) which was also previously found in #632. $\qquad \square$

Example 3.149 Let R_θ and S_θ be the rotation and reflection matrices from Example 3.133. Note, for any θ, that $\det R_\theta = \cos^2 \theta + \sin^2 \theta = 1$ and that $\det S_\theta = -\cos^2 2\theta - \sin^2 2\theta = -1$.

Example 3.150 Returning to the matrix from Example 3.145, we have

$$\begin{vmatrix} 1 & -3 & 2 \\ 0 & 7 & 1 \\ -5 & 1 & 3 \end{vmatrix}$$

$$= \underbrace{1 \times 7 \times 3}_{21} + \underbrace{(-3) \times 1 \times (-5)}_{15} + \underbrace{2 \times 0 \times 1}_{0} - \underbrace{1 \times 1 \times 1}_{1} - \underbrace{(-3) \times 0 \times 3}_{0} - \underbrace{2 \times 7 \times (-5)}_{-70}$$

$$= 105.$$

Remark 3.151 In the 2×2 and 3×3 cases, **but only in these cases,** there is a simple way to remember the determinant formula. The 2×2 formula is the product of entries on the left-to-right diagonal minus the product of those on the right-to-left diagonals. If, in the 3×3 case, we allow diagonals to 'wrap around' the vertical sides of the matrix – for example, as below:

$$\begin{pmatrix} & \searrow \\ & & \searrow \\ & \searrow & \end{pmatrix}, \quad \begin{pmatrix} \swarrow & & \\ & & \swarrow \\ & \swarrow & \end{pmatrix},$$

– then from this point of view a 3×3 matrix has three left-to-right diagonals and three right-to-left. A 3×3 determinant then equals the sum of the products of entries on the three left-to-right diagonals minus the products from the three right-to-left diagonals. This method of calculation does **not** apply to $n \times n$ determinants when $n \geqslant 4$. ∎

Definition 3.152 Let A be an $n \times n$ matrix. Given $1 \leqslant I, J \leqslant n$ the (I,J)th **cofactor** of A, denoted $C_{IJ}(A)$ or just C_{IJ}, is defined as $C_{IJ} = (-1)^{I+J} \det A_{IJ}$ and so the determinant $\det A$ can be rewritten as

$$\det A = a_{11} C_{11} + a_{21} C_{21} + \cdots + a_{n1} C_{n1}.$$

#813
#814
#815
#816

Proposition 3.153 *Let A be a triangular matrix. Then $\det A$ equals the product of the diagonal entries of A. In particular it follows that $\det I_n = 1$ for any n.*

Proof This is left to #821. □

Remark 3.154 (Summary of Determinant's Properties) As noted earlier, there are different ways to introduce determinants, each with its own particular advantages and disadvantages.

- With Definition 3.146, the determinant of an $n \times n$ matrix is at least unambiguously and relatively straightforwardly given. There are other (arguably more natural) definitions which require some initial work to show that they're well defined. For example, we shall see that det has the following *algebraic* properties:

   (i) det is linear in the rows (or columns) of a matrix (see Theorem 3.155(A)).
   (ii) If a matrix has two equal rows then its determinant is zero (see Theorem 3.155(B)).
   (iii) $\det I_n = 1$.

   In fact, these three algebraic properties uniquely characterize a function det which assigns a number to each $n \times n$ matrix (Proposition 3.167). As a consequence of this uniqueness it also follows that
   (∗) $\det A^T = \det A$ for any square matrix A (see #850).
   The problem with the above approach is that the existence and uniqueness of such a function are still moot.

- Using Definition 3.146 we avoid these issues, but unfortunately we currently have no real sense of what the determinant might convey about a matrix. The determinant of a 2×2 matrix is uniquely characterized by the two following *geometric* properties. Given a 2×2 matrix A, with associated map μ_A, it is then the case that

   (a) for any region S of the xy-plane, we have

$$\text{area of } \mu_A(S) = |\det A| \times (\text{area of } S). \tag{3.48}$$

   (b) The sense of any angle under μ_A is reversed when $\det A < 0$ but remains the same when $\det A > 0$.

These two properties best show the significance of determinants. Thinking along these lines, the following properties should seem natural enough:

(α) $\det AB = \det A \det B$ (Corollary 3.160).

(β) a square matrix is singular if and only it has zero determinant (Corollary 3.159).

However, whilst these geometric properties might better motivate the importance of determinants, they would be less useful in calculating determinants. Their meaning would also be less clear if we were working in more than three dimensions (at least until we had defined volume and sense/orientation in higher dimensions) or if we were dealing with matrices with complex numbers as entries.

- The current definition appears to lend some importance to the first column; Definition 3.146 is sometimes referred to as *expansion along the first column*. From #815 one might (rightly) surmise that determinants can be calculated by expanding along any row or column (Theorem 3.169).

- Finally, calculation is difficult and inefficient using Definition 3.146. (For example, the formula for an $n \times n$ determinant involves the sum of $n!$ separate products (Propositions 3.167 and 3.168(b).) We shall, in due course, see that a much better way to calculate determinants is via EROs. This method works well with specific examples but less well in general as too many special cases arise; if we chose to define determinants this way, even determining the general formulae for 2×2 and 3×3 determinants would become something of a chore. ∎

#817
#818   In the following we rigorously develop the theory of determinants. These proofs are
#819   often technical and not particularly illuminating. Someone looking just to understand the
#821   properties of determinants and their significance might commit (i), (ii), (iii), (∗), (a), (b), (α), (β) above to memory and move ahead to Remark 3.162, where we begin the discussion of calculating determinants efficiently.

- As in Notation 3.98 we shall write $(\mathbf{r}_1/\cdots/\mathbf{r}_n)$ for the $n \times n$ matrix with rows $\mathbf{r}_1, \cdots, \mathbf{r}_n$.

Theorem 3.155 *The map* det *defined in Definition 3.146 has the following properties:*
(A) det *is linear in each row. That is,* $\det C = \lambda \det A + \mu \det B$ *where*

$$A = (\mathbf{r}_1/\cdots/\mathbf{r}_{i-1}/\mathbf{r}_i/\mathbf{r}_{i+1}/\cdots/\mathbf{r}_n), \quad B = (\mathbf{r}_1/\cdots/\mathbf{r}_{i-1}/\mathbf{v}/\mathbf{r}_{i+1}/\cdots/\mathbf{r}_n),$$
$$C = (\mathbf{r}_1/\ldots/\mathbf{r}_{i-1}/\lambda\mathbf{r}_i + \mu\mathbf{v}/\mathbf{r}_{i+1}/\ldots/\mathbf{r}_n).$$

(B) If $A = (\mathbf{r}_1/\cdots/\mathbf{r}_n)$ *with* $\mathbf{r}_i = \mathbf{r}_j$ *for some* $i \neq j$, *then* $\det A = 0$.
(B′) If the matrix B is produced by swapping two different rows of A then $\det B = -\det A$.

Before proceeding to the main proof we will first prove the following.

Lemma 3.156 *Together, properties (A) and (B) are equivalent to properties (A) and (B′).*

Proof Suppose that det has properties (A), (B). Let $A = (\mathbf{r}_1/\cdots/\mathbf{r}_n)$ and B be produced by swapping rows i and j where $i < j$. Then

$$0 = \det(\mathbf{r}_1/\cdots/\mathbf{r}_i + \mathbf{r}_j/\cdots/\mathbf{r}_i + \mathbf{r}_j/\cdots/\mathbf{r}_n) \quad \text{[by (B)]}$$

$$= \det(\mathbf{r}_1/\cdots/\mathbf{r}_i/\cdots/\mathbf{r}_i + \mathbf{r}_j/\cdots/\mathbf{r}_n) + \det(\mathbf{r}_1/\cdots/\mathbf{r}_j/\cdots/\mathbf{r}_i + \mathbf{r}_j/\cdots/\mathbf{r}_n) \quad \text{[by (A)]}$$

$$= \left\{ \det(\mathbf{r}_1/\cdots/\mathbf{r}_i/\cdots/\mathbf{r}_i/\cdots/\mathbf{r}_n) + \det(\mathbf{r}_1/\cdots/\mathbf{r}_i/\cdots/\mathbf{r}_j/\cdots/\mathbf{r}_n) \right\}$$

$$\quad + \left\{ \det(\mathbf{r}_1/\cdots/\mathbf{r}_j/\cdots/\mathbf{r}_i/\cdots/\mathbf{r}_n) + \det(\mathbf{r}_1/\cdots/\mathbf{r}_j/\cdots/\mathbf{r}_j/\cdots/\mathbf{r}_n) \right\} \quad \text{[by (A)]}$$

$$= \{0 + \det A\} + \{\det B + 0\} \quad \text{[by (B)]}$$

$$= \det A + \det B$$

and so property (B′) follows. Conversely, if det has properties (A), (B′) and $\mathbf{r}_i = \mathbf{r}_j$ for $i \neq j$ then

$$\det(\mathbf{r}_1/\cdots/\mathbf{r}_i/\cdots/\mathbf{r}_j/\cdots/\mathbf{r}_n) = \det(\mathbf{r}_1/\cdots/\mathbf{r}_j/\cdots/\mathbf{r}_i/\cdots/\mathbf{r}_n),$$

as the two matrices are equal, but by property (B′)

$$\det(\mathbf{r}_1/\cdots/\mathbf{r}_i/\cdots/\mathbf{r}_j/\cdots/\mathbf{r}_n) = -\det(\mathbf{r}_1/\cdots/\mathbf{r}_j/\cdots/\mathbf{r}_i/\cdots/\mathbf{r}_n),$$

so that both determinants are in fact zero. □

We continue now with the proof of Theorem 3.155.

Proof (A) If $n = 1$ then (A) equates to the identity $(\lambda a_{11} + \mu v_1) = \lambda(a_{11}) + \mu(v_1)$. As an inductive hypothesis, suppose that (A) is true for $(n-1) \times (n-1)$ matrices. We are looking to show that the $n \times n$ determinant function is linear in the ith row. Note, for $j \neq i$, that $C_{j1}(C) = \lambda C_{j1}(A) + \mu C_{j1}(B)$ by our inductive hypothesis as these cofactors relate to $(n-1) \times (n-1)$ determinants. Also $C_{i1}(C) = C_{i1}(A) = C_{i1}(B)$ as C_{i1} is independent of the ith row. Hence

$$\det C = a_{11} C_{11}(C) + \cdots + (\lambda a_{i1} + \mu v_1) C_{i1}(C) + \cdots + a_{n1} C_{n1}(C)$$

$$= a_{11}(\lambda C_{11}(A) + \mu C_{11}(B)) + \cdots + \lambda a_{i1} C_{i1}(A) + \mu v_1 C_{i1}(B)$$

$$\quad + \cdots + a_{n1}(\lambda C_{n1}(A) + \mu C_{n1}(B))$$

$$= \lambda\{a_{11} C_{11}(A) + \cdots + a_{n1} C_{n1}(A)\} + \mu\{a_{11} C_{11}(B) + \cdots + v_1 C_{i1}(B) + \cdots + a_{n1} C_{n1}(B)\}$$

$$= \lambda \det A + \mu \det B.$$

By induction, property (A) holds for all square matrices.

(B) For a 2×2 matrix, if $\mathbf{r}_1 = \mathbf{r}_2$ then

$$\det A = \begin{pmatrix} a_{11} & a_{12} \\ a_{11} & a_{12} \end{pmatrix} = a_{11} a_{12} - a_{12} a_{11} = 0.$$

Assume now that (B) (or equivalently (B′)) is true for $(n-1) \times (n-1)$ matrices. Let $A = (\mathbf{r}_1/\cdots/\mathbf{r}_n)$ with $\mathbf{r}_i = \mathbf{r}_j$ where $i < j$. Then

$$\det A = a_{11} C_{11}(A) + \cdots + a_{n1} C_{n1}(A) = a_{i1} C_{i1}(A) + a_{j1} C_{j1}(A)$$

by the inductive hypothesis, as A_{k1} has two equal rows when $k \neq i, j$. Note that as $\mathbf{r}_i = \mathbf{r}_j$, with one copy of each being removed from A_{i1} and A_{j1}, then the rows of A_{i1} are the same as the rows of A_{j1} but come in a different order. The rows of A_{i1} and A_{ji} can be reordered to be the same as follows: what remains of $\mathbf{r}_j$ in A_{i1} can be moved up to the position of $\mathbf{r}_i$'s remainder in A_{j1} by swapping it $j-i-1$ times, each time with the next row above. By our inductive hypothesis

$$\det A = a_{i1} C_{i1}(A) + a_{j1} C_{j1}(A)$$
$$= (-1)^{1+i} a_{i1} \det A_{i1} + (-1)^{1+j} a_{j1} \det A_{j1} \qquad \text{[by definition of cofactors]}$$
$$= (-1)^{1+i} a_{i1} (-1)^{j-i-1} \det A_{j1} + (-1)^{1+j} a_{j1} \det A_{j1} \qquad \text{[by } j-i-1 \text{ uses of (B')]}$$
$$= (-1)^j (a_{i1} - a_{j1}) \det A_{j1} = 0 \qquad \text{[as } a_{i1} = a_{j1} \text{ because } \mathbf{r}_i = \mathbf{r}_j \text{].}$$

Hence (B) is true for $n \times n$ determinants and the result follows by induction. □

Corollary 3.157 *Let A be an $n \times n$ matrix and λ a real number.*

(a) If the matrix B is formed by multiplying a row of A by λ then $\det B = \lambda \det A$.
(b) $\det(\lambda A) = \lambda^n \det A$.
(c) If any row of A is zero then $\det A = 0$.

Proof (a) This follows from the fact that det is linear in its rows, and then if (a) is applied consecutively to each of the n rows part (b) follows. Finally if $\mathbf{r}_i = \mathbf{0}$ for some i, then $\mathbf{r}_i = 0\mathbf{r}_i$ and so (c) follows from part (a). □

Lemma 3.158 *(a) The determinants of the elementary matrices are*

$$\det M_i(\lambda) = \lambda; \qquad \det S_{ij} = -1; \qquad \det A_{ij}(\lambda) = 1.$$

In particular, elementary matrices have non-zero determinants.
(b) If E, A are $n \times n$ matrices and E is elementary then $\det EA = \det E \det A$.
(c) If E is an elementary matrix then $\det E^T = \det E$.

Proof We shall prove (a) and (b) together. If $E = M_i(\lambda)$ and then $\det EA = \lambda \det A$ by Corollary 3.157(a). If we choose $A = I_n$ then we find $\det M_i(\lambda) = \lambda$ and so we also have $\det EA = \det E \det A$ when $E = M_i(\lambda)$.

If $E = S_{ij}$ then by Theorem 3.155(B') $\det EA = -\det A$. If we take $A = I_n$ then we see $\det S_{ij} = -1$ and then we also have $\det EA = \det E \det A$ when $E = S_{ij}$.

If $E = A_{ij}(\lambda)$ and $A = (\mathbf{r}_1/\cdots/\mathbf{r}_n)$ then

$$\det(EA) = \det(\mathbf{r}_1/\cdots/\mathbf{r}_i + \lambda \mathbf{r}_j/\cdots/\mathbf{r}_j/\cdots/\mathbf{r}_n)$$
$$= \det(\mathbf{r}_1/\cdots/\mathbf{r}_i/\cdots/\mathbf{r}_j/\cdots/\mathbf{r}_n)$$
$$\quad + \lambda \det(\mathbf{r}_1/\cdots/\mathbf{r}_j/\cdots/\mathbf{r}_j/\cdots/\mathbf{r}_n) \qquad \text{[by Theorem 3.155 (A)]}$$
$$= \det A + 0 \qquad \text{[by Theorem 3.155 (B)]}$$
$$= \det A.$$

If we take $A = I_n$ then $\det A_{ij}(\lambda) = 1$ and $\det EA = \det E \det A$ follows when $E = A_{ij}(\lambda)$.

(c) Note that $M_i(\lambda)$ and S_{ij} are symmetric and so there is nothing to prove in these cases. Finally

$$\det(A_{ij}(\lambda)^T) = \det A_{ji}(\lambda) = 1 = \det A_{ij}(\lambda).$$ □

Corollary 3.159 *(Criterion for Invertibility) A square matrix A is invertible if and only if* $\det A \neq 0$, *in which case* $\det(A^{-1}) = (\det A)^{-1}$.

Proof If A is invertible then it row-reduces to the identity; that is, there are elementary matrices $E_1, \ldots, E_k$ such that $E_k \cdots E_1 A = I$. Hence, by repeated use of Lemma 3.158(b),

$$1 = \det I = \det E_k \times \cdots \times \det E_1 \times \det A.$$

In particular, $\det A \neq 0$. Further as $E_k \cdots E_1 = A^{-1}$ then $\det(A^{-1}) = (\det A)^{-1}$. If, however, A is singular then A reduces to a matrix R with at least one zero row so that $\det R = 0$. So, as before,

$$\det E_k \times \cdots \times \det E_1 \times \det A = \det R = 0$$

for some elementary matrices E_i. As $\det E_i \neq 0$ for each i, it follows that $\det A = 0$. □

Corollary 3.160 *(Product Rule for Determinants) Let* A, B *be* $n \times n$ *matrices. Then* $\det AB = \det A \det B$.

Proof If A and B are invertible then they can be written as products of elementary matrices; say $A = E_1 \ldots E_k$ and $B = F_1 \ldots F_l$. Then

$$\det AB = \det E_1 \times \cdots \times \det E_k \times \det F_1 \times \cdots \times \det F_l = \det A \det B$$

by Lemma 3.158(b). Otherwise one (or both) of A or B is singular. By Corollary 3.97, AB is singular and so $\det AB = 0$. But, also $\det A \times \det B = 0$ as one or both of A, B is singular. □

Corollary 3.161 *(Transpose Rule for Determinants) Let* A *be a square matrix. Then* $\det A^T = \det A$.

Proof A is invertible if and only if A^T is invertible. If A is invertible then $A = E_1 \ldots E_k$ for some elementary matrices E_i. Now $A^T = E_k^T \cdots E_1^T$ by the product rule for transposes and so, by Lemma 3.158(c) and the product rule in Corollary 3.160,

$$\det A^T = \det E_k^T \times \cdots \times \det E_1^T = \det E_k \times \cdots \times \det E_1 = \det A.$$

If A is singular then so is A^T and so $\det A = 0 = \det A^T$. □

Remark 3.162 Corollaries 3.159, 3.160 and 3.161 represent the most important algebraic properties of determinants. However we are still lumbered with a very inefficient way of calculating determinants in Definition 3.146. That definition is practicable up to 3×3 matrices but rapidly becomes laborious after that. A much more efficient way to calculate determinants is using EROs and ECOs, and we have been in a position to do this since showing $\det EA = \det E \times \det A$ for elementary E. An ECO involves postmultiplication by an elementary matrix but the product rule shows they will have the same effects on the determinant. Spelling this out:

- Adding a multiple of a row (resp. column) to another row (resp. column) has no effect on a determinant.
- Multiplying a row or column of the determinant by a scalar λ multiplies the determinant by λ.

- Swapping two rows or two columns of a determinant multiplies the determinant by -1. ∎

The following examples will hopefully make clear how to efficiently calculate determinants using EROs and ECOs.

Example 3.163 Use EROs and ECOs to calculate the following 4×4 determinants:

$$\begin{vmatrix} 1 & 2 & 0 & 3 \\ 4 & -3 & 1 & 0 \\ 0 & 2 & 5 & -1 \\ 2 & 3 & 1 & 2 \end{vmatrix}, \qquad \begin{vmatrix} 2 & 2 & 1 & -3 \\ 0 & 6 & -2 & 1 \\ 3 & 2 & 1 & 1 \\ 4 & 2 & -1 & 2 \end{vmatrix}.$$

#824
#825
#826
#828

Solution

$$\begin{vmatrix} 1 & 2 & 0 & 3 \\ 4 & -3 & 1 & 0 \\ 0 & 2 & 5 & -1 \\ 2 & 3 & 1 & 2 \end{vmatrix} = \begin{vmatrix} 1 & 2 & 0 & 3 \\ 0 & -11 & 1 & -12 \\ 0 & 2 & 5 & -1 \\ 0 & -1 & 1 & -4 \end{vmatrix} = \begin{vmatrix} -11 & 1 & -12 \\ 2 & 5 & -1 \\ -1 & 1 & -4 \end{vmatrix}$$

$$= \begin{vmatrix} 0 & -10 & 32 \\ 0 & 7 & -9 \\ -1 & 1 & -4 \end{vmatrix} = -1 \times \begin{vmatrix} -10 & 32 \\ 7 & -9 \end{vmatrix} = 134,$$

where, in order, we (i) add appropriate multiples of the row 1 to lower rows to clear the rest of column 1, (ii) expand along column 1, (iii) add appropriate multiples of row 3 to rows 1 and 2 to clear the rest of column 1, (iv) expand along column 1 and (v) employ the 2×2 determinant formula.

$$\begin{vmatrix} 2 & 2 & 1 & -3 \\ 0 & 6 & -2 & 1 \\ 3 & 2 & 1 & 1 \\ 4 & 2 & -1 & 2 \end{vmatrix} = \begin{vmatrix} 2 & 2 & 1 & -3 \\ 0 & 6 & -2 & 1 \\ 0 & -1 & -\frac{1}{2} & \frac{11}{2} \\ 0 & -2 & -3 & 8 \end{vmatrix} = 2 \begin{vmatrix} 6 & -2 & 1 \\ -1 & -\frac{1}{2} & \frac{11}{2} \\ -2 & -3 & 8 \end{vmatrix}$$

$$= 2 \begin{vmatrix} 0 & -5 & 34 \\ -1 & -\frac{1}{2} & \frac{11}{2} \\ 0 & -2 & -3 \end{vmatrix} = 2 \begin{vmatrix} -5 & 34 \\ -2 & -3 \end{vmatrix} = 166,$$

where, in order, we (i) add appropriate multiples of row 1 to lower rows to clear the rest of column 1, (ii) expand along column 1, (iii) add appropriate multiples of row 2 to rows 1 and 3 to clear the rest of column 1, (iv) expand along column 1 and (v) employ the 2×2 determinant formula.

Alternatively, for this second determinant, it may have made more sense to column-reduce as the third column has a helpful leading 1 and we could have instead calculated the determinant as follows:

$$
\begin{vmatrix} 2 & 2 & 1 & -3 \\ 0 & 6 & -2 & 1 \\ 3 & 2 & 1 & 1 \\ 4 & 2 & -1 & 2 \end{vmatrix} = \begin{vmatrix} 0 & 0 & 1 & 0 \\ 4 & 10 & -2 & -5 \\ 1 & 0 & 1 & 4 \\ 6 & 4 & -1 & -1 \end{vmatrix} = \begin{vmatrix} 4 & 10 & -5 \\ 1 & 0 & 4 \\ 6 & 4 & -1 \end{vmatrix}
$$

$$
= \begin{vmatrix} 0 & 10 & -21 \\ 1 & 0 & 4 \\ 0 & 4 & -25 \end{vmatrix} = - \begin{vmatrix} 10 & -21 \\ 4 & -25 \end{vmatrix} = 166.
$$

where, in order, we (i) add appropriate multiples of column 3 to other columns to clear the rest of row 1, (ii) expand along row 1, (iii) add appropriate multiples of row 2 to rows 1 and 3 to clear the rest of column 1, (iv) expand along column 1 and (v) employ the 2×2 determinant formula. □

- We will demonstrate in Theorem 3.169 the as-yet-unproven equivalence of expanding along *any* row or column.

Example 3.164 Let a, x be real numbers. Determine the following 3×3 and $n \times n$ determinants:

$$
\text{(a)} \quad \begin{vmatrix} x & a & a \\ x & x & a \\ x & x & x \end{vmatrix}, \qquad \text{(b)} \quad \begin{vmatrix} x & 1 & \cdots & 1 \\ 1 & x & \cdots & 1 \\ \vdots & \vdots & \ddots & \vdots \\ 1 & 1 & \cdots & x \end{vmatrix}.
$$

Solution (a) Subtracting row 3 from the other rows, and expanding along column 1, we obtain

$$
\begin{vmatrix} x & a & a \\ x & x & a \\ x & x & x \end{vmatrix} = \begin{vmatrix} 0 & a-x & a-x \\ 0 & 0 & a-x \\ x & x & x \end{vmatrix} = x \begin{vmatrix} a-x & a-x \\ 0 & a-x \end{vmatrix} = x(a-x)^2.
$$

Similarly, for (b), if we note that the sum of each column is the same and then add the bottom $n-1$ rows to the first row (which won't affect the determinant), we see it equals

$$
\begin{vmatrix} x+n-1 & x+n-1 & \cdots & x+n-1 \\ 1 & x & \cdots & 1 \\ \vdots & \vdots & \ddots & \vdots \\ 1 & 1 & \cdots & x \end{vmatrix} = (x+n-1) \begin{vmatrix} 1 & 1 & \cdots & 1 \\ 1 & x & \cdots & 1 \\ \vdots & \vdots & \ddots & \vdots \\ 1 & 1 & \cdots & x \end{vmatrix}
$$

$$
= (x+n-1) \begin{vmatrix} 1 & 1 & \cdots & 1 \\ 0 & x-1 & \cdots & 0 \\ \vdots & \vdots & \ddots & \vdots \\ 0 & 0 & \cdots & x-1 \end{vmatrix}
$$

where, in order, we (i) take the common factor of $x+n-1$ out of the first row, (ii) subtract the first row from each of the other rows and (iii) note the determinant is upper triangular to finally obtain a result of $(x+n-1)(x-1)^{n-1}$. □

We define now the Vandermonde [21] determinant, important to *interpolation* (#840) and *discriminants* (#843).

Example 3.165 (**Vandermonde Matrix**) For $n \geqslant 2$ and real numbers $x_1, \ldots, x_n$ we define

$$
V_n = \begin{pmatrix}
1 & x_1 & x_1^2 & \cdots & x_1^{n-1} \\
1 & x_2 & x_2^2 & \cdots & x_2^{n-1} \\
1 & x_3 & x_3^2 & \cdots & x_3^{n-1} \\
\vdots & \vdots & \vdots & \ddots & \vdots \\
1 & x_n & x_n^2 & \cdots & x_n^{n-1}
\end{pmatrix}
\qquad \text{and then} \qquad \det V_n = \prod_{i>j} (x_i - x_j).
$$

In particular, $V_n(x_1, \ldots, x_n)$ is invertible if and only if the x_i are distinct.

Solution Note that the given product is zero if and only if two (or more) of the x_i are equal. So the product is non-zero if and only if the x_i are distinct. We shall prove the determinant does indeed equal the given product by induction on n. In the case when $n = 2$ then $V_2(x_1, x_2) = x_2 - x_1$ by the 2×2 determinant formula, verifying the result for $n = 2$. Suppose now that the result holds true for the $(n-1)$th case. Then $\det V_n(x_1, \ldots, x_n)$ equals

$$
\begin{vmatrix}
1 & 0 & \cdots & 0 \\
1 & x_2 - x_1 & \cdots & x_2^{n-1} - x_1 x_2^{n-2} \\
\vdots & \vdots & \ddots & \vdots \\
1 & x_n - x_1 & \cdots & x_n^{n-1} - x_1 x_n^{n-2}
\end{vmatrix}
$$

[by adding $-x_1 \times$ (kth column) to ($k+1$)th column]

$$
= \begin{vmatrix}
x_2 - x_1 & \cdots & x_2^{n-1} - x_1 x_2^{n-2} \\
\vdots & \ddots & \vdots \\
x_n - x_1 & \cdots & x_n^{n-1} - x_1 x_n^{n-2}
\end{vmatrix}
\qquad \text{[expanding along the first row]}
$$

$$
= \left\{ \prod_{j=2}^{n} (x_j - x_1) \right\}
\begin{vmatrix}
1 & x_2 & \cdots & x_2^{n-2} \\
\vdots & \vdots & \ddots & \vdots \\
1 & x_n & \cdots & x_n^{n-2}
\end{vmatrix}
\qquad \text{[taking the factor of } x_{i+1} - x_1 \text{ from row } i]
$$

$$
= \left\{ \prod_{j=2}^{n} (x_j - x_1) \right\} V_{n-1}(x_2, \ldots, x_n)
$$

$$
= \prod_{i>j} (x_i - x_j) \qquad \text{[by the inductive hypothesis].} \qquad \square
$$

---

[21] See p.97 for some brief biographical detail on Alexandre-Théophile Vandermonde.

Exercises

#813a Calculate the following determinants:

$$\begin{vmatrix} 1 & 7 \\ -3 & 6 \end{vmatrix}, \quad \begin{vmatrix} 1 & -7 \\ -1 & 7 \end{vmatrix}, \quad \begin{vmatrix} 1 & 2 & 3 \\ 4 & 5 & 6 \\ 7 & 8 & 9 \end{vmatrix}, \quad \begin{vmatrix} a & b & c \\ 0 & d & e \\ 0 & 0 & f \end{vmatrix}.$$

#814A Give a counter-example to $\det(A+B) = \det A + \det B$.

#815a In the case of the determinant from Example 3.150, verify that the following three expressions

$$a_{21}C_{21} + a_{22}C_{22} + a_{23}C_{23}, \quad a_{12}C_{12} + a_{22}C_{22} + a_{32}C_{32}, \quad a_{13}C_{13} + a_{23}C_{23} + a_{33}C_{33},$$

all agree with the given determinant. Here C_{ij} denotes the (i,j)th cofactor.

#816A Verify directly that $\det AB = \det A \times \det B = \det BA$ where

$$A = \begin{pmatrix} 1 & 0 & 2 \\ 0 & 1 & 3 \\ 2 & 0 & 1 \end{pmatrix} \quad \text{and} \quad B = \begin{pmatrix} 0 & 0 & 1 \\ 0 & 1 & -1 \\ 2 & 0 & 3 \end{pmatrix}.$$

#817A Show that an orthogonal matrix (see #588) has determinant 1 or -1.

#818b† Let A be a square matrix. Show that if A has a real square root then $\det A \geqslant 0$. Is the converse true?

#819B Rederive Cassini's identity (#349) using #596.

#820D† Let A and B be square matrices. Show that $\det \text{diag}(A,B) = \det A \det B$.

#821B Prove, by induction, that the determinant of a triangular matrix is the product of its diagonal entries.

#822D† Let A be the $(m+n) \times (m+n)$ matrix in (3.49) where U, V, W are respectively $m \times m$, $m \times n$, $n \times n$ matrices. Show that $\det A = \det U \times \det W$.

$$A = \begin{pmatrix} U & V \\ 0_{nm} & W \end{pmatrix}. \qquad B = \begin{pmatrix} U & V \\ V & U \end{pmatrix}. \qquad (3.49)$$

#823C† Let B be as in (3.49) where U, V are square matrices of the same size. Show that $\det B = \det(U+V)\det(U-V)$.

#824B† Show, for any real number x, that

$$\begin{vmatrix} x^2 & (x+1)^2 & (x+2)^2 \\ (x+1)^2 & (x+2)^2 & (x+3)^2 \\ (x+2)^2 & (x+3)^2 & (x+4)^2 \end{vmatrix} = -8 \quad \text{and} \quad \begin{vmatrix} 1 & 1 & 1 & 1 \\ x & 1 & 1 & 1 \\ x & x & 1 & 1 \\ x & x & x & 1 \end{vmatrix} = (1-x)^3.$$

#825a Use EROs and ECOs to determine the following 3×3 determinants:

$$
\begin{vmatrix} -2 & 3 & 2 \\ 2 & -1 & 1 \\ 0 & 7 & -3 \end{vmatrix}, \quad
\begin{vmatrix} 3 & 1 & 8 \\ -2 & 1 & 2 \\ 1 & -7 & 1 \end{vmatrix}, \quad
\begin{vmatrix} 2 & 7 & 8 \\ 3 & 6 & 9 \\ 4 & 5 & 10 \end{vmatrix}.
$$

#826a Use EROs and ECOs to determine the following 4×4 determinants:

$$
\begin{vmatrix} 1 & 2 & 3 & 1 \\ -2 & 0 & 0 & 2 \\ 6 & 1 & 3 & 4 \\ 0 & 1 & 3 & -1 \end{vmatrix}, \quad
\begin{vmatrix} 1 & 2 & 4 & 2 \\ 2 & -3 & 1 & -3 \\ 7 & -1 & 0 & 1 \\ 1 & -2 & 1 & 0 \end{vmatrix}, \quad
\begin{vmatrix} 1 & 2 & 3 & 4 \\ 5 & 6 & 7 & 8 \\ 9 & 10 & 11 & 12 \\ 13 & 14 & 15 & 16 \end{vmatrix}.
$$

#827C Use EROs and ECOs to determine the following 5×5 determinants:

$$
\begin{vmatrix} 0 & 1 & 2 & 0 & 2 \\ 3 & -5 & 1 & 3 & 1 \\ 3 & 0 & 3 & 2 & 1 \\ 1 & 2 & 2 & 7 & -3 \\ 6 & -1 & 0 & 2 & -4 \end{vmatrix}, \quad
\begin{vmatrix} 2 & 3 & 1 & 0 & -4 \\ 4 & 1 & 2 & -1 & 5 \\ 3 & 4 & 1 & 0 & 3 \\ 3 & 0 & -3 & 1 & 2 \\ 2 & -1 & 3 & -2 & 3 \end{vmatrix}.
$$

#828B† Determine the determinants:

$$
\begin{vmatrix} y+z & x & x^2 \\ z+x & y & y^2 \\ x+y & z & z^2 \end{vmatrix}, \quad
\begin{vmatrix} y+z & x & x^3 \\ z+x & y & y^3 \\ x+y & z & z^3 \end{vmatrix}, \quad
\begin{vmatrix} x & a & b & c \\ a & x & b & c \\ a & b & x & c \\ a & b & c & x \end{vmatrix}.
$$

#829C† Let z_1, z_2, z_3, p, q be complex numbers and define

$$
\Delta = \begin{vmatrix} z_1 & \bar{z}_1 & 1 \\ z_2 & \bar{z}_2 & 1 \\ z_3 & \bar{z}_3 & 1 \end{vmatrix}; \quad
D = \begin{vmatrix} z & \bar{z} & 1 \\ p & \bar{p} & 1 \\ q & \bar{q} & 1 \end{vmatrix}; \quad
\delta = \begin{vmatrix} z & \bar{z} & 1 \\ p & \bar{q} & 1 \\ q & \bar{p} & 1 \end{vmatrix}.
$$

(i) Show that the points represented by z_1, z_2, z_3 are collinear if and only if $\Delta = 0$.

(ii) Deduce that the line connecting the points p and q has equation $D = 0$.

(iii) Deduce further that the perpendicular bisector of p and q has equation $\delta = 0$.

#830b Determine the first determinant in #826 by direct inductive use of Definition 3.146.

#831B† Evaluate the $n \times n$ determinants

$$
\begin{vmatrix}
2 & -1 & 0 & 0 & \cdots & 0 \\
-1 & 2 & -1 & 0 & \cdots & 0 \\
0 & -1 & 2 & -1 & \ddots & \vdots \\
0 & 0 & -1 & \ddots & \ddots & 0 \\
\vdots & \vdots & \ddots & \ddots & 2 & -1 \\
0 & 0 & \cdots & 0 & -1 & 2
\end{vmatrix},
\qquad
\begin{vmatrix}
x & a & a & a & \cdots & a \\
-a & x & a & a & \cdots & a \\
-a & -a & x & a & \ddots & \vdots \\
-a & -a & -a & \ddots & \ddots & a \\
\vdots & \vdots & \ddots & \ddots & x & a \\
-a & -a & \cdots & -a & -a & x
\end{vmatrix},
$$

$$
\begin{vmatrix}
1 & 1 & 1 & 1 & \cdots & 1 \\
x & 1 & 1 & 1 & \cdots & 1 \\
0 & x & 1 & 1 & \ddots & \vdots \\
0 & 0 & x & \ddots & \ddots & 1 \\
\vdots & \vdots & \ddots & \ddots & 1 & 1 \\
0 & 0 & \cdots & 0 & x & 1
\end{vmatrix}.
$$

#832D Define the $n \times n$ determinants D_n and Δ_n as below:

$$
D_n =
\begin{vmatrix}
1 & -1 & 0 & 0 & \cdots & 0 \\
1 & 1 & -1 & 0 & \cdots & 0 \\
0 & 1 & 1 & -1 & \ddots & \vdots \\
0 & 0 & 1 & \ddots & \ddots & 0 \\
\vdots & \vdots & \ddots & \ddots & 1 & -1 \\
0 & 0 & \cdots & 0 & 1 & 1
\end{vmatrix}
\quad \text{and} \quad
\Delta_n =
\begin{vmatrix}
x & 1 & 0 & 0 & \cdots & 0 \\
1 & x & 1 & 0 & \cdots & 0 \\
0 & 1 & x & 1 & \ddots & \vdots \\
0 & 0 & 1 & \ddots & \ddots & 0 \\
\vdots & \vdots & \ddots & \ddots & x & 1 \\
0 & 0 & \cdots & 0 & 1 & x
\end{vmatrix}.
$$

(i) Show that $D_1 = 1$ and $D_2 = 2$ and that $D_{n+2} = D_{n+1} + D_n$ for $n \geqslant 1$. Deduce that $D_n = F_{n+1}$.

(ii) Show that $\Delta_1 = x$ and $\Delta_2 = x^2 - 1$ and that $\Delta_{n+2} = x\Delta_{n+1} - \Delta_n$ for $n \geqslant 1$. Deduce that $\Delta_n = U_n(x/2)$, where U_n is a Chebyshev polynomial.

#833D† Evaluate the $n \times n$ determinants

$$
\begin{vmatrix}
x & x^2 & 0 & 0 & \cdots & 0 \\
1 & x & x^2 & 0 & \cdots & 0 \\
0 & 1 & x & x^2 & \ddots & \vdots \\
0 & 0 & 1 & \ddots & \ddots & 0 \\
\vdots & \vdots & \ddots & \ddots & x & x^2 \\
0 & 0 & \cdots & 0 & 1 & x
\end{vmatrix},
\qquad
\begin{vmatrix}
x^2 & x^3 & 0 & 0 & \cdots & 0 \\
x & x^2 & x^3 & 0 & \cdots & 0 \\
1 & x & x^2 & x^3 & \ddots & \vdots \\
0 & 1 & x & \ddots & \ddots & 0 \\
\vdots & \vdots & \ddots & \ddots & x^2 & x^3 \\
0 & 0 & \cdots & 1 & x & x^2
\end{vmatrix}.
$$

$$\begin{vmatrix} x & x^2 & x^3 & 0 & \cdots & 0 \\ 1 & x & x^2 & x^3 & \cdots & 0 \\ 0 & 1 & x & x^2 & \ddots & \vdots \\ 0 & 0 & 1 & \ddots & \ddots & x^3 \\ \vdots & \vdots & \ddots & \ddots & x & x^2 \\ 0 & 0 & \cdots & 0 & 1 & x \end{vmatrix}.$$

#834D† Let α, β, a, b, c be real numbers. Determine the $n \times n$ determinants

$$\begin{vmatrix} \alpha+\beta & \alpha & 0 & 0 & \cdots & 0 \\ \beta & \alpha+\beta & \alpha & 0 & \cdots & 0 \\ 0 & \beta & \alpha+\beta & \alpha & \ddots & \vdots \\ 0 & 0 & \beta & \ddots & \ddots & 0 \\ \vdots & \vdots & \ddots & \ddots & \alpha+\beta & \alpha \\ 0 & 0 & \cdots & 0 & \beta & \alpha+\beta \end{vmatrix} \quad \text{and} \quad \begin{vmatrix} b & a & 0 & 0 & \cdots & 0 \\ c & b & a & 0 & \cdots & 0 \\ 0 & c & b & a & \ddots & \vdots \\ 0 & 0 & c & \ddots & \ddots & 0 \\ \vdots & \vdots & \ddots & \ddots & b & a \\ 0 & 0 & \cdots & 0 & c & b \end{vmatrix}.$$

#835B† Determine $\det V_n(1,2,3,\ldots,n)$.

#836D† Let $a_1,\ldots,a_n$ be real numbers. Show that each a_i must be zero if

$$a_1 \sin x + a_2 \sin 2x + \cdots + a_n \sin nx = 0 \qquad \text{for all real } x.$$

#837D Generalize the evaluation of the second determinant in #824 to the corresponding $n \times n$ determinant.

#838E† Generalize the first determinant in #824 to the $n \times n$ case, in such a way that all the determinants are non-zero constants.

#839D† Let $z^3 + ax^2 + bx + c$ be a cubic with distinct roots α, β, γ and let

$$C = \begin{pmatrix} 0 & 0 & -c \\ 1 & 0 & -b \\ 0 & 1 & -a \end{pmatrix}.$$

Determine VCV^{-1} where $V = V_3(\alpha, \beta, \gamma)$. Deduce that $\text{trace}(C^n) = \alpha^n + \beta^n + \gamma^n$ for a natural number n.

#840D† Let $x_0, x_1, \ldots, x_n, y_0, y_1, \ldots, y_n$ be real numbers with the x_i distinct. Show that there is a unique polynomial $p(x)$ of degree n or less such that $p(x_i) = y_i$ for each $0 \leqslant i \leqslant n$. Determine this polynomial when $y_0 = 1$ and $y_i = 0$ for $i > 0$. Finding such $p(x)$ is called **Lagrangian interpolation**.

#841B Let $A = (a_{ij})$ be a 3×3 matrix. Determine the 3×3 matrix $C = (C_{ij})$ where C_{ij} is the (i,j)th cofactor of A. Compare your answer with #632.

#842B† Show that $x = y = z = 0$ is the only solution amongst the complex numbers to the equations

$$x + y + z = 0, \qquad x^2 + y^2 + z^2 = 0, \qquad x^3 + y^3 + z^3 = 0.$$

#843D† Let m, n be real numbers and let a, b, c be the roots of the cubic $z^3 + mz + n = 0$. The **discriminant** Δ of the cubic is given by

$$\Delta = (a - b)^2 (a - c)^2 (b - c)^2.$$

(i) Explain why Δ is a real number. What can be said about the roots of the cubic when $\Delta > 0$, $\Delta = 0$, $\Delta < 0$?

(ii) By considering the identity $(z - a)(z - b)(z - c) = z^3 + mz + n$, show that $a + b + c = 0$ and find expressions for m and n in terms of a, b, c.

(iii) Let $S_k = a^k + b^k + c^k$ for $k \geqslant 0$. Write down expressions for S_0, S_1, S_2 in terms of a, b, c.

(iv) Explain why $S_{k+3} + m S_{k+1} + n S_k = 0$ for $k \geqslant 0$ and hence find expressions for S_3 and S_4 in terms of m and n.

(v) Write down $\det A$ where

$$A = \begin{pmatrix} 1 & a & a^2 \\ 1 & b & b^2 \\ 1 & c & c^2 \end{pmatrix}$$

and determine the matrix $A A^T$. Deduce that $\Delta = -4m^3 - 27n^2$. (See also #90.)

#844c† With the notation of #843, show that

$$\begin{vmatrix} S_k & S_{k+1} & S_{k+2} \\ S_{k+1} & S_{k+2} & S_{k+3} \\ S_{k+2} & S_{k+3} & S_{k+4} \end{vmatrix} = a^k b^k c^k \Delta.$$

#845C† Let A be an $m \times n$ matrix. A **submatrix** of A is any matrix formed by deleting certain rows and columns of A. The **determinantal rank** of A is the greatest value of r such that there is an invertible $r \times r$ submatrix. Show that determinantal rank of a matrix equals its row rank and deduce that row rank equals column rank.

#846c Find the ranks of the following matrices. For each matrix, find all its maximal invertible submatrices.

$$\begin{pmatrix} 1 & 1 & 1 \\ 2 & 2 & 2 \\ 3 & 3 & 3 \end{pmatrix}, \qquad \begin{pmatrix} 1 & 0 & 1 \\ 2 & 1 & 2 \end{pmatrix}, \qquad \begin{pmatrix} 1 & 1 & -1 & 1 \\ 2 & 3 & 0 & 2 \\ 1 & 2 & 1 & 1 \end{pmatrix}.$$

#847D† What is the rank of the Vandermonde matrix $V_n(x_1, \ldots, x_n)$ if $x_1, \ldots, x_n$ include precisely k distinct values?

#848D† Let A be an $n \times n$ matrix which has a zero $k \times k$ submatrix where $2k > n$. Show that A is singular.

3.9 Permutation Matrices*

It was claimed in Remark 3.154 that the determinant function for $n \times n$ matrices is entirely determined by certain algebraic properties. In light of Lemma 3.156 these properties are equivalent to:

  (i) det is linear in the rows of a matrix.
  (ii) If a matrix has two equal rows then its determinant is zero.
  (ii)$'$ If the matrix B is produced by swapping two of the rows of A then $\det B = -\det A$.
  (iii) $\det I_n = 1$.

To see why these properties determine det, we first consider the $n = 2$ case. Given a 2×2 matrix $A = (a_{ij})$, we can calculate its determinant as follows.

$$
\begin{vmatrix} a_{11} & a_{12} \\ a_{21} & a_{22} \end{vmatrix} = \begin{vmatrix} a_{11} & 0 \\ a_{21} & a_{22} \end{vmatrix} + \begin{vmatrix} 0 & a_{12} \\ a_{21} & a_{22} \end{vmatrix} \qquad \text{[as det is linear in row 1]}
$$

$$
= \left\{ \begin{vmatrix} a_{11} & 0 \\ 0 & a_{22} \end{vmatrix} + \begin{vmatrix} a_{11} & 0 \\ a_{21} & 0 \end{vmatrix} \right\} + \left\{ \begin{vmatrix} 0 & a_{12} \\ a_{21} & 0 \end{vmatrix} + \begin{vmatrix} 0 & a_{12} \\ 0 & a_{22} \end{vmatrix} \right\}
$$

[as det is linear in row 2]

$$
= a_{11}a_{22} \begin{vmatrix} 1 & 0 \\ 0 & 1 \end{vmatrix} + a_{11}a_{21} \begin{vmatrix} 1 & 0 \\ 1 & 0 \end{vmatrix} + a_{12}a_{21} \begin{vmatrix} 0 & 1 \\ 1 & 0 \end{vmatrix} + a_{12}a_{22} \begin{vmatrix} 0 & 1 \\ 0 & 1 \end{vmatrix}
$$

[as det is linear in rows]

$$
= a_{11}a_{22} \begin{vmatrix} 1 & 0 \\ 0 & 1 \end{vmatrix} + a_{12}a_{21} \begin{vmatrix} 0 & 1 \\ 1 & 0 \end{vmatrix} \qquad \text{[using (ii)]}
$$

$$
= a_{11}a_{22} \begin{vmatrix} 1 & 0 \\ 0 & 1 \end{vmatrix} - a_{12}a_{21} \begin{vmatrix} 1 & 0 \\ 0 & 1 \end{vmatrix} \qquad \text{[using (ii)$'$]}
$$

$$
= a_{11}a_{22} - a_{12}a_{21} \qquad \text{[using (iii)]}.
$$

If we were to argue similarly for a 3×3 matrix $A = (a_{ij})$, we could first use linearity to expand the determinant into a linear combination of $3^3 = 27$ determinants, with entries 1 and 0, each multiplied by a *monomial* $a_{1i}a_{2j}a_{3k}$. But we can ignore those cases where i, j, k involves some repetition, as the corresponding determinant is zero. There would, in fact, be only $3! = 6$ non-zero contributions, giving us the formula

$$
a_{11}a_{22}a_{23} \begin{vmatrix} 1 & 0 & 0 \\ 0 & 1 & 0 \\ 0 & 0 & 1 \end{vmatrix} + a_{12}a_{23}a_{31} \begin{vmatrix} 0 & 1 & 0 \\ 0 & 0 & 1 \\ 1 & 0 & 0 \end{vmatrix} + a_{13}a_{21}a_{32} \begin{vmatrix} 0 & 0 & 1 \\ 1 & 0 & 0 \\ 0 & 1 & 0 \end{vmatrix}
$$

$$
+ a_{12}a_{21}a_{33} \begin{vmatrix} 0 & 1 & 0 \\ 1 & 0 & 0 \\ 0 & 0 & 1 \end{vmatrix} + a_{13}a_{22}a_{31} \begin{vmatrix} 0 & 0 & 1 \\ 0 & 1 & 0 \\ 1 & 0 & 0 \end{vmatrix} + a_{11}a_{23}a_{32} \begin{vmatrix} 1 & 0 & 0 \\ 0 & 0 & 1 \\ 0 & 1 & 0 \end{vmatrix}.
$$

The first determinant here is $\det I_3$, which we know to be 1. The other determinants all have the same rows $(1,0,0), (0,1,0), (0,0,1)$ as I_3 but appearing in some other order. In each case, it is possible (if necessary) to swap $(1,0,0)$ – which appears as some row of the determinant – with the first row, so that it is now in the correct place. Likewise the second row can be moved (if necessary) so it is in the right place. By a process of elimination the third row is now in the right place and we have transformed the determinant into $\det I_3$. We know what the effect of each such swap is, namely multiplying by -1, and so the six determinants above have values 1 or -1. For example,

$$\begin{vmatrix} 0 & 1 & 0 \\ 0 & 0 & 1 \\ 1 & 0 & 0 \end{vmatrix} = - \begin{vmatrix} 1 & 0 & 0 \\ 0 & 0 & 1 \\ 0 & 1 & 0 \end{vmatrix} = \begin{vmatrix} 1 & 0 & 0 \\ 0 & 1 & 0 \\ 0 & 0 & 1 \end{vmatrix} = 1, \begin{vmatrix} 0 & 0 & 1 \\ 0 & 1 & 0 \\ 1 & 0 & 0 \end{vmatrix} = - \begin{vmatrix} 1 & 0 & 0 \\ 0 & 1 & 0 \\ 0 & 0 & 1 \end{vmatrix} = -1.$$

So finally we have, as we found in Proposition 3.148(b), that a 3×3 determinant $\det(a_{ij})$ equals

$$a_{11}a_{22}a_{23} + a_{12}a_{23}a_{31} + a_{13}a_{21}a_{32} - a_{12}a_{21}a_{33} - a_{13}a_{22}a_{31} - a_{11}a_{23}a_{32}.$$

The general situation is hopefully now clear for an $n \times n$ matrix $A = (a_{ij})$. Using linearity to expand along each row in turn, $\det A$ can be written as the sum of n^n terms

$$\sum \det P_{i_1 \cdots i_n} a_{1i_1} \cdots a_{ni_n},$$

where $P_{i_1 \cdots i_n}$ is the matrix whose rows are $\mathbf{e}_{i_1}, \ldots, \mathbf{e}_{i_n}$ – that is the entries of $P_{i_1 \cdots i_n}$ are all zero except entries $(1, i_1), \ldots, (n, i_n)$ which are all 1. At the moment each of $i_1, \ldots, i_n$ can independently take a value between 1 and n, but most such choices lead to the determinant $\det P_{i_1 \cdots i_n}$ being zero as some of the rows $\mathbf{e}_{i_1}, \ldots, \mathbf{e}_{i_n}$ are repeated. In fact, $\det P_{i_1 \cdots i_n}$ can only be non-zero when

$$\{i_1, \ldots, i_n\} = \{1, \ldots, n\}.$$

That is $i_1, \ldots, i_n$ are $1, \ldots n$ *in some order* or equivalently the rows of $P_{i_1 \cdots i_n}$ are $\mathbf{e}_1, \ldots, \mathbf{e}_n$ *in some order*.

Definition 3.166 An $n \times n$ matrix P is said to be a **permutation matrix** if its rows are $\mathbf{e}_1, \ldots, \mathbf{e}_n$ in some order. This is equivalent to saying that each row and column contains a single entry 1 with all other entries being zero.

Thus we have shown:

Proposition 3.167 *The function* $\det$ *is entirely determined by the three algebraic properties (i), (ii) and (iii). Further, the determinant* $\det A$ *of an* $n \times n$ *matrix* $A = (a_{ij})$ *equals*

$$\det A = \sum \det P_{i_1 \cdots i_n} a_{1i_1} \cdots a_{ni_n} \tag{3.50}$$

where the sum is taken over all permutation matrices $P_{i_1 \cdots i_n} = (\mathbf{e}_{i_1} / \cdots / \mathbf{e}_{i_n})$.

We further note:

Proposition 3.168 *(a) The columns of a permutation matrix are* $\mathbf{e}_1^T, \ldots, \mathbf{e}_n^T$ *in some order.*
(b) The number of $n \times n$ *permutation matrices is* $n!$.
(c) A permutation matrix has determinant 1 or -1.
(d) When $n \geqslant 2$, *half the permutation matrices have determinant 1 and half have determinant* -1.

Proof (a) The entries in the first column of a permutation matrix P are the first entries of $\mathbf{e}_1, \mathbf{e}_2, \ldots, \mathbf{e}_n$ in some order and so are $1, 0, \ldots, 0$ in some order – that is, the first column is $\mathbf{e}_i^T$ for some i. Likewise each column of P is $\mathbf{e}_i^T$ for some i. If any of the columns of P were the same then this would mean that a row of P had two non-zero entries, which cannot occur. So the columns are all distinct. As there are n columns then each of $\mathbf{e}_1^T, \ldots, \mathbf{e}_n^T$ appears exactly once.

(b) This follows from #294.

(c) The rows of a permutation matrix P are $\mathbf{e}_1, \ldots, \mathbf{e}_n$ in some order. We know that swapping two rows of a matrix has the effect of multiplying the determinant by -1. We can create a (possibly new) matrix P_1 by swapping the first row of P with the row $\mathbf{e}_1$ (which appears somewhere); of course no swap may be needed. The matrix P_1 has $\mathbf{e}_1$ as its first row and $\det P_1 = \pm \det P$ depending on whether a swap was necessary or not. We can continue in this fashion producing matrices $P_1, \ldots, P_n$ such that the first k rows of P_k are $\mathbf{e}_1, \ldots, \mathbf{e}_k$ in *that* order and $\det P_k = \pm \det P_{k-1}$ in each case, depending on whether or not we needed to make any swap to get $\mathbf{e}_k$ to the kth row. Eventually then $P_n = I_n$ and $\det P = \pm \det P_n = 1$ or -1 depending on whether an even or odd number of swaps had to be made to turn P into I_n.

(d) Let $n \geqslant 2$ and let S_{12} be the elementary $n \times n$ matrix associated with swapping the first and second rows of a matrix. If P is a permutation matrix then $S_{12}P$ is also a permutation matrix as its rows are still $\mathbf{e}_1, \ldots, \mathbf{e}_n$ in some order; further,

$$\det S_{12}P = \det S_{12} \times \det P = -\det P.$$

For each permutation matrix P with $\det P = 1$, we have $S_{12}P$ being a permutation matrix with $\det(S_{12}P) = -1$; conversely for every permutation matrix $\tilde{P}$ with $\det \tilde{P} = -1$ we have $S_{12}\tilde{P}$ being a permutation matrix with $\det(S_{12}\tilde{P}) = 1$ As these processes are inverses of one another, because $S_{12}(S_{12}P) = P$, there are equal numbers of determinant 1 and determinant -1 permutation matrices, each separately numbering $\frac{1}{2}n!$.   □

We now prove a result already mentioned in Remark 3.154. Our inductive definition of the determinant began by expanding down the first column. In fact it is the case that we will arrive at the same answer, the determinant, whichever column or row we expand along.

Theorem 3.169 *(Equality of Determinant Expansions* [22]*) Let $A = (a_{ij})$ be an $n \times n$ matrix and let C_{ij} denote the (i,j)th cofactor of A. Then the determinant $\det A$ may be calculated by expanding along any column or row of A. So, for any $1 \leqslant i \leqslant n$, we have*

$$\det A = a_{1i}C_{1i} + a_{2i}C_{2i} + \cdots + a_{ni}C_{ni} \qquad \textit{[expansion along the ith column]} \quad (3.51)$$
$$= a_{i1}C_{i1} + a_{i2}C_{i2} + \cdots + a_{in}C_{in} \qquad \textit{[expansion along the ith row].} \qquad (3.52)$$

Proof We showed in Theorem 3.155 and Proposition 3.153 that det has properties (i), (ii), (iii), and have just shown in Proposition 3.167 that these properties uniquely determine the

---

[22] This was proved by Pierre-Simon Laplace (1749–1827) in 1772, though Leibniz had been aware of this result a century earlier. Laplace made many important contributions to applied mathematics, statistics and physics. Laplace's equation (p.432) is one of the most fundamental differential equations in mathematics and physics, being significant in analysis, fluid mechanics, electromagnetism and gravitational theory. The Laplace transform (§ 6.7) is an important tool in solving differential equations.

function det. Making the obvious changes to Theorem 3.155 and Proposition 3.153 it can similarly be shown, for any i, that the function which assigns

$$a_{1i}C_{1i} + a_{2i}C_{2i} + \cdots + a_{ni}C_{ni} \tag{3.53}$$

to the matrix $A = (a_{ij})$ also has properties (i), (ii), (iii). By uniqueness it follows that (3.53) also equals $\det A$. That is, expanding down any column also leads to the same answer of $\det A$. Then

$$
\begin{aligned}
\det A = \det A^T &= [A^T]_{1i}C_{1i}(A^T) + [A^T]_{2i}C_{2i}(A^T) + \cdots + [A^T]_{ni}C_{ni}(A^T) \\
&= a_{i1}C_{i1} + a_{i2}C_{i2} + \cdots + a_{in}C_{in}
\end{aligned}
$$

by expanding down the ith column of A^T, but this is the same sum found when expanding along the ith row of A. □

In practical terms, however, Laplace's result isn't that helpful. We have already discounted repeated expansion along rows and columns of hard-to-calculate cofactors as a hugely inefficient means to find determinants (see Remarks 3.154 and 3.162 and #883). However, it does lead us to the following theorem of interest.

Theorem 3.170 (*Existence of the Adjugate*) *Let A be an $n \times n$ matrix. Let C_{ij} denote the (i,j)th cofactor of A and let $C = (C_{ij})$ be the matrix of cofactors. Then*

$$C^T A = AC^T = \det A \times I_n.$$

In particular, if A is invertible, then

$$A^{-1} = \frac{C^T}{\det A}. \tag{3.54}$$

#870
#871
#872
#873
#874

Proof Note

$$[C^T A]_{ij} = \sum_{k=1}^{n} [C^T]_{ik}[A]_{kj} = \sum_{k=1}^{n} C_{ki}a_{kj}.$$

When $i = j$ then

$$[C^T A]_{ii} = \sum_{k=1}^{n} a_{ki}C_{ki} = \det A$$

by Theorem 3.169, as this is the determinant calculated by expanding along the ith column. On the other hand, if $i \neq j$, then consider the matrix B which has the same columns as A except for the ith column of B, which is a copy of A's jth column. As the ith and jth columns of B are equal then $\det B$ is zero. Note that the (k,i)th cofactor of B equals C_{ki} as A and B agree except in the ith column; so if expanding $\det B$ along its ith column we see

$$0 = \det B = \sum_{k=1}^{n} b_{ki}C_{ki} = \sum_{k=1}^{n} a_{kj}C_{ki} = [C^T A]_{ij}.$$

Hence $C^T A = \det A \times I_n$. That $AC^T = \det A \times I_n$ similarly follows (#869). Finally if A is invertible, then $\det A \neq 0$, and (3.54) follows. □

Definition 3.171 With notation as in Theorem 3.170 the matrix C^T is called the **adjugate** of A (or sometimes the adjoint of A) and is written adj$A$.

Corollary 3.172 *(Cramer's Rule* [23]*) Let A be an $n \times n$ matrix, $\mathbf{b}$ in $\mathbb{R}_n$ and consider the linear system $(A|\mathbf{b})$. The system has a unique solution if and only if $\det A \neq 0$, which is given by*

$$\mathbf{x} = \frac{C^T \mathbf{b}}{\det A}.$$

Proof This follows from the previous theorem, Propositions 3.93 and 3.95 and Corollary 3.159. $\qquad\square$

Remark 3.173 Writing $A = (a_{ij})$ and $\mathbf{b} = (b_1, \ldots, b_n)^T$ then Cramer's Rule with $n = 2$ expressly reads as

$$x_1 = \frac{b_1 a_{22} - b_2 a_{12}}{\det A}, \qquad x_2 = \frac{b_2 a_{11} - b_1 a_{21}}{\det A},$$

where $\det A = a_{11}a_{22} - a_{12}a_{21}$ as found in (3.25). When $n = 3$ Cramer's rule reads as

$$x_1 = \frac{b_1 \begin{vmatrix} a_{22} & a_{23} \\ a_{32} & a_{33} \end{vmatrix} - b_2 \begin{vmatrix} a_{12} & a_{13} \\ a_{32} & a_{33} \end{vmatrix} + b_3 \begin{vmatrix} a_{12} & a_{13} \\ a_{22} & a_{23} \end{vmatrix}}{\det A},$$

$$x_2 = \frac{-b_1 \begin{vmatrix} a_{21} & a_{23} \\ a_{31} & a_{33} \end{vmatrix} + b_2 \begin{vmatrix} a_{11} & a_{13} \\ a_{31} & a_{33} \end{vmatrix} - b_3 \begin{vmatrix} a_{11} & a_{13} \\ a_{21} & a_{23} \end{vmatrix}}{\det A},$$

$$x_3 = \frac{b_1 \begin{vmatrix} a_{21} & a_{22} \\ a_{31} & a_{32} \end{vmatrix} - b_2 \begin{vmatrix} a_{11} & a_{12} \\ a_{31} & a_{32} \end{vmatrix} + b_3 \begin{vmatrix} a_{11} & a_{12} \\ a_{21} & a_{22} \end{vmatrix}}{\det A},$$

where $\det A = a_{11}a_{22}a_{33} + a_{12}a_{23}a_{31} + a_{13}a_{21}a_{32} - a_{12}a_{21}a_{33} - a_{13}a_{22}a_{31} - a_{11}a_{23}a_{32}$. $\qquad\blacksquare$

Cramer's rule though is a seriously limited and impractical means of solving linear systems. The rule only applies when the matrix A is square and invertible, and the computational power required to calculate so many cofactors and $\det A$ makes it substantially more onerous than row-reduction.

We finish now by introducing a method of solving linear systems that was expressly introduced for its efficiency.

Definition 3.174 We say than a square matrix A has an **LU decomposition** if $A = LU$, where L is lower triangular with all its diagonal entries equalling 1 and U is upper triangular.

---

[23] Named after the Swiss mathematician Gabriel Cramer (1704–1752), who discovered this result in 1750.

Example 3.175 Find an LU decomposition for the following matrix A. Then solve the linear system $A\mathbf{x} = \mathbf{b}$ by separately solving the systems $L\mathbf{y} = \mathbf{b}$ and $U\mathbf{x} = \mathbf{y}$.

$$A = \begin{pmatrix} 1 & 3 & 2 & 1 \\ 2 & 0 & 1 & 1 \\ 1 & 2 & 4 & 1 \\ 0 & 2 & 1 & 3 \end{pmatrix}, \quad \mathbf{b} = \begin{pmatrix} 2 \\ 7 \\ 2 \\ 1 \end{pmatrix}.$$

Solution We begin by reducing A to an upper triangular matrix. For reasons that will become apparent in due course, the only EROs we shall use involve subtracting multiples of rows from lower rows, i.e. $A_{ij}(\lambda)$, where $i < j$. Further we shall carefully keep track of the EROs that we are using.

$$\begin{pmatrix} 1 & 3 & 2 & 1 \\ 2 & 0 & 1 & 1 \\ 1 & 2 & 4 & 1 \\ 0 & 2 & 1 & 3 \end{pmatrix} \xrightarrow[A_{13}(-1)]{A_{12}(-2)} \begin{pmatrix} 1 & 3 & 2 & 1 \\ 0 & -6 & -3 & -1 \\ 0 & -1 & 2 & 0 \\ 0 & 2 & 1 & 3 \end{pmatrix} \xrightarrow[A_{24}(1/3)]{A_{23}(-1/6)} \begin{pmatrix} 1 & 3 & 2 & 1 \\ 0 & -6 & -3 & -1 \\ 0 & 0 & 5/2 & 1/6 \\ 0 & 0 & 0 & 8/3 \end{pmatrix}.$$

$$(3.55)$$

The final matrix we have arrived at is our U and we have shown

$$A = [A_{24}(1/3)A_{23}(-1/6)A_{13}(-1)A_{12}(-2)]^{-1}U$$
$$= \underbrace{A_{12}(2)A_{13}(1)A_{23}(1/6)A_{24}(-1/3)}_{L}U = LU.$$

So

$$L = \begin{pmatrix} 1 & 0 & 0 & 0 \\ 2 & 1 & 0 & 0 \\ 1 & 1/6 & 1 & 0 \\ 0 & -1/3 & 0 & 1 \end{pmatrix} \quad \text{and} \quad (L\,|\,\mathbf{b}) = \left(\begin{array}{cccc|c} 1 & 0 & 0 & 0 & 2 \\ 2 & 1 & 0 & 0 & 7 \\ 1 & 1/6 & 1 & 0 & 2 \\ 0 & -1/3 & 0 & 1 & 1 \end{array} \right).$$

- Because of the order in which the EROs were performed, each entry of L below the diagonal is the negative of the multiplier needed to clear out that entry as we reduced A to U in (3.55).
- L is lower triangular and has diagonal entries 1 here – and in general – as the same is true of the elementary matrices $A_{ij}(\lambda)$, where $i < j$.

Thus we have $A = LU$ with L and U as above. Solving $L\mathbf{y} = \mathbf{b}$ is relatively quick as we immediately have $y_1 = 2$, then can successively find y_2, y_3, y_4 as

$$y_2 = 7 - 2y_1 = 3, \quad y_3 = 2 - y_1 - \frac{y_2}{6} = -\frac{1}{2}, \quad y_4 = 1 + \frac{y_2}{3} = 2.$$

This stage is consequently known as **forward substitution.** Solving $U\mathbf{x} = \mathbf{y}$ (as below) involves **backward substitution.** We straight away have $x_4 = 3/4$ and then

$$(U\,|\,\mathbf{y}) = \begin{pmatrix} 1 & 3 & 2 & 1 & 2 \\ 0 & -6 & -3 & -1 & 3 \\ 0 & 0 & 5/2 & 1/6 & -1/2 \\ 0 & 0 & 0 & 8/3 & 2 \end{pmatrix}$$

$$5x_3/2 + x_4/6 = -1/2 \implies x_3 = \tfrac{2}{5}\left(-\tfrac{1}{8} - \tfrac{1}{2}\right) = -\tfrac{1}{4};$$

$$-6x_2 - 3x_3 - x_4 = 3 \implies x_2 = -\tfrac{1}{6}\left(-\tfrac{3}{4} + \tfrac{3}{4} + 3\right) = -\tfrac{1}{2};$$

$$x_1 + 3x_2 + 2x_3 + x_4 = 2 \implies x_1 = 2 + \tfrac{3}{2} + \tfrac{1}{2} - \tfrac{3}{4} = \tfrac{13}{4}.$$

Note also that

$$\det A = [U]_{11}[U]_{22}[U]_{33}[U]_{44} = 1 \times (-6) \times (5/2) \times (8/3) = -40. \qquad \square$$

The merit of the LU decomposition is that we have rewritten a linear system as a combination of two triangular linear systems, which are much more efficient to solve with each stage involving (forward or backward) substitution of previously determined variables. Unfortunately a general $n \times n$ matrix need not have an LU decomposition (#877). However, this can be overcome by a simple reordering of the rows. Thus we have:

Theorem 3.176 (*Existence of a PLU Decomposition* [24]) *Given any square matrix A, there is a permutation matrix P such that PA has an LU decomposition.*

#877
#878
#880
#881

Proof By #686 it is possible to put A into row echelon form using a sequence of EROs of the form S_{ij} followed by a sequence of EROs of the form $A_{ij}(\lambda)$, where $i < j$. The combined effect of the EROs S_{ij} is a permutation matrix P. Also the elementary matrix representing $A_{ij}(\lambda)$ is lower triangular with diagonal entries equalling 1, so this is also true of their product $\tilde{L}$. As the square matrix $\tilde{L}PA = U$ is in row echelon form then U is upper triangular. If we write $L = \tilde{L}^{-1}$, then L is again lower triangular, with diagonal entries equalling 1, and we have $PA = LU$ as required. $\qquad \square$

Exercises

#849 A† (i) Show that the product of two $n \times n$ permutation matrices is a permutation matrix.

(ii) Show that the inverse of a permutation matrix is a permutation matrix.

(iii) Show that a permutation matrix is orthogonal.

(iv) Show that the trace of a permutation matrix equals the number of $1 \leqslant i \leqslant n$ such that $P\mathbf{e}_i^T = \mathbf{e}_i^T$.

#850 B† Use (3.50) to show that $\det A^T = \det A$.

---

[24] This result was demonstrated by Alan Turing in 1948. See p.411 for biographical details.

#851b† (i) Show that each elementary matrix S_{ij} is a permutation matrix; the matrices S_{ij} are called **transpositions.**
(ii) Show that any permutation matrix may be expressed as a product of transpositions.
(iii) Show that if E is a transposition and P is a permutation matrix then $P^{-1}EP$ is a transposition.

#852B† (**Parity is Well defined**) For a given permutation matrix, show that the product in #851(ii) always involves an even number of transpositions, or always an odd number. Thus we may refer to a permutation matrix as being **even** or **odd**. List the even 4×4 permutation matrices.

#853b Write the odd 4×4 permutation matrices as products of transpositions using as few transpositions as possible.

#854C† Let M and P be $n \times n$ matrices. If P is a permutation matrix, show that every entry in M appears as an entry of PM. Conversely, show that if, for all $n \times n$ matrices M, every entry of M appears in PM, then P is a permutation matrix.

#855D Let D, Q be $n \times n$ matrices, D diagonal with distinct diagonal entries and Q invertible. Show that $Q^{-1}DQ$ is diagonal if and only if $Q = \Delta P$, where Δ is diagonal and P is a permutation matrix.

#856C Rederive (3.51) by applying Definition 3.146 to AS_{i1}.

#857D† Find a new proof of #822 using (3.50).

#858b† For $r \geqslant 2$, set Σ_r to be the $r \times r$ permutation matrix to the right below, and for $r = 1$, set $\Sigma_1 = I_1$. When r is odd, find a permutation matrix P such that $(\Sigma_r)^2 = P^{-1}\Sigma_r P$. What is the situation when r is even?

#859E† (**Cycle Decomposition – Existence**) Show that for any permutation matrix P there exist positive integers $r_1,\ldots,r_k$ and a permutation matrix R such that

$$P = R^{-1}\text{diag}(\Sigma_{r_1}, \Sigma_{r_2},\ldots, \Sigma_{r_k})R.$$

$$\Sigma_r = \begin{pmatrix} 0 & 0 & \cdots & 0 & 1 \\ 1 & 0 & \cdots & 0 & 0 \\ 0 & 1 & \ddots & \vdots & \vdots \\ \vdots & \ddots & \ddots & 0 & 0 \\ 0 & \cdots & 0 & 1 & 0 \end{pmatrix}$$

#860D† (**Cycle Decomposition – Uniqueness**) With notation as in #859, show that if also $P = S^{-1}\text{diag}(\Sigma_{s_1}, \Sigma_{s_2},\ldots, \Sigma_{s_l})S$ for a permutation matrix S and positive integers $s_1,\ldots,s_l$ then $l = k$ and $s_1,\ldots,s_k$ equal $r_1,\ldots,r_k$ in some order.

#861 D† Write the following 5×5 matrices as in #859.

$$\begin{pmatrix} 0 & 0 & 0 & 0 & 1 \\ 0 & 0 & 0 & 1 & 0 \\ 0 & 0 & 1 & 0 & 0 \\ 0 & 1 & 0 & 0 & 0 \\ 1 & 0 & 0 & 0 & 0 \end{pmatrix}, \qquad \begin{pmatrix} 0 & 0 & 0 & 0 & 1 \\ 0 & 0 & 0 & 1 & 0 \\ 1 & 0 & 0 & 0 & 0 \\ 0 & 1 & 0 & 0 & 0 \\ 0 & 0 & 1 & 0 & 0 \end{pmatrix},$$

$$\begin{pmatrix} 0 & 0 & 0 & 1 & 0 \\ 0 & 0 & 0 & 0 & 1 \\ 1 & 0 & 0 & 0 & 0 \\ 0 & 1 & 0 & 0 & 0 \\ 0 & 0 & 1 & 0 & 0 \end{pmatrix}, \qquad \begin{pmatrix} 0 & 0 & 0 & 0 & 1 \\ 0 & 0 & 0 & 1 & 0 \\ 0 & 0 & 1 & 0 & 0 \\ 1 & 0 & 0 & 0 & 0 \\ 0 & 1 & 0 & 0 & 0 \end{pmatrix}.$$

#862 B† Let P and Q be $n \times n$ permutation matrices. What can you say about the cycle decompositions of PQ and of QP?

#863 D† Let P be an $n \times n$ permutation matrix expressible as in #859. What is the smallest positive integer N such that $P^N = I_n$? Deduce that $P^{n!} = I_n$.

#864 D Let $1 \leqslant i < j \leqslant r$. Show that $\Sigma_r S_{ij} = P^{-1} \mathrm{diag}\left(\Sigma_{r-i+j}, \Sigma_{j-i}\right) P$ for some $r \times r$ permutation matrix P.

#865 B Given a permutation matrix P, the **index** $\mathrm{ind}(P)$ of P is the smallest number k such that P can be expressed as a product of k transpositions. Show for permutation matrices P, Q that

$$\mathrm{ind}(PQ) \leqslant \mathrm{ind}(P) + \mathrm{ind}(Q).$$

Give an example where equality holds, and another example where equality does not hold.

#866 D† Show that $\mathrm{ind}(\Sigma_r) = r - 1$. If P is a permutation matrix expressible as in #859, show that $\mathrm{ind}(P) = n - k$.

#867 b Let A denote the 4×4 matrix in Example 3.175. Calculate the determinant of A by (i) expanding down the third column, (ii) expanding along the second row, (iii) by using EROs and ECOs.

#868 c Let A and $\mathbf{b}$ be as in Example 3.175. Solve the system $(A \,|\, \mathbf{b})$ by direct use of Cramer's rule.

#869 B In the notation of Theorem 3.170, prove that $AC^T = \det A \times I_n$.

#870 b (i) Verify that $\det A = 0$ where A is the matrix to the right.
(ii) Find $\mathrm{adj}A$ and verify directly that $A(\mathrm{adj}A) = (\mathrm{adj}A)A = 0_{33}$.
(iii) Show that the only matrices B to satisfy $BA = AB = 0_{33}$ are scalar multiples of $\mathrm{adj}A$.

$$A = \begin{pmatrix} 1 & 2 & 3 \\ 2 & 3 & 4 \\ 3 & 4 & 4 \end{pmatrix}.$$

#871 A† Let A be an invertible $n \times n$ matrix. Show that $\det(\mathrm{adj}A) = (\det A)^{n-1}$.

#872 a Let A, B be invertible $n \times n$ matrices. Show that $\mathrm{adj}(AB) = (\mathrm{adj}B)(\mathrm{adj}A)$.

#873B† Need (i) #871 hold when A is singular or (ii) #872 hold when A or B is singular?

#874B Show that if A is a symmetric matrix then so is adj$A$.

#875b (i) Show that the only 2×2 matrices which satisfy $A = $ adj$A$ are of the form λI_2 for some λ.
(ii) Find a 3×3 matrix A which satisfies $A = $ adj$A$ which is not a scalar multiple of I_3.

#876D† Let $A = (a_{ij})$ be an $n \times n$ matrix where $n \geqslant 2$. Show that adj$A = 0$ if and only if rank$A \leqslant n - 2$. What is the rank of adj$A$ if rank$A = n - 1$?

#877A† Give a 2×2 matrix which does not have an LU decomposition.

#878B† Show that if an invertible square matrix has an LU decomposition then it is unique. Show that this need not be the case for singular square matrices.

#879B Let A be a symmetric, invertible matrix with decomposition $A = LU$. Show that $L^{-1}U^T$ is diagonal.

#880b For each of the following matrices A_i, write $PA_i = LU$ as in Theorem 3.176.

$$A_1 = \begin{pmatrix} 1 & 0 & 2 \\ 1 & 2 & 1 \\ 3 & 1 & 2 \end{pmatrix}; \quad A_2 = \begin{pmatrix} 1 & 3 & 1 & 1 \\ 2 & 2 & 2 & 0 \\ 1 & 3 & 2 & 1 \\ 0 & 2 & 1 & 2 \end{pmatrix}; \quad A_3 = \begin{pmatrix} 0 & 0 & 0 & 1 \\ 0 & 1 & 2 & 0 \\ 0 & 3 & 1 & 0 \\ 2 & 1 & 0 & 3 \end{pmatrix}.$$

#881B Let $\mathbf{b} = (1,1,1,1)^T$ and A_2, A_3 be as in #880. Using the LU decompositions of A_2, A_3 solve the systems $(A_2 \,|\, \mathbf{b})$ and $(A_3 \,|\, \mathbf{b})$, using forward and backward substitution.

#882c Find all possible LU decompositions for the following singular matrices

$$\begin{pmatrix} 1 & 1 \\ 2 & 2 \end{pmatrix}, \quad \begin{pmatrix} 1 & 2 & 3 \\ 2 & 3 & 4 \\ 3 & 4 & 5 \end{pmatrix}, \quad \begin{pmatrix} 1 & 1 & 1 \\ 2 & 2 & 2 \\ 3 & 3 & 3 \end{pmatrix}.$$

#883B (i) Consider a linear system $(A \,|\, \mathbf{b})$, where A is $n \times n$ and invertible. Given $A = LU$, how many arithmetic operations (additions, subtractions, multiplications, divisions) are required to solve the systems $L\mathbf{y} = \mathbf{b}$ and $U\mathbf{x} = \mathbf{y}$ as in Example 3.175?
(ii) Suppose that an invertible $n \times n$ matrix A has an LU decomposition. How many arithmetic operations (additions, subtractions, multiplications, divisions) are required to determine that LU decomposition?
(iii) How many arithmetic operations are required to evaluate an $n \times n$ determinant using a Laplace expansion?

3.10 The Vector Product

In $\mathbb{R}^3$, *but not generally in other dimensions* [25], together with the scalar product there is also a *vector product*.

Definition 3.177 Let $\mathbf{u} = (u_1, u_2, u_3)$ and $\mathbf{v} = (v_1, v_2, v_3)$ be two vectors in $\mathbb{R}^3$. We define their **vector product** (or **cross product**) $\mathbf{u} \wedge \mathbf{v}$ as

$$\mathbf{u} \wedge \mathbf{v} = \begin{vmatrix} \mathbf{i} & \mathbf{j} & \mathbf{k} \\ u_1 & u_2 & u_3 \\ v_1 & v_2 & v_3 \end{vmatrix} = (u_2 v_3 - v_2 u_3)\mathbf{i} + (u_3 v_1 - v_3 u_1)\mathbf{j} + (u_1 v_2 - v_1 u_2)\mathbf{k}. \tag{3.56}$$

$\wedge$ is read as 'vec'. A common alternative notation is $\mathbf{u} \times \mathbf{v}$ and hence the alternative name of the cross product. The same definition of $\wedge$ can likewise be made for column vectors in $\mathbb{R}_3$.

- Note firstly that $\mathbf{u} \wedge \mathbf{v}$ is a vector (unlike $\mathbf{u} \cdot \mathbf{v}$ which is a real number).
- Note that the vector on the RHS of (3.56) appeared earlier in #509.
- In #510 we saw that $\mathbf{u} \wedge \mathbf{v}$ is perpendicular to both $\mathbf{u}$ and $\mathbf{v}$.
- In #510 we also saw that $\mathbf{u} \wedge \mathbf{v} = \mathbf{0}$ if and only if one of $\mathbf{u}$ and $\mathbf{v}$ is a multiple of the other.
- Note that $\mathbf{i} \wedge \mathbf{j} = \mathbf{k}$, $\mathbf{j} \wedge \mathbf{k} = \mathbf{i}$, $\mathbf{k} \wedge \mathbf{i} = \mathbf{j}$, whilst $\mathbf{i} \wedge \mathbf{k} = -\mathbf{j}$, $\mathbf{j} \wedge \mathbf{i} = -\mathbf{k}$, $\mathbf{k} \wedge \mathbf{j} = -\mathbf{i}$.

Example 3.178 Find $(2, 1, 3) \wedge (1, 0, -1)$. Determine all the vectors $(2, 1, 3) \wedge \mathbf{v}$ as $\mathbf{v}$ varies over $\mathbb{R}^3$.

Solution By definition we have

$$(2, 1, 3) \wedge (1, 0, -1) = \begin{vmatrix} \mathbf{i} & \mathbf{j} & \mathbf{k} \\ 2 & 1 & 3 \\ 1 & 0 & -1 \end{vmatrix} = (-1 - 0, 3 - (-2), 0 - 1) = (-1, 5, -1).$$

More generally with $\mathbf{v} = (a, b, c)$ we have

$$(2, 1, 3) \wedge \mathbf{v} = \begin{vmatrix} \mathbf{i} & \mathbf{j} & \mathbf{k} \\ 2 & 1 & 3 \\ a & b & c \end{vmatrix} = (c - 3b, 3a - 2c, 2b - a).$$

Note, as a, b, c vary, that this is a general vector in the plane $2x + y + 3z = 0$. □

Proposition 3.179 *(Characterizing Properties of the Vector Product) For* $\mathbf{u}, \mathbf{v}, \mathbf{w}$ *in* $\mathbb{R}^3$, *and reals* α, β *we have*

(a) $\mathbf{u} \wedge \mathbf{v} = -\mathbf{v} \wedge \mathbf{u}$.
(b) $\mathbf{u} \wedge \mathbf{v}$ *is perpendicular to both* $\mathbf{u}$ *and* $\mathbf{v}$.
(c) $(\alpha \mathbf{u} + \beta \mathbf{v}) \wedge \mathbf{w} = \alpha(\mathbf{u} \wedge \mathbf{w}) + \beta(\mathbf{v} \wedge \mathbf{w})$.
(d) *If* $\mathbf{u}, \mathbf{v}$ *are perpendicular unit vectors then* $\mathbf{u} \wedge \mathbf{v}$ *is a unit vector.*
(e) $\mathbf{i} \wedge \mathbf{j} = \mathbf{k}$.

---

[25] The only other space $\mathbb{R}^n$ for which there is a vector product with the properties (a)–(d) of Proposition 3.179 is $\mathbb{R}^7$. See #1228. For further reading see Fenn [13, §9.7]. This result relates to the possible dimensions of division algebras (p.53).

Proof (b) appeared as #510. The remaining parts are in the main simple calculations and are left as #890. □

Proposition 3.180 *For* **u**, **v** *in* $\mathbb{R}^3$*, we have* $|\mathbf{u} \wedge \mathbf{v}|^2 = |\mathbf{u}|^2 |\mathbf{v}|^2 - (\mathbf{u} \cdot \mathbf{v})^2$.

Proof This is a straightforward algebraic verification and is left as #887. □

Corollary 3.181 *For* **u**, **v** *in* $\mathbb{R}^3$*, we have* $|\mathbf{u} \wedge \mathbf{v}| = |\mathbf{u}| |\mathbf{v}| |\sin \theta|$ *where* θ *is an angle between* **u** *and* **v***. In particular,* $\mathbf{u} \wedge \mathbf{v} = \mathbf{0}$ *if and only if* **u** *and* **v** *are linearly dependent.*

Proof From (3.7) we have $\mathbf{u} \cdot \mathbf{v} = |\mathbf{u}| |\mathbf{v}| \cos \theta$. So by Proposition 3.180 we have

$$|\mathbf{u} \wedge \mathbf{v}|^2 = |\mathbf{u}|^2 |\mathbf{v}|^2 - (\mathbf{u} \cdot \mathbf{v})^2 = |\mathbf{u}|^2 |\mathbf{v}|^2 (1 - \cos^2 \theta) = |\mathbf{u}|^2 |\mathbf{v}|^2 \sin^2 \theta$$

and the result follows. Note, as $\sin(2\pi - \theta) = -\sin \theta$, the formula applies whichever of the two possible angles we choose between **u** and **v**. □

Corollary 3.182 *For* **u**, **v** *in* $\mathbb{R}^3$ *then* $|\mathbf{u} \wedge \mathbf{v}|$ *equals the area of the parallelogram with vertices* **0**, **u**, **v** *and* $\mathbf{u} + \mathbf{v}$.

Proof This is left as #888. □

The definition of the vector product in (3.56) is somewhat unsatisfactory as it appears to depend upon the choice of *xyz*-co-ordinates in $\mathbb{R}^3$. If the vector product represents something genuinely geometric – the way, for example, that the scalar product can be written in terms of lengths and angles as in (3.7) – then we should be able to characterize the vector product by similarly geometric means. A vector product $\mathbf{u} \wedge \mathbf{v}$ has a length specified by the geometry of **u** and **v** (Corollary 3.181); further, being perpendicular to both **u** and **v** (Proposition 3.179(b)), the geometry of two vectors determines their vector product up to a choice of a minus sign.

It can be shown (#891) that properties (a)–(d) above, none of which expressly involves co-ordinates, determine the vector product up to a choice of sign. What this essentially means is that there are two different **orientations** of three-dimensional space. The *xyz*-axes in $\mathbb{R}^3$ are *right-handed* (Figure 3.18a) in the sense that $\mathbf{i} \wedge \mathbf{j} = \mathbf{k}$ but we could easily have set up *xyz*-axes in a *left-handed* fashion instead, as in Figure 3.18b. Imagine pointing a thumb along the *x*-axis and first finger along the *y*-axis; then the middle finger of a right/left hand can point along the *z*-axis depending on whether the axes are right-/left-handed.

Say, now, that **u** and **v** are two independent vectors and **n** is a unit vector which is perpendicular to both **u** and **v**. Looking down at the plane containing **u** and **v** from the half-space containing **n**, denote the angle from **u** to **v** when measured in an anticlockwise sense as θ. The vector product $\mathbf{u} \wedge \mathbf{v}$ can then be defined as

$$\mathbf{u} \wedge \mathbf{v} = |\mathbf{u}| |\mathbf{v}| \sin \theta \, \mathbf{n}.$$

Note, if $|\mathbf{u}| = |\mathbf{v}| = 1$ and $\theta = \pi/2$, that **u**, **v** and **n** make right-handed axes. Note also that if we had chosen $-\mathbf{n}$ as our normal unit vector instead of **n**, then we would have measured $2\pi - \theta$ as the anticlockwise angle from **u** to **v**, but that the formula

$$|\mathbf{u}| |\mathbf{v}| \sin(2\pi - \theta)(-\mathbf{n}) = |\mathbf{u}| |\mathbf{v}|(-\sin \theta)(-\mathbf{n}) = |\mathbf{u}| |\mathbf{v}| \sin \theta \, \mathbf{n}$$

would still have led to the same result when calculating $\mathbf{u} \wedge \mathbf{v}$.

#884
#886
#887
#888
#890
#891

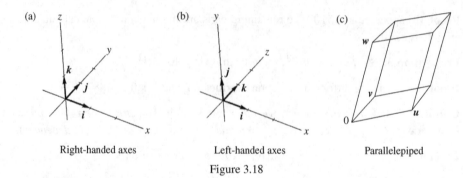

Right-handed axes · · · Left-handed axes · · · Parallelepiped

Figure 3.18

Definition 3.183 Given three vectors $\mathbf{u}, \mathbf{v}, \mathbf{w}$ in $\mathbb{R}^3$, we define the **scalar triple product** as

$$[\mathbf{u}, \mathbf{v}, \mathbf{w}] = \mathbf{u} \cdot (\mathbf{v} \wedge \mathbf{w}).$$

If $\mathbf{u} = (u_1, u_2, u_3)$, $\mathbf{v} = (v_1, v_2, v_3)$ and $\mathbf{w} = (w_1, w_2, w_3)$ then

$$[\mathbf{u}, \mathbf{v}, \mathbf{w}] = \begin{vmatrix} u_1 & u_2 & u_3 \\ v_1 & v_2 & v_3 \\ w_1 & w_2 & w_3 \end{vmatrix}. \tag{3.57}$$

Note consequently, given properties of determinants, that

$$[\mathbf{u}, \mathbf{v}, \mathbf{w}] = [\mathbf{v}, \mathbf{w}, \mathbf{u}] = [\mathbf{w}, \mathbf{u}, \mathbf{v}] = -[\mathbf{u}, \mathbf{w}, \mathbf{v}] = -[\mathbf{v}, \mathbf{u}, \mathbf{w}] = -[\mathbf{w}, \mathbf{v}, \mathbf{u}] \tag{3.58}$$

and that $[\mathbf{u}, \mathbf{v}, \mathbf{w}] = 0$ if and only if $\mathbf{u}, \mathbf{v}, \mathbf{w}$ are linearly dependent (#896).

There is a three-dimensional equivalent of a parallelogram called the **parallelepiped**, as appears in Figure 3.18c. This is a three-dimensional figure with six parallelograms for faces. Given three vectors $\mathbf{u}, \mathbf{v}, \mathbf{w}$ in $\mathbb{R}^3$, then they determine a parallelepiped with the eight vertices $\alpha\mathbf{u} + \beta\mathbf{v} + \gamma\mathbf{w}$ where each of α, β, γ is 0 or 1. If we consider $\mathbf{u}$ and $\mathbf{v}$ as determining the base of the parallelepiped then this has area $|\mathbf{u} \wedge \mathbf{v}|$. If θ is the angle between $\mathbf{w}$ and the normal to the plane containing $\mathbf{u}$ and $\mathbf{v}$, then the parallelepiped's volume is

$$\text{area of base} \times \text{height} = |\mathbf{u} \wedge \mathbf{v}| \times |\mathbf{w}| |\cos\theta| = |(\mathbf{u} \wedge \mathbf{v}) \cdot \mathbf{w}| = |[\mathbf{u}, \mathbf{v}, \mathbf{w}]|, \tag{3.59}$$

as $\mathbf{u} \wedge \mathbf{v}$ is in the direction of the normal of the plane. Similarly, the volume of the tetrahedron with vertices $\mathbf{0}, \mathbf{u}, \mathbf{v}, \mathbf{w}$ is given by $\frac{1}{6} |[\mathbf{u}, \mathbf{v}, \mathbf{w}]|$ (see #912).

Proposition 3.184 *Given three vectors* $\mathbf{u}, \mathbf{v}, \mathbf{w}$ *in* $\mathbb{R}^3$, *their* **vector triple product** *is* $\mathbf{u} \wedge (\mathbf{v} \wedge \mathbf{w})$ *and satisfies*

$$\mathbf{u} \wedge (\mathbf{v} \wedge \mathbf{w}) = (\mathbf{u} \cdot \mathbf{w})\mathbf{v} - (\mathbf{u} \cdot \mathbf{v})\mathbf{w}.$$

#885
#893
#894
#896
#897
#898
#899

Proof Both the LHS and RHS are linear in $\mathbf{u}$, so it is in fact sufficient to note that when $\mathbf{v} = (v_1, v_2, v_3)$, $\mathbf{w} = (w_1, w_2, w_3)$ then

$$(\mathbf{i} \cdot \mathbf{w})\mathbf{v} - (\mathbf{i} \cdot \mathbf{v})\mathbf{w} = w_1\mathbf{v} - v_1\mathbf{w} = (0, w_1v_2 - v_1w_2, w_1v_3 - v_1w_3) = \mathbf{i} \wedge (\mathbf{v} \wedge \mathbf{w}) \tag{3.60}$$

and two similar calculations for $\mathbf{u} = \mathbf{j}$ and $\mathbf{u} = \mathbf{k}$. The general result then follows by linearity. In fact, for those comfortable with the comments following Corollary 3.182 that the vector product is entirely determined by geometry, we can choose our x-axis to be in the direction of **u** without any loss of generality, so that the calculation in (3.60) is in fact sufficient to verify this proposition. □

Proposition 3.185 *The **scalar quadruple product** of four vectors* **a**, **b**, **c**, **d** *in* $\mathbb{R}^3$ *is* $(\mathbf{a} \wedge \mathbf{b}) \cdot (\mathbf{c} \wedge \mathbf{d})$ *and satisfies*

$$(\mathbf{a} \wedge \mathbf{b}) \cdot (\mathbf{c} \wedge \mathbf{d}) = (\mathbf{a} \cdot \mathbf{c})(\mathbf{b} \cdot \mathbf{d}) - (\mathbf{a} \cdot \mathbf{d})(\mathbf{b} \cdot \mathbf{c}).$$

Proof Set $\mathbf{e} = \mathbf{a} \wedge \mathbf{b}$. Then, by (3.58) and Proposition 3.184, $(\mathbf{a} \wedge \mathbf{b}) \cdot (\mathbf{c} \wedge \mathbf{d})$ equals

$$\mathbf{e} \cdot (\mathbf{c} \wedge \mathbf{d}) = [\mathbf{e}, \mathbf{c}, \mathbf{d}] = [\mathbf{c}, \mathbf{d}, \mathbf{e}] = \mathbf{c} \cdot (\mathbf{d} \wedge (\mathbf{a} \wedge \mathbf{b})) = \mathbf{c} \cdot \{(\mathbf{d} \cdot \mathbf{b})\mathbf{a} - (\mathbf{d} \cdot \mathbf{a})\mathbf{b}\}$$
$$= (\mathbf{a} \cdot \mathbf{c})(\mathbf{b} \cdot \mathbf{d}) - (\mathbf{a} \cdot \mathbf{d})(\mathbf{b} \cdot \mathbf{c}). \qquad \square$$

Having introduced this new product, a natural problem to address is describing the locus $\mathbf{r} \wedge \mathbf{a} = \mathbf{b}$. Note that if $\mathbf{a} \cdot \mathbf{b} \neq 0$ then this equation can have no solutions – this can be seen by dotting both sides of the equation with **a**.

Proposition 3.186 *(**Another Vector Equation for a Line**). Let* **a**, **b** *be vectors in* $\mathbb{R}^3$ *with* $\mathbf{a} \cdot \mathbf{b} = 0$. *The vectors* **r** *in* $\mathbb{R}^3$ *which satisfy* $\mathbf{r} \wedge \mathbf{a} = \mathbf{b}$ *form the line parallel to* **a** *which passes through* $(\mathbf{a} \wedge \mathbf{b}) / |\mathbf{a}|^2$.

Proof Let $\mathbf{p} = (\mathbf{a} \wedge \mathbf{b}) / |\mathbf{a}|^2$. As $\mathbf{a} \cdot \mathbf{b} = 0$ then, using the vector triple product formula,

$$\mathbf{p} \wedge \mathbf{a} = \frac{(\mathbf{a} \wedge \mathbf{b}) \wedge \mathbf{a}}{|\mathbf{a}|^2} = -\frac{\mathbf{a} \wedge (\mathbf{a} \wedge \mathbf{b})}{|\mathbf{a}|^2} = \frac{(\mathbf{a} \cdot \mathbf{a})\mathbf{b} - (\mathbf{a} \cdot \mathbf{b})\mathbf{a}}{|\mathbf{a}|^2} = \mathbf{b}.$$

We then complete the proof by noting

$$\mathbf{r} \wedge \mathbf{a} = \mathbf{b} \iff \mathbf{r} \wedge \mathbf{a} = \mathbf{p} \wedge \mathbf{a} \iff (\mathbf{r} - \mathbf{p}) \wedge \mathbf{a} = 0 \iff \mathbf{r} - \mathbf{p} = \lambda \mathbf{a} \quad \text{for some } \lambda.$$

This choice of **p** 'out of thin air' may seem unsatisfactory. For an alternative proof, note that **a** and **b** are independent, as they are perpendicular, so that by #897 we know every vector **r** in $\mathbb{R}^3$ can be written uniquely as

$$\mathbf{r} = \lambda \mathbf{a} + \mu \mathbf{b} + \nu \mathbf{a} \wedge \mathbf{b}$$

for reals λ, μ, ν. Now $\mathbf{b} = \mathbf{r} \wedge \mathbf{a}$ if and only if

$$\mathbf{b} = -\mathbf{a} \wedge (\lambda \mathbf{a} + \mu \mathbf{b} + \nu \mathbf{a} \wedge \mathbf{b}) = -\mu \mathbf{a} \wedge \mathbf{b} - \nu((\mathbf{a} \cdot \mathbf{b})\mathbf{a} - (\mathbf{a} \cdot \mathbf{a})\mathbf{b}) = -\mu \mathbf{a} \wedge \mathbf{b} + \nu |\mathbf{a}|^2 \mathbf{b}.$$

By uniqueness we may compare coefficients and see that λ may take any value, $\mu = 0$ and $\nu = 1/|\mathbf{a}|^2$. So $\mathbf{r} \wedge \mathbf{a} = \mathbf{b}$ if and only if

$$\mathbf{r} = \frac{\mathbf{a} \wedge \mathbf{b}}{|\mathbf{a}|^2} + \lambda \mathbf{a} \qquad \text{for some real } \lambda. \qquad \square$$

We claimed in (3.48), given any square matrix M, that $|\det M|$ is the area (or volume) scaling factor of the associated map μ_M. We are not in a position to prove this generally but can appreciate some of the details of this in two and three dimensions. Note that

#903
#904
#905
#906
#907
#908

premultiplication by the matrix

$$M = \begin{pmatrix} a & b \\ c & d \end{pmatrix}$$

moves the unit square with vertices

$$O = (0,0), \qquad A = (1,0), \qquad B = (1,1), \qquad C = (0,1)$$

to the parallelogram with vertices

$$O = (0,0), \qquad A' = (a,c), \qquad B' = (a+b,c+d), \qquad C' = (b,d).$$

By Corollary 3.182, the parallelogram $OA'B'C'$ has area

$$\left| OA' \wedge OC' \right| = |(a,c,0) \wedge (b,d,0)| = |(0,0,ad-bc)| = |\det M|.$$

Similarly the 3×3 matrix $M = (\mathbf{u}\,|\,\mathbf{v}\,|\,\mathbf{w})$ takes the unit cube in $\mathbb{R}^3$ to a parallelepiped with volume

$$|[\mathbf{u},\mathbf{v},\mathbf{w}]| = \left| \det M^T \right| = |\det M|.$$

Whilst we have only demonstrated that the associated map μ_M of a 2×2 or 3×3 matrix M scales area/volume by $|\det M|$ for squares and cubes, it is not hard to imagine a more complicated region R being approximated by small squares or cubes. Each such square or cube would have its area/volume scaled the same and so the image of R would have an area/volume that is $|\det M|$ times its original area/volume. What is missing here is not an algebraic issue so much as rigorous definitions for area and volume.

Exercises

#884a Determine $(1,2,3) \wedge (3,2,1)$. Determine also $(1,k,1) \wedge (1,0,1)$ and $(k,1,k) \wedge (1,0,1)$, where k is a real number.

#885A† Show, for $\mathbf{a},\mathbf{b},\mathbf{c}$ in $\mathbb{R}^3$, that $(\mathbf{a} \wedge \mathbf{b}) \wedge \mathbf{c} = (\mathbf{c} \cdot \mathbf{a})\mathbf{b} - (\mathbf{c} \cdot \mathbf{b})\mathbf{a}$.

#886A† Under what circumstances does $\mathbf{u} \wedge \mathbf{v} = \mathbf{v} \wedge \mathbf{u}$?

#887a For $\mathbf{u},\mathbf{v}$ in $\mathbb{R}^3$, show that $|\mathbf{u} \wedge \mathbf{v}|^2 = |\mathbf{u}|^2 |\mathbf{v}|^2 - (\mathbf{u} \cdot \mathbf{v})^2$.

#888A Verify Corollary 3.182.

#889A Let $\mathbf{a} = (a_1,a_2,a_3)^T$ and $\mathbf{b} = (b_1,b_2,b_3)^T$. Show that

$$\mathbf{a} \wedge \mathbf{b} = \begin{pmatrix} 0 & -a_3 & a_2 \\ a_3 & 0 & -a_1 \\ -a_2 & a_1 & 0 \end{pmatrix} \begin{pmatrix} b_1 \\ b_2 \\ b_3 \end{pmatrix}. \tag{3.61}$$

#890B Verify properties (a), (c), (d) from Proposition 3.179.

#891B† Let $\sqcap$ be a vector product which assigns to any two vectors $\mathbf{u},\mathbf{v}$ in $\mathbb{R}^3$ a vector $\mathbf{u} \sqcap \mathbf{v}$ in $\mathbb{R}^3$ and which satisfies properties (a)–(d) of Proposition 3.179. Show that $\sqcap$ is $\wedge$ or $-\wedge$. Hence only $\wedge$ satisfies each of (a)–(e).

#892D† (See also #1228.) Show that there is no vector product $\wedge$ on $\mathbb{R}^4$ with properties Proposition 3.179(a)–(d).

#893A† Let ABC be three points in space with position vectors $\mathbf{a}, \mathbf{b}, \mathbf{c}$ from an origin O. Show that O, A and B are collinear if and only if $\mathbf{a} \wedge \mathbf{b} = \mathbf{0}$. Deduce that A, B and C are collinear if and only if $\mathbf{a} \wedge \mathbf{b} + \mathbf{b} \wedge \mathbf{c} + \mathbf{c} \wedge \mathbf{a} = \mathbf{0}$.

#894B† Prove for $\mathbf{a}, \mathbf{b}, \mathbf{c}, \mathbf{d}$ in $\mathbb{R}^3$ that $(\mathbf{a} \wedge \mathbf{b}) \wedge (\mathbf{c} \wedge \mathbf{d}) = [\mathbf{a}, \mathbf{b}, \mathbf{d}]\mathbf{c} - [\mathbf{a}, \mathbf{b}, \mathbf{c}]\mathbf{d}$.

#895c† Let OAB be a triangle in the plane with centroid C. Show that the triangles OAC, OBC, ABC are each one third the area of OAB.

#896B Quoting appropriate properties of determinants:

  (i) Verify the expression (3.57).
 (ii) Verify the properties of the triple scalar product in (3.58).
(iii) Verify that $[\mathbf{u}, \mathbf{v}, \mathbf{w}] = 0$ if and only if $\mathbf{u}, \mathbf{v}, \mathbf{w}$ are dependent.

#897B† Suppose that $\mathbf{a}$ and $\mathbf{b}$ are independent vectors in $\mathbb{R}^3$. Show that $\mathbf{a}, \mathbf{b}$ and $\mathbf{a} \wedge \mathbf{b}$ are independent.

#898b† For vectors $\mathbf{u}, \mathbf{v}, \mathbf{w}$ in $\mathbb{R}^3$ show that

$$[\mathbf{u} \wedge \mathbf{v}, \mathbf{v} \wedge \mathbf{w}, \mathbf{w} \wedge \mathbf{u}] = [\mathbf{u}, \mathbf{v}, \mathbf{w}]^2.$$

Deduce that $\mathbf{u}, \mathbf{v}, \mathbf{w}$ are independent if and only if $\mathbf{u} \wedge \mathbf{v}, \mathbf{v} \wedge \mathbf{w}, \mathbf{w} \wedge \mathbf{u}$ are independent.

#899b† Find the most general solution of the vector equations $\mathbf{r} \cdot \mathbf{a} = \alpha$ and $\mathbf{r} \cdot \mathbf{b} = \beta$, where $\mathbf{a} \wedge \mathbf{b} \neq \mathbf{0}$. Under what conditions do the equations above and the equation $\mathbf{r} \cdot \mathbf{c} = \gamma$ (where $\mathbf{c} \neq \mathbf{0}$) have a unique common solution?

#900C† Let $\mathbf{a}, \mathbf{b}$ be vectors in $\mathbb{R}^3$. Under what conditions does the equation $\mathbf{r} \wedge \mathbf{a} = (\mathbf{r} \cdot \mathbf{b})\mathbf{r}$ have non-zero solutions?

#901D† Let $\mathbf{a}, \mathbf{b}$ be vectors in $\mathbb{R}^3$. Show that the equation $\mathbf{r} \wedge \mathbf{a} + \mathbf{r} = \mathbf{b}$ has a unique solution.

#902C Let $\mathbf{a}$ be a non-zero vector in $\mathbb{R}^3$. Determine all the solutions of the vector equation $(\mathbf{r} \wedge \mathbf{a}) \wedge \mathbf{r} = \mathbf{0}$.

#903b Show that the equation of the plane containing three non-collinear points with position vectors $\mathbf{a}, \mathbf{b}, \mathbf{c}$ is

$$\mathbf{r} \cdot (\mathbf{a} \wedge \mathbf{b} + \mathbf{b} \wedge \mathbf{c} + \mathbf{c} \wedge \mathbf{a}) = [\mathbf{a}, \mathbf{b}, \mathbf{c}].$$

Deduce that four points $\mathbf{a}, \mathbf{b}, \mathbf{c}, \mathbf{d}$ are coplanar if and only if

$$[\mathbf{a}, \mathbf{b}, \mathbf{c}] - [\mathbf{b}, \mathbf{c}, \mathbf{d}] + [\mathbf{c}, \mathbf{d}, \mathbf{a}] - [\mathbf{d}, \mathbf{a}, \mathbf{b}] = 0.$$

#904A Find three vectors $\mathbf{u}, \mathbf{v}, \mathbf{w}$ such that $\mathbf{u} \wedge (\mathbf{v} \wedge \mathbf{w}) \neq (\mathbf{u} \wedge \mathbf{v}) \wedge \mathbf{w}$.

#905a (i) Prove **Jacobi's identity** [26] that $\mathbf{a} \wedge (\mathbf{b} \wedge \mathbf{c}) + \mathbf{b} \wedge (\mathbf{c} \wedge \mathbf{a}) + \mathbf{c} \wedge (\mathbf{a} \wedge \mathbf{b}) = \mathbf{0}$ for any three vectors $\mathbf{a}, \mathbf{b}, \mathbf{c}$ in $\mathbb{R}^3$.

(ii) For $\mathbf{u}, \mathbf{v}, \mathbf{w}$ in $\mathbb{R}^3$, under what circumstances does $\mathbf{u} \wedge (\mathbf{v} \wedge \mathbf{w}) = (\mathbf{u} \wedge \mathbf{v}) \wedge \mathbf{w}$?

#906b† Write the equations of each of the following lines in the form $\mathbf{r} \wedge \mathbf{a} = \mathbf{b}$.

(i) The line through the points $(1, 1, 1)$ and $(1, 2, 3)$
(ii) The line with equation $(x - 1)/2 = y/3 = z + 1$
(iii) The intersection of the planes $x + y + z = 1$ and $x - y - z = 2$

#907B† Show that the distance of the point with position vector $\mathbf{c}$ from the line $\mathbf{r} \wedge \mathbf{a} = \mathbf{b}$ is $|\mathbf{c} \wedge \mathbf{a} - \mathbf{b}|/|\mathbf{a}|$.

#908B† Say $\mathbf{a} \cdot \mathbf{b} = \mathbf{c} \cdot \mathbf{d} = 0$ with $\mathbf{a} \wedge \mathbf{c} \neq \mathbf{0}$. Show that the lines $\mathbf{r} \wedge \mathbf{a} = \mathbf{b}$ and $\mathbf{r} \wedge \mathbf{c} = \mathbf{d}$ intersect if and only if $\mathbf{a} \cdot \mathbf{d} + \mathbf{b} \cdot \mathbf{c} = 0$.

#909D (From Oxford 2009 Paper A #8) Two non-parallel lines L_1 and L_2 in three-dimensional space have respective equations $\mathbf{r} \wedge \mathbf{a}_1 = \mathbf{b}_1$ and $\mathbf{r} \wedge \mathbf{a}_2 = \mathbf{b}_2$. For $i = 1, 2$, let Π_i denote the plane of the form $\mathbf{r} \cdot (\mathbf{a}_1 \wedge \mathbf{a}_2) = c_i$ which contains L_i. Show that $c_1 = \mathbf{b}_1 \cdot \mathbf{a}_2$ and find c_2. Hence show that the least distance between the lines equals

$$\frac{|\mathbf{a}_1 \cdot \mathbf{b}_2 + \mathbf{a}_2 \cdot \mathbf{b}_1|}{|\mathbf{a}_1 \wedge \mathbf{a}_2|}.$$

#910B (From Oxford 2009 Paper A #8) Let $OABC$ be a regular tetrahedron with common side length s and let $\mathbf{a}, \mathbf{b}, \mathbf{c}$ be the position vectors of A, B, C from the origin O. If

$$\mathbf{a} = (s, 0, 0), \qquad \mathbf{b} = (b_1, b_2, 0), \qquad \mathbf{c} = (c_1, c_2, c_3),$$

where $b_2, c_3 > 0$, determine b_1, b_2, c_1, c_2, c_3 in terms of s. Hence show that the shortest distance between two opposing sides of $OABC$ is $s/\sqrt{2}$.

#911B† Let A be a 3×3 matrix and $\mathbf{u}, \mathbf{v}, \mathbf{w}$ be three column vectors in $\mathbb{R}_3$. Show that $[A\mathbf{u}, A\mathbf{v}, A\mathbf{w}] = \det A \times [\mathbf{u}, \mathbf{v}, \mathbf{w}]$.

#912b† Let $0 < c < 1$. The tetrahedron T has vertices $(0, 0, 0)$, $(0, 0, 1)$, $(0, 1, 0)$ and $(1, 0, 0)$. Show that the intersection of T with the plane $z = c$ has area $(1 - c)^2/2$. Hence find the volume of T. Deduce that the volume of the tetrahedron with vertices $\mathbf{0}, \mathbf{u}, \mathbf{v}, \mathbf{w}$ is given by $|[\mathbf{u}, \mathbf{v}, \mathbf{w}]|/6$.

#913b (i) Let $\mathbf{a}$ be a fixed column vector in $\mathbb{R}_3$. Show that $T_{\mathbf{a}}(\mathbf{x}) = \mathbf{a} \wedge \mathbf{x}$ defines a linear map from $\mathbb{R}_3$ to $\mathbb{R}_3$.

(ii) What is the rank of $T_{\mathbf{a}}$? What is $\text{Null}(T_{\mathbf{a}})$? What is $\text{Col}(T_{\mathbf{a}})$?
(iii) Write down the associated matrix for $T_{\mathbf{a}}$ when $\mathbf{a} = (a_1, a_2, a_3)$.

#914B† For $\mathbf{a}$ in $\mathbb{R}^3$, define the linear map $T_{\mathbf{a}}$ by $T_{\mathbf{a}}(\mathbf{b}) = \mathbf{a} \wedge \mathbf{b}$. Determine $(T_{\mathbf{a}})^n(\mathbf{b})$.

---

[26] Named after the German mathematician Carl Jacobi (1804–1851). Jacobi made important contributions in the study of *elliptic functions* and *elliptic integrals* (see p.419). However, his name is commonly first met with reference to the *Jacobian* matrix and determinant. The Jacobian matrix represents the derivative of a multivariable function from $\mathbb{R}_n$ to $\mathbb{R}_m$ and, when $m = n$, its determinant represents a (signed) local volume-scaling factor important when changing variables with n-dimensional integrals.

#915D Let $\mathbf{n}$ be a unit vector. Show that rotation by θ about the line $\mathbf{r}(\lambda) = \lambda \mathbf{n}$ is given by

$$T(\mathbf{v}) = (\mathbf{v} \cdot \mathbf{n})\mathbf{n} + \sin\theta(\mathbf{n} \wedge \mathbf{v}) + \cos\theta(\mathbf{n} \wedge (\mathbf{v} \wedge \mathbf{n})).$$

3.11 Diagonalization

Definition 3.187 An $n \times n$ matrix A is said to be **diagonalizable** if there is an invertible matrix P such that $P^{-1}AP$ is diagonal.

Two questions immediately spring to mind: why might this be a useful definition, and how might we decide whether such a matrix P exists? In an attempt to partially answer the first question, we recall #618, where we noted

$$P^{-1}A^k P = (P^{-1}AP)^k.$$

Thus if $P^{-1}AP = D$ is diagonal then

$$A^k = PD^k P^{-1} \qquad \text{for a natural number } k$$

and so we are in a position to easily calculate the powers of A as powers of diagonal matrices can be readily calculated. So ease of calculation is clearly one advantage of a matrix being diagonalizable.

For now we will consider this reason enough to seek to answer the second question: how do we determine whether such a P exists? Suppose such a P exists and has columns $\mathbf{v}_1, \mathbf{v}_2, \ldots, \mathbf{v}_n$. As P is invertible then the $\mathbf{v}_i$ are linearly independent. Further as $AP = PD$ where $D = \text{diag}(\lambda_1, \ldots, \lambda_n)$ then we have by #563 that

$$i\text{th column of } AP = A\mathbf{v}_i \qquad \text{and} \qquad i\text{th column of } PD = P(\lambda_i \mathbf{e}_i^T) = \lambda_i \mathbf{v}_i.$$

So the columns of P are n independent vectors, each of which A maps to a scalar multiple of itself. Thus we make the following definitions.

Definition 3.188 Let A be an $n \times n$ matrix. We say that a *non-zero* vector $\mathbf{v}$ in $\mathbb{R}_n$ is an **eigenvector** [27] of A if $A\mathbf{v} = \lambda\mathbf{v}$ for some scalar λ. The scalar λ is called the **eigenvalue** of $\mathbf{v}$ and we will also refer to $\mathbf{v}$ as a λ-**eigenvector**.

Definition 3.189 n linearly independent eigenvectors of an $n \times n$ matrix A are called an **eigenbasis** (of A).

And we have partly demonstrated the following.

Theorem 3.190 *An $n \times n$ matrix A is diagonalizable if and only if A has an eigenbasis.*

#916
#917
#918
#922
#924
#952

Proof We showed earlier that if such a P exists then its columns form an eigenbasis. Conversely if $\mathbf{v}_1, \ldots, \mathbf{v}_n$ form an eigenbasis, with respective eigenvalues $\lambda_1, \ldots, \lambda_n$, we define $P = (\mathbf{v}_1 | \cdots | \mathbf{v}_n)$. Again P is invertible as its columns are linearly independent. By #563,

$$P\mathbf{e}_i^T = \mathbf{v}_i \qquad \text{and} \qquad AP\mathbf{e}_i^T = A\mathbf{v}_i = \lambda_i\mathbf{v}_i = \lambda_i P\mathbf{e}_i^T = P(\lambda_i\mathbf{e}_i^T),$$

[27] The German adjective *eigen* means 'own' or 'particular'. David Hilbert was the first to use the term in the early twentieth century. The term *proper* or *characteristic* is sometimes also used, especially in older texts.

so that $P^{-1}APe_i^T = \lambda_i e_i^T$ for each i or equivalently

$$P^{-1}AP = \mathrm{diag}(\lambda_1, \lambda_2, \ldots, \lambda_n).$$ □

Note that λ is an eigenvalue of A if and only if the equation $A\mathbf{v} = \lambda\mathbf{v}$ has a non-zero solution or equivalently if $(\lambda I_n - A)\mathbf{v} = \mathbf{0}$ has a non-zero solution. By Proposition 3.95, this is equivalent to $\lambda I_n - A$ being singular, which in turn is equivalent to $\det(\lambda I_n - A) = 0$. Thus we have shown (a) below.

Proposition 3.191 (a) *A real number λ is an eigenvalue of A if and only if $x = \lambda$ is a root of $\det(xI_n - A) = 0$.*
(b) $\det(xI_n - A)$ is a polynomial in x of degree n which is monic (i.e. leading coefficient is 1).
(c) If $\det(xI_n - A) = x^n + c_{n-1}x^{n-1} + \cdots + c_0$ then

$$c_0 = (-1)^n \det A \qquad and \qquad c_{n-1} = -\mathrm{trace}(A).$$

Proof We leave part (b) as #931. The first part of (c) follows from setting $x = 0$ and Corollary 3.157(b). The second part is left to #952.   □

Definition 3.192 Let A be a real $n \times n$ matrix. Then the **characteristic polynomial** of A is

$$c_A(x) = \det(xI_n - A).$$

Example 3.193 Find the eigenvalues of the following matrices.

$$A = \begin{pmatrix} 1 & 1 \\ 1 & 1 \end{pmatrix}; \qquad B = \begin{pmatrix} 1 & -1 \\ 1 & 1 \end{pmatrix};$$

$$C = \begin{pmatrix} 3 & 2 & -4 \\ 0 & 1 & 4 \\ 0 & 0 & 3 \end{pmatrix}; \qquad D = \begin{pmatrix} 5 & -3 & -5 \\ 2 & 9 & 4 \\ -1 & 0 & 7 \end{pmatrix}.$$

Solution By Proposition 3.191(a) this is equivalent to finding the *real* roots of the matrices' characteristic polynomials.

(a) The eigenvalues of A are 0 and 2 as

$$c_A(x) = \begin{vmatrix} x-1 & -1 \\ -1 & x-1 \end{vmatrix} = (x-1)^2 - 1 = x(x-2).$$

(b) Similarly note

$$c_B(x) = \begin{vmatrix} x-1 & 1 \\ -1 & x-1 \end{vmatrix} = (x-1)^2 + 1 = x^2 - 2x + 2.$$

Now $c_B(x)$ has no real roots (the roots are $1 \pm i$) and so B has no eigenvalues.

(c) As C is triangular then we can immediately see (#821) that $c_C(x) = (x-3)(x-1)(x-3)$. So C has eigenvalues $1, 3, 3$, the eigenvalue of 3 being a repeated root of $c_C(x)$.

(d) Finally D has eigenvalues $6, 6, 9$, the eigenvalue of 6 being repeated as $c_D(x)$ equals

$$\begin{vmatrix} x-5 & 3 & 5 \\ -2 & x-9 & -4 \\ 1 & 0 & x-7 \end{vmatrix} = (x-6) \begin{vmatrix} 1 & 1 & 1 \\ -2 & x-9 & -4 \\ 1 & 0 & x-7 \end{vmatrix}$$

$$= (x-6) \begin{vmatrix} x-7 & -2 \\ -1 & x-8 \end{vmatrix} = (x-6)^2(x-9). \qquad \square$$

Here follow some basic facts about eigenvalues, eigenvectors and diagonalizability.

Proposition 3.194 *Let A be a square matrix.*

(a) If $\mathbf{v}$ is a λ-eigenvector of A, and $c \neq 0$, then $c\mathbf{v}$ is a λ-eigenvector.

(b) If $\mathbf{v}, \mathbf{w}$ are independent λ-eigenvectors of A, and c, d are real numbers not both zero, then $c\mathbf{v} + d\mathbf{w}$ is a λ-eigenvector.

(c) For $1 \leqslant i \leqslant k$, let $\mathbf{v}_i$ be a λ_i-eigenvector of A. If $\lambda_1, \ldots, \lambda_k$ are distinct then $\mathbf{v}_1, \ldots, \mathbf{v}_k$ are independent.

Proof Parts (a) and (b) are left as #932. Part (c) may be proven by induction as follows. Note that $\mathbf{v}_1 \neq \mathbf{0}$ (as it is an eigenvector) and so $\mathbf{v}_1$ makes an independent set. Suppose, as our inductive hypothesis, that $\mathbf{v}_1, \ldots, \mathbf{v}_i$ are linearly independent vectors and that

$$\alpha_1 \mathbf{v}_1 + \cdots + \alpha_i \mathbf{v}_i + \alpha_{i+1} \mathbf{v}_{i+1} = \mathbf{0} \tag{3.62}$$

for some reals $\alpha_1, \ldots, \alpha_{i+1}$. If we apply A to both sides of (3.62), we find

$$\alpha_1 \lambda_1 \mathbf{v}_1 + \cdots + \alpha_i \lambda_i \mathbf{v}_i + \alpha_{i+1} \lambda_{i+1} \mathbf{v}_{i+1} = \mathbf{0}. \tag{3.63}$$

Now subtracting λ_{i+1} times (3.62) from (3.63) we arrive at

$$\alpha_1 (\lambda_1 - \lambda_{i+1}) \mathbf{v}_1 + \cdots + \alpha_i (\lambda_i - \lambda_{i+1}) \mathbf{v}_i = \mathbf{0}.$$

By hypothesis $\mathbf{v}_1, \ldots, \mathbf{v}_i$ are linearly independent vectors and hence $\alpha_j (\lambda_j - \lambda_{i+1}) = 0$ for $1 \leqslant j \leqslant i$. As $\lambda_1, \ldots, \lambda_i$ are distinct then $\alpha_j = 0$ for $1 \leqslant j \leqslant i$ and then by (3.62) $\alpha_{i+1} = 0$ as well. We have shown that $\mathbf{v}_1, \ldots, \mathbf{v}_{i+1}$ are linearly independent vectors and so part (c) follows by induction. $\qquad \square$

Remark 3.195 Those who read §3.6 will recognize that (b) in Proposition 3.194 shows that the λ-eigenvectors, together with $\mathbf{0}$, form a subspace of $\mathbb{R}_n$ known as an **eigenspace**, or more specifically as the λ-**eigenspace**. In fact, we see that the λ-eigenspace is the null space of $\lambda I_n - A$.

Those readers will also recognize that an eigenbasis of a diagonalizable $n \times n$ matrix is a basis for $\mathbb{R}_n$. ∎

Corollary 3.196 *If an $n \times n$ matrix has n distinct eigenvalues then it is diagonalizable.*

Proof Let $\lambda_1, \ldots, \lambda_n$ denote the distinct eigenvalues. For each i there is a λ_i-eigenvector $\mathbf{v}_i$ and by Proposition 3.194(c) $\mathbf{v}_1, \ldots, \mathbf{v}_n$ are independent. There being n of them they form an eigenbasis. $\qquad \square$

- It is important to note Corollary 3.196 is a sufficient, but not a necessary condition for diagonalizability. For example, I_n is diagonal (and so diagonalizable) but has eigenvalue 1 repeated n times.

Example 3.197 Determine the eigenvectors and diagonalizability of the matrices A, B, C, D from Example 3.193.

Solution (a) We determined that A has eigenvalues $\lambda = 0$ and 2. Note that

$$(A - 0I)\begin{pmatrix} x \\ y \end{pmatrix} = 0 \iff \begin{pmatrix} 1 & 1 \\ 1 & 1 \end{pmatrix}\begin{pmatrix} x \\ y \end{pmatrix} = 0 \iff x + y = 0;$$

$$(A - 2I)\begin{pmatrix} x \\ y \end{pmatrix} = 0 \iff \begin{pmatrix} -1 & 1 \\ 1 & -1 \end{pmatrix}\begin{pmatrix} x \\ y \end{pmatrix} = 0 \iff x = y.$$

The eigenvectors of A are $(\alpha, -\alpha)^T$ and $(\beta, \beta)^T$, where $\alpha, \beta \neq 0$. In particular, $(1, -1)^T$ and $(1, 1)^T$ form an eigenbasis and if we set

$$P = \begin{pmatrix} 1 & 1 \\ -1 & 1 \end{pmatrix} \qquad \text{then} \qquad P^{-1}AP = \begin{pmatrix} 0 & 0 \\ 0 & 2 \end{pmatrix}.$$

(b) B has no real eigenvalues and so no eigenvectors. Consequently B is not diagonalizable. (At least not using a real matrix P; however, see Example 3.198.)

(c) C has eigenvalues $1, 3, 3$. Note

$$(C - 3I)\begin{pmatrix} x \\ y \\ z \end{pmatrix} = 0 \iff \begin{pmatrix} 0 & 2 & -4 \\ 0 & -2 & 4 \\ 0 & 0 & 0 \end{pmatrix}\begin{pmatrix} x \\ y \\ z \end{pmatrix} = 0 \iff y = 2z.$$

$$(C - 1I)\begin{pmatrix} x \\ y \\ z \end{pmatrix} = 0 \iff \begin{pmatrix} 2 & 2 & -4 \\ 0 & 0 & 4 \\ 0 & 0 & 2 \end{pmatrix}\begin{pmatrix} x \\ y \\ z \end{pmatrix} = 0 \iff x + y = 0 = z.$$

The 3-eigenvectors are of the form $(\alpha, 2\beta, \beta)^T$, where α and β are not both zero and the 1-eigenvectors are of the form $(\gamma, -\gamma, 0)^T$, where $\gamma \neq 0$. Thus an eigenbasis is $(1, 0, 0)^T$, $(0, 2, 1)^T$ and $(1, -1, 0)^T$. Setting

$$P = \begin{pmatrix} 1 & 0 & 1 \\ 0 & 2 & -1 \\ 0 & 1 & 0 \end{pmatrix}, \qquad \text{then} \qquad P^{-1}CP = \text{diag}(3, 3, 1).$$

(d) D has eigenvalues $6, 6, 9$. Note that

$$(D - 6I)\begin{pmatrix} x \\ y \\ z \end{pmatrix} = 0 \iff \begin{pmatrix} -1 & -3 & -5 \\ 2 & 3 & 4 \\ -1 & 0 & 1 \end{pmatrix}\begin{pmatrix} x \\ y \\ z \end{pmatrix} = 0$$

$$\iff x - z = 0 = y + 2z.$$

$$(D - 9I)\begin{pmatrix} x \\ y \\ z \end{pmatrix} = 0 \iff \begin{pmatrix} -4 & -3 & -5 \\ 2 & 0 & 4 \\ -1 & 0 & -2 \end{pmatrix}\begin{pmatrix} x \\ y \\ z \end{pmatrix} = 0$$

$$\iff x + 2z = 0 = y - z.$$

#920
#925
#926
#927

The 6-eigenvectors are non-zero multiples of $(1, -2, 1)^T$ and the 9-eigenvectors are non-zero multiples of $(-2, 1, 1)^T$. As we can find no more than two independent eigenvectors, then there is no eigenbasis and D is not diagonalizable. $\square$

Example 3.198 Find a complex matrix P such that $P^{-1}BP$ is diagonal, where B is as given in Example 3.193.

#936
#937
#948
#949

Remark 3.199 When we defined 'diagonalizability' in Definition 3.187 we were, strictly speaking, defining 'diagonalizability over $\mathbb{R}$'. We would say that B is not diagonalizable over $\mathbb{R}$, as no such matrix P with real entries exists, but B is diagonalizable over $\mathbb{C}$ as such a complex matrix P *does* exist. In this text we will mainly be interested in real eigenvalues, real eigenvectors and diagonalizability over $\mathbb{R}$. $\blacksquare$

Solution The roots of $c_B(x) = (x-1)^2 + 1$ are $1 \pm i$. Note that

$$(A - (1+i)I)\begin{pmatrix} x \\ y \end{pmatrix} = \mathbf{0} \iff \begin{pmatrix} -i & -1 \\ 1 & -i \end{pmatrix}\begin{pmatrix} x \\ y \end{pmatrix} = \mathbf{0} \iff x = iy;$$

$$(A - (1-i)I)\begin{pmatrix} x \\ y \end{pmatrix} = \mathbf{0} \iff \begin{pmatrix} i & -1 \\ 1 & i \end{pmatrix}\begin{pmatrix} x \\ y \end{pmatrix} = \mathbf{0} \iff ix = y.$$

So $(i, 1)^T$ is a complex $(1 + i)$-eigenvector and $(1, i)^T$ is a complex $(1 - i)$-eigenvector. So we may take

$$P = \begin{pmatrix} i & 1 \\ 1 & i \end{pmatrix} \quad \text{and find} \quad P^{-1}BP = \begin{pmatrix} 1+i & 0 \\ 0 & 1-i \end{pmatrix}. \qquad \square$$

Examples 3.193 and 3.197 cover the various eventualities that may arise when investigating the diagonalizability of matrices. In summary then, the checklist when testing a matrix for diagonalizability is as follows.

Algorithm 3.200 *(Determining Diagonalizability over $\mathbb{R}$ and $\mathbb{C}$)*

(a) *Let A be an $n \times n$ matrix. Determine its characteristic polynomial $c_A(x)$.*

(b) *If any of the roots of $c_A(x)$ are not real, then A is not diagonalizable using a real invertible matrix P.*

(c) *If all the roots of $c_A(x)$ are real and distinct then A is diagonalizable using a real invertible matrix P.*

(d) *If all the roots of $c_A(x)$ are real, and for each root λ there are as many independent λ-eigenvectors as repeated factors of $x - \lambda$ in $c_A(x)$, then A is diagonalizable using a real invertible matrix P.*

(c)' *If the roots of $c_A(x)$ are distinct complex numbers then A is diagonalizable using a complex invertible matrix P.*

(d)' *If for each root λ of $c_A(x)$ and there are as many independent λ-eigenvectors in $\mathbb{C}_n$ as repeated factors of $x - \lambda$ in $c_A(x)$, then A is diagonalizable using a complex invertible matrix P.*

(c) and the same result (c′) for complex matrices, were proven in Corollary 3.196.
(d) and its complex version (d′), will be proven in Corollary 3.203.

So a square matrix can fail to be diagonalizable using a real invertible matrix P when

- not all the roots of $c_A(x)$ are real – counting multiplicities and including complex roots, $c_A(x)$ has n roots. However, we will see (Proposition 3.202) that there are at most as many independent λ-eigenvectors as repetitions of λ as a root. So if some roots are not real we cannot hope to find n independent real eigenvectors. This particular problem can be circumvented by seeking an invertible complex matrix P instead.
- some (real or complex) root λ of $c_A(x)$ has fewer independent λ-eigenvectors (in $\mathbb{R}_n$ or $\mathbb{C}_n$) than there are factors of $x - \lambda$ in $c_A(x)$.

The latter problem cannot be circumvented; however; this latter possibility is reassuringly unlikely. If a matrix's entries contain experimental data or randomly selected entries – rather than being a contrived exercise – then $c_A(x)$ will almost certainly have distinct complex roots and so A will be diagonalizable using a complex invertible matrix P.

Definition 3.201 Let A be an $n \times n$ matrix with eigenvalue λ.

(a) The **algebraic multiplicity** of λ is the number of factors of $x - \lambda$ in the characteristic polynomial $c_A(x)$.
(b) The **geometric multiplicity** of λ is the maximum number of linearly independent λ-eigenvectors. (For those familiar with §3.6 this equals the dimension of the λ-eigenspace.)

Proposition 3.202 *The geometric multiplicity of an eigenvalue is less than or equal to its algebraic multiplicity.*

Proof Let g and a respectively denote the geometric and algebraic multiplicities of an eigenvalue λ of an $n \times n$ matrix A. There are then g independent λ-eigenvectors $v_1, v_2, \ldots, v_g$ and by Corollary 3.116 we can extend these vectors to n independent vectors $v_1, \ldots, v_n$. If we put $v_1, \ldots, v_n$ as the columns of a matrix P then, arguing as in Theorem 3.190, we have

$$P^{-1}AP = \begin{pmatrix} \lambda I_g & B \\ 0_{(n-g)g} & C \end{pmatrix},$$

where B is a $g \times (n - g)$ matrix and C is $(n - g) \times (n - g)$. By the product rule for determinants and #822 we have

$$c_A(x) = \det(xI_n - A) = \det(P(xI_n - P^{-1}AP)P^{-1})$$
$$= \det(xI_n - P^{-1}AP)$$
$$= \det \begin{pmatrix} (x - \lambda)I_g & -B \\ 0_{(n-g)g} & xI_{n-g} - C \end{pmatrix} = (x - \lambda)^g c_C(x).$$

So there are at least g factors of $x - \lambda$ in $c_A(x)$ and hence $a \geqslant g$. □

Corollary 3.203 *Let A be a square matrix with all the roots of c_A being real. Then A is diagonalizable if and only if, for each eigenvalue, its geometric multiplicity equals its algebraic multiplicity.*

Proof Let the *distinct* eigenvalues of A be $\lambda_1, \ldots, \lambda_k$ with geometric multiplicities $g_1, \ldots, g_k$ and algebraic multiplicities $a_1, \ldots, a_k$. By the previous proposition

$$g_1 + \cdots + g_k \leqslant a_1 + \cdots + a_k = \deg c_A(x) = n, \tag{3.64}$$

the equalities in (3.64) following as all the roots of c_A are real. We can find g_i linearly independent λ_i-eigenvectors $\mathbf{v}_1^{(i)}, \ldots, \mathbf{v}_{g_i}^{(i)}$ for each i. If $g_i = a_i$ for each i then we have n eigenvectors in all, but if $g_i < a_i$ for any i then $g_1 + \cdots + g_k < n$ by (3.64), so we will not be able to find n independent eigenvectors and no eigenbasis exists. It remains to show that if $g_i = a_i$ for each i then these n eigenvectors are indeed independent. Say that

$$\sum_{i=1}^{k} \sum_{j=1}^{g_i} \alpha_j^{(i)} \mathbf{v}_j^{(i)} = \mathbf{0},$$

for some scalars $\alpha_j^{(i)}$. As $\lambda_1, \ldots, \lambda_k$ are distinct, arguing along the same lines as Proposition 3.194(c) it follows that

$$\mathbf{w}_i = \sum_{j=1}^{g_i} \alpha_j^{(i)} \mathbf{v}_j^{(i)} = \mathbf{0}, \qquad \text{for each } i,$$

as $\mathbf{w}_i$ is a λ_i-eigenvector (or $\mathbf{0}$). Now the vectors $\mathbf{v}_1^{(i)}, \ldots, \mathbf{v}_{g_i}^{(i)}$ are independent and so $\alpha_j^{(i)} = 0$ for each i and j, and hence these n vectors are indeed independent and so form an eigenbasis. $\qquad \square$

If we recall the matrices A, B, C, D from Example 3.193, we can now see that A meets criterion (c) and so is diagonalizable; B meets criterion (b) and so is not diagonalizable over $\mathbb{R}$ but does meet criterion (c)$'$ of Algorithm 3.200 and is diagonalizable over $\mathbb{C}$; matrix C meets criterion (d) and so is diagonalizable; matrix D fails criteria (c) and (d) and so is not diagonalizable, specifically because the eigenvalue $\lambda = 6$ has a greater algebraic multiplicity of 2 than its geometric multiplicity of 1. This problem remains true when using complex numbers and so D is also not diagonalizable over $\mathbb{C}$.

Example 3.204 Show that the matrix A below is diagonalizable and find A^n where n is a positive integer.

$$A = \begin{pmatrix} 2 & 2 & -2 \\ 1 & 3 & -1 \\ -1 & 1 & 1 \end{pmatrix}.$$

Solution Adding column 2 of $xI - A$ to column 1, we can see that $c_A(x)$ equals

$$\begin{vmatrix} x-2 & -2 & 2 \\ -1 & x-3 & 1 \\ 1 & -1 & x-1 \end{vmatrix} = \begin{vmatrix} x-4 & -2 & 2 \\ x-4 & x-3 & 1 \\ 0 & -1 & x-1 \end{vmatrix}$$

$$= \begin{vmatrix} x-4 & -2 & 2 \\ 0 & x-1 & -1 \\ 0 & -1 & x-1 \end{vmatrix} = (x-4)(x-2)x.$$

Hence the eigenvalues are $\lambda = 0, 2, 4$. That they are distinct implies immediately that A is diagonalizable. Note

$$\lambda = 0: \quad \begin{pmatrix} -2 & -2 & 2 \\ -1 & -3 & 1 \\ 1 & -1 & -1 \end{pmatrix} \begin{pmatrix} x \\ y \\ z \end{pmatrix} = 0 \iff \begin{matrix} x+y=z \\ x+3y=z \end{matrix} \iff \begin{pmatrix} x \\ y \\ z \end{pmatrix} = \begin{pmatrix} \mu \\ 0 \\ \mu \end{pmatrix};$$

$$\lambda = 2: \quad \begin{pmatrix} 0 & -2 & 2 \\ -1 & -1 & 1 \\ 1 & -1 & 1 \end{pmatrix} \begin{pmatrix} x \\ y \\ z \end{pmatrix} = 0 \iff \begin{matrix} y=z \\ x=0 \end{matrix} \iff \begin{pmatrix} x \\ y \\ z \end{pmatrix} = \begin{pmatrix} 0 \\ \mu \\ \mu \end{pmatrix};$$

$$\lambda = 4: \quad \begin{pmatrix} 2 & -2 & 2 \\ -1 & 1 & 1 \\ 1 & -1 & 3 \end{pmatrix} \begin{pmatrix} x \\ y \\ z \end{pmatrix} = 0 \iff \begin{matrix} x+z=y \\ x=y+z \end{matrix} \iff \begin{pmatrix} x \\ y \\ z \end{pmatrix} = \begin{pmatrix} \mu \\ \mu \\ 0 \end{pmatrix}.$$

#921
#929
#930
#934
#945
#955

So three independent eigenvectors are $(1,0,1)^T, (0,1,1)^T, (1,1,0)^T$. If we set

$$P = \begin{pmatrix} 1 & 0 & 1 \\ 0 & 1 & 1 \\ 1 & 1 & 0 \end{pmatrix} \quad \text{so that} \quad P^{-1} = \frac{1}{2}\begin{pmatrix} 1 & -1 & 1 \\ -1 & 1 & 1 \\ 1 & 1 & -1 \end{pmatrix},$$

then $P^{-1}AP = \mathrm{diag}(0,2,4)$ and $P^{-1}A^nP = (P^{-1}AP)^n = \mathrm{diag}(0,2^n,4^n)$. Finally,

$$A^n = \begin{pmatrix} 1 & 0 & 1 \\ 0 & 1 & 1 \\ 1 & 1 & 0 \end{pmatrix}\begin{pmatrix} 0 & 0 & 0 \\ 0 & 2^n & 0 \\ 0 & 0 & 4^n \end{pmatrix}\frac{1}{2}\begin{pmatrix} 1 & -1 & 1 \\ -1 & 1 & 1 \\ 1 & 1 & -1 \end{pmatrix}$$

$$= \begin{pmatrix} 2^{2n-1} & 2^{2n-1} & -2^{2n-1} \\ 2^{2n-1}-2^{n-1} & 2^{n-1}+2^{2n-1} & 2^{n-1}-2^{2n-1} \\ -2^{n-1} & 2^{n-1} & 2^{n-1} \end{pmatrix}. \qquad \square$$

Example 3.205 (From Oxford 2007 Paper A #4) Let

$$A = \begin{pmatrix} 6 & 1 & 2 \\ 0 & 7 & 2 \\ 0 & -2 & 2 \end{pmatrix}.$$

(a) Show that A has two eigenvalues λ_1 and λ_2. Is A diagonalizable?
(b) Show further that $A^2 = (\lambda_1 + \lambda_2)A - \lambda_1\lambda_2 I$. Are there scalars $a_0, a_1, \ldots, a_n$, for some n, such that

$$a_n A^n + a_{n-1}A^{n-1} + \cdots + a_0 I = \mathrm{diag}(1,2,3)?$$

Solution (a) We have

$$c_A(x) = \det\begin{pmatrix} x-6 & -1 & -2 \\ 0 & x-7 & -2 \\ 0 & 2 & x-2 \end{pmatrix} = (x-6)\{(x-7)(x-2)+4\} = (x-6)^2(x-3).$$

As one of the eigenvalues is repeated then we cannot immediately decide on A's diagonalizability. Investigating the repeated eigenvalue first we see

$$\lambda_1 = 6: \begin{pmatrix} 0 & -1 & -2 \\ 0 & -1 & -2 \\ 0 & 2 & 4 \end{pmatrix} \begin{pmatrix} x \\ y \\ z \end{pmatrix} = \mathbf{0} \iff y + 2z = 0 \iff \begin{pmatrix} x \\ y \\ z \end{pmatrix} = \begin{pmatrix} \mu \\ 2\nu \\ -\nu \end{pmatrix};$$

$$\lambda_2 = 3: \begin{pmatrix} -3 & -1 & -2 \\ 0 & -4 & -2 \\ 0 & 2 & 1 \end{pmatrix} \begin{pmatrix} x \\ y \\ z \end{pmatrix} = \mathbf{0} \iff \begin{matrix} x = y \\ 2y = -z \end{matrix} \iff \begin{pmatrix} x \\ y \\ z \end{pmatrix} = \begin{pmatrix} \mu \\ \mu \\ -2\mu \end{pmatrix}.$$

So an eigenbasis is $(1,0,0)^T, (0,2,-1)^T, (1,1,-2)^T$ and we see that A is diagonalizable. Further

$$A^2 - (\lambda_1 + \lambda_2)A + \lambda_1\lambda_2 I = A^2 - 9A + 18I$$

$$= \begin{pmatrix} 36 & 9 & 18 \\ 0 & 45 & 18 \\ 0 & -18 & 0 \end{pmatrix} - 9 \begin{pmatrix} 6 & 1 & 2 \\ 0 & 7 & 2 \\ 0 & -2 & 2 \end{pmatrix} + \begin{pmatrix} 18 & 0 & 0 \\ 0 & 18 & 0 \\ 0 & 0 & 18 \end{pmatrix} = 0.$$

So $A^2 = 9A - 18I$ can be written as a linear combination of A and I, and likewise we have $A^3 = 9A^2 - 18A = 81A - 180I$ can also be written as such a linear combination. More generally (say using a proof by induction), we find that any polynomial in A can be written as a linear combination of A and I. However, if $\text{diag}(1,2,3) = \alpha A + \beta I$ for some α, β then, just looking at the diagonal entries, we'd have

$$6\alpha + \beta = 1, \qquad 7\alpha + \beta = 2, \qquad 2\alpha + \beta = 3,$$

and, with a quick check, we see this system is inconsistent. Hence $\text{diag}(1,2,3)$ cannot be expressed as a polynomial in A and no such scalars $a_0, a_1, \ldots, a_n$ exist. $\qquad \square$

Example 3.206 (See also #944.) Determine x_n and y_n where $x_0 = 1$, $y_0 = 0$ and

$$x_{n+1} = x_n - y_n \qquad \text{and} \qquad y_{n+1} = x_n + y_n \qquad \text{for } n \geqslant 0.$$

Solution We can rewrite the two recurrence relations as a single recurrence relation involving a vector, namely

$$\begin{pmatrix} x_n \\ y_n \end{pmatrix} = \begin{pmatrix} 1 & -1 \\ 1 & 1 \end{pmatrix} \begin{pmatrix} x_{n-1} \\ y_{n-1} \end{pmatrix} = \begin{pmatrix} 1 & -1 \\ 1 & 1 \end{pmatrix}^n \begin{pmatrix} x_0 \\ y_0 \end{pmatrix}.$$

From Example 3.198 we have

$$P^{-1} \begin{pmatrix} 1 & -1 \\ 1 & 1 \end{pmatrix} P = \begin{pmatrix} 1+i & 0 \\ 0 & 1-i \end{pmatrix}, \qquad \text{where} \quad P = \begin{pmatrix} i & 1 \\ 1 & i \end{pmatrix}.$$

So

$$\begin{pmatrix} 1 & -1 \\ 1 & 1 \end{pmatrix}^n = P \begin{pmatrix} 1+i & 0 \\ 0 & 1-i \end{pmatrix}^n P^{-1}$$

$$= \begin{pmatrix} i & 1 \\ 1 & i \end{pmatrix} \begin{pmatrix} (1+i)^n & 0 \\ 0 & (1-i)^n \end{pmatrix} \left(-\frac{1}{2}\right) \begin{pmatrix} i & -1 \\ -1 & i \end{pmatrix}$$

$$= \frac{1}{2} \begin{pmatrix} (1+i)^n + (1-i)^n & i(1-i)^n - i(1+i)^n \\ i(1-i)^n - i(1+i)^n & (1+i)^n + (1-i)^n \end{pmatrix}$$

$$= \frac{1}{2} \begin{pmatrix} 2\operatorname{Re}(1+i)^n & 2\operatorname{Im}(1+i)^n \\ 2\operatorname{Im}(1+i)^n & 2\operatorname{Re}(1+i)^n \end{pmatrix}.$$

By De Moivre's theorem, and noting $1+i = \sqrt{2}\operatorname{cis}(\pi/4)$, we have

$$\begin{pmatrix} x_n \\ y_n \end{pmatrix} = \begin{pmatrix} \operatorname{Re}(1+i)^n & \operatorname{Im}(1+i)^n \\ \operatorname{Im}(1+i)^n & \operatorname{Re}(1+i)^n \end{pmatrix} \begin{pmatrix} 1 \\ 0 \end{pmatrix} = \begin{pmatrix} \operatorname{Re}(1+i)^n \\ \operatorname{Im}(1+i)^n \end{pmatrix}$$

$$= 2^{n/2} \begin{pmatrix} \cos(n\pi/4) \\ \sin(n\pi/4) \end{pmatrix}. \qquad \square$$

- See § 6.6 for a similar treatment of linear systems of differential equations.

We briefly return to our first question from the start of the section: why might diagonalizability be a useful definition? We have seen that it can be computationally helpful but to have a matrix in diagonal from also helps us appreciate its associated map. In §3.7 we defined, for an $n \times n$ matrix A, the associated map μ_A of $\mathbb{R}_n$. Towards the end of that section, in Example 3.143, we saw how a particular associated map might be better understood via a change of co-ordinates and, for those interested, the general theory of changing co-ordinates is described in §4.1.

For each choice of axes and co-ordinates in $\mathbb{R}_n$, the associated map μ_A is represented by a certain matrix. So a sensible question is: is there a preferential set of co-ordinates to best describe μ_A? Certainly if we can choose our matrix representative to be diagonal that will be a computational improvement but we will also better appreciate how μ_A stretches $\mathbb{R}_n$ with respect to certain axes (which is the effect of a diagonal matrix). To conclude, we recall that some matrices are not diagonalizable; this just invites the more refined question: into what preferred forms might we be able to change those matrices with a sensible choice of co-ordinates? (See #996, for example.)

Exercises

#916A Show that a square matrix is singular if and only if 0 is an eigenvalue.

#917A† Show that a square matrix has the same eigenvalues as its transpose.

#918A If A is an invertible matrix, how are the eigenvalues of A^{-1} related to those of A?

#919B† Show that similar matrices have equal characteristic polynomials, equal traces and equal determinants.

#920 b Find the eigenvalues and eigenvectors of each of the following matrices A_i. In each case, determine whether or not the matrix is diagonalizable and, if so, find an invertible matrix P such that $P^{-1}A_iP$ is diagonal.

$$A_1 = \begin{pmatrix} 2 & 1 & 2 \\ 0 & 0 & 1 \\ 0 & 1 & 0 \end{pmatrix}, \qquad A_2 = \begin{pmatrix} 1 & 1 & 0 \\ -1 & 3 & 0 \\ -1 & 4 & -1 \end{pmatrix}, \qquad A_3 = \begin{pmatrix} 2 & 1 & 1 \\ 1 & 2 & 1 \\ 1 & 1 & 2 \end{pmatrix}.$$

#921 b Find a diagonal matrix D and an invertible matrix P such that $P^{-1}MP = D$, where M is the matrix on the right. Find a matrix E such that $E^3 = D$ and a matrix N such that $N^3 = M$.

$$M = \begin{pmatrix} -5 & 3 \\ 6 & -2 \end{pmatrix}$$

#922 A† Show that the eigenvalues of a triangular matrix are its diagonal entries.

#923 c Let A be a square matrix and c be a real number. How do the eigenvalues of A relate to those of $A + cI$? Show that A is diagonalizable if and only if $A + cI$ is diagonalizable.

#924 B Let A, B be $n \times n$ matrices and k a positive integer.

(i) Show A is diagonalizable if and only if A^T is diagonalizable.
(ii) Say A is invertible. Show A is diagonalizable if and only if A^{-1} is diagonalizable.
(iii) Say B is invertible. Show A is diagonalizable if and only if $B^{-1}AB$ is diagonalizable.
(iv) Say A is diagonalizable and $A^k = 0$. Show that $A = 0$.

#925 A Show that the 2×2 matrix E_{12} is not diagonalizable, even with the use of a complex invertible matrix P.

#926 b For what values of $a, b, \alpha, \beta, \gamma$ are the following matrices diagonalizable?

$$\begin{pmatrix} 1 & a \\ b & 1 \end{pmatrix}, \qquad \begin{pmatrix} \alpha & \beta \\ \gamma & \alpha \end{pmatrix}.$$

#927 b For each of the following parts, provide an example which is not diagonal.

(i) A singular matrix which is diagonalizable.
(ii) An invertible matrix which is diagonalizable.
(iii) An upper triangular matrix which is diagonalizable.
(iv) A singular matrix which is not diagonalizable over $\mathbb{R}$ but is diagonalizable over $\mathbb{C}$.

#928 c Show that if A and B are diagonalizable then so is $\mathrm{diag}(A, B)$. (The converse is also true – see #1008.)

#929 b† (See also #998.) Let $\mathbf{v}$ be a λ-eigenvector of an $n \times n$ matrix A. Show for any polynomial $p(x)$ that $p(A)\mathbf{v} = p(\lambda)\mathbf{v}$ and hence that $c_A(A)\mathbf{v} = \mathbf{0}$. Deduce that if A is diagonalizable then $c_A(A) = 0_{nn}$.

#930B† Show that the matrix M below has three distinct eigenvalues, at least one of which is strictly negative. Deduce that there is no real matrix N such that $N^2 = M$.

$$M = \begin{pmatrix} -2 & 1 & -2 \\ 1 & 1 & 0 \\ -2 & 0 & -2 \end{pmatrix}$$

#931C Prove part (b) of Proposition 3.191.

#932C Prove parts (a) and (b) of Proposition 3.194.

#933b Find the eigenvalues and eigenvectors of the matrix A in #557. Is the matrix diagonalizable?

#934B Determine A^n for the matrix A below. Evaluate the infinite sum $S = I + A + A^2 + \cdots$ and verify that $S = (I-A)^{-1}$.

$$A = \frac{1}{12} \begin{pmatrix} 0 & -8 \\ 3 & 10 \end{pmatrix} \qquad B = \begin{pmatrix} 7 & -2 \\ 2 & 2 \end{pmatrix}$$

#935c Let $p(x)$ be a polynomial and B be the matrix above. Show that $p(B)$ is invertible unless $p(3) = 0$ or $p(6) = 0$ (or both).

#936B† Let

$$A = \begin{pmatrix} 0 & 1 & 0 & 0 \\ 0 & 0 & 1 & 0 \\ 0 & 0 & 0 & 1 \\ 1 & 0 & 0 & 0 \end{pmatrix} \qquad \text{and} \qquad B = \begin{pmatrix} x & 1 & 0 & 1 \\ 1 & x & 1 & 0 \\ 0 & 1 & x & 1 \\ 1 & 0 & 1 & x \end{pmatrix}.$$

(i) Show that A is not diagonalizable using real matrices, but is diagonalizable using complex matrices.

(ii) Show that $A^T = A^{-1}$. Hence find the eigenvalues and determinant of B.

#937B† Are any of the following matrices similar?

$$\begin{pmatrix} 2 & 6 & 0 \\ 1 & 1 & 0 \\ 1 & -2 & 3 \end{pmatrix}, \quad \begin{pmatrix} 1 & 0 & 0 \\ 0 & 1 & 0 \\ 0 & 0 & -1 \end{pmatrix}, \quad \begin{pmatrix} 0 & 1 & 0 \\ 0 & 0 & 1 \\ -1 & 1 & 1 \end{pmatrix}, \quad \begin{pmatrix} -1 & 1 & 1 \\ 0 & 3 & 0 \\ 0 & 0 & 4 \end{pmatrix}.$$

#938b† Find an invertible matrix P such that $P^{-1}MP$ is diagonal where M is the matrix in #660. Hence determine all the matrices X which commute with M.

#939C† Let A be a 2×2 matrix with $c_A(x) = (x-k)^2$ for some k. Show that $A = kI_2$ or that $P^{-1}AP = \begin{pmatrix} k & 1 \\ 0 & k \end{pmatrix}$ for some invertible matrix P.

#940c† Find necessary and sufficient conditions on a, b, c, d for $\begin{pmatrix} a & b \\ c & d \end{pmatrix}$ to be diagonalizable.

#941D† For each of the following matrices, determine under what conditions on a, b, c the matrix is diagonalizable.

$$\begin{pmatrix} a & b & c \\ c & a & b \\ b & c & a \end{pmatrix}, \qquad \begin{pmatrix} a & b & c \\ a & b & c \\ a & b & c \end{pmatrix}.$$

#942d† For the matrices A_i below, find *all* the invertible matrices P such that $P^{-1}A_iP$ is diagonal.

$$A_1 = \begin{pmatrix} 1 & 0 & 0 \\ 0 & 2 & 0 \\ 0 & 0 & 3 \end{pmatrix}; \qquad A_2 = \begin{pmatrix} 1 & 0 & 0 \\ 0 & 1 & 0 \\ 0 & 0 & 2 \end{pmatrix}; \qquad A_3 = \begin{pmatrix} 7 & 2 & -6 \\ -4 & 1 & -6 \\ -3 & -1 & -2 \end{pmatrix}.$$

#943b With A as in Example 3.205, and n an integer, determine A^n and write it as a linear combination of A and I_3.

#944C Show that $z_n = x_n + iy_n$ where x_n, y_n are as in Example 3.206, satisfies the recursion $z_{n+1} = (1+i)z_n$ and $z_0 = 1$. Hence rederive expressions for x_n and y_n.

#945B Let $J(\lambda, r)$ be as in #598. What is the algebraic multiplicity of λ? What is the geometric multiplicity of λ? What is the rank of $(J(\lambda, r) - \lambda I)^k$ for $k \geqslant 1$?

#946C† Let λ be a real number and $0 < r \leqslant s \leqslant t$ be positive integers. Let

$$M = \mathrm{diag}(J(\lambda, r), J(\lambda, s)), \qquad N = \mathrm{diag}(J(\lambda, r), J(\lambda, s), J(\lambda, t)).$$

(i) With reference to M, what is the algebraic multiplicity of λ? What is the geometric multiplicity of λ? What is the rank of $(M - \lambda I)^k$ for $k \geqslant 1$?
(ii) Repeat (i) for N instead of M.

#947C Let A be a diagonalizable $n \times n$ matrix and let $k \geqslant 0$ be an integer. Show that A^k is diagonalizable. Give an example of a square matrix B, and a positive integer k, where B^k is diagonalizable but B is not.

#948b Let $A_{\alpha, \beta}$ be the matrix as defined in #602. Find the eigenvalues of $A_{\alpha, \beta}$. Is $A_{\alpha, \beta}$ diagonalizable?

#949B† Show that a matrix is similar to $\mathrm{diag}(1, -1)$ if and only if it equals $A_{\alpha, \beta}$ for some α, β.

#950C† Show that the transpose map of M_{22} defined in Example 3.142 is diagonalizable. Is the transpose map of M_{nn} diagonalizable?

#951c Show that the 3×3 matrix in (3.61) is not diagonalizable over $\mathbb{R}$ when $\mathbf{a} \neq \mathbf{0}$ but is diagonalizable with the use of complex matrices. Determine positive integer powers of this matrix and hence rederive the solution of #914.

#952B Prove the second identity of Proposition 3.191(c).

#953D† Show that $c_{n-2} = \frac{1}{2}((\mathrm{trace}A)^2 - \mathrm{trace}(A^2))$, using the notation of Proposition 3.191(c).

#954c† Let A, B be $n \times n$ matrices. Say that A has n distinct real eigenvalues and that $AB = BA$. Show that B is diagonalizable.

#955b Solve the simultaneous recurrence relations $x_{n+1} = 2x_n + 3y_n$ and $y_{n+1} = 3x_n + y_n$ with $x_0 = y_0 = 1$.

#956c Find the solution of the simultaneous recurrence relations $x_{n+1} = 4x_n - 3y_n$ and $y_{n+1} = 6x_n + 2y_n$ with $x_0 = y_0 = 1$. Make plain that your answer is a real sequence.

#957c By introducing the vector $\mathbf{v}_n = (x_{n+1}, x_n)^T$, find the general solution of the recursion $x_{n+2} - 5x_{n+1} + 6x_n = 0$.

#958D† Solve the recurrence relation

$$x_{n+1} = \frac{1 + x_n}{4 - 2x_n} \quad \text{with} \quad x_0 = 0.$$

#959D† Let A be a 2×2 matrix. Define T from M_{22} to M_{22} by $T(X) = AX$, where X is in M_{22}. By identifying M_{22} with $\mathbb{R}_4$, T can be identified with a 4×4 matrix.

  (i) Show that λ is an eigenvalue of T if and only if λ is an eigenvalue of A. What is $c_T(x)$ in terms of $c_A(x)$?
  (ii) Show that if A is diagonalizable then T is diagonalizable.
  (iii) Does the converse of (ii) hold?

#960c Let A be the $n \times n$ matrix whose every entry equals 1. Is A diagonalizable?

3.12 Further Exercises*

Exercises on Metrics, Norms and Inner Products (See p.143.)

#961B† Show that if $\cdot$ is an inner product on $\mathbb{R}^n$ then $\|\mathbf{v}\| = \sqrt{\mathbf{v} \cdot \mathbf{v}}$ defines a norm.

#962B Show that if $\|\ \|$ is a norm on $\mathbb{R}^n$ then $d(\mathbf{x}, \mathbf{y}) = \|\mathbf{x} - \mathbf{y}\|$ defines a metric.

#963D† For what values of a and b does the following define an inner product on $\mathbb{R}^2$?

$$(x_1, x_2) \cdot (y_1, y_2) = x_1 y_1 + a x_2 y_1 + b x_1 y_2 + y_1 y_2.$$

#964B Show, for the metrics D, d, δ defined on p.143, that

$$D(\mathbf{p}, \mathbf{q}) \leqslant d(\mathbf{p}, \mathbf{q}) \leqslant \delta(\mathbf{p}, \mathbf{q}) \qquad \text{for any } \mathbf{p}, \mathbf{q} \text{ in } \mathbb{R}^n.$$

Sketch in $\mathbb{R}^2$ the loci (i) $D(\mathbf{0}, \mathbf{x}) = 1$; (ii) $d(\mathbf{0}, \mathbf{x}) = 1$; (iii) $\delta(\mathbf{0}, \mathbf{x}) = 1$.

#965B Let $\cdot$ be an inner product on $\mathbb{R}^n$ and $\|\ \|$ denote its norm. Prove the **parallelogram law**

$$\|\mathbf{v} + \mathbf{w}\|^2 + \|\mathbf{v} - \mathbf{w}\|^2 = 2\|\mathbf{v}\|^2 + 2\|\mathbf{w}\|^2.$$

#966B For $\mathbf{x}$ in $\mathbb{R}^n$ we define $\|\mathbf{x}\|_1 = D(\mathbf{0}, \mathbf{x})$ and $\|\mathbf{x}\|_\infty = \delta(\mathbf{0}, \mathbf{x})$. Show that $\|\ \|_1$ and $\|\ \|_\infty$ are norms, neither of which satisfy the parallelogram law.

#967 B† Show that the dot product is not an inner product on $\mathbb{C}^n$. Suggest a definition for the length $\|\mathbf{v}\|$ of a complex vector so that $\|\mathbf{v}\|$ is a norm. What "complex inner product" does your norm correspond to? What properties does your complex inner product have instead of IP1, IP2, IP3?

#968 B Let $A = (a_{ij})$ be an $n \times n$ matrix and $\mathbf{v}$ in $\mathbb{R}_n$. For $p = 1, 2, \infty$ we define

$$\|A\|_p = \max\{\|A\mathbf{v}\|_p : \|\mathbf{v}\|_p = 1\}.$$

(i) For each of $p = 1, 2, \infty$, show that $\|\ \|_p$ is a norm on the space M_{nn} of $n \times n$ matrices.

(ii) For each of $p = 1, 2, \infty$, show that $\|A\mathbf{v}\|_p \leqslant \|A\|_p \|\mathbf{v}\|_p$.

(iii) For each of $p = 1, 2, \infty$, show $\|AB\|_p \leqslant \|A\|_p \|B\|_p$ where A, B are $n \times n$ matrices.

#969 D† With notation as in #968, show that

$$\|A\|_1 = \max_{1 \leqslant j \leqslant n} \sum_{i=1}^n |a_{ij}| \quad \text{and} \quad \|A\|_\infty = \max_{1 \leqslant i \leqslant n} \sum_{j=1}^n |a_{ij}|.$$

#970 D† Let $A = (a_{ij})$ be an $n \times n$ matrix such that $\max |a_{ij}| < 1/n$. Show that $I - A$ is invertible. Show that if $\max |a_{ij}| = 1/n$ then $I - A$ may be singular or invertible.

#971 B For $m \times n$ matrices we can define $A \cdot B = \text{trace}(B^T A)$. Show that this defines an inner product on M_{mn}. Show that this product is the same as the usual dot product when M_{mn} is appropriately identified with $\mathbb{R}^{mn}$.

#972 C Perron's theorem [28] relates to square matrices with all positive entries. It states for such a matrix A that

(i) there is a largest positive eigenvalue r;

(ii) all other roots of c_A have modulus less than r;

(iii) $r \leqslant \|A\|_\infty$; (iv) r is not a repeated root;

(v) there is an r-eigenvector $\mathbf{v}$ with all positive co-ordinates;

(vi) any eigenvector with all positive co-ordinates is a multiple of $\mathbf{v}$;

(vii) if $\mathbf{w}$ is a similar r-eigenvector for A^T, then A^k/r^k converges to $P = (\mathbf{v}\mathbf{w}^T)/(\mathbf{w}^T\mathbf{v})$ as k becomes large;

(viii) P is a rank one projection matrix whose column space is the r-eigenspace.

Prove Perron's theorem for 2×2 matrices with positive entries.

#973 B† Let $a < b$ and let V be the space of real polynomials. Show that the following is an inner product on V.

$$p \cdot q = \int_a^b p(x)q(x)\,dx.$$

---

[28] Proved in 1907 by the German mathematician Oskar Perron (1880–1975). He was a prolific mathematician, making contributions to analysis (the Perron integral) and to partial differential equations (the Perron method). He is also known for the *Perron Paradox* which memorably demonstrates fallacies that might arise by presupposing a problem has a solution: let N be the largest integer. If $N > 1$ then $N^2 > N$, contradicting the maximality of N. Hence $N = 1$ is the largest integer.

#974D† Let $P_n(x)$ denote the nth Legendre polynomial. With $a = -1$ and $b = 1$ (in the notation of #973), show that $P_n \cdot P_m = 0$ when $n \neq m$ and determine $P_n \cdot P_n$.

#975C† Let V be the space of real polynomials of degree two or less. Show that the formula $p \cdot q = p(0)q(0) + p(1)q(1) + p(2)q(2)$ defines an inner product on V.

#976C Let n be a positive integer and $\{0,1\}^n$ denote the set of strings of 0s and 1s of length n. The **Hamming distance** [29] $d(s_1, s_2)$ between two strings of 0s and 1s equals the number of times the kth term of s_1 does not equal the kth term of s_2. So, for example, when $n = 3$ we have $d(011, 110) = 2$ as $0 \neq 1, 1 = 1, 1 \neq 0$. Show that d is a metric on $\{0,1\}^n$. How many strings lie at a distance k of a given string s?

#977C Let p be a prime number. A non-zero rational number q can uniquely be written as $q = p^a m/n$, where m and n are integers that are coprime with p. We define $|q|_p = p^{-a}$ and we set $|0|_p = 0$. The p-**adic metric** is then defined on the rationals by $d_p(q_1, q_2) = |q_1 - q_2|_p$. Show that d_p is indeed a metric. Which of the numbers $108, 144, 216, 464$ are closest using the 2-adic metric? Which using the 3-adic metric?

Exercises on Direct Sums

#978D† Let X and Y be subspaces of $\mathbb{R}^n$. By separately extending a basis for $X \cap Y$ to bases for X and Y, show that

$$\dim(X + Y) = \dim X + \dim Y - \dim(X \cap Y). \tag{3.65}$$

#979B Verify directly (3.65) when X and Y are the following subspaces of $\mathbb{R}^4$.

$$X = \langle (4,2,3,3), (1,0,1,1), (1,2,0,0) \rangle, \qquad Y = \{(y_1, y_2, y_3, y_4) : y_1 + y_2 = y_3 + y_4 = 0\}.$$

Definition 3.207 Let V be a subspace of $\mathbb{R}^n$ which contains subspaces $X_1, X_2, \ldots, X_k$. We say that V is the **direct sum** of $X_1, X_2, \ldots, X_k$ if every $\mathbf{v}$ in V can be uniquely written as $\mathbf{v} = \mathbf{x}_1 + \cdots + \mathbf{x}_k$ where $\mathbf{x}_i \in X_i$ for each i. We then write this as

$$V = X_1 \oplus X_2 \oplus \cdots \oplus X_k.$$

#980B Verify that $\mathbb{R}^4$ is the direct sum of X and Y, where

$$X = \{(x_1, x_2, x_3, x_4) : x_1 + x_2 = x_3 = 0\}, \qquad Y = \{(y_1, y_2, y_3, y_4) : y_1 + y_3 = y_2 + y_4 = 0\}.$$

Write $(1,0,0,0)$ as the sum of elements in X and Y.

#981C Determine which of the following represent $\mathbb{R}^3$ as a direct sum.

(i) $X_1 = \{(x,x,z) : x,z \in \mathbb{R}\}, X_2 = \{(x,y,y) : x,y \in \mathbb{R}\}$.
(ii) $X_1 = \{(x,0,x) : x \in \mathbb{R}\}, X_2 = \{(x,y,y) : x,y \in \mathbb{R}\}$.
(iii) $X_1 = \{(x,0,x) : x \in \mathbb{R}\}, X_2 = \{(0,x,x) : x \in \mathbb{R}\}$.
(iv) $X_1 = \{(x,0,x) : x \in \mathbb{R}\}, X_2 = \{(0,x,x) : x \in \mathbb{R}\}, X_3 = \{(x,x,x) : x \in \mathbb{R}\}$.

---

[29] After the American mathematician Richard Hamming (1915–1998) who, in a seminal paper of 1950, introduced *error-correcting* codes for transmissions. With an *error-checking* code the existence of an error might be clear; with an error-correcting code the error can be identified by the receiver and (with high probability) corrected without any need for retransmission.

#982B† Let X be a subspace of $\mathbb{R}^n$. Show that there is a subspace Y of $\mathbb{R}^n$ such that $\mathbb{R}^n = X \oplus Y$.

#983C If X, Y, Z are subspaces of $\mathbb{R}^n$ such that $\mathbb{R}^n = X \oplus Y = X \oplus Z$, must it follow that $Y = Z$?

#984B† Let V, X, Y be subspaces of $\mathbb{R}^n$. Show that $V = X \oplus Y$ if and only if $V = X + Y$ and $X \cap Y = \{\mathbf{0}\}$.

#985C† Give examples of subspaces V, X, Y, Z such that V is not a direct sum of X, Y, Z but still

$$V = X + Y + Z \qquad \text{and} \qquad X \cap Y = X \cap Z = Y \cap Z = \{\mathbf{0}\}.$$

#986B Show that vectors $\mathbf{v}_1, \ldots, \mathbf{v}_k$ in $\mathbb{R}^n$ are linearly independent if and only if

$$\langle \mathbf{v}_1, \ldots, \mathbf{v}_k \rangle = \langle \mathbf{v}_1 \rangle \oplus \cdots \oplus \langle \mathbf{v}_k \rangle.$$

#987D Let V be a subspace of $\mathbb{R}^n$ which is the direct sum of subspaces $X_1, X_2, \ldots, X_k$. Show that if $\mathcal{B}_i$ is a basis for X_i for each i, then $\mathcal{B}_1 \cup \mathcal{B}_2 \cup \cdots \cup \mathcal{B}_k$ is a basis for V. Deduce that $\dim V = \dim X_1 + \cdots + \dim X_k$.

#988D† Let V be a subspace of $\mathbb{R}^n$ and $X_1, \ldots, X_k$ be subspaces of V. Then show that $V = X_1 \oplus \cdots \oplus X_k$ if and only if

$$V = X_1 + \cdots + X_k \qquad \text{and} \qquad \dim V = \dim X_1 + \cdots + \dim X_k.$$

#989B (i) Let X, Y be as in #980. Write $\mathbf{v} = (v_1, v_2, v_3, v_4)$ in the form $\mathbf{v} = \mathbf{x} + \mathbf{y}$, where $\mathbf{x}$ is in X and $\mathbf{y}$ is in Y.
(ii) Define the maps P and Q from $\mathbb{R}^4$ to $\mathbb{R}^4$ by $P\mathbf{v} = \mathbf{x}$ and $Q\mathbf{v} = \mathbf{y}$. Show that P and Q are projections.
(iii) Find the associated matrices for P and Q (when considered as maps from $\mathbb{R}_4$ to $\mathbb{R}_4$).

#990D† Let P be a projection of $\mathbb{R}_n$. Show that $\mathbb{R}_n = \text{Null}(P) \oplus \text{Col}(P)$.

#991C Let $\mathbb{R}_n = X \oplus Y$. Define P by $P\mathbf{v} = \mathbf{x}$ where $\mathbf{v} = \mathbf{x} + \mathbf{y}$ is the decomposition of $\mathbf{v}$ in elements of X and Y.

(i) Show that P is a projection with null space Y and column space X.
(ii) Show that $\mathbf{v}$ is in X if and only if $P\mathbf{v} = \mathbf{v}$.

#992D Let A be an $n \times n$ matrix. Show that the eigenspaces of A form a direct sum. Show that A is diagonalizable if and only if this direct sum equals $\mathbb{R}_n$.

#993D Let C, D be as in Example 3.193. Show that $\mathbb{R}_3$ is the direct sum of the eigenspaces of C but the direct sum of the eigenspaces of D is a plane.

#994D† (i) Let A be a diagonalizable complex $n \times n$ matrix. Show that for every A-invariant subspace M, there exists an A-invariant subspace N such that $\mathbb{C}_n = M \oplus N$. (The converse is also true – #1009.)
(ii) Find a non-diagonalizable real matrix A for which every A-invariant subspace M has an A-invariant subspace N with $\mathbb{R}_n = M \oplus N$.

#995D† Let A be an $n \times n$ matrix. Show there is a smallest $k \geqslant 0$ with $\text{Null}(A^k) = \text{Null}(A^{k+j})$ for all $j \geqslant 0$. Show further that

$$\mathbb{R}_n = \text{Null}(A^k) \oplus \text{Col}(A^k).$$

Exercises on the Cayley–Hamilton Theorem

#996D† (**Triangularizability of Complex Matrices**) Let A be an $n \times n$ complex matrix. Show, by induction on n, that there is an invertible complex matrix P such that $P^{-1}AP$ is upper triangular.

#997D† (**Primary Decomposition Theorem**) Let

$$p(x) = (x - a_1)^{r_1}(x - a_2)^{r_2} \cdots (x - a_k)^{r_k}$$

where $a_1, \ldots, a_k$ are distinct real (or complex) numbers and $r_1, r_2, \ldots, r_k$ are positive integers. Deduce from Bézout's lemma that

$$\text{Null}(p(A)) = \text{Null}((A - a_1 I)^{r_1}) \oplus \text{Null}((A - a_2 I)^{r_2}) \oplus \cdots \oplus \text{Null}((A - a_k I)^{r_k})$$

for any square real (or complex) matrix A.

#998D (See also #635.) The **Cayley–Hamilton theorem** for a square matrix A states that $c_A(A) = 0$.

(i) Let U be an upper triangular $n \times n$ matrix with diagonal entries $\alpha_1, \alpha_2, \ldots, \alpha_n$. Show that the characteristic polynomial of U is

$$c_U(x) = (x - \alpha_1)(x - \alpha_2) \cdots (x - \alpha_n).$$

(ii) Show that $(U - \alpha_1 I)\mathbf{e}_1^T = \mathbf{0}$ and more generally, for any $i \geqslant 1$, that

$$(U - \alpha_1 I)(U - \alpha_2 I) \cdots (U - \alpha_i I)\mathbf{e}_i^T = \mathbf{0}.$$

(iii) Deduce, for any $1 \leqslant j \leqslant i$, that

$$(U - \alpha_1 I)(U - \alpha_2 I) \cdots (U - \alpha_i I)\mathbf{e}_j^T = \mathbf{0}.$$

Hence prove the Cayley–Hamilton theorem for upper triangular matrices.

(iv) Apply #996 to deduce the Cayley–Hamilton theorem for all $n \times n$ matrices.

#999D (**Alternative proof of Cayley–Hamilton Theorem**) Let A be an $n \times n$ (real or complex) matrix with

$$c_A(x) = x^n + a_{n-1}x^{n-1} + \cdots + a_1 x + a_0.$$

Explain why

$$\text{adj}(xI_n - A) = M_{n-1}x^{n-1} + M_{n-2}x^{n-2} + \cdots + M_1 x + M_0,$$

where each M_i is a (constant) $n \times n$ matrix. Deduce the Cayley–Hamilton theorem from the fact that

$$(xI_n - A)\text{adj}(xI_n - A) = c_A(x)I_n.$$

#1000E† (i) Let A and B be two $n \times n$ complex matrices such that A and $-B$ have no eigenvalue in common. Show that the only solution of the equation $AX + XB = 0$ is $X = 0$. Deduce that Sylvester's equation, $AX + XB = C$, has a unique solution for every $n \times n$ matrix C.

(ii) Say that A and $-B$ have a common eigenvalue. Find a matrix C such that $AX + XB = C$ has no solution X.

Exercises on Minimal Polynomials (See #705.)

#1001B† Let A be a square matrix with minimal polynomial $m_A(x)$. Show $p(A) = 0$ if and only if $m_A(x)$ divides $p(x)$. Consequently $m_A(x)$ divides $c_A(x)$ by the Cayley–Hamilton theorem.

#1002B† Let A be a square matrix and λ real. Without reference to the Cayley–Hamilton theorem, show that $m_A(\lambda) = 0$ if and only if $c_A(\lambda) = 0$.

#1003B† Let A be a 4×4 matrix such that $c_A(x) = (x-1)^2(x-2)^2$. Given #1001 and #1002, what are the possible minimal polynomials of A? For each such possibility, give an example of a matrix A with that minimal polynomial.

#1004B† Show that the minimal polynomial of a diagonalizable matrix is a product of distinct linear factors.

#1005B† Show that the minimal polynomial of a triangularizable matrix is a product of linear factors.

#1006D† Prove the converse of #1004.

#1007E† Prove the converse of #1005.

#1008D† Show that if $\mathrm{diag}(A,B)$ is diagonalizable then so are A and B. (This is the converse of #928.)

#1009D† Let A be a complex $n \times n$ matrix such that, for every A-invariant subspace M, there exists an A-invariant subspace N such that $\mathbb{C}_n = M \oplus N$. Show that A is diagonalizable. (This is the converse of #994.)

#1010E† Two $n \times n$ matrices A and B are said to be **simultaneously diagonalizable** if there exists an invertible matrix P such that $P^{-1}AP$ and $P^{-1}BP$ are both diagonal.

(i) Show that simultaneously diagonalizable matrices commute.
(ii) Say now that A and B commute. Show that the eigenspaces of A are B-invariant.
(iii) Deduce that commuting diagonalizable matrices are simultaneously diagonalizable.

Exercises on Graph Theory

Definition 3.208 The **adjacency matrix** of a graph with n vertices $v_1, \ldots, v_n$ is the $n \times n$ matrix whose (i,j)th entry is the number of edges connecting v_i to v_j.

Graph Theory and Euler's Formula

In graph theory, a *graph* is a collection of points called *vertices* (the singular being a *vertex*), with these vertices connected by *edges*. For our purposes we will allow an edge to begin and end in the same vertex, and for there to be more than one edge between any two vertices; other texts may focus on *simple* graphs which have at most one edge between distinct vertices and none from a vertex to itself. We will also assume a graph to be *connected*, meaning that between any two vertices there is a *walk* along the edges.

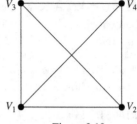

Graphs may be used to describe a wide range of physical, biological and social systems, diversely representing transport networks (such as the London underground), computer networks and website structure, evolution of words across languages and time in philology, migrations in biology and geography, etc.. The definition of a graph may be extended to include one-way edges; such graphs are known as

Figure 3.19

directed graphs or *digraphs*. We might also introduce weights to edges in a *weighted graph* to demonstrate the difficulty – say in terms of time, distance or cost – of travelling along a particular edge.

An important family of graphs is the *complete graphs* K_n. The graph K_n consists of n vertices $v_1,\ldots,v_n$ with a single edge between each pair of distinct vertices. In a similar fashion the *complete bipartite graph* $K_{m,n}$ consists of two sets of vertices $v_1,\ldots v_m$ and $w_1,\ldots,w_n$ with a single edge between each v_i and w_j. The graphs K_4 and $K_{2,3}$ are drawn in Figures 3.19 and 3.20.

Note that for each graph, some of the edges, as drawn, intersect – however, it is possible to draw both these graphs in such a way that no edges intersect. That is to say, these two graphs are *planar*. It is, however, the case that K_5 and $K_{3,3}$ aren't planar (#1016,#1017). In fact Kuratowski's theorem, named after the Polish mathematician Kazimierz Kuratowski (1896–1980), states that a graph is planar if and only if it does not 'essentially' contain K_5 or $K_{3,3}$.

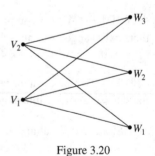

If a graph is drawn in the plane, then the edges will bound certain *faces*. For example, in Figure 3.21, there are three triangular faces and

Figure 3.20

the outside is also counted as a face. **Euler's formula** relates the numbers V, E, F of vertices, edges and faces and states that

$$V - E + F = 2.$$

For Figure 3.21 this holds as $5 - 7 + 4 = 2$. We prove Euler's formula here, introducing a few important ideas of graph theory along the way.

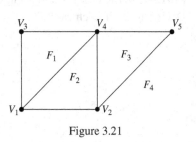

Figure 3.21

Every graph G has a *spanning tree* T; such a tree for the top graph is drawn, un-dashed, in Figure 3.22. A *tree* is a graph containing no *cycles* (walks beginning and ending in the same vertex that otherwise go through distinct vertices and edges) and is *spanning* when it includes each vertex. Also given any graph G in the plane, we can create its *dual graph*: each face of G corresponds to a vertex of the dual graph and two vertices of the dual graph have an edge between them for each edge the corresponding faces share. So G^* has as many edges as G. The dual graph of Figure 3.21 is drawn in Figure 3.23.

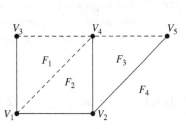

Figure 3.22

A spanning tree T of a graph G has $V - 1$ edges. The complement $G - T$ denotes those edges that are in G and not in T, together with any vertices that those edges connect.

We note now that $(G - T)^*$ is a spanning tree of G^*: to see this, note that any face (i.e. vertex of the dual graph) missing from $(G - T)^*$ would mean T contained a cycle around one or more faces of G, and a cycle amongst the vertices of $(G - T)^*$ would mean that T did not connect all the vertices of G. So as $(G - T)^*$ is a spanning tree of G^*, it has $F - 1$ edges. As the two graphs T and $(G - T)^*$ have $V + F - 2$ edges between them, and this is as many as T and $G - T$ have between them, namely E, then Euler's Formula follows. In the dual graph, Figure 3.23, $(G - T)^*$ is the tree connecting F_2 to F_1 to F_4 to F_3.

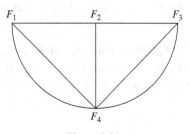

Figure 3.23

#1011B† Let A be the adjacency matrix of a graph and n be a natural number.

(i) Show that A is symmetric.
(ii) Show that $[A^n]_{ij}$ is the number of walks of length n from the ith vertex to the jth vertex.
(iii) In a simple graph, show that the number of triangles in the graph equals $\text{trace}(A^3)/6$.

#1012D† Write down the adjacency matrices A_i for each of the graphs G_i in Figure 3.24. Determine A_i^n in each case.

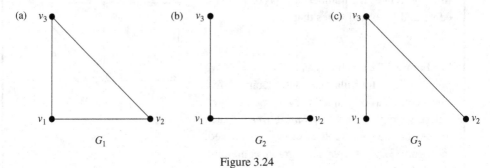

Figure 3.24

#1013B The **degree** (or **valency**) $\deg(v)$ of a vertex v in a graph is the number of edge ends at that vertex (so that an edge beginning and ending at a vertex contributes 2 to its degree). If E is the total number of edges in a graph, show that

$$\sum_{\text{vertices } v} \deg(v) = 2E.$$

#1014B Show that if $\deg(v) \geqslant 2$ for every vertex of a graph, then the graph contains a cycle.

#1015D† (i) (Euler, 1736)[30] An **Euler circuit** in a graph is a walk that begins and ends in the same vertex and which includes every edge of the graph exactly once. Show that a graph has an Euler circuit if and only if the degree of each vertex is even.
(ii) A graph is said to be **traversable** if there is a walk (not necessarily a circuit) which includes every edge exactly once. Show that a graph is traversable if and only if it has no or two vertices of odd degree.

#1016B† Show that the complete graph K_n is planar if and only if $n \leqslant 4$.

#1017C† Show that $K_{2,3}$ is planar, but that $K_{3,3}$ is not planar.[31] Show that $K_{3,3}$ can be drawn on a torus without the edges intersecting.

#1018D† A **Platonic solid** is a regular three-dimensional solid, not having any holes through the solid. In particular, this means that at each vertex m edges meet and each face has the same number of edges n. Deduce from Euler's Formula that

$$\frac{1}{m} + \frac{1}{n} > \frac{1}{2}.$$

Hence show that there are at most five regular solids. Describe five regular solids, finding V, E and F in each case.

---

[30] This is commonly considered to be the founding result of graph theory. Euler demonstrated it to show it is impossible to travel once, in a single journey, across each of the seven bridges of Königsberg (now Kaliningrad) on the Baltic coast. Euler appreciated it was the connections of the underlying graph, rather than classical geometrical notions, that were crucial. The two banks of the river and two islands in the river can be represented as four vertices having degrees 3,3,3 and 5, making for a non-transversable graph.

[31] This fact is commonly introduced as the *utilities problem*: given three houses and three utilities (gas, electric, water say), show it is not possible to connect each house with each utility without the connections intersecting.

#1019 B Prove that, between any two vertices in a tree, there is a unique walk which involves no repeated vertices.

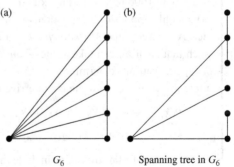

(a)   (b)

G_6 Spanning tree in G_6

Figure 3.25

#1020 D† (Adapted from MAT 2016 #6.) Let G_n denote the graph in which a vertex called the hub in Figure 3.25a is connected to each of n tips in Figure 3.25b. The n tips are connected in order 1 to 2, 2 to 3, and so on with $n-1$ connected to n.

(i)   Draw the three spanning trees that G_2 contains.
(ii) How many spanning trees does G_3 contain?
(iii) For $n \geqslant 1$, let t_n denote the number of spanning trees in G_n. By considering the possible sizes of the top group of tips and how the group is connected to the hub, give an expression for t_n in terms of t_k, where $1 \leqslant k < n$. Use the recurrence to show $t_5 = 55$.
(iv) Deduce that $t_n = F_{2n}$, the $2n$th Fibonacci number, for $n \geqslant 1$.

#1021 D† (i) Given a graph G with vertices $v_1, \ldots, v_n$, its **Laplacian matrix** is given by

$$L = \mathrm{diag}(d_1, \ldots, d_n) - A,$$

where d_i is the degree of v_i and A is the adjacency matrix of A. Show that $c_L(0) = 0$.
(ii) **Kirchoff's theorem** states that the number of spanning trees in G equals

$$(-1)^{n-1} n^{-1} c_L'(0).$$

Deduce **Cayley's theorem** from Kirchoff's theorem, which states that the number of spanning trees in K_n equals n^{n-2}. (Equivalently this is the number of trees with n vertices.)

#1022 D† Use Kirchoff's theorem to show that the number of spanning trees in $K_{m,n}$ equals $m^{n-1} n^{m-1}$.

#1023 C Use Kirchoff's theorem to determine the number of spanning trees in Figure 3.21. Now rederive this answer directly.

#1024 d Use Kirchoff's Theorem to rederive the number of spanning trees in G_5, as defined in #1020.

Definition 3.209 A graph is said to be **bipartite** if the set of vertices V can written as a disjoint union of two sets V_1 and V_2 with every edge connecting a vertex in V_1 to one in V_2.

#1025 B† Show that every tree is bipartite.

Markov Chains A Markov chain is a random process, first introduced by Andrey Markov in 1906 in an effort to extend results such as the *weak law of large numbers* – which only applies to independent events – to events that are dependent in some manner. A discrete-time *Markov chain* is a sequence of random variables $X_1, X_2, X_3, \ldots$ such that each X_n can take values from a fixed set of *states* and the probabilities for X_{n+1} are entirely determined by the previous state that X_n was in, and *not* by the previous history of how the chain arrived in that state; that is,

$$P(X_{n+1} = x_{n+1} \mid X_n = x_n) = P(X_{n+1} = x_{n+1} \mid X_1 = x_1, X_2 = x_2, \ldots, X_n = x_n).$$

for all $x_1, x_2, \ldots, x_n$

So examples of Markov chains include random walks such as the gambler's ruin problem – at any point the gambler, being in the state of having £n can gain £1 or lose £1 and move to one of the states £$(n + 1)$ or £$(n - 1)$. How the previous £n had been gained has no effect on this. Other examples include models for queues, population models, the simple weather model (#407), many board games, etc. and some processes that are not obviously Markov chains – because of a limited effect of several previous states – can be reconsidered as Markov chains (#1043).

The probabilities $p_{ij}^{(n)}$ of moving from state $X_n = i$ (that is, being in the ith state at time nth) to $X_{n+1} = j$ can then be codified in *transition matrices*. We shall consider only *homogeneous* Markov chains, which means that there is a single transition matrix $P = (p_{ij})$ whose entries do not depend on n. For such homogeneous chains, powers P^k of the transition matrix codify, for any n, the probabilities for state X_{n+k} given state X_n. If $\mathbf{x} = (p_1, \ldots, p_s)$ denotes the probabilities of X_n being in states $1, \ldots, s$ then $\mathbf{x}P$ denotes the probabilities of X_{n+1} being in states $1, \ldots, s$. (Note that, as defined, transition matrices postmultiply row vectors.)

Many Markov chains are also naturally *irreducible* with a non-zero probability of moving from any state to another, given enough time. One might then expect such a Markov chain to settle down to certain long-term probabilities of being in each state, and this is indeed the case with finitely many states (#1035). This long-term distribution, known as the *stationary distribution*, is the unique row vector $\mathbf{x}$, with entries adding to 1, that satisfies $\mathbf{x}P = \mathbf{x}$. For reducible, finite chains this vector may or may not be unique (#1038(i)) and may not exist for infinite chains (#1034).

Andrey Markov (1856–1922) was a Russian mathematician and a student of Pafnuty Chebyshev (1821–1894). Besides making significant contributions of his own within mathematics, and especially in probability, Chebyshev is considered a founding father of Russian mathematics, who had many successful students besides Markov, notably Aleksandr Lyapunov (1857–1918), who would prove the *Central Limit Theorem* (#1576).

#1026D† Show that a graph is bipartite if and only if it contains no odd length walks beginning and ending in the same vertex.

#1027D† Let A be the adjacency matrix of a bipartite graph. Show that there is a permutation matrix P such that

$$P^{-1}AP = \begin{pmatrix} 0_{mm} & B \\ B^T & 0_{nn} \end{pmatrix}.$$

Show that if λ is an eigenvalue of A then so is $-\lambda$.

#1028D† (Converse of #1027.) Let A be the adjacency matrix of a graph such that whenever λ is an eigenvalue of A (of a certain multiplicity) then $-\lambda$ is an eigenvalue (of the same multiplicity). Show that A is bipartite. [You may assume that a symmetric matrix is diagonalizable.]

Exercises on Markov Chains

#1029B† Consider the five-state Markov chain in Figure 3.26. If the system is in a given state, it then moves with equal likelihood to any neighbouring state. Explain why

$$p_A = \frac{1}{3}p_B \quad \text{and} \quad p_D = \frac{1}{3}p_B + \frac{1}{2}p_E.$$

Find the long-term probabilities p_A, p_B, p_C, p_D, p_E of being in each state. Conjecture and prove a general formula for such probabilities in terms of the degrees of the states and the total number of edges.

Definition 3.210 A **transition matrix** is a square matrix with non-negative entries that add to 1 in each row.

#1030B† Show that the product of two transition matrices is a transition matrix.

#1031B Let M denote the transition matrix of a finite-state Markov chain. Show that $e_i M^n e_j^T$ equals the probability of moving from the ith state to the jth state in n turns.

#1032B† Let M denote the transition matrix for the sunny/rainy model in #606. Calculate M^3. What is the probability that it is sunny on day 3 given it is rainy on day 0? What is the probability that it is sunny on day 11 given it is sunny on day 8? What does M^n approximate to for large values of n? (See also #606.)

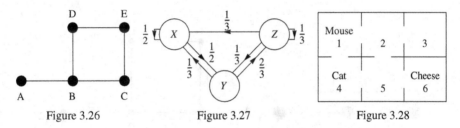

Figure 3.26   Figure 3.27   Figure 3.28

#1033D† Given the Markov chain in Figure 3.27, let x, y and z denote the long-term probabilities of being in states X, Y and Z respectively. These probabilities are known as the **stationary distribution**.

(i) Explain why $x = \frac{1}{2}x + \frac{1}{3}y + \frac{1}{3}z$. Write down two similar equations for y and z, and solve these to find x, y, z.

(ii) What is the transition matrix M for this Markov chain? Show $(x, y, z)M = (x, y, z)$.

(iii) Determine M^n where n is a positive integer. To what does M^n approximate when n is large?

#1034B†(i) Show that any finite-state Markov chain has a (not necessarily unique) stationary distribution.

(ii) Give an example of an infinite state Markov chain with no stationary distribution.

(iii) Give an example of an irreducible finite-state Markov chain with transition matrix P such that P^n does not converge to any limit.

#1035D† (i) If $\mathbf{p} = (p_1, \ldots, p_k)$ is a stationary distribution of an irreducible Markov chain, show $p_i > 0$ for each i.

(ii) Deduce that an irreducible finite-state Markov chain has a unique stationary distribution.

(iii) Show that the converse of (ii) is not true.

#1036B† As shown in Figure 3.28, a mouse moves at random about a set of six rooms (with no memory of its previous journey) picking available doors at random. A cat is in (and remains in) room 4, meaning certain death for the mouse, and some cheese is in room 6. The mouse starts in room 1; we wish to determine the chance of it finding the cheese (rather than the cat). Let p_n denote the probability of the mouse finding the cheese when in room n so that we are seeking to determine p_1.

(i) What is p_6? What is p_4?

(ii) Explain why $p_2 = \frac{1}{3}p_1 + \frac{1}{3}p_3 + \frac{1}{3}p_5$.

(iii) Write down similar equations for p_1, p_3, p_5 and solve these to find p_1.

(iv) Write down the transition matrix M for this Markov chain. What are the stationary distributions?

#1037D† With M as in #1036, determine M^n and hence rederive the values of $p_1, \ldots, p_6$.

#1038C(i) Write down the transition matrix M for the Markov chain in Figure 3.29. What is M^2? What is M^n? What row vectors $\mathbf{x}$ satisfy $\mathbf{x}M = \mathbf{x}$?

(ii) Consider the two four-state Markov chains in Figures 3.30a and 3.30b. Write down their transition matrices P and Q. What is P^2? What is P^3? Can you find an expression for the powers of P? Do the same for Q.

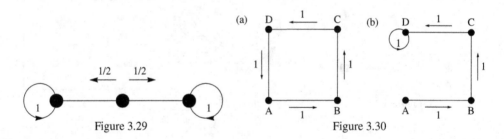

Figure 3.29   Figure 3.30

#1039D An infinite-state Markov chain models a bus queue as follows: if at time n there are $k \geqslant 0$ people in the queue, then at time $n+1$ there is probability p of one more person joining the queue and probability $q = 1 - p$ that a (suitably large) bus arrives collecting everyone.

(i) What is the long-term probability of the queue having k people?

(ii) Long term, what is the expected number of people in the queue?

#1040D† Let L denote the transition matrix in (3.66) associated with a graph of four states A, B, C, D connected in a loop. Here $0 \leqslant p \leqslant 1$ and $q = 1 - p$.

(i) If the chain starts in state A, what is expected time (number of steps) taken for it to first return to state A?

(ii) What is the expected time taken for it to first reach state B?

#1041D† Let M denote the transition matrix in (3.66) associated with a graph of four states A, B, C, D.

(i) Suppose the chain starts in state C. What is the expected time before it is first 'absorbed' into state A or D?

(ii) What is the probability that the chain is absorbed into state A if it begins in state C?

#1042D† Let N denote the transition matrix in (3.66) associated with a graph of six states A, B, C, D, E, F. If the chain starts in C, what is the probability that it is eventually absorbed in state F?

$$
L = \begin{pmatrix} 0 & p & 0 & q \\ q & 0 & p & 0 \\ 0 & q & 0 & p \\ p & 0 & q & 0 \end{pmatrix}, \quad
M = \begin{pmatrix} 1 & 0 & 0 & 0 \\ \frac{1}{3} & 0 & \frac{2}{3} & 0 \\ 0 & \frac{1}{4} & 0 & \frac{3}{4} \\ 0 & 0 & 0 & 1 \end{pmatrix}, \quad
N = \begin{pmatrix} \frac{1}{2} & 0 & 0 & \frac{1}{2} & 0 & 0 \\ \frac{1}{4} & 0 & 0 & 0 & \frac{3}{4} & 0 \\ 0 & \frac{1}{2} & \frac{1}{2} & 0 & 0 & 0 \\ \frac{1}{3} & 0 & 0 & \frac{2}{3} & 0 & 0 \\ 0 & 0 & \frac{1}{4} & \frac{1}{4} & \frac{1}{4} & \frac{1}{4} \\ 0 & 0 & 0 & 0 & 0 & 1 \end{pmatrix}.
$$

$$\tag{3.66}$$

#1043C† Consider the following weather model for sun (S) and rain (R), which gives probabilities for tomorrow's weather on the basis of the last two days' weather:

$$P(\text{R tomorrow} \mid \text{R today and R yesterday}) = 0.4;$$
$$P(\text{R tomorrow} \mid \text{S today and R yesterday}) = 0.2;$$
$$P(\text{R tomorrow} \mid \text{R today and S yesterday}) = 0.3;$$
$$P(\text{R tomorrow} \mid \text{S today and S yesterday}) = 0.1.$$

At first glance this might not seem to be a Markov chain, but can be made so if we consider the four states for the weather today and yesterday: RR, RS, SR, SS. What is the 4×4 transition matrix for this Markov chain? What are the long-term chances of (i) these four states? (ii) rain on a given day?

4

More on Matrices

4.1 Changing Bases*

• This section makes considerable use of §3.6, in particular the notion of a basis (Definition 3.109) and the co-ordinates associated with a basis (Definition 3.119). It continues the discussion begun with Example 3.143.

Let A be an $m \times n$ matrix and consider the *associated map* μ_A from $\mathbb{R}_n$ to $\mathbb{R}_m$ given by $\mu_A(\mathbf{v}) = A\mathbf{v}$. From Proposition 3.137 we know μ_A is linear and conversely that, for a linear map T from $\mathbb{R}_n$ to $\mathbb{R}_m$, there is a unique *associated matrix* A such that $T = \mu_A$. The columns of A are $T(\mathbf{e}_1^T), T(\mathbf{e}_2^T), \dots, T(\mathbf{e}_n^T)$.

To help recall this, consider the map $T = \mu_A$, where

$$A = \begin{pmatrix} 0 & -2 \\ 3 & 5 \end{pmatrix}. \tag{4.1}$$

If we multiply the standard basis vectors $\mathbf{i} = (1,0)^T$ and $\mathbf{j} = (0,1)^T$ by A we get

$$\begin{pmatrix} 0 & -2 \\ 3 & 5 \end{pmatrix} \begin{pmatrix} 1 \\ 0 \end{pmatrix} = \begin{pmatrix} 0 \\ 3 \end{pmatrix}, \quad \text{and} \quad \begin{pmatrix} 0 & -2 \\ 3 & 5 \end{pmatrix} \begin{pmatrix} 0 \\ 1 \end{pmatrix} = \begin{pmatrix} -2 \\ 5 \end{pmatrix}.$$

As claimed above, the columns of A are $T(\mathbf{i})$ and $T(\mathbf{j})$. We have seen though, particularly in the section on diagonalization, that the standard basis commonly isn't the best way to understand a linear map. The eigenvalues of A are 2 and 3 and corresponding eigenvectors are $\mathbf{v}_1 = (1, -1)^T$ and $\mathbf{v}_2 = (2, -3)^T$ (#1044). The matrix A is relatively complicated – for example, consider trying to work out A^{100} with A in its current form. This is precisely because we started with the standard xy-co-ordinate system associated with $\mathbf{i}$ and $\mathbf{j}$, which has nothing in particular to do with the map μ_A. Much more natural co-ordinates are those associated with $\mathbf{v}_1$ and $\mathbf{v}_2$; let's call those co-ordinates X and Y.

As $\mathbf{v}_1$ and $\mathbf{v}_2$ form a basis for $\mathbb{R}_2$ each vector $x\mathbf{i} + y\mathbf{j}$ can be uniquely written as

$$x\mathbf{i} + y\mathbf{j} = X\mathbf{v}_1 + Y\mathbf{v}_2. \tag{4.2}$$

In this XY-system (Figure 4.1), we have $\mathbf{v}_1 = 1\mathbf{v}_1 + 0\mathbf{v}_2$ has co-ordinates $(X, Y) = (1, 0)$ and likewise $\mathbf{v}_2$ has XY-co-ordinates $(0, 1)$. Further $T(\mathbf{v}_1) = 2\mathbf{v}_1$ and $T(\mathbf{v}_2) = 3\mathbf{v}_2$, so if we had instead started with the XY-co-ordinates we would have determined the matrix for μ_A as

$$\tilde{A} = \text{diag}(2, 3),$$

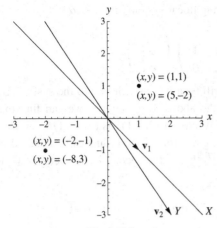

$(x,y) = (1,1)$

$(x,y) = (5,-2)$

$(x,y) = (-2,-1)$

$(x,y) = (-8,3)$

$\mathbf{v}_1$

$\mathbf{v}_2$ Y X

Figure 4.1

which is obviously much easier to calculate with than the original A. Note that the matrices A and $\widetilde{A}$ both encode the same map μ_A provided we use the correct co-ordinates. They respectively encode that

$$T(x\mathbf{i}+y\mathbf{j}) = -2y\mathbf{i}+(3x+5y)\mathbf{j},$$
$$T(X\mathbf{v}_1+Y\mathbf{v}_2) = 2X\mathbf{v}_1+3Y\mathbf{v}_2.$$

We recognize here that the diagonal entries of $\widetilde{A}$ are the eigenvalues of A. The matrix $\widetilde{A}$ would have been determined in §3.11 by noting

$$P^{-1}AP = P^{-1}\begin{pmatrix} 0 & -2 \\ 3 & 5 \end{pmatrix}P = \begin{pmatrix} 2 & 0 \\ 0 & 3 \end{pmatrix} = \widetilde{A} \qquad \text{where} \qquad P = \begin{pmatrix} 1 & 2 \\ -1 & -3 \end{pmatrix}.$$

There the matrix P would have been constructed by setting the eigenvectors as its columns. However, in this context P is better understood as a **change of basis matrix**. If x, y and X, Y are defined as in (4.2) then

$$x = X+2Y, \quad y = -X-3Y \quad \text{or equivalently} \quad \begin{pmatrix} x \\ y \end{pmatrix} = \begin{pmatrix} 1 & 2 \\ -1 & -3 \end{pmatrix}\begin{pmatrix} X \\ Y \end{pmatrix}. \quad (4.3)$$

Thus if $(\alpha, \beta)^T$ are the XY-co-ordinates of a point in $\mathbb{R}_2$ then $P(\alpha, \beta)^T$ are the xy-co-ordinates *of the same point*. In light of this we make the following definitions.

Definition 4.1 Let $\mathcal{V} = \{\mathbf{v}_1, \dots, \mathbf{v}_n\}$ be a basis for $\mathbb{R}_n$ and let $\mathbf{x}$ be in $\mathbb{R}_n$. Then there are unique numbers $\alpha_1, \dots, \alpha_n$ such that $\mathbf{x} = \alpha_1\mathbf{v}_1 + \cdots + \alpha_n\mathbf{v}_n$. Recall from Definition 3.119 that we refer to $\alpha_1, \dots, \alpha_n$ as the co-ordinates of $\mathbf{x}$ with respect to $\mathcal{V}$. We shall refer to $(\alpha_1, \dots, \alpha_n)^T$ as the $\mathcal{V}$-**vector** of $\mathbf{x}$.

Example 4.2 Find the co-ordinates of $(2,3,1)^T$ with respect to the following bases:

$$\mathcal{E} = \left\{\begin{pmatrix} 1 \\ 0 \\ 0 \end{pmatrix}, \begin{pmatrix} 0 \\ 1 \\ 0 \end{pmatrix}, \begin{pmatrix} 0 \\ 0 \\ 1 \end{pmatrix}\right\}, \quad \mathcal{F} = \left\{\begin{pmatrix} 1 \\ -1 \\ 2 \end{pmatrix}, \begin{pmatrix} 0 \\ 3 \\ -2 \end{pmatrix}, \begin{pmatrix} 2 \\ 1 \\ 0 \end{pmatrix}\right\},$$

$$\mathcal{G} = \left\{\begin{pmatrix} 1 \\ 2 \\ 3 \end{pmatrix}, \begin{pmatrix} 2 \\ 3 \\ 1 \end{pmatrix}, \begin{pmatrix} 3 \\ 1 \\ 2 \end{pmatrix}\right\}.$$

Solution As $\mathcal{E}$ is the standard basis for $\mathbb{R}_3$ then the $\mathcal{E}$-vector for $(2,3,1)^T$ is $(2,3,1)^T$. To find the $\mathcal{F}$-vector we need to find scalars $\alpha_1, \alpha_2, \alpha_3$ such that

$$(2,3,1) = \alpha_1(1,-1,2) + \alpha_2(0,3,-2) + \alpha_3(2,1,0).$$

This vector equation is equivalent to the following linear system in $\alpha_1, \alpha_2, \alpha_3$:

$$\begin{pmatrix} 2 \\ 3 \\ 1 \end{pmatrix} = \begin{pmatrix} 1 & 0 & 2 \\ -1 & 3 & 1 \\ 2 & -2 & 0 \end{pmatrix} \begin{pmatrix} \alpha_1 \\ \alpha_2 \\ \alpha_3 \end{pmatrix}.$$

As we are given that $\mathcal{F}$ is a basis then there will be a unique solution to the system. By solving the system, or equivalently inverting the above 3×3 matrix, we can find that $\alpha_1 = 7/3, \alpha_2 = 11/6, \alpha_3 = -1/6$. Hence the $\mathcal{F}$-vector of $(2, 3, 1)^T$ is $(7/3, 11/6, -1/6)^T$. Note that the 3×3 matrix has the role of changing from $\mathcal{F}$-co-ordinates to standard or $\mathcal{E}$-co-ordinates. Finally the $\mathcal{G}$-vector of $(2, 3, 1)^T$, being the second basis vector in $\mathcal{G}$, is $(0, 1, 0)^T$. $\qquad\square$

Example 4.3 For any basis $\mathcal{V} = \{v_1, \ldots, v_n\}$ for $\mathbb{R}_n$, the $\mathcal{V}$-vector of v_k is e_k^T as

$$v_k = 0v_1 + \cdots + 0v_{k-1} + 1v_k + 0v_{k+1} + \cdots + 0v_n.$$

Definition 4.4 (Matrix Representatives for Linear Maps) Let T be a linear map from $\mathbb{R}_n$ to $\mathbb{R}_m$ and let x be in $\mathbb{R}_n$. Let $\mathcal{V}$ and $\mathcal{W}$ be bases for $\mathbb{R}_n$ and $\mathbb{R}_m$ respectively. The **matrix of T with respect to initial basis $\mathcal{V}$ and final basis $\mathcal{W}$** is the $m \times n$ matrix which maps the $\mathcal{V}$-vector of x to the $\mathcal{W}$-vector of $T(x)$.

We will denote this matrix as $_\mathcal{W}T_\mathcal{V}$. To provide more detail, if $\mathcal{V} = \{v_1, \ldots, v_n\}$ and $\mathcal{W} = \{w_1, \ldots, w_m\}$ then there are unique a_{ij} such that

$$T(v_j) = a_{1j}w_1 + a_{2j}w_2 + \cdots + a_{mj}w_m \qquad (1 \leqslant i \leqslant m, 1 \leqslant j \leqslant n). \tag{4.4}$$

Then $_\mathcal{W}T_\mathcal{V} = (a_{ij})$.

- If we return to the earlier discussion, we see that $A = {_\mathcal{E}}T_\mathcal{E}$ where $\mathcal{E} = \{i, j\}$ and $\widetilde{A} = {_\mathcal{V}}T_\mathcal{V}$ where $\mathcal{V} = \{v_1, v_2\}$.
- Note that the columns of $_\mathcal{W}T_\mathcal{V}$ are the $\mathcal{W}$-vectors of $T(v_1), \ldots, T(v_n)$.

Remark 4.5 It should be stressed that, in the discussion beginning this section, we were only considering one map from $\mathbb{R}_2$ to $\mathbb{R}_2$. The matrices A and $\widetilde{A}$ both represent the *same* map T. It might help to think of x as a point of $\mathbb{R}_2$ that is moved to the point $T(x)$ by T. That these points can each be represented by different co-ordinate vectors is just a consequence of using two different sets of co-ordinates – the points x and $T(x)$ don't actually move when we change co-ordinates. The $\mathcal{E}$-vector of x (where $\mathcal{E}$ is the standard basis) contains the usual co-ordinates of x. The $\mathcal{V}$-vector of x has different entries but still represents the same point x. The matrix P takes the $\mathcal{V}$-vector of x to the $\mathcal{E}$-vector of x and so isn't actually moving the plane, just changing co-ordinates. $\qquad\blacksquare$

Example 4.6 Let $T = \mu_A$, where

$$A = \begin{pmatrix} 1 & 2 & -1 \\ 3 & 3 & 2 \end{pmatrix},$$

and set

$$v_1 = (1, 1, 1)^T, \quad v_2 = (1, 1, 0)^T, \quad v_3 = (1, 0, 0)^T, \quad w_1 = (1, 1)^T, \quad w_2 = (1, -1)^T.$$

Then

$$T(\mathbf{v}_1) = \begin{pmatrix} 1 & 2 & -1 \\ 3 & 3 & 2 \end{pmatrix} \begin{pmatrix} 1 \\ 1 \\ 1 \end{pmatrix} = \begin{pmatrix} 2 \\ 8 \end{pmatrix} = 5 \begin{pmatrix} 1 \\ 1 \end{pmatrix} - 3 \begin{pmatrix} 1 \\ -1 \end{pmatrix}.$$

So the first column of $_{\mathcal{W}}T_{\mathcal{V}}$ equals $(5, -3)^T$. Similarly,

$$T(\mathbf{v}_2) = (3, 6)^T = (9/2)\mathbf{w}_1 + (-3/2)\mathbf{w}_2; \qquad T(\mathbf{v}_3) = (1, 3)^T = 2\mathbf{w}_1 - \mathbf{w}_2.$$

Hence

$$_{\mathcal{W}}T_{\mathcal{V}} = \begin{pmatrix} 5 & 9/2 & 2 \\ -3 & -3/2 & -1 \end{pmatrix}.$$

Example 4.7 Let T be a linear map from $\mathbb{R}_n$ to $\mathbb{R}_m$ and let A be its associated matrix so that $T = \mu_A$. Then $A = {}_{\mathcal{F}}T_{\mathcal{E}}$, where $\mathcal{E}$, $\mathcal{F}$ are the standard bases for $\mathbb{R}_n$ and $\mathbb{R}_m$. That is, the associated matrix of a linear map is its matrix representative with respect to the standard initial and final bases.

Example 4.8 For any basis $\mathcal{V}$ of $\mathbb{R}_n$, the identity map I has matrix representative $I_n = {}_{\mathcal{V}}I_{\mathcal{V}}$.

The matrix representative $_{\mathcal{W}}T_{\mathcal{V}}$ for a linear map T can always be found by writing each $T(\mathbf{v}_j)$ as a linear combination of the basis $\mathcal{W}$ but often it is easiest to apply the following theorem and (4.5) in particular.

Theorem 4.9 *Let T be a linear map from $\mathbb{R}_n$ to $\mathbb{R}_m$ and let A be the associated matrix for T, so that $T = \mu_A$.*

Let $\mathcal{V} = \{\mathbf{v}_1, \ldots, \mathbf{v}_n\}$, $\mathcal{W} = \{\mathbf{w}_1, \ldots, \mathbf{w}_m\}$ be bases for $\mathbb{R}_n$ and $\mathbb{R}_m$ respectively. Set

$$P = (\mathbf{v}_1 | \mathbf{v}_2 | \cdots | \mathbf{v}_n) \qquad and \qquad Q = (\mathbf{w}_1 | \mathbf{w}_2 | \cdots | \mathbf{w}_m).$$

Then $P = {}_{\mathcal{E}}I_{\mathcal{V}}$ and $Q = {}_{\mathcal{F}}I_{\mathcal{W}}$, where $\mathcal{E}$ is the standard basis for $\mathbb{R}_n$, $\mathcal{F}$ is the standard basis for $\mathbb{R}_m$ and I denotes the relevant identity map. Also $P^{-1} = {}_{\mathcal{V}}I_{\mathcal{E}}$ and $Q^{-1} = {}_{\mathcal{W}}I_{\mathcal{F}}$. Further,

$$_{\mathcal{W}}T_{\mathcal{V}} = Q^{-1}AP. \tag{4.5}$$

Remark 4.10 Equation (4.5) may seem more natural if written as

$$_{\mathcal{W}}T_{\mathcal{V}} = ({}_{\mathcal{W}}I_{\mathcal{F}})({}_{\mathcal{F}}T_{\mathcal{E}})({}_{\mathcal{E}}I_{\mathcal{V}}),$$

especially in light of Remark 4.5. A $\mathcal{V}$-vector of $\mathbf{x}$ in $\mathbb{R}_n$ is mapped by the matrix $P = {}_{\mathcal{E}}I_{\mathcal{V}}$ to the $\mathcal{E}$-vector of $\mathbf{x}$ as the underlying map is the identity; all that's happening at this stage is that P is acting as a change of basis matrix, changing from $\mathcal{V}$-co-ordinates to $\mathcal{E}$-co-ordinates. The $\mathcal{E}$-vector of $\mathbf{x}$ is then mapped to the $\mathcal{F}$-vector of $T(\mathbf{x})$ by ${}_{\mathcal{F}}T_{\mathcal{E}}$. Finally, as the underlying map of the third matrix is again the identity, the $\mathcal{F}$-vector of $T(\mathbf{x})$ is mapped to the $\mathcal{W}$-vector of $T(\mathbf{x})$. The overall effect, of sending the $\mathcal{V}$-vector of $\mathbf{x}$ to the $\mathcal{W}$-vector of $T(\mathbf{x})$, is the same as what $_{\mathcal{W}}T_{\mathcal{V}}$ performs.

Proof (of Theorem 4.9) Let $\mathcal{E} = \{\mathbf{e}_1^T, \ldots, \mathbf{e}_n^T\}$ and $\mathcal{F} = \{\mathbf{E}_1^T, \ldots, \mathbf{E}_m^T\}$. (The different notation $\mathbf{e}_i^T$ and $\mathbf{E}_i^T$ is used as we are referring to two different standard bases of column vectors.) As $\mathbf{v}_j$ and $\mathbf{w}_j$ are the jth columns of P and Q,

$$I(\mathbf{v}_j) = \mathbf{v}_j = P\mathbf{e}_j^T = [P]_{1j}\mathbf{e}_1^T + \cdots + [P]_{nj}\mathbf{e}_n^T;$$
$$I(\mathbf{w}_j) = \mathbf{w}_j = Q\mathbf{E}_j^T = [Q]_{1j}\mathbf{E}_1^T + \cdots + [Q]_{mj}\mathbf{E}_m^T. \tag{4.6}$$

Referencing (4.4), this shows that $P = {}_\varepsilon I_\mathcal{V}$ and $Q = {}_\mathcal{F} I_\mathcal{W}$. Noting Example 4.3, we find ${}_\mathcal{V} I_\varepsilon \mathbf{v}_j = \mathbf{e}_j^T = P^{-1}\mathbf{v}_j$ as $P\mathbf{e}_j^T = \mathbf{v}_j$ and so ${}_\mathcal{V} I_\varepsilon = P^{-1}$. Similarly $Q^{-1} = {}_\mathcal{W} I_\mathcal{F}$.

Now denote the matrix ${}_\mathcal{W} T_\mathcal{V}$ as $\tilde{A}$. To determine $\tilde{A}$ we need to write $T(\mathbf{v}_j)$ as a linear combination of the $\mathbf{w}_k$. So, along similar but more general lines to the calculation in Example 4.6, we see

$$T(\mathbf{v}_j) = T\left(\sum_{i=1}^n [P]_{ij}\mathbf{e}_i^T\right) = \sum_{i=1}^n [P]_{ij}A\mathbf{e}_i^T = \sum_{i=1}^n [P]_{ij}\sum_{k=1}^m [A]_{ki}\mathbf{E}_k^T$$
$$= \sum_{i=1}^n [P]_{ij}\sum_{k=1}^m [A]_{ki}\sum_{l=1}^m [Q^{-1}]_{lk}\mathbf{w}_l,$$

where the first equality uses (4.6), the second and third that $T = \mu_A$ and $A = {}_\mathcal{F}T_\varepsilon$, and the last that $Q^{-1} = {}_\mathcal{W} I_\mathcal{F}$. By (4.4) we have

$$[\tilde{A}]_{lj} = \sum_{i=1}^n [P]_{ij}\sum_{k=1}^m [A]_{ki}[Q^{-1}]_{lk}$$
$$= \sum_{k=1}^m [Q^{-1}]_{lk}\sum_{i=1}^n [A]_{ki}[P]_{ij} = \sum_{k=1}^m [Q^{-1}]_{lk}[AP]_{kj} = [Q^{-1}AP]_{lj},$$

recalling the definition of matrix multiplication, and (4.5) follows.   □

Remark 4.11 (See #1062 and #1063.) Generally, if S is a linear map from $\mathbb{R}_m$ to $\mathbb{R}_n$ and T is a linear map from $\mathbb{R}_n$ to $\mathbb{R}_p$, and $\mathcal{V}, \mathcal{W}, \mathcal{X}$ are respectively bases for $\mathbb{R}_m, \mathbb{R}_n, \mathbb{R}_p$, then

$$\mathcal{X}(TS)_\mathcal{V} = ({}_\mathcal{X}T_\mathcal{W})({}_\mathcal{W}S_\mathcal{V}). \qquad ■ \tag{4.7}$$

Example 4.12 To verify (4.5) for Example 4.6, we note

$$Q^{-1}AP = \begin{pmatrix} 1 & 1 \\ 1 & -1 \end{pmatrix}^{-1} \begin{pmatrix} 1 & 2 & -1 \\ 3 & 3 & 2 \end{pmatrix} \begin{pmatrix} 1 & 1 & 1 \\ 1 & 1 & 0 \\ 1 & 0 & 0 \end{pmatrix}$$

$$= \frac{1}{2}\begin{pmatrix} 1 & 1 \\ 1 & -1 \end{pmatrix}\begin{pmatrix} 2 & 3 & 1 \\ 8 & 6 & 3 \end{pmatrix} = \begin{pmatrix} 5 & 9/2 & 2 \\ -3 & -3/2 & -1 \end{pmatrix}.$$

In §3.11, on diagonalization, we were dealing with square matrices and the initial and final bases were the same. In this case (4.5) reads as ${}_\mathcal{V}T_\mathcal{V} = P^{-1}AP$, where $P = {}_\varepsilon I_\mathcal{V}$. We were essentially asking: given a linear map T from $\mathbb{R}_n$ to $\mathbb{R}_n$, is there a *single* basis $\mathcal{V}$ for $\mathbb{R}_n$ such that ${}_\mathcal{V}T_\mathcal{V}$ is diagonal? We saw that this wasn't always the case, for example, with the matrix D in Examples 3.193 and 3.197. We might instead wonder what is a good alternative in such a case.

Example 4.13 Find a basis $\mathcal{V}$ with respect to which $T = \mu_D$ is upper triangular.

#1057
#1059
#1064
#1065

Solution We want to find $V = \{\mathbf{v}_1, \mathbf{v}_2, \mathbf{v}_3\}$ and a, b, c, d, e, f such that

$$_V T_V = \begin{pmatrix} a & b & d \\ 0 & c & e \\ 0 & 0 & f \end{pmatrix} \quad \text{where } T = \mu_D \text{ and } \quad D = \begin{pmatrix} 5 & -3 & -5 \\ 2 & 9 & 4 \\ -1 & 0 & 7 \end{pmatrix}.$$

This in particular means that $\mathbf{v}_1$ is an eigenvector of T with eigenvalue a. From the calculation in Example 3.197 we see we might take $\mathbf{v}_1 = (1 - 2, 1)^T$ and $a = 6$. We also require that $D\mathbf{v}_2 = b\mathbf{v}_1 + c\mathbf{v}_2$ and we can arrange this by taking $\mathbf{v}_2 = (-2, 1, 1)^T$ to be the 9-eigenvector that we found. This then means $c = 9$ and $b = 0$. For a basis we need $\mathbf{v}_3$ to be independent of $\mathbf{v}_1$ and $\mathbf{v}_2$, so let's take $\mathbf{v}_3 = (0, 0, 1)^T$.

By choice $T\mathbf{v}_1 = 6\mathbf{v}_1$, $T\mathbf{v}_2 = 9\mathbf{v}_2$ and further $T\mathbf{v}_3 = (-5, 4, 7)^T = -\mathbf{v}_1 + 2\mathbf{v}_2 + 6\mathbf{v}_3$. Hence

$$_V T_V = \begin{pmatrix} 6 & 0 & -1 \\ 0 & 9 & 2 \\ 0 & 0 & 6 \end{pmatrix}. \tag{4.8}$$

□

You might note that the matrix in (4.8) has the same characteristic polynomial as D, namely $(x - 6)^2(x - 9)$, which raises the following question: if many different square matrices can represent the same linear map, what properties must these matrix representatives have in common? For example, it would surely be weird and confusing if we determined using one basis that a linear map was invertible but came to the opposite conclusion using a different basis. Fortunately this eventuality cannot arise.

Theorem 4.14 *(Properties of Linear Maps) Let T be a linear map from $\mathbb{R}_n$ to $\mathbb{R}_n$. Let $A = {}_V T_V$ and $B = {}_W T_W$ where V and W are bases for $\mathbb{R}_n$. Then:*

(a) *There exists an invertible matrix P such that $A = P^{-1}BP$.*
(b) *$c_A(x) = c_B(x)$. In particular the eigenvalues of A and B are the same, $\det A = \det B$ and $\operatorname{trace} A = \operatorname{trace} B$.*
(c) *A is diagonalizable if and only if B is diagonalizable.*
(d) *$A^n = {}_V (T^n)_V$ for any natural number n.*
(e) *T is invertible if and only if A is invertible. In this case $A^{-1} = {}_V (T^{-1})_V$.*
(f) *$\operatorname{rank} A = \operatorname{rank} B$.*

Remark 4.15 Note that this means we can now discuss the determinant of a linear map T from $\mathbb{R}_n$ to $\mathbb{R}_n$ without any fear of ambiguity. We may define $\det T$ to be $\det A$, where $A = {}_V T_V$ is *any* matrix representative of T, knowing now that this definition does not depend on the choice of A. Likewise we can unambiguously define the trace, eigenvalues, characteristic polynomial, rank, diagonalizability and invertibility of a linear map. Note, however, that the transpose of a linear map is *not* a well-defined notion (#1061).

For a square matrix A, a non-zero vector $\mathbf{v}$ is an eigenvector if $A\mathbf{v} = \lambda\mathbf{v}$ for some scalar λ. If B is another matrix representative for μ_A then $A = P^{-1}BP$ for some change of basis matrix $P = {}_V I_\mathcal{E}$ and $P\mathbf{v}$ is a λ-eigenvector of B (#1047). Note that $P\mathbf{v}$ is the V-vector of $\mathbf{v}$. So whilst $P\mathbf{v}$ and $\mathbf{v}$ may have different co-ordinates, in the context of the bases V and $\mathcal{E}$, they represent the *same* eigenvector. ∎

Proof (of Theorem 4.14) (a) By Remark 4.11, $A = (_VI_W)B(_WI_V) = P^{-1}BP$ where $P = _WI_V$. (b) Appears as #919. (c) A is diagonalizable if there exists a matrix Q such that $Q^{-1}AQ$ is diagonal. But then

$$Q^{-1}AQ = Q^{-1}P^{-1}BPQ = (PQ)^{-1}B(PQ),$$

and we see that B is also diagonalizable. The converse follows similarly, as we have $B = (P^{-1})^{-1}AP^{-1}$. For (d) we note that $_\mathcal{E}T_\mathcal{E} = Q^{-1}AQ$ where $Q = _VI_\mathcal{E}$ and $\mathcal{E}$ is the standard basis for $\mathbb{R}_n$. Then, by Corollary 3.140,

$$\mathcal{E}(T^n)\mathcal{E} = (Q^{-1}AQ)^n = Q^{-1}A^nQ$$

and hence $A^n = (_VI_\mathcal{E})\mathcal{E}(T^n)\mathcal{E}(_\mathcal{E}I_V) = _V(T^n)_V$. (e) Say now that the linear map T is invertible – this means that there is a linear map S such that $ST = I = TS$. Then

$$(_VS_V)(_VT_V) = I_n = (_VT_V)(_VS_V),$$

so that $A^{-1} = _VS_V$. Conversely, say that the matrix $A = _VT_V$ is invertible. As in (d) we have $Q^{-1}AQ = _\mathcal{E}T_\mathcal{E}$ which has inverse $Q^{-1}A^{-1}Q$. Let $S = \mu_{Q^{-1}A^{-1}Q}$ and then we see, by Proposition 3.139, that

$$S \circ T = \mu_{Q^{-1}A^{-1}Q} \circ \mu_{Q^{-1}AQ} = \mu_{Q^{-1}A^{-1}QQ^{-1}AQ} = \mu_{I_n} = I;$$
$$T \circ S = \mu_{Q^{-1}AQ} \circ \mu_{Q^{-1}A^{-1}Q} = \mu_{Q^{-1}AQQ^{-1}A^{-1}Q} = \mu_{I_n} = I.$$

So T is invertible with inverse S and

$$_VS_V = (_VI_\mathcal{E})(_\mathcal{E}S_\mathcal{E})(_\mathcal{E}I_V) = Q(Q^{-1}A^{-1}Q)Q^{-1} = A^{-1}.$$

Part (f) is left as #1064, where a more general result for $m \times n$ matrices is proven. $\qquad\square$

Exercises

#1044 A Verify that the eigenvalues and eigenvectors of the matrix A in (4.1) are as given.

#1045 A Verify the identities giving x, y in terms of X, Y from (4.3). Determine X, Y in terms of x, y.

#1046 A With X and Y as in (4.2), sketch on the same axes (i) the region $X > 0, Y > 0$; (ii) the line $X + Y = 0$; (iii) the region $-1 \leqslant X, Y \leqslant 0$.

#1047 A† Let A, B be square matrices with $A = P^{-1}BP$. Show that $\mathbf{v}$ is a λ-eigenvector of A if and only if $P\mathbf{v}$ is a λ-eigenvector of B.

#1048 a† Let $\mathcal{F} = \{(1,1,1)^T, (1,1,0)^T, (1,0,0)^T\}$. Show that $\mathcal{F}$ is a basis for $\mathbb{R}_3$ and find the $\mathcal{F}$-vector for $(x, y, z)^T$.

#1049 b† Show that $\mathcal{B}$ below is a basis for M_{22}. Find the $\mathcal{B}$-vector for each of $E_{11}, E_{12}, E_{21}, E_{22}$.

$$\mathcal{B} = \left\{ \begin{pmatrix} 0 & 1 \\ 1 & 1 \end{pmatrix}, \begin{pmatrix} 1 & 0 \\ 1 & 1 \end{pmatrix}, \begin{pmatrix} 1 & 1 \\ 0 & 1 \end{pmatrix}, \begin{pmatrix} 1 & 1 \\ 1 & 0 \end{pmatrix} \right\}.$$

#1050C Show that $P_0(x),\ldots,P_4(x)$ form a basis $\mathcal{P}$ for the space of polynomials in x of degree four or less, where $P_n(x)$ is the nth Legendre polynomial. Determine the $\mathcal{P}$-vector for each of $1,x,\ldots,x^4$.

#1051b† Let $\mathcal{E}$ be the standard basis for $\mathbb{R}_3$ and $\mathcal{F}$ be as in #1048. Find the matrices $_{\mathcal{E}}T_{\mathcal{E}},\ _{\mathcal{F}}T_{\mathcal{E}},\ _{\mathcal{E}}T_{\mathcal{F}},\ _{\mathcal{F}}T_{\mathcal{F}}$, where T is defined by $T(x,y,z)^T = (2y+z, x-4y, 3x)^T$.

#1052c Verify directly, using the answers for #1051, that $(_{\mathcal{F}}T_{\mathcal{E}})(_{\mathcal{E}}T_{\mathcal{F}}) = (_{\mathcal{F}}T_{\mathcal{F}})^2$ and $(_{\mathcal{E}}T_{\mathcal{F}})(_{\mathcal{F}}T_{\mathcal{E}}) = (_{\mathcal{E}}T_{\mathcal{E}})^2$.

#1053B† Let Π denote the plane in $\mathbb{R}_3$ with equation $2x-3y+6z=0$. With $\mathcal{F}$ as in #1048, find the equation of Π in the form $\alpha X + \beta Y + \gamma Z = 0$, where X,Y,Z are the $\mathcal{F}$-co-ordinates of $\mathbb{R}_3$. Show that the vector with $\mathcal{F}$-vector $(\alpha,\beta,\gamma)^T$ is not normal to Π. How do you explain this apparent contradiction?

#1054b† Let T be the linear map of $\mathbb{R}_3$ defined by $T(x,y,z)^T = (x+3y, x-y+4z, 0)^T$. Find a basis $\mathcal{V}$ for $\mathbb{R}_3$ such that $_{\mathcal{V}}T_{\mathcal{V}} = \mathrm{diag}(2,-2,0)$.

#1055c† Verify by direct calculation that $_{\mathcal{V}}T_{\mathcal{V}} = P^{-1}DP$ for Example 4.13, where $P = {}_{\mathcal{E}}I_{\mathcal{V}}$.

#1056D† Repeat Example 4.13 but now finding a basis $\mathcal{V}$ for $\mathbb{R}_3$ such that $b = d = 0$. Hence determine D^n.

#1057b† Define $T(x,y)^T = (y,-x)^T$. Show that there is no basis $\mathcal{V}$ for $\mathbb{R}_2$ such that $_{\mathcal{V}}T_{\mathcal{V}}$ is upper triangular.

#1058A Let T be a linear map from $\mathbb{R}_n$ to $\mathbb{R}_n$ such that $_{\mathcal{V}}T_{\mathcal{V}}$ is triangular with respect to some basis $\mathcal{V}$ for $\mathbb{R}_n$. Show that all the roots of $c_T(x)$ are real. (The converse is also true; see #998 and #1007.)

#1059B† Let A be a 2×2 matrix with trace$A = 2$ and det$A = 1$. Show that μ_A is a shear. (See also #802.)

#1060C† Show that trace$A = 3$ and det$A = 1$ for A, the associated matrix of a shear of $\mathbb{R}_3$. Does the converse hold?

#1061b Let $T = \mu_A$ and $S = \mu_{A^T}$, where $A = \begin{pmatrix} 1 & 2 \\ 3 & 4 \end{pmatrix}$. Let $\mathcal{V} = \{(1,2)^T, (1,1)^T\}$. Show that $(_{\mathcal{V}}T_{\mathcal{V}})^T \neq {}_{\mathcal{V}}S_{\mathcal{V}}$.

#1062B† Derive (4.7) from Theorem 4.9.

#1063C Prove (4.7) directly, by mimicking the proof of (4.5).

#1064b† Let T be a linear map from $\mathbb{R}_n$ to $\mathbb{R}_m$. Let $\mathcal{A},\mathcal{B}$ be bases for $\mathbb{R}_n$ and $\mathcal{C},\mathcal{D}$ be bases for $\mathbb{R}_m$. Show that $\mathrm{rank}(_{\mathcal{C}}T_{\mathcal{A}}) = \mathrm{rank}(_{\mathcal{D}}T_{\mathcal{B}})$.

#1065B† For the matrix A below, find a basis $\mathcal{V}$ for $\mathbb{R}_3$ such that $_{\mathcal{V}}(\mu_A)_{\mathcal{V}} = \mathrm{diag}(1,B)$, where B is a 2×2 matrix. Show that $B^2 + B + I = 0$ for your choice of B.

$$A = \begin{pmatrix} 0 & 0 & 1 \\ 1 & 0 & 0 \\ 0 & 1 & 0 \end{pmatrix}$$

#1066D† With notation as in #1065, show that if $v(\mu_A)v = \text{diag}(1, B)$ then $B^2 + B + I = 0$.

#1067d† (From Oxford 2007 Paper A #2) (i) Let A denote the matrix below. Given $\mathbf{x}, \mathbf{y}$ in $\mathbb{R}_3$, find conditions that $\mathbf{y}$ must satisfy for the system $A\mathbf{x} = \mathbf{y}$ to be consistent, and find the general solution for $\mathbf{x}$ when these conditions hold.

$$A = \frac{1}{9} \begin{pmatrix} 5 & -2 & 4 \\ -2 & 8 & 2 \\ 4 & 2 & 5 \end{pmatrix}.$$

(ii) Show that if $\mathbf{y}$ is in $\text{Col}(A)$ then $A\mathbf{y} = \mathbf{y}$.
(iii) Find a basis $\mathcal{B}$ for $\mathbb{R}_3$ such that $_\mathcal{B}(\mu_A)_\mathcal{B} = \text{diag}(1, 1, 0)$.
For each of the following equations, say whether there is a real 3×3 matrix X which satisfies the equation, or show that no such X exists. (iv) $X^2 = A + 3I$. (v) $X^2 = A - 3I$.

#1068D† Let $T = \mu_A$ where A is the matrix in #1067 and B, C as in (4.9).

(i) Give as many reasons as possible why there is no basis $\mathcal{V}$ for $\mathbb{R}_3$ such that $_\mathcal{V}T_\mathcal{V} = B$.
(ii) For each reason given in part (i), show that it does not contradict there being a basis $\mathcal{W}$ for $\mathbb{R}_3$ such that $_\mathcal{W}T_\mathcal{W} = C$.
(iii) Find a basis $\mathcal{W}$ for $\mathbb{R}_3$ such that $_\mathcal{W}T_\mathcal{W} = C$.

$$B = \frac{1}{3} \begin{pmatrix} 1 & 1 & 1 \\ 2 & 2 & 2 \\ 3 & 3 & 3 \end{pmatrix}; \quad C = \frac{1}{18} \begin{pmatrix} -3 & -7 & -14 \\ 15 & 23 & 10 \\ -3 & -1 & 16 \end{pmatrix}; \quad M = \begin{pmatrix} 2 & 3 & 1 & 0 \\ 2 & -1 & 3 & 1 \\ 0 & 1 & 3 & 4 \end{pmatrix}.$$

$$(4.9)$$

#1069b Let $T = \mu_M$ with M as in (4.9). Determine $_\mathcal{W}T_\mathcal{V}$, where

$$\mathcal{V} = \{(1,0,1,0)^T, (0,1,1,1)^T, (1,0,0,0)^T, (1,1,0,0)^T\},$$
$$\mathcal{W} = \{(1,2,3)^T, (2,3,1)^T, (3,1,2)^T\}.$$

#1070c† Linear maps S, T of $\mathbb{R}_2$ are defined by $S(x, y)^T = (2x + 3y, 3x + 2y)^T$ and also by $T(x, y)^T = (2x + y, x + 2y)^T$.

(i) Show that there is a basis for $\mathbb{R}_2$ with respect to which both S and T are diagonal.
(ii) Let n be a positive integer. Determine a_n and b_n such that $T^n = a_n S + b_n I$.

#1071c† (See #691.) Let T be a linear map from $\mathbb{R}_n$ to $\mathbb{R}_m$ with $r = \text{rank}(T)$. Show that there are bases $\mathcal{V}$ for $\mathbb{R}_n$ and $\mathcal{W}$ for $\mathbb{R}_m$ such that

$$_\mathcal{W}T_\mathcal{V} = \begin{pmatrix} I_r & 0_{r,n-r} \\ 0_{m-r,r} & 0_{m-r,n-r} \end{pmatrix}.$$

#1072C† Let T be a linear map from $\mathbb{R}_n$ to $\mathbb{R}_n$ such that $\mathbb{R}_n = U \oplus V$ where U and V are T-invariant subspaces. Show that there is a basis $\mathcal{V}$ for $\mathbb{R}_n$ such that $_\mathcal{V}T_\mathcal{V} = \text{diag}(A, B)$ for some square matrices A and B.

#1073D† Let P and Q be a projection and involution of $\mathbb{R}_n$ respectively. Show that there are bases $\mathcal{V}$ and $\mathcal{W}$ for $\mathbb{R}_n$ such that $_\mathcal{V}P_\mathcal{V} = \text{diag}(I_{n-m}, 0_{mm})$ and $_\mathcal{W}Q_\mathcal{W} = \text{diag}(I_r, -I_{n-r})$.

#1074 D† (**Cyclic vectors**) Let T be a linear map from $\mathbb{R}_n$ to $\mathbb{R}_n$ such that $T^n = 0 \neq T^{n-1}$. Show that there is a vector $\mathbf{v}$ such that $\mathcal{B} = \{\mathbf{v}, T\mathbf{v}, T^2\mathbf{v}, \ldots, T^{n-1}\mathbf{v}\}$ is a basis for $\mathbb{R}_n$ and determine $_{\mathcal{B}}T_{\mathcal{B}}$. (See also p.335.)

Definition 4.16 Let $J(\lambda, r)$ be as in #598. A complex matrix

$$A = \mathrm{diag}(J(\lambda_1, r_{11}), \cdots, J(\lambda_1, r_{1g_1}), \cdots, J(\lambda_k, r_{k1}), \cdots, J(\lambda_k, r_{kg_k})) \qquad (4.10)$$

consisting of Jordan block matrices and where $r_{i1} \leqslant r_{i2} \leqslant \cdots \leqslant r_{ig_i}$ for each $1 \leqslant i \leqslant k$ is said to be in **Jordan normal form.** [1]

#1075 B† Let A be as in (4.10). Show that its minimum and characteristic polynomials are

$$m_A(x) = (x - \lambda_1)^{r_{1g_1}} \cdots (x - \lambda_k)^{r_{kg_k}} \qquad \text{and} \qquad c_A(x) = (x - \lambda_1)^{a_1} \cdots (x - \lambda_k)^{a_k}$$

where g_i is the geometric multiplicity of λ_i and the algebraic multiplicity of λ_i equals

$$a_i = r_{i1} + r_{i2} + \cdots + r_{ig_i}.$$

#1076 B† Find the Jordan normal forms of the following matrices:

$$A = \begin{pmatrix} 0 & 1 & 1 \\ 0 & 0 & 1 \\ 0 & 0 & 0 \end{pmatrix}, \qquad B = \begin{pmatrix} 0 & 1 & 0 \\ 0 & 0 & 1 \\ 1 & 0 & 0 \end{pmatrix}, \qquad C = \begin{pmatrix} 2 & 0 & 0 \\ 0 & 2 & 1 \\ 0 & 0 & 2 \end{pmatrix}.$$

#1077 D† Find the Jordan normal forms of the following matrices:

$$\begin{pmatrix} 1 & 0 & 0 \\ 1 & 1 & 0 \\ 0 & 1 & 1 \end{pmatrix}, \quad \begin{pmatrix} 1 & -1 & 1 \\ 0 & 0 & 1 \\ 0 & 1 & 0 \end{pmatrix}, \quad \begin{pmatrix} 1 & 1 & 1 & 1 \\ 1 & 1 & 1 & 1 \\ 1 & 1 & 1 & 1 \\ 1 & 1 & 1 & 1 \end{pmatrix}, \quad \begin{pmatrix} 1 & 1 & 0 & 0 \\ 0 & 1 & 0 & 0 \\ 2 & 3 & -1 & 4 \\ 1 & 1 & -1 & 3 \end{pmatrix}.$$

#1078 E† Let M be an $n \times n$ complex matrix such that $M^k = 0$ and $M^{k-1} \neq 0$. Show that M is similar to a matrix of the form

$$\mathrm{diag}(J(0, r_1), \cdots, J(0, r_m)),$$

where $r_1 \leqslant r_2 \leqslant \cdots \leqslant r_m$ and $r_m = k$.

---

[1] The name of the French mathematician Camille Jordan (1838–1922) is familiar to many undergraduates due to a number of significant foundational contributions. An easy-to-state but hard-to-prove result is the *Jordan curve theorem* which states that a simple, closed curve in the plane (that is, a curve which is a deformed circle) divides the plane into two connected pieces, one unbounded, one bounded. Jordan also shaped progress in group theory, both in the *Jordan–Hölder theorem* – an important theorem about the structure of finite groups – and also in the popularizing of Galois theory, particularly its subtle group-theoretic aspects which had largely gone unappreciated by earlier mathematicians. He is also remembered for *Jordan content* (see p.344), an early attempt to define measure in Euclidean space.

#1079D† Show that every square complex matrix is similar to a matrix in Jordan normal form which is unique (save for a possible reordering of the eigenvalues).

4.2 Orthogonal Matrices and Isometries*

We saw in the previous section (and earlier still in §3.11) that a change of co-ordinates can be very useful both in terms of computations with matrices (Example 3.204, #1056) and also in terms of appreciating the effects of linear maps (remarks at start of §4.1). We saw further in Theorem 4.14 that any conclusions about a linear map's determinant, trace, eigenvalues, diagonalizability, invertibility or rank are not affected by a change of co-ordinates associated with an invertible matrix P.

However, geometric properties – like length, area, angle – are *not* in general preserved by such a change of co-ordinates. For this we need the change of basis matrix P to be *orthogonal*. We recall (from #588) that

Definition 4.17 A square real matrix A is said to be **orthogonal** if $A^{-1} = A^T$.

Example 4.18 (Orthogonal 2 × 2 Matrices) A 2×2 orthogonal matrix has one of the forms below, for some unique θ in the range $0 \leqslant \theta < 2\pi$. From Example 3.133 we know R_θ represents rotation anticlockwise by θ about the origin and $S_{\theta/2}$ represents reflection in the line $y = x\tan(\theta/2)$.

$$R_\theta = \begin{pmatrix} \cos\theta & -\sin\theta \\ \sin\theta & \cos\theta \end{pmatrix}; \qquad S_{\theta/2} = \begin{pmatrix} \cos\theta & \sin\theta \\ \sin\theta & -\cos\theta \end{pmatrix}.$$

Solution Let A be an orthogonal 2×2 matrix. If

$$A = \begin{pmatrix} a & b \\ c & d \end{pmatrix} \qquad \text{then} \qquad I_2 = A^T A = \begin{pmatrix} a^2+c^2 & ab+cd \\ ab+cd & b^2+d^2 \end{pmatrix}.$$

So the orthogonality of A imposes three equations on its entries, namely

$$a^2 + c^2 = 1; \qquad b^2 + d^2 = 1; \qquad ab + cd = 0. \qquad (4.11)$$

Note that the first two equations imply that the columns of A are of unit length and the third equation implies they are perpendicular to one another. As $(a,c)^T$ is of unit length then there is unique θ in the range $0 \leqslant \theta < 2\pi$ such that $a = \cos\theta$ and $c = \sin\theta$. Then, as $(b,d)^T$ is also of unit length and perpendicular to $(a,c)^T$, we have two possibilities

$$(b,d)^T = (\cos(\theta \pm \pi/2), \sin(\theta \pm \pi/2)) = (\mp\sin\theta, \pm\cos\theta).$$

#1083
#1084
This means either $A = R_\theta$ or $A = S_{\theta/2}$ as desired. □

#1089
#1091
Example 4.19 In #541 an expression for Q_0, reflection in the plane $2x + y + 2z = 0$,
#1119 was found. Write down the associated matrix for Q_0 and show directly that this matrix is
#1120 orthogonal.

Solution It was found that $Q_0(x, y, z) = (x - 4y - 8z, -4x + 7y - 4z, -8x - 4y + z)/9$ which can be written in matrix form below.

$$Q_0 = \frac{1}{9} \begin{pmatrix} 1 & -4 & -8 \\ -4 & 7 & -4 \\ -8 & -4 & 1 \end{pmatrix}.$$

Showing that $Q_0^T Q_0 = I_3$ is equivalent to showing that the columns have unit length and are mutually perpendicular. Note that

$$1^2 + 4^2 + 8^2 = 4^2 + 7^2 + 4^2 = 8^2 + 4^2 + 1^2 = 9^2,$$

meaning the columns are of unit length and they are perpendicular to one another as

$$-4 - 28 + 32 = -8 + 16 - 8 = 32 - 28 - 4 = 0. \qquad \square$$

Definition 4.17 was first made in #588 and we have already seen the following properties.

Proposition 4.20 *(a) An $n \times n$ matrix A is orthogonal if and only if $A\mathbf{v} \cdot A\mathbf{w} = \mathbf{v} \cdot \mathbf{w}$ for all $\mathbf{v}, \mathbf{w}$ in $\mathbb{R}_n$. Note that this implies that the associated linear map μ_A of an orthogonal matrix is an isometry.*
(b) A square matrix is orthogonal if and only if its columns (or rows) are mutually perpendicular unit vectors.
(c) An orthogonal matrix has determinant 1 or -1. (The converse is not true.)

Proof (a) and (b) are #588(i) and (ii) and (c) is #817. $\qquad \square$

We prove here a similar result to Proposition 4.20(a) for the vector product.

Proposition 4.21 *Let A be a 3×3 orthogonal matrix with $\det A = 1$. For $\mathbf{v}, \mathbf{w}$ in $\mathbb{R}_3$ we have $A(\mathbf{v} \wedge \mathbf{w}) = A\mathbf{v} \wedge A\mathbf{w}$.*

Proof Recall from #911 that for any vectors $\mathbf{v}, \mathbf{w}, \mathbf{x}$ in $\mathbb{R}_3$ we have

$$[A\mathbf{v}, A\mathbf{w}, \mathbf{x}] = \det A \times \left[\mathbf{v}, \mathbf{w}, A^{-1}\mathbf{x}\right] = \det A \times [\mathbf{v}, \mathbf{w}, A^T\mathbf{x}].$$

If $\det A = 1$ and noting $\mathbf{a} \cdot \mathbf{b} = \mathbf{a}^T\mathbf{b}$ for any $\mathbf{a}, \mathbf{b}$ in $\mathbb{R}_3$, then

$$(A\mathbf{v} \wedge A\mathbf{w}) \cdot \mathbf{x} = (\mathbf{v} \wedge \mathbf{w}) \cdot A^T\mathbf{x} = (\mathbf{v} \wedge \mathbf{w})^T A^T\mathbf{x} = (A(\mathbf{v} \wedge \mathbf{w}))^T\mathbf{x} = (A(\mathbf{v} \wedge \mathbf{w})) \cdot \mathbf{x}.$$

As this is true for all $\mathbf{x}$ then the result follows (#501). $\qquad \square$

Remark 4.22 The corresponding result when $\det A = -1$ is $A(\mathbf{v} \wedge \mathbf{w}) = -(A\mathbf{v} \wedge A\mathbf{w})$. This ties in with comments made in Remark 3.154 about the map μ_A being sense-preserving when $\det A > 0$ and sense-reversing when $\det A < 0$. $\qquad \blacksquare$

Proposition 4.20(b) leads us to make the following definition.

Definition 4.23 Vectors $\mathbf{v}_1, \mathbf{v}_2, \ldots, \mathbf{v}_k$ in $\mathbb{R}_n$ are **orthonormal** if they are of unit length and mutually perpendicular, that is, if $\mathbf{v}_i \cdot \mathbf{v}_j = \delta_{ij}$ for $1 \leqslant i, j \leqslant k$. By the following, if $k = n$ then such a set is an **orthonormal basis**.

Proposition 4.24 *An orthonormal set* $\mathbf{v}_1, \ldots, \mathbf{v}_k$ *is linearly independent. If we have that* $\mathbf{w} = \alpha_1 \mathbf{v}_1 + \cdots + \alpha_k \mathbf{v}_k$ *then* $\alpha_i = \mathbf{w} \cdot \mathbf{v}_i$.

Proof These results are left to #1087.   □

By Proposition 4.20(a), if A is orthogonal then the angle between $A\mathbf{v}$ and $A\mathbf{w}$ is the same as the angle between $\mathbf{v}$ and $\mathbf{w}$, and also $A\mathbf{v}$ has the same length as $\mathbf{v}$. So we have:

Corollary 4.25 *Let P be an orthogonal $n \times n$ matrix. P's columns then form an orthonormal basis V for $\mathbb{R}_n$. Let $\tilde{\mathbf{x}}, \tilde{\mathbf{y}}$ be the V-vectors for $\mathbf{x}, \mathbf{y}$ in $\mathbb{R}_n$. Then $|\tilde{\mathbf{x}}| = |\mathbf{x}|$ and the angle between $\tilde{\mathbf{x}}$ and $\tilde{\mathbf{y}}$ equals the angle between $\mathbf{x}$ and $\mathbf{y}$.*

Proof This follows from Proposition 4.20(a) given that $\mathbf{x} = P\tilde{\mathbf{x}}$ and $\mathbf{y} = P\tilde{\mathbf{y}}$.   □

- Consequently, measurements of angle and length are not affected by a change of co-ordinates if and only if the change of basis matrix is orthogonal. The formulae defining length (3.1) and angle (3.6) are the same whether we use standard co-ordinates or those associated with an orthonormal basis.

Example 4.26 Let $A = R_{\pi/3} = (\mathbf{v}_1 | \mathbf{v}_2)$. Determine the co-ordinates of $P = (-1, 2)^T$ and $Q = (3, 5)^T$ with respect to $\mathbf{v}_1, \mathbf{v}_2$ and verify directly that the distance $|PQ|$ is invariant under this change of co-ordinates.

Solution Note that $|PQ| = 5$. We have

$$A = \begin{pmatrix} \cos \frac{\pi}{3} & -\sin \frac{\pi}{3} \\ \sin \frac{\pi}{3} & \cos \frac{\pi}{3} \end{pmatrix} = \begin{pmatrix} \frac{1}{2} & -\frac{\sqrt{3}}{2} \\ \frac{\sqrt{3}}{2} & \frac{1}{2} \end{pmatrix},$$

so that

$$\begin{pmatrix} X_P \\ Y_P \end{pmatrix} = A^T \begin{pmatrix} -1 \\ 2 \end{pmatrix} = \frac{1}{2} \begin{pmatrix} -1 + 2\sqrt{3} \\ \sqrt{3} + 2 \end{pmatrix};$$

$$\begin{pmatrix} X_Q \\ Y_Q \end{pmatrix} = A^T \begin{pmatrix} 3 \\ 5 \end{pmatrix} = \frac{1}{2} \begin{pmatrix} 3 + 5\sqrt{3} \\ -3\sqrt{3} + 5 \end{pmatrix}.$$

Then, as expected, $(X_P - X_Q)^2 + (Y_P - Y_Q)^2$ equals

$$\frac{1}{4} \left[\left(4 + 3\sqrt{3} \right)^2 + \left(3 - 4\sqrt{3} \right)^2 \right] = \frac{1}{4} [16 + 27 + 9 + 48] = 25 = |PQ|^2. \qquad □$$

Example 4.27 Show that μ_A is a shear of $\mathbb{R}_2$ where $A = \frac{1}{35} \begin{pmatrix} 11 & -32 \\ 18 & 59 \end{pmatrix}$.

Solution This is matrix A_2 from Example 3.134, so we already know it represents a shear. However, if we had not previously met this matrix, how might we go about appreciating this? Distance plays a crucial part in the definition of a shear and any change of co-ordinates we make needs to respect this. Note that $c_A(x) = (x - 1)^2$ and that the only 1-eigenvectors are multiples of $(4, -3)^T$ (#1080). So A is not diagonalizable but we see that $3x + 4y = 0$ is an invariant line under μ_A. We might take a unit vector $\mathbf{v}_1 = (4, -3)^T / 5$ parallel to this line and extend it to an orthonormal basis for $\mathbb{R}_2$ by choosing $\mathbf{v}_2 = (3,4)^T/5$. This change of co-ordinates will then preserve distances and angles. With $P = (\mathbf{v}_1|\mathbf{v}_2)$ we see

$$P^{-1}AP = P^T AP = \begin{pmatrix} 1 & -10/7 \\ 0 & 1 \end{pmatrix}. \tag{4.12}$$

With this change of basis we see that we have a shear with $\mathbf{v}_1$ parallel to the invariant line and points at a distance d from the invariant line moving a distance of $10d/7$ parallel to the invariant line. In particular,

$$(1,1)^T = (1/5)\mathbf{v}_1 + (7/5)\mathbf{v}_2$$

moves a distance of $(10/7) \times (7/5) = 2$, as we found in Example 3.134. □

Example 4.28 Let Q_0 be as in Example 4.19. Find an orthonormal basis V of $\mathbb{R}_3$ such that $\gamma(Q_0)\gamma = \mathrm{diag}(1,1,-1)$.

Solution If $V = \{\mathbf{v}_1, \mathbf{v}_2, \mathbf{v}_3\}$ and $\gamma(Q_0)\gamma = \mathrm{diag}(1,1,-1)$ then $Q_0\mathbf{v}_1 = \mathbf{v}_1$, $Q_0\mathbf{v}_2 = \mathbf{v}_2$ and $Q_0\mathbf{v}_3 = -\mathbf{v}_3$. This means that $\mathbf{v}_1$ and $\mathbf{v}_2$ lie in the invariant plane, $2x + y + 2z = 0$, and $\mathbf{v}_3$ is normal to it. So we might choose

$$\mathbf{v}_1 = (1,0,-1)^T/\sqrt{2}, \qquad \mathbf{v}_2 = (1,-4,1)^T/(3\sqrt{2}), \qquad \mathbf{v}_3 = (2,1,2)^T/3. \tag{4.13}$$

□

Example 4.29 Given that one of the matrices below describes a rotation of $\mathbb{R}_3$, one a reflection of $\mathbb{R}_3$, determine which is which. Determine the axis of the rotation, and the invariant plane of the reflection.

$$A = \frac{1}{25} \begin{pmatrix} 20 & 15 & 0 \\ -12 & 16 & 15 \\ 9 & -12 & 20 \end{pmatrix}; \qquad B = \frac{1}{25} \begin{pmatrix} -7 & 0 & -24 \\ 0 & 25 & 0 \\ -24 & 0 & 7 \end{pmatrix}. \tag{4.14}$$

Solution If we consider the equations $A\mathbf{x} = \mathbf{x}$ and $B\mathbf{x} = \mathbf{x}$ then the rotation will have a one-dimensional solution space (the axis of rotation) and the reflection's solution space will be two-dimensional (the plane of reflection). Reduction gives

$$A - I = \frac{1}{25} \begin{pmatrix} -5 & 15 & 0 \\ -12 & -9 & 15 \\ 9 & -12 & -5 \end{pmatrix} \xrightarrow{\text{RRE}} \begin{pmatrix} 1 & 0 & -1 \\ 0 & 1 & -1/3 \\ 0 & 0 & 0 \end{pmatrix};$$

$$B - I = \frac{1}{25} \begin{pmatrix} -32 & 0 & -24 \\ 0 & 0 & 0 \\ -24 & 0 & -18 \end{pmatrix} \xrightarrow{\text{RRE}} \begin{pmatrix} 1 & 0 & 3/4 \\ 0 & 0 & 0 \\ 0 & 0 & 0 \end{pmatrix}.$$

Therefore A is the rotation and B is the reflection. The null space of $A - I$ consists of multiples of $(3,1,3)^T$, which is parallel to the axis of rotation. We see the invariant plane of the reflection B has equation $4x + 3z = 0$. □

Example 4.30 Let A and B be the matrices in (4.14). The unit vector $\mathbf{v}_1 = (3,1,3)^T/\sqrt{19}$ is parallel to the axis of rotation of A. Extend $\mathbf{v}_1$ to an orthonormal basis V for $\mathbb{R}_3$. What is the matrix for μ_A with respect to V? Determine an orthonormal basis $\mathbf{w}_1, \mathbf{w}_2$ for the invariant plane of B and extend it to an orthonormal basis $\mathcal{W}$ for $\mathbb{R}_3$. What is the matrix for μ_B with respect to $\mathcal{W}$?

Solution (Details are left to #1082. See also Figure 4.2.) We can extend $\mathbf{v}_1$ to an orthonormal basis $\mathcal{V}$ for $\mathbb{R}_3$ by taking

$$\mathbf{v}_2 = (1, -3, 0)^T / \sqrt{10}, \qquad \mathbf{v}_3 = (9, 3, -10)^T / \sqrt{190}.$$

If we set $P = (\mathbf{v}_1 | \mathbf{v}_2 | \mathbf{v}_3)$ then P is orthogonal and we find the matrix for μ_A with respect to $\mathcal{V}$ is

$$P^{-1}AP = \begin{pmatrix} 1 & 0 & 0 \\ 0 & 31/50 & 9\sqrt{19}/50 \\ 0 & -9\sqrt{19}/50 & 31/50 \end{pmatrix} = \begin{pmatrix} 1 & 0 & 0 \\ 0 & \cos\alpha & \sin\alpha \\ 0 & -\sin\alpha & \cos\alpha \end{pmatrix}$$

$$= \mathrm{diag}\,(1, R_\alpha), \quad \text{where } \alpha = \cos^{-1}(31/50).$$

#1080
#1082
#1096
#1097
#1102

The invariant plane of B has equation $4x + 3z = 0$, and this has an orthonormal basis $\mathbf{w}_1 = (-3, 0, 4)^T / 5$, $\mathbf{w}_2 = (0, 1, 0)^T$. This can be extended to an orthonormal basis for $\mathbb{R}_3$ by taking $\mathbf{w}_3 = (4, 0, 3)^T / 5$. If we set $Q = (\mathbf{w}_1 | \mathbf{w}_2 | \mathbf{w}_3)$ then Q is orthogonal and the matrix for μ_B with respect to $\mathcal{W} = \{\mathbf{w}_1, \mathbf{w}_2, \mathbf{w}_3\}$ equals

$$Q^{-1}BQ = Q^T BQ = \mathrm{diag}(1, 1, -1). \qquad \square$$

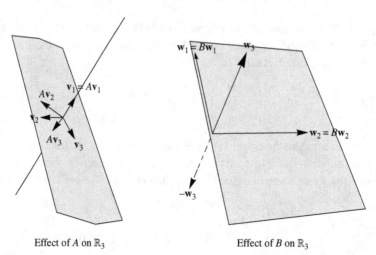

Effect of A on $\mathbb{R}_3$ Effect of B on $\mathbb{R}_3$

Figure 4.2

Matrices A and B, from Example 343, give a flavour of the general situation regarding rotations and reflections in three dimensions. Say a linear map T of $\mathbb{R}_3$ represents a rotation. By taking a unit vector parallel to the axis of rotation and extending it to an orthonormal basis $\mathcal{V}$ for $\mathbb{R}_3$ we see that $_\mathcal{V}T_\mathcal{V} = \mathrm{diag}(1, R_\theta)$, for some θ, from our knowledge of rotations in two dimensions (Example 3.133).

If a linear map S of $\mathbb{R}_3$ represents a reflection, then we may take any two orthonormal vectors in the plane of reflection and a third unit vector perpendicular to the plane. These three vectors form an orthonormal basis $\mathcal{W}$ such that $_\mathcal{W}S_\mathcal{W} = \mathrm{diag}\,(1, 1, -1)$.

• We noted in Example 4.18 that every orthogonal 2×2 matrix represents either a rotation or a reflection. Note that this is *not* the case in $\mathbb{R}_3$. For example, $-I_3$ is an orthogonal matrix representing neither a rotation nor a reflection, as it fixes only the origin.

The general situation for 3×3 orthogonal matrices is treated in Theorem 344.

Theorem 4.31 (*Classifying* 3×3 **Orthogonal Matrices**) *Let A be a 3×3 orthogonal matrix.*

(a) If $\det A = 1$ then μ_A is a rotation of $\mathbb{R}_3$ by an angle θ, where $\operatorname{trace} A = 1 + 2\cos\theta$.
(b) If $\det A = -1$ and $\operatorname{trace} A = 1$ then μ_A represents a reflection of $\mathbb{R}_3$.

Proof (a) We firstly show that when $\det A = 1$ there exists a non-zero vector $\mathbf{x}$ in $\mathbb{R}_3$ such that $A\mathbf{x} = \mathbf{x}$. Note by the product rule

$$\det(A - I) = \det((A - I)A^T) = \det(I - A^T) = \det((I - A)^T)$$
$$= \det(I - A) = (-1)^3 \det(A - I) = -\det(A - I),$$

also using $\det M^T = \det M$ and $\det(\lambda M) = \lambda^3 \det M$. Hence $\det(A - I) = 0$, and so there exists a non-zero vector $\mathbf{x}$ such that $(A - I)\mathbf{x} = \mathbf{0}$ as required. If we set $\mathbf{v}_1 = \mathbf{x}/|\mathbf{x}|$, so as to make a unit vector, then we can extend $\mathbf{v}_1$ to an orthonormal basis $\mathbf{v}_1, \mathbf{v}_2, \mathbf{v}_3$ for $\mathbb{R}_3$.

Let $P = (\mathbf{v}_1 \,|\, \mathbf{v}_2 \,|\, \mathbf{v}_3)$, noting P is orthogonal and define $B = P^{-1}AP = P^T AP$. Note B is another matrix that represents μ_A and that this change of co-ordinates preserves distances and angles. Note also that B is orthogonal, as it is the product of orthogonal matrices and that $\det B = 1$. If $\mathbf{i} = (1, 0, 0)^T$, then as $A\mathbf{v}_1 = \mathbf{v}_1$ we have

$$B\mathbf{i} = P^{-1}AP\mathbf{i} = P^{-1}A\mathbf{v}_1 = P^{-1}\mathbf{v}_1 = \mathbf{i}.$$

Hence the first column of B is $\mathbf{i}$. As the columns of B are orthonormal, B has the form

$$B = \operatorname{diag}(1, C),$$

where C is some 2×2 matrix so that $I_3 = B^T B = \operatorname{diag}(1, C^T C)$. We see C is an orthogonal 2×2 matrix and further, $\det C = 1$ as $\det B = 1$. By Example 4.18 we have $C = R_\theta$ for some $0 \leqslant \theta < 2\pi$ and that $B = \operatorname{diag}(1, R_\theta)$, so μ_A is a rotation. Finally note, by Theorem 4.14(b), that

$$\operatorname{trace} A = \operatorname{trace} B = 1 + \operatorname{trace} R_\theta = 1 + 2\cos\theta.$$

(b) Say now that $\det A = -1$ and $\operatorname{trace} A = 1$ Then $-A$ is orthogonal and we also have $\det(-A) = (-1)^3 \det A = 1$. By (a) there is an orthogonal matrix P such that

$$P^T(-A)P = \operatorname{diag}(1, R_\theta) \quad \Longrightarrow \quad P^T AP = \operatorname{diag}(-1, -R_\theta)$$

for some $0 \leqslant \theta < 2\pi$. Now $1 = \operatorname{trace} A = \operatorname{trace} P^T AP = -1 - 2\cos\theta$, showing that $\theta = \pi$, that $P^T AP = \operatorname{diag}(-1, 1, 1)$ and that μ_A is a reflection. $\qquad\square$

Example 4.32 In Example 4.29 we showed by a change of co-ordinates that the matrix A represents a rotation of angle $\alpha = \cos^{-1}(31/50)$ and that B represents a reflection. With Theorem 4.31 we might have spotted this earlier by noting that $\det A = 1$ and also that $\operatorname{trace} A = 56/25 = 1 + 2\cos\alpha$ for A and that $\det B = -1$ and $\operatorname{trace} B = 1$ for B.

In Example 4.30 and Theorem 4.31 we assumed that one or two orthonormal vectors in $\mathbb{R}_3$ can be extended to an orthonormal basis for $\mathbb{R}_3$ but never proved this. We now discuss how we might do this systematically in $\mathbb{R}_3$ (and will see that the process generalizes to higher dimensions). Our first result is to show how an orthonormal set can be constructed from a linearly independent one. Say that $\mathbf{v}_1, \mathbf{v}_2, \mathbf{v}_3$ is a basis for $\mathbb{R}_3$; we shall construct an orthonormal basis $\mathbf{w}_1, \mathbf{w}_2, \mathbf{w}_3$ for $\mathbb{R}_3$ such that

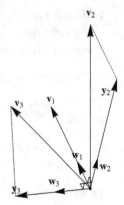

$$\langle \mathbf{w}_1 \rangle = \langle \mathbf{v}_1 \rangle;$$

$$\langle \mathbf{w}_1, \mathbf{w}_2 \rangle = \langle \mathbf{v}_1, \mathbf{v}_2 \rangle.$$

Figure 4.3

There are, in fact, only limited ways of doing this. As $\langle \mathbf{w}_1 \rangle = \langle \mathbf{v}_1 \rangle$ then $\mathbf{w}_1$ is a scalar multiple of $\mathbf{v}_1$. But as $\mathbf{w}_1$ is a unit vector then $\mathbf{w}_1 = \pm \mathbf{v}_1 / |\mathbf{v}_1|$. So there are only two choices for $\mathbf{w}_1$ and it seems most natural to take $\mathbf{w}_1 = \mathbf{v}_1 / |\mathbf{v}_1|$ (rather than needlessly introducing a negative sign). With this choice of $\mathbf{w}_1$ we then need to find a unit vector $\mathbf{w}_2$ perpendicular to $\mathbf{w}_1$ and such that $\langle \mathbf{w}_1, \mathbf{w}_2 \rangle = \langle \mathbf{v}_1, \mathbf{v}_2 \rangle$. In particular, we have

$$\mathbf{v}_2 = \alpha \mathbf{w}_1 + \beta \mathbf{w}_2 \qquad \text{for some scalars } \alpha, \beta.$$

We require $\mathbf{w}_2$ to be perpendicular to $\mathbf{w}_1$ and so $\alpha = \mathbf{v}_2 \cdot \mathbf{w}_1$. Note that

$$\mathbf{y}_2 = \beta \mathbf{w}_2 = \mathbf{v}_2 - (\mathbf{v}_2 \cdot \mathbf{w}_1) \mathbf{w}_1$$

is the component of $\mathbf{v}_2$ perpendicular to $\mathbf{v}_1$. We then have $\mathbf{w}_2 = \pm \mathbf{y}_2 / |\mathbf{y}_2|$. Again we have two choices of $\mathbf{w}_2$ but again there is no particular reason to choose the negative option.

Figure 4.3 hopefully captures the geometric nature of this process. $\mathbf{v}_1$ spans a line and so there are only two unit vectors parallel to it, with $\mathbf{w}_1 = \mathbf{v}_1 / |\mathbf{v}_1|$ being a more natural choice than its negative. $\langle \mathbf{v}_1, \mathbf{v}_2 \rangle$ is a plane divided into two half-planes by the line $\langle \mathbf{v}_1 \rangle$ and there are two choices of unit vector in this plane which are perpendicular to this line. We choose $\mathbf{w}_2$ to be that unit vector pointing into the same half-plane as $\mathbf{v}_2$ does. Continuing, $\langle \mathbf{v}_1, \mathbf{v}_2, \mathbf{v}_3 \rangle$ is a three-dimensional space divided into two half-spaces by the plane $\langle \mathbf{v}_1, \mathbf{v}_2 \rangle$. There are two choices of unit vector in this space which are perpendicular to the plane. We choose $\mathbf{w}_3$ to be that unit vector pointing into the same half-space as $\mathbf{v}_3$ does. This process is known as the *Gram–Schmidt orthogonalization process*,[2] with the rigorous details appearing below.

Theorem 4.33 (*Gram–Schmidt Orthogonalization Process*) *Let* $\mathbf{v}_1, \ldots, \mathbf{v}_k$ *be independent vectors in* $\mathbb{R}_n$ *(or* $\mathbb{R}^n$*). Then there are orthonormal vectors* $\mathbf{w}_1, \ldots, \mathbf{w}_k$ *such that, for each* $1 \leqslant i \leqslant k$*, we have*

$$\langle \mathbf{w}_1, \ldots, \mathbf{w}_i \rangle = \langle \mathbf{v}_1, \ldots, \mathbf{v}_i \rangle. \tag{4.15}$$

#1109
#1112
#1115
#1117

Proof We will prove this by induction on i. The result is seen to be true for $i = 1$ by taking

[2] Named after the Danish mathematician Jorgen Pedersen Gram (1850–1916) and the German mathematician Erhard Schmidt (1876–1959). The orthogonalization process was employed by Gram in a paper of 1883 and

$\mathbf{w}_1 = \mathbf{v}_1/|\mathbf{v}_1|$. Suppose now that $1 \leqslant I < k$ and that we have so far produced orthonormal vectors $\mathbf{w}_1, \ldots, \mathbf{w}_I$ such that (4.15) is true for $1 \leqslant i \leqslant I$. We then set

$$\mathbf{y}_{I+1} = \mathbf{v}_{I+1} - \sum_{j=1}^{I} (\mathbf{v}_{I+1} \cdot \mathbf{w}_j)\mathbf{w}_j.$$

Note, for $1 \leqslant i \leqslant I$, that

$$\mathbf{y}_{I+1} \cdot \mathbf{w}_i = \mathbf{v}_{I+1} \cdot \mathbf{w}_i - \sum_{j=1}^{I} (\mathbf{v}_{I+1} \cdot \mathbf{w}_j)\delta_{ij} = \mathbf{v}_{I+1} \cdot \mathbf{w}_i - \mathbf{v}_{I+1} \cdot \mathbf{w}_i = 0. \qquad (4.16)$$

So $\mathbf{y}_{I+1}$ is perpendicular to each of $\mathbf{w}_1, \ldots, \mathbf{w}_I$. Further, $\mathbf{y}_{I+1}$ is non-zero, for if $\mathbf{y}_{I+1} = \mathbf{0}$ then

$$\mathbf{v}_{I+1} = \sum_{j=1}^{I} (\mathbf{v}_{I+1} \cdot \mathbf{w}_j)\mathbf{w}_j \quad \text{is in} \quad \langle \mathbf{w}_1, \ldots, \mathbf{w}_I \rangle = \langle \mathbf{v}_1, \ldots, \mathbf{v}_I \rangle$$

which contradicts the linear independence of $\mathbf{v}_1, \ldots, \mathbf{v}_I, \mathbf{v}_{I+1}$. If we set $\mathbf{w}_{I+1} = \mathbf{y}_{I+1}/|\mathbf{y}_{I+1}|$, it follows from (4.16) that $\mathbf{w}_1, \ldots, \mathbf{w}_{I+1}$ form an orthonormal set. Further, by #730,

$$\langle \mathbf{w}_1, \ldots, \mathbf{w}_{I+1} \rangle = \langle \mathbf{w}_1, \ldots, \mathbf{w}_I, \mathbf{y}_{I+1} \rangle = \langle \mathbf{w}_1, \ldots, \mathbf{w}_I, \mathbf{v}_{I+1} \rangle = \langle \mathbf{v}_1, \ldots, \mathbf{v}_I, \mathbf{v}_{I+1} \rangle$$

and the proof follows by induction. $\qquad \square$

Corollary 4.34 *Every subspace of $\mathbb{R}_n$ has an orthonormal basis.*

Proof If U is a subspace of $\mathbb{R}_n$ then it has a basis $\mathbf{v}_1, \ldots, \mathbf{v}_k$ by Corollary 3.113. By applying the Gram–Schmidt orthogonalization process, an orthonormal set $\mathbf{w}_1, \ldots, \mathbf{w}_k$ can be constructed from them which is a basis for U as

$$\langle \mathbf{w}_1, \ldots, \mathbf{w}_k \rangle = \langle \mathbf{v}_1, \ldots, \mathbf{v}_k \rangle = U. \qquad \square$$

Corollary 4.35 *An orthonormal set can be extended to an orthonormal basis.*

Proof Let $\mathbf{w}_1, \ldots, \mathbf{w}_k$ be an orthonormal set in $\mathbb{R}_n$. In particular, it is linearly independent and so may be extended to a basis $\mathbf{w}_1, \ldots, \mathbf{w}_k, \mathbf{v}_{k+1}, \ldots, \mathbf{v}_n$ for $\mathbb{R}_n$ by Corollary 3.116. The Gram–Schmidt process can then be applied to construct an orthonormal basis $\mathbf{x}_1, \ldots, \mathbf{x}_n$ from this basis. The nature of the Gram–Schmidt process means that $\mathbf{x}_i = \mathbf{w}_i$ for $1 \leqslant i \leqslant k$ (#1109) and so our orthonormal basis is an extension of the original orthonormal set. $\qquad \square$

We have been primarily interested in orthogonal matrices because their associated maps preserve distances. In fact, we shall see shortly that the orthogonal matrices represent the linear isometries.

Definition 4.36 A map T from $\mathbb{R}_n$ to $\mathbb{R}_n$ is said to be an **isometry** if it preserves distances – that is, if

$$|T(\mathbf{v}) - T(\mathbf{w})| = |\mathbf{v} - \mathbf{w}| \qquad \text{for any } \mathbf{v}, \mathbf{w} \text{ in } \mathbb{R}_n.$$

by Schmidt, with acknowledgements to Gram, in a 1907 paper, but in fact the process had also been used by Laplace as early as 1812.

Example 4.37 In #541, we saw that reflection in the plane $\mathbf{r} \cdot \mathbf{n} = c$ (where $\mathbf{n}$ is a unit vector) is given by

$$Q_c(\mathbf{v}) = Q_0(\mathbf{v}) + 2c\mathbf{n} = \mathbf{v} - 2(\mathbf{v} \cdot \mathbf{n})\mathbf{n} + 2c\mathbf{n}.$$

This is an isometry as

$$|Q_0(\mathbf{v})|^2 = (\mathbf{v} - 2(\mathbf{v} \cdot \mathbf{n})\mathbf{n}) \cdot (\mathbf{v} - 2(\mathbf{v} \cdot \mathbf{n})\mathbf{n}) = |\mathbf{v}|^2 - 4(\mathbf{v} \cdot \mathbf{n})^2 + 4(\mathbf{v} \cdot \mathbf{n})(\mathbf{n} \cdot \mathbf{n}) = |\mathbf{v}|^2,$$

and more generally

$$Q_c(\mathbf{v}) - Q_c(\mathbf{w}) = (\mathbf{v} - 2(\mathbf{v} \cdot \mathbf{n})\mathbf{n} + 2c\mathbf{n}) - (\mathbf{w} - 2(\mathbf{w} \cdot \mathbf{n})\mathbf{n} + 2c\mathbf{n})$$
$$= (\mathbf{v} - \mathbf{w}) - 2((\mathbf{v} - \mathbf{w}) \cdot \mathbf{n})\mathbf{n} = Q_0(\mathbf{v} - \mathbf{w}).$$

Example 4.38 (a) If A is an orthogonal matrix then μ_A is an isometry (Proposition 4.20(a)).

(b) Given a vector $\mathbf{c}$ in $\mathbb{R}_n$ then **translation** by $\mathbf{c}$, that is the map $T(\mathbf{v}) = \mathbf{v} + \mathbf{c}$, is an isometry as

$$|T(\mathbf{v}) - T(\mathbf{w})| = |(\mathbf{v} + \mathbf{c}) - (\mathbf{w} + \mathbf{c})| = |\mathbf{v} - \mathbf{w}|.$$

The composition of two isometries is still an isometry (#1093) so any map of the form $T(\mathbf{v}) = A\mathbf{v} + \mathbf{b}$ is an isometry where A is an orthogonal matrix and $\mathbf{b}$ is in $\mathbb{R}_n$. We shall see now that all isometries of $\mathbb{R}_n$ take this form.

Proposition 4.39 *Let S be an isometry from $\mathbb{R}_n$ to $\mathbb{R}_n$ such that $S(\mathbf{0}) = \mathbf{0}$. Then*

(a) $|S(\mathbf{v})| = |\mathbf{v}|$ for any $\mathbf{v}$ in $\mathbb{R}_n$ and $S(\mathbf{u}) \cdot S(\mathbf{v}) = \mathbf{u} \cdot \mathbf{v}$ for any $\mathbf{u}, \mathbf{v}$ in $\mathbb{R}_n$.
(b) If $\mathbf{v}_1, \dots, \mathbf{v}_n$ is an orthonormal basis for $\mathbb{R}_n$ then so is $S(\mathbf{v}_1), \dots, S(\mathbf{v}_n)$.
(c) S is linear. Further, there exists an orthogonal matrix A such that $S(\mathbf{v}) = A\mathbf{v}$ for each $\mathbf{v}$ in $\mathbb{R}_n$.

Proof (a) Note $|S(\mathbf{v})| = |S(\mathbf{v}) - \mathbf{0}| = |S(\mathbf{v}) - S(\mathbf{0})| = |\mathbf{v} - \mathbf{0}| = |\mathbf{v}|$ as S is an isometry that fixes $\mathbf{0}$. We further have for any $\mathbf{u}, \mathbf{v}$ in $\mathbb{R}_n$ that

$$\mathbf{u} \cdot \mathbf{v} = \frac{1}{2}(|\mathbf{u}|^2 + |\mathbf{v}|^2 - |\mathbf{u} - \mathbf{v}|^2) = \frac{1}{2}\left(|S(\mathbf{u})|^2 + |S(\mathbf{v})|^2 - |S(\mathbf{u}) - S(\mathbf{v})|^2\right) = S(\mathbf{u}) \cdot S(\mathbf{v}).$$

(b) It follows immediately from (a) that $S(\mathbf{v}_1), \dots, S(\mathbf{v}_n)$ are orthonormal. Note that n vectors in $\mathbb{R}_n$ which are orthonormal are automatically a basis as noted in Definition 4.23.

(c) Assume for now that $S(\mathbf{e}_i^T) = \mathbf{e}_i^T$ for each i. For a given $\mathbf{v}$ in $\mathbb{R}_n$ there exist unique λ_i and μ_i such that

$$\mathbf{v} = \lambda_1 \mathbf{e}_1^T + \dots + \lambda_n \mathbf{e}_n^T \qquad \text{and} \qquad S(\mathbf{v}) = \mu_1 \mathbf{e}_1^T + \dots + \mu_n \mathbf{e}_n^T.$$

Using our assumption and (a) we have $\mu_i = S(\mathbf{v}) \cdot \mathbf{e}_i^T = S(\mathbf{v}) \cdot S(\mathbf{e}_i^T) = \mathbf{v} \cdot \mathbf{e}_i^T = \lambda_i$. Hence $S(\mathbf{v}) = \mathbf{v}$ for each $\mathbf{v}$ in $\mathbb{R}_n$.

Now, without the initial assumption that $S(\mathbf{e}_i^T) = \mathbf{e}_i^T$, let

$$A = (S(\mathbf{e}_1^T)| \cdots |S(\mathbf{e}_n^T)).$$

As $S(\mathbf{e}_1^T), \dots, S(\mathbf{e}_n^T)$ is an orthonormal basis for $\mathbb{R}_n$, then A is orthogonal. In particular A is an isometry from $\mathbb{R}_n$ to $\mathbb{R}_n$ such that $A\mathbf{e}_i^T = S(\mathbf{e}_i^T)$ for each i. Then $A^{-1}S$ is an isometry which fixes each $\mathbf{e}_i^T$ and by the previous argument $A^{-1}S(\mathbf{v}) = \mathbf{v}$ for each $\mathbf{v}$ in $\mathbb{R}_n$. The result $S(\mathbf{v}) = A\mathbf{v}$ for each $\mathbf{v}$ then follows. $\qquad \square$

Theorem 4.40 (Classifying Isometries of $\mathbb{R}_n$) *Let T be an isometry from $\mathbb{R}_n$ to $\mathbb{R}_n$. Then there is an orthogonal matrix A and a column vector $\mathbf{b}$ such that $T(\mathbf{v}) = A\mathbf{v} + \mathbf{b}$ for all $\mathbf{v}$. Further, A and $\mathbf{b}$ are unique in this regard.*

Proof The map $S(\mathbf{v}) = T(\mathbf{v}) - T(\mathbf{0})$ is an isometry which fixes $\mathbf{0}$. So there is an orthogonal matrix A such that $S(\mathbf{v}) = A\mathbf{v}$, giving $T(\mathbf{v}) = A\mathbf{v} + T(\mathbf{0})$. To show uniqueness, suppose $T(\mathbf{v}) = A_1\mathbf{v} + \mathbf{b}_1 = A_2\mathbf{v} + \mathbf{b}_2$ for all $\mathbf{v}$. Setting $\mathbf{v} = \mathbf{0}$ we see $\mathbf{b}_1 = \mathbf{b}_2$. Then $A_1\mathbf{v} = A_2\mathbf{v}$ for all $\mathbf{v}$ and hence $A_1 = A_2$. $\qquad \square$

Corollary 4.41 *Isometries preserve angles.*

Proof Let T be an isometry of $\mathbb{R}_n$ and $A, \mathbf{b}$ be as described in Theorem 4.40. Let $\mathbf{p}, \mathbf{q}, \mathbf{r}$ be points in $\mathbb{R}_n$; say that T respectively maps them to $\mathbf{p}', \mathbf{q}', \mathbf{r}'$. Note that

$$\mathbf{q}' - \mathbf{p}' = (A\mathbf{q} + \mathbf{b}) - (A\mathbf{p} + \mathbf{b}) = A(\mathbf{q} - \mathbf{p}).$$

Let θ and θ' respectively denote the angles subtended at $\mathbf{p}$ and $\mathbf{p}'$. As A is orthogonal,

$$\cos\theta' = \frac{(\mathbf{q}' - \mathbf{p}') \cdot (\mathbf{r}' - \mathbf{p}')}{|\mathbf{q}' - \mathbf{p}'| |\mathbf{r}' - \mathbf{p}'|} = \frac{A(\mathbf{q} - \mathbf{p}) \cdot A(\mathbf{r} - \mathbf{p})}{|A(\mathbf{q} - \mathbf{p})| |A(\mathbf{r} - \mathbf{p})|} = \frac{(\mathbf{q} - \mathbf{p}) \cdot (\mathbf{r} - \mathbf{p})}{|\mathbf{q} - \mathbf{p}| |\mathbf{r} - \mathbf{p}|} = \cos\theta$$

and the result follows. $\qquad \square$

Example 4.42 Let Q_k denote reflection in the plane $x + y + z = k$. Find an orthogonal matrix A_0 such that $Q_0(\mathbf{v}) = A_0\mathbf{v}$ for all $\mathbf{v}$. Show more generally that $Q_k(\mathbf{v}) = A_0\mathbf{v} + \mathbf{b}_k$ for some $\mathbf{b}_k$ to be determined.

Solution From #541, we know reflection in the plane $\mathbf{r} \cdot \mathbf{n} = c$, where $\mathbf{n}$ is a unit vector, is given by $Q_c(\mathbf{v}) = \mathbf{v} - 2(\mathbf{v} \cdot \mathbf{n})\mathbf{n} + 2c\mathbf{n}$. In this case we might use $\mathbf{n} = (1,1,1)^T/\sqrt{3}$ and rewrite the given plane as $\mathbf{r} \cdot \mathbf{n} = k/\sqrt{3}$. Note

$$Q_0(\mathbf{i}) = \mathbf{i} - \frac{2}{\sqrt{3}}\frac{1}{\sqrt{3}}(1,1,1)^T = \frac{1}{3}(1,-2,-2)^T.$$

With similar calculations for $Q_0(\mathbf{j})$ and $Q_0(\mathbf{k})$ we have

$$A_0 = \frac{1}{3}\begin{pmatrix} 1 & -2 & -2 \\ -2 & 1 & -2 \\ -2 & -2 & 1 \end{pmatrix}$$

$$Q_k(\mathbf{v}) = A_0\mathbf{v} + 2 \times \frac{k}{\sqrt{3}} \times \frac{1}{\sqrt{3}}\begin{pmatrix} 1 \\ 1 \\ 1 \end{pmatrix} = A_0\mathbf{v} + \begin{pmatrix} 2k/3 \\ 2k/3 \\ 2k/3 \end{pmatrix}. \qquad \square$$

Example 4.43 Determine the map T of $\mathbb{R}_3$ representing a rotation of π about the line $(x-1)/2 = y = (z+1)/3$.

Solution The axis of the rotation passes through $\mathbf{b} = (1,0,-1)$ and is parallel to the vector $\mathbf{v}_1 = (2,1,3)/\sqrt{14}$. We can extend $\mathbf{v}_1$ to an orthonormal basis for $\mathbb{R}_3$ by choosing

$$\mathbf{v}_2 = (1,-2,0)/\sqrt{5}, \qquad \mathbf{v}_3 = (6,3,-5)/\sqrt{70}.$$

We know that T fixes $\mathbf{b}$ and $\mathbf{b} + \mathbf{v}_1$, both being on the axis of rotation. As T is a half-turn then

$$T(\mathbf{b} + \mathbf{v}_2) = \mathbf{b} - \mathbf{v}_2, \qquad T(\mathbf{b} + \mathbf{v}_3) = \mathbf{b} - \mathbf{v}_3.$$

By Proposition 4.39, we know that $S(\mathbf{v}) = T(\mathbf{b} + \mathbf{v}) - \mathbf{b}$ is represented by an orthogonal matrix A as S is an isometry that fixes $\mathbf{0}$. From what we just noted we have that, $A(\mathbf{v}_1 \,|\, \mathbf{v}_2 \,|\, \mathbf{v}_3) = (\mathbf{v}_1 \,|\, -\mathbf{v}_2 \,|\, -\mathbf{v}_3)$ and hence

$$A = (\mathbf{v}_1 \,|\, -\mathbf{v}_2 \,|\, -\mathbf{v}_3)(\mathbf{v}_1 \,|\, \mathbf{v}_2 \,|\, \mathbf{v}_3)^T = \frac{1}{7} \begin{pmatrix} -3 & 2 & 6 \\ 2 & -6 & 3 \\ 6 & 3 & 2 \end{pmatrix}$$

with some further calculation. As $T(\mathbf{v}) = S(\mathbf{v} - \mathbf{b}) + \mathbf{b}$ then

$$T \begin{pmatrix} x \\ y \\ z \end{pmatrix} = \frac{1}{7} \begin{pmatrix} -3 & 2 & 6 \\ 2 & -6 & 3 \\ 6 & 3 & 2 \end{pmatrix} \begin{pmatrix} x-1 \\ y \\ z+1 \end{pmatrix} + \begin{pmatrix} 1 \\ 0 \\ -1 \end{pmatrix}$$

$$= \frac{1}{7} \begin{pmatrix} -3x + 2y + 6z + 16 \\ 2x - 6y + 3z + 1 \\ 6x + 3y + 2z - 11 \end{pmatrix}. \qquad \square$$

Exercises

#1080 A Verify that the characteristic polynomial, eigenvalues and eigenvectors of A in Example 4.27 are as claimed. Verify also the calculation of $P^{-1}AP$.

#1081 A Let $\mathbf{u}, \mathbf{v}$ be in $\mathbb{R}_2$. Show that there is an orthogonal matrix A such that $A\mathbf{u} = \mathbf{v}$ if and only if $|\mathbf{u}| = |\mathbf{v}|$.

#1082 a As defined in Example 4.30, verify that $\mathbf{v}_1, \mathbf{v}_2, \mathbf{v}_3$ and $\mathbf{w}_1, \mathbf{w}_2, \mathbf{w}_3$ are indeed orthonormal bases for $\mathbb{R}_3$ with $\mathbf{w}_1, \mathbf{w}_2$ being an orthonormal basis for the invariant plane of B. Verify also the expressions for $P^T AP$ and $Q^T BQ$.

#1083 b Let A be the associated matrix for Q_0 as determined in Example 4.19 and $\mathbf{v}_1, \mathbf{v}_2, \mathbf{v}_3$ be as determined in (4.13). Verify that $\mathbf{v}_1, \mathbf{v}_2, \mathbf{v}_3$ are indeed orthonormal vectors and that $P^T AP = \text{diag}(1, 1, -1)$ where $P = (\mathbf{v}_1 \,|\, \mathbf{v}_2 \,|\, \mathbf{v}_3)$.

#1084 A† Deduce directly from (4.11) that the rows of A are of unit length and are perpendicular to one another.

#1085 C If the rows and columns of a square matrix are all unit length, must the matrix be orthogonal?

#1086 A Let A be a 3×3 orthogonal matrix which represents a reflection. Show $\det A = -1$ and $\text{trace} A = 1$.

#1087 C Prove Proposition 4.24.

#1088 C Which 3×3 permutation matrices are rotations, which reflections? Find their axes and angles of rotation, or their planes of reflection.

#1089 A(i) Show that a square matrix A is orthogonal if and only A^{-1} is orthogonal.
(ii) Show that a square matrix A is orthogonal if and only if A^T is orthogonal.
(iii) Show [3] that if A and B are orthogonal $n \times n$ matrices then so is AB.

#1090 C† Let A be an invertible 3×3 matrix such that $A(\mathbf{v} \wedge \mathbf{w}) = A\mathbf{v} \wedge A\mathbf{w}$ for all $\mathbf{v}, \mathbf{w}$ in $\mathbb{R}_3$. Need A be orthogonal with $\det A = 1$?

#1091 b If an orthogonal matrix represents a reflection, show that it is symmetric. Is the converse true?

#1092 C† Let A be a 3×3 orthogonal matrix. Prove that the following are equivalent.

$$A^2 = I_3 \quad \Longleftrightarrow \quad A \text{ is symmetric} \quad \Longleftrightarrow \quad A \text{ is diagonalizable}$$

$$\Longleftrightarrow \quad A = \pm I_3, \text{ a reflection or a half-angle rotation.}$$

#1093 B Show that the composition of two isometries is an isometry. Show that an isometry of $\mathbb{R}_n$ is invertible and that its inverse is an isometry. [4]

#1094 c Show that an isometry of $\mathbb{R}_n$ maps a line to a line, a plane to a plane and a sphere to a sphere.

#1095 b† Let T be an isometry of $\mathbb{R}_3$ and P, Q, R be distinct points in $\mathbb{R}_3$. Show that the triangles PQR and $T(P)T(Q)T(R)$ have equal areas.

#1096 b† Let A and B be 2×2 matrices such that μ_A is a shear and μ_B is a stretch. Show that there are orthogonal matrices P and Q, and real numbers d and k, such that

$$P^T A P = \begin{pmatrix} 1 & d \\ 0 & 1 \end{pmatrix}; \quad Q^T B Q = \begin{pmatrix} 1 & 0 \\ 0 & k \end{pmatrix}.$$

#1097 b† Consider the orthogonal matrices

$$A = \frac{1}{3} \begin{pmatrix} 2 & 2 & -1 \\ 2 & -1 & 2 \\ -1 & 2 & 2 \end{pmatrix}, \quad B = \frac{1}{2} \begin{pmatrix} \sqrt{2} & \sqrt{2} & 0 \\ 1 & -1 & \sqrt{2} \\ 1 & -1 & -\sqrt{2} \end{pmatrix}.$$

Is either a rotation? If so, find the axis and angle of rotation. Is either a reflection? If so, find the plane of reflection.

#1098 b† Let A and B be as in #1097. Find an orthogonal matrix P such that it's the case that either $P^T A P = \text{diag}(1, R_\theta)$ or $\text{diag}(1, 1, -1)$. Do the same for B.

#1099 C† Let A be a 3×3 orthogonal matrix with $\det A = -1$. Show that there is a 3×3 orthogonal matrix B with $\det B = 1$ and such that BA represents a reflection.

---

[3] The $n \times n$ orthogonal matrices form a group $O(n)$ under matrix multiplication; this follows from (i), (iii), that I_n is orthogonal and that matrix multiplication is associative.
[4] This exercise (together with the facts that the identity map is an isometry and that composition is associative) shows the isometries to form a group, the *Euclidean group*.

#1100D (From Oxford 2007 Paper A #8) Let $A_\theta = \mathrm{diag}(1, R_\theta)$ where $0 \leqslant \theta < 2\pi$, and let B be as below.

(i) Show that B is orthogonal and that $\det B = -1$.
(ii) Show that B does not represent a reflection.
(iii) Find θ such that $A_\theta B$ represents a reflection.
(iv) For this θ, find the plane of reflection of $A_\theta B$.

$$B = \frac{1}{25} \begin{pmatrix} 15 & 0 & 20 \\ -16 & 15 & 12 \\ 12 & 20 & -9 \end{pmatrix}.$$

#1101C For an angle θ let

$$R(\mathbf{i}, \theta) = \begin{pmatrix} 1 & 0 & 0 \\ 0 & \cos\theta & -\sin\theta \\ 0 & \sin\theta & \cos\theta \end{pmatrix}, \qquad R(\mathbf{j}, \theta) = \begin{pmatrix} \cos\theta & 0 & -\sin\theta \\ 0 & 1 & 0 \\ \sin\theta & 0 & \cos\theta \end{pmatrix}.$$

Show for angles α, β, γ that $R(\mathbf{i}, \alpha) R(\mathbf{j}, \beta) R(\mathbf{i}, \gamma)$ equals

$$\begin{pmatrix} \cos\beta & -\sin\beta\sin\gamma & -\sin\beta\cos\gamma \\ -\sin\alpha\sin\beta & \cos\alpha\cos\gamma - \sin\alpha\cos\beta\sin\gamma & -\cos\alpha\sin\gamma - \sin\alpha\cos\beta\cos\gamma \\ \cos\alpha\sin\beta & \sin\alpha\cos\gamma + \cos\alpha\cos\beta\sin\gamma & -\sin\alpha\sin\theta + \cos\alpha\cos\beta\cos\gamma \end{pmatrix}.$$

$$(4.17)$$

#1102B† Let B be the matrix in #1097, and define $R(\mathbf{i}, \theta)$ and $R(\mathbf{j}, \theta)$ as in #1101. Find α, β, γ in the ranges $-\pi/2 \leqslant \alpha, \beta \leqslant \pi/2$ and $-\pi < \gamma \leqslant \pi$ such that

$$B = R(\mathbf{i}, \alpha) R(\mathbf{j}, \beta) R(\mathbf{i}, \gamma).$$

#1103D† Let R denote an orthogonal 3×3 matrix with $\det R = 1$ and let $R(\mathbf{i}, \theta)$ and $R(\mathbf{j}, \theta)$ be as in #1101.

(i) Suppose that $R\mathbf{i} = \mathbf{i}$. Show that R is of the form $R(\mathbf{i}, \theta)$ for some $-\pi < \theta \leqslant \pi$.
(ii) For general R, show that there exist α, β in the ranges $-\pi < \alpha \leqslant \pi, 0 \leqslant \beta \leqslant \pi$, and c, d such that $d \geqslant 0$ and $c^2 + d^2 = 1$, with

$$R(\mathbf{i}, \alpha)^{-1} R\mathbf{i} = c\mathbf{i} + d\mathbf{k}, \qquad R(\mathbf{j}, \beta)^{-1} R(\mathbf{i}, \alpha)^{-1} R\mathbf{i} = \mathbf{i}.$$

(iii) Deduce that $R = R(\mathbf{i}, \alpha) R(\mathbf{j}, \beta) R(\mathbf{i}, \gamma)$ for some $-\pi < \alpha \leqslant \pi, 0 \leqslant \beta \leqslant \pi, -\pi < \gamma \leqslant \pi$.

This means that every rotation about the origin can be achieved as three successive rotations about the x-, y- and x-axes respectively. It also means that (4.17) is the general form of a rotation about the origin. The angles α, β, γ are examples of **Euler angles**.

#1104B† Show that there are no isometries from $\mathbb{R}_3$ to $\mathbb{R}_2$. Generalize this result to show that there are no isometries from $\mathbb{R}_n$ to $\mathbb{R}_m$ when $n > m$.

#1105b† Find the most general form of an isometry from $\mathbb{R}_3$ to $\mathbb{R}_3$ which fixes the points $(1, 0, 0)^T, (0, 1, 0)^T, (0, 0, 1)^T$.

#1106c† An isometry from $\mathbb{R}_2$ to $\mathbb{R}_3$ maps $(1, 0)^T$ to $(1, 0, 1)^T$ and $(0, 1)^T$ to $(0, 0, 2)^T$. What are the possible images of $(0, 0)^T$?

#1107c Show that there is a unique isometry T from $\mathbb{R}_2$ to $\mathbb{R}_3$ which maps

$$\begin{pmatrix} 2 \\ 1 \end{pmatrix} \mapsto \begin{pmatrix} 3 \\ \sqrt{8} \\ -\sqrt{8} \end{pmatrix}, \quad \begin{pmatrix} 2 \\ 5 \end{pmatrix} \mapsto \begin{pmatrix} 3 \\ 0 \\ 0 \end{pmatrix}, \quad \begin{pmatrix} 5 \\ 5 \end{pmatrix} \mapsto \begin{pmatrix} 4 \\ 2 \\ 2 \end{pmatrix}.$$

#1108D† Let $n \geqslant m$. Show that any isometry from $\mathbb{R}_m$ to $\mathbb{R}_n$ has the form $T(\mathbf{v}) = A\mathbf{v} + \mathbf{b}$ where A is an $n \times m$ matrix with rank m. Show further that there is an orthogonal $n \times n$ matrix P such that

$$PA = \begin{pmatrix} I_m \\ 0_{(n-m)m} \end{pmatrix}.$$

#1109A Let $\mathbf{v}_1, \ldots, \mathbf{v}_k$ be an orthonormal set. If the Gram–Schmidt process is applied to this set to produce $\mathbf{w}_1, \ldots, \mathbf{w}_k$ show that $\mathbf{w}_i = \mathbf{v}_i$ for each i.

#1110b Apply the Gram–Schmidt process to the following sets of vectors.

(i) $\{(1,1,0),(0,1,1),(1,0,1)\}$.
(ii) $\{(1,2,3),\{3,2,1)\}$.
(iii) $\{(1,2,3,4),(2,3,4,1),(3,4,1,2)\}$.

#1111b† Find an orthonormal basis for the subspace $3x_1 + 2x_3 = x_2 + 4x_4$ in $\mathbb{R}^4$.

#1112A Let A be an $n \times n$ matrix and $\mathbf{v}_1, \ldots, \mathbf{v}_n$ be in $\mathbb{R}_n$. Show that A is orthogonal if and only if whenever $\mathbf{v}_1, \ldots, \mathbf{v}_n$ form an orthonormal basis then $A\mathbf{v}_1, \ldots, A\mathbf{v}_n$ form an orthonormal basis.

#1113B Let A be an $n \times n$ matrix and $T = \mu_A$ and $S = \mu_{A^T}$. If $\mathcal{V}$ is an orthonormal basis for $\mathbb{R}_n$, show that $(\mathcal{V}T\mathcal{V})^T = \mathcal{V}S\mathcal{V}$. (Compare this with #1061.)

#1114B† Generalize #1081 from $\mathbb{R}_2$ to $\mathbb{R}_n$.

#1115B† Let X be an invertible $n \times n$ matrix. Show that there is an upper triangular matrix U and an orthogonal matrix P such that $X = PU$.

#1116c In the notation of #1115, find P and U for the matrices A_1, A_2, A_3 in #920.

#1117b† Show there are 2^n upper triangular orthogonal $n \times n$ matrices. Deduce that there are in all 2^n different decompositions $X = PU$, as described in #1115, for an invertible matrix X.

#1118C† Let U be an upper triangular matrix which is diagonalizable. Need there exist an upper triangular matrix X such that $X^{-1}UX$ is diagonal?

#1119B† Let

$$A = \begin{pmatrix} 1 & a & b \\ c & d & 2 \\ e & 1 & f \end{pmatrix}, \quad B = \begin{pmatrix} 1 & a & b \\ c & d & -1 \\ e & \frac{1}{2} & f \end{pmatrix}, \quad C = \begin{pmatrix} \frac{1}{2} & a & b \\ c & d & \frac{1}{2} \\ e & \frac{1}{2} & f \end{pmatrix}.$$

Are there constants a,b,c,d,e,f such that (i) A is orthogonal? (ii) B is orthogonal? (iii) C is orthogonal?

#1120b† Let $M = \begin{pmatrix} a & X \\ Y & b \end{pmatrix}$. Find the conditions on the constants X, Y such that *given* X, Y, there then exist a, b such that M is orthogonal. What can be said about X, Y if such orthogonal M is unique?

#1121D Let

$$M = \begin{pmatrix} X & a & b \\ c & d & Y \\ e & Z & f \end{pmatrix}.$$

Find conditions on the constants X, Y, Z such that given X, Y, Z there then exist a, b, c, d, e, f such that M is orthogonal. What can be said about X, Y, Z if a unique orthogonal M exists?

#1122C Let P denote the associated matrix P for orthogonal projection on to the plane $\mathbf{r} \cdot \mathbf{n} = 0$, where $\mathbf{n}$ is a unit vector, as found in #808. Determine an orthogonal matrix X such that $X^{-1} P X = \mathrm{diag}(1, 1, 0)$.

#1123C Let A be a 3×3 orthogonal matrix. Show directly that the product $\sqcap$, defined by

$$\mathbf{u} \sqcap \mathbf{v} = A(A^T \mathbf{u} \wedge A^T \mathbf{v}) \qquad \text{for } \mathbf{u}, \mathbf{v} \text{ in } \mathbb{R}_3,$$

satisfies each of the properties Proposition 3.179(a)-(d) which determine $\wedge$ up to a choice of sign.

4.3 Conics

The **conics**[5] or **conic sections** are a family of planar curves. They get their name as each can be formed by intersecting the double cone $x^2 + y^2 = z^2 \cot^2 \alpha$ with a plane in $\mathbb{R}^3$. For example, intersecting the cone with a plane $z = z_0$ produces a circle. The four different possibilities are drawn in Figure 4.4.

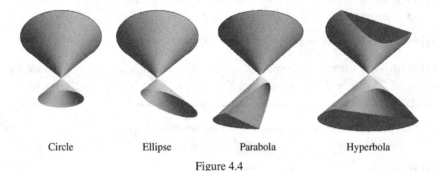

Circle   Ellipse   Parabola   Hyperbola

Figure 4.4

[5] The study of the conics dates back to the ancient Greeks. They were studied by Euclid and Archimedes, but much of their work did not survive. It is Apollonius of Perga (c. 262–190 BC), who wrote the eight-volume *Conics,* who is most associated with the curves and who gave them their modern names: ellipse, parabola, hyperbola. Later Pappus (c. 290–350) defined them via a focus and directrix (Definition 4.44). With the introduction of analytic geometry (i.e. the use of Cartesian co-ordinates), their study moved into the realm of algebra. From an algebraic point of view, the conics are a natural family of curves to study being defined by degree two equations in two variables (Theorem 4.53).

In Figure 4.5 is the cross-sectional view, in the xz-plane, of the intersection of the plane $z = (x-1)\tan\theta$ with the double cone $x^2 + y^2 = z^2 \cot^2\alpha$. We see that when $\theta < \alpha$ then the plane intersects only with bottom cone in a bounded curve (which in due course we shall see to be an *ellipse*). When $\theta = \alpha$ it intersects with the lower cone in an unbounded curve (a *parabola*), and when $\theta > \alpha$ we see that the plane intersects with both cones to make two separate unbounded curves (a *hyperbola*).

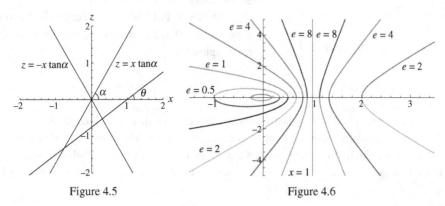

Figure 4.5   Figure 4.6

However, as a starting definition, we shall introduce conics using the idea of a *directrix* and *focus*.

Definition 4.44 Let D be a line, F be a point not on the line D and $e > 0$. Then the **conic**, with **directrix** D and **focus** F and **eccentricity** e, is the set of points P (in the plane containing F and D) which satisfy the equation

$$|PF| = e|PD|,$$

where $|PF|$ is the distance of P from the focus F and $|PD|$ is the shortest distance of P from the directrix D. That is, as the point P moves around the conic, the distance $|PF|$ is in constant proportion to the distance $|PD|$.

- If $0 < e < 1$ then the conic is called an **ellipse**.
- If $e = 1$ then the conic is called a **parabola**.
- If $e > 1$ then the conic is called a **hyperbola**.

In Figure 4.6 are sketched, for a fixed focus F (the origin) and fixed directrix D (the line $x = 1$), a selection of conics of varying eccentricity e.

Example 4.45 Find the equation of the parabola with focus $(1,1)$ and directrix $x + 2y = 1$.

Solution Using the formula from #518, we have that the distance of the point (x_0, y_0) from the line $x + 2y = 1$ is $|x_0 + 2y_0 - 1|/\sqrt{5}$. Hence the given parabola has the equation

$$\frac{|x + 2y - 1|}{\sqrt{5}} = \sqrt{(x-1)^2 + (y-1)^2}.$$

With some simplifying this becomes $4x^2 + y^2 - 8x - 6y - 4xy + 9 = 0$. $\qquad\square$

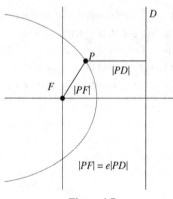

Figure 4.7

However, it is somewhat more natural to begin describing a conic's equation with polar co-ordinates whose origin is at the focus F. Let C be the conic in the plane with directrix D, focus F and eccentricity e. We may choose polar co-ordinates for the plane in which F is the origin and D is the line $r\cos\theta = k$ (or equivalently $x = k$) (Figure 4.7).

Then $|PF| = r$ and $|PD| = k - r\cos\theta$, if P is on the same side of D as F. So we have $r = e(k - r\cos\theta)$ or rearranging

$$r = \frac{ke}{1 + e\cos\theta}. \tag{4.18}$$

Note that k is purely a scaling factor here and it is e which determines the shape of the conic. Note also that when $0 < e < 1$ then r is well defined and bounded for all θ. However, when $e \geqslant 1$ then r (which must be positive) is unbounded and also undefined when $1 + e\cos\theta \leqslant 0$. If we change to Cartesian co-ordinates (u, v) using $u = r\cos\theta$ and $v = r\sin\theta$, we obtain $\sqrt{u^2 + v^2} = e(k - u)$ or equivalently

$$(1 - e^2)u^2 + 2e^2 ku + v^2 = e^2 k^2. \tag{4.19}$$

Provided $e \neq 1$, then we can complete the square to obtain

$$(1 - e^2)\left(u + \frac{e^2 k}{1 - e^2}\right)^2 + v^2 = e^2 k^2 + \frac{e^4 k^2}{1 - e^2} = \frac{e^2 k^2}{1 - e^2}.$$

Introducing new co-ordinates $x = u + e^2 k/(1 - e^2)$ and $y = v$, our equation becomes

$$(1 - e^2)x^2 + y^2 = \frac{e^2 k^2}{1 - e^2}. \tag{4.20}$$

Note that this change from uv-co-ordinates to xy-co-ordinates is simply a translation of the origin along the u-axis.

We are now in a position to write down the equations in **normal form** for the three types of conic.

- **Case 1: The Ellipse** ($0 < e < 1$).
  In this case, we can rewrite (4.20) as

$$\frac{x^2}{a^2} + \frac{y^2}{b^2} = 1, \qquad 0 < b < a,$$

where

$$a = \frac{ke}{1 - e^2}, \qquad b = \frac{ke}{\sqrt{1 - e^2}}.$$

Note that the eccentricity is $e = \sqrt{1 - b^2/a^2}$ in terms of a and b.

In the xy-co-ordinates, the focus F is at $(ae,0)$ and the directrix D is the line $x = a/e$ (Figure 4.8). However, it's clear by symmetry we could have used $F' = (-ae,0)$ and $D' : x = -a/e$ as an alternative focus and directrix and still produce the same ellipse. The area of this ellipse is $\pi\,ab$ (#1128). Further, this ellipse can be parametrized either by setting

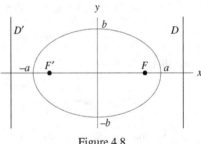

Figure 4.8

$$x = a\cos t, \qquad y = b\sin t, \qquad 0 \leqslant t < 2\pi, \tag{4.21}$$

or alternatively as

$$x = a\left(\frac{1-t^2}{1+t^2}\right), \qquad y = b\left(\frac{2t}{1+t^2}\right), \qquad (-\infty < t < \infty). \tag{4.22}$$

Note that this last parametrization omits the point $(-a,0)$, which can be thought of as corresponding to $t = \infty$.

Remark 4.46 Recall that the normal form of the ellipse is $x^2/a^2 + y^2/b^2 = 1$, where $a = ke(1 - e^2)^{-1}, b = ke(1 - e^2)^{-1/2}$. If we keep constant $l = ke$, as we let e become closer to zero we find a and b become closer to l and the ellipse approximates to the circle $x^2 + y^2 = l^2$. As a limit, then, a circle can be thought of as a conic with eccentricity $e = 0$. The two foci $(\pm ae,0)$ both move to the centre of the circle as e approaches zero and the two directrices $x = \pm a/e$ have both moved towards infinity. ∎

- **Case 2: The Hyperbola** $(e > 1)$.
  In this case, we can rewrite (4.20) as

  $$\frac{x^2}{a^2} - \frac{y^2}{b^2} = 1, \qquad 0 < a, b,$$

  where

  $$a = \frac{ke}{e^2 - 1}, \qquad b = \frac{ke}{\sqrt{e^2 - 1}}.$$

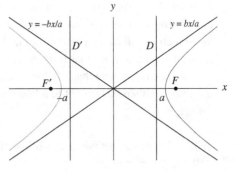

Figure 4.9

Note that the eccentricity is $e = \sqrt{1 + b^2/a^2}$ in terms of a and b.

In the xy-co-ordinates, the focus has co-ordinates $F = (ae,0)$ and the directrix D is the line $x = a/e$ (Figure 4.9). However, it is again clear from symmetry that we could have used $F' = (-ae,0)$ and $D' : x = -a/e$ as an alternative focus and directrix to produce the same hyperbola. The lines $ay = \pm bx$ are known as the **asymptotes** of the hyperbola; these are, in a sense, the tangents to the hyperbola at its two 'points at infinity'. When $e = \sqrt{2}$ (i.e. when $a = b$) then these asymptotes are perpendicular and C is known as a **right hyperbola**.

In a similar fashion to the ellipse, this hyperbola can be parametrized by

$$x = \pm a \cosh t, \qquad y = b \sinh t, \qquad (-\infty < t < \infty), \qquad (4.23)$$

or alternatively as

$$x = a\left(\frac{1+t^2}{1-t^2}\right), \qquad y = b\left(\frac{2t}{1-t^2}\right), \qquad (t \neq \pm 1). \qquad (4.24)$$

Again this second parametrization misses out the point $(-a,0)$ which in a sense corresponds to $t = \infty$. Likewise the points corresponding to $t = \pm 1$ can be viewed as the hyperbola's two 'points at infinity'.[6]

- **Case 3: The Parabola** ($e = 1$)
  In our derivation of (4.20) we assumed that $e \neq 1$. We can treat the case $e = 1$ by returning to (4.19). If we set $e = 1$ we obtain

$$2ku + v^2 = k^2.$$

  Substituting $a = k/2$, $x = a - u$, $y = v$ we obtain the normal form for a parabola

$$y^2 = 4ax.$$

  The focus is the point $(a,0)$ and the directrix is the line $x = -a$ (Figure 4.10). The *vertex* of the parabola is at $(0,0)$. In this case the conic C may be parametrized by setting $(x,y) = (at^2, 2at)$ where $-\infty < t < \infty$.

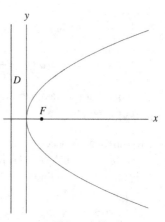

Figure 4.10

Remark 4.47 The three normal forms in Cases 1–3 are appropriately general in that *any* ellipse, parabola or hyperbola in the plane can be moved to a conic in normal form using an isometry (a distance preserving map). Equivalently we can always assume a conic to have an equation in normal form by an appropriate choice of co-ordinates. To appreciate this, we need only note that any focus can be brought to the origin by a translation and any directrix can then be brought to a line with equation $x = k$ by a rotation of the plane about the origin. ∎

Table 4.1 summarizes the details of the circle, ellipse, hyperbola and parabola.

We return now to considering conics as the curves formed by intersecting a plane with a double cone (as we started this section). We see below that, with an appropriate choice of co-ordinates, these intersections can be put into the normal forms above.

#1124
#1125
#1126
#1128
#1131
#1132

---

[6] This talk of 'points at infinity' can be made technically sound. The *real projective plane* consists of the *xy*-plane and a line at infinity. There is a point at infinity associated with each family of parallel lines. An ellipse, parabola, hyperbola intersects this line at infinity in 0,1,2 points respectively.
In a similar fashion the extended complex plane (p.59) with its one point at infinity is the *complex projective line*.

Table 4.1 *Details of the Conics' Normal Forms*

e range	$e = 0$	$0 < e < 1$	$e = 1$	$e > 1$
Conic type	Circle	Ellipse	Parabola	Hyperbola
Normal form	$x^2 + y^2 = a^2$	$x^2/a^2 + y^2/b^2 = 1$	$y^2 = 4ax$	$x^2/a^2 - y^2/b^2 = 1$
e formula	0	$\sqrt{1 - b^2/a^2}$	1	$\sqrt{1 + b^2/a^2}$
Foci	$(0,0)$	$(\pm ae, 0)$	$(a, 0)$	$(\pm ae, 0)$
Directrices	at infinity	$x = \pm a/e$	$x = -a$	$x = \pm a/e$
Notes		$0 < b < a$	Vertex: $(0,0)$	Asymptotes: $y = \pm bx/a$

Example 4.48 Let $0 < \theta, \alpha < \pi/2$. Show that the double cone $x^2 + y^2 = z^2 \cot^2 \alpha$ intersects the plane $z = \tan \theta (x - 1)$ in an ellipse, parabola or hyperbola. Determine which type of conic arises in terms of θ.

Solution We will denote as C the intersection of the cone and plane. In order to properly describe C we need to set up co-ordinates in the plane $z = \tan \theta (x - 1)$. Note that

$$\mathbf{e}_1 = (\cos \theta, 0, \sin \theta), \qquad \mathbf{e}_2 = (0, 1, 0), \qquad \mathbf{e}_3 = (\sin \theta, 0, -\cos \theta),$$

are of unit length and mutually perpendicular, with $\mathbf{e}_1, \mathbf{e}_2$ being parallel to the plane and $\mathbf{e}_3$ being perpendicular to it. Any point (x, y, z) in the plane can be written uniquely as

$$(x, y, z) = (1, 0, 0) + X\mathbf{e}_1 + Y\mathbf{e}_2 = (1 + X\cos \theta, Y, X\sin \theta)$$

for some X, Y, so that X and Y then act as the desired co-ordinates in the plane. Substituting the above expression for (x, y, z) into the cone's equation gives

$$(1 + X\cos \theta)^2 + Y^2 = (X\sin \theta)^2 \cot^2 \alpha,$$

which is the equation for C in terms of X and Y. This rearranges to

$$(\cos^2 \theta - \sin^2 \theta \cot^2 \alpha)X^2 + 2X\cos \theta + Y^2 = -1.$$

If $\theta \neq \alpha$ then we can complete the square to arrive at

$$\frac{(\cos^2 \theta - \sin^2 \theta \cot^2 \alpha)^2}{\sin^2 \theta \cot^2 \alpha} \left(X + \frac{\cos \theta}{\cos^2 \theta - \sin^2 \theta \cot^2 \alpha} \right)^2 + \frac{(\cos^2 \theta - \sin^2 \theta \cot^2 \alpha)}{\sin^2 \theta \cot^2 \alpha} Y^2 = 1.$$

If $\theta < \alpha$ then we have an ellipse as both squares have positive coefficients, whereas if $\theta > \alpha$ we have a hyperbola as the first coefficient is positive and the second is negative. If $\theta = \alpha$ then our original equation has become

$$2X\cos \alpha + Y^2 = -1,$$

which is a parabola. The eccentricity of the conic C is $\sin \theta / \sin \alpha$ (#1137). $\qquad \square$

Example 4.49 (Parabolic Mirrors) Consider the parabola C which has equation $y^2 = 4ax$ and let P be the point on C with coordinates $(at^2, 2at)$ (Figure 4.11). The focus F is at $(a, 0)$.

Let L_1 denote the line connecting P and F and let L_2 denote the horizontal line through P. Show that L_1 and L_2 make the same angle with the tangent to C at P.

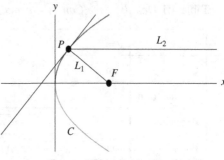

Figure 4.11

Consequently any light beam emitted from the focus F will reflect horizontally after contact with a parabolic mirror. (See also #1144.)

#1129
#1130
#1135
#1136
#1143
#1144

Solution Let θ_i denote the angle the line L_i makes with the tangent to C at P. The gradient at P can be found by the chain rule as

$$\frac{dy}{dx} = \frac{dy/dt}{dx/dt} = \frac{2a}{2at} = \frac{1}{t}.$$

Then a vector in the direction of the tangent is $(t, 1)$. A vector in the direction of L_1 is $(at^2 - a, 2at)$ and a vector in the direction of L_2 is $(1, 0)$. Hence we find

$$\cos\theta_1 = \frac{(at^2 - a, 2at) \cdot (t, 1)}{a(1 + t^2)\sqrt{1 + t^2}} = \frac{t^3 + t}{(1 + t^2)^{3/2}} = \frac{t}{\sqrt{t^2 + 1}};$$

$$\cos\theta_2 = \frac{(t, 1) \cdot (1, 0)}{\sqrt{1 + t^2}} = \frac{t}{\sqrt{t^2 + 1}},$$

and $\theta_1 = \theta_2$ as required. $\qquad\square$

Ellipses and hyperbolae have similar geometric properties, given in Proposition 4.50. Property (a) means, that if one tied a loose length of string between two fixed points and drew a curve with a pen so as to keep the string taut at all points, then the resulting curve would be an ellipse.

Proposition 4.50 *Let A, B be distinct points in the plane and k a real number.*

(a) *If $k > |AB|$ then the locus $|AP| + |BP| = k$ is an ellipse with foci at A and B.*
(b) *If $0 < |k| < |AB|$ then the locus $|AP| - |BP| = k$ is one branch of a hyperbola with foci at A and B.*

Proof These results are left as #1147 and #1149. $\qquad\square$

We introduce here an important class of equations that relates to conics.

Definition 4.51 A **degree two equation in two variables** is one of the form

$$Ax^2 + Bxy + Cy^2 + Dx + Ey + F = 0, \qquad (4.25)$$

where A, B, C, D, E, F are real constants and A, B, C are not all zero.

We begin with an example.

Example 4.52 Sketch the curve with equation

$$6x^2 + 4xy + 9y^2 - 12x - 4y - 4 = 0. \tag{4.26}$$

Explain why the curve is an ellipse and find its area.

Solution This is not the normal form of any conic. Also, because of the presence of the xy-term, it is not a simple translation away from a normal form. So we begin by making a substitution that corresponds to a rotation of the plane about the origin. We shall set

$$X = x\cos\theta + y\sin\theta, \qquad Y = -x\sin\theta + y\cos\theta;$$
$$x = X\cos\theta - Y\sin\theta, \qquad y = X\sin\theta + Y\cos\theta.$$

This corresponds to rotating our axes anticlockwise through an angle θ to new co-ordinate axes X and Y; we will choose θ in such a way as to eliminate the mixed XY-term, ultimately aiming to recognize the curve in some normal form. For ease of notation we will write $c = \cos\theta$ and $s = \sin\theta$. Our equation then becomes

$$6(Xc - Ys)^2 + 4(Xc - Ys)(Xs + Yc) + 9(Xs + Yc)^2 - 12(Xc - Ys) - 4(Xs + Yc) - 4 = 0. \tag{4.27}$$

The coefficient of the mixed XY-term is

$$-12cs - 4s^2 + 4c^2 + 18cs = 4\cos 2\theta + 3\sin 2\theta;$$

we need to choose θ so that $\tan 2\theta = -4/3$ for the above to be zero. If we take θ in the range $-\pi/2 < \theta < 0$ such that $\tan 2\theta = -4/3$, then $\sin\theta = -1/\sqrt{5}$ and $\cos\theta = 2/\sqrt{5}$ (#1153). Equation (4.27) then reads as

$$\frac{6}{5}(2X + Y)^2 + \frac{4}{5}(2X + Y)(-X + 2Y) + \frac{9}{5}(-X + 2Y)^2$$
$$- \frac{12}{\sqrt{5}}(2X + Y) - \frac{4}{\sqrt{5}}(-X + 2Y) - 4 = 0.$$

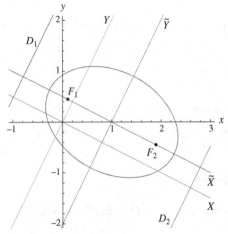

Figure 4.12

Tidying this up, we get

$$5X^2 - 4\sqrt{5}X + 10Y^2 - 4\sqrt{5}Y - 4 = 0,$$

which, when we complete the squares, simplifies further to

$$5\left(X - \frac{2}{\sqrt{5}}\right)^2 + 10\left(Y - \frac{1}{\sqrt{5}}\right)^2 = 10.$$

If we make a final change of variable

$$\tilde{X} = X - 2/\sqrt{5}, \qquad \tilde{Y} = Y - 1\sqrt{5},$$

which is a translation to change origin, then we have the normal form $\tilde{X}^2/2 + \tilde{Y}^2 = 1$, which we see to be an ellipse with area $\pi\sqrt{2}$.

The eccentricity of this ellipse is

$$e = \sqrt{1 - 1/2} = 1/\sqrt{2},$$

the foci are at $(\tilde{X}, \tilde{Y}) = (\pm 1, 0)$ and the directrices are $\tilde{X} = \pm 2$.

Note that to make these conclusions it was important that our changes of variables both preserved distance. A sketch of the curve appears in Figure 4.12. ∎

Example 4.52 contains all the ingredients for classifying any degree two equation in two variables as in (4.25). A sketch proof classifying such curves is given below (some of the details are left as exercises). The general approach is again a rotation of axes followed by a change of origin to get the equation in some normal form.

Theorem 4.53 *The solutions of the equation*

$$Ax^2 + Bxy + Cy^2 + Dx + Ey + F = 0 \tag{4.28}$$

where A, B, C, D, E, F are real constants, such that A, B, C are not all zero, is one of the following types of loci.

Case (a): If $B^2 - 4AC < 0$ then the solutions form an ellipse, a single point or the empty set.

Case (b): If $B^2 - 4AC = 0$ then the solutions form a parabola, two parallel lines, a single line or the empty set.

Case (c): If $B^2 - 4AC > 0$ then the solutions form a hyperbola or two intersecting lines.

Proof Note that we may assume $A \geqslant C$ without any loss of generality; if this were not the case we could swap the variables x and y. We begin with a rotation of the axes as before to remove the mixed term xy. Setting

$$X = xc + ys, \qquad Y = -xs + yc; \qquad x = Xc - Ys, \qquad y = Xs + Yc,$$

where $c = \cos\theta$ and $s = \sin\theta$, our equation becomes

$$A(Xc - Ys)^2 + B(Xc - Ys)(Xs + Yc) + C(Xs + Yc)^2 + D(Xc - Ys) + E(Xs + Yc) + F = 0. \tag{4.29}$$

The coefficient of the XY term is $-2Acs - Bs^2 + Bc^2 + 2Csc = B\cos 2\theta + (C - A)\sin 2\theta$ which equals zero when

$$\tan 2\theta = \frac{B}{A - C}. \tag{4.30}$$

If we now choose $-\pi/4 < \theta < \pi/4$ to be a solution of (4.30), then (4.29) simplifies to

$$(Ac^2 + Bsc + Cs^2)X^2 + (As^2 - Bsc + Cc^2)Y^2 + (Dc + Es)X + (-Ds + Ec)Y + F = 0. \tag{4.31}$$

With some further simplification this rearranges to

$$\left(\frac{A + C + H}{2}\right)X^2 + \left(\frac{A + C - H}{2}\right)Y^2 + (Dc + Es)X + (-Ds + Ec)Y + F = 0, \tag{4.32}$$

where $H = \sqrt{(A-C)^2 + B^2}$ (#1154). Note that $A+C+H$ and $A+C-H$ will have the same sign if $(A+C)^2 > H^2$ which is equivalent to the inequality

$$4AC > B^2. \tag{4.33}$$

If $4AC > B^2$ then the X^2 and Y^2 coefficients have the same sign and by completing the squares we arrive at $\tilde{X}^2/a^2 + \tilde{Y}^2/b^2 = K$ in new variables $\tilde{X}$ and $\tilde{Y}$ (resulting from a change of origin) which is either an ellipse (when $K > 0$), a single point (when $K = 0$) or empty (when $K < 0$). Details are left to #1156.

If $4AC < B^2$ then the X^2 and Y^2 coefficients have different signs and further, completing of the squares for the X and Y variables, leads to an equation of the form $\tilde{X}^2/a^2 - \tilde{Y}^2/b^2 = K$ in new variables $\tilde{X}$ and $\tilde{Y}$ (resulting from a change of origin) which is either a hyperbola (when $K \neq 0$), or a pair of intersecting lines (when $K = 0$). Details are left to #1157.

If $4AC = B^2$ then (only) one of the X^2 and Y^2 coefficients is zero. Without loss of generality, let this be the X^2-coefficient (so that $H = -A - C$). Then we have an equation of the form

$$(A+C)Y^2 + (Dc+Es)X + (-Ds+Ec)Y + F = 0.$$

Provided $Dc + Es \neq 0$ then this can be rearranged to an equation of the form $\tilde{Y}^2 = 4a\tilde{X}$, which is a parabola; if $Dc + Es = 0$ then the equation reads

$$(A+C)Y^2 + (-Ds+Ec)Y + F = 0$$

and represents two parallel lines, a single (repeated) line or the empty set depending on whether the discriminant of this quadratic in Y is positive, zero or negative. Details are left to #1158. □

Exercises

#1124a Sketch the ellipse $x^2/a^2 + y^2/b^2 = 1$, where $0 < a < b$. What is its eccentricity? What are the co-ordinates of the foci and the equations of the directrices?

#1125a Verify the parametrizations (4.21), (4.22) for the ellipse, (4.23), (4.24) for the hyperbola and the parametrization $(at^2, 2at)$ for the parabola $y^2 = 4ax$.

#1126A† Show that there are points P on the conic (4.18) on the other side of D from F if and only if $e > 1$. Show, when $e > 1$, that the other branch of the hyperbola has equation $r = ek/(e\cos\theta - 1)$.

#1127C Consider the conic with polar equation (4.18). Show that the chord of the conic, which is parallel to the directrix and which passes through the focus, has length $2ke$. This quantity is known as the **latus rectum**.

#1128B† Show that the ellipse with equation $x^2/a^2 + y^2/b^2 = 1$ has area πab.

#1129A Show that e_1, e_2, e_3, as defined in Example 4.48, are of unit length and mutually perpendicular. Show further that e_1, e_2 are parallel to the plane $z = \tan\theta(x-1)$.

#1130B Let $e_1 = (1,1)/\sqrt{2}$ and $e_2 = (-1,1)/\sqrt{2}$. Show that e_1 and e_2 are of unit length and perpendicular. Given X, Y determine x, y such that $Xe_1 + Ye_2 = x\mathbf{i} + y\mathbf{j}$. Deduce that $x^2 + xy + y^2 = 1$ is an ellipse and find its area.

#1131b Find the equation of the ellipse with foci at $(\pm 2, 0)$ which passes through $(0, 1)$.

#1132b Find the equation of the hyperbola with asymptotes $y = \pm 2x$ and with directrices $x = \pm 1$.

#1133B Find the equation of the ellipse consisting of points P such that $|AP| + |BP| = 10$, where A, B are $(\pm 3, 0)$.

#1134c† Let $A = (1, -1)$ and $B = (-1, 1)$. Find the equation of the ellipse of points P such that $|AP| + |BP| = 4$.

#1135B† Find the equation of the parabola with directrix $x + y = 1$ and focus $(-1, -1)$.

#1136a† Under what conditions on A, B, C does the line $Ax + By + C = 0$ meet the hyperbola $x^2/a^2 - y^2/b^2 = 1$?

#1137C Verify that the conic in Example 4.48 has eccentricity $\sin\theta / \sin\alpha$.

#1138b Write down the xy-co-ordinates of the foci of (4.26) and also the equations in x, y of the directrices.

#1139B Adapt the $(a\cos t, b\sin t)$ parametrization for $x^2/a^2 + y^2/b^2 = 1$ to a parametrization for (4.26). Use this to find the maximum and minimum values for x and y on the ellipse.

#1140D† Show, for the ellipse defined in (4.26), $dy/dx = (12 - 12x - 4y)/(4x + 18y - 4)$. Determine the point on the ellipse which is closest to the line $x + y = 6$.

#1141C† Rederive the second part of #1140 by using the parametrization found in #1139.

#1142b† For given a, b, what is the largest value $ax + by$ takes on the conic $x^2 + xy + y^2 = 1$? Where is this value achieved and what is the gradient of the conic at that point?

#1143b†(i) The conic C is formed by intersecting the double cone $x^2 + y^2 = z^2$ with the plane $x + y + z = 1$. Show that the point with position vector $\mathbf{r}(t)$ lies on C where

$$\mathbf{r}(t) = (1 + (\sec t - \tan t)/\sqrt{2}, \, 1 + (\sec t + \tan t)/\sqrt{2}, \, -1 - \sqrt{2}\sec t).$$

(ii) Show that the vectors $e_1 = (1/\sqrt{6}, 1/\sqrt{6}, -\sqrt{2/3})$ and $e_2 = (-1/\sqrt{2}, 1/\sqrt{2}, 0)$ are of unit length, are perpendicular to one another and are parallel to the plane $x + y + z = 1$. Show for some $a, b > 0$ that

$$\mathbf{r}(t) = (1, 1, -1) + (a\sec t)e_1 + (b\tan t)e_2.$$

(iii) Show that C has eccentricity $2/\sqrt{3}$, has foci $(1, 1, -1) \pm 2e_1$ and further has directrices $(1, 1, -1) \pm 3e_1/2 + \lambda e_2$.

#1144B† (**Reflection Property of an Ellipse**) Let E be an elliptical mirror with foci F and F'. Show that a light beam emitted from F will reflect back through F'.

#1145 B Consider the family of hyperbolae $x^2/a^2 - y^2/b^2 = 1$. What happens if we (i) fix a and let b become large? (ii) fix b and let a become large?

#1146 b† Let $a > b$. If P lies on the ellipse $x^2/a^2 + y^2/b^2 = 1$, with foci F and F', show that $|PF| + |PF'| = 2a$

#1147 D† Conversely to #1146 show that, given two points F and F' in the plane and d such that $d > |FF'|$, then the locus of points P such that $|PF| + |PF'| = d$ is an ellipse with foci F and F' and eccentricity $e = |FF'|/d$.

#1148 b Let H denote the hyperbola $x^2/a^2 - y^2/b^2 = 1$. Show points P on H satisfy $|PF'| - |PF| = \pm 2a$, where F and F' are the foci, with the sign indicating the branch of the hyperbola.

#1149 D† Given two points F and F' in the plane and $k \neq 0$, show that the locus of points P such that $|PF| - |PF'| = k$ is the branch of a hyperbola. What is the locus when $k = 0$?

#1150 b Sketch the conic $8x^2 + 4xy + 5y^2 + 8x + 2y = 7$.

#1151 b Sketch the conic $x^2 - 2xy + y^2 + 2x + 6y = 15$.

#1152 B† Use implicit differentiation to determine a formula for the gradient at the point (X, Y) of $3x^2 + 7xy + 2y^2 + x - 3y - 2 = 0$. Verify that the point $(1, -1)$ lies on the curve; what gradient does your formula give there? What possible values can your formula take for other points of the curve?

#1153 A (See Example 4.52.) Show that if $\tan 2\theta = -4/3$ and $-\pi/2 < \theta < 0$, then we have $\sin \theta = -1/\sqrt{5}$ and $\cos \theta = 2/\sqrt{5}$.

#1154 C As required for Theorem 4.53, given $\tan 2\theta = B/(A - C)$ where $-\pi/4 < \theta < \pi/4$, find expressions for $\cos 2\theta$ and $\sin 2\theta$ and verify that (4.31) simplifies to (4.32).

#1155 C Assuming that $4AC \neq B^2$ show that (4.32) rearranges to

$$\left(\frac{A+C+H}{2}\right)\left(X + \frac{Dc+Es}{A+C+H}\right)^2 + \left(\frac{A+C-H}{2}\right)\left(Y + \frac{-Ds+Ec}{A+C-H}\right)^2 = k$$

where

$$k = \left(\frac{CD^2 + AE^2 - BDE}{4AC - B^2}\right) - F. \tag{4.34}$$

#1156 C Suppose that $4AC > B^2$ and let k be as in (4.34). Deduce from #1155 that (4.28) represents (i) an ellipse when $k/(A + C + H) > 0$, (ii) a single point when $k = 0$, (iii) the empty set when $k/(A + C + H) < 0$.

#1157 C Suppose that $4AC < B^2$ and let k be as in (4.34). Deduce from #1155 that (4.28) represents (i) a hyperbola when $k \neq 0$ and (ii) two lines when $k = 0$.

#1158 C Suppose that $4AC = B^2$ Deduce, from the proof of Theorem 4.53, that (4.28) represents a parabola if $CD^2 \neq AE^2$. In the cases where $CD^2 = AE^2$, show that (4.28) represents two parallel lines, a single (repeated) line or the empty set depending on the sign of $D^2 + E^2 - 4(A + C)F$.

#1159D Given that equation (4.28) describes an ellipse, find its area as a function of A, B, C, D, E, F. Hence rederive the area of the ellipse described by (4.26).

#1160B Describe the following conics as one of ellipse, parabola, hyperbola, intersecting lines, parallel lines, single line, point, empty set. Verify, as relevant, that Theorem 4.53, #1156, #1157, #1158 hold true in each case.

   (i) $x^2 + 2x + y^2 + 4y + 6 = 0$;
   (ii) $x^2 - 4x + y^2 + 8y + 15 = 0$;
   (iii) $4x^2 + 6x - 9y^2 + 15y + 6 = 0$;
   (iv) $y^2 = 2x + 3$;
   (v) $x^2 = 1$;
   (vi) $(x + y + 1)(2x + y + 2) = 0$;
   (vii) $(2x + 3y + 1)^2 = 0$.

#1161D† Under what conditions on a, b, c, d, e, f does $(ax + by + c)(dx + ey + f) = 1$ describe a hyperbola?

#1162A Verify that equation (4.28) can be rewritten as

$$
(x, y, 1) M \begin{pmatrix} x \\ y \\ 1 \end{pmatrix} = 0 \qquad \text{where} \qquad M = \begin{pmatrix} A & \frac{B}{2} & \frac{D}{2} \\ \frac{B}{2} & C & \frac{E}{2} \\ \frac{D}{2} & \frac{E}{2} & F \end{pmatrix}. \qquad (4.35)
$$

Compare $\det M$ with (4.34).

#1163C† Using #1157 and #1158, show that if a conic contains a line then $\det M = 0$ where M is as in #1162.

#1164D† Let M be as in #1162 and let $P = (X, Y)$ be a point satisfying (4.28). Assuming the tangent line at P is well defined, show that it has equation $(x, y, 1) M (X, Y, 1)^T = 0$. How does this equation read for the curve in #1152 and $(X, Y) = (1, -1)$?

#1165C† Let $\mathbf{v}$ and $\mathbf{w}$ be independent vectors in $\mathbb{R}^2$. Show that $\mathbf{r}(t) = \mathbf{v} \cos t + \mathbf{w} \sin t$, where $0 \leqslant t < 2\pi$, is a parametrization of an ellipse.

#1166c Let $\mathbf{v}$ and $\mathbf{w}$ be independent vectors in $\mathbb{R}^2$. Show that $\mathbf{r}(t) = \mathbf{v} \tan t + \mathbf{w} \sec t$, where $-\pi/2 < t < \pi/2$, is a parametrization of a branch of a hyperbola.

#1167b Show that the point $\mathbf{p}(t) = (a \sec t, b \tan t)$ lies on the hyperbola H with equation $x^2/a^2 - y^2/b^2 = 1$. Give a domain in $\mathbb{R}$, for t, so that $\mathbf{p}$ parametrizes the entire hyperbola. Write down the equation of the tangent line to H at the point $(a \sec t, b \tan t)$. Determine the co-ordinates of the points P and Q where this tangent line meets the asymptotes of the hyperbola, and show that the area of the triangle OPQ is independent of t.

#1168D† A line l through the focus F of a conic C meets C in two points A and B. Show that the quantity

$$
\frac{1}{|AF|} + \frac{1}{|BF|}
$$

is independent of the choice of l. What is this quantity for the parabola $y^2 = 4ax$?

#1169D† Let l_θ denote the line through (a,b) making an angle θ with the x-axis. Show that l_θ is a tangent of the parabola $y = x^2$ if and only if $\tan^2 \theta - 4a\tan\theta + 4b = 0$. Show that the tangents to the parabola that pass through (a,b) subtend an angle $\pi/4$ if and only if (a,b) lies on one branch of the hyperbola $1 + 24b + 16b^2 = 16a^2$. Sketch the appropriate branch of $1 + 24y + 16y^2 = 16x^2$ and the original parabola on the same axes.

#1170D (i) Let C denote the unit circle $x^2 + y^2 = 1$ and let $P = (X,Y)$ be a point outside the circle. Show that there are two lines l_1 and l_2 through P which are tangential to C. If l_i is tangential to C at Q_i, show that the line Q_1Q_2 has equation $Xx + Yy = 1$. Show, in this way, that every point outside the circle corresponds to a line that intersects C twice. P is referred to as the **pole** and the corresponding line Q_1Q_2 as its **polar**. Are the diameters of C polars?

(ii) Let l be a line which does not intersect the unit circle C. Let P be a point on l and let l_P denote its polar. Show that, as P varies on l, the polars l_P are all concurrent at a point inside C. Show, in this way, that every point inside the circle, other than its centre, corresponds to a line (its polar) that does not intersect C and vice versa.

(iii) Show that, for points P both outside and inside C, the polar l_P is the line through the the inverse point of P which is perpendicular to OP.

#1171D† Without further calculation, explain why the properties of poles and polars, found in #1170(i) and (ii), generalize to the case when C is an ellipse. What is the equation of the polar of the point (X,Y) when C denotes the ellipse in (4.26)? Show that C consists of those points that lie on their polar.

4.4 Spectral Theory for Symmetric Matrices*

That the general degree two equation in two variables (4.25) can be put into normal forms is part of the more general theory on the diagonalizability of symmetric matrices, the so-called *spectral theorem*. (4.25) can be rewritten as

$$(x,y)M \begin{pmatrix} x \\ y \end{pmatrix} + (D,E) \begin{pmatrix} x \\ y \end{pmatrix} + F = 0, \quad \text{where} \quad M = \begin{pmatrix} A & B/2 \\ B/2 & C \end{pmatrix}. \quad (4.36)$$

Note that M is symmetric (i.e. $M^T = M$). In Theorem 4.53 we rotated the xy-axes to new XY-axes in such a way as to eliminate the mixed term XY term. So we introduced a change of variables

$$\begin{pmatrix} x \\ y \end{pmatrix} = P \begin{pmatrix} X \\ Y \end{pmatrix} \quad \text{where} \quad P = \begin{pmatrix} \cos\theta & -\sin\theta \\ \sin\theta & \cos\theta \end{pmatrix},$$

after which (4.36) became

$$(X,Y)P^T M P \begin{pmatrix} X \\ Y \end{pmatrix} + (D,E)P \begin{pmatrix} X \\ Y \end{pmatrix} + F = 0,$$

and we chose θ so that the mixed term XY vanished which is equivalent to $P^T M P$ being diagonal. As P is orthogonal then $P^T = P^{-1}$ and so we have diagonalized M using an orthogonal change of variables; it is important from a geometric viewpoint that this change of variable be orthogonal so that distances, angles and areas remain unaltered.

In these new variables X, Y, and with $P^T M P = \mathrm{diag}(\tilde{A}, \tilde{C})$ and $(D, E)P = (\tilde{D}, \tilde{E})$, our equation now reads as

$$\tilde{A}X^2 + \tilde{C}Y^2 + \tilde{D}X + \tilde{E}Y + F = 0.$$

#1172
#1173

We can now complete any squares to put this equation into a normal form.

That all this is possible is because M is symmetric and we shall see (Theorem 4.55) that symmetric matrices can generally be diagonalized by an orthogonal change of variables. It is not very hard to see that, if a square matrix N is such that $P^T N P$ is diagonal for an orthogonal matrix P, then N is symmetric (#1172); the difficulty will be in proving the converse. We begin with the following result.

Proposition 4.54 *Let A be a real $n \times n$ symmetric matrix with characteristic polynomial $c_A(x) = \det(xI - A)$.*

(a) The roots of c_A are real.
(b) If $\mathbf{v}$ and $\mathbf{w}$ are eigenvectors of A with associated eigenvalues λ and μ, where $\lambda \neq \mu$, then $\mathbf{v} \cdot \mathbf{w} = 0$.

Proof (a) Let λ be a (potentially complex) root of c_A. Then by (an appropriate complex version of) Proposition 3.191(a), there is a non-zero complex vector $\mathbf{v}$ in $\mathbb{C}_n$ such that $A\mathbf{v} = \lambda\mathbf{v}$. As the entries of A are real, when we conjugate this equation we obtain $A\bar{\mathbf{v}} = \bar{\lambda}\bar{\mathbf{v}}$. As $A = A^T$, and by the product rule for transposes, we see

$$\bar{\lambda}\bar{\mathbf{v}}^T \mathbf{v} = (\bar{\lambda}\bar{\mathbf{v}})^T \mathbf{v} = (A\bar{\mathbf{v}})^T \mathbf{v} = \bar{\mathbf{v}}^T A^T \mathbf{v} = \bar{\mathbf{v}}^T A \mathbf{v} = \bar{\mathbf{v}}^T \lambda \mathbf{v} = \lambda \bar{\mathbf{v}}^T \mathbf{v}.$$

Now for any non-zero *complex* vector $\mathbf{v} = (v_1, v_2, \ldots, v_n)^T$ we have

$$\bar{\mathbf{v}}^T \mathbf{v} = \bar{\mathbf{v}} \cdot \mathbf{v} = \overline{v_1}v_1 + \cdots + \overline{v_n}v_n = |v_1|^2 + \cdots + |v_n|^2 > 0.$$

As $(\bar{\lambda} - \lambda)\bar{\mathbf{v}}^T \mathbf{v} = 0$ then $\lambda = \bar{\lambda}$ and so λ is real.

(b) We have that $A\mathbf{v} = \lambda\mathbf{v}$ and $A\mathbf{w} = \mu\mathbf{w}$ where $\lambda \neq \mu$. Then, as A is symmetric, we have

$$\lambda\mathbf{v} \cdot \mathbf{w} = \lambda\mathbf{v}^T\mathbf{w} = (\lambda\mathbf{v})^T\mathbf{w} = (A\mathbf{v})^T\mathbf{w} = \mathbf{v}^T A^T \mathbf{w} = \mathbf{v}^T A \mathbf{w} = \mathbf{v}^T \mu\mathbf{w} = \mu\mathbf{v} \cdot \mathbf{w}.$$

As $\lambda \neq \mu$ then $\mathbf{v} \cdot \mathbf{w} = 0$. ☐

Theorem 4.55 *(Spectral Theorem[7]) Let A be a real symmetric $n \times n$ matrix. Then there exists an orthogonal $n \times n$ matrix P such that $P^T A P$ is diagonal.*

---

[7] Appreciation of this result, at least in two variables, dates back to Descartes and Fermat. But the equivalent general result was first proven by Cauchy in 1829, though independently of the language of matrices, which were yet to be invented. Rather Cauchy's result was in terms of *quadratic forms* – a quadratic form in two variables is an expression of the form $ax^2 + bxy + cy^2$.

Proof We shall prove the result by strong induction on n. When $n = 1$ there is nothing to prove as all 1×1 matrices are diagonal and so we can simply take $P = I_1$.

Suppose now that the result holds for $r \times r$ real symmetric matrices where $1 \leqslant r < n$. By the fundamental theorem of algebra, the characteristic polynomial c_A has a root λ in $\mathbb{C}$, which by Proposition 4.54(a) we in fact know to be real. Let X denote the λ-eigenspace, that is

$$X = \{\mathbf{v} \in \mathbb{R}_n : A\mathbf{v} = \lambda\mathbf{v}\}.$$

Then X is a subspace as it is $\mathrm{Null}(A - \lambda I)$ and so has an orthonormal basis $\mathbf{v}_1, \ldots \mathbf{v}_m$ which we may extend to an orthonormal basis $\mathbf{v}_1, \ldots \mathbf{v}_n$ for $\mathbb{R}_n$. Let $P = (\mathbf{v}_1 | \ldots | \mathbf{v}_n)$; then P is orthogonal and by the definition of matrix multiplication

$$\left[P^T A P \right]_{ij} = \mathbf{v}_i^T A \mathbf{v}_j = \mathbf{v}_i \cdot (A\mathbf{v}_j).$$

Note that $A\mathbf{v}_j = \lambda\mathbf{v}_j$ for $1 \leqslant j \leqslant m$ and so the first m columns of $P^T A P$ are $\lambda\mathbf{e}_1^T, \ldots, \lambda\mathbf{e}_m^T$. Also, using $A = A^T$ and the product rule for transposes, we note that $P^T A P$ is symmetric. Together this means that

$$P^T A P = \mathrm{diag}\,(\lambda I_m, M),$$

where M is a symmetric $(n - m) \times (n - m)$ matrix. By our inductive hypothesis there is an orthogonal matrix Q such that $Q^T M Q$ is diagonal. If we set $R = \mathrm{diag}\,(I_m, Q)$ then R is orthogonal, PR is orthogonal and

$$
\begin{aligned}
(PR)^T A (PR) &= R^T P^T A P R \\
&= \mathrm{diag}\,(I_m, Q^T)\,\mathrm{diag}\,(\lambda I_m, M)\,\mathrm{diag}\,(I_m, Q) \\
&= \mathrm{diag}\,(\lambda I_m, Q^T M Q)
\end{aligned}
$$

is diagonal. This concludes the proof by induction. $\qquad\square$

Example 4.56 Find an orthogonal matrix P such that $P^T M P$ is diagonal, where

$$M = \begin{pmatrix} 1 & 0 & 1 \\ 0 & -1 & 1 \\ 1 & 1 & 0 \end{pmatrix}.$$

Identify the type of quadric surface $x^2 - y^2 + 2xz + 2yz = 1$.

#1179 **Solution** With a little working we can calculate that
#1180
#1183

$$c_M(x) = \det(xI_3 - M) = \begin{vmatrix} x - 1 & 0 & -1 \\ 0 & x + 1 & -1 \\ -1 & -1 & x \end{vmatrix} = x^3 - 3x,$$

so that M has distinct eigenvalues $0, \sqrt{3}, -\sqrt{3}$. Determining (necessarily orthogonal) *unit* eigenvectors, we see

$$\lambda = 0: \quad M = \begin{pmatrix} 1 & 0 & 1 \\ 0 & -1 & 1 \\ 1 & 1 & 0 \end{pmatrix} \xrightarrow{\text{RRE}} \begin{pmatrix} 1 & 0 & 1 \\ 0 & 1 & -1 \\ 0 & 0 & 0 \end{pmatrix};$$

$$\lambda = \sqrt{3}: \quad M - \sqrt{3}I_3 = \begin{pmatrix} 1-\sqrt{3} & 0 & 1 \\ 0 & -1-\sqrt{3} & 1 \\ 1 & 1 & -\sqrt{3} \end{pmatrix} \xrightarrow{\text{RRE}} \begin{pmatrix} 1 & 0 & \frac{-1-\sqrt{3}}{2} \\ 0 & 1 & \frac{1-\sqrt{3}}{2} \\ 0 & 0 & 0 \end{pmatrix};$$

$$\lambda = -\sqrt{3}: \quad M + \sqrt{3}I_3 = \begin{pmatrix} 1+\sqrt{3} & 0 & 1 \\ 0 & \sqrt{3}-1 & 1 \\ 1 & 1 & \sqrt{3} \end{pmatrix} \xrightarrow{\text{RRE}} \begin{pmatrix} 1 & 0 & \frac{\sqrt{3}-1}{2} \\ 0 & 0 & \frac{\sqrt{3}+1}{2} \\ 0 & 0 & 0 \end{pmatrix}.$$

Take $\mathbf{v}_1 = \dfrac{1}{\sqrt{3}} \begin{pmatrix} -1 \\ 1 \\ 1 \end{pmatrix}$, $\mathbf{v}_2 = \dfrac{1}{2\sqrt{3}} \begin{pmatrix} \sqrt{3}+1 \\ \sqrt{3}-1 \\ 2 \end{pmatrix}$, $\mathbf{v}_3 = \dfrac{1}{2\sqrt{3}} \begin{pmatrix} 1-\sqrt{3} \\ -1-\sqrt{3} \\ 2 \end{pmatrix}$.

We then can take

$$P = \frac{1}{2\sqrt{3}} \begin{pmatrix} -2 & \sqrt{3}+1 & 1-\sqrt{3} \\ 2 & \sqrt{3}-1 & -1-\sqrt{3} \\ 2 & 2 & 2 \end{pmatrix}.$$

By the orthogonal change of variable $\mathbf{x} = P\mathbf{X}$, the quadric is transformed into one with equation $\sqrt{3}Y^2 - \sqrt{3}Z^2 = 1$. This is a hyperbolic cylinder (its cross-section being the same hyperbola in each plane $X = $ const.) $\square$

Example 4.57 Given the matrix A below, find orthogonal P such that $P^T A P$ is diagonal.

$$A = \begin{pmatrix} 0 & 1 & 1 & 1 \\ 1 & 0 & 1 & 1 \\ 1 & 1 & 0 & 1 \\ 1 & 1 & 1 & 0 \end{pmatrix}.$$

Solution Details of this solution are left to #1182. The characteristic polynomial of A is $c_A(x) = (x+1)^3(x-3)$. A unit length 3-eigenvector is $\mathbf{v}_1 = (1,1,1,1)^T/2$ and the -1-eigenspace is $x_1 + x_2 + x_3 + x_4 = 0$. So a basis for the -1-eigenspace is

$$(1,-1,0,0)^T, \qquad (0,1,-1,0)^T, \qquad (0,0,1,-1)^T.$$

However, to find the last three columns of P, we need an orthonormal basis for the -1-eigenspace. Applying the Gram–Schmidt process to the above three vectors, we arrive at

$$\mathbf{v}_2 = (1,-1,0,0)^T/\sqrt{2}, \qquad \mathbf{v}_3 = (1,1,-2,0)^T/\sqrt{6}, \qquad \mathbf{v}_4 = (1,1,1,-3)^T/\sqrt{12}.$$

Such a required matrix is then $P = (\mathbf{v}_1 \mid \mathbf{v}_2 \mid \mathbf{v}_3 \mid \mathbf{v}_4)$. $\square$

Algorithm 4.58 *(Orthogonal Diagonalization of a Symmetric Matrix) Let M be a symmetric matrix. The spectral theorem shows that M is diagonalizable and so has an* *eigenbasis. Setting this eigenbasis as the columns of a matrix P will yield an invertible matrix P such that $P^{-1}MP$ is diagonal (as done in §3.11) – in general though this P will not be orthogonal.*

If $\mathbf{v}$ is an eigenvector of M whose eigenvalue is not repeated, then we replace it with $\mathbf{v}/|\mathbf{v}|$. This new eigenvector is of unit size and is necessarily orthogonal to other eigenvectors with different eigenvalues by Proposition 4.54(b). If none of the eigenvalues is repeated, this is all we need do to the eigenbasis to produce an orthonormal eigenbasis.

If λ is a repeated eigenvalue then we can find a basis for the λ-eigenspace. Applying the Gram–Schmidt orthogonalization process to this basis produces an orthonormal basis for the λ-eigenspace. Again these eigenvectors are orthogonal to all eigenvectors with a different eigenvalue. We can see now that the previous non-repeated case is simply a special case of the repeated case, the Gram–Schmidt process for a single vector involving nothing other than normalizing it.

Once the given basis for each eigenspace has had the Gram–Schmidt process applied to it the entire eigenbasis has now been made orthonormal. We may put this orthonormal eigenbasis as the columns of a matrix P which will be orthogonal with $P^{-1}MP = P^T MP$ being diagonal.

Example 4.59 Find a 2×2 real symmetric matrix M such that $M^2 = A$ where

$$A = \begin{pmatrix} 3 & \sqrt{3} \\ \sqrt{3} & 5 \end{pmatrix}.$$

Solution The characteristic polynomial of A is

$$\det(xI - A) = (x-3)(x-5) - (-\sqrt{3})^2 = x^2 - 8x + 12$$

which has roots 2 and 6. Determining the eigenvectors we see

$$\lambda = 2: \quad A - 2I_2 = \begin{pmatrix} 1 & \sqrt{3} \\ \sqrt{3} & 3 \end{pmatrix} \xrightarrow{\text{RRE}} \begin{pmatrix} 1 & \sqrt{3} \\ 0 & 0 \end{pmatrix} \quad \text{so take } \mathbf{v}_1 = \frac{1}{2}\begin{pmatrix} -\sqrt{3} \\ 1 \end{pmatrix}.$$

$$\lambda = 6: \quad A - 6I_2 = \begin{pmatrix} -3 & \sqrt{3} \\ -\sqrt{3} & 1 \end{pmatrix} \xrightarrow{\text{RRE}} \begin{pmatrix} 1 & -1/\sqrt{3} \\ 0 & 0 \end{pmatrix} \quad \text{so take } \mathbf{v}_2 = \frac{1}{2}\begin{pmatrix} 1 \\ \sqrt{3} \end{pmatrix}.$$

So with $P = (\mathbf{v}_1 | \mathbf{v}_2)$ we have $P^T AP = \text{diag}(2,6)$, which has a clear square root of $\text{diag}(\sqrt{2}, \sqrt{6})$. Thus we might choose

$$M = P\text{diag}\left(\sqrt{2}, \sqrt{6}\right)P^T = \frac{1}{4}\begin{pmatrix} 3\sqrt{2}+\sqrt{6} & 3\sqrt{2}-\sqrt{6} \\ 3\sqrt{2}-\sqrt{6} & \sqrt{2}+3\sqrt{6} \end{pmatrix}. \qquad \square$$

Example 4.60 (a) Find an orthogonal matrix P such that $P^T AP$ is diagonal where

$$A = \begin{pmatrix} 1 & -1 & -1 \\ -1 & 1 & -1 \\ -1 & -1 & 1 \end{pmatrix}.$$

(b) Consider the real-valued functions f and g defined on $\mathbb{R}^3$ by

$$f(\mathbf{x}) = x^2 + y^2 + z^2 - 2xy - 2xz - 2yz, \qquad g(\mathbf{x}) = -y^2 + 2z^2 + 2\sqrt{2}xy,$$

where $\mathbf{x} = (x, y, z)^T$. Is there an invertible matrix Q such that $f(Q\mathbf{x}) = g(\mathbf{x})$? Is there an orthogonal Q?

(c) Sketch the surface $f(\mathbf{x}) = 1$.

#1186 **Solution** The details of this solution are left to #1181. (a) The characteristic polynomial
#1189 $c_A(x)$ equals $(x+1)(x-2)^2$ so that the eigenvalues are $-1, 2, 2$. The -1-eigenvectors are
#1193 multiples of $(1, 1, 1)^T$ and the 2-eigenspace is the plane $x + y + z = 0$. So an orthonormal
#1194 eigenbasis for A is

$$\frac{(1,1,1)^T}{\sqrt{3}}, \qquad \frac{(1,-1,0)^T}{\sqrt{2}}, \qquad \frac{(1,1,-2)^T}{\sqrt{6}},$$

from which we can form the required

$$P = \frac{1}{\sqrt{6}} \begin{pmatrix} \sqrt{2} & \sqrt{3} & 1 \\ \sqrt{2} & -\sqrt{3} & 1 \\ \sqrt{2} & 0 & -2 \end{pmatrix}.$$

(b) We then have that $f(P\mathbf{x}) = (P\mathbf{x})^T A (P\mathbf{x}) = \mathbf{x}^T P^T A P \mathbf{x} = -x^2 + 2y^2 + 2z^2$. Now we similarly have

$$g(\mathbf{x}) = (x, y, z) \begin{pmatrix} 0 & \sqrt{2} & 0 \\ \sqrt{2} & -1 & 0 \\ 0 & 0 & 2 \end{pmatrix} \begin{pmatrix} x \\ y \\ z \end{pmatrix}$$

and this matrix (call it B) has characteristic polynomial $c_B(x) = (x+2)(x-1)(x-2)$. This means that there is an orthogonal matrix R such that

$$g(R\mathbf{x}) = -2x^2 + y^2 + 2z^2.$$

We can then see that the (invertible but not orthogonal) map S which sends $(x, y, z)^T$ to $(x/\sqrt{2}, \sqrt{2}y, z)^T$ satisfies

$$g(RS\mathbf{x}) = -2(x/\sqrt{2})^2 + (\sqrt{2}y)^2 + 2z^2 = -x^2 + 2y^2 + 2z^2$$

and so $g(RS\mathbf{x}) = f(P\mathbf{x})$. Hence $f(Q\mathbf{x}) = g(\mathbf{x})$ where $Q = PS^{-1}R^{-1}$. This Q is invertible by not orthogonal (as S is not orthogonal). If there were an orthogonal Q such that $f(Q\mathbf{x}) = g(\mathbf{x})$ then we'd have

$$\mathbf{x}^T Q^T A Q \mathbf{x} = \mathbf{x}^T B \mathbf{x} \qquad \text{for all } \mathbf{x}.$$

It would follow (#1184) that $Q^T A Q = B$ but this is impossible as $Q^T A Q = Q^{-1} A Q$ has eigenvalues -1, $2, 2$ which are different from those of B.

(c) The surface $1 = f(\mathbf{x})$ now has the equation $1 = -x^2 + 2y^2 + 2z^2 = f(P\mathbf{x})$ in new variables.

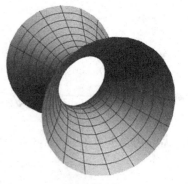

Figure 4.13

This is a hyperboloid of one sheet as sketched in Figure 4.13. The x-axis (of the new variables) is the central axis of the surface; each plane $x = const.$ intersects the surface in a circle. □

We conclude with the *singular value decomposition theorem* which applies not just to square matrices.

Theorem 4.61 (*Singular Value Decomposition*) *Let A be an $m \times n$ matrix of rank r. Then there exist an orthogonal $m \times m$ matrix P and an orthogonal $n \times n$ matrix Q such that*

$$PAQ = \begin{pmatrix} D & 0_{r,n-r} \\ 0_{m-r,r} & 0_{m-r,n-r} \end{pmatrix} \qquad (4.37)$$

where D is an invertible diagonal $r \times r$ matrix with positive entries listed in decreasing order.

#1205
#1206
#1207
Proof Note that $A^T A$ is a symmetric $n \times n$ matrix. So by the spectral theorem there is an $n \times n$ orthogonal matrix Q such that

$$Q^T A^T A Q = \begin{pmatrix} \Delta & 0_{r,n-r} \\ 0_{n-r,r} & 0_{n-r,n-r} \end{pmatrix},$$

where Δ is a diagonal $r \times r$ matrix with its diagonal entries in decreasing order. Note that $A^T A$ has the same rank r as A (#767) and that the eigenvalues of $A^T A$ are non-negative (#1185(i)), the positive eigenvalues being the entries of Δ. If we write

$$Q = \begin{pmatrix} Q_1 & Q_2 \end{pmatrix}$$

where Q_1 is $n \times r$ and Q_2 is $n \times (n-r)$ then we have in particular that

$$Q_1^T A^T A Q_1 = \Delta; \qquad Q_2^T A^T A Q_2 = 0_{n-r,n-r}; \qquad Q_1^T Q_1 = I_r; \qquad Q_1 Q_1^T + Q_2 Q_2^T = I_n,$$

the last two equations following from Q's orthogonality. From the second equation we have $A Q_2 = 0_{m,n-r}$ (again by #767).

If $\Delta = \text{diag}(\lambda_1, \ldots, \lambda_r)$ then we may set $D = \text{diag}(\sqrt{\lambda_1}, \ldots, \sqrt{\lambda_r})$ so that $D^2 = \Delta$. We then define P_1 to be the $m \times r$ matrix

$$P_1 = A Q_1 D^{-1}.$$

Note that

$$P_1 D Q_1^T = A Q_1 Q_1^T = A(I_n - Q_2 Q_2^T) = A - (A Q_2) Q_2^T = A.$$

We are almost done now as, by the transpose product rule and because D is diagonal, we have that

$$P_1^T P_1 = \left(A Q_1 D^{-1}\right)^T \left(A Q_1 D^{-1}\right) = D^{-1} Q_1^T A^T A Q_1 D^{-1} = D^{-1} \Delta D^{-1} = I_r$$

and also that

$$P_1^T A Q_1 = P_1^T P_1 D = I_r D = D.$$

That $P_1^T P_1 = I_r$ means the columns of P_1 form an orthonormal set, which may then be extended to an orthonormal basis for $\mathbb{R}_m$. We put these vectors as the columns of an orthogonal $m \times m$ matrix $P^T = \begin{pmatrix} P_1 & P_2 \end{pmatrix}$ and note that

$$P_2^T A Q_1 = P_2^T (A Q_1 D) D^{-1} = P_2^T P_1 D^{-1} = 0_{m-r,r} D^{-1} = 0_{m-r,r}$$

as the columns of P are orthogonal. Finally we have that PAQ equals

$$\begin{pmatrix} P_1^T \\ P_2^T \end{pmatrix} A \begin{pmatrix} Q_1 & Q_2 \end{pmatrix} = \begin{pmatrix} P_1^T \\ P_2^T \end{pmatrix} \begin{pmatrix} AQ_1 & 0_{m,n-r} \end{pmatrix}$$

$$= \begin{pmatrix} P_1^T AQ_1 & 0_{r,n-r} \\ 0_{m-r,r} & 0_{m-r,n-r} \end{pmatrix} = \begin{pmatrix} D & 0_{r,n-r} \\ 0_{m-r,r} & 0_{m-r,n-r} \end{pmatrix}. \qquad \square$$

Exercises

#1172A Let N be a square matrix such that $P^T NP$ is diagonal for some orthogonal matrix P. Show that N is symmetric.

#1173A Let A be an $n \times n$ matrix. Show that $A\mathbf{x} \cdot \mathbf{y} = \mathbf{x} \cdot A\mathbf{y}$ for all $\mathbf{x}, \mathbf{y}$ in $\mathbb{R}_n$ if and only if A is symmetric.

#1174B Let A be a symmetric matrix with maximum eigenvalue $\lambda_{\max}$ and minimum eigenvalue $\lambda_{\min}$. Show, for any $\mathbf{x}$ that

$$\lambda_{\max} |\mathbf{x}|^2 \geqslant \mathbf{x}^T A\mathbf{x} \geqslant \lambda_{\min} |\mathbf{x}|^2.$$

#1175A Show that every symmetric matrix has a symmetric cube root.

#1176C† Show that the only symmetric cube root of I_2 is I_2. What are the symmetric cube roots of $\operatorname{diag}(1,8)$?

#1177C Let A be a skew-symmetric (real) matrix. Show that the roots of $c_A(x)$ are purely imaginary.

#1178c Need a complex symmetric matrix have real eigenvalues?

#1179b For each of the matrices A below, find an orthogonal matrix P such that $P^T AP$ is diagonal.

$$A = \begin{pmatrix} 2 & 3 \\ 3 & 2 \end{pmatrix}, \qquad A = \begin{pmatrix} 29 & 5 & 5 \\ 5 & 53 & -1 \\ 5 & -1 & 53 \end{pmatrix}, \qquad A = \begin{pmatrix} 27 & -3 & -6\sqrt{2} \\ -3 & 19 & 2\sqrt{2} \\ -6\sqrt{2} & 2\sqrt{2} & 26 \end{pmatrix}.$$

#1180b† (i) Identify $\mathbb{R}_4$ with the space of polynomials of degree three or less. Write down the matrix for L, the differential operator in (4.38). Find, by direct calculation, the eigenvalues and eigenvectors of this matrix and an orthogonal matrix which diagonalizes it. (ii) Generalize (i) to the space of polynomials of degree n or less for a given positive integer n.

#1181C Complete the details of Example 4.60.

#1182C Complete the details of Example 4.57.

#1183b Find an orthogonal matrix P such that $P^T MP$ is diagonal, where M is as given on the right. Hence sketch the surface $x^2 + y^2 + z^2 + 4xy + 4xz + 4yz = 1$.

$$M = \begin{pmatrix} 1 & 2 & 2 \\ 2 & 1 & 2 \\ 2 & 2 & 1 \end{pmatrix}$$

Spectral Theory The nth Legendre polynomial $P_n(x)$ satisfies Legendre's equation (6.2), which can be rewritten

$$Ly = -n(n+1)y \qquad \text{where} \qquad L = \frac{\mathrm{d}}{\mathrm{d}x}\left[(1-x^2)\frac{\mathrm{d}}{\mathrm{d}x}\right]. \qquad (4.38)$$

So $P_n(x)$ can be viewed as an eigenvector of the differential operator L with eigenvalue $-n(n+1)$. Further, we saw in #974 that $P_n \cdot P_m = 0$ when $n \neq m$ (for a certain inner product), so that the Legendre polynomials are in fact orthogonal eigenvectors; further still it is true (using the same inner product) that $Ly_1 \cdot y_2 = y_1 \cdot Ly_2$ (#1524), as is similarly true of symmetric matrices (#1173).

Because of this last property, if we identify, in the natural way, $\mathbb{R}_{n+1}$ with the space of polynomials of degree n or less, then L has a symmetric matrix. This is the matrix of L with respect to the standard basis $\{1, x, \ldots, x^n\}$. We know by the spectral theorem that there is an orthonormal basis which diagonalizes L; in fact given the above comments we see that this orthonormal basis consists of scalar multiples of the Legendre polynomials $P_0(x), P_1(x), \ldots, P_n(x)$. (See #1180 for details.)

Whilst the space of polynomials is infinite dimensional, the preceding example is not at a great remove from orthogonally diagonalizing a real symmetric matrix – after all any polynomial can be written as a finite linear combination of Legendre polynomials. For contrast, *Schrödinger's equation* in quantum theory has the form

$$-\frac{\hbar^2}{2m}\frac{\mathrm{d}^2\psi}{\mathrm{d}x^2} + V(x)\psi = E\psi, \qquad \psi(0) = \psi(a) = 0. \qquad (4.39)$$

This equation was formulated in 1925 by the Austrian physicist Erwin Schrödinger (1887–1961) with (4.39) being the time-independent equation of a particle in the interval $0 \leqslant x \leqslant a$. The *wave function* ψ is a complex-valued function of x and $|\psi(x)|^2$ can be thought of as the probability density function of the particle's position. m is its mass, $\hbar$ is the (reduced) Planck constant, $V(x)$ denotes potential energy and E is the particle's energy.

A significant, confounding aspect of late nineteenth-century experimental physics was the *emission spectra* of atoms. (By the way, these two uses of the word 'spectrum' in mathematics and physics appear to be coincidental.) As an example, experiments showed that only certain discrete, quantized energies could be released by an excited atom of hydrogen. Classical physical theories were unable to explain this phenomenon.

Schrödinger's equation can be rewritten as $H\psi = E\psi$, with E being an eigenvalue of the differential operator H known as the *Hamiltonian*. One can again show that $H\psi \cdot \tilde{\psi} = \psi \cdot H\tilde{\psi}$ for a certain (complex) inner product and general wave functions $\psi, \tilde{\psi}$. And if V is constant then we can see that the only non-zero solutions of (4.39) are

$$\psi_n(x) = A_n \sin\left(\frac{n\pi x}{a}\right), \qquad \text{where} \qquad E = E_n = V + \frac{n^2\pi^2\hbar^2}{2ma^2}$$

and n is a positive integer and A_n is a constant. If $|\psi_n|^2$ is to be a pdf then we need $A_n = \sqrt{2/a}$ (see #1661) and again these ψ_n are orthonormal with respect to this inner product. Note above that the energy E can only take certain discrete values E_n.

In general though, a wave function need not be one of these *eigenstates* ψ_n and may be a finite or indeed infinite combination of them. For example we might have

$$\psi(x) = \sqrt{30/a^5}x(a-x),$$

for which $|\psi(x)|^2$ is a pdf. How might we write such a ψ as a combination of the $\psi_n(x)$? This is an infinite-dimensional version of the problem the spectral theorem solved – how in general to write a vector as a linear combination of orthonormal eigenvectors – and in the infinite-dimensional case is the subject of *Fourier analysis*, named after the French mathematician Joseph Fourier. In this case Fourier analysis shows that

$$\psi(x) = \sum_{n=0}^{\infty} \alpha_{2n+1} \psi_{2n+1}(x), \qquad \text{where} \qquad \alpha_n = \frac{8\sqrt{15}}{n^3 \pi^3}.$$

If the particle's energy was measured it would be one of the permitted energies E_n and the effect of measuring this energy is to *collapse* the above wave function ψ to one of the eigenstates ψ_n. It is the case that

$$\sum_{n=0}^{\infty} |\alpha_{2n+1}|^2 = 1$$

(this is *Parseval's identity*, which is essentially an infinite-dimensional version of Pythagoras' theorem). The probability of the particle having measured energy E_{2n+1} is $|\alpha_{2n+1}|^2$. The role of *measurement* in quantum theory is very different from that of classical mechanics; the very act of measuring some *observable* characteristic of the particle actually affects and changes the wave function.

#1184D† Let M be a symmetric $n \times n$ matrix and D be a diagonal $n \times n$ matrix with distinct entries, at least one of which is positive. If $\mathbf{x}^T M \mathbf{x} = 1$ precisely when $\mathbf{x}^T D \mathbf{x} = 1$, show that $M = D$. Show that this result need not follow if all the diagonal entries of D are negative.

#1185d† (i) Say that $M = A^T A$ where A is an $n \times n$ matrix. Show that the roots of $c_M(x)$ are real and non-negative.
(ii) Say that N is a symmetric matrix with non-negative eigenvalues. Show that $N = A^T A$ for some matrix A.

#1186A† Show that the ellipsoid with equation $x^2/a^2 + y^2/b^2 + z^2/c^2 = 1$ has volume $4\pi abc/3$.

#1187D Find all 3×3 orthogonal matrices which commute with $\text{diag}(1, 1, -1)$.

#1188D† Under what circumstances does the equation $(\mathbf{r} \wedge \mathbf{a}) \cdot (\mathbf{r} \wedge \mathbf{b}) = 1$ describe a hyperboloid of two sheets?

#1189B Let A and B be $n \times n$ symmetric square matrices with the same characteristic polynomial. Show that there exists an orthogonal matrix P such that $P^T A P = B$. Determine how many different such P there are if the eigenvalues of A (and B) are distinct.

#1190c Consider the effect on (4.28) of the euclidean change of variable

$$\begin{pmatrix} x \\ y \end{pmatrix} \longmapsto \begin{pmatrix} \cos\theta & -\sin\theta \\ \sin\theta & \cos\theta \end{pmatrix} \begin{pmatrix} x \\ y \end{pmatrix} + \begin{pmatrix} k_1 \\ k_2 \end{pmatrix}$$

on equation (4.25). Show that (4.25) transforms into another degree two equation in two variables and that $B^2 - 4AC$ remains unchanged.

#1191D Let A, B be symmetric $n \times n$ matrices and suppose that every eigenvalue of A is positive. Show that there exists an invertible matrix Q such that $Q^T A Q = I_n$. Deduce that there is an invertible matrix R such that $R^T A R$ and $R^T B R$ are both diagonal.

#1192C† Let E be an ellipse in $\mathbb{R}^2$ and consider the map $T(\mathbf{x}) = A\mathbf{x} + \mathbf{b}$, where A is an invertible 2×2 matrix and $\mathbf{b}$ is in $\mathbb{R}_2$. Show that $T(E)$ is an ellipse. Show that T likewise preserves parabolae and hyperbolae.

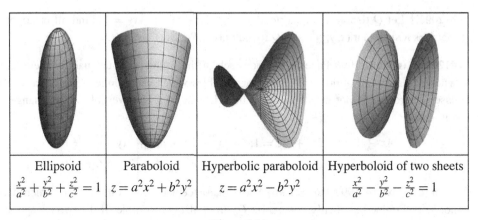

Ellipsoid	Paraboloid	Hyperbolic paraboloid	Hyperboloid of two sheets
$\frac{x^2}{a^2} + \frac{y^2}{b^2} + \frac{z^2}{c^2} = 1$	$z = a^2 x^2 + b^2 y^2$	$z = a^2 x^2 - b^2 y^2$	$\frac{x^2}{a^2} - \frac{y^2}{b^2} - \frac{z^2}{c^2} = 1$

Figure 4.14

#1193b Show that the equation $13x^2 + 13y^2 + 10z^2 + 4yz + 4zx + 8xy = 1$ defines an ellipsoid and find its volume.

#1194b Classify the quadric $4x^2 + 4y^2 - 7z^2 - 12yz - 12xz + 6xy = 1$.

#1195b† What is the maximum value of $x + y + z$ attained on the quadric in #1193?

#1196c Let $f(x, y, z) = x^2 + 4y^2 + z^2 + 4xy + 2xz + 6yz$. Classify the surfaces $f(x, y, z) = 1$ and $f(x, y, z) = -1$.

Quadrics We can analyze a degree two equation in three variables in a similar fashion to the approach of Theorem 4.53 for two variables. By an orthogonal change of variable the mixed terms xy, yz, zx can be removed and then completing squares, where possible, will deal with some or all of linear terms x, y, z. Several standard examples are drawn in Figure 4.14 and at the end of Example 4.60. There are again many special and degenerate cases such as

$$(x+y+z)(x+y+z-1) = 0, \quad x^2+y^2 = 1, \quad x^2-y^2 = 1,$$
$$x^2+y^2 = z^2, \quad x^2+y^2 = 0, \quad x^2+y^2+z^2 = -1.$$

The above respectively describe: two parallel planes, a (circular) cylinder, a hyperbolic cylinder, a (double) cone, a line and the empty set.

#1197c Classify the quadric $24y^2 - 24z^2 - 10xy + 70xz + 14yz + 6x + 12y + 10z = 0$.

#1198D† Show that the equation $3x^2 + 6y^2 + 2z^2 + 2yz + 4zx + 2xy = 1$ defines an ellipsoid and find its volume.

#1199B Let C denote the ellipse in #1130. Find all orthogonal matrices R such that $(x,y)^T$ is in C if and only if $R(x,y)^T$ is in C.

#1200E† Let Q denote the quadric $x^2 + y^2 + 2yz + 2xz - 4xy = 1$. Find all orthogonal matrices R such that $(x,y,z)^T$ is in Q if and only if $R(x,y,z)^T$ is in Q.

#1201d† (Oxford 1996 Paper 1 #6) For each of the following four quadrics, determine whether there is an orthogonal change of variables which carries one to the other. In the cases where this is not possible, can it be achieved with an invertible linear change of variables?

$$4x^2 + 15y^2 + 7z^2 + 2yz = 1; \qquad x^2 + 2y^2 - 4z^2 + 2xy - 6yz = 1;$$
$$5x^2 + 5y^2 + 2z^2 + 2xy = 1; \qquad x^2 + y^2 - 3z^2 - 12xy - 6yz = 1.$$

#1202d† (Oxford 1997 Paper 1 #6) Show that except for two values of α, which need not be determined explicitly, the matrix Q_α in (4.40) is invertible. If the two exceptional values of α are $\alpha_1 < \alpha_2$, show that the surface $x^T Q_\alpha x = 1$ is a hyperboloid of two sheets if $\alpha_1 < \alpha < \alpha_2$ and a hyperboloid of two sheets if $\alpha < \alpha_1$ or $\alpha > \alpha_2$. Describe the surfaces in the cases $\alpha = \alpha_1$ and $\alpha = \alpha_2$.

$$Q_\alpha = \begin{pmatrix} 2 & 3 & \alpha \\ 3 & 3 & \alpha^2 \\ \alpha & \alpha^2 & -2 \end{pmatrix}, \qquad M_a = \begin{pmatrix} 1 & 0 & 1 \\ 0 & a & 1 \\ 1 & 1 & 2a+2 \end{pmatrix}. \tag{4.40}$$

#1203d (i) Let a be a real number. Show that the matrix M_a in (4.40) is invertible except for two exceptional values.
(ii) Identify the type of the quadric surface $2x^2 + y^2 + 6z^2 + 4xz + 4yz = 1$.

#1204 D† (**Spectral Theorem for Hermitian Matrices**) A complex $n \times n$ matrix M is said to be **Hermitian** if $M = \overline{M}^T$ (i.e. if it equals the conjugate of its transpose), A complex $n \times n$ matrix U is said to be **unitary** if $U\overline{U}^T = I_n = \overline{U}^T U$. Adapt the proof of the spectral theorem to show that for any Hermitian matrix M there exists a unitary matrix U such that $\overline{U}^T M U$ is diagonal with real entries. Deduce that a (real or complex) skew-symmetric matrix is diagonalizable over the complex numbers and has a characteristic polynomial with purely imaginary roots .

#1205 B† Let A be an $m \times n$ matrix with two singular value decompositions

$$PAQ = \begin{pmatrix} D & 0 \\ 0 & 0 \end{pmatrix} \quad \text{and} \quad \tilde{P}A\tilde{Q} = \begin{pmatrix} \tilde{D} & 0 \\ 0 & 0 \end{pmatrix}.$$

Show that $D = \tilde{D}$. But show that it need not be the case that $P = \tilde{P}$ or $Q = \tilde{Q}$.

#1206 B† With A, P, Q, D as in (4.37), say $D = \text{diag}(\lambda_1, \ldots, \lambda_r)$ where $\lambda_1 \geqslant \lambda_2 \geqslant \cdots \geqslant \lambda_r$. Show that

$$\lambda_1 = \max\{|A\mathbf{v}| : |\mathbf{v}| = 1\}.$$

#1207 b Find singular value decompositions for the following matrices:

$$\begin{pmatrix} 1 & 1 \\ -1 & 1 \\ 3 & 0 \end{pmatrix}, \quad \begin{pmatrix} 0 & 1 & -1 \\ 1 & 0 & -2 \end{pmatrix}, \quad \begin{pmatrix} 1 & 0 & 2 & 1 \\ 0 & 2 & 1 & -1 \end{pmatrix}.$$

#1208 C† Let A be an $n \times n$ matrix. Show that there is an orthogonal matrix X such that AX is symmetric.

#1209 D† With notation as in Theorem 4.61, the **pseudoinverse** [8] of A is

$$A^+ = Q \begin{pmatrix} D^{-1} & 0_{r,m-r} \\ 0_{n-r,r} & 0_{n-r,m-r} \end{pmatrix} P.$$

(i) Show that if A is invertible then $A^{-1} = A^+$.
(ii) Show that $(A^T)^+ = (A^+)^T$.
(iii) Show further that the pseudoinverse has the following properties:

    (I) $AA^+A = A$;   (II) $A^+AA^+ = A^+$;   (III) AA^+ and A^+A are symmetric.

Deduce that AA^+ and A^+A are the orthogonal projections onto $\text{Col}(A)$ and $\text{Col}(A^T)$.

#1210 D† Show that A^+ is the only matrix to have the properties I, II, III described in #1209(iii). Deduce that if the columns of A are independent then $A^+ = (A^TA)^{-1}A^T$. (Compare with #805.)

#1211 c† Give an example to show that $(AB)^+ \neq B^+A^+$ in general.

#1212 c Find the pseudoinverses of the matrices in #1207.

---

[8] The pseudoinverse is also known as the Moore–Penrose pseudoinverse, named after E. H. Moore (1862–1932) and Roger Penrose (b. 1931).

#1213D† (i) Let A be an $m \times n$ matrix and $\mathbf{b}$ be in $\mathbb{R}_m$. If $\mathbf{x}_0 = A^+\mathbf{b}$ show that

$$|A\mathbf{x} - \mathbf{b}| \geqslant |A\mathbf{x}_0 - \mathbf{b}| \qquad \text{for all } \mathbf{x} \text{ in } \mathbb{R}_n.$$

Give examples to show that $\mathbf{x}_0$ may or may not be unique in having this property.
(ii) Let A be an $n \times n$ matrix and let P, Q be as in Theorem 4.61. Show that

$$\|X - A\| \geqslant \left\|P^T Q^T - A\right\| \qquad \text{for all orthogonal } n \times n \text{ matrices } X.$$

Here the matrix norm $\|\ \ \|$ is that associated with the inner product introduced in #971.

4.5 Further Exercises*

Geometry of the Sphere

Definition 4.62 Let S denote the unit sphere in $\mathbb{R}^3$ with equation $x^2 + y^2 + z^2 = 1$. The **great circles** of S are the intersections of planes through the origin with S. Given a point $\mathbf{p}$ on S its **antipodal point** is $-\mathbf{p}$.

#1214B (i) Show that any pair of distinct, non-antipodal points $\mathbf{p}, \mathbf{q}$ has a unique great circle through them.
(ii) $\mathbf{p}$ and $\mathbf{q}$ divide the great circle into two arcs. The **distance** $d(\mathbf{p}, \mathbf{q})$ between $\mathbf{p}$ and $\mathbf{q}$ is defined to be the length of the shorter arc. Show that $d(\mathbf{p}, \mathbf{q}) = \cos^{-1}(\mathbf{p} \cdot \mathbf{q})$.
(iii) If $0 < r < \pi$, show the points $\mathbf{q}$ such that $d(\mathbf{p}, \mathbf{q}) = r$ form a circle and find its radius.

#1215C Let $\mathbf{p}, \mathbf{q}$ and $\mathbf{x}, \mathbf{y}$ be two pairs of distinct points on S. Show that there is an orthogonal matrix A such that $A\mathbf{p} = \mathbf{x}$ and $A\mathbf{q} = \mathbf{y}$ if and only if $d(\mathbf{p}, \mathbf{q}) = d(\mathbf{x}, \mathbf{y})$. If $\mathbf{p}$ and $\mathbf{q}$ are non-antipodal, how many orthogonal matrices A are there such that $A\mathbf{p} = \mathbf{p}$ and $A\mathbf{q} = \mathbf{q}$?

#1216D† Let T be an isometry of S – that is T maps S to S and preserves distances. Show that there is an orthogonal matrix A such that $T(\mathbf{x}) = A\mathbf{x}$ for all $\mathbf{x}$.

Definition 4.63 Spherical triangles are determined by three vertices A, B, C whose position vectors $\mathbf{a}, \mathbf{b}, \mathbf{c}$ are linearly independent. The edges of the triangle are the arcs of the great circles connecting A to B, B to C, and C to A whose lengths we respectively denote c, a, b. A spherical triangle then defines six angles: the edges BC, CA, AB subtend angles a, b, c at the centre of the sphere; there are also three angles α, β, γ at the vertices between edges BA and CA, AB and CB, AC and BC (Figure 4.15). We make the requirement that all six angles lie strictly between 0 and π.

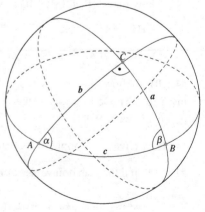

Figure 4.15

#1217D† Let $0 < a \leqslant b \leqslant c < \pi/2$ and let $A = (0,0,1)$ and $B = (\sin c, 0, \cos c)$ be points on the unit sphere.

  (i) Show that there is a point C which is at a distance b from A and at a distance a from B if and only if $a + b \leqslant c$.
 (ii) Given a spherical triangle with sides a, b, c, show that $a + b + c < 2\pi$.
(iii) Given positive a, b, c satisfying the triangle inequalities $a < b + c, b < a + c, c < a + b$, and $a + b + c < 2\pi$, show there is a spherical triangle with sides a, b, c.

#1218B† Prove the **first cosine rule**: $\cos a = \cos b \cos c + \sin b \sin c \cos \alpha$.

#1219B Prove the **sine rule** for spherical triangles: $\sin \alpha / \sin a = \sin \beta / \sin b = \sin \gamma / \sin c$.

#1220C† Show, for small spherical triangles where third and higher powers of a, b, c are negligible, that the usual cosine and sine rules from the first cosine rule and the sine rule for spherical triangles.

Definition 4.64 Every triangle ABC has a **dual triangle** $A'B'C'$ obtained as follows. The plane OBC cuts S into two hemispheres and has a normal through O which meets S in two antipodal points with position vectors $\pm \mathbf{a}'$. We take $A' = \mathbf{a}'$ to be that choice which is in the same hemisphere as A. Then B' and C' are similarly defined.

#1221D† Write down expressions for the position vectors $\mathbf{a}', \mathbf{b}', \mathbf{c}'$ in terms of $\mathbf{a}, \mathbf{b}, \mathbf{c}$. Show that if a spherical triangle has angles $a, b, c, \alpha, \beta, \gamma$ then its dual triangle has respective angles $\pi - \alpha, \pi - \beta, \pi - \gamma, \pi - a, \pi - b, \pi - c$. Deduce that

$$\alpha + \beta + \gamma > \pi$$

and deduce the **second cosine rule**: $\cos \alpha = -\cos \beta \cos \gamma + \sin \beta \sin \gamma \cos a$.

#1222D† Show that the dual triangle of the dual triangle is the original triangle.

#1223D† (**Girard's Theorem**[9]) By considering how the great circles bounding a spherical triangle divide up the sphere, show that the area of a spherical triangle equals $\alpha + \beta + \gamma - \pi$.

#1224C† Let Δ_1 and Δ_2 be two spherical triangles with sides a, b, c. Show that there is an orthogonal matrix M such that $\Delta_1 = M(\Delta_2)$. Deduce that similar spherical triangles are congruent.

#1225C Use the formula for the area of a spherical triangle to obtain a formula for the area of a spherical quadrilateral with angles $\alpha, \beta, \gamma, \delta$. Does it matter whether the quadrilateral is convex or not? Generalize the formula to an n-sided polygon on the sphere.

---

[9] Named after the mathematician Albert Girard (1595–1632). Girard was French born, but fled to the Netherlands in his youth as a religious refugee. In his 1629 work *Invention Nouvelle en l'Algebre*, he made important early observations relating to the fundamental theorem of algebra and symmetric functions of the roots of polynomials.

More on Quaternions

#1226D Given a quaternion $Q = a + b\mathbf{i} + c\mathbf{j} + d\mathbf{k}$, we write $\overline{Q} = a - b\mathbf{i} - c\mathbf{j} - d\mathbf{k}$ and set $|Q|^2 = Q\overline{Q}$.

(i) Show that any quaternion Q can be written in the form $Q = r(\cos\theta + H\sin\theta)$, where $r \geqslant 0$ and $H^2 = -1$.

(ii) A quaternion P is said to be **pure** if it is a combination of $\mathbf{i}, \mathbf{j}, \mathbf{k}$. Show that for pure quaternions P_1, P_2 we have

$$P_1 P_2 = -P_1 \cdot P_2 + P_1 \wedge P_2, \tag{4.41}$$

where $\cdot$ and $\wedge$ denote the usual scalar and vector products in $\mathbb{R}_3$.

(iii) Let $Q = \sqrt{3}/2 + (\mathbf{i} + \mathbf{j} + \mathbf{k})/(2\sqrt{3})$. Show that $|Q| = 1$. Identify the pure quaternions with $\mathbb{R}_3$ and consider the map $T(P) = QP\overline{Q}$. Determine $T(P)$ where $P = (1, 1, 1)/\sqrt{3}$ and $T(R)$ where R is a unit vector perpendicular to this P. Deduce that T represents a rotation by $\pi/3$ about the line $x = y = z$.

#1227D† (**Cayley–Dickson Construction**) Any quaternion q can be uniquely written in the form

$$q = z_1 + z_2\mathbf{j}$$

where z_1 and z_2 are complex numbers of the form $x + y\mathbf{i}$. We can then identify q with the complex matrix

$$q \leftrightarrow Q = \begin{pmatrix} z_1 & z_2 \\ -\overline{z_2} & \overline{z_1} \end{pmatrix}.$$

(i) Show that if $q_1 \leftrightarrow Q_1$ and $q_2 \leftrightarrow Q_2$ then

$$q_1 + q_2 \leftrightarrow Q_1 + Q_2, \qquad q_1 q_2 \leftrightarrow Q_1 Q_2, \qquad \overline{q} \leftrightarrow \overline{Q}.$$

Show that Q is invertible if and only if $q \neq 0$, in which case $q^{-1} \leftrightarrow Q^{-1}$.

(ii) The **octonions** are an eight-dimensional division algebra which may be constructed from the quaternions using a similar Cayley–Dickson construction. The octonions can be considered as pairs (q_1, q_2) of quaternions which add and multiply according to the rules

$$(q_1, q_2) + (r_1, r_2) = (q_1 + q_2, r_1 + r_2)$$
$$(q_1, q_2) \times (r_1, r_2) = (q_1 r_1 - \overline{r_2} q_2, r_2 q_1 + q_2 \overline{r_1}),$$

where q_1, q_2, r_1, r_2 are quaternions. Show that multiplication of octonions is not associative.

#1228E (**Vector Product in** $\mathbb{R}^7$) A vector $\mathbf{x} = (x_1, \ldots, x_7)$ can be identified with a 'pure' octonion as in #1227 by setting $q_1 = x_1\mathbf{i} + x_2\mathbf{j} + x_3\mathbf{k}$ and $q_2 = x_4 + x_5\mathbf{i} + x_6\mathbf{j} + x_7\mathbf{k}$. [This octonion is pure in the sense that the first quaternion has no real part.] Given two vectors $\mathbf{x}, \mathbf{y}$ in $\mathbb{R}^7$ show that $\mathbf{xy} + \mathbf{x} \cdot \mathbf{y}$ is a pure octonion; here the term $\mathbf{xy}$ is the product of the octonions corresponding to $\mathbf{x}$ and $\mathbf{y}$ and $\mathbf{x} \cdot \mathbf{y}$ is the dot product of the vectors $\mathbf{x}$ and $\mathbf{y}$.

Show that the vector product

$$\mathbf{x} \wedge \mathbf{y} = \mathbf{xy} + \mathbf{x} \cdot \mathbf{y}$$

on $\mathbb{R}^7$ satisfies properties (a)–(d) of Proposition 3.179. (Compare the above formula with (4.41).)

Exercises on Orthogonal Complements

Definition 4.65 Let S be a subset of $\mathbb{R}^n$ and define

$$S^\perp = \{\mathbf{v} \in \mathbb{R}^n : \mathbf{v} \cdot \mathbf{s} = 0 \text{ for all } \mathbf{s} \in S\}.$$

If S is a subspace of $\mathbb{R}^n$ then $S^\perp$ is called the **orthogonal complement** of S.

#1229 B (i) Let S be a subset of $\mathbb{R}^n$. Show that $S^\perp$ is a subspace of $\mathbb{R}^n$ (whether or not S itself is a subspace).
(ii) Show that if S is contained in T then $T^\perp$ is contained in $S^\perp$.
(iii) Let S be a subset of $\mathbb{R}^n$. Show that $\langle S \rangle^\perp = S^\perp$.

#1230 B† Let $X = \langle (0,2,1,2),(1,2,2,3) \rangle$. Find a basis for $X^\perp$.

#1231 B† (i) Let X be a subspace of $\mathbb{R}^n$. Show that $\mathbb{R}^n = X \oplus X^\perp$.
(ii) Let S be a subset of $\mathbb{R}^n$. Show that $\left(S^\perp\right)^\perp = S$ if and only if S is a subspace.

#1232 B† Let X be a subspace of $\mathbb{R}^n$. Say that $\mathbf{v} = \mathbf{x} + \mathbf{y}$ is the decomposition of $\mathbf{v}$ into $\mathbf{x}$ in X and $\mathbf{y}$ in $X^\perp$.

 (i) Show that $\mathbf{x}$ is the closest point of X to $\mathbf{v}$.
(ii) Show that the map P which sends $\mathbf{v}$ to $\mathbf{x}$ is linear, is a projection, has image X and null space $X^\perp$.

#1233 C† Show, for an $m \times n$ matrix A, we have that $(\text{Row}(A))^\perp = \text{Null}(A)$ and further that $(\text{Col}(A))^\perp = \text{Null}(A^T)$.

Exercises on Cyclic Vectors and Rational Canonical Form

Definition 4.66 Let T be a linear map from $\mathbb{R}_n$ to $\mathbb{R}_n$ and $\mathbf{v}$ be a vector in $\mathbb{R}_n$. We denote

$$Z(\mathbf{v}, T) = \{p(T)\mathbf{v} : p(x) \text{ is a real polynomial}\} = \langle \mathbf{v}, T\mathbf{v}, T^2\mathbf{v}, T^3\mathbf{v}, \ldots \rangle.$$

We say a vector $\mathbf{v}$ is **cyclic** (or T-**cyclic**) if $Z(\mathbf{v}, T) = \mathbb{R}_n$.

Definition 4.67 A subspace V of $\mathbb{R}_n$ is said to be **cyclic** (or T-**cyclic**) if there exists $\mathbf{v}$ in V such that $V = Z(\mathbf{v}, T)$.

#1234 B† Let $A = \text{diag}(1,2)$. Show that $\mathbb{R}_2$ is A-cyclic. What are the A-cyclic subspaces of $\mathbb{R}_2$?

$$_{\mathcal{V}}T_{\mathcal{V}} = \begin{pmatrix} 0 & 0 & \cdots & 0 & -c_0 \\ 1 & 0 & \cdots & 0 & -c_1 \\ 0 & 1 & \ddots & \vdots & -c_2 \\ \vdots & \ddots & \ddots & 0 & \vdots \\ 0 & \cdots & \cdots & 1 & -c_{n-1} \end{pmatrix} \tag{4.42}$$

#1235D Let T be a linear map from $\mathbb{R}_n$ to $\mathbb{R}_n$ and $\mathbf{v}$ be such that $\mathbb{R}_n = Z(\mathbf{v}, T)$. Show that $\mathcal{V} = \{\mathbf{v}, T\mathbf{v}, T^2\mathbf{v}, \ldots, T^n\mathbf{v}\}$ is a basis for $\mathbb{R}_n$ and that $_{\mathcal{V}}T_{\mathcal{V}}$ equals the matrix in (4.42) for some real scalars $c_0, c_1, \ldots, c_{n-1}$. Show further that the characteristic polynomial and minimal polynomial of T both equal

$$p(x) = x^n + c_{n-1}x^{n-1} + \cdots + c_1 x + c_0. \tag{4.43}$$

The matrix in (4.42) is known as the **companion matrix** of the polynomial $p(x)$ in (4.43) and is denoted $C(p)$.

#1236B† Find the minimum and characteristic polynomials of the matrices A, B, C in #1076.

#1237B Determine the following cyclic subspaces

$$Z((1,0,0)^T, A), \quad Z((0,1,0)^T, A), \quad Z((1,-1,0)^T, B), \quad Z((1,1,1)^T, B), \quad Z((1,1,1)^T, C),$$

where A, B, C are as in #1076.

#1238B Determine all the cyclic vectors of A and B from #1076.

#1239D†(i) With C as in #1076, write $\mathbb{R}_3$ as the direct sum of C-cyclic subspaces.
(ii) Show that if $\mathbb{R}_n$ is written as the direct sum of T-cyclic subspaces then there is a basis $\mathcal{B}$ for $\mathbb{R}_n$ such that

$$_{\mathcal{B}}T_{\mathcal{B}} = \mathrm{diag}(C(f_1), C(f_2), \ldots, C(f_k)),$$

for certain monic polynomials $f_1, \ldots, f_k$. Show that in this case $c_T = f_1 f_2 \cdots f_k$.
(iii) Show further that if f_i is a factor of f_{i+1} for $i = 1, 2, \ldots, k-1$ then $m_T = f_k$.

#1240C Let A be a 3×3 matrix, $\mathbf{v}$ and $\mathbf{w}$ non-zero vectors in $\mathbb{R}_3$ with $\mathbf{w}$ not in $Z(\mathbf{v}, A)$. For each situation below, give specific examples of $A, \mathbf{v}, \mathbf{w}$ to show that the situation is possible.

(i) $Z(\mathbf{v}, A)$ is contained in $Z(\mathbf{w}, A)$.
(ii) $\mathbf{0}$ is the only vector common to both $Z(\mathbf{v}, A)$ and $Z(\mathbf{w}, A)$.
(iii) The intersection of $Z(\mathbf{v}, A)$ and $Z(\mathbf{w}, A)$ is non-zero without either space including the other.

#1241E†(i) Let A be an $n \times n$ matrix and $\mathbf{v}, \mathbf{w}$ be in $\mathbb{R}_n$. Show there is a monic polynomial $m_{A,\mathbf{v}}(x)$ of least degree such that $m_{A,\mathbf{v}}(A)\mathbf{v} = \mathbf{0}$. Further show $\dim Z(\mathbf{v}, A) = \deg m_{A,\mathbf{v}}(x)$.
(ii) Assume that $m_{A,\mathbf{v}}(x)$ and $m_{A,\mathbf{w}}(x)$ have no common complex root. Show that

$$m_{A,\mathbf{v}+\mathbf{w}}(x) = m_{A,\mathbf{v}}(x) m_{A,\mathbf{w}}(x)$$

and deduce that $Z(\mathbf{v}, A) + Z(\mathbf{w}, A)$ is a direct sum and also a cyclic subspace.
(iii) Deduce that there is a vector $\mathbf{x}$ such that $m_{A,\mathbf{x}}(x) = m_A(x)$.
(iv) Deduce also that $c_A(x) = m_A(x)$ if and only if there is an A-cyclic vector.

#1242C† For what vectors $\mathbf{x}$ is it the case that $m_{A,\mathbf{x}}(x) = m_A(x)$ where A is as in #1076. Answer this question for B and C as well.

The above ideas ultimately lead to the following result which we shall not prove here and only state.

Definition 4.68 Let T be a linear map from $\mathbb{R}_n$ to $\mathbb{R}_n$. Then there is a basis $\mathcal{V}$ for $\mathbb{R}_n$ with respect to which

$$_\mathcal{V}T_\mathcal{V} = \text{diag}\,(C(f_1), C(f_2), \ldots, C(f_k)),$$

where $f_1(x), \ldots, f_k(x)$ are monic polynomials such that $f_i(x)$ divides $f_{i+1}(x)$ for $1 \leqslant i < k$. Further, the polynomials $f_i(x)$ are unique. Such a representation is known as the **rational canonical form** or **Frobenius normal form** [10] of T. As we saw in #1239 it follows that

$$c_T(x) = f_1(x)f_2(x)\cdots f_k(x) \qquad \text{and} \qquad m_T(x) = f_k(x).$$

#1243 B Find the rational canonical forms of the matrices A, B, C from #1076.

#1244 D† Find the rational canonical forms of the matrices in #1077.

Principles of Galilean and Special Relativity In physical theories of space and time, and especially in special relativity, an **observer** is often cast as a figure able to assign spatial and temporal co-ordinates, say x, y, z, t, to events around him/her with the observer always at the spatial origin. We are ultimately interested in what measurements and physical laws different observers will agree on. So the notion of an observer is a shorthand for an **inertial frame of reference** in which the laws of physics apply. And when we find a frame of reference where, say, Newton's laws don't apply – for example, within a rotating frame which seemingly introduces extra fictitious, not real forces, such as the Coriolis force – we exclude such frames as being *non-inertial* rather than questioning Newton's laws.

In Galilean relativity, two different observers O and O' may be moving at some constant relative speed to one another. As seen in #1245 these two observers will agree when two events are simultaneous and also on the distance between such events. They should also agree on physical laws, but **Maxwell's equations** in electromagnetism (1861) provided an important example of physical laws that were not invariant under Galilean transformations. For some considerable time physicists sought to resolve this issue until Einstein proposed, in his 1905 paper on special relativity, that the speed of light c should be constant for all observers. This has significant consequences for notions of space–time and for what two different observers can agree on – in particular, observers can no longer agree on simultaneity, length, velocity, colour, mass and energy (#1248). Rather than using Galilean transformations, inertial frames in special relativity are connected via a Lorentz transformation (#1246) – under which Maxwell's equations remain invariant. Whilst such transformations approximate Galilean transformations for low relative speeds (#1247(i)), those differences become significant when the relative speed of the observers is comparable with the speed of light c.

---

[10] The German mathematician Georg Frobenius (1849–1917) made significant contributions across mathematics and especially in group theory, with his development of *character theory*, which is an important tool in the understanding of a group's structure.

Exercises on Special Relativity

#1245B† (i) Two observers O and O' are moving along the real line. Observer O' is moving with velocity v relative to O. Explain why the time and space co-ordinates t, x of O are related to those t', x' of O' by

$$\begin{pmatrix} t \\ x \end{pmatrix} = \begin{pmatrix} 1 & 0 \\ v & 1 \end{pmatrix} \begin{pmatrix} t' \\ x' \end{pmatrix} + \begin{pmatrix} c_1 \\ c_2 \end{pmatrix}.$$

(ii) The above transformation between t, x and t', x' is known as a **Galilean transformation**. Show that the composition of two Galilean transformations is a Galilean transformation, as is the inverse of one.

(iii) Show that the two observers agree on whether two events are simultaneous. Show further that they agree on the distance between two simultaneous events.

#1246B A 2×2 **Lorentz matrix** [11] is one of the form

$$L(u) = \gamma(u) \begin{pmatrix} 1 & u/c \\ u/c & 1 \end{pmatrix} \qquad \text{where} \qquad \gamma(u) = \frac{1}{\sqrt{1 - u^2/c^2}}.$$

(i) Show that we may rewrite $L(u)$ in the form below, saying how ϕ and u are related.

$$\begin{pmatrix} \cosh\phi & \sinh\phi \\ \sinh\phi & \cosh\phi \end{pmatrix}.$$

Deduce that the product of two Lorentz matrices is a Lorentz matrix, as is the inverse of a Lorentz matrix.

(ii) Let $g = \operatorname{diag}(1, -1)$ and let L be a Lorentz matrix. Show that $L^T g L = g$.

(iii) Let M be a 2×2 matrix such that

$$M^T g M = g, \qquad [M]_{11} > 0, \qquad \det M > 0.$$

Show that M is a Lorentz matrix.

#1247B (i) The co-ordinates of two observers O and O' are related by

$$\begin{pmatrix} ct \\ x \end{pmatrix} = \gamma(u) \begin{pmatrix} 1 & u/c \\ u/c & 1 \end{pmatrix} \begin{pmatrix} ct' \\ x' \end{pmatrix}, \tag{4.44}$$

where $|u| < c$. Show that if $u/c \approx 0$ then t, x and t', x' are approximately related by a Galilean transformation.

(ii) Show that the velocity of O' as measured by O is u.

(iii) Say now that O' is moving with velocity u relative to O and O'' is moving with velocity v relating to O'. Show that O'' is moving with velocity

$$\frac{u + v}{1 + uv/c^2}$$

relative to O. What is this velocity when $v = c$?

---

[11] Hendrik Lorentz (1853–1928) was a Dutch physicist, Nobel Prize winner and professor at Leiden for much of his career. His investigations and conjectures provided much of the ground work for Einstein's theory of special relativity, including time dilation, the Lorentz contraction and Lorentz transformations.

#1248D† (i) The co-ordinates of two observers O and O' are related as in (4.44). Show that the two observers may disagree on whether two events E_1 and E_2 are simultaneous.

(ii) Say that O' is at $(ct',x') = (ct_1',0)$ and $(ct',x') = (ct_2',0)$ at two different events. What is the time difference as measured by O'? What is the time difference as measured by O? This phenomenon is known as **time dilation**.

(iii) (**Twin Paradox**) Two twins O and O' meet at $(ct,x) = (ct',x') = (0,0)$ on Earth. Twin O', an astronaut, then travels away from O at speed v to a planet distance d away and immediately returns to Earth at the same speed. Show that the two twins are no longer of the same age, the difference in ages being approximately dv/c^2.

(iv) Observer O' has a rod of length L with its ends having co-ordinates $(ct',0)$ and (ct',L). Noting that O measures distances between simultaneous events, show that O measures the rod as having length $L/\gamma(u)$. This phenomenon is known as **Lorentz contraction**.

#1249D† Two observers O, O' are travelling along the real line, O' moving with velocity v relative to O. The first sends out two light signals separated by an interval τ as measured by O'. What is the interval between the times according to O' when these signals are received by O' if both light signals are emitted (i) before O' passes O; (ii) after O' passes O? If the light signal has frequency ω as measured by O, show that the frequency as measured by O' equals

$$\text{(i)} \quad \omega\sqrt{\frac{c+u}{c-u}}, \qquad \text{(ii)} \quad \omega\sqrt{\frac{c-u}{c+u}}.$$

This means in (i) that O' sees the light as bluer than O or **blue shift**. The phenomenon in (ii) of diminishing frequency is **red shift**.

Exercises on Angular Velocity

#1250D† (i) For each t, let $A(t)$ denote an orthogonal matrix which is differentiable with respect to t. Show that $A'(t)A(t)^T$ is skew-symmetric.

(ii) For a fixed vector $\mathbf{v}_0$ define $\mathbf{v}(t) = A(t)\mathbf{v}_0$. Show that there is a vector $\boldsymbol{\omega}(t)$ such that

$$\mathbf{v}'(t) = \boldsymbol{\omega}(t) \wedge \mathbf{v}(t).$$

The vector $\boldsymbol{\omega}(t)$ is known as the **angular velocity**.

#1251B Set $A(t) = R_{\theta(t)}$ in the notation of Example 3.133. Show that the angular velocity equals $\boldsymbol{\omega}(t) = \theta'(t)\mathbf{k}$.

#1252B Let

$$A(t) = \begin{pmatrix} \cos^2 t & \sin^2 t & \sqrt{2}\sin t\cos t \\ \sin^2 t & \cos^2 t & -\sqrt{2}\sin t\cos t \\ -\sqrt{2}\sin t\cos t & \sqrt{2}\sin t\cos t & \cos^2 t - \sin^2 t \end{pmatrix}.$$

Given that $A(t)$ is an orthogonal matrix, find its angular velocity $\boldsymbol{\omega}(t)$. Find a right-handed orthonormal basis for $\mathbb{R}_3$ with respect to which $A(t)$ equals the matrix below

$$\begin{pmatrix} 1 & 0 & 0 \\ 0 & \cos\omega t & -\sin\omega t \\ 0 & \sin\omega t & \cos\omega t \end{pmatrix},$$

where $\omega = |\boldsymbol{\omega}|$.

#1253C Let

$$A(t) = \frac{1}{9} \begin{pmatrix} 4+5\cos t & -4+4\cos t+3\sin t & 2-2\cos t+6\sin t \\ -4+4\cos t-3\sin t & 4+5\cos t & -2+2\cos t+6\sin t \\ 2-2\cos t-6\sin t & -2+2\cos t-6\sin t & 1+8\cos t \end{pmatrix}.$$

Given that the matrices $A(t)$ are orthogonal, find the angular velocity.

#1254C When a rigid body (such as a spinning top) **precesses**, the fixed-in-space and fixed-in-body axes are related by

$$A(t) = \begin{pmatrix} \sin\alpha\cos\omega t & \cos\alpha\cos nt\cos\omega t - \sin nt\sin\omega t & -\cos\alpha\sin nt\cos\omega t - \cos nt\sin\omega t \\ \sin\alpha\sin\omega t & \cos\alpha\cos nt\sin\omega t + \sin nt\cos\omega t & -\cos\alpha\sin nt\sin\omega t + \cos nt\cos\omega t \\ \cos\alpha & -\sin\alpha\cos nt & \sin\alpha\sin nt \end{pmatrix},$$

where α and ω are constants. Show that the angular velocity satisfies

$$\begin{pmatrix} n\sin\alpha\cos\omega t \\ n\sin\alpha\sin\omega t \\ \omega+n\cos\alpha \end{pmatrix} = \omega \begin{pmatrix} 0 \\ 0 \\ 1 \end{pmatrix} + n \begin{pmatrix} \sin\alpha\cos\omega t \\ \sin\alpha\sin\omega t \\ \cos\alpha \end{pmatrix}.$$

This conveys that the top is spinning around the vertical axis at an angular speed ω and about the axis of the top at an angular speed n.

#1255C Let

$$\mathbf{e}_r = (\sin\theta\cos\phi, \sin\theta\sin\phi, \cos\theta)$$

where θ and ϕ are functions of time t.

(i) Show that $\mathbf{e}_r$ has unit length and that

$$\dot{\mathbf{e}}_r = \dot{\theta}\mathbf{e}_\theta + \dot{\phi}\sin\theta\mathbf{e}_\phi$$

for two unit vectors $\mathbf{e}_\theta$ and $\mathbf{e}_\phi$ which you should determine. Find similar expressions for $\dot{\mathbf{e}}_\theta$ and $\dot{\mathbf{e}}_\phi$.

(ii) Show that $\mathbf{e}_r$, $\mathbf{e}_\theta$, $\mathbf{e}_\phi$ form a right-handed orthonormal basis.

(iii) Find the angular velocity $\boldsymbol{\omega}$, in terms of $\mathbf{e}_r$, $\mathbf{e}_\theta$, $\mathbf{e}_\phi$, such that

$$\dot{\mathbf{e}}_r = \boldsymbol{\omega}\wedge\mathbf{e}_r, \qquad \dot{\mathbf{e}}_\theta = \boldsymbol{\omega}\wedge\mathbf{e}_\theta, \qquad \dot{\mathbf{e}}_\phi = \boldsymbol{\omega}\wedge\mathbf{e}_\phi.$$

5

Techniques of Integration

Remark 5.1 This chapter is aimed at those who have met integration to some extent before. The opening section seeks to define, somewhat rigorously, the notion of what an integral is, and the later sections are dedicated to slightly more advanced techniques of integration. Readers should be comfortable with the notion of integration as the inverse of differentiation, that a definite integral represents a signed area and with the integrals of some standard functions, especially polynomials.

Those wishing solely to improve their range of integration methods may wish to omit the first section. Those interested in investigating yet more advanced integration techniques might further consider Boros [5] and Nahin [24]. ∎

Notation 5.2 Let a, b be real numbers with $a \leqslant b$. We introduce here the following notation for **intervals** of the real line. We will write

$$[a, b] = \{x \in \mathbb{R} : a \leqslant x \leqslant b\}; \qquad (a, b) = \{x \in \mathbb{R} : a < x < b\};$$
$$(a, b] = \{x \in \mathbb{R} : a < x \leqslant b\}; \qquad [a, b) = \{x \in \mathbb{R} : a \leqslant x < b\}.$$

The interval $[a, b]$ is referred to as a **closed** interval and the interval (a, b) as **open**. Note by these definitions that the empty set is an interval.

5.1 History and Foundations*

Calculus is arguably humankind's single greatest invention from the last five hundred years. Put simply, calculus is the study of change and is typically split into two branches: *differential calculus* – the study of rates of change – and *integral calculus* – the study of accumulated changes. These two processes, differentiation and integration, are essentially inverses of one another, a fact made explicit in the *fundamental theorem of calculus* (Theorem 5.18). But such a description does little to warrant this paragraph's opening claim; it is the pervasiveness of calculus in mathematics, the physical sciences, and beyond, that has ultimately made calculus so important to our understanding of the world about us. The language of calculus and the use of differential equations (Chapter 6) are central to most mathematical descriptions of the real world including gravity (Newtonian and Einsteinian), electromagnetism (Maxwell's equations, wave equation), classical mechanics (Lagrange's and Hamilton's equations), quantum theory (Schrödinger's equation), fluid dynamics (Navier–Stokes equations), economics and finance (Black–Scholes), thermodynamics (heat equation) and mathematical biology (predator–prey and epidemiology models). "The calculus has served for three centuries as the principal quantitative language of Western science." (Edwards [12]).

Below we will give a reasonably formal introduction to the Riemann integral. In most of our examples, the integrals we consider will represent (signed) areas, which is certainly an important application of integration. However, integrals can much more generally represent quantities other than area. A natural generalization is in calculating volumes – in the same way that an integral of a function $y = f(x)$ can represent the area under that curve, a double integral of $z = f(x,y)$ can represent the (signed) volume under that surface. Integration is generally a means of summing up infinitesimal contributions. So if the *integrands* (the functions being integrated) represent other quantities then so will their integrals: if the integrand represents density, its integral will represent mass; if an integrand represents (infinitesimal) chances, its integral will represent a probability (see p.413); if the integrand represents temperature at a point then its integral will represent heat energy [1]; if the integrand represents an element of arc length then its integral will represent the length of a curve (see p.416).

Still more generally, it is very natural to integrate over regions that are not simply intervals or rectangles (in the case of repeated integrals – p.395). For example, if a particle were moving in a gravitational field then a *line integral* along its journey would calculate the work done by/against gravity. A *surface integral* might be used to calculate the surface area of a sphere. Or if the integrand represented the velocity of a fluid, then a *flux integral* over a surface would represent the rate at which fluid crossed that surface. We also noted on p.60 that integrals of complex functions can be determined and complex analysis is a rich and highly applicable area of mathematics.

We will spend a little time now looking to define the integral of a function $f(x)$ on the interval $[a,b]$. Without the rigorous language of analysis it will be impossible to 'dot every i' but the following includes all the main ideas involved in defining the **Riemann integral.** As a starting point for defining integrals, it is hopefully uncontroversial to define the area of a rectangle with sides x and y to be xy. Given we also wish our integral to be linear, this leads us naturally to a family of functions, the *step functions*.

Definition 5.3 (a) By a **subinterval** of the interval $[a,b]$ we shall mean a subset which is itself an interval. That is, a subinterval has one of the following forms:

$$[c,d], \qquad (c,d), \qquad (c,d], \qquad [c,d),$$

where $a \leqslant c \leqslant d \leqslant b$. Note again that this includes the empty set as a subinterval. For each of the subintervals I above we define its **length** to be $l(I) = d - c$.

(b) Given a subinterval I, its **indicator function** (or **characteristic function**) is the function $\mathbf{1}_I$ defined by

$$\mathbf{1}_I(x) = \begin{cases} 1 & \text{if } x \text{ is in } I, \\ 0 & \text{if } x \text{ is not in } I. \end{cases}$$

---

[1] More precisely, the heat energy will be the integral of the product of temperature, density and specific heat of the material.

(c) A function ϕ is said to be a **step function** if it can be written as a linear combination of indicator functions of subintervals. That is, if ϕ can be written

$$\phi = \sum_{k=1}^{n} y_k \mathbf{1}_{I_k} \tag{5.1}$$

for some real constants $y_1, \ldots, y_n$ and some subintervals $I_1, \ldots, I_n$.

A Brief History of Calculus: Integration

The invention of calculus is traditionally credited to Isaac Newton (1643–1727) and Gottfried Leibniz (1646–1716), though their progress was very much built on the work of others (Pierre de Fermat and John Wallis, being notable examples) and a rigorous conclusion to the story would take us at least into the nineteenth century (with the work of Karl Weierstrass (1815–1897) or later still with the work of Henri Lebesgue (1875–1941)).

However whilst calculus may only be three centuries old, the questions it addresses are as old as mathematics itself, in particular the calculation of areas and volumes. The ancient Greeks' *method of exhaustion*, albeit more limited in scope, was in fact more rigorous than anything the nascent calculus could muster until the nineteenth century. Exhaustion involves inscribing regions of known area within an unknown area until all strictly lower possibilities for the area are eliminated or 'exhausted'; the method was used to great effect by Archimedes.

In fact, during the eighteenth century, integration was viewed as little more than the inverse process of differentiation, and commonly its development was stymied not by lack of imagination in solving problems, rather by a perceived lack of need to ask the right questions. Much of the problem related to the limited notion of a function during the eighteenth century. Euler himself, in his *Introduction to Analysis of the Infinite*, had defined a function as a single analytical expression. What had been seminal in the eighteenth century, the setting of functions (rather than curves and other geometric objects), as a centrepiece of mathematics, was due a radical upgrade a century later; a more inclusive notion of function was becoming necessary.

Cauchy defined in 1823 a definite integral for continuous functions and gave a (not wholly rigorous) proof of the fundamental theorem of calculus for such functions. It was more the limited concept of function in the early nineteenth century, rather than any technical difficulties, that meant his integral dealt solely with continuous functions. However, the work of Joseph Fourier (1768–1830) (see p.328), expressing solutions to the heat equation in terms of infinite sums of trigonometric functions, would significantly extend the need for a broader concept of function.

Riemann's integral embraced arbitrary bounded functions, though it technically is not at much of a remove from Cauchy's integral. And in fact the (equivalent) formulation given in Definition 5.7 in terms of lower and upper integrals is due to Jean-Gaston

Darboux (1842–1917) et al. This reformulation was important in that it naturally generalizes – for example, to Jordan's notion of *content* where, say, regions of the plane have defined area if the areas of polygons contained in the set and areas of polygons that contain the set can be made arbitarily close.

In the latter half of the nineteenth century, mathematicians were becoming aware of pathological examples stretching (or breaking) certain self-evident notions that had gone too long without formal definition. Previous ideas of area, dimension, sets, even proof, were all being tested. The theory of integration, and related notions of *measure* (e.g. length, area, volume), largely reached a conclusion with the work of Henri Lebesgue (1875–1941). The **Lebesgue integral** has major advantages over the Riemann integral: proper integrals on $(0, \infty)$ can be defined, rather than considered as limits of other integrals; there are powerful convergence theorems (see below right) for dealing with sequences of functions; the theory encompasses a much greater range of functions than Riemann's theory – for example, the Dirichlet function (Example 5.15) is Lebesgue integrable, its integral being zero. In fact any function that can be constructively written down is *Lebesgue measurable* and can be handled within the theory; there are non-measurable functions and sets, but their existence can be demonstrated only by using assumptions beyond standard set theory, such as the *axiom of choice*.

We conclude with a statement of the Banach–Tarski paradox proven in 1924 by two Polish mathematicians, Stefan Banach (1892–1945) and Alfred Tarski (1901–1983). This gives a real sense of the foundational issues that mathematicians were facing in the late nineteenth and early twentieth centuries. It states that it is possible to decompose a solid sphere into as few as five parts and to rearrange those five parts, using only rigid motions (isometries), to reproduce two solid spheres of the same size as the original. The result is paradoxical as we cannot envision how volume might be doubled, but the pieces that the sphere is decomposed into are so pathological as to have no defined volume and consequently there need be no invariance in the volume. Perhaps surprisingly, this is not true in two dimensions – this is because when a disc is decomposed into finitely many pieces, each piece can be assigned an area and so area is preserved overall. For a very readable history of the calculus see Edwards [12].

Monotone Convergence Theorem Let f_n be a sequence of (Lebesgue) integrable functions with $f_{n+1}(x) \geqslant f_n(x)$, for all x and all n. Say there is a K such that

$$\int_{-\infty}^{\infty} f_n(x)\,dx < K \qquad \text{for all } n.$$

Then there is an integrable function $f(x)$ such that $f_n(x) \to f(x)$ as $n \to \infty$ (except possibly on some *null* set which doesn't affect the value of the integrals) and such that

$$\int_{-\infty}^{\infty} f(x)\,dx = \lim_{n \to \infty} \int_{-\infty}^{\infty} f_n(x)\,dx.$$

Bernhard Riemann (Bettmann/ Getty Images)

Bernhard Riemann (1826–1866) Despite a relatively short life spent in poor health, Riemann made lasting, revolutionary contributions to mathematics, particularly in the fields of geometry, analysis and number theory. He was arguably 'the first modern mathematician' (James [20, p.188]) introducing profound, almost visionary, alternative ways of addressing fundamental areas of mathematics – for example, Riemannian geometry is the language of general relativity and of curved space more generally. In number theory Riemann wrote only one paper but it included what is currently the most celebrated open problem in

mathematics, the *Riemann hypothesis*. This is important to understanding the distributions of the primes, and made both Hilbert's list of 23 problems in 1900 and the list of Clay Institute's list of Millennium Problems in 2000. The *Riemann zeta function*

$$\zeta(s) = \sum_{n=1}^{\infty} \frac{1}{n^s} = \prod_{p \text{ prime}} \left(1 - \frac{1}{p^s}\right)^{-1}$$

had previously been studied by Euler. This infinite series converges for $s > 1$ and can also be made sense of when s is a complex number with $\text{Re}(s) > 1$. Now considering $\zeta(s)$ as a function of a complex variable, Riemann was able to extend its definition to all complex $s \neq 1$, and showed that

$$\zeta(s) = 2^s \pi^{2-1} \sin\left(\frac{\pi s}{2}\right) \Gamma(1-s)\zeta(1-s)$$

where Γ denotes the gamma function. One can then see that $\zeta(s) = 0$ for the values $s = -2, -4, -6, \ldots$, these being the so-called *trivial zeros*. (Note that it does not follow that $\zeta(2n)$ is zero at even positive integers as $\Gamma(1-2n)$ is infinite.) Using the original definition and above functional relation, one can show all other zeros are in the strip $0 < \text{Re}(s) < 1$. Riemann conjectured that in fact all these *non-trivial zeros* lie on the *critical line* $\text{Re}(s) = 1/2$. These zeros are important to number theory as Riemann gave an explicit formula for $\pi(n)$, the numbers of primes less than n, in terms of the zeros of the zeta function. Hardy and Littlewood showed in 1921 that there are infinitely many zeros on the critical line and computationally the first trillion zeros have been found to be on the critical line, but the hypothesis remains unproven.

Another significant contribution was the idea of a *Riemann surface*. In the same way that $f(x,y) = 0$ typically defines a curve in $\mathbb{R}^2$ when x and y are real variables, then such an equation defines a two-dimensional surface in four-dimensional $\mathbb{C}^2$ when x and y are *complex* variables, It may well be that for given x there are many y that solve the equation; one alternative is to take a set of principal values for y, a so-called *branch*. So, as seen in #73, the equation $y^2 = x$ has two choices y_1, y_2 of y, for any non-zero x, which can be defined in a cut-plane and as we cross the cut we move from one choice to the other. We might instead seek to consider all choices together which is the idea behind the Riemann surface $\{(x,y) \in \mathbb{C}^2 : y^2 = x\}$. The set of points (x, y_1) forms half the Riemann surface, the other branch forming the other half. As we move from one branch to another we simply continue moving around the Riemann surface. Essentially the Riemann surface is the more fundamental object, and the two branches are just arbitrary ways to carve up the surface into two parts. The theory of Riemann surfaces is a rich area of mathematics connecting algebra, analysis, topology and geometry with many applications including to string theory.

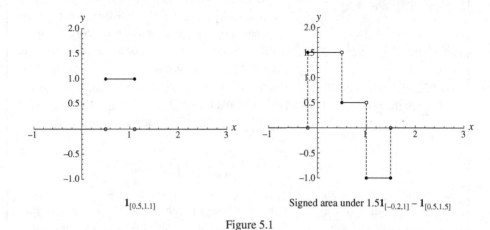

$$\mathbf{1}_{[0.5,1.1]}$$ Signed area under $1.5\mathbf{1}_{[-0.2,1]} - \mathbf{1}_{[0.5,1.5]}$

Figure 5.1

Given our earlier comment on the area of a rectangle, we define the indicator function $\mathbf{1}_I$ to have integral $l(I)$, If our integral is to be linear this next definition necessarily follows.

Definition 5.4 Given the step function ϕ in (5.1), we define its integral as

$$\int_a^b \phi(x)\,dx = \sum_{k=1}^n y_k l(I_k). \tag{5.2}$$

#1256
#1257
#1259
#1260
So the first step function in Figure 5.1 has integral $1 \times (1.1 - 0.5) = 0.6$ and the second has integral $1.5 \times 1.2 - 1 \times 1 = 0.8$. Note that the second step function can also be written as

$$1.5\mathbf{1}_{[-0.2,0.5)} + 0.5\mathbf{1}_{[0.5,1)} - \mathbf{1}_{[1,1.5]}$$

which has integral

$$1.5 \times 0.7 + 0.5 \times 0.5 - 1 \times 0.5 = 0.8.$$

So there is, unfortunately, a potential problem with the above definition. A given step function ϕ will be expressible as a linear combination of indicator functions in a number of ways; for each such expression, the RHS of (5.2) leads to a sum defining the integral of ϕ. If the integral of the given step function is to be well defined, all these sums must be equal. We state this result here though we leave the proof to #1263.

Proposition 5.5 (*Integral of a Step Function is Well Defined*) *Let ϕ be a step function expressible as*

$$\phi = \sum_{k=1}^{n} y_k 1_{I_k} = \sum_{k=1}^{N} Y_k 1_{J_k}$$

for real numbers $y_1, \ldots, y_n, Y_1, \ldots, Y_N$ and subintervals $I_1, \ldots, I_n, J_1 \ldots, J_N$, Then

$$\sum_{k=1}^{n} y_k l(I_k) = \sum_{k=1}^{N} Y_k l(J_k).$$

The following properties are also true of the integral we have just defined.

Proposition 5.6 *Let ϕ and ψ be step functions on the interval $[a,b]$, and $a < c < b$.*

(a) *For real numbers λ, μ then $\lambda \phi + \mu \psi$ is a step function and*

$$\int_a^b [\lambda \phi(x) + \mu \psi(x)] \, dx = \lambda \int_a^b \phi(x) \, dx + \mu \int_a^b \psi(x) \, dx.$$

(b) *If $\phi(x) \leqslant \psi(x)$ for all x then*

$$\int_a^b \phi(x) \, dx \leqslant \int_a^b \psi(x) \, dx.$$

(c) *The restriction of ϕ to the subinterval $[a,c]$ is a step function on that subinterval and*

$$\int_a^b \phi(x) \, dx = \int_a^c \phi(x) \, dx + \int_c^b \phi(x) \, dx.$$

Proof (a) By #1256(ii) there are subintervals $I_1, \ldots, I_n$ and reals $y_1, \ldots, y_n, Y_1, \ldots, Y_n$ such that

$$\phi = \sum_{k=1}^{n} y_k 1_{I_k} \qquad \text{and} \qquad \psi = \sum_{k=1}^{n} Y_k 1_{I_k}. \tag{5.3}$$

We see $\lambda \phi + \mu \psi = \sum_{k=1}^{n} (\lambda y_k + \mu Y_k) 1_{I_k}$ is a step function and

$$\int_a^b [\lambda \phi(x) + \mu \psi(x)] \, dx = \sum_{k=1}^{n} (\lambda y_k + \mu Y_k) l(I_k)$$

$$= \lambda \sum_{k=1}^{n} y_k l(I_k) + \mu \sum_{k=1}^{n} Y_k l(I_k) = \lambda \int_a^b \phi(x) \, dx + \mu \int_a^b \psi(x) \, dx.$$

(b) Writing ϕ and ψ as in (5.3), as $\phi \leqslant \psi$ then $y_k \leqslant Y_k$ for each k and hence

$$\int_a^b \phi(x) \, dx = \sum_{k=1}^{n} y_k l(I_k) \leqslant \sum_{k=1}^{n} Y_k l(I_k) = \int_a^b \psi(x) \, dx.$$

(c) Note that each subinterval I_k can be written as the union of the two subintervals

$$J_k = \{x \in I_k : x \leqslant c\} \qquad \text{and} \qquad K_k = \{x \in I_k : x \geqslant c\},$$

and note that $l(I_k) = l(J_k) + l(K_k)$. Then the restriction of ϕ to $[a,c]$ is $\sum_{k=1}^{n} y_k \mathbf{1}_{J_k}$, and so is a step function. Further, we have

$$\int_a^c \phi(x)\,dx + \int_c^b \phi(x)\,dx = \sum_{k=1}^{n} y_k l(J_k) + \sum_{k=1}^{n} y_k l(K_k)$$

$$= \sum_{k=1}^{n} y_k(l(J_k) + l(K_k)) = \sum_{k=1}^{n} y_k l(I_k) = \int_a^b \phi(x)\,dx. \qquad \square$$

We are finally in a position to define what it means for a bounded function f on an interval to be *integrable*. For such f we are seeking to define a number $\int_a^b f(x)\,dx$ and so far we have only defined integrals for step functions. However, based on Proposition 5.6(b) we would expect, for step functions ϕ and ψ with $\phi(x) \leqslant f(x) \leqslant \psi(x)$, that

$$\int_a^b \phi(x)\,dx \leqslant \int_a^b f(x)\,dx \leqslant \int_a^b \psi(x)\,dx. \tag{5.4}$$

This leads us to our definition.

Definition 5.7 Let f be a bounded function on $[a,b]$. We say that f is **integrable** [2] if the requirement (5.4) uniquely specifies $\int_a^b f(x)\,dx$ as a real number, when the inequalities are considered over all step functions ϕ and ψ satisfying $\phi \leqslant f \leqslant \psi$ on $[a,b]$.

Definition 5.8 For $a > b$ and f integrable on $[a,b]$, we **define** $\int_a^b f(x)\,dx = -\int_b^a f(x)\,dx$.

Example 5.9 Show, for $a > 0$, that

$$\int_0^a x^2\,dx = \frac{a^3}{3}. \tag{5.5}$$

Solution

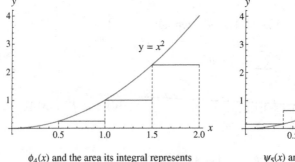

$\phi_4(x)$ and the area its integral represents

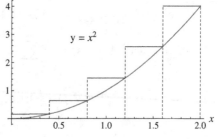

$\psi_5(x)$ and the area its integral represents

Figure 5.2

#1282
#1283

Let n be a positive integer and define the step functions ϕ_n and ψ_n for $0 \leqslant x < a$ by

$$\phi_n(x) = \left(\frac{(k-1)a}{n}\right)^2 \quad \text{on} \quad \left[\frac{(k-1)a}{n}, \frac{ka}{n}\right);$$

$$\psi_n(x) = \left(\frac{ka}{n}\right)^2 \quad \text{on} \quad \left[\frac{(k-1)a}{n}, \frac{ka}{n}\right),$$

where $k = 1, 2, \ldots n$. (Such step functions are graphed in Figure 5.2.) Note that we have $\phi_n(x) \leqslant x^2 \leqslant \psi_n(x)$ for each n and each x in $[0, a]$. As each subinterval has length a/n then by Definition 5.4 we have

$$\int_0^a \phi_n(x)\,dx = \sum_{k=1}^{n}\left(\frac{(k-1)a}{n}\right)^2 \frac{a}{n} = \frac{a^3}{n^3}\sum_{k=0}^{n-1}k^2 = \frac{(n-1)n(2n-1)a^3}{6n^3}$$

$$= \left(1 - \frac{1}{n}\right)\left(2 - \frac{1}{n}\right)\frac{a^3}{6}, \tag{5.6}$$

using the summation formula from #243. Likewise

$$\int_0^a \psi_n(x)\,dx = \sum_{k=1}^{n}\left(\frac{ka}{n}\right)^2 \frac{a}{n} = \left(1 + \frac{1}{n}\right)\left(2 + \frac{1}{n}\right)\frac{a^3}{6}.$$

If the integral in (5.5) exists then we would have, for all n, that

$$\left(1 - \frac{1}{n}\right)\left(2 - \frac{1}{n}\right)\frac{a^3}{6} \leqslant \int_0^a x^2\,dx \leqslant \left(1 + \frac{1}{n}\right)\left(2 + \frac{1}{n}\right)\frac{a^3}{6}. \tag{5.7}$$

Note that the LHS of (5.7) is less than $a^3/3$ but approximates to $a^3/3$ as n increases and the RHS exceeds $a^3/3$ but approximates ever more closely to that value as n increases. The only real number satisfying these inequalities for all n is $a^3/3$. □

Remark 5.10 Note that the functions ϕ_n and ψ_n were defined on $[0, a)$ rather than on $[0, a]$. We could have extended them by setting $\phi_n(a) = \psi_n(a) = a^2$, but this is not really an 'i worth dotting'. We instead might recognize that the notions of being integrable on $[0, a]$, $[0, a)$, $(0, a]$, $(0, a)$ each lead *essentially* to the same functions. See #1264. ■

Remark 5.11 (Dummy Variables) In (5.5) the variable x is a *dummy variable* and can be thought of as ranging over the interval $0 \leqslant x \leqslant a$. As x varies the integral is (loosely put) a continuous sum of all the infinitesimal contributions $x^2\,dx$. It is just as true that

$$\int_0^a u^2\,du = \frac{a^3}{3}$$

with the new dummy variable u 'counting off' these contributions in the same manner as x does as it varies over the same interval. This is entirely comparable with the sums

$$\sum_{k=1}^{n} k = \frac{n(n+1)}{2} = \sum_{j=1}^{n} j.$$

The dummy variables (k and j) in these sums have the same role varying over the integers $1, 2, \ldots, n$. The kth term might refer to any of the terms in the first sum, whilst the nth term can only refer to the last one. It is therefore wrong to ever write sums or integrals such as

$$\sum_{k=1}^{k} k \qquad \text{or} \qquad \int_0^x x^2 \, dx,$$

as the dummy variables (k and x) cannot denote both a general number in a particular range and the upper limit of that range. ∎

Reassuringly, the properties of the integral which we proved earlier still hold for integrable functions, namely:

Proposition 5.12 *(Properties of the Integral) Let f and g be integrable functions on $[a,b]$, and $a < c < b$.*

(a) For real numbers λ, μ, then $\lambda f + \mu g$ is integrable and

$$\int_a^b [\lambda f(x) + \mu g(x)] \, dx = \lambda \int_a^b f(x) \, dx + \mu \int_a^b g(x) \, dx.$$

(b) If $f(x) \leqslant g(x)$ for all x then $\int_a^b f(x) \, dx \leqslant \int_a^b g(x) \, dx$.

(c) The restriction of f to $[a,c]$ is integrable, and $\int_a^b f(x) \, dx = \int_a^c f(x) \, dx + \int_c^b f(x) \, dx$.

Notation 5.13 If we are being pedantic then we should note that f is not a function on $[a,c]$ but rather a function on $[a,b]$. When we write $\int_a^c f(x) \, dx$, this is a shorthand for $\int_a^c \check{f}(x) \, dx$ where $\check{f}$ denotes the restriction of f to $[a,c]$.

Remark 5.14 Several times in the next proof, we will make use of the following facts:

- The only real number x, which satisfies $0 \leqslant x < \varepsilon$ for all $\varepsilon > 0$, is $x = 0$.
- The only real number y, satisfying $x - \varepsilon < y < x + \varepsilon$ for all $\varepsilon > 0$, is $y = x$.

So an alternative equivalent criterion for a function f being integrable on $[a,b]$ is this: for all $\varepsilon > 0$ there are step functions ϕ and ψ such that

$$\phi(x) \leqslant f(x) \leqslant \psi(x) \text{ for all } x \text{ in } [a,b] \qquad \text{and} \qquad 0 \leqslant \int_a^b (\psi(x) - \phi(x)) \, dx < \varepsilon. \qquad \blacksquare$$

#1265
#1266
#1268
#1274
#1276
#1277

Proof (a) For any required accuracy $\varepsilon > 0$, there are step functions ϕ and ψ such that $\phi \leqslant f \leqslant \psi$ and

$$\left(\int_a^b f(x) \, dx \right) - \varepsilon \leqslant \int_a^b \phi(x) \, dx \leqslant \int_a^b f(x) \, dx \leqslant \int_a^b \psi(x) \, dx \leqslant \left(\int_a^b f(x) \, dx \right) + \varepsilon. \quad (5.8)$$

We can find similar step functions $\tilde{\phi}$ and $\tilde{\psi}$ for g. Note then that

$$\phi(x) + \tilde{\phi}(x) \leqslant f(x) + g(x) \leqslant \psi(x) + \tilde{\psi}(x),$$

with $\phi + \tilde{\phi}$ and $\psi + \tilde{\psi}$ being step functions. So if any real number I satisfies

$$\int_a^b \left[\phi(x) + \tilde{\phi}(x) \right] dx \leqslant I \leqslant \int_a^b \left[\psi(x) + \tilde{\psi}(x) \right] dx,$$

then

$$\left(\int_a^b f(x) \, dx \right) + \left(\int_a^b g(x) \, dx \right) - 2\varepsilon \leqslant I \leqslant \left(\int_a^b f(x) \, dx \right) + \left(\int_a^b g(x) \, dx \right) + 2\varepsilon.$$

The only real number meeting these inequalities for all $\varepsilon > 0$ is $I = \int_a^b f(x)\,dx + \int_a^b g(x)\,dx$. This uniqueness means that $f(x) + g(x)$ is integrable and that

$$\int_a^b [f(x) + g(x)]\,dx = \int_a^b f(x)\,dx + \int_a^b g(x)\,dx,$$

showing the additivity of the integral. Now suppose that $\lambda > 0$. Then $\lambda\phi$ and $\lambda\psi$ are step functions such that

$$\lambda\phi(x) \leqslant \lambda f(x) \leqslant \lambda\psi(x)$$

and by (5.8) and the linearity of the integral for step functions

$$\lambda\left(\int_a^b f(x)\,dx\right) - \lambda\varepsilon \leqslant \int_a^b \lambda\phi(x)\,dx \leqslant \int_a^b \lambda\psi(x)\,dx \leqslant \lambda\left(\int_a^b f(x)\,dx\right) + \lambda\varepsilon.$$

The only real number J that, for all $\varepsilon > 0$, satisfies

$$\lambda\left(\int_a^b f(x)\,dx\right) - \lambda\varepsilon \leqslant J \leqslant \lambda\left(\int_a^b f(x)\,dx\right) + \lambda\varepsilon,$$

is $J = \lambda\int_a^b f(x)\,dx$ and so we have shown that λf is integrable and that

$$\int_a^b \lambda f(x)\,dx = \lambda\int_a^b f(x)\,dx. \tag{5.9}$$

Now (5.9) is clear when $\lambda = 0$ and it can be shown to hold for $\lambda < 0$ in a similar fashion to the $\lambda > 0$ argument above, noting carefully the inequality reversing effect of multiplying by a negative number (#1266).

(b) Say now that f and g are integrable functions with $f \leqslant g$. Then $g - f$ is non-negative and integrable by (a). So 0, the zero function, is a step function such that $0 \leqslant g - f$. By the definition of being integrable and by part (a), we have

$$0 = \int_a^b 0\,dx \leqslant \int_a^b (g(x) - f(x))\,dx = \int_a^b g(x)\,dx - \int_a^b f(x)\,dx$$

and the result follows.

(c) Given any function ψ on $[a,b]$, we might consider its restriction $\check{\psi}$ to $[a,c]$. Note that if ψ is a step function then so is $\check{\psi}$. We aim to show that $\check{f}$ is integrable. From (5.8) we have

$$0 \leqslant \int_a^b \psi(x)\,dx - \int_a^b \phi(x)\,dx < 2\varepsilon,$$

and, as $\phi \leqslant \psi$, it follows from Proposition 5.6(b) and (c) that

$$0 \leqslant \int_a^c \check{\psi}(x)\,dx - \int_a^c \check{\phi}(x)\,dx < 2\varepsilon. \tag{5.10}$$

So we have that $\check{\phi} \leqslant \check{f} \leqslant \check{\psi}$ and, because of (5.10), there is only one real number satisfying

$$\int_a^c \check{\phi}(x)\,dx \leqslant I \leqslant \int_a^c \check{\psi}(x)\,dx$$

for all such restrictions of step functions $\check{\phi}$ and $\check{\psi}$. So $\check{f}$ is integrable and this number I is $\int_a^c \check{f}(x)\,dx$ by definition. Finally, given a function g on $[a,b]$ we can define

$$g_1(x) = \begin{cases} g(x) & \text{if } a \leqslant x \leqslant c, \\ 0 & \text{if } c < x \leqslant b, \end{cases} \qquad g_2(x) = \begin{cases} 0 & \text{if } a \leqslant x \leqslant c, \\ g(x) & \text{if } c < x \leqslant b, \end{cases}$$

noting $g = g_1 + g_2$. It is then the case that g_1 is integrable on $[a,b]$ and

$$\int_a^b g_1(x)\,dx = \int_a^c \check{g}(x)\,dx = \int_a^c g(x)\,dx \tag{5.11}$$

– the details are left to #1267 – with a similar result holding for g_2. By the additivity of the integral we then have

$$\int_a^b g(x)\,dx = \int_a^b g_1(x)\,dx + \int_a^b g_2(x)\,dx = \int_a^c g(x)\,dx + \int_c^b g(x)\,dx. \qquad \square$$

For a given bounded function f on $[a,b]$, there will always be at least one real number [3] satisfying (5.4) but it need not be *uniquely* specified, as we shall see with the following example. Consequently the following function is *not* integrable. [4]

Example 5.15 (Dirichlet Function) Define the function f on $[0,1]$ by

$$f(x) = \begin{cases} 1 & x \text{ is rational}, \\ 0 & x \text{ is irrational}. \end{cases}$$

If ϕ and ψ are step functions satisfying $\phi \leqslant f \leqslant \psi$ then

$$\int_a^b \phi(x)\,dx \leqslant 0 \qquad \text{and} \qquad 1 \leqslant \int_a^b \psi(x)\,dx. \tag{5.12}$$

Consequently the inequalities in (5.4) are satisfied by all real numbers between 0 and 1.

Solution Let $\phi = \sum_{k=1}^n y_k \mathbf{1}_{I_k}$ and assume that the subintervals I_k are disjoint (#1256). If the subinterval I_k has positive length then it will contain an irrational number and hence $y_k \leqslant 0$ and so $y_k l(I_k) \leqslant 0$. If the subinterval I_k has zero length then $y_k l(I_k) = 0$. The first inequality in (5.12) follows.

Similarly let $\psi = \sum_{k=1}^N Y_k \mathbf{1}_{J_k}$ and assume again that the subintervals J_k are disjoint. If the subinterval J_k has positive length then it will contain a rational number and in particular $Y_k \geqslant 1$. Consequently $\psi(x) < 1$ is only possible on finitely many subintervals J_k of zero length (i.e. at single points) and we can set $\psi(x) = 1$ at these points without changing the integral of ψ. So we may assume $\psi \geqslant 1$ on the entirety of $[0,1]$ and the second inequality in (5.12) follows. $\square$

So not every bounded function on $[a,b]$ is (Riemann) integrable. However, the definition does apply to a wide range of functions. In particular, we have:

Theorem 5.16 *A continuous function on $[a,b]$ is integrable.*

---

[3] This is a consequence of the *completeness axiom* for real numbers.
[4] Dirichlet (see p.122 for a brief biographical note) demonstrated that this function is not integrable in 1829.

Proving this result is not possible without some proper analysis.[5] Very loosely speaking, a function is continuous if its graph can be drawn without taking one's pen off the paper. Somewhat more rigorously a function $f(x)$ is continuous at a point c if

$$f(x) \to f(c) \quad \text{as } x \to c \qquad \text{which is also written} \qquad \lim_{x \to c} f(x) = f(c), \qquad (5.13)$$

conveying a sense that as x better approximates c then $f(x)$ better approximates $f(c)$. But in most ways this is just hiding important details behind notation. (See p.358 for further details on the rigorous definition of a limit.) Importantly, any function that we can differentiate is continuous. Thus we immediately know that a wide variety of functions – polynomials, exponentials, logarithms, trigonometric – are integrable, as will be sums, products, quotients, compositions of these when the product, quotient and chain rules of differentiation apply.

The most important theorem for us in determining integrals will be the *fundamental theorem of calculus*. We begin first with the following result that shows differentiation and integration are inverse processes.

Proposition 5.17 *Let $f(x)$ be a continuous function on $[a,b]$. Then*

$$\frac{d}{dX} \int_a^X f(x)\,dx = f(X) \qquad \text{for } a < X < b.$$

Proof *Sketch Proof*: (See Binmore [4, Theorem 13.12] for a rigorous proof.) We define

$$F(X) = \int_a^X f(x)\,dx$$

for X in the interval $a < X < b$. If we consider a small positive number ε such that $X + \varepsilon \leqslant b$ then

$$F(X + \varepsilon) - F(X) = \int_a^{X+\varepsilon} f(x)\,dx - \int_a^X f(x)\,dx = \int_X^{X+\varepsilon} f(x)\,dx.$$

Rather loosely speaking, as f is continuous at $x = X$ and if ε is suitably small, then we can approximate $f(x)$ as $f(X)$ for x in the small interval $X \leqslant x \leqslant X + \varepsilon$. Hence

$$F(X + \varepsilon) - F(X) \approx \int_X^{X+\varepsilon} f(X)\,dx = \varepsilon f(X),$$

and hence

$$\frac{F(X + \varepsilon) - F(X)}{\varepsilon} \approx f(X),$$

with the approximation becoming exact in the limit. Thus as ε approaches 0 we have

$$\frac{d}{dX} \int_a^X f(x)\,dx = F'(X) = \lim_{\varepsilon \to 0} \left(\frac{F(X + \varepsilon) - F(X)}{\varepsilon} \right) = f(X).$$

See p.429 for a discussion of the rigorous definition of a derivative. In fact, what the above sketch proof shows is that the right-hand derivative of F at $x = X$ is $f(X)$. An almost identical proof shows that the left-hand derivative of F at $x = X$ is also $f(X)$, and so it is correct to write $F'(X) = f(X)$. □

---

[5] Most texts covering a first course in analysis – such as Binmore [4, Chapter 13] – contain such material.

Thus we may finally conclude:

Theorem 5.18 *(**Fundamental Theorem of Calculus** or **FTC**) Let $f(x)$ be a differentiable function on $[a,b]$ with continuous derivative $f'(x)$. Then*

$$\int_a^b f'(x)\,dx = f(b) - f(a).$$

Proof We define

$$F(X) = \int_a^X f'(x)\,dx,$$

so that, by Proposition 5.17, $F'(X) = f'(X)$ for $a < X < b$. Thus

$$\frac{d}{dX}(F(X) - f(X)) = 0 \quad \text{for} \quad a < X < b.$$

The only functions with zero derivative are constant functions Binmore [4, Thm 11.7]. So there is a real number c such that

$$F(X) - f(X) = c \quad \text{for} \quad a \leqslant X \leqslant b.$$

As $F(a) = 0$ by definition, then $c = -f(a)$. Finally, setting $X = b$ we have

$$\int_a^b f'(x)\,dx = F(b) = f(b) + c = f(b) - f(a),$$

as required. □

We shall see this theorem applied throughout the following sections. By way of an example, as $\sin'(x) = \cos x$, we have for any a that

$$\int_a^X \cos x\,dx = \sin X - \sin a.$$

In fact, as we may well have no particular preference for a choice of a, then we often write

$$\int \cos x\,dx = \sin x + \text{const.}$$

which conveys the fact that the only functions which differentiate to $\cos x$ are of the form $\sin x + c$ for a constant c. This is the so-called **indefinite integral**. Such a function that differentiates to $\cos x$ is called an **antiderivative** or **primitive** of $\cos x$. In general, we may choose to leave the constant term unspecified or can provide information about the value $f(x)$ takes at a specific value of x in order to determine this constant. By contrast, the integrals which we were previously discussing, where the limits a and b were specified, are known as **definite integrals**.

Definition 5.7 relates only to bounded functions on an interval $[a,b]$ and so does not include integrals on an unbounded interval or of unbounded functions on a bounded interval. So, for example, we cannot readily address integrals such as

$$\int_0^\infty e^{-x}\,dx, \qquad \int_1^\infty \frac{dx}{x^2}, \qquad \int_0^1 \frac{dx}{\sqrt{x}}.$$

From the fundamental theorem of calculus we have, for $a > 0$, that

$$\int_0^a e^{-x}\,dx = 1 - e^{-a} \qquad \text{and} \qquad \int_1^a \frac{dx}{x^2} = 1 - \frac{1}{a}. \tag{5.14}$$

If a increases without bound then both RHSs approximate to 1 and so we might feel we can reasonably write

$$\int_0^\infty e^{-x}\,dx = 1 \qquad \text{and} \qquad \int_1^\infty \frac{dx}{x^2} = 1. \tag{5.15}$$

The above are examples of **improper (Riemann) integrals**. Definition 5.7 does not extend to the integrals in (5.15), but we might reasonably consider them as the limits of the proper integrals in (5.14). The values assigned the integrals in (5.15) are the only values that might sensibly be given to those integrals and there are more general theories of integration – most notably the Lebesgue integral – which treat the integrals in (5.15) as well-defined integrals in themselves, rather than as limits of such integrals. (See #1282 for a treatment of the third integral.)

Exercises

#1256A(i) Let ϕ, ψ be step functions on $[a,b]$. Show that there are disjoint subintervals $I_1,\ldots,I_n$ and real numbers $y_1,\ldots,y_n$ such that $\phi = \sum_{k=1}^n y_k \mathbf{1}_{I_k}$.
(ii) Show that there are disjoint subintervals $J_1,\ldots,J_N$ and scalars $y_1,\ldots,y_N,Y_1,\ldots,Y_N$ such that

$$\phi = \sum_{k=1}^N y_k \mathbf{1}_{J_k} \qquad \text{and} \qquad \psi = \sum_{k=1}^N Y_k \mathbf{1}_{J_k}.$$

#1257a† Let ϕ and ψ be step functions on $[a,b]$. Show that the following are also step functions

$$\phi + \psi, \qquad \phi - \psi, \qquad \phi\psi, \qquad |\phi|, \qquad \max\{\phi,\psi\}, \qquad \min\{\phi,\psi\}.$$

#1258C† Show that if $\phi(x)$ is a step function then so is $\phi(x^2)$.

#1259a Show that the following step functions are equal. Verify that their integrals agree.

$$\phi = 2 \times \mathbf{1}_{[0,3]} - \mathbf{1}_{(1,2]}, \qquad \psi = 2 \times \mathbf{1}_{[0,1]} + \mathbf{1}_{(1,2]} + 2 \times \mathbf{1}_{(2,3]}.$$

#1260B† Which of the following functions are step functions?

$$\mathbf{1}_{(0,\infty)} - \mathbf{1}_{(1,\infty)}, \qquad \sum_{k=1}^\infty \frac{(-1)^k}{k}\mathbf{1}_{[1/(k+1),1/k)}, \qquad \sum_{k=1}^\infty \frac{1}{2^k}\mathbf{1}_{[0,1]}.$$

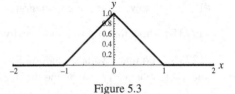

Figure 5.3

#1261D† Let $f(x)$ be the function graphed in Figure 5.3 (with $f(x) = 0$ for $|x| > 2$).
(i) Find an antiderivative $F(x)$ such that $F'(x) = f(x)$ for all x.
(ii) Determine the following integrals:

$$\int_{-1}^1 f(x)\,dx, \quad \int_{-1}^1 f(x)^2\,dx, \quad \int_{-1}^1 f(x^2-1)\,dx, \quad \int_{-\infty}^\infty f(x^2+1)\,dx, \quad \int_{-\infty}^\infty f(x^2)\,dx.$$

#1262c† Prove that a step function on an interval is integrable – according to Definition 5.7 – and that its integral agrees with Definition 5.4.

#1263D† Prove Proposition 5.5.

#1264C In a like manner to Definition 5.7, we can say what it is for a bounded function on $[a,b)$ or $(a,b]$ or (a,b) to be integrable. Show that a function f is integrable on $[a,b]$ if and only if its restriction to (a,b) is integrable.

#1265b† Let $I_1, \cdots, I_n, J_1, \ldots, J_N$ be subintervals of $[a,b]$ with the I_k being disjoint. If the union of the I_k is contained in the union of the J_k, show that

$$\sum_{k=1}^{n} l(I_k) \leqslant \sum_{k=1}^{N} l(J_k).$$

#1266A Complete the proof of (5.9) for $\lambda < 0$.

#1267C Let g be an integrable function on $[a,b]$ and $a < c < b$. Define g_1 and $\check{g}$ as in the proof of Proposition 5.12. Show that g_1 is integrable on $[a,b]$ and prove (5.11).

#1268b† In light of Definition 5.8, show that the identity in Proposition 5.12(c) holds generally, irrespective of the relative ordering of a,b,c.

#1269A† Find an integrable function $f(x)$ on $[0,1]$ such that $\int_0^x f(t)\,dt$ is not differentiable (i.e. does not have a well-defined gradient for some x).

#1270b† A function $f(x)$ from $\mathbb{R}$ to $\mathbb{R}$ is said to be **periodic** with period $a > 0$ if, for all x, we have $f(x+a) = f(x)$.
(i) Show that if $f(x)$ is differentiable with period a, then $f'(x)$ is periodic with period a.
(ii) If $f(x)$ is periodic with period a, under what conditions is $F(x) = \int_0^x f(t)\,dt$ periodic with period a?

#1271D† Let f,g be integrable functions on $[a,b]$. Show that $|f|$ is integrable and that

$$\left| \int_a^b f\,dx \right| \leqslant \int_a^b |f|\,dx. \tag{5.16}$$

How does this inequality relate to the triangle inequality? Deduce that $\max\{f,g\}$ and $\min\{f,g\}$ are also integrable.

#1272D† Show that if f,g are integrable on $[a,b]$ then so is their product fg.

#1273E† Show that if f is a non-negative integrable function on $[a,b]$ then so is $\sqrt{f}$.

#1274b A complex-valued function f on $[a,b]$ is said to be integrable if $\operatorname{Re}f$ and $\operatorname{Im}f$ are integrable. For such f, we define the integral of f by

$$\int_a^b f\,dx = \int_a^b \operatorname{Re}f\,dx + i\int_a^b \operatorname{Im}f\,dx.$$

Show that if f, g are integrable complex-valued functions and α,β are complex numbers

$$\int_a^b (\alpha f + \beta g)\,dx = \alpha \int_a^b f\,dx + \beta \int_a^b g\,dx.$$

#1275 D† Let f be a complex-valued integrable function on $[a,b]$. Show $|f|$ is integrable and that (5.16) remains true.

#1276 B Let f,g be integrable on $[a,b]$. Modify the proof of the Cauchy–Schwarz inequality (3.3) to show that

$$\left(\int_a^b f(x)g(x)\,dx \right)^2 \leqslant \left(\int_a^b f(x)^2\,dx \right)\left(\int_a^b g(x)^2\,dx \right).$$

#1277 B Let f,g be integrable functions on $[a,b]$. Say $m \leqslant f(x) \leqslant M$ and $g(x) \geqslant 0$ for all x. Show that

$$m \int_a^b g(x)\,dx \leqslant \int_a^b f(x)g(x)\,dx \leqslant M \int_a^b g(x)\,dx.$$

Give an example to show the result does not hold without the requirement that $g \geqslant 0$.

#1278 B† **(Integral Test)** Let $f(x)$ be a decreasing positive function defined for $x \geqslant 0$. Show, for any integer $n > 0$, that

$$\sum_{k=0}^{n-1} f(n) \geqslant \int_0^n f(x)\,dx \geqslant \sum_{k=1}^{n} f(k).$$

#1279 C† Let f,g be continuous functions on $[a,b]$. Show that

$$\int_{x=a}^{x=b} \left(\int_{y=a}^{y=b} f(x)g(y)\,dy \right) dx = \int_{x=a}^{x=b} \left(\int_{y=a}^{y=x} f(x)g(y)\,dy \right) dx + \int_{y=a}^{y=b} \left(\int_{x=a}^{x=y} f(x)g(x)\,dx \right) dy.$$

#1280 B† Let f,g be increasing, positive integrable functions on $[a,b]$. Show that

$$\left(\int_a^b f(x)\,dx \right)\left(\int_a^b g(x)\,dx \right) \leqslant (b-a) \int_a^b f(x)g(x)\,dx.$$

How does this result relate to Chebyshev's inequality (#430)?

#1281 c† Let k be a natural number and $0 < a < b$. Show from first principles that

$$\int_0^a x^k\,dx = \frac{a^{k+1}}{k+1} \qquad \text{and deduce that} \qquad \int_a^b x^k\,dx = \frac{b^{k+1} - a^{k+1}}{k+1}.$$

#1282 B Let $0 < \varepsilon < 1 < R$ and α, β be real. Define

$$I = \int_\varepsilon^1 x^\alpha\,dx, \qquad J = \int_1^R x^\beta\,dx.$$

For what values of α does the integral I remain bounded as ε becomes arbitrarily small? For what values of β does the integral J remain bounded as R becomes arbitrarily large?

#1283 b Let $r \neq 0$ and $a < b$. Carefully show, using Definition 5.7, that

$$\int_a^b e^{rx}\,dx = \frac{e^{rb} - e^{ra}}{r}.$$

#1284 c† Carefully show, using Definition 5.7, that $\int_0^{\pi/2} \sin x\,dx = 1$.

The Formal Definition of a Limit The notion of a **limit** is a relatively intuitive one. It is fundamental to calculus and analysis, and limits have been informally worked with since the time of the ancient Greeks. So it is perhaps surprising that the notion of a limit was not put on a rigorous footing until the nineteenth century with the work of Bernard Bolzano (1781–1848) and Karl Weierstrass (1815–1897). We begin this brief summary with the notion of the limit of a sequence. A (real) **sequence** is an infinite list of real numbers such as

$$1/2, 2/3, 3/4, 4/5, \ldots, \qquad 1, -1, 1, -1, 1, -1, \ldots.$$

Here the nth term of the first sequence is $n/(n+1)$ and becomes ever closer to 1 as n increases. It seems reasonable to say that 1 is the limit of this sequence, and to say that the second sequence has no limit, but what formally does this entail? One important element is that we are discussing the long-term behaviour of the sequence – if a sequence appears to play around at random, even for a billion terms, but eventually exhibits behaviour like the first sequence then its limit will still be 1. Another important element is that the sequence becomes arbitarily close to the limit. This not mean that each successive term is getting closer to the limit (though that is the case with the sequence above), but rather that any accuracy can be achieved. The more accuracy needed, the further we will typically need to go down the sequence, but that should always be achievable. So if we require the sequence to be within ten decimal places of the limit then that should occur at some point; further by some point this should become normal with all subsequent terms being of the required accuracy. So the formal definition reads as follows: the sequence with nth term x_n converges to the **limit** L if

for any $\varepsilon > 0$, there is a natural number N such that $|x_n - L| < \varepsilon$ for all $n \geqslant N$.

We would write this $x_n \to L$ as $n \to \infty$ or equally $\lim_{n \to \infty} = L$. We say a sequence is **convergent** if it has a limit and is **divergent** otherwise. In this definition, ε represents the arbitrary accuracy we earlier referred to. The expression $|x_n - L|$ is the distance of the nth term from the limit. So, for the particular example above, for any given $\varepsilon > 0$ we would first need to find a natural number N such that

$$\left| \frac{N}{N+1} - 1 \right| < \varepsilon \qquad \text{or equally} \qquad N > \frac{1}{\varepsilon} - 1.$$

Note that N depends on the choice of ε and will need to be larger for smaller ε; this is typical. But as the natural numbers are not bounded above then we can always find such an N. In this case, for such an N, then any $n > N$ will also lead to an nth term of the required accuracy. For other convergent sequences, a first N might be just a fluke and we will need to find later N after which the required accuracy is *always* met. (See Figure 5.4.)

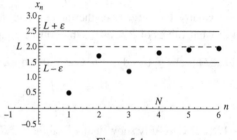

Figure 5.4

We can also define what it means for a function to have a limit. This idea is central to the study of continuity and in formally defining a derivative. It is claimed in #1286 that $\sin x/x \to 1$ as $x \to 0$. What *formally* does this mean? Taking pointers from our previous definition, we might suggest that $\sin x/x$ can be made as close to 1 as we desire (within any given ε) for some small value of x. Further, we don't wish this to be some fluke of an x but that this accuracy is again achieved for all smaller positive values of x. So given a real-valued function $f(x)$ defined for $x > 0$, we say that $f(x)$ has a limit of L at $x = 0$ if

for any $\varepsilon > 0$, there is $\delta > 0$ such that $|f(x) - L| < \varepsilon$ whenever $0 < x < \delta$.

Technically this says that $f(x)$ has a limit of L at 0 *from the right* and we write this as $\lim_{x \to 0^+} f(x) = L$.

We write $\lim_{x \to 0} f(x) = L$ to denote a function having a limit of L at $x = 0$, when a function has a limit of L from both the left and the right. And we say $f(x)$ is **continuous** at $x = 0$ if it has a limit at $x = 0$ which agrees with the value $f(0)$ that the function takes there, written $f(x) \to f(0)$ as $x \to 0$.

For example, if we consider the following function

$$f(x) = \begin{cases} 1 & \text{if} \quad x \geqslant 0 \\ 0 & \text{if} \quad x < 0 \end{cases}$$

at $x = 0$, we see $f(x)$ has a limit of 1 from the right, and 0 from the left and so in particular is not continuous at $x = 0$ (Figure 5.5).

The definition of a limit extends readily to any real a (not just 0), is central to analysis, and is also part of the definition of a derivative (p.429). Much of the fundamental theory of analysis concerns itself with showing rigorously when limits exist

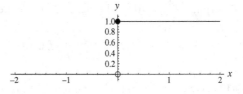

Figure 5.5

and proving theorems relating to the handling and algebra of limits, for example providing conditions when the order of limiting processes can be swapped.

#1285 b† Let f be a continuous function on an interval $[a, b]$ with $a < x < x+1 < b$. Define

$$A(x) = \int_x^{x+1} f(t)\, dt.$$

(i) Show that if $A(x)$ is maximal or minimal, then $f(x+1) = f(x)$.
(ii) For what values of x is $A(x)$ minimal where $f(x) = |x^2 - 1|$?
(iii) For what value of x is $A(x)$ maximal where $f(x) = 2^{-x^2}$?

#1286B† Consider the triangles and sector in Figure 5.6, where OA and OC have unit length. Let $0 < x < \pi/2$ denote the angle COA. Show that

$$\sin x < x < \tan x.$$

Deduce that

$$\frac{\sin x}{x} \to 1 \qquad \text{as} \qquad x \to 0.$$

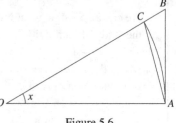

Figure 5.6

#1287B†(i) Let x be a real number such that $|x| < 1$. Show that $x^n \to 0$ as $n \to \infty$.
(ii) Let a be a real number. Show that $a^n/n! \to 0$ as $n \to \infty$.
(iii) Let $a > 0$. Show that $a^{1/n} \to 1$ as $n \to \infty$.

#1288C Assuming that $\sin\theta < \theta < \tan\theta$ for $0 < \theta < \pi/2$, rank the integrals below in order of size without directly calculating any of them.

$$\int_0^{\pi/4} x^3 \cos x \, dx, \qquad \int_0^{\pi/4} x^3 \cos^2 x \, dx, \qquad \int_0^{\pi/4} x^2 \sin x \cos x \, dx, \qquad \int_0^{\pi/4} x^3 \cos 2x \, dx.$$

#1289B For $n \geqslant 2$, we define the function $f_n(x)$ on the interval $[0,1]$ by

$$f_n(x) = nx\mathbf{1}_{[0,1/n]}(x) + (2 - nx)\mathbf{1}_{(1/n,2/n]}(x).$$

Determine $\int_0^1 f_n(x)\,dx$ and $\lim_{n\to\infty} f_n(x)$ and deduce that

$$\lim_{n\to\infty} \int_0^1 f_n(x)\,dx \neq \int_0^1 \lim_{n\to\infty} f_n(x)\,dx.$$

#1290D† Give examples of continuous functions f, g, h on the interval $(0,1)$ such that

(i) $\int_a^1 f(x)\,dx$ has a finite limit as $a \to 0$.
(ii) $\int_a^1 g(x)\,dx$ increases without bound as $a \to 0$.
(iii) $\int_a^1 h(x)\,dx$ has no limit (finite or infinite) as $a \to 0$.

#1291D Give an example of an unbounded, positive function f on $\mathbb{R}$ for which $\lim_{R\to\infty} \int_{-R}^R f(x)\,dx$ exists.

#1292D† Let f be a positive function on $\mathbb{R}$ for which $\lim_{R\to\infty} \int_{-R}^R f(x)\,dx$ exists. Show that the limit

$$\lim_{S\to\infty} \left(\lim_{R\to\infty} \int_{-S}^R f(x)\,dx \right)$$

exists. Show that this need not be the case without the assumption that f is positive.

#1293D† [6] For a positive integer n, let $f_n(x)$ be an integrable function on $[a, b]$. Let $f(x)$ be a function such that

$$|f_n(x) - f(x)| \leq \frac{1}{n} \qquad \text{for all } x \text{ in } [a, b].$$

Show that $\lim_{n \to \infty} f_n(x) = f(x)$ for each x. Show further that $f(x)$ is integrable and that

$$\int_a^b f(x)\,dx = \lim_{n \to \infty} \int_a^b f_n(x)\,dx.$$

#1294C Let $f(x)$ denote Dirichlet's function from Example 5.15. Find a sequence of integrable functions $f_n(x)$ on $[0, 1]$ such that $\lim_{n \to \infty} f_n(x) = f(x)$ for each x.

#1295D† (**Riemann–Lebesgue Lemma**) Suppose that the function f is integrable on the interval $[-\pi, \pi]$. Prove that

$$\int_{-\pi}^{\pi} f(x) \cos nx\,dx \to 0 \qquad \text{and} \qquad \int_{-\pi}^{\pi} f(x) \sin nx\,dx \to 0 \qquad \text{as } n \to \infty.$$

#1296D† [7] The **Thomae function** $T(x)$ is defined on $(0, 1)$ as follows:

$$T(x) = \begin{cases} \frac{1}{n} & x \text{ is rational and } x = \frac{m}{n} \text{ in simplest form;} \\ 0 & x \text{ is irrational.} \end{cases}$$

Sketch the function $T(x)$. Show that $T(x)$ is integrable and find $\int_0^1 T(x)\,dx$.

5.2 Standard Functions

In this section we will put the fundamental theorem of calculus (Theorem 5.18) to use in determining a range of integrals and demonstrating some functions' algebraic properties. As a consequence of the fundamental theorem, we know the indefinite integral of a function $f(x)$ to represent the antiderivatives of $f(x)$ – i.e. those functions that differentiate to $f(x)$.

Proposition 5.19 (*Properties of the Natural Logarithm*) *Define the* **natural logarithm** $\ln x$ *for* $x > 0$ *by*

$$\ln x = \int_1^x \frac{dt}{t}.$$

Show that

(a) $\ln(xy) = \ln x + \ln y;$ (b) $\ln(x^{-1}) = -\ln x;$ (c) $\dfrac{d}{dx}(\ln x) = \dfrac{1}{x}.$

Show further that $\ln x$ *is a strictly increasing function and unbounded.*

Proof (a) For any positive x, y we have

$$\ln(xy) = \int_1^{xy} \frac{dt}{t} = \int_1^x \frac{dt}{t} + \int_x^{xy} \frac{dt}{t} \qquad \text{[\#1268]}$$

$$= \int_1^x \frac{dt}{t} + \int_{u=1}^{u=y} \frac{x\,du}{xu} \qquad \text{[substituting } t = xu \text{ in second integral]}$$

$$= \int_1^x \frac{dt}{t} + \int_1^y \frac{du}{u} = \ln x + \ln y.$$

---

[6] This exercise (essentially) shows that the *uniform* limit of integrable functions is integrable. This need not be the case for *pointwise* limits, as Dirichlet's function shows (#1294).

[7] Named after the German mathematician Johannes Thomae (1840–1921).

(For those unfamiliar with substitution, see §5.4.) For (b) we note

$$0 = \ln 1 = \ln(x^{-1}x) = \ln(x^{-1}) + \ln x$$

and so $\ln(x^{-1}) = -\ln x$ and (c) follows from Proposition 5.17. Now if $0 < x < y$, and noting $t^{-1} \geqslant y^{-1}$ for $x \leqslant t \leqslant y$, then

$$\ln y = \int_1^y \frac{dt}{t} = \int_1^x \frac{dt}{t} + \int_x^y \frac{dt}{t} \geqslant \int_1^x \frac{dt}{t} + \frac{y-x}{y} > \int_1^x \frac{dt}{t} = \ln x,$$

and so $\ln x$ is a strictly increasing function. Further, for any natural number n, we have $\ln 2^n = n \ln 2$, which increases without bound as n becomes large. $\qquad\square$

Remark 5.20 Note for $x < 0$ that the function $\ln(-x)$ differentiates, by the chain rule, to $-1/(-x) = 1/x$ and so $\ln|x|$ is an antiderivative of $1/x$ for all non-zero values of x. For example, by the fundamental theorem of calculus, we have

$$\int_{-4}^{-2} \frac{dx}{x} = [\ln|x|]_{-4}^{-2} = \ln|-2| - \ln|-4| = \ln 2 - \ln 4 = -\ln 2. \qquad\blacksquare$$

Note $\ln 1 = 0$ and $\ln x$ is strictly increasing and unbounded for $x > 0$, so it follows for every $k \geqslant 0$ there is a unique value $x \geqslant 1$ satisfying $\ln x = k$. Similarly, as $\ln(x^{-1}) = -\ln x$, it follows that for every $k < 0$ there exists a unique value $0 < x < 1$ such that $\ln x = k$. So for every real k there is a unique value x such that $\ln x = k$. We will denote this value of x as $\exp k$. As $\ln x$ is itself differentiable with positive derivative then $\exp x$ must also be differentiable. [8]

Proposition 5.21 *(**Properties of the Exponential Function**) The exponential function $\exp x$ is the unique map from $\mathbb{R}$ to the positive reals $(0, \infty)$ satisfying*

$$\ln(\exp x) = x. \tag{5.17}$$

It follows that

$$\exp'(x) = \exp x \qquad and \qquad \exp(0) = 1. \tag{5.18}$$

In fact (5.18) may be taken as an alternative definition of $\exp x$. Also for real s, t we have

$$\exp(s + t) = \exp s \exp t. \tag{5.19}$$

Proof (5.17) follows from the definition of the exponential function. Differentiating (5.17) with the chain rule we get

$$\ln'(\exp x)\exp'(x) = 1 \quad\Longrightarrow\quad \frac{1}{\exp x}\exp' x = 1 \quad\Longrightarrow\quad \exp'(x) = \exp x.$$

Also, as $\ln 1 = 0$ (by definition) then $\exp 0 = 1$. That there is a unique differentiable function satisfying (5.18) is left to #1301. Finally for a constant real a, we define

$$f_a(x) = \exp(x + a)\exp(-x).$$

Differentiating using the chain and product rules, and recalling $\exp' = \exp$, we find

$$f_a'(x) = \exp(x + a)\exp(-x) - \exp(x + a)\exp(-x) = 0.$$

---

[8] Again a rigorous proof of this paragraph's claims would require various theorems of analysis, especially the *Intermediate Value Theorem* (Binmore [4, Corollary 9.10]), but hopefully these claims are intuitively clear.

It follows that $f_a(x)$ is a constant function. Note that $f_a(0) = \exp(a)$. Thus we have for all real x and a that

$$\exp(x+a)\exp(-x) = \exp(a).$$

If we set $a = s+t$ and $x = -t$ then we arrive at (5.19). ☐

Definition 5.22 (General Powers and Logarithms) (a) Given $a > 0$ and a real number x, we **define**

$$a^x = \exp(x\ln a).$$

By (5.17) it follows that $\ln(a^x) = x\ln a$. (b) If $a \neq 1$ then for any $k > 0$ there exists a unique value of x such that $a^x = k$. We write the value of x as $\log_a k$.

#1301
#1305
#1310
#1311

    This definition of a^x may seem rather surprising, but consider how else one might define such powers for a *general* exponent x. The notion of a^x being 'a multiplied by itself x times' may work for natural numbers and can be extended to a negative integer x via $a^x = (a^{-x})^{-1}$. When $x = m/n$ is rational, then we can also define $a^{m/n}$ as the unique positive number satisfying $(a^{m/n})^n = a^m$. However, we need some definition, like the one above, for just what a^x means when x is irrational.

    Now the number $e = \exp(1)$ satisfies $\ln e = 1$ so that, by definition, we have

$$e^x = \exp(x\ln e) = \exp x$$

and so we will typically write e^x instead of $\exp x$ from now on.

Proposition 5.23 (*Power Rules*) *Let $a > 0$ and x, y be real. Then*

    (a) $\quad a^{x+y} = a^x a^y;$ *(b)* $\quad (a^x)^y = a^{xy};$ *(c)* $\quad \dfrac{\mathrm{d}}{\mathrm{d}x}(a^x) = (\ln a)a^x;$

and (d) for $x > 0$ and r real

$$\frac{\mathrm{d}}{\mathrm{d}x}(x^r) = rx^{r-1}.$$

Proof (a) By (5.19) we have

$$a^{x+y} = \exp((x+y)\ln a) = \exp(x\ln a)\exp(y\ln a) = a^x a^y.$$

(b) Applying Definition 5.22(a) and (b) we see

$$(a^x)^y = \exp(y\ln(a^x)) = \exp(y(x\ln a)) = \exp(xy\ln a) = a^{xy}.$$

(c) That $a^x = e^{x\ln a}$ differentiates to $(\ln a)a^x$ follows from the chain rule. Finally, for (d), we have $x^r = \exp(r\ln x)$ is differentiable by the chain rule as $\ln$ and $\exp$ are both differentiable. So by the chain rule

$$\frac{\mathrm{d}}{\mathrm{d}x}(x^r) = \frac{r}{x}\exp(r\ln x) = r\exp(-\ln x)\exp(r\ln x) = r\exp((r-1)\ln x) = rx^{r-1},$$

using Proposition 5.19(b) and (5.19). ☐

Proposition 5.24 (*Logarithm Rules*) *For $a, b, c > 0$, $a \neq 1$ and real x we have*

    (a) $\quad \log_a(bc) = \log_a b + \log_a c;$ *(b)* $\quad \log_a(b^x) = x\log_a b;$ *(c)* $\quad \log_a b = \dfrac{\ln b}{\ln a}.$

#1298

Table 5.1 *Some Standard Indefinite Integrals*

#1299
#1306
#1308
#1316
#1319
#1321

$f(x)$	x^a $(a \neq -1)$	x^{-1}	e^{ax} $(a \neq 0)$	$\frac{f'(x)}{f(x)}$
$\int f(x)\,dx$	$\frac{x^{a+1}}{a+1}+c$	$\ln|x|+c$	$e^{ax}/a+c$	$\ln|f(x)|+c$
$f(x)$	$\sin x$	$\cos x$	$\tan x$	$\sec x$
$\int f(x)\,dx$	$-\cos x+c$	$\sin x+c$	$-\ln|\cos x|+c$	$\ln|\sec x+\tan x|+c$
$f(x)$	$\cot x$	$\sec^2 x$	$\tan x\sec x$	$\csc x$
$\int f(x)\,dx$	$\ln|\sin x|+c$	$\tan x+c$	$\sec x+c$	$-\ln|\csc x+\cot x|+c$

c denotes an arbitrary constant.

#1309
#1312
#1313
#1322

Proof These are left to #1309. $\qquad\qquad\square$

We now move on to the trigonometric functions. In #1286 we showed that $(\sin\delta)/\delta \to 1$ as $\delta \to 0$. So for real x and $\varepsilon \neq 0$ we have

$$\frac{\sin(x+\varepsilon)-\sin x}{\varepsilon} = \frac{2\cos(x+\varepsilon/2)\sin(\varepsilon/2)}{\varepsilon} = \cos\left(x+\frac{\varepsilon}{2}\right)\left(\frac{\sin(\varepsilon/2)}{\varepsilon/2}\right) \to \cos x$$

as $\varepsilon \to 0$ using the continuity of cosine. This shows that

$$\frac{d}{dx}(\sin x) = \cos x.$$

As $\cos x = \sin(\pi/2 - x)$ then by the chain rule

$$\frac{d}{dx}(\cos x) = \frac{d}{dx}(\sin(\pi/2-x)) = -\sin'(\pi/2-x) = -\cos(\pi/2-x) = -\sin x.$$

Using variously the product, quotient and chain rules we note

$$\frac{d}{dx}(\tan x) = \frac{d}{dx}\left(\frac{\sin x}{\cos x}\right) = \frac{\cos^2 x+\sin^2 x}{\cos^2 x} = \sec^2 x; \qquad \frac{d}{dx}(\ln\cos x) = -\frac{\sin x}{\cos x} = -\tan x;$$

$$\frac{d}{dx}(\cot x) = \frac{d}{dx}\left(\frac{\cos x}{\sin x}\right) = \frac{-\sin^2 x-\cos^2 x}{\sin^2 x} = -\csc^2 x; \qquad \frac{d}{dx}(\ln\sin x) = \frac{\cos x}{\sin x} = \cot x;$$

$$\frac{d}{dx}(\ln(\sec x+\tan x)) = \frac{\tan x\sec x+\sec^2 x}{\tan x+\sec x} = \sec x; \qquad \frac{d}{dx}(\sec x) = -\frac{(-\sin x)}{\cos^2 x} = \tan x\sec x;$$

$$\frac{d}{dx}(\ln(\csc x+\cot x)) = \frac{-\cot x\csc x-\csc^2 x}{\cot x+\csc x} = -\csc x; \qquad \frac{d}{dx}(\csc x) = -\frac{\cos x}{\sin^2 x} = -\cot x\csc x.$$

Thus, applying the fundamental theorem of calculus, we have demonstrated the standard indefinite integrals shown in Table 5.1.

Definition 5.25 (Hyperbolic Functions) Given a real number x then its **hyperbolic cosine** $\cosh x$ and **hyperbolic sine** $\sinh x$ are defined by

$$\cosh x = \frac{e^x+e^{-x}}{2}; \qquad \sinh x = \frac{e^x-e^{-x}}{2}.$$

In a similar fashion to the trigonometric functions the following are also defined

$$\tanh x = \frac{\sinh x}{\cosh x}; \qquad \coth x = \frac{\cosh x}{\sinh x}; \qquad \operatorname{sech} x = \frac{1}{\cosh x}; \qquad \operatorname{csch} x = \frac{1}{\sinh x}.$$

The graphs of the sinh, cosh and tanh are sketched in Figure 5.7.

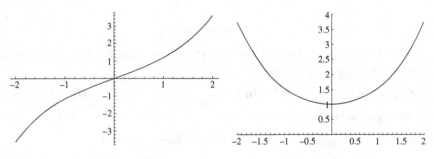

The graph of $y = \sinh x$ The graph of $y = \cosh x$

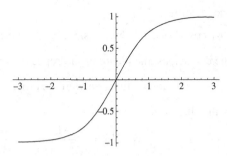

The graph of $y = \tanh x$

Figure 5.7

Note that $\cosh^2 x = 1 + \sinh^2 x$ for all x, which means that as t varies over the reals $(\pm \cosh t, \sinh t)$ parametrize the two branches of the hyperbola $x^2 - y^2 = 1$, which explains the name *hyperbolic* functions.

#1300
#1303
#1304
#1307

Proposition 5.26 (*Properties of the Hyperbolic Functions*) *Let x and y be real numbers. Then*

(a) $\cosh^2 x = 1 + \sinh^2 x,$ $\operatorname{sech}^2 x + \tanh^2 x = 1,$ $\coth^2 x = 1 + \operatorname{csch}^2 x.$

(b) $\sinh(x+y) = \sinh x \cosh y + \cosh x \sinh y;$ $\sinh 2x = 2\sinh x \cosh x;$

$\cosh(x+y) = \cosh x \cosh y + \sinh x \sinh y;$ $\cosh 2x = 2\cosh^2 x - 1$

$= 2\sinh^2 x + 1;$

$$\tanh(x+y) = \frac{\tanh x + \tanh y}{1 + \tanh x \tanh y}; \qquad \tanh 2x = \frac{2\tanh x}{1 + \tanh^2 x}.$$

(c) $\cosh' x = \sinh x;$ $\sinh' x = \cosh x;$ $\tanh' x = \operatorname{sech}^2 x.$

(d) *Their inverses, and their derivatives, are given by*

$$\sinh^{-1} x = \ln\left(x + \sqrt{x^2 + 1}\right) \text{ for real } x; \qquad \frac{d}{dx}\sinh^{-1} x = \frac{1}{\sqrt{x^2 + 1}};$$

$$\cosh^{-1} x = \ln\left(x + \sqrt{x^2 - 1}\right) \text{ for } x \geqslant 1; \qquad \frac{d}{dx}\cosh^{-1} x = \frac{1}{\sqrt{x^2 - 1}};$$

$$\tanh^{-1} x = \frac{1}{2}\ln\left(\frac{1+x}{1-x}\right) \text{ for } |x| < 1; \qquad \frac{d}{dx}\tanh^{-1} x = \frac{1}{1 - x^2}.$$

Proof These are left as #1300, #1303 and #1304.   □

Exercises

#1297B The functions $s(x) = \sin x$ and $c(x) = \cos x$ can be defined as the unique differentiable functions satisfying

$$s''(x) = -s(x), \qquad s(0) = 0, s'(0) = 1; \qquad c''(x) = -c(x), \qquad c(0) = 1, c'(0) = 0.$$

Use these definitions to show that $\sin'(x) = \cos(x)$ and $\cos'(x) = -\sin(x)$.

#1298a Use differentiation to prove that $\sin^2 x + \cos^2 x = 1$ for all x.

#1299A By differentiating $\cos c - \cos x \cos(x - c) - \sin x \sin(x - c)$ where c is a constant, prove the trigonometric identities

$$\cos(a + b) = \cos a \cos b - \sin a \sin b; \qquad \sin(a + b) = \sin a \cos b + \cos a \sin b.$$

#1300b Prove from the definitions, the identities in Proposition 5.26(a) and (b).

#1301B† Show that $\exp x$ is the unique function with the properties in (5.18).

#1302c† By means of the chain rule show that

$$\frac{d}{dx} \sin^{-1} x = \frac{1}{\sqrt{1 - x^2}}; \qquad \frac{d}{dx} \cos^{-1} x = \frac{-1}{\sqrt{1 - x^2}}; \qquad \frac{d}{dx} \tan^{-1} x = \frac{1}{1 + x^2}.$$

#1303a Verify the derivatives given in Proposition 5.26(c).

#1304b† Verify the identities in Proposition 5.26(d) for the inverse hyperbolic functions and their derivatives.

#1305b† Let a be a real number. Evaluate the following definite integrals.

$$\int_{-1}^{2} 3^x \, dx, \qquad \int_0^{\infty} e^{-2x-1} \, dx, \qquad \int_{-3}^{-2} e^{|x|} dx,$$

$$\int_{-\infty}^{\infty} e^{-|x|} dx, \qquad \int_{-\infty}^{\infty} e^{-|x-a|} dx, \qquad \int_0^{\infty} e^{-|x-a|} dx.$$

#1306a Evaluate the following definite integrals.

$$\int_0^{\pi/4} \cos 2x \, dx, \qquad \int_0^{\pi} \sin^2 x \, dx, \qquad \int_{-1}^{0} \tan(2x + 1) \, dx, \qquad \int_0^{\pi/4} \tan^2 x \, dx.$$

#1307a Evaluate the following definite integrals.

$$\int_0^1 \cosh(2x + 3) \, dx, \qquad \int_{-1}^{1} \sinh^2 x \, dx, \qquad \int_0^{\infty} \operatorname{sech} x \, dx,$$

$$\int_0^1 \frac{\sinh^2 x}{\cosh x} \, dx, \qquad \int_0^{\infty} \operatorname{csch}^2(3x + 1) \, dx, \qquad \int_0^1 \tanh^2 x \, dx.$$

#1308A Show that $\ln|\sec x + \tan x|$, which is an antiderivative of $\sec x$, can also be written as each of

$$\ln|(1 + \tan(x/2))/(1 - \tan(x/2))|, \qquad \ln|\tan(x/2 + \pi/4)|, \qquad 2\tanh^{-1}\tan(x/2).$$

For what values of x is the third function an antiderivative of $\sec x$?

#1309 A Prove the identities in Proposition 5.24.

#1310 B† Use calculus to demonstrate a more general form of Bernoulli's inequality (#227), namely

$$(1+x)^\alpha \geqslant 1+\alpha x \qquad \text{for real } x \geqslant -1 \text{ and real } \alpha \geqslant 1.$$

#1311 A† Show that $1-x \leqslant e^{-x} \leqslant (1+x)^{-1}$ and $x \geqslant \ln(1+x)$ for $x \geqslant 0$.

#1312 a By differentiating the RHS, show that $\int \ln x \, dx = x \ln x - x + \text{const}$.

#1313 B† Let $0 < \varepsilon < 1$. Determine

$$\int_\varepsilon^1 \ln x \, dx \qquad \text{and} \qquad \int_0^1 \ln x \, dx.$$

#1314 C† Sketch $y = \ln x/(x-1)$ for $x > 0$. For values of y near $x = 1$, it may help to treat $y(1)$ as the limit associated with a derivative.

#1315 B† Find all the solutions x to the following equations.

(i) $\displaystyle\int_x^{x^2} e^{-t^2} \, dt = 0;$ (ii) $\displaystyle\int_0^x \sin(\sin t) \, dt = 0;$ (iii) $\displaystyle\int_{-x}^x \sin^3(t \cos t) \, dt = 0.$

#1316 B† (i) A regular m-sided polygon I_m is inscribed in a circle S of radius 1, and a regular n-sided polygon C_n is circumscribed about S. By considering the perimeter of I_m and the area of C_n prove, for all $m, n \geqslant 3$, that

$$m \sin\left(\frac{\pi}{m}\right) < \pi < n \tan\left(\frac{\pi}{n}\right). \tag{5.20}$$

(ii) Archimedes showed (using this method) that $3\frac{10}{71} < \pi < 3\frac{1}{7}$. Use a calculator to work out the smallest values of m and n needed to verify Archimedes' inequality. (Archimedes himself used 96-sided polygons.)

(iii) Use integration to prove that $\sin\theta < \theta < \tan\theta$ for $0 < \theta < \pi/2$, and show (5.20) for all real numbers $m, n > 2$.

#1317 D† (i) Use the inequality $\sin y \leqslant y$ for $y \geqslant 0$ to show $\cos y \geqslant 1 - y^2/2$ for $y \geqslant 0$.

(ii) Show more generally, for any positive integer n and $y \geqslant 0$, that

$$y - \frac{y^3}{3!} + \frac{y^5}{5!} - \cdots - \frac{y^{4n-1}}{(4n-1)!} \leqslant \sin y \leqslant y - \frac{y^3}{3!} + \frac{y^5}{5!} - \cdots - \frac{y^{4n-1}}{(4n-1)!} + \frac{y^{4n+1}}{(4n+1)!};$$

$$1 - \frac{y^2}{2!} + \frac{y^4}{4!} - \cdots - \frac{y^{4n-2}}{(4n-2)!} \leqslant \cos y \leqslant 1 - \frac{y^2}{2!} + \frac{y^4}{4!} - \cdots - \frac{y^{4n-2}}{(4n-2)!} + \frac{y^{4n}}{(4n)!}.$$

(iii) Use these inequalities to determine $\sin 1$ to within 0.01. Could you (in principle at least) use such inequalities to find $\sin 100$ to within 0.01?

#1318 D† Show for n a natural number and $x \geqslant 0$ that

$$1 + x + \frac{x^2}{2!} + \cdots + \frac{x^n}{n!} \leqslant e^x.$$

Hence deduce for $\alpha, \beta > 0$ that

$$\lim_{x \to \infty} \left(x^\alpha e^{-x}\right) = 0 \qquad \text{and} \qquad \lim_{x \to 0} \left(x^\beta \ln x\right) = 0.$$

#1319b By writing the denominator as $R\cos(x - \alpha)$ for some α, determine $\int (4\cos x + 3\sin x)^{-1} \, dx$.

#1320D† Let k be a non-zero integer. Show that

$$\int_0^{\pi/2} \operatorname{cis}(kx) \, dx = \frac{i^k - 1}{ik}, \qquad \text{and} \qquad \int_0^{\pi/2} \cos^n x \cos nx \, dx = \frac{\pi}{2^{n+1}}.$$

#1321B Let

$$I_1 = \int \frac{\sin x \, dx}{\sin x + \cos x} \qquad \text{and} \qquad I_2 = \int \frac{\cos x \, dx}{\sin x + \cos x}.$$

By considering $I_1 + I_2$ and $I_2 - I_1$, find I_1 and I_2. Generalize your method to calculate

$$\int \frac{\sin x \, dx}{a\sin x + b\cos x} \qquad \text{and} \qquad \int \frac{\cos x \, dx}{a\sin x + b\cos x}.$$

#1322b Evaluate

$$\int_3^\infty \frac{dx}{(x-1)(x-2)} \qquad \text{and} \qquad \int_3^\infty \frac{dx}{(x-1)^2(x-2)^2}.$$

#1323C Show

$$\int \cosh^{2n+1} x \, dx = \sum_{k=0}^n \binom{n}{k} \frac{\sinh^{2k+1} x}{2k+1} + \text{const.}.$$

#1324B† (See also #1401.) The **error function** $\operatorname{erf}(x)$ is defined by

$$\operatorname{erf}(x) = \frac{2}{\sqrt{\pi}} \int_0^x e^{-t^2} \, dt.$$

Show that

$$\int \operatorname{erf}(x) \, dx = \frac{e^{-x^2}}{\sqrt{\pi}} + x\operatorname{erf}(x) + \text{const.}.$$

Show further, for real a, b, c with $a > 0$, that

$$\int e^{-(ax^2 + 2bx + c)} \, dx = \frac{e^{(b^2/a) - c}}{2} \sqrt{\frac{\pi}{a}} \operatorname{erf}\left(\sqrt{a}x + \frac{b}{\sqrt{a}}\right) + \text{const.}.$$

#1325C (**Harmonic Series**) By considering rectangular areas which lie above the graph of $y = 1/x$ for $x \geqslant 1$, show that the infinite series below increases without bound:

$$1 + \frac{1}{2} + \frac{1}{3} + \frac{1}{4} + \cdots + \frac{1}{n} + \cdots.$$

#1326D†(i) Let n be a positive integer. Show that $0 \leqslant \gamma_{n+1} \leqslant \gamma_n$ for each n where

$$\gamma_n = 1 + \frac{1}{2} + \frac{1}{3} + \cdots + \frac{1}{n} - \ln n.$$

It follows that the sequence γ_n converges to a limit γ, known as **Euler's constant**. [9]
(ii) Show that $s_n = \gamma_{2n} - \gamma_n + \ln 2$, and deduce that $s_n \to \ln 2$ as $n \to \infty$, where

$$s_n = 1 - \frac{1}{2} + \frac{1}{3} - \frac{1}{4} + \cdots + \frac{1}{2n-1} - \frac{1}{2n}.$$

(iii) Define t_n and u_n by

$$t_n = 1 + \frac{1}{3} - \frac{1}{2} + \frac{1}{5} + \frac{1}{7} - \frac{1}{4} + \frac{1}{9} + \frac{1}{11} - \frac{1}{6} + \cdots + \frac{1}{4n-3} + \frac{1}{4n-1} - \frac{1}{2n}.$$

$$u_n = 1 - \frac{1}{2} - \frac{1}{4} + \frac{1}{3} - \frac{1}{6} - \frac{1}{8} + \frac{1}{5} - \frac{1}{10} - \frac{1}{12} + \cdots + \frac{1}{2n-1} - \frac{1}{4n-2} - \frac{1}{4n}.$$

Find an expression for t_n involving $\gamma_n, \gamma_{2n}, \gamma_{4n}$ and show that $t_n \to \frac{3}{2}\ln 2$ as $n \to \infty$. Find the limit of u_n.

#1327D (See also #241.) By considering rectangular areas that lie beneath the graph of $y = 1/x^2$, show that the infinite sum [10]

$$S = 1 + \frac{1}{4} + \frac{1}{9} + \frac{1}{16} + \cdots + \frac{1}{n^2} + \cdots$$

is finite. Show further that $S \leqslant 2$.

#1328D† Let n be a positive integer. With $\lfloor x \rfloor$ denoting the integer part of x, show that

$$\ln n! = \int_1^n \frac{n - \lfloor x \rfloor}{x}\, dx.$$

#1329B Let $a, b > 0$. Using the fact that

$$\int_{y=a}^{y=b} \left(\int_{x=0}^{x=\infty} e^{-xy}\, dx \right) dy = \int_{x=0}^{x=\infty} \left(\int_{y=a}^{y=b} e^{-xy}\, dy \right) dx,$$

show that

$$\int_0^\infty \frac{e^{-ax} - e^{-bx}}{x}\, dx = \ln\left(\frac{b}{a}\right).$$

#1330B† Let $0 < y < 1$. Show that

$$\int_0^1 \frac{x-y}{(x+y)^3}\, dx = \frac{-1}{(y+1)^2}.$$

Deduce that

$$\int_{y=0}^{y=1} \left(\int_{x=0}^{x=1} \frac{x-y}{(x+y)^3}\, dx \right) dy = -\frac{1}{2} \neq \frac{1}{2} = \int_{x=0}^{x=1} \left(\int_{y=0}^{y=1} \frac{x-y}{(x+y)^3}\, dy \right) dx.$$

---

[9] Euler's constant γ has approximate value $0.5772156649\ldots$. Arguably it is the third most important mathematical constant (after π and e) appearing in a wide range of identities relating it to special functions, yet relatively little is known about it; for example, it is still an open problem as to whether γ is irrational or not. However, by computing the convergents of the continued fraction of γ it has been shown that if γ is rational its denominator is astronomically large (Havil [18, p.97]).

[10] The sum of this series was famously determined by Euler to be $\pi^2/6$ in 1734 after generations of mathematicians had failed to evaluate it. The problem is often referred to as the *Basel Problem*, after Euler's hometown.

Absolute and Conditional Convergence Notice that the two infinite series from #1326

$$1 - \frac{1}{2} + \frac{1}{3} - \frac{1}{4} + \frac{1}{5} - \frac{1}{6} + \cdots \quad \text{and} \quad 1 + \frac{1}{3} - \frac{1}{2} + \frac{1}{5} + \frac{1}{7} - \frac{1}{4} + \cdots$$

contain exactly the same terms. They are admittedly summed in different orders but we might still have expected their sums to be equal, and yet the first sums to $\ln 2$ whilst the second sum is 50% bigger. This certainly could not happen with any finite sum and it's perhaps disconcerting that this can happen at all with infinite sums. These two series are consequently called **conditionally convergent**.

The same phenomenon can also occur with integrals. In #1420 we show

$$\int_0^\infty \frac{\sin x}{x}\, dx = \lim_{X \to \infty} \int_0^X \frac{\sin x}{x}\, dx = \frac{\pi}{2}.$$

(See Figure 5.8.) This is the value achieved when we let the upper limit of the integral tend to ∞. It might seem odd to do so, but we might also seek to determine the signed area represented by this integral by taking the area of two 'positive' humps at a time followed by a single 'negative' hump (in a manner similar to the second series above). All the humps would still contribute the same signed area as they had before but again the answer would be greater than $\pi/2$. So this integral also exists only in some conditional sense.

Reassuringly, these problems can be resolved if a series or integral is **absolutely convergent**. This means that

$$\sum_{k=1}^{\infty} |a_k|, \quad \text{or} \quad \int_0^\infty |f(x)|\, dx,$$

also converges. An absolutely convergent series is unconditionally convergent in the sense that the sum

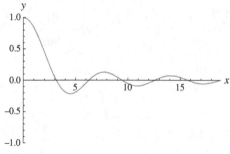

Figure 5.8

is always the same, no matter what order the terms are summed in. You may note that the infinite series above are not absolutely convergent, as they lead to the divergent harmonic series. In a similar manner one can show that $|\sin x / x|$ is not integrable on $(0, \infty)$ (see #1421).

In a similar fashion repeated integrals of a function $f(x, y)$ may lead to different answers if the order of integration is swapped (see #1330), but this cannot happen if either repeated integral of $|f(x, y)|$ exists. This follows from two theorems of the Italian mathematicians Guido Fubini (1879–1943) and Leonida Tonelli (1885–1946).

We conclude with a surprising and perhaps perturbing result of Riemann. Given a conditionally convergent series, like either of the previous two series, and given any real number S, then the series can be summed in some rearranged order to produce S as its sum. The relevant property of such series is that the positive terms alone sum to ∞ and the negative terms alone sum to $-\infty$. So if one wished to achieve a sum of 100, say, then one could start adding the positive terms until 100 was exceeded, then sum the negative terms until below 100, back to the positive terms, and so on. As their respective sums are $\pm\infty$ then this process can always continue, but the process will also lead to every term of the series being included in the sum. Such an argument is not possible with an absolutely convergent series, as the separate positive and negative terms both have finite sums.

#1331D [11] For n a positive integer and $x \geqslant 0$ we define $f_n(x) = n/(n+x)$. For fixed $X > 0$ and $0 \leqslant x \leqslant X$, show that $|f_n(x) - 1| \leqslant X/n$. By considering the integrals $\int_0^X f_n(x)\,dx$, show

$$\left(1 + \frac{X}{n}\right)^n \to e^X \qquad \text{as } n \to \infty. \tag{5.21}$$

Prove (5.21) also holds for $X \leqslant 0$.

#1332D Determine the limits of the following sequences as n becomes large.

$$(1 + n^{-2})^n, \qquad (1 + n^{-1})^{n^2}, \qquad (1 - n^{-2})^n, \qquad (1 - 1/n)^{n^2}.$$

#1333D† (**Jensen's Inequality** [12]) A function $f(x)$ is said to be **convex** if $f''(x) \geqslant 0$ for all x. For a convex function f, and any a, b, show that

$$f\left(\frac{a+b}{2}\right) \leqslant \frac{f(a) + f(b)}{2}.$$

#1334D† Prove the more general form of Jensen's inequality: for $0 \leqslant t \leqslant 1$ and convex $f(x)$, show that

$$f(ta + (1-t)b) \leqslant tf(a) + (1-t)f(b).$$

What does this say about the graph of $y = f(x)$ and the line connecting $(a, f(a))$ to $(b, f(b))$? Deduce that

$$f\left(\frac{x_1 + \cdots + x_n}{n}\right) \leqslant \frac{f(x_1) + \cdots + f(x_n)}{n},$$

for real $x_1, \ldots, x_n$. By making an appropriate choice of f, rederive the AM–GM inequality.

#1335D†(i) (**Weighted AM–GM Inequality**) Let $w_1, \ldots, w_n, x_1, \ldots, x_n$ be non-negative with $w_1 + \cdots + w_n = 1$. Show that

$$w_1 x_1 + w_2 x_2 + \cdots + w_n x_n \geqslant x_1^{w_1} x_2^{w_2} \cdots x_n^{w_n}.$$

(ii) Deduce **Young's inequality,** [13] which states, for $p, q > 0$ such that $1/p + 1/q = 1$, we have

$$xy \leqslant \frac{x^p}{p} + \frac{y^q}{q} \qquad \text{where } x, y \geqslant 0.$$

---

[11] The exponential function and e were first discovered as in (5.21) by Jacob Bernoulli (see p.124) in 1683 when investigating compound interest.

[12] After the Danish mathematician and engineer Johan Jensen (1859–1925).

[13] Named after the English mathematician William Young (1863–1942). Young was an analyst and independently of Lebesgue constructed a theory of integration essentially equivalent to Lebesgue's. In 1910 he wrote *The Fundamental Theorems of the Differential Calculus*, which was subsequently highly influential on the teaching of multivariable calculus.

(iii) Further deduce **Hölder's inequality**,[14] which states, for $p, q > 0$ with $1/p + 1/q = 1$, that

$$\sum_{k=1}^{n} x_k y_k \leqslant \left(\sum_{k=1}^{n} (x_k)^p \right)^{1/p} \left(\sum_{k=1}^{n} (y_k)^q \right)^{1/q}, \qquad \text{where } x_k, y_k \geqslant 0.$$

#1336D† (Minkowski's Inequality [15]) Let $p \geqslant 1$. Show that

$$\left(\sum_{k=1}^{n} |x_k + y_k|^p \right)^{1/p} \leqslant \left(\sum_{k=1}^{n} |x_k|^p \right)^{1/p} + \left(\sum_{k=1}^{n} |y_k|^p \right)^{1/p}$$

for vectors $\mathbf{x} = (x_1, \ldots, x_n)$, $\mathbf{y} = (y_1, \ldots, y_n)$ in $\mathbb{R}^n$. Deduce that

$$\|\mathbf{x}\|_p = \left(\sum_{k=1}^{n} |x_k|^p \right)^{1/p}$$

defines a norm on $\mathbb{R}^n$. Is this still the case when $p < 1$?

#1337E† Let f be an integrable function on $[0, 1]$ such that the limit $\lim_{x \to 0} f(x) = L$ exists. Show that

$$\lim_{n \to \infty} \int_0^{1/\sqrt{n}} f(x) n e^{-nx} \, \mathrm{d}x = L \qquad \text{and deduce that} \qquad \lim_{n \to \infty} \int_0^1 f(x) n e^{-nx} \, \mathrm{d}x = L.$$

Sketch the functions ne^{-nx} on $[0, 1]$ for various n. What properties do these functions have that are of relevance to this question?

5.3 Integration by Parts. Reduction Formulae

Integration by parts (IBP) can be used to tackle integrals of products, but not just any product. Suppose we have an integral

$$\int f(x) g(x) \, \mathrm{d}x$$

in mind. This will be approachable with IBP if one of these functions $f(x), g(x)$ integrates, or differentiates, perhaps after repeated efforts, to something simpler, whilst the other function differentiates, or integrates to something similar and as tractable. Typically then, $f(x)$ might be a polynomial which, after differentiating enough times, will become a constant; $g(x)$, on the other hand, could be something like e^x, $\sin x$, $\cos x$, $\sinh x$, $\cosh x$, all of which are functions which continually integrate to something similar. This remark reflects the nature of the formula for IBP, which is:

---

[14] Named after the German mathematician Otto Hölder (1859–1937). Hölder's name appears variously in mathematics and especially within *group theory*. Hölder was the first to fully describe quotient groups and the Jordan–Hölder theorem is a significant result about the structure of groups.

[15] Named after Hermann Minkowski (1864–1909), a Lithuanian–German mathematician and tutor of Einstein. In 1907 he showed that special relativity is most readily understood in terms of a four-dimensional space–time, *Minkowski space–time*, in which there is interaction between the spatial and temporal co-ordinates. It was also at Minkowski's encouragement that Hilbert gave his celebrated 1900 International Congress of Mathematics lecture in which he listed 23 open problems of mathematics for the attention of mathematicians as the new century dawned.

Proposition 5.27 (*Integration by Parts*) *Let F and G be integrable functions with derivatives f and g. Then*

$$\int F(x)g(x)\,dx = F(x)G(x) - \int f(x)G(x)\,dx.$$

IBP takes the integral of the product Fg and leaves us with the integral of another product fG – but as we commented above, the point is that $f(x)$ should be a 'simpler' function than $F(x)$ was whilst $G(x)$ should be no 'worse' a function than $g(x)$ was.

Proof The proof is straightforward – integrate the product rule of differentiation which states $(FG)' = Fg + fG$, and rearrange. □

Example 5.28 Determine

$$\int x^2 \sin x\,dx \qquad \text{and} \qquad \int_0^1 x^3 e^{2x}\,dx.$$

Solution For the first integral x^2 is a function that simplifies with differentiation, and $\sin x$ is a function that will integrate without complications. So we have, with *two* applications of IBP:

$$\int x^2 \sin x\,dx = x^2(-\cos x) - \int 2x(-\cos x)\,dx \qquad \text{[IBP]}$$

$$= -x^2\cos x + \int 2x\cos x\,dx \qquad \text{[rearranging]}$$

$$= -x^2\cos x + \left(2x\sin x - \int 2\sin x\,dx\right) \qquad \text{[IBP]}$$

$$= -x^2\cos x + 2x\sin x - 2(-\cos x) + \text{const.}$$

$$= (2 - x^2)\cos x + 2x\sin x + \text{const.}$$

In a similar fashion, now with three applications of IBP:

$$\int_0^1 x^3 e^{2x}\,dx = \left[x^3\frac{e^{2x}}{2}\right]_0^1 - \int_0^1 3x^2\frac{e^{2x}}{2}\,dx \qquad \text{[IBP]}$$

$$= \frac{e^2}{2} - \left(\left[3x^2\frac{e^{2x}}{4}\right]_0^1 - \int_0^1 6x\frac{e^{2x}}{4}\,dx\right) \qquad \text{[IBP]}$$

$$= \frac{e^2}{2} - \frac{3e^2}{4} + \left[6x\frac{e^{2x}}{8}\right]_0^1 - \int_0^1 6\frac{e^{2x}}{8}\,dx \qquad \text{[IBP]}$$

$$= \frac{-e^2}{4} + \frac{3e^2}{4} - \left[\frac{6e^{2x}}{16}\right]_0^1$$

$$= \frac{e^2}{8} + \frac{3}{8}. \qquad \square$$

This is by far the main use of IBP, the idea of eventually differentiating out one of the two functions. There are other important uses of IBP which don't quite fit into this type. These next two examples fall into the original class, but are a little unusual: in these cases we choose to integrate the polynomial factor instead, as it is easier to differentiate the other factor. In particular, this is the case when we have a logarithm or an inverse trigonometric/hyperbolic function as the second factor.

Example 5.29 Find

$$\int (2x+1)\ln(x+1)\,dx \qquad \text{and} \qquad \int_0^1 x^2 \tan^{-1} x\,dx.$$

Solution In both cases, integrating the second factor looks rather daunting, but each factor differentiates nicely; recall

$$\frac{d}{dx}\ln(x+1) = \frac{1}{x+1} \qquad \text{and that} \qquad \frac{d}{dx}\tan^{-1}x = \frac{1}{1+x^2}.$$

(See Example 5.40 if the second derivative is new to you.) So if we apply IBP to the first integral, integrating the factor of $2x+1$, we get

$$\int (2x+1)\ln(x+1)\,dx = (x^2+x)\ln(x+1) - \int (x^2+x)\frac{1}{x+1}\,dx$$

$$= (x^2+x)\ln(x+1) - \int x\,dx$$

$$= (x^2+x)\ln(x+1) - \frac{x^2}{2} + \text{const.}$$

and likewise, for the second integral, again integrating the polynomial factor, we find

$$\int_0^1 x^2 \tan^{-1} x\,dx = \left[\frac{1}{3}x^3 \tan^{-1}x\right]_0^1 - \frac{1}{3}\int_0^1 \frac{x^3}{x^2+1}\,dx$$

$$= \frac{\pi}{12} - \frac{1}{3}\int_0^1 \left(x - \frac{x}{x^2+1}\right)\,dx$$

$$= \frac{\pi}{12} - \frac{1}{3}\left[\frac{x^2}{2} - \frac{1}{2}\ln(x^2+1)\right]_0^1 = \frac{\pi}{12} - \frac{1}{6} + \frac{\ln 2}{6}. \qquad \square$$

In the same vein as this, we can use IBP to integrate functions which, at first glance, don't seem to be products – this is done by treating a function $F(x)$ as the product $F(x) \times 1$.

Example 5.30 Determine

$$\int \ln x\,dx \qquad \text{and} \qquad \int \tan^{-1} x\,dx.$$

Solution With IBP we see (integrating 1 and differentiating $\ln x$) that

$$\int \ln x\,dx = \int \ln x \times 1\,dx = (\ln x)x - \int \frac{1}{x}x\,dx = x\ln x - \int dx = x\ln x - x + \text{const.}$$

#1338
#1339
#1341
#1342

(as already seen in #1312) and similarly

$$\int \tan^{-1} x \, dx = \int \tan^{-1} x \times 1 \, dx = (\tan^{-1} x)x - \int \frac{x}{1+x^2} \, dx$$

$$= x \tan^{-1} x - \frac{1}{2} \ln(1+x^2) + \text{const.} \qquad \square$$

Sometimes both functions don't become complicated as we integrate, but also don't simplify – in these cases we might hope to return (in due course) to the original integral, as with the integrals below.

Example 5.31 (See also #1347 and #1348.) Find

$$I = \int e^x \sin x \, dx \qquad \text{and} \qquad J = \int_0^{\pi/2} \cos 2x \sin 3x \, dx.$$

Solution Neither of the functions e^x, $\sin x$ in I simplify with differentiation or integration, but if we apply IBP twice, integrating the e^x factor each time, then we see I equals

$$e^x \sin x - \int e^x \cos x \, dx = e^x \sin x - \left(e^x \cos x - \int e^x (-\sin x) \, dx \right) = e^x (\sin x - \cos x) - I.$$

We see that we have returned to the original integral of I and so we can rearrange this equality to get

$$I = \int e^x \sin x \, dx = \frac{1}{2} e^x (\sin x - \cos x) + \text{const.}$$

We can approach J similarly with IBP, twice integrating one of the factors and twice differentiating the other. So

$$J = \left[\frac{\sin 2x}{2} \sin 3x \right]_0^{\pi/2} - \int_0^{\pi/2} \left(\frac{\sin 2x}{2} \right) (3\cos 3x) \, dx \qquad \text{[IBP]}$$

$$= -\frac{3}{2} \left\{ \left[-\frac{\cos 2x}{2} \cos 3x \right]_0^{\pi/2} - \int_0^{\pi/2} \left(-\frac{\cos 2x}{2} \right) (-3\sin 3x) \, dx \right\} \qquad \text{[IBP]}$$

$$= -\frac{3}{2} \left\{ \frac{1}{2} - \frac{3}{2} J \right\} = -\frac{3}{4} + \frac{9}{4} J.$$

Rearranging this last equation we see that $3/4 = 5J/4$ and hence $J = 3/5$. $\qquad \square$

We now move on to discuss the systematic use of IBP in dealing with families of integrals. This introduces the idea of *reduction formulae* (which we met in a limited way in Example 2.8 and #262, #263). When faced with a family of integrals such as

$$I_n = \int \cos^n \theta \, d\theta, \qquad n \geqslant 0,$$

our aim will be to write I_n in terms of other I_k where $k < n$, eventually reducing the problem to calculating I_0, or I_1 say, which hopefully are straightforward integrals.

For this particular integral, using IBP we see

$$I_n = \int \cos^{n-1}\theta \times \cos\theta \; d\theta$$

$$= \cos^{n-1}\theta \sin\theta - \int (n-1)\cos^{n-2}\theta \, (-\sin\theta)\sin\theta \; d\theta$$

$$= \cos^{n-1}\theta \sin\theta + (n-1)\int \cos^{n-2}\theta \left(1 - \cos^2\theta\right) d\theta$$

$$= \cos^{n-1}\theta \sin\theta + (n-1)(I_{n-2} - I_n).$$

Rearranging this we see

$$I_n = \frac{\cos^{n-1}\theta \sin\theta}{n} + \frac{n-1}{n}I_{n-2}. \tag{5.22}$$

With repeated use of this reduction formula, I_n can be rewritten in terms of simpler and simpler integrals until we are left only needing to calculate I_0, if n is even, or I_1, if n is odd.

Example 5.32 Calculate $I_7 = \int \cos^7\theta \; d\theta$.

Solution Using the reduction formula above

$$I_7 = \frac{\cos^6\theta \sin\theta}{7} + \frac{6}{7}I_5$$

$$= \frac{\cos^6\theta \sin\theta}{7} + \frac{6}{7}\left(\frac{\cos^4\theta \sin\theta}{5} + \frac{4}{5}I_3\right)$$

$$= \frac{\cos^6\theta \sin\theta}{7} + \frac{6\cos^4\theta \sin\theta}{35} + \frac{24}{35}\left(\frac{\cos^2\theta \sin\theta}{3} + \frac{2}{3}I_1\right)$$

$$= \frac{\cos^6\theta \sin\theta}{7} + \frac{6\cos^4\theta \sin\theta}{35} + \frac{8\cos^2\theta \sin\theta}{35} + \frac{16}{35}\sin\theta + \text{const.} \qquad \square$$

Example 5.33 Calculate $\int_0^1 x^3 e^{2x} \, dx$.

Solution This is an integral we previously calculated in Example 5.28. We can approach this, in a more systematic and notationally clearer fashion, by setting up a reduction formula. For a natural number n, let

$$J_n = \int_0^1 x^n e^{2x} \, dx.$$

We can then use IBP to show

$$J_n = \left[x^n \frac{e^{2x}}{2}\right]_0^1 - \int_0^1 nx^{n-1}\frac{e^{2x}}{2} \, dx = \frac{e^2}{2} - \frac{n}{2}J_{n-1} \qquad \text{if } n \geqslant 1,$$

and so the calculation in Example 5.28 simplifies enormously (at least on the eye). We first note

$$J_0 = \int_0^1 e^{2x} \, dx = \left[\frac{e^{2x}}{2}\right]_0^1 = \frac{e^2 - 1}{2},$$

#1350
#1351
#1352
#1353

and then applying the reduction formula, we find

$$J_3 = \frac{e^2}{2} - \frac{3}{2}J_2 = \frac{e^2}{2} - \frac{3}{2}\left(\frac{e^2}{2} - \frac{2}{2}J_1\right) = \frac{e^2}{2} - \frac{3e^2}{4} + \frac{3}{2}\left(\frac{e^2}{2} - \frac{1}{2}J_0\right) = \frac{e^2}{8} + \frac{3}{8}. \qquad \square$$

Some families of integrands may involve two variables, such as the following integral of Euler's.

Example 5.34 The **beta function** $B(a,b)$ is defined for positive reals a,b by

$$B(a,b) = \int_0^1 x^{a-1}(1-x)^{b-1}\ dx.$$

Calculate $B(m,n)$ for positive integers m,n.

#1357
#1358
#1359
#1360
#1363

Solution Note that

$$B(m,1) = \int_0^1 x^{m-1}\ dx = \frac{1}{m}. \tag{5.23}$$

So it would seem best to find a reduction formula moving us towards such an integral. Using IBP, if $b > 1$ we have

$$B(a,b) = \left[\frac{x^a}{a}(1-x)^{b-1}\right]_0^1 - \int_0^1 \frac{x^a}{a} \times (b-1) \times (-1)(1-x)^{b-2}\ dx$$

$$= 0 + \frac{b-1}{a}\int_0^1 x^a(1-x)^{b-2}\ dx$$

$$= \frac{b-1}{a}B(a+1,b-1).$$

So if $n \geqslant 2$ we can apply this result repeatedly to see

$$B(m,n) = \frac{n-1}{m}B(m+1,n-1)$$

$$= \frac{n-1}{m} \times \frac{n-2}{m+1}B(m+2,n-2)$$

$$= \left(\frac{n-1}{m}\right)\left(\frac{n-2}{m+1}\right)\cdots\left(\frac{1}{m+n-2}\right)B(m+n-1,1)$$

$$= \left(\frac{n-1}{m}\right)\left(\frac{n-2}{m+1}\right)\cdots\left(\frac{1}{m+n-2}\right)\frac{1}{m+n-1} \qquad \text{[by (5.23)]}$$

$$= \frac{(m-1)!(n-1)!}{(m+n-1)!}.$$

Equation (5.23) shows this formula also holds for $n = 1$. $\qquad \square$

Example 5.35 Define $I_n = \int \sec^n \theta\ d\theta$ for an integer n. Determine a reduction formula for I_n and hence find I_6.

Solution Recalling that the derivatives of $\tan\theta$ and $\sec\theta$ are respectively $\sec^2\theta$ and $\tan\theta\sec\theta$, we have that

$$I_n = \int \sec^{n-2}\theta \sec^2\theta \, d\theta$$

$$= \sec^{n-2}\theta \tan\theta - \int \left\{ (n-2)\sec^{n-3}\theta \sec\theta \tan\theta \right\} \tan\theta \, d\theta \qquad \text{[IBP]}$$

$$= \sec^{n-2}\theta \tan\theta - (n-2)\int \sec^{n-2}\theta \tan^2\theta \, d\theta$$

$$= \sec^{n-2}\theta \tan\theta - (n-2)\int \sec^{n-2}\theta (\sec^2\theta - 1) d\theta$$

$$= \sec^{n-2}\theta \tan\theta - (n-2)(I_n - I_{n-2}).$$

Solving for I_n we have, for $n \neq 1$, that

$$I_n = \frac{\sec^{n-2}\theta \tan\theta}{n-1} + \left(\frac{n-2}{n-1} \right) I_{n-2}.$$

Recall, we have as standard integrals from §5.2, that

$$I_1 = \ln|\sec\theta + \tan\theta| + \text{const.} \qquad \text{and} \qquad I_2 = \tan\theta + \text{const.}.$$

So to determine I_6 we find

$$I_6 = \frac{\sec^4\theta \tan\theta}{5} + \frac{4}{5}I_4$$

$$= \frac{\sec^4\theta \tan\theta}{5} + \frac{4}{5}\left(\frac{\sec^2\theta \tan\theta}{3} + \frac{2}{3}I_2 \right)$$

$$= \frac{\sec^4\theta \tan\theta}{5} + \frac{4}{15}\sec^2\theta \tan\theta + \frac{8}{15}\tan\theta + \text{const.}. \qquad \square$$

Exercises using Integration by Parts

#1338A Given integrable functions F and g, then IBP may be applied to the integrals $\int Fg \, dx$ and $\int_a^b Fg \, dx$. Show that this application of IBP does not depend on the choice of G such that $G' = g$.

#1339B Determine

$$\int x \sec^2 x \, dx, \qquad \int (x^3 + 2x - 1)\cos 2x \, dx, \qquad \int (x^2 + 2x - 3)e^x \, dx.$$

#1340c† Determine

$$\int_0^\pi x \sin^2 x \, dx, \qquad \int_0^1 (x^2 + 1)\sinh x \cosh x \, dx, \qquad \int_0^\infty (2x^2 - 2)e^{-2x-3} \, dx.$$

#1341B By treating the following as products of the form $F(x) \times 1$, determine

$$\int \sin^{-1}x \, dx, \qquad \int_0^1 \tan^{-1}x \, dx, \qquad \int \cosh^{-1}x \, dx, \qquad \int \ln(x^2 + 1) \, dx.$$

#1342b† Using IBP to integrate the polynomial factor, determine

$$\int_0^1 x^6 \ln x \, dx, \qquad \int_1^\infty x \sec^{-1}x \, dx, \qquad \int (x+2)\ln(x^2 + 1) \, dx.$$

#1343 D† Determine

$$\int \ln\left(\frac{x^2}{x^2+1}\right)\,dx \qquad \text{and show that} \qquad \int_0^\infty \ln\left(\frac{x^2}{x^2+1}\right)\,dx = -\pi.$$

#1344 b Let a, t be real with $t > 0$. Show that

$$\int_0^\infty e^{-tx}\cos ax\,dx = \frac{t}{t^2+a^2} \qquad \text{and} \qquad \int_0^\infty e^{-tx}\sin ax\,dx = \frac{a}{t^2+a^2}.$$

#1345 C Let a, b, t be real with $t > 0$. Evaluate

$$I = \int_0^\infty e^{-tx}\sin ax\sin bx\,dx \qquad \text{and} \qquad J = \int_0^1 \sinh(2x-1)\cos(3x+2)\,dx.$$

#1346 c† Let a be real and n be a natural number. Evaluate

$$I_n = \int_0^\infty x^n e^{-x}\mathrm{cis}\,ax\,dx, \qquad J_n = \int_0^\infty x^n e^{-x}\cos ax\,dx, \qquad K_n = \int_0^\infty x^n e^{-x}\sin ax\,dx.$$

#1347 C Use Euler's identity to write $e^x\sin x = \mathrm{Im}(e^{(1+i)x})$ and so rederive the integral I from Example 5.31.

#1348 c Rewrite $2\sin 3x\cos 2x = \sin A + \sin B$ and so rederive the integral J from Example 5.31.

#1349 C Show, for any polynomial $P(x)$, and $a \neq 0$, that

$$\int P(x)e^{ax}\,dx = e^{ax}\sum_{k\geq 0}(-1)^k\frac{P^{(k)}(x)}{a^{k+1}} + \text{const}.$$

Exercises on Reduction Formulae

#1350 B Let m, n be natural numbers. Show that

$$\int x^n(\ln x)^m\,dx = \frac{x^{n+1}(\ln x)^m}{n+1} - \frac{m}{n+1}\int x^n(\ln x)^{m-1}\,dx.$$

Hence find $\int x^3(\ln x)^2\,dx$.

#1351 B Use Example 5.35 to determine $\int \sec^9\theta\,d\theta$.

#1352 B† Determine a reduction formula for $\int \tan^n\theta\,d\theta$.

#1353 B Show that

$$\int \cos^5 x\,dx = \sin x - \frac{2}{3}\sin^3 x + \frac{1}{5}\sin^5 x + \text{const}.$$

#1354 C† Rederive Example 5.32 by writing $\cos^7\theta$ as a linear combination of functions of the form $\cos n\theta$.

#1355 C Show that the reduction formula found in Example 5.35 where n is negative, is the same as the reduction formula found in (5.22).

#1356 d† Show, for a natural number n, that

$$\int_0^1 (1-x^2)^n\,dx = \frac{2^{2n}(n!)^2}{(2n+1)!}.$$

#1357B† Show that

$$\int_0^\pi \sec x \cos(2n+1)x\, dx = (-1)^n \pi.$$

#1358B Show, using integration by parts, that

$$I_n = n\left(\frac{\pi}{2}\right)^{n-1} - n(n-1)I_{n-2} \qquad \text{where} \qquad I_n = \int_0^{\pi/2} x^n \sin x\, dx.$$

Evaluate I_0 and I_1, and hence find evaluate I_5 and I_6.

#1359B† (For a proof that $I_0 = \sqrt{\pi}/2$, see #1401.) Show, for $n \geqslant 2$, that

$$I_n = \frac{1}{2}(n-1)I_{n-2} \qquad \text{where} \qquad I_n = \int_0^\infty x^n e^{-x^2}\, dx.$$

Find an expression for I_n.

#1360B Euler's **gamma function** [16] $\Gamma(a)$ is defined for all $a > 0$ by the integral

$$\Gamma(a) = \int_0^\infty x^{a-1} e^{-x}\, dx.$$

Show that $\Gamma(a+1) = a\Gamma(a)$ for $a > 0$, and deduce that $\Gamma(n+1) = n!$ for a natural number n. Explain how the relation $\Gamma(a+1) = a\Gamma(a)$ can be used to defines $\Gamma(a)$ for all real a except $0, -1, -2, \ldots$.

#1361C Let m be a positive integer and α a positive real. Evaluate $B(m, \alpha)$ where B is the beta function.

#1362C Find reduction formulae for

$$I_n = \int_0^\infty e^{-x} \sin^n x\, dx \qquad \text{and} \qquad J_n = \int_0^\infty e^{-x} \cos^n x\, dx.$$

#1363B† For $-1 < \alpha < 1$, and n a positive integer, find reduction formulae involving the integrals below and evaluate I_5. [Note that J_1 is evaluated in #1410(iv).]

$$I_n = \int_0^1 \left(1+x^2\right)^n \sqrt{x}\, dx. \qquad J_n = \int_0^\infty \frac{x^\alpha}{\left(1+x^2\right)^n}\, dx.$$

#1364C Find a reduction formula for the integrals

$$I_n = \int_0^1 \frac{dx}{(4-x^2)^n}.$$

#1365C For a natural number n, we define

$$I_n = \int_1^\infty (x^2 - 1)^n e^{-x}\, dx.$$

Find a reduction formula involving I_n and hence determine I_5. Show for all n that eI_n is a positive integer.

---

[16] Euler introduced the gamma function in 1729–1730 as a means of extending the factorial function beyond the domain of the natural numbers.

#1366C† Let n be a natural number. Find reduction formulae for each of the following. Hence evaluate I_n and J_n.

$$I_n = \int_{-1}^{1} \frac{x^n}{\sqrt{1-x^2}}\,dx. \qquad J_n = \int_{-1}^{1} x^n \cos^{-1} x\,dx. \qquad K_n = \int_{0}^{1} x^n \sinh^{-1} x\,dx.$$

Exercises involving Infinite Series

#1367D† (See also #1374.) Find a reduction formula for the integrals $I_n = \int_0^1 \frac{x^n}{1+x}\,dx$ and deduce that

$$\ln 2 = 1 - \frac{1}{2} + \frac{1}{3} - \frac{1}{4} + \cdots + \frac{(-1)^{n+1}}{n} + \cdots.$$

#1368D† Find a reduction formula for the integrals $I_n = \int_0^1 \frac{x^n}{1+x^2}\,dx$ and deduce that [17]

$$\frac{\pi}{4} = 1 - \frac{1}{3} + \frac{1}{5} - \frac{1}{7} + \cdots + \frac{(-1)^n}{2n+1} + \cdots.$$

#1369D† (**Irrationality of** e) Find a reduction formula for the integrals

$$I_n = \int_0^1 (1-x)^n e^x\,dx.$$

Show that $0 \leqslant I_n \leqslant e/(n+1)$, where n is a positive integer, and deduce that

$$e = 1 + 1 + \frac{1}{2!} + \frac{1}{3!} + \cdots + \frac{1}{k!} + \cdots.$$

Prove the inequalities below and deduce that e is irrational:

$$0 < e - \sum_{k=0}^{n} \frac{1}{k!} < \frac{1}{n!\,n}.$$

#1370d† (**Exponential series**) Let a be a real number. Adapt #1369 to show that

$$e^a = 1 + a + \frac{a^2}{2!} + \frac{a^3}{3!} + \cdots + \frac{a^k}{k!} + \cdots.$$

#1371D† Let n be a positive integer. Show that

$$\left(1 + \frac{1}{n}\right)^n \leqslant \sum_{k=0}^{n} \frac{1}{k!} \leqslant \left(1 + \frac{1}{n}\right)^{n+1}.$$

#1372C Show that

$$-\int_0^1 \left(\frac{1-x^{4n}}{1+x^2}\right) \ln x\,dx = \sum_{k=0}^{n-1} \frac{(-1)^k}{(2k+1)^2}.$$

Deduce that

$$\int_0^1 \frac{-\ln x}{1+x^2}\,dx = \sum_{k=0}^{\infty} \frac{(-1)^k}{(2k+1)^2}.$$

The common value of this integral and sum is known as **Catalan's constant** and is commonly denoted G.

---

[17] This result was known to the Indian mathematician Madhave of Sangamagrama (c.1340–1425), as a special case of the inverse tangent series. In the west, three centuries later, the series was independently found in 1671 by the Scottish mathematician James Gregory (1638–1675) and in 1674 by Leibniz.

#1373D† **(Taylor's theorem** [18]**)** Show, for a natural number n and a $(n+1)$-times differentiable function $f(x)$, that

$$f(x) = f(0) + f'(0)x + \frac{f''(0)}{2!}x^2 + \cdots + \frac{f^{(n)}(0)}{n!} + \frac{1}{n!}\int_0^x (x-t)^n f^{(n+1)}(t)\,dt.$$

#1374D† [19] (See also #1367.) Let n be a positive integer and $f(x) = \ln(1+x)$ for $|x| < 1$. Show that

$$\ln(1+x) = x - \frac{x^2}{2} + \frac{x^3}{3} - \frac{x^4}{4} + \cdots + (-1)^{n+1}\frac{x^n}{n} + (-1)^n \int_0^x \frac{(x-t)^n}{(1+t)^{n+1}}\,dt.$$

Show that the integral in the above identity tends to 0 as $n \to \infty$.

#1375D† Show that

$$\int_0^1 \ln x \ln(1-x)\,dx = 2 - \sum_{n=1}^{\infty}\frac{1}{n^2}.$$

#1376D† (Euler, 1748) For real x and a natural number n, define

$$I_n(x) = \int_0^{\pi/2} \cos^n t \cos xt\,dt.$$

(i) Find a reduction formula for $I_n(x)$ and deduce that

$$\frac{I_{n-2}(x)}{I_{n-2}(0)} = \left(1 - \frac{x^2}{n^2}\right)\frac{I_n(x)}{I_n(0)}.$$

(ii) Show that

$$0 \leqslant I_n(0) - I_n(x) \leqslant \frac{x^2}{2}(I_{n-2}(0) - I_n(0))$$

and deduce, for any x, that $I_n(x)/I_n(0)$ converges to 1 as n becomes large.

(iii) Deduce that

$$\sin x = x \prod_{k=1}^{\infty}\left(1 - \frac{x^2}{k^2\pi^2}\right).$$

---

[18] The theorem is named after the English mathematician Brook Taylor (1685–1731), who stated a version of the theorem in 1712. Taylor wasn't, though, the first to be aware of the theorem – or at least earlier variants had been known to Newton in 1691 and to Johann Bernoulli in 1694 – and it was only later in the century applied widely by the Scottish mathematician Colin MacLaurin (1698–1746) and Lagrange, who was the first to produce an expression for the remainder term (though different to the one given in #1373).

[19] This series was first published by Nicholas Mercator (1620–1687) in 1668 and independently discovered by Newton and Gregoire de Saint-Vincent (1584–1667). The infinite series also converges for $x = 1$.

#1377 D†[20] (Irrationality of π)(i) For a positive integer n, set

$$I_n(x) = x^{2n+1} \int_{-1}^{1} (1-t^2)^n \cos(xt)\,dt.$$

Use integration by parts to show that $I_n(x) = 2n(2n-1)I_{n-1}(x) - 4n(n-1)x^2 I_{n-2}(x)$.

(ii) Show that $I_0(x) = 2\sin x$ and $I_1(x) = -4x\cos x + 4\sin x$. Deduce that there are polynomials $P_n(x)$ and $Q_n(x)$, of degree at most $2n$ and with integer coefficients, such that

$$I_n(x) = n!\,(P_n(x)\sin x + Q_n(x)\cos x).$$

(iii) Let $x = \pi/2$ and suppose for a contradiction that $\pi/2 = a/b$ where a and b are positive integers. Show that

$$\frac{a^{2n+1}}{n!} \int_{-1}^{1} \left(1-t^2\right)^n \cos(\pi t/2)\,dt$$

is an integer for all n. Explain how this leads to a contradiction and conclude that π is irrational.

5.4 Substitution

Integration is not something that should be treated as a table of formulae, for at least two reasons: one is that many of the formulae would be far from memorable, and the second is that each technique is more flexible and general than any memorized formula ever could be. If you can approach an integral with a range of techniques at hand you will find the subject less confusing and not be fazed by new and different integrands.

In many ways the aspect of integration that most exemplifies this is *substitution*, a technique which can almost become an art form. Substitution is such a varied and flexible approach that it is impossible to classify its uses, though we shall discuss later some standard trigonometric substitutions useful in integrating rational functions. We begin here with some examples of substitution correctly used and highlighting two examples that show some care is needed when using substitution, before stating our formal result.

Given an integral involving a variable x, say $\int f(x)\,dx$, the idea behind substitution is to introduce a new variable $u(x)$ which simplifies the integral. It is a recognition that the original integral is, in some way at least, much more naturally expressed in terms of u rather than x. Without hints, there is of course something of a skill in deciding what makes an effective substitution. Once we have decided on our choice of u, the function $f(x)$ will now be a new function of u and we also have that $du = u'(x)\,dx$ and the function $u'(x)$ will also need writing in terms of the new variable u. Should we be faced with a definite integral $\int_{x=a}^{x=b} f(x)\,dx$ then the limits $x = a$ and $x = b$ correspond to new limits $u = u(a)$ and $u = u(b)$. Hopefully the following examples will be illustrative.

Example 5.36 Determine

$$\int \frac{dx}{1+e^x}, \qquad \int_{-1}^{1} \sqrt{1-x^2}\,dx, \qquad \int_{-\infty}^{\infty} \frac{dx}{(1+x^2)^2}.$$

---

[20] The irrationality of π was first proved in 1761 by the Swiss polymath Johann Heinrich Lambert (1728–1777). The proof here is due to Mary Cartwright (see p.384 for a biographical note), who set the proof in 1945 as part of an examination at the University of Cambridge.

Emmy Noether

Sophie Germain

Sofia Kovalevskaya
(Photo by Fine Art
Images/Heritage
Images/Getty Images)

Dame Mary Lucy
Cartwright. © National
Portrait Gallery, London

Women in Mathematics The organization and practice of mathematics has changed much over the centuries. In the time of Tartaglia and Cardano secrecy overrode any desire to share knowledge; in the time of Newton and Leibniz there were no professional journals to publish in, a practice that seems almost unfathomably distant given today's internet and the desire for open access research. Prior to the twentieth century the number of professional mathematicians was relatively small by today's measure. This was most keenly seen in the general lack of opportunity for female mathematicians (often excluded from higher education or not permitted to graduate) and the sometimes entrenched hostility towards women of proven mathematical ability who were not allowed to join academic faculties. This was, of course, one aspect of a wider problem that riddled society at large. **Emmy Noether** (1882–1935) was a significant figure in the rise of modern algebra (specifically her three isomorphism theorems), also remembered for her theorem showing to every symmetry of a physical system there is a corresponding conservation law (e.g. time invariance leads to energy conservation). But despite such results, and despite the backing of mathematical heavyweights David Hilbert and Felix Klein, it would take four years before Noether was able to join the teaching faculty in Göttingen in 1919. Historically, other female mathematicians, including Sophie Germain (1776–1831) and Sofia Kovalevskaya (1850–1891), also found it hard to study at university or enter academia despite their mathematical successes. Germain was fortunate, though, in the correspondences she was able to maintain during her life with other mathematicians including Gauss, Legendre and Lagrange. Others pictured here are from the twentieth century. Mary Cartwright (1900–1998) was the first female president of the London Mathematical Society (LMS) (1961–1963);

Frances Kirwan
(Photograph
by Peter Jackson)

Frances Kirwan became the second; and Caroline Series became the third in 2017. Ulrike Tillmann is a noted algebraic topologist and FRS. Maryam Mirzakhani in 2014 became the first woman to win the Fields medal, the nearest equivalent to the Nobel Prize for mathematics. In the United Kingdom it remains the case that maths is somewhat male-dominated. In some European countries (e.g. Italy) this has not been the case for some decades. The situation is improving with the percentage of female academics increasing steadily across Europe over recent decades. There are various initiatives promoting opportunities for female students and researchers, such as *UK Mathematics Trust* summer schools for girls, the LMS *Women in Mathematics* committee and *European Women in Mathematics*.

Maryam Mirzakhani

Caroline Series (Photograph by Michelle Tennison)

Ulrike Tillmann (Photograph by Marc Atkins)

Solution For the first integral no obvious antiderivative springs to mind. So what might be a good substitution? We might try $u = 1 + e^x$ say. (Other substitutions would also work, such as $u = e^x$.) Setting $u = 1 + e^x$ we also have

$$\mathrm{d}u = e^x \, \mathrm{d}x = (u - 1) \, \mathrm{d}x.$$

Using partial fractions, the first integral equals

$$\int \frac{1}{u}\frac{du}{u-1} = \int \left(\frac{1}{u-1}-\frac{1}{u}\right)du = \ln|u-1|-\ln|u|+c$$

$$= \ln|e^x|-\ln|1+e^x|+c = -\ln(1+e^{-x})+c,$$

where c is a constant. One can of course verify that $-\ln(1+e^{-x})$ differentiates to $(1+e^x)^{-1}$, and by particularly insightful inspection this antiderivative might have been spotted. The aim of substitution is to simplify an integrand to a point where such leaps of intuition are not necessary.

For the second integral, we will set $x = \sin u$ (or $u = \sin^{-1} x$). Why is this a sensible substitution? The rationale for this substitution is the identity

$$1-x^2 = 1-\sin^2 u = \cos^2 u.$$

The entire purpose of the substitution is to turn the expression within the square root sign into a square. (Note $x = \cos u$ would also have worked.) Setting $x = \sin u$, we then have $dx = \cos u\,du$, and so the second integral becomes

$$\int_{x=-1}^{x=1} \sqrt{1-x^2}\,dx = \int_{u=-\pi/2}^{u=\pi/2} \sqrt{1-\sin^2 u}\,(\cos u\,du)$$

$$= \int_{u=-\pi/2}^{u=\pi/2} \cos^2 u\,du = \frac{1}{2}\int_{u=-\pi/2}^{u=\pi/2}(1+\cos 2u)\,du = \frac{\pi}{2}.$$

This is not surprising when we note the graph of $y = \sqrt{1-x^2}$ is that of a semicircle with radius 1.

Finally, for the third integral, we will set $x = \tan u$ (or $u = \tan^{-1} x$). Again this is a sensible substitution because of the identity

$$1+x^2 = 1+\tan^2 u = \sec^2 u,$$

and so we have

$$\int_{x=-\infty}^{x=\infty} \frac{dx}{(1+x^2)^2} = \int_{u=-\pi/2}^{u=\pi/2} \frac{\sec^2 u\,du}{(\sec^2 u)^2} = \int_{u=-\pi/2}^{u=\pi/2} \cos^2 u\,du = \frac{\pi}{2}. \qquad \square$$

By way of contrast, here are two examples where substitution appears not to work. This shows that, at the very least, some care is needed using substitution.

Example 5.37 Evaluate

$$\int_0^\pi \frac{dx}{1+\tan^2 x}, \qquad \int_{-\infty}^\infty \frac{e^{-1/x^2}}{1+x^2}\,dx.$$

Solution If we set $u = \tan x$ in the first integral we have that $du = \sec^2 x\,dx = (1+u^2)\,dx$ and seemingly find that

$$\int_{x=0}^{x=\pi} \frac{dx}{1+\tan^2 x} = \int_{u=0}^{u=0} \frac{du/(1+u^2)}{(1+u^2)} = 0.$$

This clearly cannot be correct, as the original integrand is positive on the interval $[0,\pi]$. In fact, a more sensible approach to the integral is to note

$$\int_0^\pi \frac{dx}{1+\tan^2 x} = \int_0^\pi \frac{dx}{\sec^2 x} = \int_0^\pi \cos^2 x\,dx = \frac{1}{2}\int_0^\pi (1+\cos 2x)\,dx = \frac{\pi}{2},$$

which is the correct answer; so just what did go wrong with our attempt with substitution?

The problem is the behaviour of the function $\tan x$ on the interval $[0,\pi]$. In particular it becomes positively infinite as x approaches $\pi/2$ returning from negative infinity. If instead we had split the integral into two integrals on $[0,\pi/2]$ and $[\pi/2,\pi]$, then the same substitution would given

$$\int_{x=0}^{x=\pi/2} \frac{dx}{1+\tan^2 x} + \int_{x=\pi/2}^{x=\pi} \frac{dx}{1+\tan^2 x} = \int_{u=0}^{u=\infty} \frac{du}{(1+u^2)^2} + \int_{u=-\infty}^{u=0} \frac{du}{(1+u^2)^2}$$

$$= \int_{u=-\infty}^{u=\infty} \frac{du}{(1+u^2)^2}$$

which we know to equal $\pi/2$ from Example 5.36.

In a similar fashion, with the second integral, we might have tried substituting $u = 1/x$, so that $du = -dx/x^2 = -u^2\,dx$, and argued that

$$\int_{x=-\infty}^{x=\infty} \frac{e^{-1/x^2}}{1+x^2}\,dx = \int_{u=0}^{u=0} \frac{e^{-u^2}}{1+u^{-2}} \left(\frac{-du}{u^2}\right) = 0.$$

Again this is impossible, as the integrand is positive (for $x \neq 0$). And again the issue is the discontinuity of $u = 1/x$ (from $-\infty$ to ∞ at $x = 0$) and the correct answer can be found by splitting the integral into two integrals on $(-\infty,0]$ and $[0,\infty)$. We then find

$$\int_{-\infty}^0 \frac{e^{-1/x^2}}{1+x^2}\,dx + \int_0^\infty \frac{e^{-1/x^2}}{1+x^2}\,dx = -\int_0^{-\infty} \frac{e^{-u^2}}{u^2+1}\,du - \int_\infty^0 \frac{e^{-u^2}}{u^2+1}\,du$$

$$= \int_{-\infty}^\infty \frac{e^{-u^2}}{u^2+1}\,du. \tag{5.24}$$

Evaluation of the final integral, which equals $e\pi(1-\text{erf}\,1)$, is left to #1419. $\square$

Now we state our formal result.

Proposition 5.38 *(Integration by Substitution) Let f be a continuous function on $[a,b]$ and g be a differentiable function from $[c,d]$ to $[a,b]$ with $g(c) = a$ and $g(d) = b$. Then*

$$\int_a^b f(x)\,dx = \int_c^d f(g(t))g'(t)\,dt.$$

Remark 5.39 Note that the substitutions we made in the two examples that went awry were $u = \tan x$ on $[0,\pi]$ and $u = 1/x$ on $(-\infty,\infty)$ and each had discontinuities on the given interval. $\blacksquare$

#1378 **Proof** For $a \leqslant u \leqslant b$, we define the antiderivative
#1379
#1382
#1383
$$F(u) = \int_a^u f(x)\,dx$$

of f. Then the function $F(g(t))$ is defined and differentiable for $c \leqslant t \leqslant d$ and by the chain rule

$$\frac{d}{dt}(F(g(t))) = F'(g(t))g'(t) = f(g(t))g'(t).$$

Hence, by the FTC,

$$\int_c^d f(g(t))g'(t)\,dt = F(g(d)) - F(g(c)) = F(b) - F(a) = \int_a^b f(x)\,dx. \qquad \square$$

Here are some standard integrals you may have met before and which can be demonstrated using substitution.

$$\int \frac{dx}{x^2+1} = \tan^{-1}x + \text{const.}, \qquad \int \frac{dx}{\sqrt{x^2+1}} = \sinh^{-1}x + \text{const.}$$

As ever, the purpose of substitution is to simplify the integrand and these two particular integrals are well suited to trigonometric and hyperbolic substitutions because of the identities [21] below.

Trigonometric identities:	Hyperbolic identities:
$\sin^2 u + \cos^2 u = 1$;	$\cosh^2 u - \sinh^2 u = 1$;
$\tan^2 u + 1 = \sec^2 u$;	$1 - \tanh^2 u = \text{sech}^2 u$;
$1 + \cot^2 u = \csc^2 u$;	$\coth^2 u - 1 = \text{csch}^2 u$.

Once such techniques become comfortable, and their rationale understood, then the above integrals (and similar) become more approachable as part of a general method rather than integrals to be rote-learned.

We begin with three such integrals.

Example 5.40 Determine

$$\int \frac{dx}{x^2+1}, \qquad \int \frac{dx}{\sqrt{x^2+1}}, \qquad \int_2^\infty \frac{dx}{(x^2+4)^{3/2}}.$$

#1384
#1385
#1405
#1406
#1407

Solution We need some identity that will simplify $x^2 + 1$ and $x = \tan u$ does this to some extent. (You can check that $x = \cot u$ works just as well.) We then have $dx = \sec^2 u\,du$ and find

$$\int \frac{dx}{x^2+1} = \int \frac{\sec^2 u\,du}{\tan^2 u + 1} = \int \frac{\sec^2 u\,du}{\sec^2 u} = u + \text{const.} = \tan^{-1}x + \text{const.}$$

For the second integral, we again need to simplify the expression $x^2 + 1$, ideally into a square. If we again set $x = \tan u$ then we find

$$\int \frac{dx}{\sqrt{x^2+1}} = \int \frac{\sec^2 u\,du}{\sqrt{\tan^2 u + 1}} = \int \sec u\,du = \ln|\sec u + \tan u| + c$$
$$= \ln\left|\sqrt{x^2+1}+x\right| + \text{const.}$$

---

[21] The three trigonometric identities are essentially the same (being multiples of one another), as are the hyperbolic identities. In fact, as $\cos(it) = \cosh t$ and $\sin(it) = i\sinh t$, all six identities are essentially the same identity.

This is the right answer, though perhaps not in its most palatable form. Instead we might have tried $x = \sinh u$, seeking to use the first hyperbolic identity. Then we would find

$$\int \frac{dx}{\sqrt{x^2+1}} = \int \frac{\cosh u\, du}{\sqrt{\sinh^2 u + 1}} = \int \frac{\cosh u\, du}{\sqrt{\cosh^2 u}} = \int du = u + \text{const.} = \sinh^{-1} x + \text{const.},$$

which is a rather cleaner way to write the answer. (Note Proposition 5.26(d) for the equality of these answers.)

For the third integral, we need to simplify the expression $x^2 + 4$. We can no longer use $x = \sinh u$, but we might try $x = 2\sinh u$ on the basis of $4 + (2\sinh u)^2 = (2\cosh u)^2$. This gives

$$\int_{x=2}^{x=\infty} \frac{dx}{(x^2+4)^{3/2}} = \int_{u=\sinh^{-1} 1}^{u=\infty} \frac{2\cosh u\, du}{(4\sinh^2 u + 4)^{3/2}} = \frac{1}{4}\int_{u=\sinh^{-1} 1}^{u=\infty} \text{sech}^2 u\, du$$

$$= \frac{1}{4}[\tanh u]_{\sinh^{-1} 1}^{\infty} = \frac{1}{4}\left(1 - \tanh \sinh^{-1} 1\right) = \frac{1}{4}\left(1 - \frac{1}{\sqrt{2}}\right). \qquad \square$$

In a like manner to Table 5.1, we might consider the following as standard integrals of this form and worth committing to memory.

Here are some further examples; recall the aim of substitution is to simplify $f(x)\,dx$, essentially recognizing that it is more naturally expressed in terms of another variable.

Example 5.41 Determine

$$\int \frac{x^2\, dx}{1+x^6}, \qquad \int_0^1 (x+3)\sqrt[3]{x^2+6x+2}\,dx, \qquad \int_0^\infty \frac{x\, dx}{(x^4+2x^2+2)^{3/2}}.$$

Solution Various substitutions might suggest themselves to us for the first integral. We might consider $u = x^2$, as the integrand's numerator and denominator can easily be written in terms of u but this would be forgetting that it is $x^2\, dx$ that needs to simplify; continuing that thought, $u = x^3$ might be better as the denominator can be expressed in terms of it and $du = 3(x^2\, dx)$. So instead with this latter substitution

$$\int \frac{x^2\, dx}{1+x^6} = \int \frac{du/3}{1+u^2} = \frac{1}{3}\tan^{-1} u + \text{const.} = \frac{1}{3}\tan^{-1}\left(x^3\right) + \text{const.}$$

For the second integral, anything simplifying the expression under the root might be helpful. We set $u = x^2 + 6x + 2$, and then note $du = 2(x+3)\,dx$, so that the second integral becomes

$$\int_{x=0}^{x=1} \sqrt[3]{x^2+6x+2}\,(x+3)\,dx = \int_{u=2}^{u=9} \sqrt[3]{u}\,\frac{du}{2} = \left[\frac{3}{8}u^{4/3}\right]_{u=2}^{u=9} = \frac{3}{8}\left(9^{4/3} - 2^{4/3}\right).$$

Table 5.2 *Some Further Standard Integrals*

$f(x)$	$\frac{1}{1+x^2}$	$\frac{1}{\sqrt{1-x^2}}$	$\frac{1}{\sqrt{1+x^2}}$	$\frac{1}{\sqrt{x^2-1}}$
$\int f(x)\,dx$	$\tan^{-1} x + c$	$\sin^{-1} x + c$	$\sinh^{-1} x + c$	$\cosh^{-1} x + c$

#1391
#1393
#1394
#1401
#1408
#1409

For the final integral, a substitution of $u = x^4 + 2x^2 + 2$ won't help as nothing related to du appears elsewhere in the integral. However, if we complete the square in the denominator, we see

$$x^4 + 2x^2 + 2 = (x^2 + 1)^2 + 1$$

and so we set $u = x^2 + 1$ as then $du = 2x\,dx$ which also appears in the integral. The third integral then becomes

$$\int_0^\infty \frac{x\,dx}{(x^4 + 2x^2 + 2)^{3/2}} = \int_{u=1}^{u=\infty} \frac{du/2}{(u^2 + 1)^{3/2}} = \frac{1}{2} \int_{t=\sinh^{-1} 1}^{t=\infty} \frac{\cosh t\, dt}{(\sinh^2 t + 1)^{3/2}}$$

$$= \frac{1}{2} \int_{t=\sinh^{-1} 1}^{t=\infty} \operatorname{sech}^2 t\, dt = \frac{1}{2} [\tanh t]_{t=\sinh^{-1} 1}^{t=\infty} = \frac{1}{2}\left(1 - \frac{1}{\sqrt{2}}\right)$$

when we make a second substitution $x = \sinh t$, and as

$$\sinh t = 1 \implies \cosh t = \sqrt{2} \implies \tanh t = \frac{1}{\sqrt{2}}. \qquad \square$$

Exercises on Substitution

#1378b† Determine

$$\int \frac{\ln x}{x}\,dx, \qquad \int \frac{dx}{1 + \sqrt{x}}, \qquad \int \frac{\ln(\tan x)}{\cos^2 x}\,dx.$$

#1379b Evaluate

$$\int_0^{\pi/2} \cos x \sqrt{\sin x}\,dx, \qquad \int_2^\infty \frac{dx}{x\sqrt{x-1}}, \qquad \int_0^1 \exp(\sin^{-1} x)\,dx.$$

#1380c† Determine

$$\int_0^\infty \frac{\tan^{-1} x}{1 + x^2}\,dx, \qquad \int \frac{\tan^{-1} x}{(1+x)^2}\,dx, \qquad \int \frac{\tan^{-1} x}{\left(1 + x^2\right)^{3/2}}\,dx.$$

#1381B Determine

$$\int \frac{dx}{\sinh x}, \qquad \int \frac{dx}{\cosh x}, \qquad \int_0^\infty \frac{dx}{\cosh x}.$$

#1382B Evaluate

$$\int_0^{\pi/2} \sin^{10} x \cos x\,dx, \qquad \int_0^{\pi/2} \cos^{11} x \sin^3 x\,dx,$$

$$\int_0^{\pi/2} \sin^{11} x \cos^5 x\,dx, \qquad \int_0^{\pi/2} \sin^6 x \cos^6 x\,dx.$$

#1383B† Let m, n be natural numbers. Show that $\int_0^1 x^n (\ln x)^m\,dx = (-1)^m m!\,(n+1)^{-m-1}$.

#1384b Determine, using trigonometric and/or hyperbolic substitutions,

$$\int \frac{dx}{\sqrt{4 - x^2}}, \qquad \int_2^\infty \frac{dx}{(x^2 - 1)^{3/2}}, \qquad \int_0^1 \frac{x+1}{\sqrt{x^2 + 4}}\,dx.$$

#1385 B By completing the square in the denominator, and by making the substitution $x = (\sqrt{2}\tan\theta - 1)/3$, determine

$$\int \frac{dx}{3x^2 + 2x + 1}.$$

#1386 D† Show, for a positive integer n, that

$$\int_0^\infty \frac{dx}{(1+x^2)^n} = \frac{\pi n}{2^{2n}(2n-1)}\binom{2n}{n}.$$

#1387 D† A curve has equation $y = f(x)$ where $f(x)$ is a decreasing function such that $f(0) > 0$ and $f(X) = 0$ as in Figure 5.9. Say that the curve, in polar co-ordinates, has equation $r = R(\theta)$. Show that the area $A(\alpha)$ of the sector bounded by the curve and the half-lines $\theta = 0$, $\theta = \alpha$ equals

$$A(\alpha) = \frac{1}{2}\int_{\theta=0}^{\theta=\alpha} R(\theta)^2\, d\theta.$$

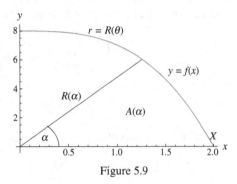

Figure 5.9

#1388 d† Let $a > 0$. Show that

$$\int_0^1 \frac{\ln x}{1+x^2}\, dx = -\int_1^\infty \frac{\ln x}{1+x^2}\, dx \quad \text{and find} \quad I(a) = \int_0^\infty \frac{\ln x}{x^2 + a^2}\, dx.$$

#1389 C† Let $a, b > 0$. Show that

$$\int_{-\infty}^\infty \frac{x}{ae^x + be^{-x}}\, dx = \frac{\pi}{4\sqrt{ab}}\ln\left(\frac{b}{a}\right).$$

#1390 d† Show that

$$\int_0^\infty \frac{1}{x}\sin\left(x - \frac{1}{x}\right) dx = 0.$$

What is the relevant property of sin that leads to this integral equalling 0?

#1391 B† By making the substitution $t = -x$ determine the integral

$$I = \int_{-\pi/2}^{\pi/2} \frac{\cos x}{e^x + 1}\, dx.$$

#1392 c† Evaluate

$$\int_{-\infty}^\infty \frac{dx}{(1+x^2)(e^x + 1)} \quad \text{and} \quad \int_{-\pi/4}^{\pi/4} \frac{dx}{1 + \sec x + \tan x}.$$

#1393 b† Given that $\int_{-\infty}^\infty \frac{\sin x}{x}\, dx = \pi$ (see #1420), evaluate

$$\int_{-\infty}^\infty \frac{\cos \pi x}{2x - 1}\, dx \quad \text{and} \quad \int_{-\infty}^\infty \frac{\cos \pi x}{4x^2 - 1}\, dx.$$

#1394B† Evaluate

$$\int_0^{\pi/2} \cos x \ln(\sin x)\,dx, \qquad \int_0^{\pi/2} \sin x \ln(\cos x)\,dx,$$

$$\int_0^{\pi/2} \sin x \ln(\sin x)\,dx, \qquad \int_0^{\pi/2} \cos x \ln(\cos x)\,dx.$$

#1395D† Show that

$$\int_0^{\pi/2} \ln(\sin x)\,dx = \int_0^{\pi/2} \ln(\cos x)\,dx = \int_0^{\pi/2} \ln(\sin 2x)\,dx.$$

Deduce that the above integrals equal $-(\pi/2)\ln 2$. Hence evaluate

$$\int_0^\infty \frac{\ln(x^2+1)}{x^2+1}\,dx \qquad \text{and} \qquad \int_0^\infty \frac{x}{\sqrt{e^x-1}}\,dx.$$

#1396d† Show that

$$\int_0^\pi \frac{x^2\,dx}{1-\cos x} = 4\pi \ln 2; \qquad \int_0^\pi \frac{x-\sin x}{1-\cos x}\,dx = 2.$$

#1397D† (See also #505) Determine $\int \cosh^2 t\,dt$ and hence show that

$$\frac{n\sqrt{1+n^2}}{2} + \frac{1}{2}\sinh^{-1} n \leqslant \sum_{k=1}^n \sqrt{k^2+1} \leqslant \frac{(n+1)\sqrt{2+2n+n^2}}{2} + \frac{1}{2}\sinh^{-1}(n+1).$$

#1398c† Let $a,b > 0$. Determine

$$\int \frac{\sinh x\,dx}{\sqrt{a^2-\cosh^2 x}} \qquad \text{and} \qquad \int \frac{\sinh x\,dx}{\sqrt{b^2-\sinh^2 x}}.$$

#1399D† Let $a > 0$. Show that

$$\int_0^{\pi/2} \frac{\sin x}{\sqrt{1+a^2\sin^2 x}}\,dx = \frac{\tan^{-1} a}{a}.$$

Exercises involving Special Functions

#1400c† Let $a,b > 0$. Determine the following integrals in terms of the gamma function.

$$\int_0^\infty x^{a-1} e^{-bx}\,dx, \qquad \int_0^\infty x^{a-1} e^{-bx^2}\,dx, \qquad \int_1^\infty \frac{(\ln x)^{a-1}}{x^2}\,dx \qquad \int_0^\infty e^{-ax^b}\,dx.$$

#1401B† Let $a > 0$. We define

$$I_a = \int_0^\infty e^{-ax^2}\,dx \quad \text{and} \quad J = \int_0^\infty \frac{e^{-y}}{\sqrt{y}}\,dy.$$

(i) Show that $J = 2I_1$ and that $I_a = a^{-1/2}I_1$.
(ii) Using the fact that

$$\int_0^\infty \left(\int_0^\infty e^{-y(1+x^2)}\,dy \right) dx = \int_0^\infty e^{-y} \left(\int_0^\infty e^{-yx^2}\,dx \right) dy,$$

determine I_a and J and show that

$$\int_{-\infty}^\infty e^{-x^2}\,dx = \sqrt{\pi} \qquad \text{and} \qquad \Gamma\left(\frac{1}{2}\right) = \sqrt{\pi}.$$

Remark 5.42 In #1401 we showed that $\int_0^\infty e^{-x^2} dx = \sqrt{\pi}/2$ which accounts for the normalizing factor in the definition of erf$x$ in #1324. It then follows that erf$x \to \pm 1$ as $x \to \pm\infty$. ∎

#1402 d† (i) For real a, b, c with $a > 0$, evaluate

$$\int_{-\infty}^\infty e^{-(ax^2+bx+c)} dx \qquad \text{and} \qquad \int_0^\infty e^{-(ax^2+bx+c)} dx.$$

(ii) Let $a, b > 0$. Prove that

$$\int_a^\infty \frac{e^{-bx^2}}{x^2} dx = \frac{e^{-ba^2}}{a} - \sqrt{b\pi}\left(1 - \text{erf}\left(a\sqrt{b}\right)\right).$$

(iii) For a, b real with $a > 0$, show that

$$\int_0^\infty xe^{-ax^2-2bx} dx = \frac{1}{2a} - \frac{b}{2a}\sqrt{\frac{\pi}{a}}\exp\left(\frac{b^2}{a}\right)\left(1 - \text{erf}\left(\frac{b}{\sqrt{a}}\right)\right).$$

#1403 E† Let $a, b > 0$. Show that $\int_{-\infty}^\infty e^{-ax^2-b/x^2} dx = \sqrt{\frac{\pi}{a}}e^{-2\sqrt{ab}}$.

#1404 d† The **logarithmic integral** li(x) is defined for $x > 0$ as

$$\text{li}(x) = \int_0^x \frac{dt}{\ln t}.$$

For $a > -1$, evaluate $\int_0^x \frac{t^a dt}{\ln t}$ in terms of li, and further prove that

$$\text{li}(x) = \int_{u=0}^{u=1} \frac{x\,du}{\ln x + \ln u}.$$

#1405 b Let $a, b > 0$. Show that $B(a, b) = B(b, a)$ where B denotes the beta function from Example 5.34.

#1406 B Use the substitution $x = (1 - \sin\theta)/2$ to show that $B(3/2, 3/2) = \pi/8$.

#1407 b For $a < b$, determine

$$\int_a^b \frac{dx}{\sqrt{(x-a)(b-x)}}.$$

#1408 B† By splitting the interval $0 \leqslant x \leqslant 1$ of the integral in Example 5.34 in two, show that

$$B(a, b) = \int_0^1 \frac{t^{a-1} + t^{b-1}}{(1+t)^{a+b}} dt.$$

#1409 b† Show for $-1 < a < 0$ that

$$B(a+1, -a) = \int_0^\infty \frac{x^a}{1+x} dx = -a\int_0^\infty x^{a-1}\ln(1+x)\,dx = 2\int_0^{\pi/2} \tan^{2a+1} x\,dx.$$

Find this value when $a = -1/2$.

#1410d† Given that $B(a+1, -a) = -\pi \csc(\pi a)$, something we will show in #1416, evaluate the following integrals.

(i) $\displaystyle \int_0^\infty \frac{dx}{1+x^\alpha} \quad \alpha > 1;$ (ii) $\displaystyle \int_0^\infty \frac{x^\beta \, dx}{(1+x)^2} \quad -1 < \beta < 1;$

(iii) $\displaystyle \int_0^\infty \frac{x^\gamma \, dx}{(1+x)^3} \quad -1 < \gamma < 2;$ (iv) $\displaystyle \int_0^\infty \frac{x^\delta}{1+x^2} \, dx \quad -1 < \delta < 1.$

#1411D† (See also #1709.) Let $a, b > 0$. Show that

$$B(a,b) = \int_0^\infty \frac{x^{a-1}}{(1+x)^{a+b}} \, dx \quad \text{and} \quad \frac{1}{(1+x)^{a+b}} = \frac{1}{\Gamma(a+b)} \int_0^\infty e^{-(1+x)t} t^{a+b-1} \, dt.$$

By swapping the order of integration in a repeated integral, deduce that

$$B(a,b) = \frac{\Gamma(a)\Gamma(b)}{\Gamma(a+b)}.$$

#1412C† Write the integrals below in terms of the beta function.

(i) $\int_0^\infty \frac{dx}{(x^n+1)^m} \qquad m, n > 0, mn > 1.$

(ii) $\int_0^\infty \frac{x^m \, dx}{\sqrt{1+x^n}} \qquad n > 2(m+1) > 0.$

#1413E† Let $x > 0$. Show that

$$B(x, n+1)n^x \to \Gamma(x) \qquad \text{as } n \to \infty.$$

Deduce that when $x \neq 0, -1, -2, \ldots$ then

$$\Gamma(x) = \lim_{n \to \infty} \frac{n! \, n^x}{x(x+1)(x+2)\cdots(x+n)}.$$

#1414D† (See also #263 and #278.) Show that

$$\binom{2n}{n} \frac{\sqrt{\pi n}}{2^{2n}} \to 1 \qquad \text{as } n \to \infty.$$

#1415D† Let x be a real number such that $2x$ is not a non-positive integer. Show that

$$\Gamma(2x) = \frac{\Gamma(x)\Gamma(x+\frac{1}{2})}{\Gamma\left(\frac{1}{2}\right)} 2^{2x-1}.$$

#1416D† (Reflection Property of the Gamma Function) Let a be a real number which is not an integer. Show that

$$\Gamma(a)\Gamma(1-a) = \frac{\pi}{\sin \pi a}.$$

#1417C† (Raabe's Formula. [22]**)** Prove that

$$\int_0^1 \ln \Gamma(x) \, dx = \frac{1}{2} \ln 2\pi.$$

Hence, for $a \geqslant 0$, evaluate $\int_a^{a+1} \ln \Gamma(x) \, dx$.

---

[22] Proved by Joseph Raabe (1801–1859) in 1840.

#1418C† Let $-1 < a < 1$. Use #1409 and #1416 to show that

$$\int_{-\infty}^{\infty} \frac{e^{ax}}{\cosh x} \, dx = \frac{\pi}{\cos(\pi a/2)}$$

and hence evaluate

$$\int_{-\infty}^{\infty} \frac{\cosh ax}{\cosh bx} \, dx \qquad \text{where} \qquad 0 < a < b.$$

Exercises involving Repeated Integrals (See also #1329 and #1330.)

#1419D Use the equality of the repeated integrals

$$\int_{x=-\infty}^{x=\infty} \left(\int_{y=0}^{y=\infty} e^{-y-x^2-yx^2} \, dy \right) dx = \int_{y=0}^{y=\infty} \left(\int_{x=-\infty}^{x=\infty} e^{-y-x^2-yx^2} \, dx \right) dy$$

to evaluate the final integral in (5.24).

#1420D† (See also #1554.) Let $a \geqslant 0$. By applying the method of #1329 to $e^{-yx} \sin x$, show that

$$\int_0^{\infty} \left(1 - e^{-ax}\right) \frac{\sin x}{x} \, dx = \tan^{-1} a.$$

Deduce that $\int_0^{\infty} \frac{\sin x}{x} \, dx = \pi/2$. What is $\int_0^{\infty} \frac{\sin cx}{x} \, dx$ where c is a real number?

#1421D† Show that the integrals $\int_0^{n\pi} \left| \frac{\sin x}{x} \right| \, dx$ increase without bound as n becomes large.

#1422D† By applying the method of #1329, determine

$$\int_{y=0}^{y=\infty} \left(\int_{x=0}^{x=\infty} \frac{dx}{(1+y)(1+x^2 y)} \right) dy \qquad \text{and evaluate} \qquad \int_0^{\infty} \frac{\ln x}{x^2 - 1} \, dx.$$

#1423D† Let $a, b > 0$. Use the fact that

$$\int_{y=0}^{y=a} \left(\int_{x=0}^{x=\infty} \frac{\sin(yx)}{x} \, dx \right) dy = \int_{x=0}^{x=\infty} \left(\int_{y=0}^{y=a} \frac{\sin(yx)}{x} \, dy \right) dx$$

to show

$$\int_0^{\infty} \frac{1 - \cos ax}{x^2} \, dx = \frac{\pi}{2} a.$$

Deduce that

$$\int_0^{\infty} \frac{\sin ax \sin bx}{x^2} \, dx = \frac{\pi}{2} \min\{a, b\}.$$

#1424D† Show that

$$\int_0^{\infty} \frac{x - \sin x}{x^3} \, dx = \frac{\pi}{4}.$$

#1425E† Let $A, a_1, \ldots, a_n$ be positive real numbers such that $A \geqslant a_1 + a_2 + \cdots + a_n$. Show that

$$\int_0^{\infty} \frac{\sin Ax \sin a_1 x \sin a_2 x \cdots \sin a_n x}{x^{n+1}} \, dx = \frac{\pi}{2} a_1 a_2 \cdots a_n.$$

#1426E† Let $a, b > 0$. Show that

$$\int_0^\infty \frac{(1 - \cos ax) \sin bx}{x^3} \, dx = \begin{cases} \frac{1}{4}\pi a^2 & a \leqslant b, \\ \frac{1}{4}\pi b(2a - b) & a > b. \end{cases}$$

Deduce, for $a \geqslant b > 0$ and $c > 0$, that

$$\int_0^\infty \frac{\sin ax \sin bx \sin cx}{x^3} \, dx$$

$$= \begin{cases} \frac{1}{2}\pi ab & a + b \leqslant c; \\ \frac{\pi}{8}\left(2ab + 2bc + 2ca - a^2 - b^2 - c^2\right) & a - b < c < a + b; \\ \frac{1}{2}\pi bc & c < a - b. \end{cases}$$

5.5 Rational and Algebraic Functions

With a little imagination we can extend the trigonometric and hyperbolic substitutions of the previous section to determine the integrals of a wide range of rational and algebraic functions. To begin we will see how to determine integrals of the form

$$\int \frac{Ax + B}{Cx^2 + Dx + E} \, dx \quad \text{and} \quad \int \frac{Ax + B}{\sqrt{Cx^2 + Dx + E}} \, dx.$$

The first integral is not particularly problematic if the quadratic in the denominator has real roots; we may then use partial fractions. However, this method won't work if the quadratic has complex roots. In such cases we can complete the square in the integrand to obtain

$$C\left\{ \left(x + \frac{D}{2C}\right)^2 + E - \frac{D^2}{4C^2} \right\}.$$

At this point the trigonometric substitution

$$x + \frac{D}{2C} = \sqrt{E - \frac{D^2}{4C^2}} \tan t,$$

will resolve the integral in a like manner to what we saw in the previous section. We might deal with the quadratic in the denominator under the square root sign in the second integral similarly. If

$$E - \frac{D^2}{4C^2} > 0 \quad \text{substitute} \quad x + \frac{D}{2C} = \sqrt{E - \frac{D^2}{4C^2}} \sinh t;$$

$$E - \frac{D^2}{4C^2} < 0 \quad \text{substitute} \quad x + \frac{D}{2C} = \sqrt{\frac{D^2}{4C^2} - E} \cosh t,$$

these various substitutions helping because of the respective identities

$$1 + \tan^2 t = \sec^2 t; \quad \sinh^2 t + 1 = \cosh^2 t; \quad \cosh^2 t - 1 = \sinh^2 t.$$

Note that the answers to such integrals become increasingly complicated, so simply memorizing them isn't much of an option. Rather, it is much better to remember the general approaches and rationale behind the substitutions outlined above.

Example 5.43 Determine

$$\int \frac{dx}{2x^2 + 4x + 3}, \qquad \int_0^\infty \frac{1}{x^2 + 5x + 6}\, dx, \qquad \int_{-1}^1 \frac{x}{\sqrt{5 + 2x - 3x^2}}\, dx.$$

#1427
#1428

Solution If we complete the square in the denominator of the first integral we have

$$\int \frac{dx}{2x^2 + 4x + 3} = \int \frac{dx}{2(x+1)^2 + 1} = \int \frac{dx/2}{(x+1)^2 + 1/2}.$$

We set $x + 1 = (\tan u)/\sqrt{2}$ so the denominator becomes $(\tan^2 u + 1)/2 = (\sec^2 u)/2$. Then we find

$$\frac{1}{2}\int \frac{(\sec^2 u/\sqrt{2})\, du}{(\sec^2 u)/2} = \int \frac{du}{\sqrt{2}} = \frac{u}{\sqrt{2}} + \text{const.} = \frac{\tan^{-1}(\sqrt{2}(x+1))}{\sqrt{2}} + \text{const.}$$

The denominator in the second integral has real roots and factorizes as $(x+2)(x+3)$. So substitutions aren't needed and instead we can use partial fractions to see

$$\int_0^\infty \frac{dx}{x^2 + 5x + 6} = \int_0^\infty \left(\frac{1}{x+2} - \frac{1}{x+3}\right) dx = \left[\ln\left(\frac{x+2}{x+3}\right)\right]_0^\infty = \ln 1 - \ln\frac{2}{3} = \ln\frac{3}{2}.$$

We can again complete the square in the third integral. We have

$$5 + 2x - 3x^2 = 3\left(\frac{5}{3} + \frac{2}{3}x - x^2\right) = \frac{16}{3} - 3\left(x - \frac{1}{3}\right)^2.$$

So we might set $x - 1/3 = (4/3)\sin t$ so that the RHS above equals $16\cos^2 t/3$. With this substitution we see

$$\int_{x=-1}^{x=1} \frac{x}{\sqrt{5 + 2x - 3x^2}}\, dx = \int_{u=-\pi/2}^{u=\pi/6} \frac{(1/3 + 4\sin u/3)}{(4/\sqrt{3})\cos u} \left(\frac{4}{3}\cos u\, du\right)$$

$$= \frac{1}{3\sqrt{3}}\int_{u=-\pi/2}^{u=\pi/6} (1 + 4\sin u)\, du = \frac{2\pi}{9\sqrt{3}} - \frac{2}{3}. \qquad \square$$

More generally, a **rational function** is one of the form

$$\frac{a_m x^m + a_{m-1} x^{m-1} + \cdots + a_0}{x^n + b_{n-1} x^{n-1} + \cdots + b_0},$$

where the a_i and b_i are real numbers – that is, a rational function is the quotient of two polynomials. The fundamental theorem of algebra shows that the roots of the denominator can all be found amongst the complex numbers, and it is also the case that complex roots come in conjugate pairs. So it is possible (in principle at least – the fundamental theorem provides no help in determining these roots) to rewrite the denominator as

$$x^n + b_{n-1} x^{n-1} + \cdots + b_0 = p_1(x)p_2(x)\cdots p_k(x),$$

where the polynomials $p_i(x)$ are either real linear factors of the form $x + A$ or real quadratic factors of the form $x^2 + Ax + B$ with negative discriminant. From here we can use partial fractions to simplify the function further.

Given a rational function

$$\frac{a_m x^m + a_{m-1} x^{m-1} + \cdots + a_0}{p_1(x)p_2(x)\cdots p_k(x)},$$

where the factors in the denominator are linear or quadratic terms as above, we follow the steps below to put it into a form we can integrate.

Algorithm 5.44 (*Integrating Rational Functions*)

1. *If the degree m of the numerator is at least that of the denominator n, then we divide the denominator into the numerator (using polynomial long division) until we have an expression of the form*

$$P(x) + \frac{A_{n-1}x^{n-1} + A_{n-1}x^{n-1} + \cdots + A_0}{p_1(x)p_2(x)\cdots p_k(x)}, \tag{5.25}$$

   where P(x) is a polynomial, and $A_{n-1}, \ldots, A_0$ are real numbers. Integrating the polynomial part P(x) will not cause us any difficulty so we will ignore that term from now on.

2. *Let's suppose, for now, that none of the factors in the denominator are repeated. In this case we can use partial fractions to rewrite this new rational function as*

$$\frac{A_{n-1}x^{n-1} + A_{n-2}x^{n-2} + \cdots + A_0}{p_1(x)p_2(x)\cdots p_k(x)} = \frac{\alpha_1(x)}{p_1(x)} + \frac{\alpha_2(x)}{p_2(x)} + \cdots + \frac{\alpha_k(x)}{p_k(x)},$$

   where each polynomial $\alpha_i(x)$ is of smaller degree than $p_i(x)$. This means that we have rewritten the rational function in terms of rational functions of the form

$$\frac{A}{x+B} \qquad and \qquad \frac{Cx+D}{x^2+Ex+F}, \qquad where\ E^2 < 4F.$$

   The former integrates to $A\ln|x+B|$ and we showed how to integrate the second at the beginning of this section.

3. *If, however, a factor $p_i(x)$ is repeated N times in the denominator then the best we can do with partial fractions is to reduce it to an expression of the form*

$$\frac{\beta_1(x)}{p_i(x)} + \frac{\beta_2(x)}{(p_i(x))^2} + \cdots + \frac{\beta_N(x)}{(p_i(x))^N},$$

   where the polynomials $\beta_i(x)$ have smaller degree than $p_i(x)$. This means our final rewriting of (5.25) may include functions of the form

$$\frac{A}{(x+B)^n} \qquad and \qquad \frac{Cx+D}{(x^2+Ex+F)^n}, \qquad where\ E^2 < 4F. \tag{5.26}$$

   The former will not prove problematic to integrate and the latter can be simplified by completing the square. In particular, when integrating $(x^2 + Ex + F)^{-n}$ then a trigonometric substitution transforms such an integral into one of the form

$$\int \cos^m \theta \, d\theta,$$

   which we know can be found using reduction formulae. Fuller details follow after the next example.

Example 5.45 Determine the following integral by using partial fractions to simplify the integrand.

$$\int \frac{x^5}{(x-1)^2(x^2+1)} \, dx.$$

#1429
#1431
Solution The numerator has degree five whilst the denominator has degree four, so we will need to divide the denominator into the numerator first. The denominator expands out to

$$(x-1)^2(x^2+1) = x^4 - 2x^3 + 2x^2 - 2x + 1.$$

Using polynomial long division we see that

$$
\begin{array}{r}
x \quad +2 \\
x^4 - 2x^3 + 2x^2 - 2x + 1 \,\overline{\big)\, x^5 \quad +0x^4 \quad +0x^3 \quad +0x^2 \quad +0x \quad +0} \\
x^5 \quad -2x^4 \quad +2x^3 \quad -2x^2 \quad +x \\
\hline
2x^4 \quad -2x^3 \quad +2x^2 \quad -x \quad +0 \\
2x^4 \quad -4x^3 \quad +4x^2 \quad -4x \quad +2 \\
\hline
2x^3 \quad -2x^2 \quad +3x \quad -2
\end{array}
$$

So we have that

$$\frac{x^5}{(x-1)^2(x^2+1)} = x + 2 + \frac{2x^3 - 2x^2 + 3x - 2}{(x-1)^2(x^2+1)},$$

which leaves us to find the constants A, B, C, D, in the identity

$$\frac{2x^3 - 2x^2 + 3x - 2}{(x-1)^2(x^2+1)} = \frac{A}{x-1} + \frac{B}{(x-1)^2} + \frac{Cx+D}{x^2+1}.$$

Multiplying through by the denominator, we find

$$2x^3 - 2x^2 + 3x - 2 = A(x-1)(x^2+1) + B(x^2+1) + (Cx+D)(x-1)^2.$$

As this holds for all values of x, then we can set $x = 1$ to find $B = 1/2$. If we set $x = 0$, compare x^3 coefficients and compare x coefficients we find

$$-2 = -A + 1/2 + D; \qquad 2 = A + C; \qquad 3 = A + C - 2D.$$

Hence $3 = 2 - 2D$, so that $D = -1/2$. From the first equation $A = 2$ and so $C = 0$. Finally then we have

$$\frac{x^5}{(x-1)^2(x^2+1)} = x + 2 + \frac{2x^3 - 2x^2 + 3x - 2}{(x-1)^2(x^2+1)} = x + 2 + \frac{2}{x-1} + \frac{1/2}{(x-1)^2} - \frac{1/2}{x^2+1}.$$

The above has now been rewritten as a sum of functions we can straightforwardly integrate. We see

$$\int \frac{x^5 \, dx}{(x-1)^2(x^2+1)} = \frac{x^2}{2} + 2x + 2\ln|x-1| - \frac{1/2}{x-1} - \frac{1}{2}\tan^{-1}x + \text{const..} \qquad \Box$$

Returning to the most general form of a rational function, we were able to reduce (using partial fractions) the problem to integrands of the form (5.26). Integrating functions of the first type causes us no difficulty as

$$\int \frac{A \, dx}{(x+B)^n} = \begin{cases} \frac{A}{(1-n)}(x+B)^{1-n} + \text{const.} & n \neq 1; \\ A\ln|x+B| + \text{const.} & n = 1. \end{cases}$$

The second integrand in (5.26) can be simplified, first by completing the square and then with a trigonometric substitution. Note that

$$x^2 + Ex + F = \left(x + \frac{E}{2F}\right)^2 - \frac{E^2}{4F^2}.$$

If we make a substitution of the form $u = x + E/(2F)$ then we can simplify this integral to something of the form

$$\int \frac{(au+b)\,du}{(u^2+k^2)^n} \qquad \text{for new constants } a,b \text{ and } k > 0.$$

Part of this we can integrate directly

$$\int \frac{u\,du}{(u^2+k^2)^n} = \begin{cases} \frac{1}{2(1-n)}(u^2+k^2)^{1-n} + \text{const.} & n \neq 1; \\ \frac{1}{2}\ln(u^2+k^2) + \text{const.} & n = 1. \end{cases}$$

The remaining part

$$\int \frac{du}{\left(u^2+k^2\right)^n}$$

can be simplified with a trigonometric substitution $u = k\tan\theta$, the integral becoming

$$\int \frac{du}{(u^2+k^2)^n} = \int \frac{k\sec^2\theta\,d\theta}{(k^2\tan^2\theta+k^2)^n} = \frac{1}{k^{2n-1}}\int \frac{\sec^2\theta\,d\theta}{(\sec^2\theta)^n} = \frac{1}{k^{2n-1}}\int \cos^{2n-2}\theta\,d\theta.$$

We have already considered such integrals via reduction formulae in §5.3.

Here then is a final example. Its solution surely demonstrates the importance of remembering the method and not the final formula!

Example 5.46 Determine

$$I = \int \frac{dx}{(3x^2+2x+1)^2}.$$

Solution The first step is to complete the square, so we find

$$I = \int \frac{dx}{(3x^2+2x+1)^2} = \frac{1}{9}\int \frac{dx}{(x^2+2x/3+1/3)^2} = \frac{1}{9}\int \frac{dx}{((x+1/3)^2+2/9)^2}.$$

Our first substitution is a translation – let $u = x+1/3$ noting that $du = dx$, giving

$$I = \frac{1}{9}\int \frac{du}{(u^2+2/9)^2}.$$

Then setting $u = \frac{\sqrt{2}}{3}\tan\theta$ to further simplify the integral, we have

$$I = \frac{1}{9}\int \frac{(\sqrt{2}/3)\sec^2\theta\,d\theta}{(2\sec^2\theta/9)^2} = \frac{1}{9}\times\frac{2}{\sqrt{3}}\times\left(\frac{9}{2}\right)^2\int \cos^2\theta\,d\theta$$

$$= \frac{3}{2\sqrt{2}}\int \frac{1}{2}(1+\cos 2\theta)\,d\theta \qquad [\text{using } \cos 2\theta = 2\cos^2\theta - 1]$$

$$= \frac{3}{4\sqrt{2}}\left(\theta + \frac{1}{2}\sin 2\theta\right) + \text{const.}$$

$$= \frac{3}{4\sqrt{2}}\left(\theta + \sin\theta\cos\theta\right) + \text{const.} \qquad [\text{using } \sin 2\theta = 2\sin\theta\cos\theta]$$

$$= \frac{3}{4\sqrt{2}}\left(\tan^{-1}\frac{3u}{\sqrt{2}} + \left(\sin\tan^{-1}\frac{3u}{\sqrt{2}}\right)\left(\cos\tan^{-1}\frac{3u}{\sqrt{2}}\right)\right) + \text{const.}$$

by undoing the substitution $u = \frac{\sqrt{2}}{3}\tan\theta$. We note that

$$\sin\tan^{-1}x = \frac{x}{\sqrt{1+x^2}} \qquad \text{and} \qquad \cos\tan^{-1}x = \frac{1}{\sqrt{1+x^2}},$$

and hence

$$
I = \frac{3}{4\sqrt{2}} \left(\tan^{-1} \left(\frac{3u}{\sqrt{2}} \right) + \frac{3u/\sqrt{2}}{\sqrt{1+9u^2/2}} \times \frac{1}{\sqrt{1+9u^2/2}} \right) + \text{const.}
$$

$$
= \frac{3}{4\sqrt{2}} \left(\tan^{-1} \left(\frac{3u}{\sqrt{2}} \right) + \frac{6u}{\sqrt{2}\left(2+9u^2\right)} \right) + \text{const.}
$$

$$
= \frac{3}{4\sqrt{2}} \left(\tan^{-1} \left(\frac{3}{\sqrt{2}} \left(x + \frac{1}{3} \right) \right) + \frac{6x+2}{\sqrt{2}\left(9x^2+6x+3\right)} \right) + \text{const.}
$$

$$
= \frac{3}{4\sqrt{2}} \tan^{-1} \left(\frac{3x+1}{\sqrt{2}} \right) + \frac{3x+1}{4\left(3x^2+2x+1\right)} + \text{const.} \qquad \square
$$

Thus we are in a position, in principle, to integrate any rational function. Algebraic functions that involve nothing worse than the square root of a quadratic can similarly be approached by completing the square and using a trigonometric or hyperbolic substitution. However, an algebraic function involving the square roots of a cubic cannot usually be approached in this way and their solution will commonly involve special functions, known as *elliptic integrals* – see p.419.

Finally we introduce a trigonometric substitution

$$
t = \tan\left(x/2\right)
$$

that will prove of use handling integrands that are rational functions of trigonometric functions.

Proposition 5.47 *Let* $t = \tan\left(x/2\right)$. *Then*

$$
\sin x = \frac{2t}{1+t^2}; \qquad \cos x = \frac{1-t^2}{1+t^2}; \qquad \tan x = \frac{2t}{1-t^2}; \qquad dx = \frac{2\,dt}{1+t^2}. \qquad (5.27)
$$

Proof Note that

$$
\frac{2t}{1+t^2} = \frac{2\tan\left(x/2\right)}{1+\tan^2\left(x/2\right)} = \frac{2\tan\left(x/2\right)}{\sec^2\left(x/2\right)} = 2\sin\left(\frac{x}{2}\right)\cos\left(\frac{x}{2}\right) = \sin x;
$$

$$
\frac{1-t^2}{1+t^2} = \frac{1-\tan^2\left(x/2\right)}{1+\tan^2\left(x/2\right)} = \frac{1-\tan^2\left(x/2\right)}{\sec^2\left(x/2\right)} = \cos^2\left(\frac{x}{2}\right) - \sin^2\left(\frac{x}{2}\right) = \cos x,
$$

making use of the $\sin 2\theta$ and $\cos 2\theta$ formulae. The third identity is simply an expression of the $\tan 2\theta$ formula. The final identity comes from differentiating $x = 2\tan^{-1}t$. $\qquad \square$

The t-substitution is generally useful for integrating rational functions of trigonometric functions, converting such integrals into ones involving rational functions.

Example 5.48 Determine

$$
\int \sec x\,dx, \qquad \int \frac{dx}{2+\cos x}, \qquad \int_0^{2\pi} \frac{dx}{2+\cos x}.
$$

Solution We already saw the first integral treated in §5.2 though our approach was admittedly ad hoc. Now the t-substitution gives us a more natural means of addressing this

#1440
#1441
#1444

integral. By (5.27)

$$\int \sec x \, dx = \int \frac{dx}{\cos x} = \int \left(\frac{1+t^2}{1-t^2}\right)\left(\frac{2\,dt}{1+t^2}\right) = \int \frac{2\,dt}{1-t^2} = \int \left(\frac{1}{1-t} + \frac{1}{1+t}\right) dt$$

$$= \ln|1+t| - \ln|1-t| + \text{const.} = \ln\left|\frac{1+t}{1-t}\right| + \text{const.}$$

If we make the t-substitution in the second integral, we find

$$\int \frac{dx}{2+\cos x} = \int \frac{\frac{2\,dt}{1+t^2}}{2 + \left(\frac{1-t^2}{1+t^2}\right)} = \int \frac{2\,dt}{2(1+t^2) + (1-t^2)} = \int \frac{2\,dt}{3+t^2}.$$

We met such integrals earlier in the previous section and know that $t = \sqrt{3}\tan u$ works well, the integral becoming

$$\int \frac{2\,dt}{3+t^2} = 2\int \frac{\sqrt{3}\sec^2 u \, du}{3\sec^2 u} = \frac{2u}{\sqrt{3}} + \text{const.} = \frac{2}{\sqrt{3}}\tan^{-1}\left(\frac{t}{\sqrt{3}}\right) + \text{const.}$$

$$= \frac{2}{\sqrt{3}}\tan^{-1}\left(\frac{1}{\sqrt{3}}\tan\left(\frac{x}{2}\right)\right) + \text{const..}$$

If we substitute the limits $x = 0$ and $x = 2\pi$ into this antiderivative we find

$$\int_0^{2\pi} \frac{dx}{2+\cos x} = \left[\frac{2}{\sqrt{3}}\tan^{-1}\left(\frac{1}{\sqrt{3}}\tan\left(\frac{x}{2}\right)\right)\right]_0^{2\pi} = \frac{2}{\sqrt{3}}\tan^{-1}(0) - \frac{2}{\sqrt{3}}\tan^{-1}(0) = 0.$$

However, this answer is clearly wrong, as the integrand is positive on all of $0 \leqslant x \leqslant 2\pi$. The problem here is the same as that in Example 5.37 with $\tan(x/2)$ having a discontinuity at $x = \pi$. However, if we separate the given integral as two integrals on $(0, \pi)$ and $(\pi, 2\pi)$, integrals which are equal, we see

$$\int_0^{2\pi} \frac{dx}{2+\cos x} = 2\int_0^{\pi} \frac{dx}{2+\cos x} = 2\left[\frac{2}{\sqrt{3}}\tan^{-1}\left(\frac{1}{\sqrt{3}}\tan\left(\frac{x}{2}\right)\right)\right]_0^{\pi}$$

$$= \frac{4}{\sqrt{3}}\left(\tan^{-1}(\infty) - \tan^{-1}(0)\right) = \frac{2\pi}{\sqrt{3}},$$

which is the correct answer.   □

Exercises

#1427 B By completing the square in the denominator, or otherwise, show that

$$\int_0^1 \frac{x\,dx}{x^2+2x+2} = \frac{\pi}{4} + \frac{1}{2}\ln\left(\frac{5}{2}\right) - \tan^{-1} 2.$$

#1428 b Determine

$$\int \sqrt{x^2+1}\,dx, \qquad \int \left(2x^2+3x+2\right)^{3/2} dx, \qquad \int \sqrt{x^2-3x+2}\,dx.$$

#1429 b Determine

$$\int \frac{x^3-4x}{x^2+1}\,dx \qquad \text{and} \qquad \int \frac{x^5}{x^3-1}\,dx.$$

#1430 b† Show that
$$\int_0^1 \frac{x^4(1-x)^4}{1+x^2}\,dx = \frac{22}{7} - \pi.$$
Is $22/7$ an overestimate or underestimate for π?

#1431 b† Determine
$$\int \frac{9x+16}{x(x^2+3x+4)}\,dx, \qquad \int \frac{1}{1+x^3}\,dx, \qquad \int_0^\infty \frac{dx}{(2x^2+12x+13)^{3/2}}.$$

#1432 c† Determine
$$\int \frac{dx}{\sqrt{x^2+2x+5}}, \qquad \int_0^\infty \frac{dx}{4x^2+4x+5}, \qquad \int_0^\infty \frac{dx}{(x^2+1)\sqrt{x^2+4}}.$$

#1433 d† Let $a > 0$. Verify the first integral below and evaluate the remaining three.
$$\int_0^\infty \frac{dx}{x^4+a^4} = \frac{\pi}{2\sqrt{2}a^3}, \qquad \int_0^\infty \frac{x^2\,dx}{x^4+a^4}, \qquad \int_0^1 \frac{dx}{x^4+1}, \qquad \int_0^1 \frac{x^2\,dx}{x^4+1}.$$
Verify that your answers to the first two integrals agree with those found in #1410.

#1434 D† (Newton, 1676) Show that
$$\frac{\pi}{2\sqrt{2}} = \int_0^1 \frac{1+x^2}{1+x^4}\,dx = 1 + \frac{1}{3} - \frac{1}{5} - \frac{1}{7} + \frac{1}{9} + \frac{1}{11} - \cdots.$$

#1435 D† Let m,n be natural numbers such that $m+2 \leqslant 2n$. Find a reduction formula for the integrals
$$I_{m,n} = \int_{-\infty}^\infty \frac{(3x+4)^m}{(x^2+2x+3)^n}\,dx.$$
Hence evaluate $I_{3,4}$.

#1436 D† Let $0 < a < b$. Show that
$$\int_0^\infty \frac{dx}{a+b\cosh x} = \frac{2}{\sqrt{b^2-a^2}}\tan^{-1}\left(\sqrt{\frac{b-a}{b+a}}\right).$$

#1437 d† Let $a > 0$. Show that
$$\int_{-\infty}^\infty \frac{dx}{\cosh^2 x + \sinh^2 a} = \frac{4a}{\sinh 2a}.$$

#1438 D† Let $a > 0$. Evaluate
$$\int_0^\pi \frac{dx}{\cosh a - \cos x}, \qquad \int_0^\pi \frac{dx}{\cosh^2 a - \cos^2 x}, \qquad \int_0^\pi \frac{x\,dx}{\cosh^2 a - \cos^2 x}.$$

#1439 c Use the $t = \tan(x/2)$ substitution to re-evaluate the second integral in #1392.

#1440 b Use the t-substitution to evaluate $\int_0^{\pi/2} \frac{d\theta}{(1+\sin\theta)^2}$.

#1441 B Let $-1 < \alpha < 1$. Show that
$$\int_0^{2\pi} \frac{d\theta}{1+\alpha\cos\theta} = \frac{2\pi}{\sqrt{1-\alpha^2}}.$$

#1442d Let $a > 1$. Show that

$$\int_0^{2\pi} \frac{d\theta}{(a+\cos\theta)^2} = \frac{2\pi a}{(a^2-1)^{3/2}},$$

and deduce the value of $\int_0^{2\pi} \frac{d\theta}{(a-\cos\theta)^2}$.

#1443d Show that

$$\int_0^{\pi/3} \frac{1-2\cos^2 x}{1+2\cos^2 x}\,dx = \frac{\pi}{6}\left(\sqrt{3}-2\right).$$

#1444B Let $t = \tanh(x/2)$. Show that

$$\sinh x = \frac{2t}{1-t^2}; \qquad \cosh x = \frac{1+t^2}{1-t^2}; \qquad \tanh x = \frac{2t}{1+t^2}.$$

Deduce that

$$\int \frac{dx}{\sinh x} = \ln\left|\tanh\left(\frac{x}{2}\right)\right| + \text{const.}$$

#1445d† By making the substitution $x = \tan^{-1} u^2$, or otherwise, find $\int \sqrt{\tan x}\,dx$.

#1446E† Determine $\int \sqrt[3]{\tanh x}\,dx$.

5.6 Numerical Methods

Of course it's not always possible to evaluate integrals exactly and there are numerical rules that will (typically) provide approximate values for integrals – approximate values, which by 'sampling' the function more and more times, can commonly be made more and more accurate.

Suppose that $f(x)$ is a real-valued function on the interval $[a,b]$ that we wish to integrate. Our idea will be to sample the function at $n+1$ evenly spread points through the interval

$$x_k = a + k\left(\frac{b-a}{n}\right) \quad \text{for } k = 0, 1, 2, \dots, n,$$

so that $x_0 = a$ and $x_n = b$. The corresponding y-value we will denote as

$$y_k = f(x_k).$$

For ease of notation the width between each sample we will denote as

$$h = \frac{b-a}{n}.$$

There are various rules and approaches for making an estimate of $\int_a^b f(x)\,dx$ based on this data. We will consider two: the *Trapezium Rule* and *Simpson's Rule*.

- **Trapezium Rule** This estimates the area as

$$\int_a^b f(x)\,dx \approx \frac{h}{2}(y_0 + 2y_1 + 2y_2 + \cdots + 2y_{n-1} + y_n). \tag{5.28}$$

This estimate is arrived at (as you might guess from the name) by approximating the area under the graph with trapezia. We assume that the graph behaves approximately linearly between (x_k, y_k) and (x_{k+1}, y_{k+1}) – not that unreasonable an assumption for small h – and take the area under the line segment connecting these points as our contribution to the integral. That (5.28) is the correct formula for this approximate area is left to #1447.

• **Simpson's Rule** [23] This requires that n be even and estimates the area as

$$\int_a^b f(x)\,dx \approx \frac{h}{3}(y_0 + 4y_1 + 2y_2 + 4y_3 + 2y_4 + \cdots + 2y_{n-2} + 4y_{n-1} + y_n). \qquad (5.29)$$

Simpson's rule works on the assumption that between the three points (x_k, y_k), (x_{k+1}, y_{k+1}), (x_{k+2}, y_{k+2}) (where k is even) the function $f(x)$ changes quadratically and it calculates the area contributed beneath each of these quadratic curves. That (5.29) is the correct formula for this approximate area is left to #1448.

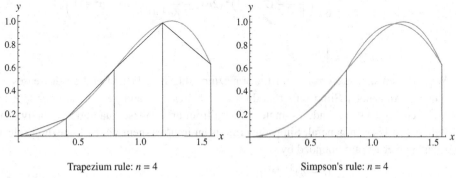

Trapezium rule: $n = 4$ Simpson's rule: $n = 4$

Figure 5.10

Figure 5.10 shows applications of the trapezium rule and Simpson's rule in calculating

$$\int_0^{\pi/2} \sin(x^2)\,dx$$

with $n = 4$ subintervals.

Example 5.49 Estimate the integral

$$\int_0^1 x^3\,dx$$

using both the trapezium rule and Simpson's rule using $2n$ subintervals.

#1447
#1448
#1449
#1450

Solution This is, of course, an integral we can calculate exactly as $1/4$. The two rules above yield:

$$\text{Trapezium estimate} = \frac{1}{4n}\left(0^3 + 2\left(\frac{1}{2n}\right)^3 + \cdots + 2\left(\frac{2n-1}{2n}\right)^3 + 1^3\right)$$

$$= \frac{1}{4n}\left(\frac{1}{4n^3}\sum_{k=1}^{2n-1} k^3 + 1\right)$$

---

[23] Named after the English mathematician Thomas Simpson (1710–1761), though this was in fact a result Simpson himself learnt from Newton.

$$= \frac{1}{4n} \left(\frac{1}{4n^3} \times \frac{1}{4} (2n-1)^2 (2n)^2 + 1 \right) \qquad \text{[using \#245]}$$

$$\frac{1}{4} + \frac{1}{16n^2},$$

and we also have

$$\text{Simpson's estimate} = \frac{1}{6n} \left(0^3 + 4 \left(\frac{1}{2n} \right)^3 + 2 \left(\frac{2}{2n} \right)^3 + \cdots + 1^3 \right)$$

$$= \frac{1}{6n} \left(0 + \frac{4}{(2n)^3} \sum_{k=1}^{2n-1} k^3 - \frac{2}{(2n)^3} \sum_{k=1}^{n-1} (2k)^3 + 1 \right)$$

$$= \frac{1}{6n} \left(\frac{4}{8n^3} \frac{1}{4} (2n-1)^2 (2n)^2 - \frac{2}{8n^3} \frac{8}{4} (n-1)^2 n^2 + 1 \right)$$

$$= \frac{1}{4}. \qquad\qquad \square$$

We see then that the error from the trapezium rule is $1/(16n^2)$ and so decreases very quickly. Amazingly Simpson's rule does even better here and gets the answer spot on – the overestimates and underestimates of area from under these quadratics actually cancel out (see #1454). In general, Simpson's rule is an improvement on the trapezium rule; the rules' two errors are bounded by

$$\left| E_{\text{trapezium}} \right| \leqslant \frac{(b-a)^3}{12n^2} \max\{ |f''(x)| : a \leqslant x \leqslant b \} \qquad (n \text{ subintervals}); \qquad (5.30)$$

$$\left| E_{\text{Simpson}} \right| \leqslant \frac{(b-a)^5}{2880n^4} \max\{ |f''''(x)| : a \leqslant x \leqslant b \} \qquad (2n \text{ subintervals}). \quad (5.31)$$

Note that the error decreases as a function of $1/n^4$ for Simpson's rule but only as a function of $1/n^2$ for the trapezium rule. These error bounds are left to #1455 and #1456 to determine.

#1452
#1453

We now apply the trapezium rule to derive an important approximation for factorials.

#1455
#1458
#1460
#1464

Proposition 5.50 (***Stirling's Approximation*** [24]) *As n becomes large then*

$$\frac{n!}{\sqrt{2\pi n} \left(\frac{n}{e} \right)^n} \quad \text{approximates to} \quad 1.$$

This same result is often written as

$$n! \sim \sqrt{2\pi n} \left(\frac{n}{e} \right)^n.$$

Proof Firstly we note that

$$\ln n! = \ln 2 + \ln 3 + \cdots + \ln n.$$

We can find a good approximation to the sum on the RHS by applying the trapezium rule to $\ln x$ on the interval $[1, n]$. Let $f(x)$ denote the approximating function to $\ln x$ whose integral the trapezium rule determines using $n-1$ steps – that is $f(x)$ satisfies $f(k) = \ln k$ for

---

[24] This approximation is due to the Scottish mathematician James Stirling (1692–1770), who published the formula in his *Methodus Differentialis* in 1730.

each integer $k = 1, 2, \ldots, n$ and is linear between those values. Note that for x in the range $k \leqslant x \leqslant k + 1$ we have

$$\frac{1}{k+1} \leqslant \frac{1}{x} \leqslant \frac{1}{k}$$

and so integrating we have

$$\int_k^x \frac{dt}{k+1} \leqslant \int_k^x \frac{dt}{t} \leqslant \int_k^x \frac{dt}{k}$$

or equivalently

$$\ln k + \frac{x-k}{k+1} \leqslant \ln x \leqslant \ln k + \frac{x-k}{k}.$$

Further, as $\ln x$ is concave (that is, a chord connecting two points of the graph lies under the graph), we have $f(x) \leqslant \ln x$ on the interval $[k, k+1]$, and as $f'(x) \geqslant (k+1)^{-1}$ on the interval (that being the minimum gradient of $\ln x$ whilst $f'(x)$ has the average gradient) we have

$$\ln k + \frac{x-k}{k+1} \leqslant f(x) \leqslant \ln x \qquad \text{for } k \leqslant x \leqslant k+1.$$

So we have the inequalities

$$0 \leqslant \ln x - f(x) \leqslant \left(\frac{1}{k} - \frac{1}{k+1} \right)(x-k) \qquad \text{for } k \leqslant x \leqslant k+1,$$

and integrating on the interval $[k, k+1]$ we find

$$0 \leqslant \int_k^{k+1} (\ln x - f(x))\, dx \leqslant \left(\frac{1}{k} - \frac{1}{k+1} \right) \int_k^{k+1} (x-k)\, dx = \frac{1}{2} \left(\frac{1}{k} - \frac{1}{k+1} \right).$$

Summing up the contributions from the intervals $[1, 2], [2, 3], \ldots, [n-1, n]$ we find

$$0 \leqslant \int_1^n (\ln x - f(x))\, dx \leqslant \frac{1}{2} \sum_{k=1}^{n-1} \left(\frac{1}{k} - \frac{1}{k+1} \right) = \frac{1}{2} \left(1 - \frac{1}{n} \right),$$

as most of the terms in the above sum cancel consecutively.

Recalling an antiderivative of $\ln x$ to be $x \ln x - x$ and using the formula for the trapezium rule we then have

$$\int_1^n (\ln x - f(x))\, dx = [x \ln x - x]_1^n - 1 \left(\frac{\ln 1}{2} + \ln 2 + \ln 3 + \cdots + \ln(n-1) + \frac{\ln n}{2} \right)$$

$$= n \ln n - n + 1 - \left(\ln n! - \frac{1}{2} \ln n \right)$$

$$= \left(n + \frac{1}{2} \right) \ln n - n + 1 - \ln n!.$$

So we have

$$0 \leqslant \int_1^n (\ln x - f(x))\, dx = \left(n + \frac{1}{2} \right) \ln n - n + 1 - \ln n! \leqslant \frac{1}{2} \left(1 - \frac{1}{n} \right).$$

These integrals form an increasing sequence of numbers which we see are bounded above by $1/2$. This means that the sequence must converge to some limit L say [25]. Applying the exponential function we find

$$\frac{n!}{\sqrt{n}\left(\frac{n}{e}\right)^n}$$

converges to some limit M say (explicitly $M = e^{1-L}$). We have in fact already determined this limit twice before. If the above converges to M, then

$$\binom{2n}{n}\frac{\sqrt{n}}{2^{2n}} = \frac{(2n)!}{n!\,n!}\frac{\sqrt{n}}{2^{2n}} = \sqrt{2}\times\frac{(2n!)}{\sqrt{2n}\left(\frac{2n}{e}\right)^{2n}}\times\left(\frac{\sqrt{n}\left(\frac{n}{e}\right)^n}{n!}\right)^2$$

converges to $\sqrt{2}\times M/M^2 = \sqrt{2}/M$. But we know from #279 and #1414 the above to have limit $1/\sqrt{\pi}$ and so $M = \sqrt{2\pi}$ and Stirling's approximation follows. □

Remark 5.51 It is the case that Stirling's approximation is always an underestimate. A more refined argument can also give an overestimate so as to bound $n!$ from above and below. Specifically,

$$\sqrt{2\pi n}\left(\frac{n}{e}\right)^n < n! < \sqrt{2\pi n}\left(\frac{n}{e}\right)^n\exp\left(\frac{1}{12n}\right) \qquad \text{for } n \geqslant 1. \blacksquare$$

#1447A Verify that (5.28) is the correct expression for the estimate produced by the trapezium rule.

#1448B† Verify that (5.29) is the correct expression for the estimate produced by Simpson's rule.

#1449b Estimate, via the trapezium rule using four subintervals, the integral $\int_0^1 \sin(x^2)\,dx$.

#1450b† Calculate $\int_0^1 e^x\,dx$ using the trapezium rule and Simpson's rule, both with $2n$ steps.

#1451D† Show that the estimates found in #1450 approach the correct value as n becomes large.

#1452b† Let $a > 0$. Find the estimates for $\int_0^a x^2\,dx$ using $2n$ intervals with the trapezium and Simpson's rules. Show that both these estimates approach the correct answer as n becomes large.

#1453B† If $f(x)$ is a convex function, i.e. $f''(x) \geqslant 0$ for all x, show that the trapezium rule overestimates an integral.

#1454C Let $f(x)$ be a polynomial of degree 3 or less. Show directly for any a, b, n that Simpson's rule estimates the definite integral exactly.

---

[25] That a bounded increasing sequence converges is a standard early result of any analysis course. It is hoped that this seems a reasonable fact to assume here without clouding the proof of Stirling's approximation. For those wishing to review a proof, see Binmore [4, Theorem 4.17].

#1455 B† Show, for a twice differentiable function $f(x)$, that

$$\int_a^b f(x)\,dx - \frac{(b-a)\,(f(b)+f(a))}{2} = \frac{1}{2}\int_a^b (x-a)(x-b)f''(x)\,dx.$$

Hence deduce the error bound for the trapezium rule given in (5.30).

#1456 E† Define the function $p(x)$ on $[a,b]$ by

$$p(x) = \begin{cases} \frac{1}{3}(x-a)^3\,(3x-a-2b) & \text{for } a \leqslant x \leqslant \frac{a+b}{2}, \\ \frac{1}{3}(x-b)^3\,(3x-2a-b) & \text{for } \frac{a+b}{2} \leqslant x \leqslant b. \end{cases}$$

Let $f(x)$ be a four-times differentiable function on $[a,b]$ such that $f''''(x)$ is continuous. Show that

$$\int_a^b p(x)f''''(x)\,dx = 4(a-b)\left(f(a)+4f\left(\frac{a+b}{2}\right)+f(b)\right) + 24\int_a^b f(x)\,dx.$$

Hence deduce the error bound for the trapezium rule given in (5.31).

#1457 C Determine the estimate from the trapezium rule of $\int_{1/n}^1 \ln x\,dx$, when it is applied with $n-1$ subintervals. As n becomes large, does the estimate converge to $\int_0^1 \ln x\,dx$?

#1458 B† Catalan's constant $G = 0.915965594\ldots$ was defined in #1372. Show that

$$G = \frac{1}{2}\int_0^{\pi/2} \frac{t}{\sin t}\,dt.$$

(i) With the aid of a computer, determine how many steps the trapezium rule takes to approximate G to within an accuracy of 10^{-4}. (ii) Do the same for Simpson's rule.

#1459 C Give four examples to show the trapezium rule and Simpson's rule can separately overestimate/underestimate a given integral.

#1460 B† If the trapezium rule, using n subintervals, overestimates an integral $\int_{-1}^1 f(x)\,dx$ what, if anything, can be said for its estimates of

$$\int_{-1}^1 (1+f(x))\,dx, \qquad \int_{-1}^1 (-f(x))\,dx, \qquad \int_{-1}^1 f(-x)\,dx,$$

$$\int_0^2 f(x-1)\,dx, \qquad \int_{-2}^2 f(x/2)\,dx,$$

using the same number of subintervals?

#1461D† (Adapted from MAT 2014 #3) The function $f(x)$ is defined for all real numbers and has the following properties, valid for all x and y:

$\quad$ (A)$\quad$ $f(x+y)=f(x)f(y)$;$\quad\quad$ (B)$\quad$ $df/dx=f(x)$;$\quad\quad$ (C)$\quad$ $f(x)>0$.

Throughout this question, these should be the only properties of f that you use; no use of the exponential function should be made in your answer.

(i) Show that $f(0)=1$ and that $I=\int_0^1 f(x)\,dx=f(1)-1$.
(ii) The trapezium rule with n steps is used to produce an estimate I_n for the integral I. Show that

$$I_n=\frac{1}{2n}\left(\frac{b+1}{b-1}\right)(f(1)-1)\quad\quad\text{where }b=f\left(\frac{1}{n}\right).$$

(iii) Show that $I_n\geqslant I$ for all n, and deduce that

$$f(1)\leqslant\left(1+\frac{2}{2n-1}\right)^n.$$

#1462C Let $f(x)$ denote Dirichlet's function. What estimate for $\int_0^1 f(x)\,dx$ does the trapezium rule with n subintervals give?

#1463D† Let $T(x)$ denote Thomae's function. What does the trapezium rule with 2^n intervals estimate $\int_0^1 T(x)\,dx$ to equal? What is the answer when using pq intervals where p and q are distinct primes?

#1464A With the aid of a calculator, determine the smallest n such that Stirling's formula approximates $n!$ to (i) 10%, (ii) 1%, (iii) 0.1%, (iv) 0.01%.

#1465D† Consider a point, with position X_t at time t, on a random walk along the integers which begins with $X_0=0$ and has probability p of increasing by 1 and probability $q=1-p$ of decreasing by 1.

(i) Write down $p_t=P(X_t=0)$.
(ii) Show that $\sum_{t=0}^{\infty}p_t$, the expected number of times the point returns to 0, is finite if $p\neq q$ and infinite if $p=q=1/2$.

#1466D† Consider a point P with position (X_t,Y_t) at time t, on a two-dimensional random walk of the integer grid $\{(x,y):x,y\in\mathbb{Z}\}$, which begins with $X_0=Y_0=0$. Let $0<a,b,c,d<1$ with $a+b+c+d=1$. If $(X_t,Y_t)=(x,y)$ then (X_{t+1},Y_{t+1}) equals, with respective probabilities a,b,c,d, one of

$$(x+1,y),\quad\quad(x-1,y),\quad\quad(x,y+1),\quad\quad(x+1,y-1).$$

(i) Let p_t denote the probability that $(X_t,Y_t)=(0,0)$. Show that $p_t=0$ when t is odd and determine p_t as a sum when t is even.
(ii) Use #278 to show that $\sum_{t=0}^{\infty}p_t$ becomes infinite when $a=b=c=d=1/4$.
(iii) What can be said of this sum when $a\neq b$ or $c\neq d$?

#1467E† What can be said of $\sum_{t=0}^{\infty}p_t$, in the notation of #1466, when $a=b\neq c=d$?

Statue of Alan Turing at Bletchley Park (Creative Commons license. Credit Antoine Taveneaux)

Computers in Mathematics. Numerical Analysis
A significant advance in the mathematical theory of computers came in a 1936 paper of Alan Turing's (1912–1954), in which he introduced a 'universal machine' (now known as a *Turing machine*) capable of generating any computable sequence of numbers, and gave a negative answer to Hilbert's *decision problem* of 1928. His paper was in fact a highly intuitive reformulation of results of Alonzo Church (1903–1995) (of which Turing was unaware) who had a year earlier shown the problem to be false.

That the decision problem is false means there is a function (on the natural numbers) that no algorithm can generate. This is a result in *computability*, a topic studying the theoretical limits of computers, and at the pure end of computer science. Turing would become a central figure in the development of computers over the next two decades, not just because of his contributions cracking the German *Enigma code* at Bletchley during World War II. In 1946 he provided the first detailed description of a stored-program computer. Also in 1950 he described the *Turing test*, a means for deciding on the presence of artificial intelligence.

In pure mathematics, functions are employed with little regard as to how, in the real world, such functions might be arrived at or how quickly or accurately calculations might be executed. In practice, following some experiment, such a 'function' would at best be present as a large set of data from recorded measurements, which might be considered as outputs of the function for certain inputs. Techniques from *statistics* or *numerical analysis* can then be used to interpolate or extrapolate that data to other values. Issues might arise during long repeated calculations, such as efficiency – how quickly the calculations can be done – and accuracy – how good an estimation of the actual function we have; there may be issues internal to the computer with the very real possibility that repeated rounding errors become so problematic as to make the ultimate answer nonsensical. For such cases we would wish our methods to be *numerically stable*, avoiding functions that exaggerate initially small errors of measurement.

5.7 A Miscellany of Definite Integrals

Evaluate the following definite integrals. The necessary techniques may be any of those described in this chapter. Some answers will have to remain in terms of special functions such as the beta, gamma and error functions; others are dependent on integrals calculated in the chapter or in earlier exercises. Each exercise comes with a hint, but you are particularly encouraged to attempt any integral in one or more ways before looking at the hint.

#1468 B† $$\int_0^\infty e^{-\sqrt{x}}\,dx.$$	**#1469 B†** $$\int_{-2}^2 |x^2-3x+1|\,dx.$$	**#1470 B†** $a,b>0,$ $$\int_0^\infty \frac{dx}{(x+a)^2+b^2}.$$	**#1471 A†** $a,b>0,$ $$\int_{-\infty}^\infty e^{-ax^2}\sinh bx\,dx.$$
#1472 B† $$\int_0^\infty \frac{\sinh x}{(1+\cosh x)^2}\,dx.$$	**#1473 D†** $$\int_0^{2\pi} \frac{\sin^2 x}{2+\cos x}\,dx.$$	**#1474 B†** $$\int_0^{\pi/2} \frac{2x-\sin x}{x^2+\cos x}\,dx.$$	**#1475 B†** $$\int_0^1 \frac{x-1}{\ln x}\,dx.$$
#1476 B† $a,b>0,$ $$\int_{-\infty}^\infty e^{-ax^2}\cosh bx\,dx.$$	**#1477 D†** $$\int_0^{\pi/3} \sqrt{\cos x+\cos(2x)}\,dx.$$	**#1478 D†** Natural $n,$ $$\int_0^1 \frac{x^n}{\sqrt{1-x}}\,dx.$$	**#1479 B†** $$\int_0^1 \frac{dx}{\sqrt{-\ln x}}.$$
#1480 B† $$\int_0^{\pi/2} x\cot x\,dx.$$	**#1481 D†** $a,b>0,$ $$\int_0^\infty \frac{\tan^{-1}ax-\tan^{-1}bx}{x}\,dx.$$	**#1482 B†** Integers $m,n,$ $$\int_0^{2\pi} \sin(mx)\cos(nx)\,dx.$$	**#1483 D†** $$\int_0^1 \frac{dx}{\sqrt{1-x^4}}.$$
#1484 B† $$\int_0^1 \frac{dx}{\sqrt{1+3x+2x^2}}.$$	**#1485 D†** $a,b>0,$ $$\int_0^\infty e^{-ax^2}\sinh bx\,dx.$$	**#1486 D†** $$\int_0^1 (\cos^{-1}x)(\ln x)\,dx.$$	**#1487 D†** $$\int_0^\pi \sin(\sqrt[3]{2x+1})\,dx.$$
#1488 D† $b>a>0,$ $$\int_{-\infty}^\infty \frac{\cosh(2ax)}{(\cosh x)^{2b}}\,dx.$$	**#1489 D†** $$\int_{-\pi}^\pi x^2\left(\frac{1}{1-e^x}-\frac{1}{2}\right)\,dx.$$	**#1490 B†** Integers $m,n,$ $$\int_0^{2\pi} \cos(mx)\cos(nx)\,dx.$$	**#1491 B†** $$\int_0^\pi xe^{-x}\sin x\,dx.$$
#1492 D† $a,b>0,$ $$\int_0^{\pi/2} \sin^a x\cos^b x\,dx.$$	**#1493 D†** $$\int_0^{\pi/4} \frac{x\,dx}{(\cos x+\sin x)\cos x}.$$	**#1494 B†** $$\int_0^{\pi/2} \sqrt{\sec x}\,dx.$$	**#1495 D†** $$\int_0^\infty e^{-x}\ln(\sinh x)\,dx.$$
#1496 B† $$\int_{-\infty}^\infty \ln\left(\frac{x^2+2}{x^2+1}\right)\,dx.$$	**#1497 B†** $a>0,$ $$\int_0^{\pi/4} \frac{dx}{a^2+\tan^2 x}.$$	**#1498 D†** $$\int_0^1 \frac{\ln(1+x)}{1+x^2}\,dx.$$	**#1499 B†** Natural $n,$ $$\int_0^1 \frac{x^n}{\sqrt{x-x^2}}\,dx.$$
#1500 D† Integer $n\geqslant 2,$ $$\int_0^\infty \left(\sqrt{x^2+1}-x\right)^n\,dx.$$	**#1501 B†** $a,b,c>0,$ $$\int_0^\infty \frac{e^{-ax}}{\sqrt{bx+c}}\,dx.$$	**#1502 D†** $0<a<1,$ $$\int_{-\infty}^\infty \frac{dx}{(x^2+1)\sqrt{x^2+a^2}}.$$	**#1503 D†** $a>1,$ $$\int_{-\infty}^\infty \frac{dx}{(x^2+1)\sqrt{x^2+a^2}}.$$
#1504 B† $$\int_0^1 \frac{\ln x}{\sqrt{1-x^2}}\,dx.$$	**#1505 D†** Natural $n,$ $$\int_0^1 \frac{x^n}{1+x}\,dx.$$	**#1506 D†** $a,b>0,$ $$\int_0^\infty \frac{e^{-ax}-e^{-bx}}{x^{3/2}}\,dx.$$	**#1507 D†** $a>0,$ $$\int_0^1 \sqrt{1-x^a}\,dx.$$
#1508 B† $$\int_0^1 \frac{x}{\sqrt{1-x^4}}\,dx.$$	**#1509 D†** $a>0,$ $$\int_0^{\pi/2} \frac{\tan x}{\cos^a x+\sec^a x}\,dx.$$	**#1510 D†** $$\int_0^{\pi/4} \ln(\cos x-\sin x)\,dx.$$	**#1511 D†** $$\int_0^{\pi/2} \frac{\sqrt{\sec x}-1}{\tan x}\,dx.$$

5.8 Further Exercises*

Exercises on Continuous Distributions

Definition 5.52 A real random variable X is said to be **continuous** if its **cumulative distribution function (cdf)**

$$F_X(x) = P(X \leqslant x) = \text{probability that } X \leqslant x$$

is a continuous function of x and X has **probability density function (pdf)** f_X if

$$F_X(x) = \int_{-\infty}^{x} f_X(t)\,dt \quad \text{for all } x \quad \text{and} \quad f_X(t) \geqslant 0 \quad \text{for all } t.$$

The **expectation** of X, or **mean**, written $E(X)$ is defined by

$$E(X) = \int_{-\infty}^{\infty} t f_X(t)\,dt$$

and the **variance** of X, written $\text{Var}(X)$, is defined by

$$\text{Var}(X) = E(X^2) - E(X)^2 = \left(\int_{-\infty}^{\infty} t^2 f_X(t)\,dt \right) - E(X)^2,$$

provided these are defined (i.e. the integrals are finite).

#1512A Show that, if f_X is a pdf, then $\int_{-\infty}^{\infty} f_X(t)\,dt = 1$.

#1513B Let $a < b$. The pdf of the **uniform distribution on** $[a,b]$ is $(b-a)^{-1}\mathbf{1}_{[a,b]}(x)$. Find and sketch the cdf, and determine the expectation and variance.

#1514D Say that X has uniform distribution on $[a,b]$. What are the cdf and pdf of X^2 when (i) $0 < a < b$, (ii) $a < 0 < a+b < b$, (iii) $a < a+b < 0 < b$? Sketch the cdf and pdf in each case.

#1515B The pdf of an **exponential distribution** is $\lambda e^{-\lambda x}\mathbf{1}_{(0,\infty)}(x)$ where $\lambda > 0$. Find and sketch the cdf, and determine the expectation and variance.

#1516B A **normal** (or **Gaussian**) **distribution** has pdf

$$c \exp\left(-\frac{(x-\mu)^2}{2\sigma^2} \right), \quad \text{where} \quad \mu, \sigma, c \text{ are real and } \sigma \neq 0.$$

Show that $c = 1/\sqrt{2\pi\sigma^2}$. Show further that the expectation is μ and the variance is σ^2. Determine the cdf in terms of the error function.

#1517E† Let X and Y be independent normally distributed random variables with means μ_X and μ_Y, and variances σ_X^2 and σ_Y^2. Let $Z = X+Y$. Explain why

$$F_Z(z) = \int_{-\infty}^{\infty} F_Y(z-x) f_X(x)\,dx.$$

Deduce that $Z = X+Y$ is normally distributed with mean $\mu_X + \mu_Y$ and variance $\sigma_X^2 + \sigma_Y^2$.

#1518D(i) The **gamma distribution**, with parameters $a > 0, \lambda > 0$, has pdf

$$\frac{1}{\Gamma(a)} \lambda^a x^{a-1} e^{-\lambda x} \mathbf{1}_{(0,\infty)}(x).$$

(Note that for $a = 1$ this is just the exponential distribution.) Find the expectation and variance.

(ii) Let X be a normally distributed random variable with $\mu = 0$ and $\sigma^2 = 1$. Show that X^2 has a gamma distribution with $a = \lambda = 1/2$. (This special case of the gamma distribution is an example of the χ^2-**distribution**, pronounced 'chi-squared distribution'.)

#1519C The **Cauchy distribution** has pdf $c(1 + x^2)^{-1}$ where c is a constant. Determine c. What is the cdf? Show that the expectation and variance are undefined.

#1520B Let $f_X(x)$ be the pdf of a continuous random variable. We say that m is a **median** if

$$P(X \leqslant m) = P(X \geqslant m) = \frac{1}{2}.$$

Find the median of the (i) uniform, (ii) exponential, (iii) normal, (iv) Cauchy distributions.

#1521C Give an example of a continuous distribution which has a non-unique median.

#1522D† (i) The subsets C_n of $[0, 1]$ are defined by

$$C_0 = [0, 1] \quad \text{and} \quad C_{n+1} = \frac{1}{3}C_n \cup \left(\frac{1}{3}C_n + \frac{2}{3}\right) \quad \text{for } n \geqslant 0,$$

so that

$$C_1 = [0, 1/3] \cup [2/3, 1], \qquad C_2 = [0, 1/9] \cup [2/9, 1/3] \cup [2/3, 7/9] \cup [8/9, 1].$$

What is $\int_0^1 \mathbf{1}_{C_n}(x)\,dx$? The **Cantor set** [26] C consists of those points x such that x is in C_n for each n. Show that $\mathbf{1}_C(x)$ is integrable and that $\int_0^1 \mathbf{1}_C(x)\,dx = 0$.

(ii) The **Cantor distribution** (or **devil's staircase**) has cdf $C(x)$ which is defined on $[0, 1]$ as follows: if x is not in C then there is a first n such that x is not in C_n. The set C_{n-1} consists of 2^{n-1} subintervals and say x is in the kth of these. We define

$$C(x) = \frac{2k-1}{2^n}.$$

Sketch a graph of $C(x)$. (We have defined $C(x)$ for x not in the Cantor set C. It is possible to define $C(x)$ on C as well so that $C(x)$ is continuous but we will not concern ourselves with the details of this.)

(iii) Evaluate $\int_0^1 C(x)\,dx$.

(iv) Show that the Cantor distribution does not have a pdf.

---

[26] Georg Cantor (1845–1918) was a significant figure in nineteenth-century mathematics, single-handedly introducing set theory, defining what it means for a set to be infinite and showing that there are infinite sets of different cardinality (for example, showing that the infinity of the real numbers is greater than that of the integers). He raised issues that had long-standing philosophical significance for mathematics, such as the *continuum hypothesis*, which asks whether there is an infinity strictly between those of the integers and the reals. The surprising answer was found to be that this isn't in fact a theorem of standard set theory but rather something to be assumed as a further axiom or denied.

#1523D† The prime number theorem (p.79) implies that the number of primes less than x is approximately

$$\int_2^x \frac{dt}{\ln t}.$$

Consequently the probability of a positive integer $n \geqslant 2$ being prime can be seen to be approximately $1/\ln n$. Such thinking can be a guide to the plausibility of open problems relating to prime numbers by considering whether certain sums or integrals converge. *Of course any such analysis does not constitute a proof.*

(i) Show that the integral

$$\int_2^x \frac{dt}{\ln(t^2 + 1)}$$

increases without bound as x becomes large. To what extent does this lend plausibility to Landau's problem: there being infinitely many prime numbers of the form $n^2 + 1$?

By consideration of similar integrals and investigation of their convergence, can you argue for the plausibility of

(ii) there being infinitely many Mersenne primes of the form $2^n - 1$?

(iii) there being infinitely many Fibonacci numbers that are prime?

(iv) there being infinitely many Fermat primes of the form $2^{2^n} + 1$?

(v) Legendre's conjecture: there always being a prime between n^2 and $(n+1)^2$?

(vi) twin prime conjecture: there are infinitely many pairs of primes that differ by 2?

(vii) Can you make an argument for the plausibility of Goldbach's conjecture, which states that every positive even integer other than 2 can be written as the sum of two primes?

Exercises on Orthogonal Polynomials

#1524B With the inner product as defined in #973 and $a = -1, b = 1$, and L as given in (4.38), show that

$$Ly_1 \cdot y_2 = y_1 \cdot Ly_2.$$

#1525D†(i) Let $P_k(x)$ denote the kth Legendre polynomial. Show that for a polynomial $p(x)$ of degree n

$$p(x) = \alpha_0(p \cdot P_0)P_0(x) + \alpha_1(p \cdot P_1)P_1(x) + \cdots + \alpha_n(p \cdot P_n)P_n(x)$$

where $\cdot$ is the inner product of #1524 and where $\alpha_0, \alpha_1, \ldots, \alpha_n$ are constants to be determined.

(ii) Now let $f(x)$ be a continuous function defined on $[-1, 1]$. Show that

$$f_n(x) = \alpha_0(f \cdot P_0)P_0(x) + \alpha_1(f \cdot P_1)P_1(x) + \cdots + \alpha_n(f \cdot P_n)P_n(x)$$

is the polynomial of degree n or less which minimizes the integral

$$\int_{-1}^1 |f(x) - f_n(x)|^2 \, dx.$$

#1526B† Let $L_n(x)$ denote the nth Laguerre polynomial (#470). Show that the Laguerre polynomials are orthogonal in the sense that

$$L_m \cdot L_n = \int_0^\infty L_m(x)L_n(x)e^{-x} \, dx = 0,$$

when $m \neq n$. Are the polynomials orthonormal with respect to this inner product?

#1527D† Let $H_n(x)$ denote the nth Hermite polynomial (#469). Show that the Hermite polynomials are orthogonal in the sense that

$$H_m \cdot H_n = \int_{-\infty}^{\infty} H_m(x)H_n(x)e^{-x^2}\, dx = 0,$$

when $m \neq n$. Are the polynomials orthonormal with respect to this inner product?

#1528B Let $T_n(x)$ denote the nth Chebyshev polynomial (#473). Show that the Chebyshev polynomials are orthogonal in the sense that

$$T_m \cdot T_n = \int_{-1}^{1} \frac{T_m(x)T_n(x)}{\sqrt{1-x^2}}\, dx = 0,$$

when $m \neq n$. Are the polynomials orthonormal with respect to this inner product?

#1529B(i) Chebyshev's equation is

$$\left(1-x^2\right)\frac{d^2y}{dx^2} - x\frac{dy}{dx} + n^2y = 0, \qquad (5.32)$$

where n is a non-negative integer. By making the substitution $x = \cos\theta$, show that Chebyshev's equation has two independent solutions:

$$T_n(x) = \cos\left(n\cos^{-1}x\right), \qquad V_n(x) = \sin\left(n\cos^{-1}x\right).$$

(ii) Show that

$$T_n(x) + iV_n(x) = \left(x + i\sqrt{1-x^2}\right)^n$$

and deduce that $T_n(x)$ is a polynomial of degree n.
(iii) We define the differential operator C by

$$Cy = \sqrt{1-x^2}\frac{d}{dx}\left(\sqrt{1-x^2}\frac{dy}{dx}\right).$$

Show that $Cy_1 \cdot y_2 = y_1 \cdot Cy_2$ using the inner product of #1528 and that (5.32) is equivalent to $Cy = -n^2y$.

#1530B† Let n be a positive integer. Say that

$$x^n = \sum_{k=0}^{n} a_k T_k(x).$$

Show that

$$a_k = \frac{2}{\pi}\int_0^{\pi} \cos^n\theta \cos k\theta\, d\theta \qquad \text{when } k \geqslant 1$$

and find a_0.

Exercises on Arc Length, Surface Area and Volume

Definition 5.53 Let $\mathbf{r}(t) = (x(t), y(t))$, where $a \leqslant t \leqslant b$, be a parametrization of a curve in $\mathbb{R}^2$. The **arc length** of the curve is defined to be

$$s = \int_{t=a}^{t=b} \left|\frac{d\mathbf{r}}{dt}\right|\, dt = \int_{t=a}^{t=b} \sqrt{x'(t)^2 + y'(t)^2}\, dt. \qquad (5.33)$$

So the graph of $y = f(x)$ where $a \leqslant x \leqslant b$ has length

$$\int_a^b \sqrt{1 + f'(x)^2}\, dx.$$

#1531b† (i) Show that a circle of radius a has circumference $2\pi a$.
(ii) Sketch the curve $y^2 = x^3$. Find the arc length of the curve from $(1, -1)$ to $(4, 8)$.
(iii) Find the arc length of the curve given parametrically by (t^2, t^3) where $-1 \leqslant t \leqslant 2$.
(iv) Let $a > 0$. The curve $r = a(1 + \cos\theta)$ is known as a **cardioid**. Find its arc length.

#1532b† Show that the arc length of a curve described in polar co-ordinates $r = R(\theta)$, where $\alpha \leqslant \theta \leqslant \beta$, equals

$$\int_\alpha^\beta \sqrt{R(\theta)^2 + R'(\theta)^2}\, d\theta.$$

#1533B† Let C be a curve parametrized as $\mathbf{r}(t) = (x(t), y(t))$, where $a \leqslant t \leqslant b$. Let $\mathbf{s}(u)$, where $\alpha \leqslant u \leqslant \beta$, be a second parametrization of the same curve C with say $\mathbf{r}(t) = \mathbf{s}(f(t))$ for some map f (a bijection) from $[a, b]$ to $[\alpha, \beta]$. Show that the arc length of C as defined by (5.33) is independent of the choice of parametrization.

#1534B Let $\mathbf{r}(t) = (\cos\theta(t), \sin\theta(t), z(t))$, where $a \leqslant t \leqslant b$, be a parametrized curve on the cylinder $x^2 + y^2 = 1$. Show that the curve has arc length

$$\int_a^b \sqrt{\theta'(t)^2 + z'(t)^2}\, dt.$$

Deduce that the map $(\theta, z) \mapsto (\cos\theta, \sin\theta, z)$ from $\mathbb{R}^2$ to the cylinder is an isometry (i.e. it preserves the lengths of curves). Find the shortest distance from $(1, 0, 0)$ to $(0, 1, 1)$ when measured on the cylinder.

#1535D† Find the shortest distance on the double cone $x^2 + y^2 = z^2$ between the points (i) $(1, 0, 1)$ and $(0, 1, 1)$; (ii) $(1, 0, 1)$ and $(0, 2, 2)$; (iii) $(0, 2, 2)$ and $(1, 0, -1)$.

#1536B† Let $\mathbf{r}(t)$ be a parametrization of a curve in the xy-plane where $t \geqslant 0$. Let $s(t)$ denote the arc length along the curve from $\mathbf{r}(0)$ to $\mathbf{r}(t)$. We could instead **parametrize the curve by arc length** and use s as the parameter instead; that is we'd set

$$\mathbf{R}(s(t)) = \mathbf{r}(t).$$

(i) Reparametrize the following curves using arc length

$$\mathbf{r}_1(t) = (a\cos t, a\sin t); \qquad \mathbf{r}_2(t) = (t, t); \qquad \mathbf{r}_3(t) = (t, \cosh t).$$

(ii) Show that $\mathbf{t}(s) = \mathbf{R}'(s)$ is a unit vector.
(iii) The **curvature** is defined to be $\kappa(s) = |\mathbf{t}'(s)|$. Show that the curvature of the graph $y = f(x)$ equals

$$\kappa = \frac{|f''(x)|}{(1 + f'(x)^2)^{3/2}}.$$

(iv) If $\kappa(s) > 0$, show that $\mathbf{t}'(s) = \kappa\mathbf{n}$ where $\mathbf{n}$ is a unit vector perpendicular to $\mathbf{t}$. Show further that $\mathbf{n}'(s) = -\kappa\mathbf{t}$.
(v) The **evolute** of a curve consists of those points with position vectors $\mathbf{e} = \mathbf{R} + \mathbf{n}/\kappa$. (The quantity $1/\kappa$ is called the **radius of curvature**.) Find and sketch the evolute of the parabola $y = x^2$.

#1537D† Show that a planar curve of constant curvature is a line segment or an arc of a circle.

#1538B† Let $\mathbf{r}(s)$ be a curve, parametrized by arc length, in xyz-space. Define $\mathbf{r}'(s) = \mathbf{t}(s)$ and $\mathbf{t}'(s) = \kappa\mathbf{n}(s)$ as in #1536. We define the **binormal** vector $\mathbf{b}(s)$ by

$$\mathbf{b}(s) = \mathbf{t}(s) \wedge \mathbf{n}(s).$$

  (i) [27] Show that $\mathbf{b}'(s) = -\tau(s)\mathbf{n}(s)$ for some scalar function $\tau(s)$, called **torsion.**
 (ii) Show that $\mathbf{n}'(s) = -\kappa(s)\mathbf{t}(s) + \tau(s)\mathbf{b}(s).$
(iii) Show that the curve $\mathbf{r}(s)$ lies in a plane if and only if τ is identically zero.
(iv) A curve of the form

$$\mathbf{r}(t) = (R\cos t, R\sin t, Vt), \qquad -\infty < t < \infty$$

is a **helix.** Determine the arc length $s(t)$ and show that the helix has constant curvature and torsion.

#1539D† Show that a curve of constant curvature and constant torsion is an arc of a helix.

#1540B† Let $f(x)$ be a positive function defined for $a \leqslant x \leqslant b$. A **surface of revolution** S can be formed by rotating the graph of $y = f(x)$ about the x-axis.
(i) Show that the surface S has equation $y^2 + z^2 = f(x)^2$ (where $a \leqslant x \leqslant b$).

The **surface area of** S and **volume within** S are defined by the formulas

$$A = 2\pi \int_{x=a}^{x=b} y\frac{ds}{dx}\,dx; \qquad V = \pi \int_{x=a}^{x=b} y^2\,dx.$$

Use these formulae to determine the surface area and volume of the following:
(ii) A sphere of radius a
(iii) A cone of height h and base radius a
(iv) The ellipsoid $a^2x^2 + y^2 + z^2 = 1$ where $a > 1$

#1541B The **cycloid** is the curve given parametrically by the equations

$$x(t) = t - \sin t \quad \text{and} \quad y(t) = 1 - \cos t \quad \text{for } 0 \leqslant t \leqslant 2\pi.$$

  (i) A fixed point on the circumference of a unit disc begins at $(0,0)$. Show that the curve $(x(t), y(t))$ is the trace of that point as the disc rolls along the x-axis. Sketch the cycloid.
 (ii) Find the arc length of the cycloid.
(iii) Find the area bounded by the cycloid and the x-axis.
(iv) Find the area of the surface of revolution generated by rotating the cycloid around the x-axis.
 (v) Find the volume enclosed by the surface of revolution generated by rotating the cycloid around the x-axis.

---

[27] The three equations in #1536(iv) and #1538(i), (ii) giving formulae for $\mathbf{t}'(s)$, $\mathbf{n}'(s)$ and $\mathbf{b}'(s)$ are known as the Serret–Frenet formulae, named after the French mathematicians Joseph Serret (1819–1885) and Jean Frenet (1816–1900).
   The vectors $\mathbf{t}, \mathbf{n}, \mathbf{b}$ form an orthonormal basis moving with the curve. Curvature is a measure of the extent to which the curve is departing from the tangent line, and torsion is a measure of the extent to which the curve is moving out of the plane through $\mathbf{r}(s)$ parallel to $\mathbf{t}$ and $\mathbf{n}$. The *fundamental theorem of the local theory of curves* shows that curvature and torsion uniquely determine a curve (up to an isometry of $\mathbb{R}^3$).

#1542 C† Consider an upside-down cycloid with parametrization

$$x(u) = u - \sin u \quad \text{and} \quad y(u) = \cos u \quad \text{for } 0 \leqslant u \leqslant \pi.$$

A smooth wire in the shape of this cycloid is fashioned, and a particle of mass m is released from the point $(0, 1)$, descending under gravity g.

(i) Find the time taken for the particle to travel from $(0, 1)$ to $(\pi, 0)$.
(ii) Show that the time taken in (i) equals the time taken by the particle from rest at $(x(u_0), y(u_0))$ to reach $(\pi, 0)$. This shows that this cycloid is the solution of the **tautochrone** (or **isochrone**) problem.[28]
(iii) Show that this time in (i) is less than the time for the particle to travel than a straight linear wire between these two points. In fact, the time for the cycloid is the least for any connecting curve between these two points.[29]

#1543 D† An engineer builds himself a unicycle with a square wheel of side $2a$, and wonders now what type of road surface he should make in order that he can ride his unicycle while remaining at a constant height. A mathematician friend of his says a road surface with the equation

$$y = a\left(1 - \cosh\frac{x}{a}\right)$$

would work and gives her reasoning as follows: call the road surface $(x(s), y(s))$, parametrized by arc length and without loss of generality start the middle of the base of the wheel at $(0, 0)$ with the base parallel to the x-axis.

(i) Let's now say the square wheel has rolled distance s. Find the co-ordinates of the centre of the wheel and hence show that a rider's height will remain constant if

$$y(s) - sy'(s) + a\sqrt{1 - y'(s)^2} = a.$$

(ii) By differentiating this equation with respect to s, or otherwise, find $y(s)$ and $x(s)$, and hence show the mathematician's answer to be correct.

Exercises on Elliptic Integrals

Definition 5.54 Let $0 \leqslant k \leqslant 1$. The **(complete) elliptic integrals** $K(k)$ and $E(k)$ are defined by

$$K(k) = \int_0^{\pi/2} \frac{dx}{\sqrt{1 - k^2 \sin^2 x}}, \quad E(k) = \int_0^{\pi/2} \sqrt{1 - k^2 \sin^2 x}\, dx.$$

#1544 B Show that the arc length of the ellipse $x^2/a^2 + y^2/b^2 = 1$ equals

$$\int_0^{2\pi} \sqrt{a^2 \cos^2 x + b^2 \sin^2 x}\, dx.$$

Show further, that if $a \geqslant b$, the above equals $4aE\left(\sqrt{1 - b^2/a^2}\right)$.

---

[28] As demonstrated by the Dutch mathematician Christiaan Huygens (1629–1695) in 1673.
[29] This is the *brachistochrone* problem which was solved by Johann Bernoulli in 1696.

#1545D† Let $a, b > 0$. Evaluate the following in terms of elliptic integrals.

$$\int_0^1 \frac{dx}{\sqrt{1 - x^4}}, \qquad \int_0^\infty \frac{dx}{\sqrt{(x^2 + a^2)(x^2 + b^2)}}.$$

#1546C† Let $a > b > 0$. Evaluate the following in terms of elliptic integrals.

$$\int_0^\pi \sqrt{a + b \cos x}\, dx, \qquad \int_0^\pi \frac{dx}{\sqrt{a + b \cos x}}.$$

#1547D By setting $x = 2 \tan^2 t$ in both of the following integrals, show that

$$\int_0^\infty \frac{dx}{\sqrt{x(x+2)(x+3)}} = \frac{2}{\sqrt{3}} K\left(\frac{1}{\sqrt{3}}\right); \qquad \int_0^\infty \frac{dx}{\sqrt{x(x^2 + 2x + 4)}} = \sqrt{2} K\left(\frac{1}{2}\right).$$

#1548D† Show that

$$\int_0^{\pi/2} \frac{dx}{\left(1 - k^2 \sin^2 x\right)^{3/2}} = \frac{E(k)}{1 - k^2}.$$

#1549E† Let $0 \leqslant k \leqslant 1$ and define $k' = \sqrt{1 - k^2}$. By substituting $\tan(x - t) = k' \tan t$, show that

$$K\left(\frac{1 - k'}{1 + k'}\right) = \left(\frac{1 + k'}{2}\right) K(k).$$

#1550E† Let $0 \leqslant k \leqslant 1$. By making the substitution $\sin x = (1 + k) \sin t / (1 + k \sin^2 t)$, show that

$$K\left(\frac{2\sqrt{k}}{1 + k}\right) = (1 + k) K(k).$$

#1551D† For $a, b > 0$ define

$$I(a, b) = \int_0^{\pi/2} \frac{dx}{\sqrt{a^2 \cos^2 x + b^2 \sin^2 x}}.$$

(i) Write $I(a, b)$ in terms of the elliptic integral K.

(ii) Let $c > 0$. Show that

$$I(a, b) = I(b, a) \qquad \text{and} \qquad I(ca, cb) = c^{-1} I(a, b).$$

What is $I(a, a)$?

(iii) Show that

$$I(a, b) = I\left(\frac{a + b}{2}, \sqrt{ab}\right).$$

(iv) (See #432.) Deduce that

$$\mathrm{agm}(a, b) I(a, b) = \frac{\pi}{2}.$$

Special Functions Polynomials, logarithms, exponentials, trigonometric and hyperbolic functions and their inverses, are generally referred to as *elementary functions*. Various *special functions* have been introduced in this text including the gamma function, error function, elliptic integrals and Bessel functions. Other orthogonal polynomials – such as those due to Legendre, Chebyshev, Laguerre and Hermite – have also been introduced and we have seen in the exercises on p.415 the algebraic properties that make them useful. It is clear though that they, being polynomials, are also elementary functions.

A harder question though is as to whether $\Gamma, \mathrm{erf}, K, E, J_n$ can be expressed in terms of elementary functions, which proves not to be the case. A first result in this direction is *Liouville's theorem* which addresses the problem of when antiderivatives can be expressed in terms of elementary functions – see Conrad [7] for an approachable discussion of the details. Loosely put, the theorem shows that the elementary antiderivatives of a given family of elementary functions are the elementary functions themselves and logarithms thereof, in a like manner to which integrals of rational functions need not be rational but can be expressed as rational functions and logarithms of such.

More generally, *differential Galois theory* is the study of which differential equations can be solved by the repeated introduction of antiderivatives – this theory began with the work of Émile Picard and Ernest Vessiot in the late nineteenth century. In a manner analogous to Galois theory (see p.33), if a differential equation can be solved by the repeated introduction of integrals, then its 'differential Galois group' will be solvable in a similar manner to when a polynomial can be solved by the repeated introduction of radicals.

Before the advent of computers, and in particular scientific computing packages such as *Mathematica*, many texts were solely dedicated to the integration and evaluation of special functions, the 1000+ page tome (Gradshteyn, Ryzhik [14]) being a notable example. The need for tables of their values has certainly passed and relatively little research into special functions, for their own sake, is now done compared with that done by nineteenth-century mathematicians – for example Legendre and Jacobi devoted much study to elliptic integrals – but the more important special functions, such as Γ and erf, retain a common presence in modern mathematics.

#1552 B In #1667 it is shown that a pendulum of length l in gravity g has period

$$T = 4\sqrt{\frac{l}{2g}} \int_0^\alpha \frac{\mathrm{d}\theta}{\sqrt{\cos\theta - \cos\alpha}}.$$

Make the substitution $\sin(\theta/2) = \sin(\alpha/2)\sin u$ to rewrite T in terms of an elliptic integral.

Exercises on Differentiation under the Integral Sign

Definition 5.55 By **differentiation under the integral sign** we mean the identity

$$\frac{\mathrm{d}}{\mathrm{d}\alpha}\int_a^b f(x,\alpha)\,\mathrm{d}x = \int_a^b \frac{\partial f}{\partial \alpha}(x,\alpha)\,\mathrm{d}x.$$

This identity can be shown to hold under a wide range of reasonable hypotheses (essentially ensuring that the integrals and functions in the identity are well defined), and is another instance where a theorem of analysis shows that the order of two limit processes – integration with respect to x and differentiation with respect to α – can be interchanged. To explain the above identity a little further, note that as x is the dummy variable in the first integral, then

$$\int_a^b f(x,\alpha)\,dx$$

is actually a function of α alone. The expression on the left is then the usual derivative of this function. However, $f(x,\alpha)$ is itself a function of two variables and it is possible to form two partial derivatives

$$\frac{\partial f}{\partial \alpha}(x,\alpha) \quad \text{and} \quad \frac{\partial f}{\partial x}(x,\alpha).$$

To calculate the former (which appears in the identity) we treat x as a constant and differentiate with respect to α, and to calculate the latter we treat α as a constant and differentiate with respect to x. See p.431 for further details about partial differentiation.

#1553B†(i) With $a,b > 0$, rederive $\int_0^\infty \frac{e^{-ax}-e^{-bx}}{x}\,dx$ which was determined in #1330.
(ii) Explain how the integral in #1442 can be deduced from that in #1441.
(iii) Use #1418 to determine $\int_{-\infty}^\infty \frac{xe^{ax}}{\cosh x}\,dx$.

#1554D† (See also #1420.) For $a \geqslant 0$ define

$$I(a) = \int_0^\infty e^{-ax}\frac{\sin x}{x}\,dx.$$

Show that $I'(a) = -(1+a^2)^{-1}$. Evaluate $I(a)$ and deduce that

$$\int_0^\infty \frac{\sin x}{x}\,dx = \frac{\pi}{2}.$$

#1555C† Let $a \geqslant 0$. Evaluate

$$\int_0^\infty e^{-ax}\frac{\sin ax}{x}\,dx, \qquad \int_0^\infty \frac{e^{-x}(1-\cos x)}{x^2}\,dx.$$

#1556D† Evaluate

$$\int_0^\infty \frac{\ln x}{(a^2+x^2)^2}\,dx \quad \text{and} \quad \int_0^\infty \frac{x^{1/2}\ln x}{(1+x)^2}\,dx.$$

#1557D For real a define

$$I(a) = \frac{2}{\pi}\int_0^{\pi/2} \cos(a\cos x)\,dx.$$

Show that

$$aI''(a) + I'(a) + aI(a) = 0, \qquad I(0) = 1, I'(0) = 0.$$

This shows that $I(a) = J_0(a)$, the Bessel function defined in Example 6.67.

#1558 E† For real a, show that

$$\int_{-\infty}^{\infty} e^{-x^2} \cos 2ax \, dx = \sqrt{\pi} e^{-a^2}.$$

By considering

$$J(c) = \int_{-\infty}^{\infty} e^{-c^2(x^2+1)} \frac{\cos ax}{x^2+1} \, dx$$

show, for $a, b > 0$, that

$$\int_{-\infty}^{\infty} \frac{\cos ax}{x^2+1} \, dx = \frac{\pi}{e^a} \qquad \text{and} \qquad \int_{-\infty}^{\infty} \frac{\cos ax}{x^2+b^2} \, dx = \frac{1}{b} \frac{\pi}{e^{a/b}}.$$

#1559 C Evaluate

$$\int_{-\infty}^{\infty} e^{-x^2} \cos^2 x \, dx, \qquad \int_{-\infty}^{\infty} e^{-x^2} \sin^2 x \, dx, \qquad \int_{-\infty}^{\infty} e^{-x^2} \cos^3 x \, dx, \qquad \int_{-\infty}^{\infty} e^{-x^2} \sin^4 x \, dx.$$

- For #1560, #1561 see §6.5 and #1562 see §6.3 for the relevant material on solving differential equations if necessary.

#1560 D† Let $a > 0$ and define

$$I(a) = \int_{-\infty}^{\infty} \frac{\cos ax}{x^2+1} \, dx.$$

Carefully show that $I''(a) = I(a)$ and hence show that $I(a) = \pi e^{-a}$.

#1561 E† Let $a > 0$ and define

$$I(a) = \int_{-\infty}^{\infty} \frac{\cos ax}{x^4+1} \, dx.$$

Show that $I''''(a) = -I(a)$ and hence evaluate $I(a)$.

#1562 D† Evaluate

$$\int_{0}^{\infty} \frac{e^{-x^2}}{x^2+1} \, dx.$$

#1563 E† Let X, Y be independent normally distributed random variables with mean 0 and variance 1. Show that $Z = X/Y$ has the Cauchy distribution.

#1564 D† Let $K(k)$ and $E(k)$ denote the elliptic integrals from Definition 5.54. Show that

$$E'(k) = \frac{E(k) - K(k)}{k}, \qquad \text{and} \qquad K'(k) = \frac{E(k)}{k(1-k^2)} - \frac{K(k)}{k}.$$

Hence evaluate

$$\int_{0}^{\pi/2} \frac{dx}{\left(1 - k^2 \sin^2 x\right)^{5/2}}.$$

#1565 E† Let a, b, c be real with $c > 0$. Show that

$$\int_{-\infty}^{\infty} e^{-cx^2} \operatorname{erf}(ax+b) \, dx = \sqrt{\frac{\pi}{c}} \operatorname{erf}\left(\frac{b\sqrt{c}}{\sqrt{a^2+c}}\right).$$

Exercises on Fractional Harmonic Numbers and the Digamma Function

Definition 5.56 For $n \geqslant 1$ the **harmonic number** H_n is defined by

$$H_n = 1 + \frac{1}{2} + \frac{1}{3} + \cdots + \frac{1}{n}.$$

We have seen that H_n increases without bound as n increases (#1325) and that $H_n - \ln n$ converges to Euler's constant γ as $n \to \infty$ (#1326).

#1566D† (i) For a positive integer n, show that $H_n = \int_0^1 (1 - x^n)/(1 - x)\,dx$. Thus we may extend this to define the **fractional harmonic number** H_α for $\alpha > -1$ by

$$H_\alpha = \int_0^1 \frac{1 - x^\alpha}{1 - x}\,dx.$$

(ii) Show for $\alpha > 0$ that

$$H_\alpha = H_{\alpha-1} + \frac{1}{\alpha} \quad \text{and} \quad H_\alpha = \sum_{k=1}^{\infty} \frac{\alpha}{k(k+\alpha)}.$$

#1567D† Show, for a positive integer n, that

$$\int_0^n H_x\,dx = n\gamma + \ln n!.$$

#1568B Show that

$$H_{1/2} = 2 - 2\ln 2; \qquad H_{1/3} = 3 - \frac{\pi}{2\sqrt{3}} - \frac{3}{2}\ln 3; \qquad H_{1/4} = 4 - \frac{\pi}{2} - 3\ln 2.$$

#1569D† The **digamma function** $\psi(x)$ is defined for $x \neq 0, -1, -2, \ldots$ as

$$\psi(x) = \frac{\Gamma'(x)}{\Gamma(x)} = \frac{d}{dx}\ln|\Gamma(x)|.$$

(i) Show that $\psi(x+1) = \psi(x) + \frac{1}{x}$.
(ii) When x is not an integer, show that $\psi(1-x) - \psi(x) = \pi\cot\pi x$.
(iii) Show, for $\alpha > -1$, that $\psi(\alpha+1) = H_\alpha - \gamma$. Hence generalize #1567 to real $n > 0$.

#1570C† Show that

$$\int_0^\infty e^{-x}\ln x\,dx = \int_0^1 \ln(-\ln x)\,dx = -\gamma.$$

#1571D† Show that $\psi(a)$, or equally $\Gamma'(a)$, has a unique positive root which is in the interval $1 < a < 2$.

#1572D† Let $b > a > 0$. Evaluate

$$\int_0^\infty \frac{\sinh ax}{\sinh bx}\,dx.$$

Exercises on Moment Generating Functions

#1573B The **moment generating function** $M_X(t)$ of a continuous random variable with pdf f_X equals

$$M_X(t) = E\left(e^{tX}\right) = \int_{-\infty}^{\infty} e^{tx} f_X(x) \, dx.$$

Determine the moment generating function of the (i) uniform, (ii) exponential, (iii) gamma, (iv) Cauchy distributions, stating carefully for what t the above integral converges.

#1574D† (i) Say that the random variable X has moment generating function $M(t)$ defined for all t. Show that

$$E(X) = M'(0) \qquad \text{and that} \qquad \text{Var}(X) = M''(0) - M'(0)^2.$$

(ii) Given real numbers a, b, show that $M_{aX+b}(t) = e^{bt} M_X(at)$.

#1575D† (i) Let U and V be independent random variables. Show $M_{U+V}(t) = M_U(t) M_V(t)$.
(ii) Show that the moment generating function of the normal distribution with expectation μ and variance σ^2 equals

$$\exp\left(\mu t + \frac{1}{2}\sigma^2 t^2\right).$$

(iii) Assuming that $M_R(t) = M_S(t)$ for all t implies R and S are identically distributed, rederive the result of #1517.

#1576E† (**Central Limit Theorem**)(i) Say that the random variable X has moment generating function $M_X(t)$ defined for all t. Using the monotone convergence theorem (p.344), show that

$$M_X(t) = \sum_{k=0}^{\infty} M_k \frac{t^k}{k!}$$

where $M_k = E(X^k)$ is the kth **moment** of X.
(ii) Denoting $\mu = E(X)$ and $\sigma^2 = \text{Var}(X)$, show that

$$\frac{M_X(t) - 1 - \mu t - \frac{1}{2}(\mu^2 + \sigma^2)t^2}{t^2} \to 0 \qquad \text{as } t \to 0.$$

(iii) Let $X_1, X_2, \ldots$ be independent, identically distributed random variables with expectation μ and variance σ^2. Let

$$Z_n = \frac{X_1 + X_2 + \cdots + X_n - n\mu}{\sigma^2 \sqrt{n}}.$$

Show, for any t, that

$$M_{Z_n}(t) \to e^{t^2/2} = M_N(t) \qquad \text{as } n \to \infty$$

where N denotes the normal distribution with expectation 0 and variance 1. The central limit theorem then follows: Z_n converges to N in distribution, which means for any x that

$$P(Z_n \leqslant x) \to P(N \leqslant x) = \frac{1}{\sqrt{2\pi}} \int_{-\infty}^{x} e^{-t^2/2} \, dt \qquad \text{as } n \to \infty.$$

6

Differential Equations

6.1 Introduction and History

The study of differential equations (DEs) is as old as calculus itself and dates back to the seventeenth century. At that time, most of the interest in DEs came from applications in physics and astronomy – one of Newton's greatest achievements, in his *Principia Mathematica* (1687), was to show that a force (a second derivative) between a planet and the sun, which is inversely proportional to the square of the distance between them, would lead to an elliptical orbit (see p.485).

The study of DEs grew as increasingly varied mathematical and physical situations led to DEs, and as more and more sophisticated techniques were found to solve them. Besides in astronomy, DEs began appearing naturally in applied areas such as fluid dynamics, heat flow, wave motion, magnetism, in determining the curve a chain between two points will make under its own weight, and equally in pure mathematics, finding the shortest path between two points on a surface, the surface across a given boundary of smallest area (such as the shape of a soap film), the largest area a curve of fixed length can bound, etc.

Still today, much of applied mathematics is concerned with the solution of DEs that arise from modelling real world situations. In this chapter, we will be studying **ordinary differential equations** (ODEs) rather than **partial differential equations** (PDEs). This means that the solution (often denoted y) will be a function of just one variable (commonly x) rather than of several variables; these ODEs will involve *full* derivatives, such as dy/dx and d^2y/dx^2 etc. (which we will also denote $y'(x)$ and $y''(x)$ etc.).

Example 6.1 Examples and types of ODEs.

- A **kth order inhomogeneous linear differential equation** is one of the form

$$a_k(x)\frac{d^k y}{dx^k} + a_{k-1}(x)\frac{d^{k-1}y}{dx^{k-1}} + \cdots + a_1(x)\frac{dy}{dx} + a_0(x)y = f(x),$$

and the equation is called **homogeneous** if $f(x) = 0$.
- The DE

$$y'(x) = y(x), \qquad y(0) = 1$$

uniquely characterizes the exponential function $y = e^x$ (see #1301).
- (See Example 6.27 and #1667.) The equation for simple harmonic motion (SHM) is

$$\frac{d^2 y}{dt^2} = -\omega^2 y, \tag{6.1}$$

426

where $\omega > 0$ and the DE governing the swinging of a pendulum (of length l, under gravity g) is

$$\frac{d^2\theta}{dt^2} = -\frac{g}{l}\sin\theta.$$

- Let n be a natural number. The DE

$$(1-x^2)\frac{d^2y}{dx^2} - 2x\frac{dy}{dx} + n(n+1)y = 0 \tag{6.2}$$

is known as **Legendre's equation** and amongst the solutions are the so-called **Legendre polynomials** $P_n(x)$. See #470, #471, #974 and #1582.

We give here, and solve, a simple example which involves some of the key ideas of ODEs; the example here is the movement of a particle P under gravity in one vertical dimension. Suppose that we write $h(t)$ for the height (in metres m, say) of P over the ground at time t (measured in seconds s, say). If we ignore air resistance and assume gravity (denoted as g) to be constant then h satisfies the DE

$$h''(t) = -g. \tag{6.3}$$

The **velocity** of the particle is the quantity dh/dt – the rate of change of displacement (here, height) with time, measured in ms^{-1}. The rate of change of velocity with time is called **acceleration** and is the quantity d^2h/dt^2 on the LHS of (6.3), having units ms^{-2}. The acceleration here is entirely due to gravity. Note the need for a minus sign here, as gravity is acting downwards.

Equation (6.3) is not a difficult DE to solve; we can integrate first once, to obtain

$$h'(t) = -gt + K_1 \tag{6.4}$$

and then again, to find

$$h(t) = -gt^2/2 + K_1 t + K_2, \tag{6.5}$$

where K_1 and K_2 are constants. Currently we don't know enough about the specific case of the particle P to be able to say anything more about these constants. Note, though, that whatever the values of K_1 and K_2, the graph of h against t is a parabola.

Definition 6.2 An **ordinary differential equation** (ODE) is an equation relating a function, say y, in one variable, say x, and finitely many of its derivatives. i.e. something that can be written in the form

$$f\left(y, \frac{dy}{dx}, \frac{d^2y}{dx^2}, \cdots, \frac{d^ky}{dx^k}\right) = 0$$

for some function f and some positive integer k. Here x is the **independent variable** and the DE governs how the **dependent variable** y varies with x. The equation may have no or many functions $y(x)$ which satisfy it; the problem usually is to find the most general form of **solution** $y(x)$ which satisfies the DE.

Definition 6.3 A derivative of the form d^ky/dx^k is said to be of order k and we say that a DE has **order** k if it involves derivatives of order k and less.

We return now to (6.3), which is a second-order DE. In some loose sense, solving a DE of order k involves integrating k times, though not usually in such an obvious fashion as in the case of (6.3). So we might expect the solution of an order k DE to have k undetermined constants in it, and this will be the case in most of the examples that we meet here. However, this is not generally the case and we will see other examples where more than k constants, or fewer, are present in the solution (see Example 6.4).

The expression (6.5) is the **general solution** of the DE (6.3), that is, an expression which encompasses, by means of the indeterminates K_1 and K_2, all solutions of the DE. At the moment then (6.5) is not a unique function, but rather depends on two undetermined constants. And this isn't unreasonable as a particle under gravity could follow many a path; at the moment we don't have enough information to characterize that path uniquely.

One way of filling in the missing info would be to say how high P was at $t = 0$ and how fast it was going at that point. For example, suppose P began at a height of 100m and was propelled upwards at a speed of 10ms^{-1}, that is,

$$h(0) = 100 \qquad \text{and} \qquad h'(0) = 10. \tag{6.6}$$

Putting these values into (6.4) and (6.5) we get two simultaneous equations in K_1 and K_2 which have solutions $K_1 = 10$ and $K_2 = 100$. So the height of P at time t has been uniquely determined and is given by

$$h(t) = 100 + 10t - gt^2/2. \tag{6.7}$$

The extra bits of information given in (6.6) are called **initial conditions** – the DE (6.3) with the initial conditions (6.6) is called an **initial-value problem.**

Alternatively, suppose we were told that P was thrown at time $t = 0$ from a height of 100m and was subsequently one second later caught at 105m. That is,

$$h(0) = 100 \qquad \text{and} \qquad h(1) = 105. \tag{6.8}$$

Putting these values into (6.5) gives $K_2 = 100$ and $-g/2 + K_1 + K_2 = 105$ giving that $K_1 = 5 + g/2$ and so

$$h(t) = -gt^2/2 + (5 + g/2)t + 100.$$

Again we have uniquely characterized the trajectory of P by saying where the particle is at two times. The conditions (6.8) are called **boundary conditions** and the DE (6.3) together with (6.8) is called a **boundary-value problem.**

#1577
#1578
#1579
#1585

Having solved in (6.7) the earlier initial-value problem and found an equation for h, we can readily answer other questions about P's behaviour such as:

- What is the greatest height P achieves? The maximum height will be a stationary value for $h(t)$ and so we need to solve the equation $h'(t) = 0$, which has solution $t = 10/g$. At this time the height is

$$h(10/g) = 100 + 100/g - (100g)/(2g^2) = 100 + 50/g.$$

- What time does P hit the ground? To solve this we see that $0 = h(t) = 100 + 10t - gt^2/2$ has two real solutions. However, one of these times is meaningless (being negative, and so before our experiment began) and so we take the other (positive) solution and see that P hits the ground at $t = (10 + 10\sqrt{1 + 2g})/g$.

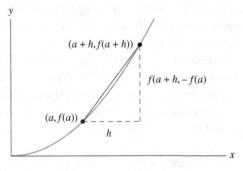

Chord's gradient approximating $f'(a)$

Figure 6.1

A Brief History of Calculus: Differentiation.

Isaac Newton (1643–1727) and Gottfried Leibniz (1646–1716) are generally regarded as the founders of calculus. This certainly does not mean that a rigorous appreciation of what even a derivative or integral means was possible during the seventeenth century, nor that substantial groundwork towards an understanding of the calculus had not been made earlier. The result that the derivative at a local maximum or minimum equals 0 is commonly referred to as *Fermat's theorem*, after Pierre de Fermat (see also p.93) who was a predecessor of Newton and Leibniz. That Fermat's name is associated with this result is a consequence of his 1638 letter *Method for Determining Maxima and Minima and Tangents to Curved Lines* to Descartes, though the work dated back to 1629. One of course does not need the full toolkit of calculus to appreciate (or even determine) a maximum of a curve with a suitably nice equation. That said, Fermat had shown that the gradient of the curve $y = x^n$ equals na^{n-1} when $x = a$ and that the area under this curve for $0 \leqslant x \leqslant a$ equals $a^{n+1}/(n+1)$.

The formal definition of the *derivative* of $f(x)$, when $x = a$, is

$$f'(a) = \lim_{h \to 0} \frac{f(a+h) - f(a)}{h}.$$

From Figure 6.1 we see that the above quotient is the approximating gradient of a short chord. With $f(x) = x^2$ we have

$$f'(a) = \lim_{h \to 0} \frac{(a+h)^2 - a^2}{h}$$
$$= \lim_{h \to 0} (2a+h) = 2a.$$

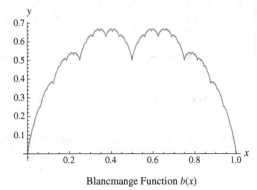

Blancmange Function $b(x)$

Figure 6.2

For the polynomials that Fermat was working with there was no need to let h approach 0; rather, with appropriate algebraic simplification this approximating gradient could be expressed in such a way that h could just be set to 0. Such algebraic manipulation would not, however, resolve the same limit for a general function.

Newton's early work on calculus dates to 1664–1666, with Leibniz's independent work dating to 1672–1676. However, Leibniz would be the first to publish in 1684 whilst Newton's ideas would first appear in print in his *Philosophiæ Naturalis Principia Mathematica* in 1687. What separates the work of Newton and Leibniz from that of Fermat is the recognition that a limiting process is taking place, though it would be wrong to suggest that seventeenth-century mathematicians were in a position to fully understand or rigorously handle the limit process, as would become clear in the subsequent criticisms of George Berkeley (1685–1783) in *The Analyst* (1734). Newton and Leibniz had intriguingly different emphases in their work. Newton shows himself in the *Principia* – which also includes determining the power series for sine and cosine – to be the technically proficient problem solver. Leibniz's prorities were more aligned to generality of approach and the benefits of clear notation, his being the more suggestive dy/dx notation.

The tale of the discovery of the calculus ends sadly with a period of hateful controversy that would sour the relationship between British and European mathematics for centuries. Around the end of the seventeenth century, various supporters of Newton began claiming that Leibniz had plagiarized (or at least been inspired by) the earlier ideas of Newton without giving due credit. Leibniz would in 1711 seek exoneration from the Royal Society for these claims but, with Newton as president of the society, Leibniz's case was poorly treated and the society found in favour of Newton in 1713. The controversy would lead to the separation of British and European mathematicians for years and an effective isolation of British mathematics. See Hellman [19, Chapter 3].

In first discussions of the calculus, students are often left with the impression that all functions are differentiable, or given no more than the example of $f(x) = |x|$ as a function with no derivative when $x = 0$. The *blancmange function* $b(x)$ in Figure 6.2 is an easily described example of a continuous function which has no derivative *anywhere*. If $a(x)$ is the minimum distance of a real number x from an integer, then $b(x) = \sum_{n=1}^{\infty} 2^{-n} a(2^n x)$.

The next example is designed to show that we should not be cavalier when solving DEs.

Example 6.4 Find the general solution of the DE

$$\left(\frac{dy}{dx}\right)^2 = 4y. \tag{6.9}$$

Solution Given this equation we might argue as follows – taking square roots and rearranging we get

$$\frac{1}{2\sqrt{y}} \frac{dy}{dx} = 1. \tag{6.10}$$

From the chain rule we recognize the LHS as the derivative of $\sqrt{y}$ and so, integrating we have $\sqrt{y} = x + K$, where K is a constant. Squaring, we might think that the general solution has the form $y = (x + K)^2$. Certainly these are solutions of (6.9).

What, if anything, could have gone wrong with this argument? We could have been more careful to include positive and negative square roots at the (6.10) stage, but actually

Partial Differentiation. In most real-world applications, quantities vary with more than one variable. For example, the temperature around you, or air pressure, depend on three spatial co-ordinates and time; height $h(\theta,\phi)$ above sea-level is a function of latitude θ and longitude ϕ; the gravitational field about you is a vector-valued function which again depends on spatial and temporal co-ordinates.

So when it comes to studying variation in a function $f(x,y)$, say, which is dependent on two variables, then we may be interested in how $f(x,y)$ varies as x changes, or as y changes, or as both change in some manner. For such an f we can immediately consider two *partial* derivatives

$$f_x = \frac{\partial f}{\partial x}, \qquad f_y = \frac{\partial f}{\partial y},$$

the first one being the change with f as x varies and y remains constant, the second where y varies and x remains constant. So with the earlier example of $h(\theta,\phi)$, the partial derivative $\partial h/\partial \theta$ detemines the change in height as we walk along a meridian (ϕ remains constant) in the direction of increasing θ. The partial derivative $\partial h/\partial \phi$ would measure change in height as we move along a line of latitude (like the equator or tropics) in the direction of increasing ϕ.

We can then likewise form higher order partial derivatives

$$f_{xx} = \frac{\partial^2 f}{\partial x^2}, \qquad f_{yy} = \frac{\partial^2 f}{\partial y^2}, \qquad f_{yx} = \frac{\partial^2 f}{\partial x \partial y}, \qquad f_{xy} = \frac{\partial^2 f}{\partial y \partial x}.$$

The first and second come from partially differentiating with respect to x or y. The last two respectively come from partially differentiating with respect to y then x, or x then y. Typically (for appropriately nice functions) we find that these two expressions agree. This result (on the basis of different hypotheses) is due to Alexis Clairault (1713–1765), Hermann Schwarz (1843–1921) and William Young (1863–1942).

By way of an example, let $g(x,y) = \sin(xy)$. We then have that $g_x = y\cos(xy)$, $g_y = x\cos(xy)$ and

$$g_{xx} = -y^2 \sin(xy), \qquad g_{xy} = g_{yx} = \cos(xy) - xy\sin(xy), \qquad g_{yy} = -x^2 \sin(xy).$$

So we see here that the mixed derivatives g_{xy} and g_{yx} agree as expected. In a like manner to functions of a single variable, we say that $f(x,y)$ has a *stationary* or *critical point* if $f_x = 0 = f_y$ and we can similarly look to the second derivatives to classify such a stationary point. The three functions

$$f_1(x,y) = x^2 + y^2, \qquad f_2(x,y) = -x^2 - y^2, \qquad f_3(x,y) = x^2 - y^2$$

each have a stationary value at $(0,0)$ and between them give a flavour of what a (non-degenerate) stationary point can look like. Note that f_1 has a *minimum* at $(0,0)$ - however we move from the origin, f_1 increases. Likewise, f_2 has a *maximum* there. f_3, though, is neither a minimum nor a maximum and instead is a *saddle point*; f_3 increases if we move along the x-axis and decreases along the y-axis. We similarly see that g has a saddle point at $(0,0)$ as g increases if we move into the two quadrants where $xy > 0$ and decreases in the other two quadrants.

Partial Differential Equations (PDEs). The study of DEs involving partial derivatives has some results analogous to those for ODEs. Say a function $f(x,y)$ of two variables satisfies the PDE

$$\frac{\partial f}{\partial x} = 0.$$

For a function of one variable, the corresponding ODE would have constant f as the general solution; in a similar way the above PDE has functions which are constant *with respect to x* but can still vary with any other variables. So the above PDE has general solution $f(x,y) = a(y)$, where a is an arbitrary function. Likewise the PDE $f_{xx} = 0$ has general solution $f(x,y) = a(y)x + b(y)$, where a and b are arbitrary functions. This corresponds to $f''(x) = 0$ having the general solution $f(x) = Ax + B$. So the rule of thumb that the general solution of an ODE of order k includes k arbitrary constants is replaced with an order k PDE's solution including k arbitrary functions.

Before this starts seeming like a simple step up from ODEs, consider the PDE $f_x + f_y = 0$. It is relatively straightforward to show that $f(x,y) = a(x-y)$ solves the PDE, where a is an arbitrary function, but how might you show that *all* solutions take this form, or have conceived of such solutions in the first place?

Second-order linear PDEs can be grouped as one of three forms, with important examples of each arising in nature. Three such examples (where $a,b > 0$) are

$$f_{xx} + f_{yy} = 0, \quad f_x = af_{yy}, \quad f_{xx} - bf_{yy} = 0.$$

The first is *Laplace's equation* (which arises naturally in complex analysis, electromagnetism and gravitational theory); the second is the *diffusion equation* (which arises when modelling heat flow and also in biological systems) and the third is the *wave equation* (arising when studying vibrations and light). Respectively they are important examples of *elliptical, parabolic and hyperbolic* PDEs whose solutions and theory differ significantly.

---

we don't lose any solutions by this oversight. Thinking a little more, we might realize that we have missed an obvious solution: the zero function, $y = 0$, which isn't present in our 'general' solution. At this point we might scold ourselves for dividing by 0 at stage (6.10), rather than treating $y = 0$ as a separate case. But we have missed many more solutions, the general solution of (6.9) being

$$y(x) = \begin{cases} (x-a)^2 & x \leqslant a \\ 0 & a \leqslant x \leqslant b, \\ (x-b)^2 & b \leqslant x \end{cases} \tag{6.11}$$

where a and b are constants satisfying $-\infty \leqslant a \leqslant b \leqslant \infty$. Note also that the general solution requires *two* constants in its description even though the DE is only first order. $\qquad\Box$

It is possible to gain some qualitative appreciation of DEs and their solutions pictorially. Consider the first-order DE, $dy/dx = x - y$. By the end of §6.3 we will be able to solve such equations using integrating factors. The general solution is of the form $y = x - 1 + Ae^{-x}$. But even without this knowledge we can perform some qualitative analysis of what the solutions look like.

In the Figure 6.3a is a plot of the direction field of this DE – that is at each point (x, y) in the diagram is drawn a short vector whose gradient equals $x - y$. The graph of any solution to the DE is tangential to this vector field at each point.

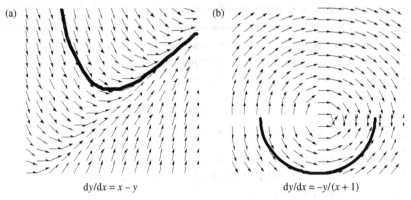

$dy/dx = x - y$ $dy/dx = -y/(x + 1)$

Figure 6.3

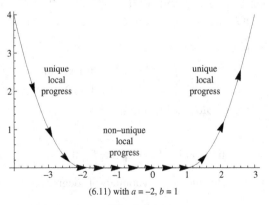

(6.11) with $a = -2$, $b = 1$

Figure 6.4

The direction field is plotted for $-3 < x, y < 3$ and the solution $y = x - 1 + e^{-x}$ is represented in bold. Even though we are not currently able to solve the DE we can see that solutions typically remain on one side of the line $y = x - 1$ and tend towards this line as x increases.

The second direction field (Figure 6.3b) is for the DE, $dy/dx = -x/(y + 1)$, which we will solve in Example 6.9. The solution that is plotted is $y = -1 - \sqrt{4 - x^2}$, which is valid for $-2 < x < 2$. This is the solution to the DE with initial condition $y(0) = -3$. Again, even though we have not yet solved the DE, the solutions' semicircular nature is clear and also that the solutions stay on one side of the line $y = -1$. As each solution approaches the line $y = -1$ we see that there is nowhere further that we can extend the solution.

Finally, Figure 6.4 is a plot of the solution (6.11) with $a = -2$ and $b = 1$. When we are at a point with $y > 0$ the vector field for (6.9) would include two vectors of gradients $\pm 2\sqrt{y}$. If our solution is decreasing (as when $x < -2$) or if our solution is increasing (as when

#1580
#1581
#1582
#1583
#1588

$x > 1$) there is only one way to smoothly continue our solution. However, at a point $(x_0, 0)$, we can see that we have the option of continuing flat along the x-axis or taking off on the half-parabola $y = (x - x_0)^2$, where $x \geqslant x_0$. A theorem of Picard's,[1] a proof of which is well beyond the scope of these notes, guarantees a unique local solution to the ODE

$$\frac{dy}{dx} = f(x, y)$$

when the partial derivative $\partial f / \partial y$ is defined. Note in (6.10) that $f(x, y) = 2\sqrt{y}$ and we see $\partial f / \partial y = 1/\sqrt{y}$ is undefined on the x-axis and only there. We will not have to worry about the concerns of Picard's theorem in this chapter's exercises; we highlight this example here by way of introducing some need for care when solving ODEs and as a flavour of the subject at a more advanced level.

Exercises

#1577a Show that the function $y(x)$ in (6.11) is indeed a solution to (6.9).

#1578a Show that the function $y(t) = A\cos(\omega t) + B\sin(\omega t)$ (where A and B are constants) is a solution of (6.1). Show further that there is a unique solution of this form meeting the initial conditions $y(0) = C$ and $y'(0) = D$.

#1579b Solve the initial-value problem $y''(x) = (1 + x^2)^{-1}$, $y(0) = y'(0) = 0$, and confirm that $y(1) = \pi/4 - (\ln 2)/2$.

#1580B† Verify that

$$y(x) = \begin{cases} \frac{1}{2}e^x - 1 & \text{for } x \leqslant \ln 2, \\ 1 - 2e^{-x} & \text{for } \ln 2 \leqslant x, \end{cases}$$

is a solution of the initial-value problem $dy/dx = 1 - |y|$, $y(0) = -1/2$. Sketch a graph of the solution.

#1581B Sketch the direction field in the xy-plane for the DE: $dy/dx = 1 - |y|$.

#1582b For $1 \leqslant n \leqslant 3$, find a polynomial of degree n which solves Legendre's equation (6.2).

#1583B† Find the solution $y(x)$ of the DE, $(2y - x)y' = y$, subject to the initial condition $y(1) = 3$. Now find the solution of the same DE subject to the initial condition $y(1) = -2$.

#1584C Where possible, find all solutions to (6.9) satisfying the following initial-value and boundary-value problems.

$\quad$ (i) $\;y(0) = 0$; $\qquad$ (ii) $\;y(0) = 0$, $y'(0) = 1$; $\qquad$ (iii) $\;y(0) = 1$, $y(2) = 1$;

$\;\;$ (iv) $\;y(0) = 1$, $y(1) = 1$; $\qquad$ (v) $\;y(1) = 1$, $y'(1) = 2$; $\qquad$ (vi) $\;y(1) = 1$, $y(5) = 4$.

#1585C Find conditions that K, L, α, β must satisfy for (6.3) to have any solutions to the boundary conditions $h(\alpha) = K$, $h(\beta) = L$.

---

[1] Émile Picard (1856–1941) was a French mathematician who made significant contributions to complex analysis and DEs.

#1586 C In each case below, find conditions that K, L, α, β must satisfy for (6.9) to have any solutions when (i) $y(\alpha) = K$, (ii) $y(\alpha) = K$, $y(\beta) = L$. If solutions are possible, find all possible solutions.

#1587 C (Regarding the discussion following Example 6.4.) One of the conditions required by Picard's theorem for the DE $dy/dx = f(x, y)$ to have a unique local solution is that there exists $K > 0$ such that

$$|f(x, y_1) - f(x, y_2)| \leqslant K |y_1 - y_2| \qquad \text{for all } x, y_1, y_2.$$

Show that the function $f(x, y) = \sqrt{y}$ does meet this condition for $y \geqslant 1$, but not for $y \geqslant 0$.

#1588 b Write the LHS of the DE

$$(2x + y) + (x + 2y)y' = 0,$$

in the form $\frac{d}{dx}(F(x, y)) = 0$, where $F(x, y)$ is a polynomial in x and y. Hence find the general solution of the equation.

Use this method to find the general solution of

$$(y \cos x + 2x e^y) + (\sin x + x^2 e^y - 1)y' = 0.$$

#1589 C Show that there is no polynomial $F(x, y)$ in x and y such that the DE

$$(x^2 - xy + y^2)y' + 3x^2 = 0$$

can be rewritten in the form $\frac{d}{dx}(F(x, y)) = 0$. However, show that this is possible if we first multiply the DE by $x + y$.

#1590 D† **Clairaut's equation** has the form

$$y = xy' + F(y').$$

(i) Show that, by differentiating with respect to x, the equation can be reduced to a first-order DE in y'. Hence solve the DE

$$y = xy' + (y')^2. \tag{6.12}$$

(ii) Transform (6.12) into an ODE in z and x by setting $y = z - x^2/4$. Now review your solutions from (i).

6.2 Separable Equations

Recall that a first-order DE has the general form $dy/dx = f(x, y)$. A special class of first-order DEs is one of the form

$$\frac{dy}{dx} = a(x)b(y),$$

known as **separable equations**. Such equations can, in principle, be easily rearranged and solved as follows: firstly we have

$$\frac{1}{b(y)} \frac{dy}{dx} = a(x) \tag{6.13}$$

and then, integrating with respect to x, we find

$$\int \frac{dy}{b(y)} = \int \frac{1}{b(y)} \frac{dy}{dx} \, dx = \int a(x) \, dx. \tag{6.14}$$

Remark 6.5 In some texts, or previously at A-level, (6.13) may be written

$$\frac{1}{b(y)}\,dy = a(x)\,dx \tag{6.15}$$

and subsequently integrated. At first glance, though, this makes no rigorous sense. dy/dx exists as the limit of an approximating gradient $\delta y/\delta x$ whereas in the limit δx and δy are both 0. So (6.15) seems to say nothing more than $0 = 0$. However, the equation can be made rigorous with knowledge of differentials (which are beyond the scope of this text) and no error will result from an equation like (6.15), but we should recognize, with our current definition of dy/dx, that the equation has no rigorous meaning. ∎

Remark 6.6 Equation (6.14) may offer some practical problems as a solution to the DE. Firstly the functions $1/b$ and a may not have known integrals. But further, even if they both are readily integrated, to end with a solution of the form $B(y) = A(x) + $const., this isn't really solving the DE as asked. We originally said a solution was a function $y(x)$ which satisfies the DE and our solution is not in this form, and it may be difficult to determine y so explicitly. ∎

Example 6.7 (Exponential Growth and Decay) In many examples from science the rate of change of a variable is proportional to the variable itself, e.g. in the growth in a bacterial sample, in a capacitor discharging, in radioactive decay. That the rate of change with time t of a quantity, say y, is proportional to y is encoded in the DE

$$\frac{dy}{dt} = ky \tag{6.16}$$

where k is a constant. Note that $y = 0$ is clearly a solution. Assuming then that $y \neq 0$ we can separate the variables to obtain $y^{-1}dy/dt = k$ and integrating with respect to t we have, for some constant C,

$$\ln|y| = kt + C \Longrightarrow |y| = e^C e^{kt} \implies y = \pm e^C e^{kt}.$$

As $\pm e^C$ can be an arbitrary non-zero constant, together with the solution $y = 0$, we see that

$$y = A e^{kt} \qquad \text{where } A \text{ is any real number}$$

is the general solution of (6.16). Note that we can characterize our solution uniquely with an initial condition.

Remark 6.8 Our solution to (6.16) is rather cumbersome in its treatment of the different cases. With knowledge of the general solution we might instead have started by introducing a new function $z(t) = y(t)e^{-kt}$, our aim then being to show that z is constant. This is left as #1591. ∎

Example 6.9 Find the solution of the initial-value problem

$$\frac{dy}{dx} = \frac{-x}{y+1}, \qquad y(0) = -3. \tag{6.17}$$

Solution The equation is separable. Rearranging to $(y+1)dy/dx = -x$ and integrating gives $(y+1)^2/2 = -x^2/2 + C$. As $y = -3$ when $x = 0$ then $C = 2$ and we have

$$x^2 + (y+1)^2 = 4, \tag{6.18}$$

which is the equation of a circle. However, by definition, a solution of a DE is a function y of x whereas for each x in the range $-2 < x < 2$ there are two y which satisfy (6.18). But any solution $y(x)$ of (6.17) must satisfy (6.18). So we can solve for y obtaining

$$y = -1 - \sqrt{4 - x^2}$$

noting we take the negative root as $y(0) = -3$. Note also that the solution is valid for $-2 < x < 2$. □

Example 6.10 Find the general solution to the separable DE

$$\sin x \frac{dy}{dx} = y \ln y. \tag{6.19}$$

Solution Separating variables we find $(y \ln y)^{-1} dy/dx = \csc x$ and integrating with respect to x we get

$$\ln |\ln y| = \ln |\tan (x/2)| + C,$$

where C is a constant, which rearranges to

$$y = \exp \{A \tan (x/2)\}, \tag{6.20}$$

where A is another constant. The solution would be valid on, say, $-\pi < x < \pi$. □

Example 6.11 Find the solution of the initial-value problem

$$(5y^4 - 1)\frac{dy}{dx} = 1, \qquad y(0) = 1.$$

Solution We can straight away integrate both sides and obtain $y^5 - y = x + C$. Putting $x = 0$ and $y = 1$ we see that $C = 0$ and so

$$y^5 - y = x. \tag{6.21}$$

This equation does not give us a simple expression for the solution y in terms of x, but the solution y has to satisfy (6.21).

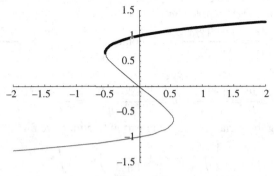

Figure 6.5

If we sketch the curve $y^5 - y = x$ (as in Figure 6.5) we see that the solution y, which goes through the point $(0, 1)$, can be extended arbitrarily to greater x but can only be extended down to $(-4 \times 5^{-5/4}, -5^{1/4})$, where dy/dx becomes infinite; there are no values we can assign to y for $x < -4 \times 5^{-5/4}$ without having a discontinuity in the solution y.

#1591
#1592
#1593
#1594
#1598
#1599

On the other hand, we cannot solve (6.21) explicitly for given x and it remains our only way to describe the solution. It should be appreciated though that the whole curve is not the solution but just the part which appears in bold in Figure 6.5.   □

There are many DEs which aren't separable, but which can be adjusted to a separable equation with a sensible change of variable. By a **homogeneous polar** DE we will mean one of the form

$$\frac{dy}{dx} = f\left(\frac{y}{x}\right). \tag{6.22}$$

The term 'polar' here is appropriate, as $y/x = \tan\theta$ where θ is the polar co-ordinate. So the RHS is a function solely of θ.

- *Note these DEs are often simply called homogeneous, but we will use the term homogeneous polar here to distinguish them from other DEs which we will meet later and which are also referred to as homogeneous.*

These can be solved with a substitution of the form

$$y(x) = xv(x) \tag{6.23}$$

to get another equation in terms of x and the new dependent variable v. Note from the product rule of differentiation that $dy/dx = v + x\,dv/dx$ and so making the substitution (6.23) into the DE (6.22) gives us the new DE

$$x\frac{dv}{dx} = f(v) - v,$$

which is a separable DE.

Example 6.12 Find the general solution of the DE

$$\frac{dy}{dx} = \frac{x + 3y}{x + y}.$$

Solution At first glance this may not look like a homogeneous polar DE, but we can see this to be the case by dividing both the numerator and denominator on the RHS by x. If we make the substitution $y(x) = xv(x)$ then

$$\frac{dy}{dx} = \frac{1 + 3(y/x)}{1 + (y/x)} \qquad \text{becomes} \qquad v + x\frac{dv}{dx} = \frac{1 + 3v}{1 + v}.$$

Rearranging gives

$$x\frac{dv}{dx} = \frac{1 + 3v}{1 + v} - v = \frac{1 + 2v - 3v^2}{1 + v} = \frac{(1 - v)(1 + 3v)}{1 + v}$$

and so, separating the variables and using partial fractions, we find

$$\frac{1}{2}\int\left(\frac{1}{1 - v} + \frac{1}{1 + 3v}\right)dv = \int\frac{dx}{x}.$$

Integrating and recalling $v = y/x$ leads us to the general solution

$$\frac{1}{2}\left\{-\ln|1 - y/x| + \frac{1}{3}\ln|1 + 3y/x|\right\} = \ln|x| + C,$$

where C is a constant. This can be simplified somewhat to $x + 3y = Ax^4(x - y)^3$ where A is a constant.   □

Example 6.13 Solve the initial-value problem

$$\frac{dy}{dx} = \frac{y}{x+y+2}, \qquad y(0) = 1.$$

#1601
#1605
#1606
#1607
#1609
#1610

Solution This DE is not homogeneous polar, but it can easily be made into such a DE with a suitable change of variables. We introduce new variables

$$X = x + a \quad \text{and} \quad Y = y + b,$$

with the aim of choosing a and b to make the DE homogeneous polar. Making these substitutions, the RHS becomes

$$\frac{Y-b}{X+Y+2-a-b},$$

which is homogeneous polar if $b = 0$ and $a = 2$. With these values of a and b, noting that $dY/dX = dy/dx$ and that the initial condition has become $Y(X = 2) = y(x = 0) = 1$, our initial-value problem now reads as

$$\frac{dY}{dX} = \frac{Y}{X+Y}, \qquad Y(2) = 1,$$

which now is homogeneous polar. Substituting in $Y = VX$ gives us

$$V + X\frac{dV}{dx} = \frac{VX}{X+VX} = \frac{V}{1+V}, \qquad V(2) = \frac{1}{2}.$$

Rearranging and separating the variables gives

$$\frac{1}{V} - \ln|V| = \int \left(-\frac{1}{V^2} - \frac{1}{V}\right) dV = \int \frac{dx}{X} = \ln|X| + K.$$

Substituting in our initial condition, $V(2) = 1/2$, we see that $K = 2$. Recalling $V = Y/X$, the above becomes

$$X - Y\ln Y = 2Y, \qquad X, Y > 0.$$

Finally, as $X = x + 2$ and $Y = y$, our solution to the initial-value problem has become

$$x + 2 = 2y + y\ln y, \qquad x > -2, y > 0. \qquad \square$$

Example 6.14 By means of a substitution, transform the following into a separable DE. Find its general solution.

$$\frac{dy}{dx} = \cos(x+y).$$

Solution This is neither separable nor homogeneous polar, but the substitution $z = x + y$ would seem a sensible one to simplify the RHS. We might then hope to get a separable DE in x and z. As

$$y = z - x, \qquad \frac{dy}{dx} = \frac{dz}{dx} - 1 \qquad \text{and the DE becomes} \qquad \frac{dz}{dx} - 1 = \cos z.$$

This is separable and we find

$$\int \frac{dz}{\cos z + 1} = x + C.$$

Integrating the LHS is made easier with the trigonometric identity $\cos z + 1 = 2\cos^2(z/2)$, giving

$$\tan\frac{z}{2} = \frac{1}{2}\int \sec^2\left(\frac{z}{2}\right)\,dz = x + C.$$

So $\tan(x+y)/2 = x + C$ and rearranging gives the general solution

$$y = 2\tan^{-1}(x+C) - x.$$

$\square$

Exercises

#1591A Introduce a new dependent variable $z(t) = y(t)e^{-kt}$ to turn (6.16) into a DE in z and t. Solve this new DE to find the general solution of (6.16).

#1592A† Which of the solutions in (6.20) satisfy the initial condition $y(0) = 1$? How many solutions satisfy the initial condition $y(0) = 2$? Why are these answers unsurprising when we look at the original DE (6.19)?

#1593b Find the solutions of the DEs (i) $y'(x) = xe^x$, $y(0) = 0$. (ii) $y'(x) = -(x+1)y^3$, $y(0) = 1/2$.

#1594b Find the general solutions of the following separable DEs.

$$\frac{dy}{dx} = \frac{x^2}{y}; \qquad \frac{dy}{dx} = \frac{\cos^2 x}{\cos^2 2y}; \qquad \frac{dy}{dx} = e^{x+2y}.$$

#1595B Show that the solution of the initial-value problem

$$\frac{dy}{dx} = \sqrt{\frac{k}{y} - 1}, \quad y(0) = 0,$$

where k is a positive constant can be parametrized as

$$x(\theta) = \frac{k}{2}(\theta - \sin\theta), \quad y(\theta) = \frac{k}{2}(1 - \cos\theta).$$

#1596c Find all solutions of the following separable DEs.

$$\frac{dy}{dx} = \frac{y - xy}{xy - x}; \qquad \frac{dy}{dx} = \frac{\sin^{-1} x}{y^2\sqrt{1 - x^2}}.$$

#1597C Let $x(t)$ be the solution of the initial-value problem

$$\frac{dx}{dt} = (1+x)e^x, \quad x(0) = 0,$$

and let T be the value of t for which $x(T) = 1$. Express T as an integral and show that $(1 - e^{-1})/2 < T < 1 - e^{-1}$.

#1598b† By making a substitution $z = dy/dx$, solve the initial-value problem

$$\frac{d^2y}{dx^2} = (1 + 3x^2)\left(\frac{dy}{dx}\right)^2 \quad \text{where } y(1) = 0 \text{ and } y'(1) = \frac{-1}{2}.$$

#1599B† Solve the initial-value problem $y'' = (y')^2 + 1$, $y(0) = y'(0) = 0$.

#1600c Find the solution of the following initial-value problems. On separate axes sketch the solution to each.

$$\frac{dy}{dx} = \frac{1-2x}{y}, \quad y(1) = -2; \qquad \frac{dy}{dx} = \frac{x(x^2+1)}{4y^3}, \quad y(0) = \frac{-1}{\sqrt{2}}; \qquad \frac{dy}{dx} = \frac{1+y^2}{1+x^2}, \quad y(0) = 1.$$

#1601B By making the substitution $y(x) = xv(x)$ as in (6.23), solve the following DEs.

$$\frac{dy}{dx} = \frac{x^2+y^2}{xy}; \qquad x\frac{dy}{dx} = y + \sqrt{x^2+y^2}.$$

#1602c Solve the initial-value problems.

$$\frac{dy}{dx} = \frac{2xy^2+x}{x^2y-y}, \quad y(\sqrt{2}) = 0; \qquad \sin x \sin y \frac{dy}{dx} = \cos x \cos y, \qquad y(\pi/2) = \pi/4.$$

#1603c Solve the DEs: (i) $x^2y' + xy = y^2$; (ii) $x^2y' = y^2 - 2x^2$.

#1604c Rederive the solution of #1589.

#1605B By introducing the new dependent variable $z = y^2$, solve $(y^4 - 3x^2)y' + xy = 0$.

#1606b Make substitutions, of the form $x = X + a$, $y = Y + b$, to turn the DE

$$\frac{dy}{dx} = \frac{x+y-3}{x-y-1}$$

into a homogeneous polar DE in X and Y. Hence find the general solution of the above equation.

#1607B Show that the DE

$$\frac{dy}{dx} = \frac{x+y-1}{2x+2y-1}$$

cannot be transformed into a homogeneous polar DE by means of substitutions $x = X + a$, $y = Y + b$. By means of the substitution $z = x + y$, find the general solution of the equation.

#1608c Find the solution of the initial-value problem $\quad y' = (y-x+1)/(y-x+5)$, $y(0) = 0$.

#1609B† A particle P moves in the xy-plane with time t. Its co-ordinates $x(t)$ and $y(t)$ satisfy the equations

$$dy/dt = x + y \qquad \text{and} \qquad dx/dt = x - y,$$

and at $t = 0$ the particle is at $(1,0)$. Find, and solve, a homogeneous polar DE relating x and y. By changing to polar co-ordinates, or otherwise, sketch the particle's journey for $t \geqslant 0$.

#1610b† Find the general solutions of the DEs (i) $\quad y' = \sin^2(x-2y)$. (ii) $\quad xy' + y = e^{xy}$.

#1611C† Find the general solutions of the following DEs.

$$\sin x \left(\frac{dy}{dx} + 1\right) + 2(x+y)\cos x = 0; \qquad e^y \frac{dy}{dx} = x + e^y - 1.$$

#1612c Find the general solutions of the DEs.

$$\text{(i)} \quad \frac{dy}{dx} = (x+y-4)^2. \qquad \text{(ii)} \quad \frac{dy}{dx} = \sqrt{2x+3y-2}.$$

#1613c† Solve the initial-value problem $4y^3y' = 1 + y^8$, $y(0) = 1$.

#1614B Find a family of solutions of the DE $(dy/dx)^2 + y^2 = 1$, depending on an arbitrary constant. Note also that $y = 1$ and $y = -1$ are solutions. Given Example 6.4, what is the general solution of this DE?

#1615d (See also #1580.) Find the solutions of the initial-value problems consisting of the DE $dy/dx = 1 - |y|$ and the initial conditions

(i) $y(0) = 2$,   (ii) $y(0) = 1/2$,   (iii) $y(0) = -2/3$,   (iv) $y(0) = -2$.

Sketch these four solutions on the same axes.

#1616C† Let α and β be distinct real numbers. Say that y satisfies

$$\frac{d^2y}{dx^2} - (\alpha + \beta)\frac{dy}{dx} + \alpha\beta y = 0.$$

Let $f(x) = y'(x) - \alpha y(x)$ and $g(x) = y'(x) - \beta y(x)$. Find and solve a first-order DE that $f(x)$ satisfies. Do the same for $g(x)$ and hence determine the general solution for $y(x)$.

#1617C (**Logistic Equation**[2]) Consider a population $N(t)$, varying with time t, which satisfies the DE

$$\frac{dN}{dt} = rN\left(1 - \frac{N}{K}\right)$$

where $r > 0$ is the growth rate of the population when resources are not an issue and $K > 0$ represents population capacity. Solve the DE and show that $N(t)$ approaches K as t becomes large.

#1618d (**SIR Epidemic Model**) An influenza spreads through a fixed population according to the following model. Initially the entire population is susceptible S, but increasingly with time t becomes infected I and then eventually recovers R. (Here S, I, R represent the fractions of the population in those categories so that $S = 1, I = R = 0$ when $t = 0$ and $S + I + R = 1$ for all t). The DEs governing this spread are

$$\frac{dS}{dt} = -\alpha SI; \qquad \frac{dI}{dt} = \alpha SI - \beta I; \qquad \frac{dR}{dt} = \beta I,$$

where α, β are positive constants.

(i) What assumptions about the spread of the influenza have led to this model? What does a high α mean about the influenza? What about a low β?
(ii) Find S and I as functions of R.
(iii) Find the maximum value of $I(t)$, showing that it is an increasing function of β/α.

6.3 Integrating Factors

An important class of DEs consists of the linear DEs. These are important because of their rich theory, because a great many important ODEs are linear and also because a linear DE can often be used to approximate a non-linear equation.

---

[2] The Belgian mathematician Pierre-François Verhulst (1804–1849) published this equation in 1838 as a model for population growth, naming it the *logistic equation*. He argued that a population's growth is limited by availability of resources; in his model the effective growth rate is $r(1 - N/K)$. When N is small this approximates exponential growth but the growth rate decreases as the population approaches its capacity K.

Definition 6.15 An **inhomogeneous linear** DE of order k is one of the form

$$a_k(x)\frac{d^k y}{dx^k} + a_{k-1}(x)\frac{d^{k-1} y}{dx^{k-1}} + \cdots + a_1(x)\frac{dy}{dx} + a_0(x)y = f(x),$$

where $a_k(x) \neq 0$. If $f(x) = 0$ then the DE is called **homogeneous linear**.

So a first-order homogeneous linear DE is one of the form $P(x)y'(x) + Q(x)y = 0$, which we can see is separable – we have already tackled such DEs in §6.2. A first order *inhomogeneous* linear DE is of the form

$$P(x)\frac{dy}{dx} + Q(x)y = R(x), \tag{6.24}$$

and these can be approached by using **integrating factors**.

- *The idea is to multiply the LHS of (6.24) by a factor which makes it the derivative of a product $A(x)y$.*

Consider the DE in (6.24). In general, the LHS of (6.24) won't be expressible as the derivative of a product $A(x)y$. However, if we multiply both sides of the DE by an appropriate integrating factor $I(x)$ then we can turn the LHS into the derivative of such a product, as shown below. Let's first of all simplify the equation by dividing through by $P(x)$, and then multiply by an integrating factor $I(x)$ (which we have yet to determine) to get

$$I(x)\frac{dy}{dx} + I(x)\frac{Q(x)}{P(x)}y = I(x)\frac{R(x)}{P(x)}. \tag{6.25}$$

We would like the LHS to be the derivative of a product $A(x)y$ for some function $A(x)$. By the product rule $A(x)y$ differentiates to

$$A(x)\frac{dy}{dx} + A'(x)y \tag{6.26}$$

and so, equating the coefficients of y and y' in (6.25) and (6.26), we have

$$A(x) = I(x) \qquad \text{and} \qquad A'(x) = I(x)Q(x)/P(x),$$

from which we deduce $I'(x)/I(x) = Q(x)/P(x)$. The LHS is the derivative of $\ln I(x)$ and so we see

$$I(x) = \exp \int \frac{Q(x)}{P(x)}\,dx.$$

(We are only looking for one such $I(x)$ with this property and so don't need to concern ourselves with a constant of integration.) For this $I(x)$, (6.25) now reads as

$$\frac{d}{dx}(I(x)y) = \frac{I(x)R(x)}{P(x)},$$

which has the general solution

$$y(x) = \frac{1}{I(x)}\left(\int \frac{I(x)R(x)}{P(x)}\,dx + \text{const.}\right).$$

Example 6.16 Find the general solution of the DE $\quad xy' + (x-1)y = x^2$.

Solution If we divide through by x we get $dy/dx + (1 - x^{-1})y = x$, which we see has integrating factor

$$I(x) = \exp\left\{\int\left(1 - \frac{1}{x}\right)dx\right\} = \exp(x - \ln x) = \frac{1}{x}e^x.$$

Multiplying through by the integrating factor gives

$$\frac{d}{dx}\left(\frac{1}{x}e^x y\right) = \frac{1}{x}e^x\frac{dy}{dx} + \left(\frac{1}{x} - \frac{1}{x^2}\right)e^x y = e^x.$$

Integrating gives $x^{-1}e^x y = e^x + C$, where C is a constant, and rearranging gives the general solution

$$y(x) = x + Cxe^{-x}. \qquad \square$$

Example 6.17 Solve the initial-value problem $y' - 2xy = 1$, $y(0) = 0$.

#1619
#1620
#1621
#1622
#1625
#1626

Solution The integrating factor here is $I(x) = \exp\int(-2x)\,dx = \exp(-x^2)$. Multiplying through we get

$$\frac{d}{dx}(e^{-x^2}y) = e^{-x^2}\frac{dy}{dx} + 2xe^{-x^2}y = e^{-x^2}.$$

Noting that $y(0) = 0$, when we integrate this we arrive at $e^{-x^2}y = \int_0^x e^{-t^2}\,dt$, and rearranging gives

$$y(x) = e^{x^2}\int_0^x e^{-t^2}\,dt = \frac{\sqrt{\pi}}{2}e^{x^2}\operatorname{erf}(x). \qquad \square$$

Example 6.18 Solve the initial-value problem $yy' + \sin x = y^2$, $y(0) = 1$.

Solution This DE is neither linear nor separable. However, if we note that $2yy' = (y^2)'$ then we see that the substitution $z = y^2$ turns the given DE into

$$z' - 2z = -2\sin x, \qquad z(0) = 1^2 = 1$$

which is solvable using integrating factors. In this case the integrating factor is e^{-2x} and we get

$$\frac{d}{dx}(ze^{-2x}) = -2e^{-2x}\sin x.$$

Integrating the RHS by parts as in Example 5.31 we get

$$ze^{-2x} = e^{-2x}(4\sin x + 2\cos x)/5 + C.$$

As $z = 1$ when $x = 0$ then $C = 3/5$ and so, recalling that $z = y^2$, we have

$$y = \sqrt{(4\sin x + 2\cos x + 3e^{2x})/5},$$

taking the positive root, and with the solution valid on the interval containing 0 for which $4\sin x + 2\cos x + 3e^{2x} > 0$. $\qquad \square$

Exercises

#1619a Use the method of integrating factors to solve the following equations with initial conditions.

$$\frac{dy}{dx} + xy = x, \quad y(0) = 0; \quad 2x^3\frac{dy}{dx} - 3x^2y = 1, \quad y(1) = 0; \quad \frac{dy}{dx} - y\tan x = 1, \quad y(0) = 1.$$

#1620 B Solve the following DEs.

$$(1-x^2)\frac{dy}{dx} + 2xy = (1-x^2)^{3/2}; \qquad \frac{dy}{dx} - (\cot x)y + \csc x = 0.$$

#1621 b Solve the initial-value problem $xy' - 2y = x^3 e^x$, $y(1) = 0$.

#1622 B† Find the general solution of the DE $y'(x) = (x+y)/(x+1)$.

#1623 b By multiplying by e^y, solve the DE below and sketch the solution.

$$e^y\frac{dy}{dx} = x\frac{dy}{dx} + 1, \quad y(0) = 0.$$

#1624 c Solve the following DEs.

$$\frac{dy}{dx} = \frac{1}{\cos y - x\tan y}; \qquad \frac{dy}{dx} = \frac{3y}{3y^{2/3} - x}.$$

#1625 b By making the substitution $y = 1/z$, find the solution $y(x)$ of the initial-value problem

$$12xy' + 24y = (12 + 5x)y^2, \quad y(1) = 1.$$

#1626 B By treating y as the independent variable, solve $(x+y^3)dy/dx = y$.

#1627 C Bernoulli's equation [3] has the form

$$\frac{dy}{dx} + P(x)y = Q(x)y^n \qquad \text{where } n \neq 1. \tag{6.27}$$

Show that (6.27) can be reduced to an inhomogeneous linear first-order DE by the substitution $z = y^{1-n}$.

#1628 c Solve the following DEs:

$$y' + y = xy^{2/3}; \qquad xy' + y = 2x^{5/2}y^{1/2}; \qquad 3xy^2y' + 3y^3 = 1.$$

6.4 Linear Differential Equations

Recall that a homogeneous linear DE of order k is one of the form

$$a_k(x)\frac{d^k y}{dx^k} + a_{k-1}(x)\frac{d^{k-1}y}{dx^{k-1}} + \cdots + a_1(x)\frac{dy}{dx} + a_0(x)y = 0. \tag{6.28}$$

The space of solutions to such a DE has some nice algebraic properties.

Theorem 6.19 *Let y_1 and y_2 be solutions of a homogeneous linear DE and α_1, α_2 be real numbers. Then $\alpha_1 y_1 + \alpha_2 y_2$ is also a solution of the DE. Note also that the zero function is always a solution.*

Proof As y_1 and y_2 are solutions of (6.28) then

$$a_k(x)\frac{d^k y_1}{dx^k} + a_{k-1}(x)\frac{d^{k-1}y_1}{dx^{k-1}} + \cdots + a_1(x)\frac{dy_1}{dx} + a_0(x)y_1 = 0; \tag{6.29}$$

$$a_k(x)\frac{d^k y_2}{dx^k} + a_{k-1}(x)\frac{d^{k-1}y_2}{dx^{k-1}} + \cdots + a_1(x)\frac{dy_2}{dx} + a_0(x)y_2 = 0. \tag{6.30}$$

[3] Named after Jacob Bernoulli. See p.124 for a biographical note on the Bernoulli family.

If we add α_1 times (6.29) to α_2 times (6.30) and rearrange we find

$$a_k(x)\frac{d^k(\alpha_1 y_1 + \alpha_2 y_2)}{dx^k} + \cdots + a_1(x)\frac{d(\alpha_1 y_1 + \alpha_2 y_2)}{dx} + a_0(x)(\alpha_1 y_1 + \alpha_2 y_2) = 0,$$

which shows that $\alpha_1 y_1 + \alpha_2 y_2$ is also a solution of (6.28) also.   □

Remark 6.20 The theorem's statement says that the solutions of the DE form a real vector space – see Remark 3.11. The details of the proof largely boil down to the following rule of differentiation

$$\frac{d}{dx}(\alpha f + \beta g) = \alpha\frac{df}{dx} + \beta\frac{dg}{dx}$$

for functions f, g and real numbers α, β – i.e. that differentiation $D = d/dx$ is a linear map (Definition 3.136) and we then see that the space of solutions of (6.28) is the null space (Definition 3.122) of the linear map

$$a_k(x)D^k + a_{k-1}(x)D^{k-1} + \cdots + a_1(x)D + a_0(x)I.$$ ■

We focus for now on the second-order case. A second-order homogeneous linear DE has the form

$$P(x)\frac{d^2 y}{dx^2} + Q(x)\frac{dy}{dx} + R(x)y = 0. \tag{6.31}$$

We shall consider the situation where, either by inspection or other means, we already know of a non-zero solution $Y(x)$. Knowing this it is possible to transform (6.31) into a first-order DE.

We begin by making the substitution

$$y(x) = Y(x)z(x),$$

thus turning (6.31) into a DE involving z and x. Note

$$\frac{dy}{dx} = Y\frac{dz}{dx} + \frac{dY}{dx}z, \qquad \frac{d^2 y}{dx^2} = Y\frac{d^2 z}{dx^2} + 2\frac{dY}{dx}\frac{dz}{dx} + \frac{d^2 Y}{dx^2}z.$$

Substituting these expressions into (6.31) and rearranging, we find

$$PY\frac{d^2 z}{dx^2} + \left(2P\frac{dY}{dx} + QY\right)\frac{dz}{dx} + \left(P\frac{d^2 Y}{dx^2} + Q\frac{dY}{dx} + RY\right)z(x) = 0. \tag{6.32}$$

Now the bracket $PY'' + QY' + RY$ equals 0 as Y is a solution to (6.31). Further, if we let $w(x) = dz/dx$ then we can see (6.32) is a first-order separable DE in w, namely

$$PY\frac{dw}{dx} + \left(2P\frac{dY}{dx} + QY\right)w = 0, \tag{6.33}$$

which is separable and having found w, we may integrate it to find z and so y.

Example 6.21 Show that $Y(x) = 1/x$ is a solution of $xy'' + 2(1 - x)y' - 2y = 0$. Hence find this DE's general solution.

Solution We see that $Y(x) = 1/x$ is a solution as $x(2x^{-3}) + 2(1 - x)(-x^{-2}) - 2x^{-1} = 0$. From the above discussion we know to make the substitution $y(x) = x^{-1}z(x)$ and we then have

$$y' = x^{-1}z' - x^{-2}z, \qquad y'' = x^{-1}z'' - 2x^{-2}z' + 2x^{-3}z.$$

Our original DE then becomes the following DE in z and x:

$$x(x^{-1}z'' - 2x^{-2}z' + 2x^{-3}z) + 2(1-x)(x^{-1}z' - x^{-2}z) - 2x^{-1}z = 0,$$

which simplifies to $z'' - 2z' = 0$ and so $z'(x) = Ae^{2x}$ for some constant A. Integrating this, we find $z = ae^{2x} + b$ (where $a = A/2$ and b is another constant) and hence the original DE's general solution is

$$y(x) = z(x)Y(x) = (ae^{2x} + b)/x. \qquad \square$$

Example 6.22 Show that Legendre's equation (see Example 6.1) with $n = 1$,

$$(1-x^2)\frac{d^2y}{dx^2} - 2x\frac{dy}{dx} + 2y = 0,$$

has a solution of the form $Y(x) = ax + b$ and hence determine its general solution.

Solution The function $Y(x) = ax + b$ is a solution of the DE if

$$(1-x^2) \times 0 - 2xa + 2(ax + b) = 0.$$

Equating coefficients we see that $-2a + 2a = 0$ and $2b = 0$. Hence $b = 0$ and we see a can take any value; in particular, $Y(x) = x$ is a solution. This time we need to make the substitution $y(x) = xz(x)$ and find

$$y' = xz' + z, \qquad y'' = xz'' + 2z'.$$

Our original DE then becomes the following DE in z and x,

$$(1-x^2)(xz'' + 2z') - 2x(xz' + z) + 2xz = 0$$

which simplifies to $(x^3 - x)z'' = (2 - 4x^2)z'$. Separating variables and integrating gives

$$\ln z' = \int \frac{2 - 4x^2}{x^3 - x} \, dx = \int \left(\frac{-2}{x} + \frac{-1}{x-1} + \frac{-1}{x+1} \right) dx = -2\ln|x| - \ln|x-1| - \ln|x+1| + C,$$

for a constant C and so

$$z = \int \frac{A \, dx}{x^2(x-1)(x+1)} = A\left(\frac{1}{x} + \frac{\ln|x-1|}{2} - \frac{\ln|x+1|}{2} \right) + B,$$

where $A = e^C$ and B are constants. Hence the general solution is

$$y(x) = xz(x) = A + \frac{Ax}{2}\ln\left|\frac{x-1}{x+1}\right| + Bx. \qquad \square$$

In the case when the DE is linear, but inhomogeneous, solving the inhomogeneous equation still strongly relates to the solution of the associated homogeneous equation, as Theorem 6.23 makes explicit.

Theorem 6.23 *Let $Y(x)$ be a solution, known as a **particular solution**, or **particular integral**, of the inhomogeneous linear DE*

$$a_k(x)\frac{d^ky}{dx^k} + a_{k-1}(x)\frac{d^{k-1}y}{dx^{k-1}} + \cdots + a_1(x)\frac{dy}{dx} + a_0(x)y = f(x). \qquad (6.34)$$

That is, $y = Y$ satisfies (6.34). Then a function $y(x)$ is a solution of the inhomogeneous linear DE (6.34) if and only if $y(x)$ can be written as

$$y(x) = z(x) + Y(x),$$

where $z(x)$ is a solution of the corresponding homogeneous linear DE

$$a_k(x)\frac{\mathrm{d}^k z}{\mathrm{d}x^k} + a_{k-1}(x)\frac{\mathrm{d}^{k-1}z}{\mathrm{d}x^{k-1}} + \cdots + a_1(x)\frac{\mathrm{d}z}{\mathrm{d}x} + a_0(x)z = 0. \tag{6.35}$$

The general solution $z(x)$ to the corresponding homogeneous DE (6.35) is known as the **complementary function**.

Proof If $y(x) = z(x) + Y(x)$ is a solution of (6.34) then

$$a_k(x)\frac{\mathrm{d}^k (Y+z)}{\mathrm{d}x^k} + a_{k-1}(x)\frac{\mathrm{d}^{k-1}(Y+z)}{\mathrm{d}x^{k-1}} + \cdots + a_1(x)\frac{\mathrm{d}(Y+z)}{\mathrm{d}x} + a_0(x)(Y+z) = f(x).$$

Rearranging the brackets we get

$$\left(a_k(x)\frac{\mathrm{d}^k z}{\mathrm{d}x^k} + \cdots + a_1(x)\frac{\mathrm{d}z}{\mathrm{d}x} + a_0(x)z\right) + \left(a_k(x)\frac{\mathrm{d}^k Y}{\mathrm{d}x^k} + \cdots + a_1(x)\frac{\mathrm{d}Y}{\mathrm{d}x} + a_0(x)Y\right) = f(x).$$

Now the second bracket equals $f(x)$, as $Y(x)$ is a solution of (6.34). Hence the first bracket must equal 0 – that is, $z(x)$ is a solution of the corresponding homogeneous DE (6.35). □

Remark 6.24 In practice, a particular solution is usually found by educated guess work and trial and error with functions that are roughly of the same type as $f(x)$. ∎

Remark 6.25 The space of solutions of (6.34) is not a vector space. They form what is known as an **affine space**. A homogeneous linear DE always has 0 as a solution, whereas this is not the case for the inhomogeneous equation.

Compare this with the vector geometry we have seen (Definition 3.24): a plane through the origin is a vector space (#495) and if vectors $\mathbf{a}$ and $\mathbf{b}$ span it then every point will have a position vector $\lambda\mathbf{a} + \mu\mathbf{b}$; points on a plane parallel to it will have position vectors $\mathbf{p} + \lambda\mathbf{a} + \mu\mathbf{b}$, where $\mathbf{p}$ is a particular point on the plane. The point $\mathbf{p}$ acts as a choice of origin in the plane, playing the same role as the particular solution Y in the inhomogeneous case above. Likewise the general solution of the DE

$$y'' - y = x \qquad \text{is} \qquad y = \lambda e^x + \mu e^{-x} - x$$

with such solutions forming a plane; with the given choice of parameters $-x, e^x, e^{-x}$ take on the roles of $\mathbf{p}, \mathbf{a}, \mathbf{b}$ respectively.

This same aspect of linear algebra is at work in the proof of Theorem 2.37, which deals with inhomogeneous linear recurrence relations, and the solutions of homogeneous and inhomogeneous linear systems is more generally treated in #646. ∎

Example 6.26 Find the general solution of

$$x\frac{\mathrm{d}^2 y}{\mathrm{d}x^2} + 2(1-x)\frac{\mathrm{d}y}{\mathrm{d}x} - 2y = 12x. \tag{6.36}$$

Solution We already showed in Example 6.21 that the general solution of the corresponding homogeneous equation is $y(x) = (ae^{2x} + b)/x$. So we just need to find a particular solution of (6.36). A reasonable first attempt would be to see if there is a solution of the form $Y(x) = Ax + B$. Such a Y is a solution if

$$x \times 0 + 2(1-x)A - 2(Ax+B) = 12x.$$

Rearranging this becomes $(2A - 2B) - 4Ax = 12x$ so, comparing coefficients, we see that $A = -3$ and $B = -3$. That is $Y(x) = -3x - 3$ is a particular solution. The general solution of (6.36) is

$$y(x) = (ae^{2x} + b)x^{-1} - 3x - 3. \qquad \square$$

Exercises

#1629 b Show that $y(x) = x$ satisfies $x^2y'' - xy' + y = 0$ and hence find the general solution.

#1630 b Show that $y(x) = e^x$ satisfies $xy'' - (1+x)y' + y = 0$ and hence find the general solution.

#1631 b Show that $y(x) = x^2$ satisfies the DE below and hence find its general solution.

$$x^2y'' + (2x^3 - 5x)y' + (8 - 4x^2)y = 0.$$

#1632 b† Find the general solution of $x^2y'' - xy' + y = x^2 + 1$.

#1633 b† Find the general solution of $xy'' - (1+x)y' + y = 1 + x^3$.

#1634 d† Find the general solution of $xy'' - (1+x)y' + y = x^2 \sin x$.

#1635 d (i) Show that $y(x) = x^2$ is a solution of $xy'' - (x+1)y' + 2y = 0$.
(ii) Find the DE's unique solution, in terms of an integral, given that $y(1) = y'(1) = 1$.

#1636 D† Find a particular solution of $(x^2 + 1)y'' = 2y$ and so find the general solution.

#1637 c Show that $y(x) = x^4$ satisfies $x^2y'' - 7xy' + 16y = 0$ and hence find the general solution.

#1638 c Show that $y(x) = x^{-2}$ satisfies $xy'' + (3 - x)y' - 2y = 0$ and hence find the general solution.

#1639 c (i) Find a solution of the form $y(x) = e^{ax}$ to $(x+2)y'' + (2x+5)y' + 2y = 0$.
(ii) Write, as an integral, the solution to the DE in (i) that satisfies the initial conditions $y(0) = y'(0) = 1$.

#1640 c Find a particular solution of the DE below, and hence find its general solution.

$$x^2y'' - x(x+2)y' + (x+2)y = 0.$$

#1641 D (i) Given a homogeneous linear DE of order k in $y(x)$, and knowledge of a non-zero solution $Y(x)$, show that the substitution $y(x) = Y(x)z(x)$ gives a homogeneous linear DE of order $k - 1$ in $z'(x)$.
(ii) Say now that two independent solutions $Y_1(x)$ and $Y_2(x)$ are known to a homogeneous linear DE of order k in $y(x)$. Explain how to reduce the DE to a homogeneous linear DE of order $k - 2$.

6.5 Linear Constant Coefficient Differential Equations

We now turn our attention to solving linear DEs

$$a_k \frac{d^k y}{dx^k} + a_{k-1}\frac{d^{k-1}y}{dx^{k-1}} + \cdots + a_1\frac{dy}{dx} + a_0 y = f(x)$$

where the functions $a_0, a_1, \ldots, a_k$ are **constants**. We have already seen in Theorem 6.23 that the difference between solving the inhomogeneous and homogeneous DEs is in finding a particular solution, so for now we will focus on the homogeneous case, when $f = 0$.

Mainly we shall treat examples when the equations are second order, though the theory extends naturally to similar higher order equations. We begin with the example of **simple harmonic motion** (SHM), the equation describing the vibrating of a spring or the swinging of a pendulum through small oscillations. The DE governing such motions is

$$\frac{d^2 y}{dx^2} = -\omega^2 y \qquad \text{where } \omega > 0. \tag{6.37}$$

Example 6.27 Show that the general solution of (6.37) is of the form

$$y(x) = A\cos\omega x + B\sin\omega x.$$

The constant ω is the **angular frequency** of these oscillations, with the solutions having **period** $2\pi/\omega$.

Solution We firstly set $v = dy/dx$. By the chain rule

$$\frac{d^2 y}{dx^2} = \frac{dv}{dx} = \frac{dy}{dx}\frac{dv}{dy} = v\frac{dv}{dy}$$

and the DE (6.37) becomes

$$v\frac{dv}{dy} = -\omega^2 y,$$

which is a separable one. If we separate the variables and integrate we find $v^2 = -\omega^2 y^2 + K$, for a non-negative constant K. Recalling $v = dy/dx$ we have

$$\frac{dy}{dx} = \sqrt{K - \omega^2 y^2},$$

which is again separable and we may solve this to find

$$x = \int \frac{dy}{\sqrt{K - \omega^2 y^2}}.$$

We know from §5.4 that such integrals can be tackled with trigonometric substitutions, a sensible one here being $y = (\sqrt{K}/\omega)\sin t$, which simplifies the integral to

$$x = \int \frac{(\sqrt{K}/\omega)\cos t\, dt}{\sqrt{K}\cos t} = \int \frac{dt}{\omega} = \frac{t}{\omega} + L,$$

for some constant L. Recalling $y = (\sqrt{K}/\omega)\sin t$ we have

$$y = (\sqrt{K}/\omega)\sin\omega(x - L),$$

or, by using the $\sin(\alpha + \beta)$ formula, we can alternatively write the above expression as $y = A\cos\omega x + B\sin\omega x$ for constants A and B. □

The SHM equation is a special case of the more general DE we are interested in, treated in the following theorem.

Theorem 6.28 *Let Q, R be real constants. The DE*

$$\frac{d^2y}{dx^2} + Q\frac{dy}{dx} + Ry = 0, \tag{6.38}$$

*has **auxiliary equation** (AE)*

$$m^2 + Qm + R = 0. \tag{6.39}$$

The general solution of (6.38) is

- $y = Ae^{\alpha x} + Be^{\beta x}$ *when the AE has two distinct real solutions $m = \alpha$ and $m = \beta$;*
- $y = (Ax + B)e^{\alpha x}$ *when the AE has a repeated real solution $m = \alpha$;*
- $y = e^{\alpha x}(A\cos \beta x + B\sin \beta x)$ *when the AE has complex conjugate roots $m = \alpha \pm i\beta$.*

Remark 6.29 Note, in Example 6.27, we tackled the case when the auxiliary equation's roots are $m = \pm\omega i$. ∎

Proof *First and second cases*: Let's call the roots of (6.39) α and β, and presume for the moment that they are real roots, but not necessarily distinct. We can rewrite the original DE (6.38) as

$$\frac{d^2y}{dx^2} - (\alpha + \beta)\frac{dy}{dx} + \alpha\beta y = 0. \tag{6.40}$$

Firstly note that $Y(x) = e^{\alpha x}$ is a solution, as substituting this into (6.40) gives

$$\alpha^2 e^{\alpha x} - (\alpha + \beta)\alpha e^{\alpha x} + \alpha\beta e^{\alpha x} = 0.$$

Following §6.4, when a solution is known, we set $y(x) = z(x)e^{\alpha x}$ to obtain

$$y' = e^{\alpha x}z' + \alpha e^{\alpha x}z,$$
$$y'' = e^{\alpha x}z'' + 2\alpha e^{\alpha x}z' + \alpha^2 e^{\alpha x}z, \tag{6.41}$$

and (6.40) ultimately simplifies to

$$z'' = (\beta - \alpha)z'. \tag{6.42}$$

We now have two cases to consider: when $\alpha = \beta$ we can integrate (6.42) twice to obtain $z(x) = Ax + B$ for constants A and B and then

$$y(x) = z(x)e^{\alpha x} = (Ax + B)e^{\alpha x},$$

as claimed for the theorem's second case. In the case when $\alpha \neq \beta$ then (6.42) has the solution $z'(x) = c_1 e^{(\beta - \alpha)x}$ for a constant c_1 (Example 6.7), and integrating this gives

$$z(x) = \frac{c_1}{\beta - \alpha}e^{(\beta - \alpha)x} + c_2 \qquad \text{for a second constant } c_2,$$

to finally find the solution to the theorem's first case:

$$y(x) = z(x)e^{\alpha x} = \frac{c_1}{\beta - \alpha}e^{\beta x} + c_2 e^{\alpha x}.$$

Third case: Suppose now that the roots of the equation are conjugate complex numbers $\alpha \pm i\beta$. For those who are aware of Euler's identity $e^{i\theta} = \cos\theta + i\sin\theta$ (p.27), we can simply treat this case the same as the first case. Allowing the constants A and B to be complex numbers, then (6.38) has a general solution of the form

$$y(x) = Ae^{(\alpha+i\beta)x} + Be^{(\alpha-i\beta)x} = e^{\alpha x}(\tilde{A}\cos\beta x + \tilde{B}\sin\beta x),$$

where $\tilde{A} = A + B$ and $\tilde{B} = i(A - B)$.

Alternatively, here is a proof which does not rely on Euler's identity. When the AE has roots $\alpha \pm i\beta$, then the original DE (6.38) reads as

$$\frac{d^2y}{dx^2} - 2\alpha\frac{dy}{dx} + (\alpha^2 + \beta^2)y = 0. \tag{6.43}$$

If we first make the substitution $y(x) = z(x)e^{\alpha x}$ and use the formulae from (6.41), then (6.43) becomes

$$(e^{\alpha x}z'' + 2\alpha e^{\alpha x}z' + \alpha^2 e^{\alpha x}z) - 2\alpha(e^{\alpha x}z' + \alpha e^{\alpha x}z) + (\alpha^2 + \beta^2)e^{\alpha x}z = 0.$$

This simplifies to $z'' + \beta^2 z = 0$, the SHM equation from Example 6.27, which we know to have general solution $z(x) = A\cos\beta x + B\sin\beta x$ and so we conclude as required that

$$y(x) = e^{\alpha x}(A\cos\beta x + B\sin\beta x). \qquad \square$$

Example 6.30 Find the general solution of the DE $\quad y'' - 6y' + 9y = 0$.

#1642
#1644
#1659
#1661
Solution This has auxiliary equation $0 = m^2 - 6m + 9 = (m - 3)^2$, which has a repeated root of 3. Hence the general solution is $y = (Ax + B)e^{3x}$. $\qquad\square$

Example 6.31 Solve the initial-value problem $\quad y'' - 3y' + 2y = 0, \quad y(0) = 1, \quad y'(0) = 0$.

Solution This has auxiliary equation $0 = m^2 - 3m + 2 = (m - 1)(m - 2)$, which has roots $m = 1$ and $m = 2$. So the general solution of the equation is

$$y(x) = Ae^x + Be^{2x}.$$

Now the initial conditions imply

$$1 = y(0) = A + B; \quad 0 = y'(0) = A + 2B.$$

Hence $A = 2$ and $B = -1$ and the unique solution of this DE is $y(x) = 2e^x - e^{2x}$. $\qquad\square$

Example 6.32 Find all solutions of the boundary-value problem

$$\frac{d^2y}{dx^2} - 2\frac{dy}{dx} + 2y = 0, \quad y(0) = 0 = y(\pi).$$

Solution This has auxiliary equation $0 = m^2 - 2m + 2 = (m - 1)^2 + 1$, which has roots $m = 1 \pm i$. So the general solution of the equation is

$$y(x) = e^x(A\cos x + B\sin x).$$

Now the boundary conditions imply

$$0 = y(0) = A; \quad 0 = y(\pi) = -e^\pi A.$$

Hence $A = 0$ and there are no constraints on B. So the general solution is $y(x) = Be^x\sin x$ for any B. $\qquad\square$

The theory behind the solving of homogeneous linear DEs with constant coefficients extends to such DEs of all orders, not just to second-order DEs, provided suitable adjustments are made.

Example 6.33 Write down the general solution of the following DE

$$\frac{d^7y}{dx^7} + \frac{d^6y}{dx^6} - \frac{d^5y}{dx^5} - 5\frac{d^4y}{dx^4} + 4\frac{d^2y}{dx^2} + 4\frac{dy}{dx} - 4y = 0.$$

Solution This has auxiliary equation

$$m^7 + m^6 - m^5 - 5m^4 + 4m^2 + 4m - 4 = 0.$$

We can see (with a little effort) that this factorizes as $(m-1)^3(m^2 + 2m + 2)^2 = 0$ which has 1 as a *triple* root and $-1 \pm i$ as *double* roots. So the general solution of this DE is

$$y(x) = (Ax^2 + Bx + C)e^x + (Dx + E)e^{-x}\cos x + (Fx + G)e^{-x}\sin x. \qquad \square$$

We now turn our attention to constant coefficient linear DEs which are *inhomogeneous* – i.e., those where the RHS is non-zero. These can occur quite naturally; for example, an inhomogeneous version of SHM, such as

$$\frac{d^2x}{dt^2} + \omega^2 x = A\sin\Omega\, t,$$

governs the behaviour of a spring with natural frequency ω which is being forced at another frequency Ω (see Example 6.37 for the solution of this DE).

More generally, as has already been noted in Theorem 6.23:

- **The solutions $y(x)$ of an inhomogeneous linear DE are of the form $z(x) + Y(x)$ where $z(x)$ is a solution of the corresponding homogeneous equation, and $Y(x)$ is a particular solution of the inhomogeneous equation.**

A particular solution $Y(x)$ is usually found by a mixture of educated guesswork and trial and error.

Example 6.34 Find the general solution of $\quad y'' - 3y' + 2y = x.$

Solution As the function on the right is $f(x) = x$ then it would seem sensible to *try* to find a particular solution of the form $Y(x) = Cx + D$ where C and D are, as yet, undetermined constants. There is no presumption that such a solution exists, but this seems a sensible family of functions amongst which we might well find a particular solution. Note that $Y' = C$ and $Y'' = 0$. So *if* $Y(x)$ is a particular solution of the DE then we would find

$$0 - 3C + 2(Cx + D) = x$$

and this is an identity which must hold for all values of x. So comparing coefficients we see that $C = 1/2$ and $D = 3/4$. What this means is that

$$Y(x) = x/2 + 3/4$$

is a particular solution of the DE. Having already found in Example 6.31 the *complementary function,* that is the general solution of the corresponding homogeneous DE, to be

$Ae^x + Be^{2x}$, then by Theorem 6.23 we know the general solution of our inhomogeneous DE is

$$y(x) = Ae^x + Be^{2x} + \frac{x}{2} + \frac{3}{4}, \qquad \text{for arbitrary constants } A \text{ and } B. \qquad \square$$

Example 6.35 Solve the initial-value problem $\quad y'' - 3y' + 2y = \sin x, \quad y(0) = y'(0) = 0.$

Solution Our first thoughts might be to seek a particular solution of the form $Y(x) = C \sin x$. However, it can be quite quickly seen that no such solution exists; for such a function, Y and Y'' are multiples of $\sin x$ but Y' is a multiple of $\cos x$. Instead we would be better off seeking a particular solution in the family of functions

$$Y(x) = C \sin x + D \cos x. \tag{6.44}$$

This highlights a sensible approach when proposing a family of functions in order to find a particular solution: **the family of functions should be closed under differentiation.** So if Y is in the family so should Y', Y'', etc. Here $Y'(x) = C \cos x - D \sin x$ is also in the suggested family. *If* (6.44) is indeed a solution of the given DE, then

$$(-C \sin x - D \cos x) - 3(C \cos x - D \sin x) + 2(C \sin x + D \cos x) = \sin x$$

and comparing coefficients of $\sin x$ and $\cos x$ we have $3D + C = 1$ and $D = 3C$ so that $C = 1/10$ and $D = 3/10$, and a particular solution is $y(x) = (1/10) \sin x + (3/10) \cos x$. For the general solution of the DE we need to add the complementary function (found in Example 6.31) – so the general solution is

$$Y(x) = Ae^x + Be^{2x} + \frac{1}{10} \sin x + \frac{3}{10} \cos x.$$

The constants A and B are specified by the initial conditions $y(0) = 0 = y'(0)$; we have

$$y(0) = 0 = A + B + 3/10; \quad y'(0) = 0 = A + 2B + 1/10,$$

so that $A = -1/2$ and $B = 1/5$. Hence the unique solution of the initial-value problem is

$$y(x) = -\frac{1}{2}e^x + \frac{1}{5}e^{2x} + \frac{1}{10} \sin x + \frac{3}{10} \cos x. \qquad \square$$

Example 6.36 Solve the initial-value problem $\quad y'' - 6y' + 9y = e^{3x}, \quad y(0) = y'(0) = 1.$

Solution From Example 6.30 we know that the complementary function is $y = (Ax + B)e^{3x}$. This means that trying neither e^{3x} nor xe^{3x} as a particular solution would be worthwhile, as substituting either of them into the LHS of the DE would give 0. Instead we will try a particular solution of the form $Y(x) = Cx^2 e^{3x}$. For this Y we find

$$Y'' - 6Y' + 9Y = \{(2 + 12x + 9x^2) - 6(2x + 3x^2) + 9x^2\}Ce^{3x} = 2Ce^{3x}.$$

Hence $C = 1/2$ gives a particular solution and we see the general solution of the given inhomogeneous DE is

$$y(x) = (x^2/2 + Ax + B)e^{3x}.$$

The initial conditions then specify A and B. Note $y(0) = B = 1$; also as

$$y'(x) = (x + A + 3x^2/2 + 3Ax + 3B)e^{3x}$$

then $y'(0) = A + 3B = 1$ and so $A = -2$. Hence the initial-value problem has solution

$$y(x) = (x^2/2 - 2x + 1)e^{3x}.$$ □

Example 6.37 Show that the solutions of the DE $x'' + \omega^2 x = \cos \Omega t$ are bounded when $\Omega \neq \omega$, but become unbounded when $\Omega = \omega$.

#1653 **Solution** By Example 6.27, the complementary function is $x(t) = A \cos \omega t + B \sin \omega t$. If
#1654 $\Omega \neq \omega$ we can find a particular solution of the form $X(t) = C_0 \cos \Omega t + D_0 \sin \Omega t$ for some
#1656 C_0, D_0 (details are left to #1665) and hence the general solution is

$$x(t) = A \cos \omega t + B \sin \omega t + C_0 \cos \Omega t + D_0 \sin \Omega t,$$

each function of which is bounded whatever the values of A, B. But if $\Omega = \omega$ then we will not be able to find a particular solution of the above form as it is part of the complementary function (so substituting it into the LHS of the DE yields 0). Instead we can find a particular solution of the form $X(t) = \gamma_0 t \cos \Omega t + \delta_0 t \sin \Omega t$ (again see #1665). Each function in the general solution

$$x(t) = \alpha \cos \omega t + \beta \sin \omega t + \gamma_0 t \cos \Omega t + \delta_0 t \sin \Omega t$$

is unbounded, whatever the values of α, β. (We recognize the DE in question as governing a system (e.g. a stretched spring) with natural frequency ω being forced at a frequency Ω. If $\Omega \neq \omega$ then the driving motion works with the system as commonly as it impedes it. When $\Omega = \omega$ then the driving motion is synchronized with the natural frequency of the system and **resonance** occurs. See #1665.) □

Example 6.38 (**Finding particular solutions**) For each of the given $f(x)$, describe a family of functions $y(x)$ which contains a particular solution $Y(x)$ of the following DE:

$$y'' - 3y' + 2y = f(x).$$

Note that the complementary function of the above DE is $y(x) = Ae^x + Be^{2x}$.

- $f(x) = \sin 2x$ – simply trying $Y(x) = C \sin 2x$ would do no good as $Y'(x)$ would contain $\cos 2x$ terms whilst $Y(x)$ and $Y''(x)$ would only contain $\sin 2x$ terms. Instead we need to try the more general $Y(x) = C \sin 2x + D \cos 2x$.
- $f(x) = e^{3x}$ – this causes few problems and, as we might expect, we can find a solution of the form $Y(x) = Ce^{3x}$.
- $f(x) = e^x$ – this is different from the previous case, as e^x is part of the complementary function; setting $y(x) = Ce^x$ in the LHS of the DE would yield 0. Instead we can successfully try a solution of the form $Y(x) = Cxe^x$.
- $f(x) = xe^{2x}$ – again e^{2x} is part of the solution to the homogeneous DE. Also, as with the previous function, we can see that Cxe^{2x} would only help us with an e^{2x} term on the RHS. So we need to 'move up' a further power and try $Y(x) = (Cx^2 + Dx)e^{2x}$.
- $f(x) = e^x \sin x$ – though this may look more complicated a particular solution can be found of the form

$$Y(x) = e^x(C \sin x + D \cos x).$$

- $f(x) = \sin^2 x$ – making use of the identity $\sin^2 x = (1 - \cos 2x)/2$, we can see a solution can be found amongst the functions

$$Y(x) = C + D\sin 2x + E\cos 2x. \qquad \square$$

Exercises

#1642a Find the most general solution of the following DEs.

$$y'' - y = 0; \quad y'' + 4y = 0; \quad y'' + 3y' + 2y = 0; \quad y'' + 2y' + 3y = 0.$$

#1643c Find the most general solution of the following DEs.

$$y'' + 2y = 0; \quad y'' + 2y' + 2y = 0; \quad y'' - 5y' - 6y = 0; \quad y'' = 0.$$

#1644b Find the most general solution of the following DEs.

$$\frac{d^4 y}{dx^4} - y = 0; \quad \frac{d^3 y}{dx^3} - y = 0; \quad \frac{d^3 y}{dx^3} + \frac{d^2 y}{dx^2} + \frac{dy}{dx} + y = 0; \quad \frac{d^4 y}{dx^4} + \frac{d^2 y}{dx^2} = 0.$$

#1645c Find the most general solution of the following DEs.

$$y''' + 3y'' + 3y' + y = 0; \quad y'''' - 2y'' + y = 0;$$
$$y'''' + 2y'' + y = 0; \quad y''' + 5y'' - 3y' - 3y = 0.$$

#1646a Find the most general solution of the following DEs.

$$y'' - y = x; \qquad y'' + 4y = 3x^2; \qquad y'' + 3y' + 2y = \sin x; \qquad y'' + 2y' + 3y = x + \cos x.$$

#1647b† Find the most general solution of the following inhomogeneous DEs.

$$y'' + 5y' + 6y = x; \quad y'' - 6y' + 9y = \sin x; \quad y'' + 2y' + 2y = e^x; \quad y'' + 3y' + 2y = e^{-x}.$$

#1648B Find the most general solution of the following DEs.

$$y'' - y = e^x; \quad y'' + 4y = \sin 2x; \quad y'' + 3y' + 2y = x^2 + e^{-x}; \quad y'' + 3y' + 2y = xe^{-2x}.$$

#1649c Find the general solution of the DE $\quad y'' - (a+b)y' + aby = e^{cx}$, when (i) a, b, c are distinct, (ii) $a = c \neq b$.

#1650c Find a particular solution of $\quad y'' + 3y' + 2y = f(x)$, for each of the following different choices of $f(x)$.

(i) $f(x) = x^2$.   (ii) $f(x) = e^x$.   (iii) $f(x) = \sin x$.   (iv) $f(x) = e^{-x}$.

#1651c Find a particular solution of the DE $\quad y''' - 2y'' - y' + 2y = 6x + \sin x$.

#1652d† Find a particular solution of the DE $\quad y''' - 2y'' + y' - 2y = 6x + \sin x$.

#1653B Find a particular solution of $\quad y'' + y = f(x)$ where (i) $f(x) = \sin^2 x$, (ii) $f(x) = \sin x$, (iii) $f(x) = x\sin 2x$.

#1654B† For each of the following $f(x)$, write down a family of functions $y(x)$ which contains a particular solution of the DE $\quad y'' + 4y' + 4y = f(x)$, where
(i) $f(x) = x^2$, (ii) $f(x) = xe^x$, (iii) $f(x) = xe^{-2x}$, (iv) $f(x) = x^2\sin x$, (v) $f(x) = \sin^3 x$.

#1655c Determine a particular solution for each of the six choices of $f(x)$ in Example 6.38.

#1656B† Solve the following initial-value problems.

(i) $y'' + y = e^x \cos x$, $y(0) = 1$, $y'(0) = 0$. (ii) $y'' + 4y = \cos^2 x$, $y(0) = 0$, $y'(0) = 2$.

#1657b† Solve the initial-value problem $y'' - 5y' + 6y = 2^x$, $y(0) = 1$, $y'(0) = 0$.

#1658d† Solve the initial-value problem $y'' + 2y' + y = \cosh x$, $y(0) = y'(0) = 0$.

#1659B Find *all* the solutions (if any) of the following boundary-value problems.

(i) $y'' = \pi^2 y$, $y(0) = 1 = -y(1)$. (ii) $y'' = -\pi^2 y$, $y(0) = 1 = -y(1)$.

(iii) $y'' = -\pi^2 y$, $y(0) = 1$, $y(1) = 0$.

#1660C Show that the boundary-value problem $y'' + y = x + c \sin x$, $y(0) = y(\pi) = 0$ has no solution unless $c = -2$. Find the general solution when $c = -2$.

#1661B† (i) Show the only solutions $\psi(x)$ to (4.39), where V is constant, and which satisfy $\int_0^a |\psi(x)|^2 \, dx = 1$ are

$$\psi_n(x) = c_n \sqrt{\frac{2}{a}} \sin \frac{n \pi x}{a} \qquad \text{where} \quad E = V + \frac{n^2 \pi^2 \hbar^2}{2ma^2},$$

where n is a positive integer and c_n is a complex number with $|c_n| = 1$.

(ii) Show that the ψ_n are orthonormal in the sense that

$$\int_0^a \left(\sqrt{\frac{2}{a}} \sin \frac{m \pi x}{a} \right) \left(\sqrt{\frac{2}{a}} \sin \frac{n \pi x}{a} \right) dx = \delta_{mn}.$$

#1662D† By means of the substitution $z = \ln x$, find the general solution of the DE

$$x^2 y'' + xy' + y = 0.$$

#1663D† By making the substitution $t = e^x$ and setting $z(t) = y(x)$, rewrite the DE

$$y'' - y' + e^{2x} y = xe^{2x} - 1,$$

as one in terms of z and t. Hence find the DE's general solution.

#1664d† Find the general solution of $(x + 1)^2 y'' + 3(x + 1)y' + y = x^2$.

#1665D Find the values of $C_0, D_0, \gamma_0, \delta_0$ from Example 6.37. Further, find A, B, α, β given $x(0) = x'(0) = 0$. Show that $x(t)$ is bounded when $\Omega \neq \omega$ and unbounded when $\Omega = \omega$. What range of values can $x(t)$ take when $\omega \neq \Omega$?

#1666D† Solve the simultaneous DEs subject to initial conditions $f(0) = g(0) = 0$,

$$f'(x) - g'(x) - 2f(x) = e^{-2x}; \qquad 3f'(x) - 2g'(x) - g(x) = 0.$$

#1667D† (See also #1552.) A mass swinging on the end of a light rod is governed by the DE

$$\frac{d^2\theta}{dt^2} = -\frac{g}{l}\sin\theta.$$

Suppose that the maximum swing of the mass is α. Show that the time T of an oscillation equals

$$T = 4\sqrt{\frac{l}{2g}}\int_0^\alpha \frac{d\theta}{\sqrt{\cos\theta - \cos\alpha}}.$$

If α is small enough that the approximation $\cos\theta \approx 1 - \theta^2/2$ applies for $|\theta| \leqslant \alpha$, show that $T \approx 2\pi\sqrt{l/g}$.

#1668D Show that, for any values of α and β, the function

$$y(x) = \sin x \left(\int_\alpha^x f(t)\cos t\, dt\right) - \cos x \left(\int_\beta^x f(t)\sin t\, dt\right)$$

is a solution to the DE $\quad y'' + y = f(x)$. Hence solve the boundary-value problem

$$y'' + y = \csc x, \quad y(0) = y(\pi/2) = 0.$$

#1669D† Let $a_0, a_1, \ldots, a_k$ be real constants and suppose that the polynomial

$$z^k + a_{k-1}z^{k-1} + \cdots + a_1 z + a_0 = 0$$

has k distinct real roots $\gamma_1, \gamma_2, \ldots, \gamma_k$. What is the general solution of the DE

$$y^{(k)}(x) + a_{k-1}y^{(k-1)}(x) + \cdots + a_1 y'(x) + a_0 y(x) = 0? \tag{6.45}$$

Let $c_0, c_1, c_2, \ldots, c_{k-1}$ be real numbers. Show that there is a unique solution to the DE (6.45) which meets the initial conditions

$$y(0) = c_0, \quad y'(0) = c_1, \quad \ldots \quad y^{(k-1)}(0) = c_{k-1}.$$

6.6 Systems of Linear Differential Equations*

In this section we shall be interested in solving simultaneous linear DEs. We begin with the following specific example.

Example 6.39 Solve the following simultaneous equations involving the functions $x(t)$ and $y(t)$.

$$\frac{dx}{dt} = 3x + y, \quad \frac{dy}{dt} = 6x + 4y, \quad x(0) = y(0) = 1. \tag{6.46}$$

Solution *Method One*: we might argue as follows. From the first equation $y = x' - 3x$, and using this to eliminate y and y' from the second equation we find with a little rearranging that

$$x'' - 7x' + 6x = 0, \quad x(0) = 1 \quad \text{and} \quad x'(0) = 3x(0) + y(0) = 4. \tag{6.47}$$

Using the methods of §6.5 (details being left to #1670) we then have that

$$x(t) = (2e^t + 3e^{6t})/5, \quad y(t) = x'(t) - 3x(t) = (9e^{6t} - 4e^t)/5. \tag{6.48}$$

This is all fine and satisfactory for the above example. However, we know from our study of linear systems (§3.5) that we need to be much more systematic when faced with many equations in many variables. Here then is an alternative method which we will see generalizes more easily.

#1670
#1671
#1675
#1676

Method Two: It is in fact easier to think of (6.46) as a single initial-value problem in a vector quantity $\mathbf{r}(t) = (x(t), y(t))^T$ rather than as two separate equations in two variables $x(t)$ and $y(t)$. Rewriting (6.46) as such we see

$$\frac{d\mathbf{r}}{dt} = \begin{pmatrix} dx/dt \\ dy/dt \end{pmatrix} = \begin{pmatrix} 3x+y \\ 6x+4y \end{pmatrix} = \begin{pmatrix} 3 & 1 \\ 6 & 4 \end{pmatrix} \mathbf{r}, \qquad \mathbf{r}(0) = \begin{pmatrix} x(0) \\ y(0) \end{pmatrix} = \begin{pmatrix} 1 \\ 1 \end{pmatrix}.$$

(6.49)

Let's denote the 2×2 matrix above as A. The eigenvalues of A are 1 and 6 and two corresponding eigenvectors are $(1, -2)^T$ and $(1, 3)^T$ (see #1671). The matrix A is diagonalizable and if we set

$$P = \begin{pmatrix} 1 & 1 \\ -2 & 3 \end{pmatrix}, \qquad \text{then} \qquad \frac{d\mathbf{r}}{dt} = A\mathbf{r} = P \underbrace{\begin{pmatrix} 1 & 0 \\ 0 & 6 \end{pmatrix} P^{-1}\mathbf{r}}_{P^{-1}AP}.$$

Because the entries of P^{-1} are constant (i.e. independent of t), we can rewrite this as

$$\frac{d(P^{-1}\mathbf{r})}{dt} = \begin{pmatrix} 1 & 0 \\ 0 & 6 \end{pmatrix} (P^{-1}\mathbf{r}).$$

Finally, if we set $(X, Y)^T = P^{-1}\mathbf{r}$, then the above system is equivalent to the two (scalar) DEs

$$dX/dt = X \qquad \text{and} \qquad dY/dt = 6Y \tag{6.50}$$

with the initial condition now reading

$$\begin{pmatrix} X(0) \\ Y(0) \end{pmatrix} = P^{-1} \begin{pmatrix} x(0) \\ y(0) \end{pmatrix} = \frac{1}{5} \begin{pmatrix} 3 & -1 \\ 2 & 1 \end{pmatrix} \begin{pmatrix} 1 \\ 1 \end{pmatrix} = \begin{pmatrix} 2/5 \\ 3/5 \end{pmatrix}.$$

We can solve the DEs in (6.50) to find $X(t) = 2e^t/5$ and $Y(t) = 3e^{6t}/5$ and hence

$$\begin{pmatrix} x(t) \\ y(t) \end{pmatrix} = P \begin{pmatrix} X(t) \\ Y(t) \end{pmatrix} = \begin{pmatrix} 1 & 1 \\ -2 & 3 \end{pmatrix} \begin{pmatrix} \frac{2}{5}e^t \\ \frac{3}{5}e^{6t} \end{pmatrix} = \begin{pmatrix} \frac{2}{5}e^t + \frac{3}{5}e^{6t} \\ -\frac{4}{5}e^t + \frac{9}{5}e^{6t} \end{pmatrix}.$$

A sketch of the solution

$$\mathbf{r}(t) = \frac{2}{5}e^t \begin{pmatrix} 1 \\ -2 \end{pmatrix} + \frac{3}{5}e^{6t} \begin{pmatrix} 1 \\ 3 \end{pmatrix} \tag{6.51}$$

appears in Figure 6.6 with the eigenvectors $\mathbf{v}_1 = (1, -2)^T$ and $\mathbf{v}_2 = (1, 3)^T$ also shown. For $t \ll 0$ we see $\mathbf{r}(t)$ is near the origin, moving approximately in the direction $\mathbf{v}_1$ and for $t \gg 0$ we see $\mathbf{r}(t)$ is moving approximately in the direction $\mathbf{v}_2$. The path of the particle lies entirely in the sector between the vectors $\mathbf{v}_1$ and $\mathbf{v}_2$ and the turning point of the curve is at

$$\sqrt[5]{2/27}(4, -6)/9.$$

By eliminating t from the equations for $x(t)$ and $y(t)$ we can see (details left to #1673) that the curve has equation

$$64(2x+y) = 3(3x-y)^6. \qquad \square$$

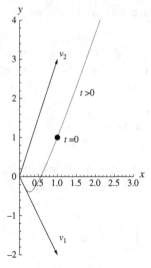

Figure 6.6

More generally, suppose that we have a system of n linear constant coefficient first-order DEs in n variables $x_1, x_2, \ldots, x_n$. These equations

$$\frac{dx_1}{dt} = a_{11}x_1 + \cdots + a_{1n}x_n; \quad \frac{dx_2}{dt} = a_{12}x_1 + \cdots + a_{2n}x_n; \quad \cdots \quad \frac{dx_n}{dt} = a_{n1}x_1 + \cdots + a_{nn}x_n,$$

(6.52)

can be more succinctly written as a single DE

$$\frac{d\mathbf{r}}{dt} = A\mathbf{r}$$

(6.53)

in a vector $\mathbf{r} = (x_1, x_2, \ldots, x_n)^T$, where $A = (a_{ij})$. The following theorem is a straightforward generalization of the second method used in solving (6.46).

Theorem 6.40 *Let $x_1(t), \ldots, x_n(t)$ be n variables satisfying the simultaneous DEs in (6.52) with $A = (a_{ij})$. If A is diagonalizable with eigenbasis $\mathbf{v}_1, \mathbf{v}_2, \ldots, \mathbf{v}_n$ and corresponding eigenvalues $\lambda_1, \lambda_2, \ldots, \lambda_n$ then the general solution of (6.52) is*

$$\mathbf{r}(t) = A_1 e^{\lambda_1 t}\mathbf{v}_1 + A_2 e^{\lambda_2 t}\mathbf{v}_2 + \cdots + A_n e^{\lambda_n t}\mathbf{v}_n,$$

(6.54)

where $\mathbf{r} = (x_1, x_2, \ldots, x_n)^T$ and $A_1, A_2, \ldots, A_n$ are arbitrary constants.

Proof The proof is a generalization of the working in 'Method Two' used to solve (6.46) and is left to #1674.   □

Example 6.41 Solve the following initial-value problem in the variables $x(t)$, $y(t)$, $z(t)$.

$$x' = 6x + y + 2z; \quad y' = 7y + 2z; \quad z' = -2y + 2z,$$

(6.55)

where $x(0) = 1$, $y(0) = 1$, $z(0) = 1$.

Solution The corresponding matrix A for this system was discussed in Example 3.205 and was found there to have an eigenbasis $(1, 1, -2)^T$, $(1, 0, 0)^T$, $(0, 2, -1)^T$ and corresponding eigenvalues $3, 6, 6$. By Theorem 6.40 we know that the general solution of (6.55) is

$$(x, y, z) = Ae^{3t}(1, 1, -2) + Be^{6t}(1, 0, 0) + Ce^{6t}(0, 2, -1),$$

and the initial conditions at $t = 0$ give us

$$(1, 1, 1) = A(1, 1, -2) + B(1, 0, 0) + C(0, 2, -1).$$

Solving, we find $A = -1$, $B = 2$, $C = 1$. Hence the desired solution is

$$x = 2e^{6t} - e^{3t}, \quad y = 2e^{6t} - e^{3t}, \quad z = 2e^{3t} - e^{6t}. \qquad \square$$

Example 6.42 Solve

$$x' = 3x + y + e^{2t}; \quad y' = 6x + 4y + e^{3t}, \tag{6.56}$$

subject to the initial conditions $x(0) = -1$ and $y(0) = 1$.

Solution This is an inhomogeneous version of the system in (6.46), which we've already solved, so we need only find a particular solution to this new system. Looking at the two RHSs, we might reasonably *try* a solution of the form

$$x(t) = Ce^{2t} + De^{3t}, \quad y(t) = Ee^{2t} + Fe^{3t}.$$

Substituting these into (6.56) we obtain equations

$$2C = 3C + E + 1; \quad 3D = 3D + F; \quad 2E = 6C + 4E; \quad 3F = 6D + 4F + 1,$$

which have solution $C = 1/2$, $D = -1/6$, $E = -3/2$, $F = 0$. Hence the general solution of (6.56) is

$$x(t) = Ae^{t} + Be^{6t} + e^{2t}/2 - e^{3t}/6, \quad y(t) = -2Ae^{t} + 3Be^{6t} - 3e^{2t}/2.$$

To complete the problem we see that the initial conditions then specify $A = -13/10$ and $B = -1/30$. $\qquad \square$

Example 6.43 Solve the linear system

$$x' = x + y; \quad y' = -x + y; \quad x(0) = y(0) = 1.$$

Solution The matrix

$$A = \begin{pmatrix} 1 & 1 \\ -1 & 1 \end{pmatrix}$$

is not diagonalizable over the reals, but does have complex eigenvalues $1 \pm i$ and corresponding eigenvectors $(1, i)^T$ and $(1, -i)^T$. The formula (6.54) holds but – in light of Theorem 6.28 and Euler's identity – is best interpreted as

$$\begin{pmatrix} x(t) \\ y(t) \end{pmatrix} = Ae^{t} \operatorname{cis} t \begin{pmatrix} 1 \\ i \end{pmatrix} + Be^{t} \operatorname{cis}(-t) \begin{pmatrix} 1 \\ -i \end{pmatrix}.$$

As $x(0) = y(0) = 1$ then $A + B = 1$ and $i(A - B) = 1$, so that $A = (1 - i)/2$ and $B = (1 + i)/2$. Hence

$$x(t) = \frac{(1 - i)}{2} e^{t} \operatorname{cis} t + \frac{(1 + i)}{2} e^{t} \operatorname{cis}(-t) = e^{t}(\cos t + \sin t);$$

$$y(t) = \frac{(i + 1)}{2} e^{t} \operatorname{cis} t + \frac{(-i + 1)}{2} e^{t} \operatorname{cis}(-t) = e^{t}(\cos t - \sin t). \qquad \square$$

To conclude this section, we note that the solution to the linear system

$$\frac{dx}{dt} = 3x + y; \quad \frac{dy}{dt} = 6x + 4y; \quad x(0) = \alpha; \quad y(0) = \beta, \tag{6.57}$$

which are the DEs in (6.46) with arbitrary initial conditions, is

$$x(t) = \frac{1}{5}[(3\alpha - \beta)e^t + (2\alpha + \beta)e^{6t}], \quad y(t) = \frac{1}{5}[(2\beta - 6\alpha)e^t + (6\alpha + 3\beta)e^{6t}]. \tag{6.58}$$

We can rewrite (6.57) in terms of $\mathbf{r}(t) = (x(t), y(t))^T$ as

$$\frac{d\mathbf{r}}{dt} = A\mathbf{r}, \quad \mathbf{r}(0) = \begin{pmatrix} \alpha \\ \beta \end{pmatrix} \quad \text{where} \quad A = \begin{pmatrix} 3 & 1 \\ 6 & 4 \end{pmatrix},$$

and *if* it made sense to approach this system as a single separable DE (as we did in Example 6.7) we might expect a solution of the form $\mathbf{r}(t) = e^{At}(\alpha, \beta)^T$; the obvious worry is what the *exponential of a matrix* might mean. We know from (6.58) that the solution is

$$\mathbf{r}(t) = \begin{pmatrix} x(t) \\ y(t) \end{pmatrix} = \frac{1}{5}\begin{pmatrix} 3e^t + 2e^{6t} & e^{6t} - e^t \\ 6e^{6t} - 6e^t & 2e^t + 3e^{6t} \end{pmatrix}\begin{pmatrix} \alpha \\ \beta \end{pmatrix}.$$

Does it therefore make any sense to define

$$\exp\left(\begin{pmatrix} 3 & 1 \\ 6 & 4 \end{pmatrix}t\right) = \frac{1}{5}\begin{pmatrix} 3e^t + 2e^{6t} & e^{6t} - e^t \\ 6e^{6t} - 6e^t & 2e^t + 3e^{6t} \end{pmatrix} \tag{6.59}$$

as the exponential of the matrix At? This may seem still more convincing when we note that

$$\frac{d}{dt}(e^{At}) = Ae^{At} \quad \text{and} \quad e^{At} = I_2 \text{ when } t = 0, \tag{6.60}$$

as these correspond to the defining characteristics of the real exponential function. See p.486 for further exercises.

Exercises

#1670A† Verify that the solution to the (6.46) is as given in (6.48). From (6.48), find an equation relating x and y.

#1671a Verify that the eigenvalues and eigenvectors the matrix A in (6.49) are as given.

#1672C† From (6.46) produce and solve a homogeneous polar DE involving x and y.

#1673C (i) Verify for the solution given in (6.51) that $\mathbf{r}(t) \approx \mathbf{0}$ and $\mathbf{r}'(t)$ is approximately parallel to $\mathbf{v}_1$ for $t \ll 0$ and that $\mathbf{r}'(t)$ is approximately parallel to $\mathbf{v}_2$ for $t \gg 0$.
(ii) Show that the turning point of the path of $\mathbf{r}(t)$ is at $\frac{1}{9}\sqrt[5]{2/27}(4, -6)$.
(iii) By eliminating t, show that the path of $\mathbf{r}(t)$ has equation $64(2x + y) = 3(3x - y)^6$.
(iv) Use implicit differentiation and the identity in (iii) to find a homogeneous polar DE involving x and y.

#1674C Generalize 'Method Two' of Example 6.39 to provide a proof of Theorem 6.40.

#1675b Find the general solution of the system

$$dx/dt = x + 4y; \qquad dy/dt = x + y.$$

Find the solution $(x(t), y(t))$ to the above system for which $x(0) = 0$ and $y(0) = 2$. Sketch this curve in the xy-plane, and include in in your sketch two independent eigenvectors of the matrix associated with the system.

#1676b† Show that the 2×2 matrix associated with the system below is not diagonalizable.

$$dx/dt = 2x + y; \qquad dy/dt = -4x + 6y; \qquad x(0) = y(0) = 1.$$

In spite of this, determine and sketch the system's solution.

#1677D† Solve the linear system

$$dx/dt = -2x + 3y; \qquad dy/dt = -3x - 2y; \qquad x(0) = y(0) = 1.$$

Show that $(x(t), y(t))$ approaches $(0, 0)$ as t becomes large. Find the distance travelled by $(x(t), y(t))$ on its journey from $(1, 1)$ to $(0, 0)$.

Example 6.44 The **Lotka–Volterra** equations (see Remark 6.45(a)) are

$$dF/dt = -mF + aFR, \qquad dR/dt = bR - kFR,$$

where a, b, k, m are positive constants, and they model competing numbers of predators F (foxes, say) and prey R (rabbits, say) – see Remarks 6.45(b) and (c). Dividing one equation by the other (and applying the chain rule) we obtain

$$\frac{dF}{dR} = \frac{F(-m + aR)}{R(b - kF)},$$

which is separable. Separating the variables and solving the equation we arrive at

$$F^b R^m e^{-aR - kF} = A \qquad \text{for some constant } A > 0. \tag{6.61}$$

Figures 6.7 and 6.8 are sketches of possible solutions where $a = b = m = 2$ and $k = 3$. Figure 6.7 shows, on FR-axes, curves of the form (6.61) where A is constant. We see that these are somewhat egg-shaped curves centred roughly around the point $(2/3, 1)$. These are the curves on which the point $(F(t), R(t))$ moves with time but Figure 6.7 gives no sense of where this point will be at a particular time t. Figure 6.8 shows the fox and rabbit populations changing with time as the point (F, R) moves around one of these egg-shaped curves. Notice how the fox population grows/declines when the rabbit population is high/low.

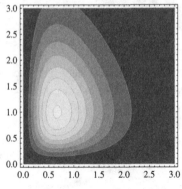

Lines of constant $F^b R^m e^{-aR-kF}$

Figure 6.7

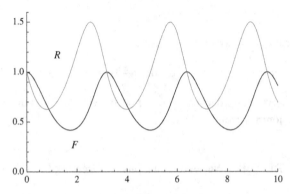

$F(t)$ and $R(t)$ plotted against time

Figure 6.8

The point $(2/3, 1)$ in Figure 6.7 is where the system is in equilibrium (for the given values of a, b, k, m). An **equilibrium** is a constant solution of the system; if (f, r) is an equilibrium of the Lotka–Volterra system this means that $dF/dt = 0$ and $dR/dt = 0$ at (f, r). So at such an equilibrium we have

$$0 = (-m + ar)f \qquad \text{and} \qquad 0 = (b - kf)r.$$

Our two equilibria are then $(0, 0)$ – where both populations have become extinct – and $(b/k, m/a)$. This agrees with the point $(2/3, 1)$ for the given values. To analyze further the egg-shaped curves near the equilibrium we move our origin there by a change of variables $\phi = F - b/k$ and $\rho = R - m/a$. The Lotka–Volterra equations then become

$$\phi' = (ab/k)\rho + a\phi\rho; \quad \rho' = (-km/a)\phi - k\phi\rho. \tag{6.62}$$

Imagine that the system was in equilibrium and ϕ and ρ represent the introduction of relatively small numbers of foxes and rabbits. What effect does this small change have on the system? The second-order terms involving $\phi\rho$ are then going to be small and negligible compared with the linear terms involving ϕ and ρ (see Remark 6.45(d)). If we ignore these small terms we arrive at a system which we can solve exactly

$$\phi' = (ab/k)\rho; \qquad \rho' = (-km/a)\phi. \tag{6.63}$$

These equations form a linear system of equations that was discussed earlier in this section; or we can readily note that

$$\phi'' = (ab/k)\rho' = (ab/k)(-km/a)\phi = -bm\phi;$$
$$\rho'' = (-km/a)\phi' = (-km/a)(ab/k)\rho = -bm\rho,$$

which are SHM equations and ϕ and ρ are given by

$$\phi = \alpha_1 \cos(\sqrt{bm}t + \beta_1); \quad \rho = \alpha_2 \cos(\sqrt{bm}t + \beta_2),$$

for constants $\alpha_1, \alpha_2, \beta_1, \beta_2$. Note that the period of these oscillations is $2\pi/\sqrt{bm} = \pi$ for the values of b, m in the above plots. It is also easy to see that $k^2 m\phi^2 + a^2 b\rho^2$ is a constant of the system in (6.63) which explains the approximately elliptical nature of the curves (see #1682). We would say that the equilibrium at $(b/k, m/a)$ is **stable** as a small nudge of the system from its equilibrium position moves the system into a nearby motion; this is in stark contrast to the equilibrium at $(0, 0)$ which is **unstable** (see #1684). A perturbation from $(0, 0)$, introducing small numbers of rabbits and foxes, would see their populations move on a large egg-shaped curve, rather than remaining in the vicinity of $(0, 0)$.

Remark 6.45 (a) These equations are named after the American mathematical biologist Alfred Lotka (1880–1949) and the Italian mathematician Vito Volterra (1860–1940) who independently studied them.

(b) This model is arrived at from the following assumptions: the rabbits breed at a certain rate $b > 0$ (which also incorporates death from disease and old age) and their population would grow exponentially were it not for the foxes; the number of rabbits being killed is proportional to both the rabbit population (the more rabbits there are, the more will get caught) and the foxes (the more predators, the more rabbits killed), so this is an *FR* term; the foxes, on the other hand, rely on the rabbits as a food source to multiply and, for the same reasons as just given, their growth term is an *FR* term countered by a term proportional to F due to death from disease and old age.

(c) Clearly the number of foxes or rabbits alive at any given time is a natural number, but if numbers are large enough these populations might still be modelled well as real variables. And if the populations are measured in thousands or millions then it makes sense for the actual numbers not to be integers.

(d) Following on from point (c), the small perturbations that ϕ and ρ represent from the equilibrium values might represent handfuls of rabbits and foxes relative to a population of thousands. So if ϕ, ρ are of the order 10^{-3} then terms like $\phi^2, \phi\rho, \rho^2$ will be of the order 10^{-6} and higher order terms will be still smaller. The process of neglecting the higher order terms in (6.62) to arrive at those in (6.63) is known as *linearization*. ∎

#1678d Find the general solution of the simultaneous DEs

$$dx/dt = 5x - y - z; \qquad dy/dt = x + 3y + z; \qquad dz/dt = -2x + 2y + 4z.$$

#1679B† Solve the inhomogeneous linear system

$$dx/dt = 2x + 3y + 2e^{2t}; \qquad dy/dt = x + 4y + 3e^{2t}; \qquad x(0) = -2/3, \qquad y(0) = 1/3.$$

#1682
#1686
#1687
#1688

#1680b Solve the system

$$x' = 3x + y + 2; \qquad y' = 6x + 4y + e^{2t}; \qquad x(0) = y(0) = 1.$$

#1681c† Find the general solution of the simultaneous equations

$$dx/dt = -2x + 2y + 2e^t; \qquad dy/dt = 3x - y + e^t.$$

#1682A Show that $k^2 m\phi^2 + a^2 b\rho^2$ is a constant of the system in (6.63). Explain why this means the paths travelled near the equilibrium point $(2/3, 1)$ in Figure 6.7 are roughly elliptical.

#1683D† Solve the following simultaneous DEs for x, y, z as functions of t.

$$x' = x - 2y + 2z; \qquad y' = x + y; \qquad z' = 2y - z; \qquad (x(0), y(0), z(0)) = (-2, 2, 3).$$

#1684C Show that (i) $F(t) = 0$, $R(t) = \varepsilon e^{bt}$ and (ii) $F(t) = \varepsilon e^{-mt}$, $R(t) = 0$ are solutions of the Lotka–Volterra equations. What does this say about the stability of the equilibrium point $(0, 0)$?

#1685c Find the solution of the system

$$dx/dt = -3x + \sqrt{2}y; \qquad dy/dt = \sqrt{2}x - 2y,$$

where $x(0) = \varepsilon_1$ and $y(0) = \varepsilon_2$. Is the system's equilibrium at $(0, 0)$ stable?

#1686b Find, and determine the stability of, the equilibria of the systems: (i) $x' = x$; $y' = y$. (ii) $x' = -x$; $y' = -y$. (iii) $x' = x$; $y' = -y$. (iv) $x' = y$; $y' = x$. (v) $x' = y$; $y' = -x$.

#1687B (i) Find, and determine the stability of, the equilibria of the system

$$x' = y; \qquad y' = 1 - x^4 - y^2.$$

(ii) By writing $x = 1 + u$ and $y = v$, where u and v are assumed to remain suitably small, show that the system approximately moves near the point $(1, 0)$ on ellipses with equations

$$4(x - 1)^2 + y^2 = \text{const.}$$

#1688b† Find, and determine the stability of, the equilibria of the system

$$x' = x + 2y, \qquad y' = -2x - y - x^3/2.$$

Show that the system moves along the curves $x^2 + xy + y^2 + x^4/8 = \text{const.}$ and sketch these curves near the origin.

#1689d Find, and determine the stability of, the equilibria of the system

$$x' = x(y - x + 1); \qquad y' = y(2y - x - 2).$$

#1690D Show that the system

$$x' = x + e^{x+y}; \qquad y' = x + 2y + 1$$

has a unique equilibrium $(-a, (a - 1)/2)$ where $a^2 e^{a+1} = 1$. Is this equilibrium stable?

6.7 Laplace Transform*

This section is about the *Laplace transform* which is one of a number of important integral transforms in mathematics. We introduce the Laplace transform in the chapter on DEs as it is (commonly) a means of transforming DEs into purely algebraic ones: ideally a DE in $f(x)$ is transformed into an algebraic equation in its transform $\bar{f}(s)$, an equation which we might solve, our remaining problem being to recognize this solution $\bar{f}(s)$ and so find the solution $f(x)$ of the original DE that transforms to $\bar{f}(s)$.

Definition 6.46 Let $f(x)$ be a real-valued function defined when $x > 0$. Then the **Laplace transform** [4] $\bar{f}(s)$ of $f(x)$ is defined to be

$$\bar{f}(s) = \int_0^\infty f(x)e^{-sx}\,dx, \tag{6.64}$$

for those s where this integral exists. $\bar{f}$ is also commonly denoted as $\mathcal{L}f$ and the Laplace transform itself as $\mathcal{L}$.

Example 6.47 Let $f(x) = e^{ax}$. Then

$$\bar{f}(s) = \int_0^\infty e^{ax}e^{-sx}\,dx = \int_0^\infty e^{-(s-a)x}\,dx = \left[\frac{e^{-(s-a)x}}{a-s}\right]_0^\infty = \frac{1}{s-a},$$

provided $s > a$. Note that $\bar{f}(s)$ is undefined when $s \leqslant a$ as the defining integral does not converge.

Example 6.48 Let $f_n(x) = x^n$ where $n \geqslant 0$ is an integer. Then, provided $s > 0$,

$$\bar{f_n}(s) = \int_0^\infty x^n e^{-sx}\,dx = \frac{-1}{s}\left[x^n e^{-sx}\right]_0^\infty + \frac{n}{s}\int_0^\infty x^{n-1}e^{-sx}\,dx = \frac{n}{s}\bar{f_{n-1}}(s),$$

which is a reduction formula. As

$$\bar{f_0}(s) = \int_0^\infty e^{-sx}\,dx = \frac{1}{s},$$

then

$$\bar{f_n}(s) = \frac{n}{s}\times\bar{f_{n-1}}(s) = \frac{n}{s}\times\frac{n-1}{s}\times\cdots\times\frac{1}{s}\times\bar{f_0}(s) = \frac{n!}{s^{n+1}}.$$

Again the integral in the definition of $\bar{f_n}(s)$ is undefined when $s \leqslant 0$.

Remark 6.49 See #1695 for the Laplace transform of x^α where $\alpha > -1$ is not necessarily an integer. ∎

Example 6.50 Let $f(x) = \cos ax$ and $g(x) = \sin ax$. For $s > 0$ we have

$$\bar{f}(s) = \frac{s}{s^2+a^2}; \qquad \bar{g}(s) = \frac{a}{s^2+a^2}. \tag{6.65}$$

---

[4] See p.240 for some brief biographical details of Laplace.

Solution By integration by parts, and provided $s > 0$,

$$\bar{f}(s) = \int_0^\infty e^{-sx} \cos ax \, dx = \frac{-1}{s}\left[e^{-sx}\cos ax\right]_0^\infty - \frac{a}{s}\int_0^\infty e^{-sx}\sin ax \, dx = \frac{1}{s} - \frac{a}{s}\bar{g}(s)$$

$$\bar{g}(s) = \int_0^\infty e^{-sx} \sin ax \, dx = \frac{-1}{s}\left[e^{-sx}\sin ax\right]_0^\infty + \frac{a}{s}\int_0^\infty e^{-sx}\cos ax \, dx = \frac{a}{s}\bar{f}(s).$$

Solving the simultaneous equations

$$\bar{f}(s) + \frac{a}{s}\bar{g}(s) = \frac{1}{s}, \qquad \bar{g}(s) = \frac{a}{s}\bar{f}(s),$$

gives the expressions in (6.65). □

Example 6.51 (Heaviside Function) Let $H(x) = \mathbf{1}_{(0,\infty)}(x)$. For $a > 0$ determine the Laplace transform of

$$H(x - a) = \begin{cases} 0 & 0 < x \leqslant a, \\ 1 & a < x. \end{cases}$$

Solution [5] We have for $s > 0$ that $H(x - a)$ has Laplace transform

$$\int_0^\infty H(x-a)e^{-sx}\,dx = \int_a^\infty e^{-sx}\,dx = \left[\frac{e^{-sx}}{-s}\right]_a^\infty = \frac{e^{-as}}{s}. \qquad □$$

We now note the following properties of the Laplace transform, which you may have already have noted as applying to the preceding examples.

Proposition 6.52 *Let $f(x)$ be a real-valued function defined when $x > 0$, such that the integral (6.64) in the definition of $\bar{f}(s_0)$ exists for some real value s_0. Then*
  (a) $\bar{f}(s)$ exists for all $s \geqslant s_0$.
  (b) $\bar{f}(s)$ tends to 0 as s tends to infinity.

Proof *Sketch proof.* (a) Not having developed a rigorous theory of integration on $(0, \infty)$ in this text, we shall only give a sketch proof of the proposition assuming somewhat more restrictive hypotheses. We shall assume that $f(x)$ is continuous and that there exists constants a, M such that

$$|f(x)| \leqslant Me^{ax} \qquad \text{for all } x > 0.$$

This means for all $t > 0$ that the integral

$$\bar{f}(a+t) = \int_0^\infty f(x)e^{-(a+t)x}\,dx = \int_0^\infty f(x)e^{-ax}e^{-tx}\,dx$$

converges as $f(x)e^{-ax}$ is a continuous function, satisfying $\left|f(x)e^{-(a+t)x}\right| \leqslant Me^{-tx}$ and Me^{-tx} is integrable on $(0, \infty)$ having integral M/t. Further (b) follows as

$$\left|\bar{f}(a+t)\right| \leqslant M/t, \qquad \text{implies} \qquad \bar{f}(a+t) \to 0 \quad \text{as} \quad t \to \infty. \qquad □$$

For the Laplace transform to be of use treating DEs, it needs to handle derivatives well and this is indeed the case.

---

[5] The function $H(x) = \mathbf{1}_{(0,\infty)}(x)$ is the *Heaviside function*, named after the mathematician and physicist Oliver Heaviside (1850–1925).

Proposition 6.53 *Provided the Laplace transforms of $f'(x)$ and $f(x)$ converge, then*

$$\bar{f}'(s) = s\bar{f}(s) - f(0).$$

Proof We have, by integration by parts, that

$$\bar{f}'(s) = \int_0^\infty f'(x)e^{-sx}\,dx = \left[f(x)e^{-sx}\right]_0^\infty - \int_0^\infty f(x)\left(-se^{-sx}\right)\,dx = (0 - f(0)) + s\bar{f}(s). \quad \square$$

Corollary 6.54 *Provided that the Laplace transforms of $f(x), f'(x), f''(x)$ converge then*

$$\bar{f}''(s) = s^2\bar{f}(s) - sf(0) - f'(0).$$

Proof If we write $g(x) = f'(x)$ then

$$\bar{f}''(s) = \bar{g}'(s) = s\bar{g}(s) - g(0) = s\bar{f}'(s) - f'(0) = s^2\bar{f}(s) - sf(0) - f'(0). \quad \square$$

Example 6.55 Determine $\bar{f}(s)$ where the function $f(x)$ is the solution of the DE

$$f''(x) - 3f'(x) + 2f(x) = 0, \qquad f(0) = f'(0) = 1.$$

Solution Applying the Laplace transform to the DE, and noting $f(0) = f'(0) = 1$ we see

$$\left(s^2\bar{f}(s) - s - 1\right) - 3(s\bar{f}(s) - 1) + 2\bar{f}(s) = 0.$$

Rearranging this equation, we find

$$\left(s^2 - 3s + 2\right)\bar{f}(s) = s - 2$$

and hence

$$\bar{f}(s) = \frac{s-2}{(s-1)(s-2)} = \frac{1}{s-1}. \qquad \square$$

We recognize $\bar{f}(s)$ as the Laplace transform of e^x but can we reasonably now claim that $f(x) = e^x$? Reassuringly the answer is positive, as the Laplace transform has an inverse. A full statement of the inverse theorem, let alone a proof, would take us well beyond the scope of this text, as it is properly a topic of complex analysis, contour integration and residue theory (see p.60). It is also possible to invert an infinite series in powers of $1/s$ term by term noting that the inverse of $1/s^{n+1}$ is $x^n/n!$. Here though we will simply state a version of the inverse theorem without further comment and so in future – when faced with inverting a Laplace transform – we shall have to rely on recognizing the Laplace transforms of standard functions and a range of techniques for combining those standard transforms.

Theorem 6.56 *(Inverse Theorem)* *Let $f(x)$ and $g(x)$ be two continuous functions defined for $x > 0$ such that $\bar{f}(s) = \bar{g}(s)$ for all $s \geqslant s_0$. Then $f(x) = g(x)$.*

Example 6.57 Find the Laplace inverses of

$$\bar{f}(s) = \frac{1}{s^2(s+1)}, \qquad \bar{g}(s) = \frac{1}{s^2 + 2s + 4}.$$

Solution Both the given transforms are not in instantly recognizable forms, given the examples we have seen so far, but with some simple algebraic rearrangement we can quickly circumvent this. Firstly, using partial fractions, we see

$$\bar{f}(s) = \frac{1}{s^2(s+1)} = \frac{1}{s^2} - \frac{1}{s} + \frac{1}{s+1},$$

and inverting the Laplace transform we have $f(x) = x - 1 + e^{-x}$.

If we complete the square in the denominator of $\bar{g}(s)$ we also see

$$\bar{g}(s) = \frac{1}{s^2 + 2s + 4} = \frac{1}{(s+1)^2 + 3}.$$

Now we know by Example 6.50 that $\sin(\sqrt{3}x)$ transforms to $\sqrt{3}/(s^2+3)$. Given the next proposition, and the inverse theorem, we can now conclude that

$$g(x) = \frac{1}{\sqrt{3}} e^{-x} \sin \sqrt{3}x. \qquad \square$$

◆

Proposition 6.58 *Assuming that the Laplace transform of $f(x)$ converges for $s \geqslant s_0$, then $g(x) = e^{ax}f(x)$ has Laplace transform*

$$\bar{g}(s) = \bar{f}(s-a) \qquad for\ s \geqslant a + s_0.$$

Proof For $s \geqslant a + s_0$ we have

$$\bar{g}(s) = \int_0^\infty e^{ax}f(x)e^{-sx}\,dx = \int_0^\infty f(x)e^{-(s-a)x}\,dx = \bar{f}(s-a). \qquad \square$$

Example 6.59 Find the Laplace inverse of $s^{-2}e^{-s}$.

Solution We know that $H(x-1)$ has transform $s^{-1}e^{-s}$ and we also know that x transforms to s^{-2}, so perhaps we can combine these facts somehow. If we consider $f(x) = xH(x-1)$ then we get

$$\bar{f}(s) = \int_0^\infty xH(x-1)(x)e^{-sx}\,dx = \int_1^\infty xe^{-sx}\,dx = \left[-s^{-2}(sx+1)e^{-sx}\right]_1^\infty = (s^{-1} + s^{-2})e^{-s},$$

which is close to what we want. In fact we can see we are wrong precisely by $\mathcal{L}(H(x-1))$. So we see that the inverse transform of $s^{-2}e^{-s}$ is

$$xH(x-1) - H(x-1) = (x-1)H(x-1). \qquad \square$$

This is a particular instance of the following result.

Proposition 6.60 *Let $f(x)$ be defined for $x > 0$ with Laplace transform $\bar{f}(s)$ defined for $s \geqslant s_0$. For $a > 0$,*

$$f(x-a)H(x-a) \overset{\mathcal{L}}{\longmapsto} e^{-as}\bar{f}(s).$$

Note that the function $f(x-a)H(x-a)$ is essentially the function $f(x)$ delayed by a. The graph of $f(x-a)H(x-a)$ is the graph of $f(x)$ shifted to the right by a and with value 0 for $0 < x \leqslant a$.

Proof Note that $f(x-a)H(x-a)$ has Laplace transform

$$\int_0^\infty f(x-a)H(x-a)e^{-sx}\,dx = \int_a^\infty f(x-a)e^{-sx}\,dx = \int_0^\infty f(u)e^{-s(u+a)}\,du = e^{-as}\bar{f}(s),$$

where we make the substitution $u = x - a$ for the second equality. $\qquad\square$

We introduce here a further result that will help us broaden our growing library of standard transforms

Proposition 6.61 *Assuming that the Laplace transforms of $f(x)$ and $g(x) = xf(x)$ converge for $s \geqslant s_0$ then*

$$\bar{g}(s) = -\frac{d\bar{f}}{ds}.$$

#1697
#1698
#1705
#1706

Proof For this result we will need to apply *differentiation under the integral sign* – see p.421 for further details of this. Note that

$$\frac{d\bar{f}}{ds} = \frac{d}{ds}\int_0^\infty f(x)e^{-sx}\,dx = \int_0^\infty \frac{\partial}{\partial s}\left(f(x)e^{-sx}\right)\,dx = -\int_0^\infty xf(x)e^{-sx}\,dx = -\bar{g}(s). \qquad\square$$

Example 6.62 Find the inverse Laplace transform of $(s-a)^{-n}$.

Solution From Example 6.48 we know that the Laplace transform of $x^{n-1}/(n-1)!$ is s^{-n}. Hence by Proposition 6.58 we see that the $x^{n-1}e^{ax}/(n-1)!$ has transform $(s-a)^{-n}$. Alternatively, we could make use of Proposition 6.61 and Example 6.47 to note

$$(s-a)^{-n} = \frac{1}{(n-1)!}\left(-\frac{d}{ds}\right)^{n-1}(s-a)^{-1} = \frac{1}{(n-1)!}\left(-\frac{d}{ds}\right)^{n-1}$$

$$\mathcal{L}(e^{ax}) = \frac{1}{(n-1)!}\mathcal{L}(x^{n-1}e^{ax}) = \mathcal{L}\left(\frac{x^{n-1}e^{ax}}{(n-1)!}\right). \qquad\square$$

Proposition 6.63 *The inverse Laplace transform of*

$$\bar{f}(s) = \frac{As+B}{(s+\alpha)^2+\beta^2},$$

is

$$f(x) = Ae^{-\alpha x}\cos\beta x + \frac{(B-A\alpha)}{\beta}e^{-\alpha x}\sin\beta x.$$

Proof Using Example 6.50 and Proposition 6.58 we know that

$$e^{-bx}\cos ax \xrightarrow{\mathcal{L}} \frac{s+b}{(s+b)^2+a^2}, \qquad e^{-bx}\sin ax \xrightarrow{\mathcal{L}} \frac{a}{(s+b)^2+a^2}.$$

Hence

$$\frac{As+B}{(s+\alpha)^2+\beta^2} = \frac{A(s+\alpha)+(B-A\alpha)}{(s+\alpha)^2+\beta^2} \xrightarrow{\mathcal{L}^{-1}} Ae^{-\alpha x}\cos\beta x + \frac{(B-A\alpha)}{\beta}e^{-\alpha x}\sin\beta x. \qquad\square$$

The above inverse is not particularly memorable and it is probably better to remember the method of solution and the relevant properties of the Laplace transform for inverting such rational functions. In fact, we now have (in principle at least) the means of inverting any rational function (where the denominator has greater degree than the numerator). By the fundamental theorem of algebra the denominator can be factorized into linear factors and irreducible quadratic factors. Using partial fractions this means that any such rational function can be written as a sum of functions of the form

$$\frac{A}{(s-a)^n} \quad \text{and} \quad \frac{Bs+C}{(s^2+Ds+E)^n}.$$

We have already seen how to invert the former (Example 6.62) and the latter can be approached as in Proposition 6.63 and further employing Proposition 6.61.

Example 6.64 Find the inverse Laplace transform of $(s^2+2s+2)^{-2}$.

Solution We know that

$$\sin x \overset{\mathcal{L}}{\longmapsto} \frac{1}{s^2+1}, \qquad \cos x \overset{\mathcal{L}}{\longmapsto} \frac{s}{s^2+1}.$$

So by Proposition 6.61

$$x\sin x \overset{\mathcal{L}}{\longmapsto} -\frac{d}{ds}\left(\frac{1}{s^2+1}\right) = \frac{2s}{(s^2+1)^2}; \qquad x\cos x \overset{\mathcal{L}}{\longmapsto} -\frac{d}{ds}\left(\frac{s}{s^2+1}\right) = \frac{s^2-1}{(s^2+1)^2}.$$

So we might write

$$\frac{1}{(s^2+1)^2} = \frac{1}{2}\left\{\frac{1}{s^2+1} - \frac{s^2-1}{(s^2+1)^2}\right\} \overset{\mathcal{L}^{-1}}{\longmapsto} \frac{1}{2}(\sin x - x\cos x).$$

Thus

$$\frac{1}{(s^2+2s+2)^2} = \frac{1}{((s+1)^2+1)^2} \overset{\mathcal{L}^{-1}}{\longmapsto} \frac{e^{-x}}{2}(\sin x - x\cos x). \qquad \square$$

We return now to applying the Laplace transform to DEs.

Example 6.65 Solve the DE

$$f''(x) - 3f'(x) + 2f(x) = x, \qquad f(0) = 1 = f'(0).$$

Solution Applying the Laplace transform we see

$$(s^2\bar{f}(s) - s - 1) - 3(s\bar{f}(s) - 1) + 2\bar{f}(s) = \frac{1}{s^2}.$$

Rearranging we see $(s^2 - 3s + 2)\bar{f}(s) = s^{-2} + s - 2$ and hence that

$$\bar{f}(s) = \frac{1 + s^3 - 2s^2}{s^2(s-1)(s-2)}.$$

Using partial fractions we can rewrite this as

$$\bar{f}(s) = \frac{3/4}{s} + \frac{1/2}{s^2} + \frac{0}{s-1} + \frac{1/4}{s-2}.$$

Hence, inverting the Laplace transform, we have

$$f(x) = \frac{3}{4} + \frac{x}{2} + \frac{1}{4}e^{2x}. \qquad \square$$

Example 6.66 (See also #1666.) Solve the following simultaneous DEs subject to the initial conditions $f(0) = g(0) = 0$,

$$f'(x) - g'(x) - 2f(x) = e^{-2x}; \qquad 3f'(x) - 2g'(x) - g(x) = 0.$$

Solution (Details are left to #1714.) Applying the Laplace transform to both equations, and noting $f(0) = g(0) = 0$, we get

$$s\bar{f} - s\bar{g} - 2\bar{f} = \frac{1}{s+2} \qquad \text{and} \qquad 3s\bar{f} - 2s\bar{g} - \bar{g} = 0.$$

Solving for $\bar{f}(s)$ and $\bar{g}(s)$ we find

$$\bar{f}(s) = \frac{-(2s+1)}{(s+1)(s+2)^2}; \qquad \bar{g}(s) = \frac{-3s}{(s+1)(s+2)^2}.$$

Using partial fractions and recognizing standard transforms we find

$$f(x) = e^{-x} - e^{-2x} - 3xe^{-2x}; \qquad g(x) = 3e^{-x} - 3e^{-2x} - 6xe^{-2x}. \qquad \square$$

Example 6.67 **Bessel's function** $J_0(x)$ satisfies the initial-value problem

$$x\frac{d^2 J_0}{dx^2} + \frac{dJ_0}{dx} + xJ_0 = 0, \qquad J_0(0) = 1, \quad J_0'(0) = 0.$$

Show that $\overline{J_0}(s) = (1 + s^2)^{-1/2}$.

Solution By Propositions 6.53 and 6.61, when we apply the Laplace transform to both sides of the above DE we get

$$-\frac{d}{ds}(s^2\overline{J_0} - s) + (s\overline{J_0}(s) - 1) - \frac{d\overline{J_0}}{ds} = 0.$$

Simplifying we see

$$(s^2 + 1)\frac{d\overline{J_0}}{ds} + s\overline{J_0} = 0.$$

This equation is separable and we may solve it to find

$$\overline{J_0}(s) = A(1 + s^2)^{-1/2}$$

where A is some constant. We might try to determine A by recalling that $\overline{J_0}(s)$ approaches 0 as s becomes large; however, this is the case for all values of A. Instead we can note that

$$\overline{J_0'}(s) = s\overline{J_0}(s) - J_0(0) = As(1 + s^2)^{-1/2} - 1 = A(1 + s^{-2})^{-1/2} - 1,$$

must also approach 0 as s becomes large. So $A - 1 = 0$ and $\overline{J_0}(s) = (1 + s^2)^{-1/2}$. $\qquad \square$

Example 6.68 Let $L_n(x)$ denote the nth Laguerre polynomial (see #476). Show that $\overline{L_n}(s) = (s - 1)^n s^{-n-1}$ and that $L_n(x)$ is a solution of the DE

$$xy'' + (1 - x)y' + ny = 0. \tag{6.66}$$

Solution Recall

$$L_n(x) = \frac{e^x}{n!}\frac{d^n}{dx^n}(x^n e^{-x}).$$

By Proposition 6.61 we have that the transform of $x^n e^{-x}$ equals

$$\mathcal{L}\left(x^n e^{-x}\right) = \left(-\frac{d}{ds}\right)^n \mathcal{L}\left(e^{-x}\right) = \left(-\frac{d}{ds}\right)^n \frac{1}{s+1} = \frac{n!}{(s+1)^{n+1}}.$$

Note that $x^n e^{-x}$ takes value 0 at $x = 0$, as do its first $n - 1$ derivatives. So by repeated applications of Proposition 6.53 we see

$$\mathcal{L}\left(\frac{d^n}{dx^n}(x^n e^{-x})\right) = \frac{n! s^n}{(s+1)^{n+1}}.$$

Finally, by Proposition 6.58, it follows that $\overline{L}_n(s) = (s-1)^n s^{-n-1}$.

Now the DE (6.66) transforms to

$$-\frac{d}{ds}\left(s^2\bar{y} - sy(0) - y'(0)\right) + (s\bar{y} - y(0)) + \frac{d}{ds}(s\bar{y} - y(0)) + n\bar{y} = 0,$$

which rearranges to

$$s(1-s)\frac{d\bar{y}}{ds} + (n-s+1)\bar{y} = 0.$$

If we set $\bar{y}(s) = (s-1)^n s^{-n-1}$ then we see the LHS equals

$$s(1-s)\left\{n(s-1)^{n-1}s^{-n-1} - (n+1)(s-1)^n s^{-n-2}\right\} + (n-s+1)(s-1)^n s^{-n-1}$$
$$= (s-1)^n s^{-n-1}[-\{ns - (n+1)(s-1)\} + (n-s+1)] = 0.$$

Hence $L_n(x)$ is indeed a solution to (6.66). ☐

Example 6.69 Find a non-zero solution to the DE

$$xf''(x) + 2f'(x) + xf(x) = 0. \tag{6.67}$$

Solution Applying the Laplace transform we obtain

$$-\frac{d}{ds}\left(s^2\overline{f}(s) - sf(0) - f'(0)\right) + 2(s\overline{f}(s) - f(0)) - \frac{d\overline{f}}{ds} = 0,$$

which rearranges to

$$-2s\overline{f}(s) - s^2\frac{d\overline{f}}{ds} + f(0) + 2(s\overline{f}(s) - f(0)) - \frac{d\overline{f}}{ds} = 0,$$

and eventually to $-d\overline{f}/ds = \frac{A}{s^2+1}$, where A is a constant. We know that $-d\overline{f}/ds$ is the Laplace transform of $xf(x)$ and the RHS is the Laplace transform of $A\sin x$ and hence inverting we obtain

$$f(x) = A\frac{\sin x}{x}.$$

We were only asked for a non-zero solution so can choose $A = 1$. Note that $A = f(0)$ exists only as a limit. The function $\sin x/x$ is undefined at $x = 0$ but does have a limit of 1 there (#1286). ☐

Example 6.70 Find an independent second solution of (6.67). Why was this solution not found by Laplace transform methods?

Solution From §6.4 we know to make the substitution $f(x) = (\sin x / x)g(x)$. We then have

$$f' = \frac{\sin x}{x}g' + \frac{\cos x}{x}g - \frac{\sin x}{x^2}g;$$

$$f'' = \frac{\sin x}{x}g'' + 2\left(\frac{\cos x}{x} - \frac{\sin x}{x^2}\right)g' + \left(\frac{2\sin x}{x^3} - \frac{2\cos x}{x^2} - \frac{\sin x}{x}\right)g.$$

Substituting these expressions into (6.67) and rearranging, $(\sin x)g'' + (2\cos x)g' = 0$. This is a separable DE and we find

$$\frac{g''}{g'} = -2\cot x \quad \Longrightarrow \quad \ln g' = -2\ln \sin x + \text{const.} \quad \Longrightarrow \quad g' = \frac{A}{\sin^2 x} \quad \Longrightarrow \quad g = A\cot x,$$

where A is a constant. A second independent solution is therefore

$$f(x) = \frac{\sin x}{x} \times \cot x = \frac{\cos x}{x}.$$

The reason that this second solution $f(x)$ was not found in the first place is that its Laplace transform does not converge. For any s, the area under $e^{-sx}x^{-1}\cos x$ is infinite on the interval $0 < x < 1$. This is because near 0 the functions e^{-sx} and $\cos x$ both have a limit of 1 and so $f(x)$ is of the order of x^{-1}. We saw in #1282 that the integral of x^{-1} on the interval $\varepsilon < x < 1$ is unbounded as ε becomes small. □

Example 6.71 Find a non-zero solution to the DE

$$x^2 f''(x) = \left(x^2 - 4x + 2\right)f(x). \tag{6.68}$$

Solution Applying the Laplace transform and making use of Proposition 6.61 we obtain

$$\frac{d^2}{ds^2}\left(s^2\bar{f}(s) - sf(0) - f'(0)\right) = \frac{d^2\bar{f}}{ds^2} + 4\frac{d\bar{f}}{ds} + 2\bar{f}.$$

This rearranges and simplifies further to

$$\frac{d^2\bar{f}}{ds^2} = \frac{-4}{s+1}\frac{d\bar{f}}{ds}.$$

Either by inspection, or by separating variables, we can then see that $d\bar{f}/ds = A(s+1)^{-4}$ for some constant A and integrating we find

$$\bar{f}(s) = B(s+1)^{-3} + C$$

for constants B, C (where $B = -A/4$). As $\bar{f}(s)$ becomes 0 as s becomes large then we further have $C = 0$. Now we know that the transform of e^{-x} is $(s+1)^{-1}$ and hence the Laplace transform of $x^2 e^{-x}$ is

$$\left(-\frac{d}{ds}\right)^2\left(\frac{1}{s+1}\right) = \frac{2}{(s+1)^3}.$$

Hence we conclude that $f(x) = Dx^2 e^{-x}$ solves of the original DE (where $D = B/2$). □

We introduce here a final property of the Laplace transform which again will be helpful in expanding our library of transforms we can recognize and invert. This is the notion of the *convolution* of two functions.

Definition 6.72 Given two functions f, g defined on $(0, \infty)$, we define their **convolution** $f * g$ by

$$(f * g)(x) = \int_0^x f(t)g(x-t)\,dt \qquad \text{for } x \geqslant 0.$$

Note that by setting $u = x - t$ we can see

$$(f * g)(x) = \int_0^x f(t)g(x-t)\,dt = \int_x^0 f(x-u)g(u)(-du) = \int_0^x g(u)f(x-u)\,du = (g * f)(x).$$

Example 6.73 Let $f(x) = \sin x$ and $g(x) = \sin x$. Then we have

$$(f * g)(x) = \int_0^x \sin t \sin(x-t)\,dt = \frac{1}{2}\int_0^x \{\cos(2t-x) - \cos x\}\,dt$$

$$= \frac{1}{2}\left[\frac{1}{2}\sin(2t-x) - t\cos x\right]_0^x = \frac{1}{2}(\sin x - x\cos x).$$

Recall we met this in Example 6.64 as the Laplace inverse of $(s^2+1)^{-2} = \bar{f}(s)\bar{g}(s)$.

Example 6.74 Let $f(x) = e^{ax}$ and $g(x) = e^{bx}$ where $a \neq b$. Then

$$(f * g)(x) = \int_0^x e^{at}e^{b(x-t)}\,dt = e^{bx}\int_0^x e^{(a-b)t}\,dt = e^{bx}\left[\frac{e^{(a-b)t}}{a-b}\right]_0^x = \frac{e^{ax} - e^{bx}}{a-b}.$$

This transforms to

$$\overline{f * g}(s) = \frac{1}{a-b}\left\{\frac{1}{s-a} - \frac{1}{s-b}\right\} = \frac{1}{(s-a)(s-b)} = \bar{f}(s)\bar{g}(s).$$

Consequently the following theorem may not come as that great a surprise.

Theorem 6.75 *Given two functions f and g whose Laplace transforms $\bar{f}$ and $\bar{g}$ exist for $s \geqslant s_0$. Then*

$$\overline{f * g}(s) = \bar{f}(s)\bar{g}(s) \qquad \text{for } s \geqslant s_0.$$

#1708
#1709
#1718
Proof We will write $h = f * g$ and prove the above by using the equality of the repeated integrals of the function

$$f(x)g(y-x)H(y-x)e^{-sy} \qquad \text{where } x, y > 0 \text{ and } s \geqslant s_0.$$

One of the repeated integrals is

$$\int_{y=0}^{y=\infty}\int_{x=0}^{x=\infty} f(x)g(y-x)H(y-x)e^{-sy}\,dx\,dy = \int_{y=0}^{y=\infty} e^{-sy}\int_{x=0}^{x=y} f(x)g(y-x)\,dx\,dy$$

$$= \int_{y=0}^{y=\infty} e^{-sy}h(y)\,dy = \bar{h}(s),$$

as $H(y-x) = 0$, where $x > y$. The other repeated integral is

$$\int_{x=0}^{x=\infty}\left(\int_{y=0}^{y=\infty} f(x)g(y-x)H(y-x)e^{-sy}\,dy\right)dx$$

$$= \int_{x=0}^{x=\infty} f(x)\left(\int_{y=x}^{y=\infty} g(y-x)e^{-sy}\,dy\right)dx \qquad [\text{as } H(y-x) = 0 \text{ when } y < x]$$

$$= \int_{x=0}^{x=\infty} f(x)\left(\int_{u=0}^{u=\infty} g(u)e^{-s(u+x)}\,du\right)dx \qquad [\text{setting } u = y - x]$$

$$= \int_{x=0}^{x=\infty} f(x) e^{-sx} \left(\int_{u=0}^{u=\infty} g(u) e^{-su} \, du \right) dx$$

$$= \int_{x=0}^{x=\infty} f(x) e^{-sx} \bar{g}(s) \, dx$$

$$= \bar{g}(s) \int_{x=0}^{x=\infty} f(x) e^{-sx} \, dx = \bar{g}(s) \bar{f}(s)$$

as required. □

Example 6.76 Find the Laplace inverse of

$$\bar{f}(s) = \frac{s}{(s^2 + 1)^2}.$$

Solution Note that $\bar{f}(s)$ is the product of the transforms of $\cos x$ and $\sin x$. Hence by the convolution theorem

$$f(x) = \int_0^x \sin t \cos(x - t) \, dt$$

$$= \cos x \left(\int_0^x \sin t \cos t \, dt \right) + \sin x \left(\int_0^x \sin^2 t \, dt \right)$$

$$= \cos x \left[\frac{1}{4} - \frac{1}{4} \cos 2x \right] + \sin x \left[\frac{1}{2} x - \frac{1}{4} \sin 2x \right]$$

$$= \frac{1}{4} \left\{ \cos x - \cos x \left(1 - 2 \sin^2 x \right) + 2x \sin x - 2 \sin^2 x \cos x \right\} = \frac{1}{2} x \sin x.$$

Having determined this answer we see we could also have realized by Proposition 6.61 that

$$\frac{s}{(s^2 + 1)^2} = \frac{1}{2} \left(-\frac{d}{ds} \right) \left(\frac{1}{s^2 + 1} \right) = \frac{1}{2} \left(-\frac{d}{ds} \right) (\overline{\sin}(s)) = \frac{1}{2} \mathcal{L} (x \sin x). \qquad \square$$

Example 6.77 Determine the solution of the DE

$$y''(x) + 3y'(x) + 2y(x) = f(x), \qquad y(0) = y'(0) = 1,$$

writing your solution involving a convolution.

#1710
#1711
#1712
#1713
#1715
Solution Applying the Laplace transform we find

$$(s^2 \bar{y} - s - 1) + 3(s\bar{y} - 1) + 2\bar{y} = \bar{f},$$

which rearranges to

$$\bar{y}(s) = \frac{s + 4}{(s + 2)(s + 1)} + \frac{\bar{f}(s)}{(s + 2)(s + 1)} = \frac{3}{s + 1} - \frac{2}{s + 2} + \left(\frac{1}{s + 1} - \frac{1}{s + 2} \right) \bar{f}(s).$$

Hence we can write the answer to include a convolution as

$$\bar{y}(s) = 3e^{-x} - 2e^{-2x} + \int_0^x (e^{-t} - e^{-2t}) f(x - t) \, dt. \qquad \square$$

We now move on to a brief discussion of the *Dirac delta function*, which the Laplace transform handles particularly well. Mathematics often seeks to model the real world about us using simplifying assumptions, omitting extraneous and unimportant details. Such examples of simplifications are point masses, point electrical charges and instantaneous impulses. And there are often good reasons for these simplifications, as we wish, essentially, to consider the radii of the mass and charge, or the contact time of the impulse, as negligible. But these simplifications can also lead to difficulties and singularities, at least if we are trying to be rigorous. For example, if we were considering a point unit mass at the origin of the real line, then the density of matter $\delta(x)$ would need to have the following two properties

$$\delta(x) = 0 \quad \text{when } x \neq 0; \qquad \int_{-\infty}^{\infty} \delta(x)\,dx = 1, \qquad (6.69)$$

the first property saying that the mass is solely present at the origin and the second saying that the total mass present is 1. Note that if we could make sense of such a function $\delta(x)$ then we would have

$$\int_{-\infty}^{x} \delta(t)\,dt = \begin{cases} 0 & x < 0 \\ 1 & x > 0 \end{cases} = H(x),$$

and so we might hope, in some sense, to view $\delta(x)$ as the derivative of the (non-differentiable!) Heaviside function.

A major drawback, though, is that no such function $\delta(x)$ actually exists with the properties in (6.69), as the density would need to be infinite at $x = 0$. In a similar fashion an infinite force would be needed to impart an instantaneous impulse. However, such simplifying ideas are plentiful and of great use, and many mathematicians including Cauchy, Poisson[6], Heaviside and Helmholtz[7] sought to resolve this lack of rigour and precision. It was Dirac[8] who popularized the notation $\delta(x)$ in his influential text *The Principles of Quantum Mechanics*; he chose the notation δ, as $\delta(x-y)$ can be thought of as a continuous version of the Kronecker delta δ_{ij} and indeed has a similar sifting property (#1727). Ultimately it was the work in the late 1940s of Laurent Schwartz (1915–2002) on *distributions*, also known as *generalized functions*, that influentially resolved the matter and for which he would win the Fields Medal in 1950.

We begin by seeking to resolve the introduction of an instantaneous impulse for the SHM problem below.

Example 6.78 (**Kick Start**) Let $T > 0$. Consider the DE below governing the extension x at time t of a mass m on a spring with spring constant k, and say that an instantaneous impulse I is applied to the mass at time T.

$$mx''(t) + kx(t) = 0 \quad \text{for } t \neq T, \qquad x(0) = x'(0) = 0.$$

---

[6] The French applied mathematician and theoretical physicist Siméon-Denis Poisson (1781–1849), is today remembered in probability for his distribution and for *Poisson processes* (essentially a type of Markov chain in continuous time), and in applied mathematics for his solution of Laplace's equation in a half-plane and for *Poisson's equation*, his correction of Laplace's equation to describe (say) the evolution of gravitational potential inside (as well as outside) matter.

[7] Hermann von Helmholtz (1821–1894) was a German physiologist, physicist and philosopher of science who variously worked on mathematical theories of sound, vision, resonance and electromagnetism.

[8] Paul Dirac (1902–1984) was an English theoretical physicist who made significant early advances in quantum theory. He jointly won the 1933 Nobel Prize in physics with Schrödinger. The Dirac equation (1928) is a relativistic version of Schrödinger's equation for electrons and predicted anti-matter, the existence of which would be experimentally verified in 1932.

Determine the extension $x(t)$.

Solution From Example 6.27 we know the solution to the problem is of the form

$$x(t) = \begin{cases} A\cos\omega t + B\sin\omega t & t < T; \\ C\cos\omega t + D\sin\omega t & t > T, \end{cases}$$

where $\omega^2 = k/m$. From the initial conditions we have that $A = B = 0$ and the system sits at rest for $t < T$. By the continuity of $x(t)$ we also have that

$$C\cos\omega T + D\sin\omega T = 0.$$

However, there is a discontinuity of I in the momentum $mx'(t)$ of the particle at $t = T$. So

$$m\omega(-C\sin\omega T + D\cos\omega T) = I.$$

The equations can be combined as

$$\begin{pmatrix} \cos\omega T & \sin\omega T \\ -\sin\omega T & \cos\omega T \end{pmatrix}\begin{pmatrix} C \\ D \end{pmatrix} = \begin{pmatrix} 0 \\ \frac{I}{m\omega} \end{pmatrix} \implies \begin{pmatrix} C \\ D \end{pmatrix} = \begin{pmatrix} \cos\omega T & -\sin\omega T \\ \sin\omega T & \cos\omega T \end{pmatrix}\begin{pmatrix} 0 \\ \frac{I}{m\omega} \end{pmatrix}$$

and so

$$x(t) = \frac{I}{m\omega}\{-\sin\omega T\cos\omega t + \cos\omega T\sin\omega t\} = \frac{I\sin\omega(t-T)}{m\omega} \qquad \text{for } t \geqslant T.$$

We can use the Heaviside function to rewrite this solution as

$$x(t) = \frac{I\sin\omega(t-T)}{m\omega}H(t-T). \qquad \square \tag{6.70}$$

#1717
#1721
#1722
#1723
#1724
#1727

Note that we can rewrite the DE of Example 6.78 as

$$mx''(t) + kx(t) = I\delta(t-T) \tag{6.71}$$

to capture both the DE that applies when $t \neq T$, and that there is a discontinuity of size I in the momentum $mx'(t)$ at time T. If we apply the Laplace transform to (6.70), then by Example 6.50 and Proposition 6.60, we find its Laplace transform is

$$\bar{x}(s) = \frac{I}{m\omega}e^{-sT}\frac{\omega}{s^2+\omega^2} = \frac{I}{m}e^{-sT}\frac{1}{s^2+\omega^2}.$$

If we applied the Laplace transform to (6.71) we get

$$m(s^2\bar{x}(s) - sx(0) - x'(0)) + k\bar{x}(s) = I\mathcal{L}(\delta(t-T)),$$

which we can solve for $\bar{x}(s)$ to find

$$\bar{x}(s) = \frac{I\mathcal{L}(\delta(t-T))}{ms^2+k} = \frac{I}{m}\frac{\mathcal{L}(\delta(t-T))}{s^2+\omega^2},$$

so that the Laplace transform of $\delta(t-T)$ appears to be e^{-sT}.

For $a > 0$, this also ties in with our notion of $\delta(x-a)$ being the derivative of $H(x-a)$. The Laplace transform of $H(x-a)$ is e^{-as}/s and so its derivative should have Laplace transform

$$s\left(\frac{e^{-as}}{s}\right) - H(-a) = e^{-as}.$$

We conclude with a table of transforms that we have so far determined.

Table 6.1 *Standard Laplace Transforms*

$f(x)$	$\bar{f}(s)$	$f(x)$	$\bar{f}(s)$
x^n	$n!/s^{n+1}$	$f'(x)$	$s\bar{f}(s) - f(0)$
e^{ax}	$(s-a)^{-1}$	$f''(x)$	$s^2\bar{f}(s) - sf(0) - f'(0)$
$\cos ax$	$s/(s^2+a^2)$	$xf(x)$	$-d\bar{f}/ds$
$\sin ax$	$a/(s^2+a^2)$	$f(x-a)H(x-a)$	$e^{-as}\bar{f}(s)$
$\delta(x-a)$	e^{-as}	$e^{-ax}f(x)$	$\bar{f}(s+a)$

Exercises

#1691a† Determine the Laplace transforms of the following functions.
$$(x+1)^2, \quad \sinh x, \quad \cosh(2x+1), \quad \cos(x+1), \quad xe^{-2x}.$$

#1692b Write down the inverses of the following Laplace transforms.
$$\frac{1}{s^2+s}, \quad \frac{1-e^{-s}}{s}, \quad \frac{e^{-s-1}}{s^2}, \quad \frac{e^{-2s+3}}{s^2+1}.$$

#1693B† Give examples of two *distinct* functions $f(x)$ and $g(x)$ such that $\bar{f}(s) = \bar{g}(s)$.

#1694C Give an example of a function $f(x)$ defined for $x > 0$, such that the Laplace transform $\bar{f}(s)$ converges for some value s_0, but there are no constants a, M with $M > 0$ such that $|f(x)| \leqslant Me^{ax}$ for all x. Show directly that Proposition 6.52 still applies.

#1695B Let $\alpha > -1$ and $f(x) = x^\alpha$ for $x > 0$. Show, for $s > 0$, $\bar{f}(s) = \Gamma(\alpha + 1)s^{-\alpha-1}$.

#1696d† Find the Laplace transform of e^{-rx^2}, where $r > 0$.

#1697b† Find the Laplace transform of $(1 - \cos(ax))/x$.

#1698b† Find the Laplace transform of $\int_0^x \frac{\sin t}{t}\, dt$.

#1699D† Let $a > 0$. Show that the Laplace transform of $x^{-1/2}e^{-a/x}$ equals $\sqrt{\pi/s}\exp\left(-\sqrt{4as}\right)$.

#1700c† Find the Laplace transform of $\ln x$.

#1701D† Show that the Laplace transform of $\text{erf}\, x$ equals $s^{-1}\exp(s^2/4)(1 - \text{erf}(s/2))$.

#1702c Let $a > b > c$. Find the Laplace inverse of $(s-a)^{-1}(s-b)^{-1}(s-c)^{-1}$ and $(s-a)^{-2}(s-c)^{-1}$. Compare your answers as b approaches a.

#1703c Using partial fractions, find the inverse Laplace transform of $(s^3+1)^{-1}$.

#1704D Use the convolution to find the inverse Laplace transform of $(s^3+1)^{-1}$.

#1705b Find the Laplace inverse of $(s^4 - a^4)^{-1}$, where $a > 0$.

#1706b Find the Laplace inverse of $(s^2 - a^2)^{-2}$, where $a > 0$.

#1707D† Find the Laplace inverse of $s^{-1/2}(s-1)$.

#1708B Find the convolution of $f(x) = e^x$ and $g(x) = \sin x$. Verify directly the convolution theorem $\overline{f * g}(s) = \bar{f}(s)\bar{g}(s)$.

#1709B† Let $\alpha, \beta > 0$. Apply the convolution theorem to $f(x) = x^\alpha$ and $g(x) = x^\beta$ to rederive #1411.

#1710B Solve the DE $\quad f'(x) + f(x) = x, \qquad f(0) = 0$.

#1711b Solve the DE $\quad f''(x) - f(x) = 4e^x, \qquad f(0) = f'(0) = 1$.

#1712b Solve the DE $\quad f''(x) + 4f'(x) + 8f(x) = x, \qquad f(0) = 1, \quad f'(0) = 0$.

#1713b Find the general solution of the DE $\quad xf''(x) + (x-1)f'(x) - f(x) = 0$.

#1714C Complete the details of Example 6.66.

#1715D Find a second solution to (6.68), expressed as an integral, which is independent of the one found in Example 6.71. Comment on why this second solution was not found using the Laplace transform.

#1716c Solve the DE in (6.66) by applying the Laplace transform. Find $L_n(x)$, for the values $n = 0, 1, 2$.

#1717b Solve the DE $\quad f'(x) + f(x) = 1_{[0,1]}(x), \qquad f(0) = 0$.

#1718B Solve the integral equation

$$f'(x) - 2\int_0^x f(t)e^{t-x}dt = e^{2x}, \qquad f(0) = 0.$$

#1719C Find the solution of the integral equation

$$f'(x) + \int_0^x f(t)\cos(x-t)dt = e^{2x}, \qquad f(0) = 0.$$

#1720c Find the solution of the integral equation

$$f'(x) + \int_0^x (x-t)f(t)dt = x + \frac{x^2}{2}, \qquad f(0) = 1.$$

#1721B† Let $0 < a < 1$. Solve directly the boundary-value problem

$$f''(x) = \delta(x-a), \qquad f(0) = f(1) = 0.$$

#1722B Solve the DE $\quad f''(x) + \omega^2 f(x) = k(x)$, where $f(0) = f'(0) = 0$, writing your solution as a convolution. Explicitly determine your solution when $k(x) = \delta(x-a)$ where $a > 0$.

#1723B Let k be real and $a > 0$. Solve directly the initial-value problem

$$f''(x) - 3f'(x) + 2f(x) = k\delta(x-a), \qquad f(0) = f'(0) = 1.$$

Then rederive your result using the Laplace transform.

Dirac Delta Function. Distributions. As previously noted there is no function which has the properties

$$\delta(x) = 0 \quad \text{for } x \neq 0; \qquad \int_{-\infty}^{\infty} \delta(x)\,dx = 1,$$

that we would like $\delta(x)$ to have. Nonetheless this chimera was being applied to great effect across mathematics and physics and a rigorous way of handling such a 'function' was clearly needed . A rich theory addressing $\delta(x)$, and other *distributions* or *generalized functions*, was developed by the French mathematician Laurent Schwartz (1915–2002), extending earlier work of the Russian mathematician Sergei Sobolev (1908–1989).

The traditional way of describing a continuous function $f(x)$ on $\mathbb{R}$ is to provide a rule for $f(x)$ in terms of x. However, there are other ways of conveying this same information, and one of these other approaches generalizes to non-standard functions like $\delta(x)$. For example, if the values of the integrals

$$\int_{-\infty}^{\infty} f(x)\phi(x)\,dx, \qquad \phi \text{ is an indicator function on some interval,}$$

were all known, then it would be possible to reconstruct the function $f(x)$, as it is continuous. By limiting the indicator functions to small intervals around any a we could determine $f(a)$. Each *traditional* function $f(x)$ would now be associated with the rule

$$\phi \mapsto \int_{-\infty}^{\infty} f(x)\phi(x)\,dx. \qquad (6.72)$$

Given the *sifting property* of the delta function (#1727), it makes sense for $\delta(x - a)$ to correspond to the rule $\phi \mapsto \phi(a)$. No traditional function has this property, but we now have a framework for describing $\delta(x)$. In fact the sifting property isn't so much something to be proven as the definition of $\delta(x - a)$. From this point of view functions aren't so much defined on points but on some choice of *test functions* $\phi(x)$, for example $H(x - a)$ corresponds to

$$\phi \mapsto \int_{-\infty}^{\infty} H(x - a)\phi(x)\,dx = \int_{a}^{\infty} \phi(x)\,dx.$$

However, we can be yet more ambitious. If we hope to differentiate our new generalized functions, then it makes sense to require that the test functions $\phi(x)$ be differentiable. If so and $f(x)$ is a differentiable traditional function then $f'(x)$ corresponds to

$$\phi \mapsto \int_{-\infty}^{\infty} f'(x)\phi(x)\,dx = \left[f(x)\phi(x)\right]_{-\infty}^{\infty} - \int_{-\infty}^{\infty} f(x)\phi'(x)\,dx \qquad \text{by IBP.}$$

If we further require that the test functions equal 0 outside some bounded interval, so $\phi(\infty) = \phi(-\infty) = 0$, then the above defines a way to differentiate distributions. In fact we can see that the derivative of $H(x - a)$ corresponds to

$$\phi \mapsto -\int_{-\infty}^{\infty} H(x - a)\phi'(x)\,dx = -\int_{a}^{\infty} \phi'(x)\,dx = \phi(a),$$

and so the derivative of $H(x - a)$ is indeed $\delta(x - a)$. However, this does *not* show that $H(x - a)$ is a differentiable function; it is not even continuous at $x = a$ and its gradient is not defined there. Its derivative exists *only* as a distribution, and distibutions need not be defined at individual points – rather distributions are best thought of as having well-defined local averages. From this point of view you might require the test functions $\phi(x)$ to be probability density functions (without restricting the theory of distributions), and (6.72) is then an expectation of f.

#1724 B (Kick Stop) Consider a mass on a spring where the extension of the spring $x(t)$ satisfies
$$m\ddot{x} + kx = I\delta(t - T),$$
where m is the mass, and $k > 0$ is the spring constant. Suppose initially $x(0) = a$ and $x'(0) = 0$ and that at time $t = T > 0$ an instantaneous impulse I is applied to the mass. Obtain the motion of the mass for $t > 0$, and find conditions on I and T such that the impulse completely stops the motion. Explain the result physically.

#1725 C Let $\tau_2 > \tau_1 > 0$. Solve the ODE
$$x''(t) + x(t) = \delta(t - \tau_1) - \delta(t - \tau_2), \qquad x(0) = 0, \quad x'(0) = 0.$$
Find conditions on τ_1 and τ_2 such that $x(t) = 0$ for $t > \tau_2$.

#1726 D† For $a, r > 0$, define the function
$$f_{r,a}(x) = \sqrt{r/\pi} \exp(-r(x - a)^2).$$
Show that
$$\int_{-\infty}^{\infty} f_{r,a}(x)\,dx = 1 \qquad \text{and} \qquad \lim_{r \to \infty} f_{r,a}(x) = 0 \quad \text{for } x \neq a.$$
Determine the limit of $\overline{f_{r,a}}(s)$ as $r \to \infty$.

#1727 B† (Sifting Property of the Dirac Delta) Let $a > 0$ and let $f(x)$ be a differentiable function defined on $(0, \infty)$. Show that
$$\int_0^{\infty} f(x)\delta(x - a)\,dx = f(a).$$

#1728 C† The lifetime T of a particular brand of light bulb is modelled as follows. There is a probability p of the light bulb blowing immediately (so that $T = 0$); given that the light bulb does not blow immediately, the probability of it having lifetime $\tau > 0$ or less is $1 - e^{-\lambda\tau}$ (where $\lambda > 0$).
  (i) Write down the cumulative distribution function $F_T(t)$ of T.
  (ii) Write down the (generalized) probability density function $f_T(t)$ of T.
  (iii) What is the expectation of T?

#1729 D Bessel's function $J_1(x)$ satisfies the initial-value problem
$$x^2\frac{d^2 J_1}{dx^2} + x\frac{dJ_1}{dx} + (x^2 - 1)J_1 = 0. \qquad J_1(0) = 0, \quad J_1'(0) = \frac{1}{2}.$$
Show that
$$\overline{J_1}(s) = 1 - \frac{s}{\sqrt{1 + s^2}}.$$

#1730 C† Show that
$$\int_0^x tJ_0(t)\,dt = xJ_1(x).$$

6.8 Further Exercises*

Exercises on Euler's Method

Definition 6.79 (Euler's Method) Euler's method is a means of finding an approximate solution to the first-order initial-value problem

$$\frac{dy}{dx} = f(x,y), \qquad y(a) = y_0.$$

We use the iteration

$$y_{n+1} = y_n + hf(a+nh, y_n)$$

where $y_k = y(a + kh)$ and h is the length of increment in x. The method works on the assumption that y grows linearly on the interval $a+nh \leqslant x \leqslant a+(n+1)h$ at the same rate as it was growing at $x = a + nh, y = y_n$. As h grows smaller we might hope that the approximate solution y_n becomes closer to the actual solution $y(a+nh)$.

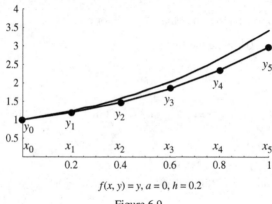

$f(x, y) = y, a = 0, h = 0.2$

Figure 6.9

#1731B† Find, using Euler's method, an approximation to $y(X)$ where y satisfies the initial-value problem

$$\frac{dy}{dx} = y + x, \qquad y(0) = 1.$$

Use an increment of X/N and show that y_N approaches $y(X)$ as N becomes large. What value of N do we need to get $y(1)$ to within 5%? To within 1%?

#1732B Find the solution $y(x)$ of the initial-value problem

$$\frac{dy}{dx} = x + 2y, \qquad y(0) = 0.$$

A numerical approximation to $y(x)$ is calculated using Euler's method by setting

$$x_n = nh, \qquad \frac{y_{n+1} - y_n}{h} = x_n + 2y_n.$$

Show that there is a solution of the form $y_n = A(1 + 2h)^n + Bn + C$ to the above recurrence relation which satisfies $y_0 = 0$, and confirm that, as h becomes small, the numerical approximation converges to the solution $y(x)$ of the initial-value problem.

#1733 C Find, using Euler's method, an approximate solution $y(x)$ of the initial-value problem $y'(x) = x^2$ with $y(0) = 0$ and show that, as the interval size reduces, the approximate solution converges to the actual solution.

#1734 D Consider the initial-value problem $y'' + 2y' + y = 0$, $y(0) = 1$, $y'(0) = 0$. Use the approximations

$$y''(x_{i+1}) \approx \frac{y_{i+2} - 2y_{i+1} + y_i}{h^2}, \qquad y'(x_{i+1}) \approx \frac{y_{i+2} - y_i}{2h}$$

to make an estimate of $y(x)$, where $x = nh$. Show your estimate approaches the correct answer as h becomes small.

#1735 B Find approximate solutions to the linear system of DEs in (6.57) by solving the recurrence relations

$$x_{n+1} = x_n + h(3x_n + y_n), \qquad y_{n+1} = y_n + h(6x_n + 4y_n), \qquad x_0 = \alpha, \, y_0 = \beta.$$

Show that the approximate solutions approach the actual solutions as h becomes small.

Exercises on the Gravitational Two-Body Problem

#1736 B† (i) The two vectors $\mathbf{e}$ and $\mathbf{f}$ in $\mathbb{R}^2$ are given by

$$\mathbf{e} = (\cos\theta, \sin\theta) \qquad \text{and} \qquad \mathbf{f} = (-\sin\theta, \cos\theta),$$

where θ is a function of time t. Show that

$$\dot{\mathbf{e}} = \dot{\theta}\mathbf{f} \qquad \text{and} \qquad \dot{\mathbf{f}} = -\dot{\theta}\mathbf{e},$$

where the dot denotes differentiation with respect to time.
(ii) Let $\mathbf{r}(t) = r(t)\mathbf{e}(t)$. Show that

$$\dot{\mathbf{r}} = \dot{r}\mathbf{e} + r\dot{\theta}\mathbf{f} \qquad \text{and} \qquad \ddot{\mathbf{r}} = (\ddot{r} - r\dot{\theta}^2)\mathbf{e} + \frac{1}{r}\frac{d}{dt}(r^2\dot{\theta})\mathbf{f}.$$

#1737 D† We can model[9] the solar system by placing the sun at the origin and having a planet (or similar) at position $\mathbf{r}(t) = r(t)\mathbf{e}(t)$. According to Newton's law of gravitation the acceleration on the planet is given by

$$\ddot{\mathbf{r}} = \frac{-GM}{r^2}\mathbf{e}, \qquad\qquad (6.73)$$

where M is the mass of the sun and G is the universal gravitational constant.[10]
(i) Explain why $r^2\dot{\theta} = h$ is constant during the planet's motion.
(ii) Set $u = 1/r$. Use the chain rule to show that

$$\frac{dr}{dt} = -h\frac{du}{d\theta} \qquad \text{and} \qquad \frac{d^2r}{dt^2} = -h^2u^2\frac{d^2u}{d\theta^2}.$$

---

[9] Note that other interplanetary forces are assumed to be negligible.
[10] G has approximate value and units 6.67408×10^{-11} m^{3}kg^{-1}s^{-2}.

(iii) By considering the component of the acceleration parallel to **e**, show that

$$\frac{d^2 u}{d\theta^2} + u = \frac{GM}{h^2},$$

and find its general solution. Explain why the planet traces a conic section.

(iv) Suppose initially that $\theta = 0$, $\mathbf{r} = (R, 0)$ and $\dot{\mathbf{r}} = (0, V)$. Show that the planet's path is an ellipse if $RV^2 < 2GM$.

#1738D† (**Kepler's Three Laws**) Johannes Kepler observed the following laws in 1609 and 1619 using data from the Danish astronomer Tycho Brahe.

(I) Planets move in ellipses, with the sun at one focus.

(II) The area of the sector swept out by a planet in a given time is constant.

(III) If a planet takes time T to orbit the sun and R is the semi-major axis of its orbit, then T^2 is proportional to R^3. Moreover, the constant of proportion is the same for each planet (in the solar system).

(I) was proved above in #1737. Show that (II) follows from #1387. For (III) show that if

$$r = \frac{1}{A\cos\theta + GM/h^2},$$

where $A > 0$, then

$$R = \frac{a}{A(a^2 - 1)} \quad \text{and} \quad T = \frac{1}{A^2 h} \times \frac{2\pi a}{(a^2 - 1)^{3/2}},$$

where $a = GM/(Ah^2) > 1$. Deduce Kepler's third law.

Exercises on the Exponential of a Matrix

#1739B Verify that the matrix in (6.59) does indeed have the properties claimed of it in (6.60).

#1740B What is the determinant of the matrix in (6.59)? What are its eigenvalues and eigenvectors?

Definition 6.80 We shall say that a square matrix A has **exponential** e^A if we can define a matrix e^{At} for each t satisfying (6.60), namely (i) $\frac{d}{dt}\left(e^{At}\right) = Ae^{At}$ and (ii) $e^{At} = I$ at $t = 0$. We then set $t = 1$ to define e^A.

We will assume, for now, that for any square matrix A there exist matrices e^{At} such that (i) and (ii) above apply and further that e^{At} commutes with A. We shall prove this in Remark 6.81 but assume this result for the preceding exercises.

#1741D† Let A be a square matrix, and $M(t)$ be matrices such that $M'(t) = AM(t)$ and $M(0) = I$. Show that $M(t) = e^{At}$. That is, e^{At} are the unique matrices having properties (i) and (ii).

#1742D† Let A be a square matrix and s, t real numbers. Show that

$$e^{As} e^{At} = e^{A(s+t)}.$$

Deduce that e^A is invertible. For every invertible matrix B, is there a matrix A such that $e^A = B$?

#1743 B (See also #597.) Let t be a real number and define

$$A(t) = 3^{t-1} \begin{pmatrix} 2t+3 & -t \\ 4t & 3-2t \end{pmatrix}.$$

Show, for real numbers s, t, that $A(s+t) = A(s)A(t)$. Is $A(t) = e^{Mt}$ for any matrix M?

#1744 B† Let A and B be square matrices.
(i) Show that $e^{A^T} = (e^A)^T$.
(ii) Show that $e^{\mathrm{diag}(A,B)} = \mathrm{diag}(e^A, e^B)$.

#1745 D The matrix $A(t)$ below is orthogonal and has determinant 1.

$$A(t) = \begin{pmatrix} \cos \omega t & -\sin \omega t \\ \sin \omega t & \cos \omega t \end{pmatrix}.$$

(i) Show that $A'(t) = WA$ where W is a constant skew-symmetric matrix. Show that $A(t) = e^{Wt}$.
(ii) Show, in general, that if X is a skew-symmetric matrix then e^X is an orthogonal matrix.

#1746 D (i) Let D be a diagonal matrix. What is e^D?
(ii) Let A, P be $n \times n$ matrices with P invertible. Show that

$$e^{P^{-1}AP} = P^{-1}e^A P.$$

Hence rederive the exponential given in (6.59).

#1747 B† Show that the 2×2 matrix E_{12} is not diagonalizable. Determine its exponential nonetheless.

#1748 D† Let a, b, c be real and $a \neq c$. For

$$A = \begin{pmatrix} a & b \\ 0 & c \end{pmatrix}, \qquad \text{show that} \qquad e^A = \begin{pmatrix} e^a & b\left(\frac{e^a - e^c}{a-c}\right) \\ 0 & e^c \end{pmatrix}.$$

How should this result be interpreted when $a = c$?

#1749 B Determine e^A, e^B, e^{A+B} and show that $e^{A+B} \neq e^A e^B$, where

$$A = \begin{pmatrix} 0 & 1 \\ 0 & 1 \end{pmatrix}, \qquad B = \begin{pmatrix} 1 & 1 \\ 0 & 0 \end{pmatrix}.$$

#1750 B Let A be a square matrix such that $A^k = 0$. Show that

$$e^A = I + A + \frac{A^2}{2!} + \frac{A^3}{3!} + \cdots + \frac{A^{k-1}}{(k-1)!}.$$

#1751 B† The square matrix A satisfies $A^2 = 3A - 2I$. Show that

$$e^{At} = (2e^t - e^{2t})I + (e^{2t} - e^t)A.$$

#1752D† Let $J(\lambda, r)$ be as in #598 where λ is real. Show that

$$e^{J(\lambda,r)t} = e^{\lambda t} \begin{pmatrix} 1 & t & t^2/2! & \cdots & t^{r-1}/(r-1)! \\ 0 & 1 & t & \ddots & \vdots \\ 0 & 0 & 1 & \ddots & t^2/2! \\ \vdots & \ddots & \ddots & \ddots & t \\ 0 & \cdots & 0 & 0 & 1 \end{pmatrix}.$$

Let A be a square matrix. Show that $\det e^A = e^{\text{trace} A}$.

Remark 6.81 (Existence of the Matrix Exponential) We are now in a position to demonstrate in several ways the existence of the exponential of a matrix. By #1741, if such an exponential exists it is unique.

(a) By #1746(ii), if A has an exponential then so does any matrix similar to A. And by #1744(ii) and #1752 any matrix in Jordan normal form has an exponential. Finally, by #1079 we know that every square matrix is similar to one in Jordan normal form and so therefore any square matrix has an exponential.

(b) The above route to the existence of the exponential is circuitous at best and also includes exponentials of complex numbers. Given #1750 we might instead hope to define the exponential of a square matrix A to be

$$e^A = I + A + \frac{1}{2!}A^2 + \frac{1}{3!}A^3 + \cdots + \frac{1}{n!}A^n + \cdots . \tag{6.74}$$

As we have a definition of limit (p.358) then we can even rigorously define what we mean by the above, in that it entails for each entry (that is for each i,j) the finite sums

$$[I]_{ij} + [A]_{ij} + \frac{1}{2!}\left[A^2\right]_{ij} + \frac{1}{3!}\left[A^3\right]_{ij} + \cdots + \frac{1}{n!}\left[A^n\right]_{ij}$$

have a limit as $n \to \infty$. One could now either show that in the limit the above sum is absolutely convergent and so convergent (see p.370), or we could show, using any of the norms defined in #968, that

$$\left\| \sum_{k=m}^{n} \frac{A^k}{k!} \right\| \to 0 \qquad \text{as } m, n \to \infty,$$

thus showing that the sequence of partial sums is *Cauchy* and so convergent. We clearly have not developed enough analytical theory yet to be able to argue rigorously down this route. Nonetheless the infinite sum in (6.74) is a very natural definition of the exponential of a matrix.

(c) A third way involves minimal polynomials. Any square matrix A satisfies some monic polynomial $m_A(x)$. This means that we can write

$$e^{At} = \sum_{k=0}^{\infty} \frac{A^k}{k!} t^k = a_0(t)I + a_1(t)A + a_2(t)A^2 + \cdots + a_{r-1}(t)A^{r-1}$$

for some functions $a_0(t), \ldots, a_{r-1}(t)$ where $r = \deg m_A(x)$. As $\left(e^{At}\right)' = Ae^{At}$ then we have

$$a_0'(t)I + a_1'(t)A + a_2'(t)A^2 + \cdots + a_{r-1}'(t)A^{r-1}$$
$$= a_0(t)A + a_1(t)A^2 + a_2(t)A^3 + \cdots + a_{r-1}(t)A^r$$
$$= a_0(t)A + a_1(t)A^2 + a_2(t)A^3 + \cdots + a_{r-1}(t)\left(m_0I + m_1A + \cdots + m_{r-1}A^{r-1}\right)$$

where $m_A(x) = x^r - m_{r-1}x^{r-1} - \cdots - m_0$. Comparing coefficients in the above (the powers of A being linearly independent by the minimality of m_A), we see

$$a_0'(t) = m_0 a_{r-1}(t), \qquad a_1'(t) = a_0(t) + m_1 a_{r-1}(t), \qquad \cdots \qquad a_{r-1}'(t) = a_{r-2}(t) + m_{r-1} a_{r-1}(t).$$

If we use these equations to eliminate everything except $a_{r-1}(t)$ and its derivatives we find

$$a_{r-1}^{(r)}(t) - m_{r-1}a_{r-1}^{(r-1)}(t) - m_{r-2}a_{r-1}^{(r-2)}(t) - \cdots - m_1 a_{r-1}'(t) - m_0 a_{r-1}(t) = 0.$$

We know how to solve such linear constant coefficient equations from §6.5, and so may determine $a_{r-1}(t)$ and then ultimately $a_0(t), a_1(t), \ldots, a_{r-2}(t)$ also. Here then is a third method of showing the exponential of a matrix does exist.

(d) The exponential of a matrix is more than a novelty and actually plays a key role in the study of Lie groups (transformation groups) and their associated Lie algebras, [11] which is a rich area of theory connecting algebraic and geometric structures, with many applications to DEs and physics. #1745 gives a small flavour of such connections. ∎

#1753 B Let A be a square matrix such that $A^2 = 2A - I$. Use the method of Remark 6.81(c) to determine e^{At}.

#1754 B Let A be a square matrix such that $A^2 = 2A - 2I$. Use the method of Remark 6.81(c) to determine e^{At}.

#1755 D Let A be a square matrix such that $(A - \alpha I)(A - \beta I)(A - \gamma I) = 0$ where α, β, γ are distinct reals. Determine e^{At} as a linear combination of I, A, A^2.

#1756 D† Let A and B be commuting $n \times n$ matrices. Show that $e^{A+B} = e^A e^B$.

#1757 C† Find the exponential of

$$A = \begin{pmatrix} 1 & 2 & -1 & 3 \\ 0 & 1 & 1 & -2 \\ 0 & 0 & 1 & 1 \\ 0 & 0 & 0 & 1 \end{pmatrix} \quad \text{and} \quad B = \begin{pmatrix} 2 & 0 & 0 & 0 \\ 0 & 1 & 2 & -1 \\ 0 & 0 & 1 & 3 \\ 0 & 0 & 0 & 1 \end{pmatrix}.$$

---

[11] Named after the Norwegian mathematician Sophus Lie (1842–1899). His surname is pronounced 'lee'.

Hints to Selected Exercises

Complex Numbers

#5 The different cases relate to the y-co-ordinates of the stationary points of $y = p(x)$.

#7 Consider the range of values the LHS and RHS can achieve in each case.

#14 Consider $x = i$ in rule (B).

#16 Simplify and compare real and imaginary parts.

#17 One root is not hard to spot, or set $z = x + yi$.

#19 Note $A = B \neq 0 = C$ is a special case.

#22 Set $w = z^2$.

#23 Note $z - 3/2 - 5i$ must be a factor of the cubic.

#26 For each part, find the modulus first.

#29 Remember the identity $a\bar{a} = |a|^2$.

#30 For the second part, relate the equation $|z|^2 + |w|^2 = |z - w|^2$ to Pythagoras' theorem.

#31 Note $a^2 + b^2 + c^2 = (a + b + c)^2 - 2(ab + bc + ca)$ and $a\bar{a} = b\bar{b} = c\bar{c}$.

#33 The points $z, z^2, \ldots, z^{628}$ all lie on a circle centred at 0. Consider dividing this circle into appropriate arcs.

#35 For (iii), take the modulus first.

#37 Any complex roots will come in conjugate pairs.

#38 A half-plane has the form $Ax + By > C$, where A and B aren't both 0.

#40 Suppose for a contradiction that $\operatorname{Im}\beta \leqslant 0$. What can you then say about $P'(\beta)/P(\beta)$?

#41 How can the disc $|z| < R$ be thought of as the intersection of a set of half-planes?

#42 If $k \geqslant 2|\alpha| > 0$ seek a solution of the form $z = \lambda\alpha$.

#44 For (ix), use polar representation $z = r\operatorname{cis}\theta$.

#45 If $z = r\operatorname{cis}\theta$, show that $\left|z + z^{-1}\right| = k \iff r^2 + r^{-2} = k^2 - 2\cos 2\theta$.

#47 Recall Example 1.8.

#51 You should find #28 helpful.

#57 Show that if $z^n + a_{n-1}z^{n-1} + \cdots + a_0 = (z - z_1)\cdots(z - z_n)$ for all z, then $-a_{n-1} = z_1 + \cdots + z_n$.

#58 Note that $\zeta^5 - 1 = 0$ and factorize this equation.

#59 Note that $\zeta^7 = -1$.

#61 Consider the roots of $64x^7 - 112x^5 + 56x^3 - 7x + 1$ and explain why their product equals $-1/64$.

#63 In each case, the roots of the polynomials solve $z^n = 1$ for some value of n.

#64 In each case, rearrange the equation to one of the form $w^n = 1$, where w is a function of z.

#66 The complex roots of $z^{2n} - 1$ come in conjugate pairs.

#67 Make use of (1.28) for a choice of z.

#68 Consider factorizing $(z^{4n} - 1)/(z^4 - 1)$.

#73 Show that $-1 + \varepsilon i$ has modulus 1 and argument $\pi - \varepsilon$ (when ε^2 is considered negligible).

#74 The quintic factorizes as a linear factor and a repeated quadratic factor.

#77 De Moivre's theorem shows that summing $\text{cis}kx$ is the real part of a geometric series.

#79 This is the real part of an infinite geometric series that converges.

#83 $D = 1, t = 4, u = -2$. For the second cubic consider $x \mapsto x + 1/2$.

#84 Try setting $x \mapsto ax + b$ for appropriately chosen a and b.

#85 We are seeking x and y that satisfy $x^3 + y^3 = n$ and $-3xy = m$.

#90 Note that a change in q translates the graph of $y = x^3 + px + q$ up and down.

#91 The cubic equation $ax + b = x^3 - x$ will have a repeated root when the line is tangential.

#93 Choose k so that $z^3 - mz$ becomes a multiple of $\cos 3\theta$.

#100 Explain why $\angle a0b$ is a right angle if and only if $\bar{a}b$ is purely imaginary.

#102 Show that the given condition rearranges to $(z_2 - z_1)/(z_3 - z_1) = (z_1 - z_3)/(z_2 - z_3)$.

#105 The maps h and k can be chosen to have invariant lines that pass through the origin.

#107 Applying the same reflection twice produces the identity.

#108 Begin by showing that the fixed points of the map form the given line.

#110 Find the general form of a glide reflection that begins with the reflection in #108.

#112 Use polar representation for f_1 and f_2. Complete the square in f_4.

#117 Write $z = 1 + \text{cis}\alpha$ and use #26.

#120 The map $z \mapsto (z - z_1)/(z_2 - z_1)$ takes z_1 to 0 and z_2 to 1.

#125 Common inverse points exist if and only if $\left|R^2 - 1 - c^2\right| > 2c$.

#127 (ii) You may find it helpful to assume, without loss of generality, that C has centre 0 and radius 1.

#128 The farthest point from the origin on the circle $z\bar{z} + \bar{B}z + B\bar{z} = 0$ is $-2B$.

#130 Proposition 1.43 and #108 should help.

#131 The normal form of an ellipse is $x^2/a^2 + y^2/b^2 = 1$.

#132 If w is not in the interval, show that the equation $f(z) = w$ has one root inside and one outside the unit disc.

#135 If $a \neq 1$, determine the fixed point of the isometry and use this as a new origin.

#138 You may wish to introduce barycentric co-ordinates associated with A, B, O or $A, B, C.$.

#139 Note that the points on the line through a and b have the form $\lambda a + \mu b$, where $\lambda + \mu = 1$.

#140 Make use of Example 1.50.

#144 The identity $\sin \arg z = (\text{Im}z)/|z|$ may be of use.

#145 Without any loss of generality, you can assume $A = 0, B = 1, C = \text{cis}(\pi/3)$.

#147 c is equidistant from $0, a, b$. Encode this as two simultaneous equations in c and $\bar{c}$.

#148 Proposition 1.43 may help.

#149 The map $1/z$ takes the second diagram to the first, when we take the intersection of C_3 and C_4 as the origin.

#150 If $1/z$ maps B, C, D to B', C', D', then $\left|B'C'\right| + \left|C'D'\right| \geqslant \left|B'D'\right|$.

#152 Show that both conditions are equivalent to $[(d - c)(b - a)]/[(d - a)(c - b)]$ being positive, where $A = a$, etc.

#153 The maps taking C_1 to C_2 have the form $z \mapsto az + (a_2 - aa_1)$ where $|a| = r_2/r_1$.

#154 Show that $\overrightarrow{AE} = \omega\overrightarrow{CD} = \omega^2\overrightarrow{BF}$.

#156 The first isogonic centres will correspond under the map $w \mapsto Aw + B$.

#158 Start with $P = \bar{P}$, where P is as in (1.44). Then note from #157 the conditions for P being at 0 or 1.

#159 Show that $\text{Im} P > 0$ if and only if $\text{Im}\left(\omega^2 z\right) \text{Im}\left(\omega (z-1)\right) > 0$.

#161 Use barycentric co-ordinates to show that $d(Z) > d(P)$ when Z lies outside the triangle.

#163 1 is the only non-primitive pth root of unity. Also note k is coprime with p^n if and only if k is coprime with p.

#164 Each nth root of unity is a primitive dth root of unity for some d. For the second part, consider $x^{2n} - 1$.

#165 Try a strong form of induction and use the results of #164.

#167 Use #164 and #162.

#171 Explain why a repetition in the remainders must occur. For (ii) take $m = 10$.

#173 For (iv) consider $z_0 = \text{cis}(\alpha \pi i)$, where α is irrational. Consider the binary representation of α.

#175 Note that $(f(x_1) - f(\alpha))/(x_1 - \alpha) \approx f'(\alpha)$.

#179 It is possible to focus on the cases $0 \leqslant x_0 \leqslant \pi$.

#181 $z_n(1/4)$ is an increasing sequence that converges to $1/2$.

#183 Consider the image of the circle $|w| = 1$ under the map $w \mapsto (w/2)(1 - w/2)$.

#184 Show with a sketch that $f(f(x))$ has the same fixed points as $f(x)$.

#185 Factorize $P_c(P_c(z))$.

#188 Π's closest distance to the origin is $|D| / \sqrt{A^2 + B^2 + C^2}$. For (ii) find when $f(x + yi)$ is in the plane Π.

#192 (i) If $c \neq 0$ then $(az + b)/(cz + d) = a/c + (bc - ad)/(c^2 z + dc)$.

#193 Don't forget to consider when ∞ is fixed.

#195 For the last part apply #192(ii) and (iii).

#197 For (i), (iii), (iv) make sure firstly that the intersections correspond.

#199 Show that if $T(z_1)\overline{T(z_2)} = -1$ then $T(z_1)\overline{T(z_2)} = -1$.

#201 Show that $\pi \circ R = \tilde{R} \circ \pi$.

#202 Let $M_{A,B}$ be the map T from #199. Can each such map be written $M_{\text{cis}\alpha,0} \circ M_{1,ti} \circ M_{\text{cis}\beta,0}$?

#204 #192(i) should help with (ii). For (iii) recall #195.

#205 Use a Möbius transformation to take $g(0), g(1), g(\infty)$ to $0, 1, \infty$.

#206 Take one of the circles to be $|z| = 1$ and the other circle to have a diameter on the real line.

#208 Note that f_1 and f_2 commute if and only if $g^{-1} f_1 g$ and $g^{-1} f_2 g$ commute; then use #207.

#210 For (i) try parallel lines. For (ii) try concentric circles.

#212 Using #192(i) will help keep the algebra straightforward.

#215 For such a Möbius transformation g, find a and θ such that $f_{a,\theta} \circ g^{-1}$ fixes 0 and 1.

#217 Use #216. Points p, q being on the line gives simultaneous equations in A and B.

#219 $f_{a,\theta}$ is distance-preserving and a Möbius transformation.

#221 Without loss of generality, C can be taken to be at the origin and B on the positive real axis.

#222 Find an expression for $\sinh^2 b \sinh^2 c \sin^2 \hat{A}$ which is symmetric in a, b, c.

Induction

#229 For (ii) notice that a lot of the terms cancel out.

#230 Use the strong form of induction.

#232 Note that $f(z) = z^2 + c$, where $-\beta \leqslant z \leqslant \beta$ has its maxima at $z = \pm\beta$ and its minimum at $z = 0$.

#235 Note that $2^{10} \approx 10^3$.

#236 It may help to write $n = -k$.

#239 For the inductive step, you may wish to consider the n numbers $x_1, x_2, \ldots, x_{n-1}, x_n + x_{n+1}$.

#242 Prove each inequality separately using induction.

#246 Expand the LHS of (2.4) and use the strong form of induction.

#247 For (ii), note that $\lfloor \sqrt{k} \rfloor = m$ for $m^2 \leqslant k \leqslant m^2 + 2m$.

#248 As a hypothesis, say that every number less than b^n has an expansion and take x in the range $b^n \leqslant x < b^{n+1}$.

#251 Show $Q_n \geqslant A_n$ is equivalent to $\sum_{i<j} (x_i - x_j)^2 \geqslant 0$.

#252 (iii) Generate sequences x_n, y_n from the values found in (i). For (iv) note $(x_n/y_n)^2 - 2 = 1/(y_n)^2$.

#253 For uniqueness in (i) note that $\sqrt{2}$ is irrational.

#254 Note that $\cos(n+2)\theta = 2\cos(n+1)\theta \cos\theta - \cos(n\theta)$.

#255 Note $\tan(x/2) = 2\cot(x/2) - 2\cot x$ is a rearrangement of the $\tan 2\theta$ formula.

#258 Show that $76^n - 76$ is divisible by 100.

#259 Set $A_n = 3 \times 2^{2n} + 2 \times 3^{2n}$ and consider $A_{n+1} - 4A_n$.

#261 The remainders when $7^m - 11^n$ is divided by 19 have a period of 3 in both m and n.

#262 Note $\sin nx - \sin(n-2)x = 2\cos(n-1)x \sin x$.

#263 Show first that $I_{n+1} = n(I_{n-1} - I_{n+1})$.

#269 Compare the expansion with $1 + 1 + 1/2 + 1/4 + \cdots + 1/2^{n-1}$.

#271 Consider $(z+7)^3$ and $(z+3)^4$.

#273 If c_k is the coefficient of x^k, show that $c_{k+1}/c_k = (1000 - k)/(3k + 3)$.

#275 The c_k increase while $k \leqslant (na - 1)/(1 + a)$.

#277 Recall that the nth row adds to 2^n and use #276.

#278 Treat each inequality separately. At the inductive step work backwards from the desired inequalities.

#279 Note that I_n is decreasing so that $I_{2n-2}/I_{2n} \geqslant I_{2n-1}/I_{2n} \geqslant I_{2n}/I_{2n} = 1$.

#280 Express $1 - k^{-2}$ as $(k-1)(k+1)/k^2$ and cancel terms.

#282 Consider $\binom{n}{k}$ with $2k \leqslant n$ and $k \leqslant 6$ and note $\binom{14}{7} > 3003$.

#283 Write $\binom{2^k l}{2^k}$ as a fraction and consider the factors of 2 in the numerator and denominator.

#284 Consider whether a subset includes $n+1$ or not.

#285 How many triples (i, j, k) of non-negative integers are there such that $i + j + k = n$?

#288 The two numbers differ by a factor of $3!$.

#290 Take note of the repeated letters.

#291 Note how $n!/(k!l!m!)$ changes when k is increased by 1 and l reduced by 1.

#294 Having chosen a first element there are $n - 1$ choices for the second.

#296 Split $\{1, 2, \ldots, 2n\}$ as n pairs.

#297 Note firstly that there are $\binom{52}{5}$ possible poker hands.

#300 Count the ways that heads can win by focusing on the winning heads toss.

#304 A standard inductive hypothesis implies the result for $p'(x)$.

#305 The converse follows from #304.

#306 When n is odd, a subset has an odd number of elements if and only if its complement has an even number.

#307 Determine $(1+i)^n$ using De Moivre's theorem and the binomial theorem.

#308 Note #307 and Example 2.20(b).

#310 The second sum can be arrived at using the first sum and #309.

#312 Use De Moivre's theorem and the binomial theorem.

#313 Compare the coefficients of x^n.

#314 Consider the product $(1+x)^n(1-x)^n$. Treat n odd/even as separate cases.

#317 Use the identity from expanding $(1+x)^n$ and also differentiate that identity.

#318 Consider the identity $(x+x^{-1}+y+y^{-1})^{2n} = (x+y)^{2n}(1+x^{-1}y^{-1})^{2n}$.

#319 Use the binomial theorem to write $\cos^{2n+1}\theta$ as a sum of functions of the form $\cos m\theta$.

#320 It may help with the induction to note $\cos(n+2)\theta = 2\cos(n+1)\theta\cos\theta - \cos n\theta$.

#321 Try integrating a certain binomial expansion.

#323 Consider the cases $0 \leqslant p \leqslant 1/3, 1/3 \leqslant p \leqslant 2/3, 2/3 \leqslant p \leqslant 1$.

#324 Show in general that $E(X)^2 \leqslant E(X^2)$ with equality only if X is constant.

#328 Argue as in #275.

#330 Consider expanding $(q+p)^n$ and $(q-p)^n$.

#332 Let l be the number of ys at the start of the word. There are $\binom{n+m-l}{m}$ such words.

#333 Note $\binom{k}{r}\binom{l}{s}$ equals the coefficient of $x^r y^s$ in $(1+x)^k(1+y)^l$,

#334 Let k denote the $(r+1)$th element of a subset when ordered.

#336 In how many ways can a given subset of r elements be contained in a larger subset?

#337 Compositions such as $4 = 1+1+2$ might be represented as $(*|*|**)$.

#338 When $n = 7$ and $k = 4$ then $7 = 2+3+0+2$ might be represented as $(**|***||**)$.

#340 Use $(1+x)^{2^n} = ((1+x)^{2^{n-1}})^2$ to show that the entries in the 2^rth row are all even except for the first and last.

#342 Assuming the result for row n, prove it for rows $2n$ and $2n+1$.

#346 For the second task, note $F_{2n} = F_{2n+1} - F_{2n-1}$.

#353 Should a plausible expression remain elusive, apply Proposition 2.30 to determine the sum directly.

#355 The binomial theorem should help.

#357 For the inductive step, focus on whether a subset contains the largest element or not.

#358 Note that, when there is more than one row, the second row and above form a block fountain.

#360 For (i), use the strong form of induction and #347.

#361 Apply Lemma 2.17 and induction.

#362 Note when for $k = 1$ this is Proposition 2.32.

#364 Use Proposition 2.30.

#370 Show, for $n \geqslant 3$, that L_n lies between consecutive Fibonacci numbers.

#374 Prove that $F_{n-k} + F_{n+k} = F_k L_n$ for odd k.

#376 Show that L_m divides L_{mn} and note that $L_k \geqslant 3$ when $k \geqslant 3$.

#377 To show uniqueness, show that any decomposition of N must include the largest $F_K \leqslant N$.

#380 Note $\beta^n = F_n\beta + F_{n-1}$.

#382 Note $(4^{\circ n})_\beta = 5^{n/2}\beta^{3n}$.

#383 Rearrange $1/(F_n F_{n+1}) - 1/(F_{n+1}F_{n+2})$.

#384 Show that the Nth partial sum equals $\left[\sqrt{5}(\alpha^{2N+2} - 1)\right] / \left[2(\alpha^{2N+2} + 1)\right]$.

#385 Show that the Nth partial sum equals $3 - (F_{2^N-1})/(F_{2^N})$.

#388 For the last part consider $\arg((0.6+0.8i)^3)$.

#390 Use Theorem 2.35 for this and subsequent exercises. The auxiliary equation has roots 1 and 2.

#393 Note that $|x|^n \approx 0$ if $|x| < 1$ and n is very large.

#394 Begin by calculating x_4 and x_5 for general a.

#395 Note 4 and 9 will be the roots of the auxiliary equation.

#396 For the second part, express the solution x_n as $x_n = $ constant $+$ terms involving powers of ε.

#397 The auxiliary equation has roots $2, i, -i$. There is a one-parameter family of solutions.

#398 The auxiliary equation factorizes as two quadratics with complex roots.

#400 For (iii) try a particular solution of the form $x_n = (An^2 + Bn)(-3)^n$.

#402 Try a particular solution of the form $x_n = A\sin n + B\cos n$.

#405 Take care to show the sum converges. Take note of the auxiliary equation's largest root.

#407 If r and s denote the long-term chance of rain on a day, explain why $s = 0.5r + 0.9s$.

#408 For (ii), if the player has £n, consider what may happen next and the subsequent chance of success.

#410 Note that X_t is distributed as $2B(t, 1/2) - t$.

#412 Consider the cases: the first toss is T, the first two tosses are HT, the first two tosses are HH.

#414 HT appears for the first time as $T^t H^h T$, where $t \geqslant 0$ and $h \geqslant 1$.

#416 For Person A, introduce expectations associated with H or HT being the first tosses.

#417 For Person A, introduce probabilities of finishing HTH having started H or HT.

#420 Consider the different ways the rightmost two tiles of a $2 \times n$ grid might be covered.

#423 How might the three leftmost tiles of a $3 \times n$ grid be covered? Show $D_n = C_{n-1} + D_{n-2}$.

#426 The square of the cuboid's volume equals the product of the areas of three of its faces.

#429 Show that $x + x^{-1} \geqslant 2$ for positive x.

#430 Show that if the b_i are not ordered then the sum $a_1 b_1 + \cdots + a_n b_n$ can be increased by reordering.

#432 For (ii) show that $b_{n+1} - a_{n+1} \leqslant (b_n - a_n)/2$.

#433 In (i) show that $a_n b_n = ab$ for all n. For (ii) consider the sequences found in calculating $\text{agm}(1/a, 1/b)$.

#434 In (ii) the mouse is one move from the hole. In (iii) when $m = 1$ the mouse must move first.

#435 Note that the numbers $g(h, m)$ are entirely determined by the properties proven in (ii), (iii), (iv) of #434.

#437 For (iv) proceed by induction, assuming every word of length $2n$ with the desired properties is an Oxword.

#438 Explain why the numbers C_n satisfy the same recursion as in (2.27).

#440 For (ii) note that $[1, 2, 2, \ldots, 2] = [1, 1 + [1, 2, \ldots 2]]$. For (iv) note $a_{n+2} - 2a_{n+1} - a_n = 0$.

#441 Note that q_0 is an integer and that $0 < [q_1, \ldots, q_m]^{-1} < 1$.

#442 For (ii), if α is irrational then each α_i is irrational.

#446 $\sqrt{7}$ ultimately repeats with period 4 and $\sqrt{13}$ with period 5.

#447 For (i) note $\left[\lfloor \alpha \rfloor, \lfloor \alpha_1 \rfloor, \ldots, \lfloor \alpha_n \rfloor, \lfloor \alpha_{n+1} \rfloor, x\right] = \left[\lfloor \alpha \rfloor, \lfloor \alpha_1 \rfloor, \ldots, \lfloor \alpha_n \rfloor, \lfloor \alpha_{n+1} \rfloor + x^{-1}\right]$.

#448 If $q = a/b$ define $r_0 = a - \lfloor \alpha \rfloor b$, $r_1 = b - \lfloor \alpha_1 \rfloor r_0$, $r_2 = r_0 - \lfloor \alpha_2 \rfloor r_1$, $r_3 = r_1 - \lfloor \alpha_3 \rfloor r_2, \ldots$

#450 Note that $\alpha = [a, b, \alpha]$.

#451 If the repeat is at q_{m+1} then $\alpha = \left[q_0, \ldots, q_m, \beta\right] = (a_m \beta + a_{m-1})/(b_m \beta + b_{m-1})$.

#452 Write $x^2 - y^2 = (x - y)(x + y)$ and consider the factors of a.

#455 Note firstly that $a_{2n+2} = 2a_{2n+1} + a_{2n}$ and $a_{2n+3} = a_{2n+2} + a_{2n+1}$.

#456 Note that $a_{4n+1} = a_{4n} + a_{4n-1}; a_{4n+2} = a_{4n+1} + a_{4n}; a_{4n+3} = a_{4n+2} + a_{4n+1}$ and that $a_{4n+4} = 4a_{4n+3} + a_{4n+2}$.

#457 Show first that the sequence $(a_n)^2 - 4(b_n)^2$ has period 4.

#459 For (v), if $x^2 - dy^2 = 1$ with $x > 1$, then consider ζ^k where $\zeta = x + y\sqrt{d}$.

#460 Given $x, y > 0$ there are finite $X, Y > 0$ such that $1 < X + Y\sqrt{d} < x + y\sqrt{d}$. Then consider $(X + Y\sqrt{d})^k$.

#461 For (i), consider where 0, 1 and the fractional parts of $\alpha, 2\alpha, 3\alpha, \ldots, (Q-1)\alpha$ lie in the interval $0 \leqslant x \leqslant 1$.

#464 If $x_0^2 - 2y_0^2 = -1$ and $a^2 - 2b^2 = 1$ then consider $(x_0 + y_0\sqrt{2})(a + b\sqrt{2})$.

#465 Consider the cases $x = 3n - 1, 3n, 3n + 1$ where n is an integer.

#466 Rearrange as $(x+y)^2 - 2(x-y)^2 = -1$ and use #464.

#467 For (ii), show the coefficient of k^s in $B_m(k+1) - B_m(k)$ equals $\sum_{r=0}^{m-s-1} \binom{m}{r}\binom{m-r}{s} B_r$. What is this for $s < m - 1$?

#468 For (iii) note that $B_n(0) = B_n(1) = B_n$ for $n \geqslant 2$ and $B_n(1-x)$ and $(-1)^n B_n(x)$ have equal second derivatives.

#471 Show that the polynomials $\frac{1}{2^n n!} \frac{\mathrm{d}}{\mathrm{d}x^n} \left[\left(x^2 - 1 \right)^n \right]$ satisfy the same recurrence that defines $P_n(x)$.

#472 Write $1 + t + t^2 = (t + \alpha)(t + \beta)$ for some α, β and then employ #471.

#474 Prove (ii) and (iii) first with $x = \cos\theta$; note distinct polynomials can only agree for a finite number of values.

#477 Show that $a_{t+1} = a_t/2 + a_{t-1}/2$.

#478 Note X_t has the distribution $2B(t,p) - t$.

#481 Show that $P_a = pP_{a+1} + qP_{a-1}$ for $a \geqslant 1$.

#482 For (i), note that the possible directions on alternate days.

#483 For (ii), note that there is a half chance of X_t changing at each step and an equal chance of it increasing/decreasing.

#484 For (ii) consider the coefficient of $x^k y^l$ in $6^{-t}(x + x^{-1} + y + y^{-1} + x^{-1}y^{-1} + xy)^t$.

#485 For the 'only if' part consider decomposing $a_{n+1} - 1$. For the 'if' part use the strong form of induction.

#487 (i) Use Zeckendorf's theorem. If F_N is missing, use F_{N-1}, F_{N-2} instead. For (ii) recall #347.

Vectors and Matrices

#491 Write x, y, z in terms of α, β and eliminate α, β.

#492 For (ii) write $\mathbf{w} = c_1\mathbf{u} + c_2\mathbf{v}$ for some real c_1, c_2

#495 Consider when $S = \{\mathbf{0}\}$, or contains a non-zero vector, or contains two which aren't multiples of one another.

#501 Take $\mathbf{x} = \mathbf{e}_i$ for each i, or take $\mathbf{x} = \mathbf{v} - \mathbf{w}$ and rearrange.

#502 (iii) is a version of the triangle inequality.

#505 Choose $\mathbf{v}_k$ so that $|\mathbf{v}_k|^2 = 1 + k^2$.

#506 Apply Cauchy–Schwarz to $2x + 3y - 4z = (2, 3, -4) \cdot (x, y, z)$.

#507 Why is there $0 \leqslant \theta \leqslant \pi$ such that $\sqrt{x^2 + y^2} = \sin\theta$ and $z = \cos\theta$?

#511 Find x, y, z in terms of λ and eliminate λ.

#512 Set $\lambda = (x-2)/3 = (y+1)/2$.

#513 Write $\mathbf{r} = \mathbf{c} + \lambda\mathbf{d}$ and use the given information to work out $\mathbf{c}$ and $\mathbf{d}$.

#516 Write $\mathbf{r} = \mathbf{p} + \lambda\mathbf{q}$.

#517 Determine $|\mathbf{r}(\lambda) - (1, 1, 6)|^2$ and use calculus, or determine when $\mathbf{r}(\lambda) - (1, 1, 6)$ is perpendicular to the line.

#518 Note the vector (a, b) is perpendicular to the line.

#520 Determine when $\mathbf{r}(\lambda) - \mathbf{s}(\mu)$ is perpendicular to both lines.

#524 For (iii) write $\lambda = \cos t, \mu = \sin t$ and the circle's centre $\mathbf{p}$ in the from $\mathbf{p} = \alpha\mathbf{a} + \beta\mathbf{b}$.

#527 Explain why we may assume $a_1 \neq 0$ and $a_1 b_2 - a_2 b_1 \neq 0$ without loss of generality.

#531 Apply #530.

#534 Find two independent vectors perpendicular to the plane's normal.

#536 $\mathbf{n}$ must be perpendicular to $(1,0,3)$ and $(2,3,-1)$.

#537 Note Example 3.25.

#539 To begin, note $\mathbf{r}_1(e,f) = \mathbf{r}_2(0,0)$ for some e,f.

#540 Write $P(\mathbf{v}) = \mathbf{v} + \lambda\mathbf{n}$ and $Q(\mathbf{v}) = \mathbf{v} + 2\lambda\mathbf{n}$.

#541 Find the orthogonal projection of $\mathbf{v}$ on to the plane $\mathbf{r} \cdot \mathbf{n} = c$ first.

#545 Set $\mathbf{p} = \lambda\mathbf{a}, \mathbf{q} = \mathbf{a} + \mu(\mathbf{b} - \mathbf{a}), \mathbf{r} = (1-\nu)\mathbf{b}$ and show concurrence if $\lambda\mu\nu = (1-\lambda)(1-\mu)(1-\nu)$.

#546 Note $\frac{1}{4}(\mathbf{a}+\mathbf{b}+\mathbf{c}+\mathbf{d}) = \frac{1}{2}\frac{1}{2}(\mathbf{a}+\mathbf{b}) + \frac{1}{2}\frac{1}{2}(\mathbf{c}+\mathbf{d}) = \frac{1}{4}\mathbf{a} + \frac{3}{4}\frac{1}{3}(\mathbf{b}+\mathbf{c}+\mathbf{d})$.

#547 (ii) Altitudes from $\mathbf{a}$ and $\mathbf{b}$ intersect at $\mathbf{d}$. (iii) The perpendicular bisector from A in XYZ is the altitude of A in ABC.

#548 (ii) Note that $\mathbf{a} \cdot (\mathbf{c} - \mathbf{b}) = 0$ and $\mathbf{b} \cdot (\mathbf{c} - \mathbf{a}) = 0$.

#550 Write x_1, x_2, x_3, x_4 in terms of λ, μ and eliminate λ, μ.

#551 (ii) Note E is on OA and in BCD. (iii) Note $\alpha/(\beta + \gamma + \delta) > 0$.

#560 Assume n to be the smallest k such that $A^k = 0$. Consider the cases $m \geqslant n$ and $m < n$.

#563 $\mathbf{e}_i$ is the $1 \times m$ matrix with $[\mathbf{e}_i]_{1j} = \delta_{ij}$.

#564 Use #563(ii).

#566 Use (3.13).

#567 Use the sifting property.

#570 Try finding 3×3 matrices which do not commute and such that $A^2 = 0 = B^2$.

#573 Show first that $P_1 P_2 = -P_2 P_1$ and then consider $P_1 P_2 P_1$.

#574 Show that $p(A) = \frac{1}{2}((p(1) - p(-1))A + \frac{1}{2}((p(1) + p(-1))I_2$.

#578 Prove first for $n = 2$ and then use induction.

#582 Note that $[A]_{ij} = [A]_{ii}\delta_{ij}$ for a diagonal matrix A.

#585 Show by induction that $[A^r]_{ij} = 0$ when $i - j \geqslant 1 - r$.

#588 (i) Use #587. (iii) Note $A(\mathbf{v} + \mathbf{w}) \cdot A(\mathbf{v} + \mathbf{w}) = (\mathbf{v} + \mathbf{w}) \cdot (\mathbf{v} + \mathbf{w})$.

#591 For (i) show that $|A\mathbf{v}|^2 = 0$.

#593 For (ii), use the notation from (3.13) to express trace(CD) as a double sum.

#594 Set $B = E_{IJ}$.

#597 For the second part, try setting $n = 1/2$ and check your answer works.

#603 Show a solution of $a^2 + bc = 1$ has the form $a = \sec\alpha\cos\beta, b = \sec\alpha\sin\beta + \tan\alpha$ and also $c = \sec\alpha\sin\beta - \tan\alpha$.

#604 Show a solution of $a^2 + bc = 0$ has the form $a = z\cos\alpha, b = z\sin\alpha + z, c = z\sin\alpha - z$.

#605 Proceed by induction on n and writing a square root of D in the form $\begin{pmatrix} B & \mathbf{v} \\ \mathbf{w}^T & \alpha \end{pmatrix}$.

#607 For (v) note that $Y(a_n, b_n, c_n)^T = (a_{n+1}, b_{n+1}, c_{n+1})^T$.

#610 $ax + by + cz = d$ and $Ax + By + Cz = D$ are parallel if and only if (a,b,c) and (A,B,C) are dependent.

#612 Expressions for x and y found in (i) apply just as well in (ii) for matrices and in (iii) for polynomials.

#613 For the converse, pick e,f to show that $ad - bc$ must divide a.

#615 Apply the product rule for transposes to $(A^{-1})^T A^T$ and $A^T (A^{-1})^T$.

#616 We already know that $A^r A^s = A^{r+s}$ from #571 when r,s are natural numbers.

#619 For the last part, find first a matrix C such that $C^3 = P^{-1}AP$.

#622 If A has a zero row find a non-zero row vector $\mathbf{v}$ such that $\mathbf{v}A = \mathbf{0}$.

#623 For (ii) consider $B(k_1, \ldots, k_n)^T$.

#626 Use the transpose product rule.

#628 For (ii) recall that $\text{trace}(CD) = \text{trace}(DC)$.

#630 Recall a factorization of $1 - x^n$.

#631 Find an inverse of the form $I + kA$ where k is a constant.

#635 For the last part note $A^n = (\text{trace}A)^{n-2}A^2$ for $n \geqslant 2$.

#636 $\left[A^n\right]_{ij}$ satisfies the same recurrence relation for each i,j with different initial values.

#638 For (v) note $\mathbf{b} = A\mathbf{x}$ for some $\mathbf{x}$ is equivalent to the system being consistent.

#639 Add $\mathbf{x}$ to a row or rows of B_1 or B_2.

#640 Consider the equation $BA = I_2$ as a linear system in the entries of B.

#645 Note that if $A = \mathbf{v}\mathbf{w}^T$ then the rows of A are multiples of one another.

#650 Show that $(A\,|\,\mathbf{0})$ reduces to $(I_3\,|\,\mathbf{0})$ if $(a-2)(1-a^2) \neq 0$.

#651 Show that the linear system is consistent if and only if $(a+4)(a-3) = 0$.

#655 Note that the columns of B satisfy the same consistency conditions required of $\mathbf{b}$.

#656 For (i) consider the form $(\text{RRE}(A)\,|\,\mathbf{0})$ takes.

#657 For (iv) note that the columns of A_2X are linear combinations of the columns of A_2.

#658 Row-reduce the system $(A\,|\,B)$.

#659 For (iv) note how the calculations in (3.24) relied on the commuting of real numbers.

#661 Treat Sylvester's equation as a linear system in the entries of X.

#663 For (ii) argue as in #662(ii).

#666 Row-reduce $(A\,|\,I_3)$ to $\left(I_3\,|\,A^{-1}\right)$, keeping a record of the EROs used.

#667 Row-reduce $(B\,|\,C)$.

#668 Note that $AX = B$ implies $X = CB$ but not the converse.

#670 Row-reduction yields a 2×3 matrix B such that $BA = I_2$ and a row vector $\mathbf{v}$ such that $\mathbf{v}A = \mathbf{0}$.

#672 Recall #578.

#673 Note that an invertible triangular matrix can be reduced using triangular elementary matrices.

#674 Use EROs and ECOs to partially reduce A_n into $\text{diag}(A_{n-3}, I_3)$.

#678 The matrices have rank 2 and 3 respectively.

#679 Determine the number of arithmetic operations needed to get each column into RRE form in turn.

#681 Recall that a matrix's rowspace is invariant under EROs.

#682 Consider the effects on a matrix's rows and recall #565.

#683 Consider the effects on a matrix's rows.

#684 Separate out the cases based on the overlap of the sets $\{i,j\}$ and $\{k,l\}$.

#686 Proceed by induction on the number of rows of the matrix.

#690 Applying ECOs is equivalent to postmultiplication by elementary matrices.

#691 What is the form of a matrix that is both row-reduced and column-reduced?

#692 Row-reduce $(A\,|\,I_3)$ to $(PA\,|\,P)$ and then column-reduce $\left(\dfrac{PA}{I_4}\right)$ to $\left(\dfrac{PAQ}{Q}\right)$.

#694 If $\text{Row}(A)$ is unaffected by ECOs then $\text{Row}(A) = \text{Row}(AQ) = \text{Row}(PAQ)$.

#695 What would happen if Corollary 3.90 were applied as a test of independence?

#702 Use the method of Corollary 3.90.

#703 Without loss of generality assume the leading 1s of A to be in the first m columns.

#704 If $A^T A \mathbf{x} = \mathbf{0}$ then $|A\mathbf{x}|^2 = \mathbf{x}^T A^T A \mathbf{x} = 0$.

#705 Apply #695 to these $n^2 + 1$ vectors.

#706 Recall #618.

#708 Note $(A - \lambda_1 I)\mathbf{e}_1^T = \mathbf{0}$ and $(A - \lambda_1 I)(A - \lambda_2 I)\mathbf{e}_2^T = \mathbf{0}$.

#709 Recall Row(BA) is contained in Row(A) and $A(k_1, \ldots, k_n)^T = k_1\mathbf{c}_1 + \cdots + k_n\mathbf{c}_n$.

#710 Reduce the matrix with rows $\mathbf{v}_1, \ldots, \mathbf{v}_k$, focusing on the columns with no leading 1.

#711 Show for each ERO E that if A has k independent rows then EA has k independent rows.

#712 If $\mathbf{v}$ is dependent on $\mathbf{v}_1, \ldots, \mathbf{v}_k$ show how the last row of B can be reduced to $\mathbf{0}$.

#713 Recall every invertible matrix can be written as a product of elementary matrices.

#714 If Row$(A) = $ Row(B) consider the columns of RRE(A) and RRE(B) containing leading 1s.

#715 Note that the sum of A and B's first rows can be written as a combination of the rows of A.

#716 Consider reducing the matrix which is A over $m_2 - m_1$ zero rows.

#717 Apply the EROs that reduce A and B to diag(A, B).

#718 Note that a leading 1 of RRE(A) could be used to eliminate all other entries in a column.

#722 What if $\alpha_1 \mathbf{v}_1 + \alpha_2 \mathbf{v}_2 + \cdots + \alpha_k \mathbf{v}_k + \alpha \mathbf{v} = \mathbf{0}$ and $\alpha \neq 0$?

#724 Begin by removing a vector $\mathbf{v}_i$ which is dependent on the other vectors.

#725 Recall Proposition 3.88(c).

#726 To find the hyperplane's dimension consider $\mathbf{a} \cdot \mathbf{x} = 0$ as a linear system.

#730 To show two spans are equal show each is contained in the other.

#731 $\mathbf{e}_i$ is in Row$(\mathbf{v}_1 / \cdots / \mathbf{v}_k)$ if and only if the ith column of RRE$(\mathbf{v}_1 / \cdots / \mathbf{v}_k)$ has a leading 1.

#733 $\mathbf{w}_1, \ldots, \mathbf{w}_4$ is a basis if and only if RRE$(\mathbf{w}_1 / \cdots / \mathbf{w}_4) = I_4$.

#734 Find linear dependencies involving I_2, A, A^2 and I_3, B, B^2, B^3.

#738 H_1 and H_2 are defined by two independent homogeneous linear equations.

#739 Recall #589(iii).

#740 Find homogeneous linear equations that characterize being in X and Y.

#743 A counter-example can be found using subsets X, Y of $\mathbb{R}$.

#744 If $A = $ diag$(1, 2, 3)$ and $\mathbf{v} = (a, b, c)^T$ consider the span of $\mathbf{v}, A\mathbf{v}, A^2\mathbf{v}$ when $abc \neq 0$.

#745 For (ii), take the ith column of P to be in V_i and not V_{i-1}.

#746 Extend a basis for Null(C), to one for Null(C^2), to one for Null(C^3), etc.

#748 Recall Proposition 3.88(d) and that column rank equals row rank.

#750 Find the general solution of the system.

#754 Recall the linear dependencies involving I_2, A and I_3, B, B^2 found in #734.

#755 Write down the general form of a matrix in each such subspace.

#756 Begin with a basis for $X \cap Y$.

#757 For (ii), show $\mathbf{v}_k$ must be of the form $(x_1, \ldots, x_k, 0, \ldots 0)^T$, where $x_k \neq 0$.

#758 Write down the general form of polynomial in each such subspace.

#759 Note $x^k = (x - 1 + 1)^k$ and use the binomial theorem.

#760 With relabelling, prove by induction $\langle \mathbf{w}_1, \ldots, \mathbf{w}_i, \mathbf{v}_{i+1}, \ldots, \mathbf{v}_k \rangle = \mathbb{R}^n$ with $1 \leqslant i \leqslant l$.

#762 Show an $m \times n$ matrix with $m < n$ cannot have a left inverse.

#763 The columns of a matrix span its column space, and therefore contain a basis for it.

#766 Note much of the detailed calculation for (ii) has already been done in (i).

#767 Note that $\mathbf{v}^T A^T A\mathbf{v} = (A\mathbf{v})^T(A\mathbf{v}) = |A\mathbf{v}|^2$ to show the nullities are equal.

#768 Note $AX = 0_{mp}$ if and only if each column of X is in the null space of A.

#769 Consider vectors $A\mathbf{v}, A^2\mathbf{v}, A^3\mathbf{v}, \ldots, A^{n-1}\mathbf{v}$ where $A^{n-1}\mathbf{v} \neq \mathbf{0}$.

#771 Recall Proposition 3.125.

#772 Recall Proposition 3.95(d)

#773 Note if $B\mathbf{v} = \mathbf{0}$ then $AB\mathbf{v} = \mathbf{0}$ and so nullity$(B) \leqslant$ nullity(AB).

#774 For (ii) recall Proposition 3.88(c). For (iii) find a right inverse and use the basis found in (i).

#777 (i) Recall Proposition 3.88(c). (ii) If $P = \text{ERRE}(P)$ remove the zero rows of RRE(P) and same columns of E.

#778 Recall #711. Show a linear dependency in the rows is unaffected by postmultiplication by an elementary matrix.

#780 Note every $\mathbf{v}$ in $\mathbb{R}_n$ can be written in the form $T\mathbf{w}$, where $\mathbf{w}$ is in $\mathbb{R}_n$.

#783 Note $(4, -3)^T$ is in the invariant line and that $A_1 (1, 1)^T$ remains at the same distance from the line as $(1, 1)^T$.

#784 Write $(3, -1)^T$ as a linear combination of $(2, 1)^T$ and $(1, 2)^T$.

#786 Write $(0, 1, 2)^T$ as a linear combination of $(2, 3, 4)^T$ and $(1, 2, 3)^T$.

#788 Note $(\mathbf{v}_1 \mid \ldots \mid \mathbf{v}_n)$ is invertible.

#789 Find a $\mathbf{v}$ which is independent of $(2, 3, 4)^T$ and $(1, 2, 3)^T$. What are the possible images of $\mathbf{v}$?

#791 Note that $\mathbf{e}_1^T \leftrightarrow E_{11}, \mathbf{e}_2^T \leftrightarrow E_{12} + E_{21}, \mathbf{e}_3^T \leftrightarrow E_{12} - E_{21}, \mathbf{e}_4^T \leftrightarrow E_{22}$ now.

#792 If $\mathbf{v}$ is parallel to the line consider the cases $A\mathbf{v} = \mathbf{0}$ and $A\mathbf{v} \neq \mathbf{0}$.

#793 (ii), (iii) If A is singular but non-zero, in what ways can the columns of A be scalar multiples of one another?

#794 For the last part, a general point of S_3 can be written $(\cos\theta, \sin\theta)^T$.

#796 (iii) The area of the ellipse $x^2/a^2 + y^2/b^2 = 1$ is πab.

#799 The point $(0, 0, 1)^T$ moves parallel to the plane $z = 0$.

#802 Consider the effect of the shear on units vector parallel and perpendicular to the invariant line.

#803 (ii) Without loss of generality, assume the invariant line of S to be the x-axis.

#804 Note $\mathbf{e}_1^T, \mathbf{e}_2^T, \mathbf{e}_3^T, \mathbf{e}_4^T$ are identified with $1, x, x^2, x^3$.

#805 For (vi) show $\mathbf{v}^T X^T (\mathbf{y} - P\mathbf{y}) = 0$ for $\mathbf{v}$ in $\mathbb{R}_n$, $\mathbf{y}$ in $\mathbb{R}_m$. Any element of Col(X) has the form $X\mathbf{v}$.

#806 The columns of X should be a basis of the given plane.

#807 Note $a^2 + bc = a$ can be rewritten as $(2a - 1)^2 + (b + c)^2 = 1 + (b - c)^2$.

#812 For S, T use co-ordinates associated with the matrices E_{ii} and $(E_{ij} \pm E_{ji})/2$ where $i < j$.

#818 Use the product rule for determinants.

#820 Prove by induction on the columns of A.

#822 Prove by induction on the columns of U in a manner similar to #820.

#823 Use EROs and ECOs and then apply #822.

#824 For the second determinant, begin by subtracting the first row from the others.

#828 For the first determinant, begin by adding the second column to the first.

#829 Note z_1, z_2, z_3 are collinear if $w = \overline{w}$ where $w = (z_3 - z_1)/(z_2 - z_1)$.

#831 Call the first determinant a_n. Expand along the first row to show $a_n = 2a_{n-1} - a_{n-2}$.

#833 Call the first determinant a_n. Show that $a_n = xa_{n-1} - x^2 a_{n-2}$ and deduce that $a_{n+3} = -x^3 a_n$.

#834 Show $D_n - (\alpha + \beta) D_{n-1} + \alpha\beta D_{n-2} = 0$ and $\Delta_n = b\Delta_{n-1} - ca\Delta_{n-2}$.

#835 Use Example 3.165.

#836 Differentiate the given identity twice, then twice more, and so on.

#838 Write the $n \times n$ determinant with (i, j)th entry $(x + i + j - 1)^{n-1}$ in terms of Vandermonde determinants.

#839 Show $VC = \text{diag}(\alpha, \beta, \gamma)V$.

#840 Consider $V(x_0, \ldots, x_n)(a_0, \ldots, a_n)^T = (y_0, \ldots, y_n)^T$, where $a_0, \ldots, a_n$ are the coefficients of the polynomial $p(x) = a_0 + a_1 x + \cdots + a_n x^n$.

#842 Write the system as $(x, y, z)V(x, y, z) = 0$.

#843 (ii) Expand and compare coefficients. (iv) Show $a^{k+3} + ma^{k+1} + na^k = 0$.

#844 Let P_k denote the given determinant. Use #843(iv) to show $P_k = abcP_{k-1}$.

#845 Use #711 to show that the determinantal rank does not exceed r.

#847 Show that the determinantal rank is at least $n - k + 1$.

#848 Use EROs and ECOs to place the zero submatrix more conveniently.

#849 (i) Explain why every e_i^T is still present among the columns of the product. (iii) Use #588.

#850 Rewrite the product $a_{i_1 1} a_{i_2 2} \cdots a_{i_n n}$ as $a_{1 j_1} a_{2 j_2} \cdots a_{n j_n}$ for a different permutation (specifically the inverse).

#851 (ii) Use induction on n. (iii) Say $Pe_I^T = e_i^T$ and $Pe_J^T = e_j^T$.

#852 Focus on the determinant of P.

#854 For the converse if P is not a permutation matrix find M such that PM and M have different entries.

#857 To make a non-zero contribution to (3.50) P must permute $1, \ldots, m$ (and so also $m + 1, \ldots, m + n$).

#858 If r is even show that $(\Sigma_r)^2$ has two cycles of length $r/2$.

#859 Use strong induction on n. Show that the first repetition in the list $e_1^T, Pe_1^T, P^2 e_1^T, P^3 e_1^T, \ldots$ is a repeat of e_1^T.

#860 If $P = R^{-1} \text{diag}(\Sigma_{r_1}, \Sigma_{r_2}, \ldots, \Sigma_{r_k})R$ consider when P^m fixes any of the standard basis, where $m \geqslant 1$.

#861 The first matrix swaps e_1^T and e_5^T, e_2^T and e_4^T, fixes e_3^T. So let R relabel $1, 5$ to $1, 2$ and $2, 4$ to $3, 4$ and 3 to 5.

#862 Note that $PQ = Q^{-1}(QP)Q$.

#863 Note $(P^{-1} \Sigma_r P)^m = I_r$ if and only m is a multiple of r.

#866 Show by induction on m, that a product of m transpositions has at least $n - m$ cycles.

#871 Use Theorem 3.170 and Corollary 3.157(b).

#873 (i) Replace A with $A - xI$, noting distinct polynomials can only agree at finitely many x.

#876 If the bottom $n - 1$ rows of A are independent, show $C_{11} \neq 0$ or e_1 is in their span.

#877 Try $E_{12} + E_{21}$.

#878 Show that $L_1 U_1 = L_2 U_2$ implies $L_2^{-1} L_1 = U_2 U_1^{-1}$.

#885 Use Proposition 3.179(a) and Proposition 3.184.

#886 When is $\mathbf{u} \wedge \mathbf{v} = \mathbf{0}$?

#891 What are the possibilities for $\mathbf{i} \sqcap \mathbf{j}$, $\mathbf{j} \sqcap \mathbf{k}$ and $\mathbf{k} \sqcap \mathbf{i}$?

#892 What are the possibilities for $e_1 \wedge e_2$, $e_1 \wedge e_3$ and $e_1 \wedge e_4$? Consider $e_1 \wedge (e_2 + e_3)/\sqrt{2}$.

#893 For second part, consider $(\mathbf{b} - \mathbf{a}) \wedge (\mathbf{c} - \mathbf{a})$.

#894 Set $\mathbf{e} = \mathbf{a} \wedge \mathbf{b}$ and use the vector triple product.

#895 The triangle OAB has area $|\mathbf{a} \wedge \mathbf{b}|/2$.

#897 Dot a zero linear combination of $\mathbf{a}, \mathbf{b}, \mathbf{a} \wedge \mathbf{b}$ with $\mathbf{a} \wedge \mathbf{b}$.

#898 First work out $(\mathbf{v} \wedge \mathbf{w}) \wedge (\mathbf{w} \wedge \mathbf{u})$. Recall also #896(iii).

#899 Write $\mathbf{r} = X\mathbf{a} + Y\mathbf{b} + Z(\mathbf{a} \wedge \mathbf{b})$ for some X, Y, Z.

#900 Begin by dotting the given equation with $\mathbf{a}$.

#901 If **a** and **b** are independent write $\mathbf{r} = \alpha\mathbf{a} + \beta\mathbf{b} + \gamma\mathbf{a} \wedge \mathbf{b}$.

#906 Recall that $\mathbf{r} \wedge \mathbf{a} = \mathbf{p} \wedge \mathbf{a}$ is a line parallel to **a** through **p**.

#907 If **p** is on the line the shortest distance is $|\mathbf{c} - \mathbf{p}| \sin\theta$ where θ is the angle between $\overrightarrow{PC}$ and the line.

#908 If $\mathbf{a} \cdot \mathbf{d} + \mathbf{b} \cdot \mathbf{c} = 0$ note a general point of $\mathbf{r} \wedge \mathbf{a} = \mathbf{b}$ is $\mathbf{r} = (\mathbf{a} \wedge \mathbf{b})/|\mathbf{a}|^2 + \lambda\mathbf{a}$.

#911 Recall the transpose and product rules for determinants.

#912 The volume of T is then $\int_0^1 \frac{1}{2}(1 - c)^2 \, dc$. Consider where $(\mathbf{u}\,|\,\mathbf{v}\,|\,\mathbf{w})$ maps T.

#914 You will need to treat even and odd n separately.

#917 How do the characteristic polynomials of A and A^T relate?

#919 Note $xI - P^{-1}AP = P^{-1}(xI - A)P$.

#922 Recall #821.

#929 Note $A^k\mathbf{v} = \lambda^k\mathbf{v}$. If A is diagonalizable then A has an eigenbasis $\mathbf{v}_1, \ldots, \mathbf{v}_n$.

#930 Note that $c_M(-2) > 0 > c_M(0)$.

#936 For (ii), note that $B = A + A^{-1} + xI$.

#937 Similar matrices have the same characteristic polynomial. The converse is not true.

#938 Note $XM = MX$ if and only if $*P^{-1}XP)(P^{-1}MP) = (P^{-1}MP)(P^{-1}XP)$. Recall #579.

#939 If A is upper triangular with respect to **v**, **x**, try a basis of the form $\mathbf{V} = \alpha\mathbf{v}, \mathbf{X} = \mathbf{x} + \beta\mathbf{v}$.

#940 Consider the discriminant of $c_A(x)$.

#941 Set $X = x - a$ in $c(x)$ for the first matrix. Note the second matrix is singular.

#942 There is a permutation matrix Π such that $\Pi^{-1}A_1\Pi = \mathrm{diag}(2, 1, 3)$.

#946 For (i) consider the cases $1 \leqslant k < r, r \leqslant k < s, s \leqslant k$.

#949 Recall #948 and #603.

#950 The eigenvectors are the symmetric and skew-symmetric matrices.

#953 Show the x^{n-2} term has coefficient $\sum_{i<j}(a_{ii}a_{jj} - a_{ij}a_{ji})$.

#954 Recall #579.

#958 Set up simultaneous linear recurrence relations in some a_n and b_n such that $x_n = a_n/b_n$.

#959 If A has eigenbasis **v**, **w**, consider the matrices $(\mathbf{v}\,|\,\mathbf{0}), (\mathbf{0}\,|\,\mathbf{v}), (\mathbf{w}\,|\,\mathbf{0}), (\mathbf{0}\,|\,\mathbf{w})$.

#961 For N3 mimic the proof of the Cauchy–Schwarz inequality.

#963 Pick specific values of (x_1, x_2) so that $\|(x_1, x_2)\| > 0$ imposes constraints on a and b.

#967 Note we can have $\mathbf{z}^T\mathbf{z} = 0$ for complex $\mathbf{z} \neq \mathbf{0}$.

#969 $\|A\|_1 \leqslant$ maximum absolute column sum – use triangle inequality. Then find **v** where this maximum is achieved.

#970 By #968(iii) $\|A^k\|_1$ tends to 0 as k becomes large.

#973 Note if $p(x_0) > 0$ at some point, there is an interval I of non-zero length containing x_0 on which $p(x) > 0$.

#974 Legendre's equation can be rewritten as $\left((1 - x^2)y'\right)' = -n(n + 1)y$.

#975 A non-zero polynomial of degree two can have at most two real roots.

#978 Given bases $\mathcal{B}_1$ for $X \cap Y$, $\mathcal{B}_1 \cup \mathcal{B}_2$ for X and $\mathcal{B}_1 \cup \mathcal{B}_3$ for Y, consider $\mathcal{B}_1 \cup \mathcal{B}_2 \cup \mathcal{B}_3$.

#982 Begin by taking a basis for X and extend it to one for $\mathbb{R}^n$.

#984 Any **v** in $X \cap Y$ can be decomposed as $\mathbf{v} = \mathbf{v} + \mathbf{0} = \mathbf{0} + \mathbf{v}$.

#985 A counter-example can be found in $\mathbb{R}^2$.

#988 It may help to recall Proposition 3.112(d).

#990 For **v** in $\mathbb{R}_n$ write $\mathbf{v} = (\mathbf{v} - P\mathbf{v}) + P\mathbf{v}$.

#994 $\mathbb{C}_n = E_1 \oplus \cdots \oplus E_k$ is a direct sum of its eigenspaces. Show $M = (M \cap E_1) \oplus \cdots \oplus (M \cap E_k)$.

#995 Show $\text{Null}(A^k) = \text{Null}(A^{k+1})$ implies $\text{Null}(A^k) = \text{Null}(A^{k+j})$ for all $j \geqslant 0$.

#996 Show A has an eigenvector $\mathbf{v}$. Extend $\mathbf{v}$ to a basis of $\mathbb{C}_n$ which you should set as the columns of a matrix.

#997 Apply Bézout's lemma to the polynomials $(x - a_1)^{r_1}$ and $(x - a_2)^{r_2} \cdots (x - a_k)^{r_k}$.

#1000 If $c_A(x)$ and $c_{-B}(x)$ have no common root, apply Bézout's lemma and use the Cayley–Hamilton theorem.

#1001 If $p(A) = 0$, write $p(x) = q(x)m_A(x) + r(x)$ where $r(x)$ has degree less than $m_A(x)$.

#1002 If $m_A(\lambda) = 0$ write $m_A(x) = (x - \lambda)q(x)$. If $c_A(\lambda) = 0$ and $A\mathbf{v} = \lambda\mathbf{v}$ note $A^k\mathbf{v} = \lambda^k\mathbf{v}$.

#1003 Use #1001, #1002 and the Cayley–Hamilton theorem.

#1004 Prove this first for diagonal matrices. Note similar matrices have the same minimal polynomials.

#1005 Show a triangular matrix A with diagonal entries $\alpha_1, \alpha_2, \ldots, \alpha_n$ satisfies the polynomial $(x - \alpha_1)(x - \alpha_2) \cdots (x - \alpha_n)$.

#1006 Use the primary decomposition theorem.

#1007 Use the primary decomposition theorem and #746.

#1008 Find $m_C(x)$, where $C = \text{diag}(A, B)$ in terms of factorized versions of $m_A(x)$ and $m_B(x)$.

#1009 The eigenspaces form a direct sum which is also A-invariant.

#1010 For (iii) if the columns of P are an eigenbasis of A explain why $P^{-1}BP$ has the form $\text{diag}(B_1, B_2, \ldots, B_k)$.

#1011 (iii) Note that a triangle corresponds to six different cycles.

#1012 There is permutation matrix P such that $A_3 = P^{-1}A_2P$.

#1015 (i) Use #1014 and then induct on E. (ii) Add an edge between two odd-degree vertices.

#1016 Note that a face is bounded by at least three edges.

#1017 (ii) Note that a face is bounded by at least four edges. (iii) A torus can be formed by gluing opposite edges of a square.

#1018 Show that $mV = 2E = nF$.

#1020 Exercises from §2.4, involving sums of Fibonacci numbers, may prove helpful.

#1021 (i) Find a 0-eigenvector. (ii) Use Example 3.164(b).

#1022 Show $c_L(x) = x(x - m)^{n-1}(x - n)^{m-1}(x - m - n)$.

#1025 Put alternate vertices into V_1 and V_2. Why can this be done unambiguously?

#1026 For the converse, fix a vertex and let V_1 be all vertices connected to it by an even-length walk.

#1027 If $(\mathbf{v}, \mathbf{w})^T$ is a λ-eigenvector when $\mathbf{v}$ is in $\mathbb{R}_m$ and $\mathbf{w}$ is in $\mathbb{R}_n$, then find a $-\lambda$-eigenvector.

#1028 Consider $\text{trace}(A^n)$, where n is odd.

#1029 The system can be in state A only by having been in B before with a 1-in-3 chance of moving from B to A.

#1030 A transition matrix has non-negative entries and $(1, 1, \ldots, 1)^T$ as a 1-eigenvector.

#1032 Note that M^3 is the transition matrix from day n to day $n + 3$ for any n.

#1033 Note $x + y + z = 1$ must also hold.

#1034 Consider the row and column rank of $M - I$, where M is the transition matrix.

#1035 (ii) If $\mathbf{p}$ and $\mathbf{P}$ are independent stationary distributions, consider $\lambda\mathbf{p} + (1 - \lambda)\mathbf{P}$.

#1036 Note that p_n here is not the long-term probability of being in state n and the p_n need not sum to 1.

#1037 Note that $(p_1, p_2, p_3, p_4, p_5, p_6)^T = \lim_{n \to \infty} M^n \mathbf{e}_6^T$.

#1040 (i) After an even number of moves the chain can only be at A or C. Work out L^2.

#1041 To avoid absorption the chain must alternate between B and C.

#1042 Absorption at F is via a route $(CBE)^n F$ including possible prolonged stays at C or E.

#1043 Work out the 16 probabilities $P(AB|CD)$ for weather tomorrow A, today B and C, and yesterday D.

More on Matrices

#1047 Note $\mathbf{v} \neq \mathbf{0}$ if and only if $P\mathbf{v} \neq \mathbf{0}$.

#1048 Put the vectors in $\mathcal{F}$ as the rows of a 3×3 matrix and row-reduce it.

#1049 Expressing each E_{ij} in terms of $\mathcal{B}$ shows $\mathcal{B}$ to span M_{22}.

#1051 Note that $_{\mathcal{E}}I_{\mathcal{F}}$ has the vectors of $\mathcal{F}$ as its columns.

#1053 Note that this change of basis matrix does not respect the dot product.

#1054 So $\mathcal{V}$ is an eigenbasis.

#1055 Note that $_{\mathcal{E}}I_{\mathcal{V}}$ has the vectors of $\mathcal{V}$ as its columns.

#1056 Note that $(D - 6I)\mathbf{v}_3 = e\mathbf{v}_2$.

#1057 Recall #922.

#1059 Take a vector parallel to the invariant line and another perpendicular to it.

#1060 Knowing $\det A = 1$ and $\text{trace} A = 3$ does not completely specify $c_A(x)$.

#1062 Note $_{\mathcal{W}}S_{\mathcal{V}} = Q^{-1}AP$ and $_{\mathcal{X}}T_{\mathcal{W}} = R^{-1}BQ$ where $P = {}_{\mathcal{E}}I_{\mathcal{V}}, Q = {}_{\mathcal{F}}I_{\mathcal{W}}$ and $R = {}_{\mathcal{G}}I_{\mathcal{X}}$.

#1064 Recall Proposition 3.88(d) and #748.

#1065 Note that $\mathbf{v}_1$ is a 1-eigenvector.

#1066 Note $\mathbf{v}_2$ and $\mathbf{v}_3$ span a plane that is invariant under T. Show the plane must be $x + y + z = 0$.

#1067 For (iv) and (v), answer this question first when using the $\mathcal{B}$-co-ordinates.

#1068 Recall Proposition 4.14 and also note $A^2 = A$.

#1070 For (ii) use the co-ordinates associated with the basis found in (i).

#1071 This is a rewording of #691 with $P = {}_{\mathcal{V}}I_{\mathcal{F}}$ and $Q = {}_{\mathcal{E}}I_{\mathcal{W}}$.

#1072 Take the union of bases for U and V.

#1073 Recall #771 and apply the rank-nullity theorem.

#1074 Note that $T(T^i\mathbf{v}) = T^{i+1}\mathbf{v}$.

#1075 Recall #820. Also note $(\lambda I_r - J(\lambda, r))^n = 0$ if and only if $n \geqslant r$.

#1076 A has one eigenvalue with a geometric multiplicity of 1.

#1077 For the fourth matrix, work out the minimal and characteristic polynomials.

#1078 Begin with, then adapt, a basis for $\mathbb{C}_n$ by extending a basis for $\text{Null}(M^{l-1})$ to $\text{Null}(M^l)$ and do this for $2 \leqslant l \leqslant k$.

#1079 Apply the fundamental theorem of algebra, primary decomposition theorem and #1078.

#1084 Note a and c cannot both be zero. If $a \neq 0$ then $b = -cd/a$.

#1090 Consider $A\mathbf{i}, A\mathbf{j}$ and $A\mathbf{k}$.

#1092 You may wish to show their equivalence in the order (i) $\implies$ (ii) $\implies$ (iv) $\implies$ (iii) $\implies$ (i).

#1095 The area of the triangle PQR equals $\frac{1}{2}\left|\overrightarrow{QP} \wedge \overrightarrow{RP}\right|$.

#1096 In each case choose an orthonormal basis $\mathbf{v}$ and $\mathbf{w}$ with $\mathbf{v}$ parallel to the invariant line of the shear/stretch.

#1097 Use Theorem 4.31.

#1098 Apply the method of Example 4.30.

#1099 By Theorem 4.31 there is a rotation matrix P such that $P^T AP = \text{diag}(-1, R_\theta)$.

#1102 Note that $R(\mathbf{i}, -\alpha)B\mathbf{i}$ must be a linear combination of $\mathbf{i}$ and $\mathbf{k}$.

#1103 (i) If $R\mathbf{i} = \mathbf{i}$, what are the first row and first column of R?

#1104 Assume $T(\mathbf{0}) = \mathbf{0}$ and consider $T(\mathbf{i})$, $T(\mathbf{j})$, $T(\mathbf{k})$.

#1105 Apply Theorem 4.40.

#1106 For each possible image point, you still need to exhibit an isometry that sends the origin there.

#1108 Let $S(\mathbf{v}) = T(\mathbf{v}) - \mathbf{b}$ and consider $A = (S(\mathbf{e}_1^T) \,|\, S(\mathbf{e}_2^T) \,|\, \ldots \,|\, S(\mathbf{e}_m^T))$.

#1111 Find the general solution for $3x_1 + 2x_3 = x_2 + 4x_4$ so as to find a basis.

#1114 If $\mathbf{u} \neq \mathbf{0}$ then extend $\mathbf{u}/|\mathbf{u}|$ to an orthonormal basis.

#1115 Note that the columns of X form a basis, and apply the Gram–Schmidt process to those columns.

#1117 Show that there are 2^n triangular, orthogonal matrices.

#1118 #1115 may prove helpful.

#1119 For C show first that $b^2 = d^2 = e^2$ and $a^2 = c^2 = f^2$. Show then that a, say, satisfies a quartic.

#1120 Note the special cases $|X| = |Y| = 1$ and $X = Y = 0$.

#1126 Review the reasoning leading to (4.18).

#1128 Consider the determinant of the stretch $(x, y) \mapsto (ax, by)$.

#1134 The answer is not a conic with a normal form.

#1135 #518 may prove helpful.

#1136 Eliminate x to form a quadratic in y. Note the special case when this 'quadratic' is in fact degree one.

#1140 Differentiate the ellipse's equation implicitly.

#1141 Recall the chain rule $dy/dx = (dy/dt)/(dx/dt)$.

#1142 Use #1130 to find a parametrization for $x^2 + xy + y^2 = 1$.

#1143 Every point of the plane can be written in the form $(1, 1, -1) + X\mathbf{e}_1 + Y\mathbf{e}_2$ for some X, Y.

#1144 Mimic the solution of Example 4.49.

#1146 Note $|PF| + |PF'| = e|DP| + e|PD'|$.

#1147 We can choose Cartesian co-ordinates so that $F = (c, 0)$ and $F' = (-c, 0)$.

#1149 Need only consider the cases $k \geqslant 0$ as $k < 0$ is the same as the $-k$ case with F and F' swapped.

#1152 Note that the curve's equation factorizes.

#1161 Apply Theorem 4.53 and #1157.

#1163 First show that the matrix has determinant $(B^2 - 4AC)k/4$ where k is as given in (4.34).

#1164 Implicitly differentiate (4.28) to find the gradient at (X, Y).

#1165 Without loss of generality assume $\mathbf{v} = (a, 0)$ and then eliminate the parameter t.

#1168 Use the polar equation for a conic, $r = ke/(1 + e\cos\theta)$.

#1169 Parametrize l_θ as $(a + t\cos\theta, b + t\sin\theta)$.

#1171 Consider a map of the plane that takes the unit circle to an ellipse.

#1176 For the second part, #579 may be of some help.

#1180 (ii) Note the Legendre polynomials are eigenvectors of L.

#1184 Say $\lambda_1 > 0$. Begin by setting $\mathbf{x} = (\lambda_1^{-1/2}, 0, \ldots 0)$. What does this imply for the entries of M?

#1185 (i) Note $\mathbf{v}^T A^T A \mathbf{v} = |A\mathbf{v}|^2$. (ii) Use the spectral theorem.

#1186 The matrix $\text{diag}(a, b, c)$ maps the unit sphere to the given ellipsoid.

#1188 Choose co-ordinates so that $\mathbf{a} = (1, 0, 0)$, $\mathbf{b} = (b\cos\theta, b\sin\theta, 0)$. Take care of special cases of θ.

#1192 Note $\mathbf{x}$ is in $T(E)$ if and only if $T^{-1}(\mathbf{x})$ is in E. Apply Theorem 4.53.

#1195 Make an orthogonal change of co-ordinates using calculations from #1193 and apply Cauchy–Schwarz.

#1198 To find the volume, relate the product of a cubic's roots to its constant coefficient.

#1200 #1184 and #579 should help.

#1201 Quadrics are related by an orthogonal/invertible variable change if their matrices' eigenvalues are the same/same signs.

#1202 Show that $\det Q_\alpha = 0$ has two real roots (which you need not identify).

#1204 Last part: if X is skew-symmetric then $M = iX$ is Hermitian.

#1205 Note that $\tilde{P}P^T$ and $Q^T\tilde{Q}$ are both orthogonal and so preserve length.

#1206 Note that $|A\mathbf{v}| = |PAQ\mathbf{v}|$ for any vector $\mathbf{v}$ in $\mathbb{R}_n$.

#1208 As PAQ is diagonal then $PAQ = (PAQ)^T$.

#1209 (iii) Find an expression for AA^+ and show it is orthogonal projection onto $P^T\langle \mathbf{e}_1^T, \ldots, \mathbf{e}_r^T\rangle$.

#1210 If M satisfies I,II,III show first that $AM = AA^+$ and $MA = A^+A$.

#1211 Examples can be found using singular 2×2 matrices.

#1213 AA^+ is orthogonal projection onto $\mathrm{Col}(A)$. #1232 may help. $\|MN\| = \|N\|$ if M is orthogonal.

#1216 First find an orthogonal matrix P such that PT fixes $(1,0,0)$.

#1217 (i) Every point at a distance b from A has the form $(\sin b \cos\phi, \sin b \sin\phi, \cos b)$ for some ϕ. Seek to solve for ϕ.

#1218 Show that $\cos a = \mathbf{b}\cdot\mathbf{c}$ and $\cos\alpha = (\mathbf{a}\wedge\mathbf{b})\cdot(\mathbf{a}\wedge\mathbf{c})/(\sin c \sin b)$.

#1220 Use the approximations $\cos x \approx 1 - x^2/2$ and $\sin x \approx x$.

#1221 With the vertices ABC going clockwise around the outward-pointing normal, $\mathbf{b}\wedge\mathbf{c}$ points into $A's$ hemisphere.

#1222 Start by showing $\mathbf{b}'\wedge\mathbf{c}' = [\mathbf{a},\mathbf{b},\mathbf{c}]\mathbf{a}/(\sin b \sin c)$.

#1223 Consider the area of the lunes, the two slices of the sphere made by the great circles meeting at a vertex of the triangle.

#1224 Use the first cosine rule to note their angles are also equal. For the last part use the second cosine rule.

#1227 (ii) Octonion multiplication need not be associative, as quaternions need not commute.

#1230 Find an orthonormal basis for X and extend it to one for $\mathbb{R}^4$.

#1231 Extend an orthonormal basis for X to one for $\mathbb{R}^n$. #987 may be helpful.

#1232 (i) If $\mathbf{v} = \mathbf{x} + \mathbf{y}$ and $\mathbf{x}$ is in X show that $|\mathbf{v} - \mathbf{x}'|^2 \geqslant |\mathbf{y}|^2$ with equality when $\mathbf{x}' = \mathbf{x}$.

#1233 For the second inequality apply the first to A^T.

#1234 The one-dimensional subspace $\langle\mathbf{v}\rangle$ is A-cyclic if and only if $\mathbf{v}$ is an eigenvector.

#1236 Recall that the minimal polynomial divides the characteristic polynomial by the Cayley–Hamilton theorem.

#1239 (ii) Recall #987 and #820.

#1241 (ii) By Bézout's lemma there are $v(x)$ and $w(x)$ with $v(x)m_{A,\mathbf{v}}(x) + w(x)m_{A,\mathbf{w}}(x) = 1$.

#1242 If $m(x) = c(x)$ then the question is equivalent to finding the cyclic vectors.

#1244 Begin by finding $m(x)$ and $c(x)$ in each case. (iv) Note $D - I$ has rank 2.

#1245 The t', x' co-ordinates of O' are always such that $x' = 0$.

#1248 (iii) O' measures the time to the planet as $d/(v\gamma(v))$ whilst O measures it as d/v.

#1249 (i) If O sends out a light signal at time t_1 as measured by O the t, x co-ordinates of the light signal are $(ct, c(t_1 - t))$.

#1250 (ii) Note $\mathbf{v}'(t) = A'(t)A(t)^T A(t)\mathbf{v}_0$.

Techniques of Integration

#1257 Make use of #1256(ii).

#1258 For a subinterval I define J (which may be a union of subintervals) such that $1_I(x) = 1_J(x^2)$.

#1260 A step function may take at most finitely many values.

#1261 If $-1 \leqslant x \leqslant 1$ then $-1 \leqslant x^2 - 1 \leqslant 0$. Note $x^2 + 1 \geqslant 1$ for all real x.

#1262 Let the step function ϕ play the role of each of ϕ, f, ψ noting $\phi \leqslant \phi \leqslant \phi$.

#1263 If $\phi = \sum y_k 1_{I_k} = \sum Y_l 1_{J_l}$, consider an ordered list of endpoints of all the subintervals I_k and J_l.

#1265 Consider the indicator function on the union of the I_k.

#1268 Consider carefully each of the remaining five cases such as $c < b < a$.

#1269 Choose f to have a discontinuity, for example, a step function.

#1270 (ii) Note $F(a) = F(0) = 0$ if F is periodic. Now write $na \leqslant x < (n+1)a$ for an integer n.

#1271 If $\phi \leqslant f \leqslant \psi$ then $\min\{|\phi|, |\psi|\} \leqslant |f| \leqslant \max\{|\phi|, |\psi|\}$. Also $\max\{f, g\} = \frac{1}{2}(f + g + |f - g|)$.

#1272 Assume initially that $f, g \geqslant 0$.

#1273 Say $\phi \leqslant f \leqslant \psi$. If $\sqrt{\psi}$ is small it contributes little to $\int_a^b (\sqrt{\psi} - \sqrt{\phi}) \, dx$. Else use the identity $\sqrt{\psi} - \sqrt{\phi} = (\psi - \phi)/(\sqrt{\psi} + \sqrt{\phi})$.

#1275 Use #1272 and #1273 applied to f and $\bar{f}$. Take λ with $|\lambda| = 1$ such that $\int_a^b \lambda f(x) \, dx$ is real.

#1278 Consider rectangular areas above and below the graph $y = f(x)$.

#1279 Introduce antiderivatives F and G for f and g.

#1280 Note that $(f(x) - f(y))(g(x) - g(y)) \geqslant 0$ for all x, y.

#1281 Use #246.

#1284 Use #257.

#1285 Note that $A'(x) = 0$ at an internal maximum or minimum.

#1286 For the limit, note that $\cos x < x^{-1} \sin x < 1$.

#1287 (i) If $0 < x < 1$ write $x^{-1} = 1 + t$ and apply Bernoulli's inequality to $(1 + t)^n$.

#1290 For (iii) define h with appropriate signed areas so that $\int_a^1 h(x) \, dx$ oscillates as $a \to 0$.

#1292 You may assume that an increasing function which is bounded above converges to a limit.

#1293 Use integrable f_N sufficiently close to f and step function ϕ sufficiently close to f_N.

#1295 Prove this first for indicator and step functions.

#1296 Find a step function with arbitrarily small integral and such that $\psi \geqslant T$.

#1301 If $f'(x) = f(x)$ and $f(0) = 1$, consider $g(x) = f(x)e^{-x}$.

#1302 If $f(x) = \sin^{-1} x$ then differentiate $\sin f(x) = x$ with respect to x.

#1304 Rearrange $\sinh y = x$ as a quadratic equation in e^y.

#1305 Separate the fifth integral into two integrals on $(-\infty, a)$ and (a, ∞). Do the same for the final integral.

#1310 Let $f(x) = (1 + x)^\alpha - 1 - \alpha x$. Show that $f(x)$ is decreasing for $-1 < x < 0$ and increasing for $0 < x$.

#1311 For the second inequality consider $g(x) = (1 + x)e^{-x}$.

#1313 Note $\varepsilon \ln \varepsilon \to 0$ as $\varepsilon \to 0$. This is proved in #1318.

#1314 Consider the derivative of $\ln x$ at 1, formally defined as a limit.

#1315 (i) The integrand is positive. (ii) What is the integral over one period? (iii) The integrand is odd.

#1316 For (iii) begin with $\cos \theta < 1 < \sec^2 \theta$ and $\cos 0 = 1 = \sec^2 0$.

#1317 (ii) Use careful induction. (iii) Note $100^n/n!$ converges to 0 as n becomes large.

#1318 Use induction and integration for the first part. For the last limit try setting $x = e^{-t}$.

#1320 For the second integral use the binomial theorem and Proposition 1.15.

#1324 Differentiate the RHSs using Proposition 5.17.

#1326 (i) Compare x^{-1} on $[n, n+1]$ with $(n+1)^{-1}$. (ii) Include once and remove twice the even terms in s_n.

#1328 $\lfloor x \rfloor = k$ on the interval $[k, k+1)$.

#1330 Having calculated the first repeated integral, the second repeated integral should be clear.

#1333 Show that $f'(x)$ is an increasing function. Rewrite the inequality in terms of integrals of $f'(x)$.

#1334 The average value of $f'(x)$ on $[a, c]$ is less than it is on $[c, b]$. For the last part set $f(x) = -\ln x$.

#1335 (iii) Explain why you may assume $\sum_{k=1}^{n} (x_k)^p = 1 = \sum_{k=1}^{n} (y_k)^q$.

#1336 Begin with $|x_k + y_k|^p \leqslant |x_k| \, |x_k + y_k|^{p-1} + |y_k| \, |x_k + y_k|^{p-1}$ and use Hölder's inequality.

#1337 For any $\varepsilon > 0$ there is n such that $|f(x) - L| < \varepsilon/2$ for $0 < x < 1/\sqrt{n}$.

#1340 (ii) Recall $2 \sinh x \cosh x = \sinh 2x$.

#1342 Note $\sec^{-1} x = \cos^{-1}(1/x)$.

#1343 Care is needed with the limits at 0 and ∞.

#1346 Note $I_n = J_n + iK_n$ and use De Moivre's theorem to find J_n, K_n.

#1352 Use $\tan^2 \theta = \sec^2 \theta - 1$ and recall $\tan \theta$ differentiates to $\sec^2 \theta$.

#1354 Recall $2 \cos \theta = z + z^{-1}$ where $z = \operatorname{cis} \theta$.

#1356 $2n \times (2n-2) \times (2n-4) \times \cdots \times 2 = 2^n n!$ may be a helpful rearrangement.

#1357 Note $\cos(2n+1)x + \cos(2n-1)x = 2 \cos(2nx) \cos x$.

#1359 Write $x^n e^{-x^2} = x^{n-1}(x e^{-x^2})$ and use IBP.

#1363 Use IBP. In each case writing $x^2 = (1 + x^2) - 1$ may help yield a reduction formula.

#1366 Use IBP to relate J_n to I_{n+1}.

#1367 Write $\ln 2 - I_{2n}$ as a finite sum. Note that $0 \leqslant I_{2n} \leqslant (2n+1)^{-1}$.

#1368 Write $\pi/4 - I_{4n}$ as a finite sum. Note that $0 \leqslant I_{4n} \leqslant (4n+1)^{-1}$.

#1369 For the last part assume $e = m/n$ and rearrange the inequality to produce a contradiction.

#1370 Consider the integrals of $(1 - x)^n e^{ax}$ for $0 \leqslant x \leqslant 1$.

#1371 Use the binomial theorem for the first inequality; for the second adapt #1331.

#1373 Use induction and integration by parts.

#1374 Use Taylor's theorem and show that the remainder term does not exceed $1/(n+1)$.

#1375 Use #1374 on $\ln(1 - x)$. Some care is needed in swapping the order of the sum and integral.

#1376 (ii) Recall $1 - \cos xt \leqslant x^2 t^2/2$ (#1317) and $t \leqslant \tan t$ (#1286).

#1377 For the contradiction, find a bound for the integral and note $a^n/n!$ tends to 0 as n becomes large.

#1378 (i) Set $x = e^u$. (ii) Set $x = u^2$. (iii) Set $u = \tan x$.

#1380 Set $x = \tan u$ for each of the three integrals.

#1383 Set $x = e^{-t}$ and relate the integral to the gamma function.

#1386 The reduction formula from #1353 may help.

#1387 Treat the area as a triangle and remainder. Recall $f(x) = R(\theta) \sin \theta$, $x = R(\theta) \cos \theta$.

#1388 Set $u = 1/x$ for the first part.

#1389 Relate this integral to the integral in #1388.

#1390 Begin by splitting the integral into two on intervals $(0, 1)$ and $(1, \infty)$.

#1391 Set $x = -t$ first to find an alternative expression for the integral. Then add the two integrals.

#1392 Use the method of #1391.

#1393 For the second integral note $2/(4x^2 - 1) = 1/(2x-1) - 1/(2x+1)$.

#1394 For the third integral note $\ln \sin x = \frac{1}{2} \ln(1 - \cos x) + \frac{1}{2} \ln(1 + \cos x)$.

#1395 Note $\sin 2x$ is even about $x = \pi/4$ and recall $\sin 2x = 2 \sin x \cos x$.

#1396 For the first integral, note $1 - \cos x = 2 \sin^2(x/2)$ and apply IBP.

#1397 Note $\cosh^2 t = (\cosh 2t + 1)/2$ and that $\sqrt{x^2 + 1}$ is an increasing function.

#1398 For the second integral write $b^2 - \sinh^2 x = b^2 + 1 - \cosh^2 x$.

#1399 Write $1 + a^2 \sin^2 x = (1 + a^2) - a^2 \cos^2 x$.

#1400 Use the substitutions $u = bx, u = bx^2, x = e^u, u = ax^b$.

#1401 Substituting $y = x^2$ into J and substitute $u = \sqrt{a}x$ into I_a.

#1402 For (i) and (iii) begin by completing the square. For (ii) use IBP to begin.

#1403 Note $-ax^2 - b/x^2 = -ac^2/u^2 - bu^2/c^2 = -au^2 - b/u^2$ when $u = c/x$ and $c = \sqrt{b/a}$.

#1404 Begin with the substitution $t = u^k$ for an appropriate choice of k.

#1408 For $0 \leqslant x \leqslant 1/2$ write $x = u/(1+u)$.

#1409 Recall the definition of $B(a+1, -a)$ and set $u = x/(1-x)$ or equally $x = u/(u+1)$.

#1410 For (ii) assume $0 < \beta < 1$ initially. For $-1 < \beta < 0$ use the substitution $u = 1/x$.

#1411 Set $x = u/(1+u)$ in the defining integral for $B(a, b)$.

#1412 For each integral set $u = x^n$.

#1413 Show $B(x, n+1)n^x = \int_{u=0}^{u=n} u^{x-1}(1 - u/n)^n \, du$. Care is needed taking the limit through the integral sign.

#1414 Apply #1413 to $\Gamma(1/2) = \sqrt{\pi}$.

#1415 Apply #1414 to the $\Gamma(x)$ and $\Gamma(x + 1/2)$ terms.

#1416 Use #1414 and #1376.

#1417 Use the gamma reflection property and #1395. For the last part, differentiate the desired integral.

#1418 Make a substitution and use #1409.

#1420 For the 'deduce' part, let a tend to infinity.

#1421 Find a lower bound for the integral of $|\sin x/x|$ between $k\pi$ and $(k+1)\pi$.

#1422 If integrating y first, use partial fractions. If integrating x first, set $x = y^{-1/2} \tan u$.

#1423 Use #1420 and for the second integral $2 \sin ax \sin bx = \cos(a - b)x - \cos(a + b)x$.

#1424 Consider a repeated integral of $(1 - \cos yx)/x^2$.

#1426 Apply #1424 and $2 \cos ax \cos yx = \cos((a + y)x) + \cos((a - y)x)$.

#1425 Use induction and the identity $2 \sin a_n x \cos yx = \sin(a_n + y)x + \sin(a_n - y)x$.

#1430 For the second part, note that the integrand is positive for $0 < x < 1$.

#1431 Use partial fractions.

#1432 For the third integral, set $x = 2 \sinh t$.

#1433 Note that $x^4 + a^4 = (x^2 + \sqrt{2}ax + a^2)(x^2 - \sqrt{2}ax + a^2)$.

#1434 Write $(1 + x^4)^{-1}$ as an infinite geometric series.

#1435 Note $(3x + 4)^2 = 9(x^2 + 2x + 3) + 2(3x + 4) - 19$.

#1436 Begin by setting $u = e^x$.

#1437 Begin by setting $u = e^{2x}$.

#1438 For the third integral, set $u = \pi - x$ and add your two expressions for that integral.

#1445 Set $x = \tan^{-1} u^2$, and note $u^4 + 1 = (u^2 + 1 + \sqrt{2}u)(u^2 + 1 - \sqrt{2}u)$.

#1446 Set $x = \tanh^{-1} u^3$ and note $1 - u^6 = (1 - u)(1 + u)(u^2 + u + 1)(u^2 - u + 1)$.

#1448 Find the equation of the quadratic through $(x_{2k}, y_{2k}), (x_{2k+1}, y_{2k+1}), (x_{2k+2}, y_{2k+2})$.

#1450 Write the estimates as geometric sums.

#1451 Note that as $n \to \infty$ then $\delta = 1/(2n) \to 0$. Write the limits of the estimates in terms of derivatives.

#1452 Recall the summation formula from #243.

#1453 Use the first part of #1334.

#1455 Use IBP twice and recall #1277.

#1456 Show that $p(c_-) = p(c_+), p'(c_-) = p'(c_+), p''(c_-) = p''(c_+)$ where $c = (a+b)/2$.

#1458 Begin by setting $x = \tan t$ and then use IBP.

#1460 Consider the geometrical transformation that takes the graph $y = f(x)$ to the graphs of the new integrands.

#1461 (i) Use (A) and use (B). (iii) Use #1453.

#1463 Use $y_0 = \lim_{x \to 0} T(x) = 0$ and $y_n = \lim_{x \to 1} T(x) = 0$.

#1465 (ii) You may find the integral test helpful showing expected returns are infinite when $p = q = 1/2$.

#1466 (ii) #318 should prove useful. (iii) If $a \neq b$, ignore up/down movements and argue as in #1465.

#1467 It is enough to show that $\sum_{n=1}^{\infty} p_{4n}$ is infinite.

#1468 Substitute $x = u^2$.

#1469 Note $x^2 - 3x + 1 = 0$ when $x = \left(3 \pm \sqrt{5}\right)/2$ and split the integral accordingly.

#1470 Set $x + a = b \tan t$.

#1471 Note that the integrand is odd.

#1472 Set $u = \cosh x$.

#1473 Split the integral into two on $(0, \pi)$ and $(\pi, 2\pi)$ and substitute $t = \tan(x/2)$.

#1474 Note that the numerator is the derivative of the denominator

#1475 Substitute $x = e^{-t}$ and recall #1329.

#1476 Complete the squares in the exponentials and recall #1401.

#1477 Use the cosine double angle formula twice to write the integrand in terms of $x/2$.

#1478 Substitute $x = \cos^2 t$.

#1479 Recall J from #1401.

#1480 Use IBP, taking care with the limits at $x = 0$. Recall #1395.

#1481 Consider the repeated integral of $(1 + x^2 y^2)^{-1}$ where $x > 0$ and $a \leq y \leq b$.

#1482 Use $2 \sin mx \cos nx = \sin(m+n)x + \sin(m-n)x$.

#1483 The answer is in terms of the beta function.

#1484 Complete the square and make a hyperbolic substitution.

#1485 Complete the squares in the exponentials. The answer is in terms of the error function.

#1486 Use IBP and then a trigonometric substitution (or vice versa).

#1487 Set $u^3 = 2x + 1$.

#1488 Set $u = e^{2x}$ and recall #1411. The answer is in terms of the beta function.

#1489 Show that the integrand is odd.

#1490 Note $2 \cos mx \cos nx = \cos(m+n)x + \cos(m-n)x$. Take some care when $m = n$.

#1491 Use IBP.

#1492 Set $u = \sin^2 x$ to relate the integral to the beta function.

#1493 Note $(\cos x + \sin x) \cos x = 1 + \cos 2x + \sin 2x = \sqrt{2} \cos(2x - \pi/4)$.

#1494 Set the t-substitution $t = \tan(x/2)$. The answer is in terms of the beta function.

#1495 Set $u = e^{-x}$

#1496 Use $\ln(a/b) = \ln a - \ln b$. Take care with the integral's limits.

#1497 Set $u = \tan x$.

#1498 Begin by setting $x = \tan u$.

#1499 Write the integral in terms of the beta function, but give your answer explicitly.

#1500 Set $x = \sinh u$.

#1501 Begin with $u^2 = bx + c$. Give the answer in terms of the error function.

#1502 Set $x = a \tan u$.

#1503 Set $x = a \tan u$.

#1504 Set $x = \sin u$. Recall #1395.

#1505 Treat separately the cases of odd/even n. Give the answer in terms of harmonic numbers.

#1506 Start with IBP. Take care with the integral's limits.

#1507 Set $u = x^a$. Give the answer in terms of the beta function.

#1508 Set $u = x^2$.

#1509 Begin by setting $u = \cos x$.

#1510 Note $\cos x - \sin x = \sqrt{2}\cos(x + \pi/4)$. Give the answer in terms of Catalan's constant.

#1511 Set $u^2 = \sec x$ and use partial fractions.

#1517 Differentiate under the integral sign the expression for $F_Z(z)$.

#1522 (iv) Explain why any pdf $f(x)$ must satisfy $f(x) = 0$ for all points not in C.

#1523 The integral test should prove helpful for most parts.

#1525 (ii) Show that $\|f - g_n\|^2 = \|f - f_n\|^2 + \|f_n - g_n\|^2$ for any polynomial g_n of degree n.

#1526 Replace $L_m(x)$ with the expression from (2.28) and apply IBP repeatedly.

#1527 Use the expression for $H_m(x)$ from #469(v) and apply IBP repeatedly.

#1530 Make use of the inner product in #1528.

#1531 Note that your answers to (ii) and (iii) should agree.

#1532 Parametrize the curve as $x(\theta) = R(\theta)\cos\theta$, $y(\theta) = R(\theta)\sin\theta$ where $\alpha \leqslant \theta \leqslant \beta$.

#1533 Note f is an increasing function if $\mathbf{s}$ parametrizes C in the same direction as $\mathbf{r}$, and otherwise is decreasing.

#1535 Wrap the sector of the plane $r > 0, 0 < \theta < \sqrt{2}\pi$ isometrically on to the cone.

#1536 (iv) Differentiate $\mathbf{t} \cdot \mathbf{t} = 1$ with respect to s.

#1537 If κ is constant and non-zero, consider $\mathbf{c} = \mathbf{r} + \mathbf{n}/\kappa$.

#1538 (i) Differentiate $\mathbf{t} \cdot \mathbf{b} = 0$ and $\mathbf{b} \cdot \mathbf{b} = 1$ with respect to s. (ii) Differentiate $\mathbf{t} \cdot \mathbf{n} = 0$ and $\mathbf{b} \cdot \mathbf{n} = 0$.

#1539 Show that $\mathbf{n}''(s) = -(\kappa^2 + \tau^2)\mathbf{n}(s)$.

#1540 (ii) Use $f(x) = \sqrt{a^2 - x^2}$. (iii) Use $f(x) = ax/h$.

#1542 The particle's energy $m((dx/dt)^2 + (dy/dt)^2)/2 + mgy$ is constant throughout the motion.

#1543 Show first that the midpoint of the wheel's base is at $(x(s), y(s)) - s(x'(s), y'(s))$.

#1545 (i) Set $x = \sin u$. (ii) Set $x = a \tan u$.

#1546 Use $\cos x = 1 - 2\sin^2(x/2)$.

#1548 Write the derivative of $\sin x \cos x(1 - k^2 \sin^2 x)^{-1/2}$ in terms of $1 - k^2 \sin^2 x$.

#1549 Show that $\sin x = (1 + k')\sin t \cos t(1 - k^2 \sin^2 t)^{-1/2}$.

#1550 Show that $\cos t / \cos x = (1 + k \sin^2 t)(1 - k^2 \sin^2 t)^{-1/2}$.

#1551 (iv) Get sequences a_n (or b_n) of arithmetic (or geometric) means decreasing (or increasing) to $\mathrm{agm}(a, b)$.

#1553 (i) Write the integral as $I(a)$, find $I'(a)$ and note $I(b) = 0$ to find the constant of integration.

#1554 Find $I'(a)$ by differentiating under the integral. When integrating consider $I(a)$ as $a \to \infty$.

#1555 For the second integral, consider integrating $e^{-x}(1 - \cos ax)/x^2$ on $(0, \infty)$.

#1556 Recall #1388. For the second integral set $x = u^2$.

#1558 #1403 may help.

#1560 Be careful that $x^{\alpha} \sin ax/(x^2 + 1)^{\beta}$ and $x^{\alpha} \cos ax/(x^2 + 1)^{\beta}$ are integrable on $(0, \infty)$ when $\alpha \leqslant 2\beta - 1$.

#1561 Apply a mixture of IBP and differentiation under the integral.

#1562 Recall #1401. Show $I(a) - I'(a) = \frac{1}{2}\sqrt{\pi/a}$.

#1563 First note $P(X/Y \leqslant z) = P(X \leqslant Yz \text{ and } Y \geqslant 0) + P(X > Yz \text{ and } Y < 0)$.

#1564 Recall #1548.

#1565 Assume to begin that $c = 1$. Differentiate with respect to a.

#1566 (ii) Consider the integral of $(1 - x^{\alpha})(1 + x + x^2 + \cdots + x^N)$ over $(0, 1)$. Then let $N \to \infty$.

#1567 Use #1566(ii) to work out $\int_0^1 H_x \, dx$.

#1569 (i) Use #1360. (ii) Use #1416. (iii) Use #1413.

#1570 Differentiate under the integral. Note $\Gamma'(1) = \psi(1) = -\gamma$ by #1569(iii).

#1571 Show $\Gamma''(a) > 0$ so that $\Gamma'(a)$ is an increasing function for $a > 0$.

#1572 Set $u = e^{-x}$ initially. Use #1569(ii) and (iii).

#1574 (i) Differentiate under the integral sign.

#1575 Note $E(f(U)g(V)) = E(f(U))E(g(V))$ for independent variables U, V.

#1576 (i) Apply the monotone convergence theorem to $f_n(x) = \sum_{k=0}^{n} \frac{(tx)^k}{k!} f_X(x)$.

Differential Equations

#1580 Check $y(x)$ is also differentiable at $x = \ln 2$.

#1583 Rearrange the DE as $2yy' = y + xy'$ and integrate.

#1590 (ii) Compare the transformed equation with Example 6.4.

#1592 Note that setting $x = 0$ and $y = 1$ into the DE gives $0 = 0$, and says nothing about $y'(0)$.

#1598 Recall the second integral in #1431.

#1599 Treat the DE as a first-order DE in $z(x) = y'(x)$.

#1609 Note by the chain rule that $dy/dx = (dy/dt)/(dx/dt)$.

#1610 (i) Set $z = x - 2y$. (ii) Set $z = xy$.

#1611 (i) Set $z = x + y$. (ii) Set $z = e^y$.

#1613 Set $z = y^4$.

#1616 Show $f'(x) - \beta f(x) = 0$.

#1622 Rearrange the DE as an inhomogeneous linear first-order DE.

#1632 Try a particular solution of the form $y = Ax^2 + B$.

#1633 Try a particular solution of the form $y = Ax^3 + Bx^2 + Cx + D$.

#1634 As an alternative, rewrite the DE in terms of $z = y' - y$.

#1636 Find a particular solution of the form $y = Ax^2 + Bx + C$.

#1647 (iv) Note e^{-x} is part of the complementary function. Try Axe^{-x}.

#1652 Note that $\sin x$ is part of the complementary function.

#1654 (v) Recall $\sin^3 x = (3/4)\sin x - (1/4)\sin 3x$.

#1656 (ii) Recall $\cos^2 x = (\cos 2x + 1)/2$.

#1657 It may help to write $2^x = e^{x \ln 2}$.

#1658 Note $\cosh x = (e^x + e^{-x})/2$ but e^{-x} and xe^{-x} are both parts of complementary function.

#1661 Consider separately the cases $E > V, E = V, E < V$.

#1662 Show $x (dy/dx) = dy/dz$; $x^2 \left(d^2y/dx^2\right) = d^2y/dz^2 - dy/dz$.

#1663 Show $y'(x) = tz'(t)$ and $y''(x) = tz'(t) + t^2 z''(t)$.

#1664 Make the substitution $z = \ln(x+1)$.

#1666 Eliminate $g(x)$ to obtain a second-order inhomogeneous DE in $f(x)$.

#1667 Begin by setting $\omega = d\theta/dt$ and showing $d^2\theta/dt^2 = \omega(d\omega/d\theta)$.

#1669 Recall the invertibility of the Vandermonde matrix.

#1670 Solve for e^t and e^{6t} in terms of x and y and use $(e^t)^6 = e^{6t}$.

#1672 Note that $dy/dx = (dy/dt)/(dx/dt)$ by the chain rule.

#1676 Eliminate y to obtain a second order homogeneous DE in $x(t)$.

#1677 Review the solution of Example 6.43.

#1679 Try a particular solution of the form $x = \alpha e^{2t}$, $y = \beta e^{2t}$.

#1681 As e^t is in the complementary function, try a particular solution $x = (Ct+D)e^t$, $y = (Et+F)e^t$.

#1683 The characteristic polynomial has a real root and two imaginary ones. Argue as in Example 6.43.

#1688 #1130 should prove helpful with the sketches.

#1691 For (v) note that the Laplace transform of x is s^{-2}.

#1693 The functions need only differ at a point to be distinct.

#1696 Give your answer in terms of the error function.

#1697 If $f(x) = (1 - \cos ax)/x$, apply the Laplace transform to $xf(x)$.

#1698 With $f(x)$ being the given function, apply the Laplace transform to $\sin x = xf'(x)$.

#1699 Make a substitution and then use #1403 .

#1700 #1570 may prove useful.

#1701 Write the error function as a convolution.

#1707 Use the convolution theorem.

#1709 Recall #1695. Show $(f * g)(x)$ is a multiple of $x^{\alpha+\beta+1}$.

#1721 The DE is $f''(x) = 0$ for $x \neq a$ and $f'(x)$ has a discontinuity of 1 at $x = a$.

#1726 The Laplace transform $\overline{f_{r,a}}(s)$ involves the error function.

#1727 Use IBP and recall that $\delta(x - a)$ is the derivative of $H(x - a)$.

#1728 Note $F_T(t)$ has a discontinuity of size p at $t = 0$. Use the sifting property to evaluate $E(T)$.

#1729 It may be helpful to note both $\overline{J_1(s)}$ and $\overline{J_1''(s)}$ both converge to 0 as s becomes large.

#1730 Note that $\int_0^x tJ_0(t)\,dt$ is the convolution of 1 and $xJ_0(x)$.

#1731 Recall the limit from #1331.

#1736 Use the chain rule.

#1737 (i) There is no component of acceleration in the transverse direction **f**.

#1738 (iii) If r_+ and r_- denote the extremes of the planet's distance from the sun then $R = (r_+ + r_-)/2$.

#1741 Consider the matrices $N(t) = e^{-At}M(t)$.

#1742 Consider $M(s + t)M(-t)$ treating s as fixed and t as the variable.

#1744 The exponential of a matrix is characterized by properties (i) and (ii) in Definition 6.80.

#1747 Rewrite $M'(t) = E_{12}M(t)$ as separate DEs in the entries of $M(t)$.

#1748 The limit of $(e^a - e^c)/(a - c)$ as a approaches c can be considered as a derivative.

#1751 Show that $M(t) = (2e^t - e^{2t})I + (e^{2t} - e^t)A$ has the property $M'(t) = AM(t)$.

#1752 Prove the desired result first for $A = J(\lambda, r)$ and then apply #1079.

#1756 Note that if B commutes with A then B commutes with e^{At}.

#1757 Note that $(A - I)^4 = 0$.

Bibliography

[1] Arnoux, Pierre. Some Remarks About Fibonacci Multiplication. *Applied Mathematics Letters*, 2:4, 319–320, 1989.

[2] Baker, Alan. *A Comprehensive Course in Number Theory*. Cambridge: Cambridge University Press, 2012.

[3] Berger, Marcel. *Geometry I and II*. Berlin and Heidelberg: Springer, 2008.

[4] Binmore, K. G. *Mathematical Analysis: A Straightforward Approach,* 2nd ed. Cambridge: Cambridge University Press, 1982.

[5] Boros, George, Moll, Victor H. *Irresistible Integrals*. Cambridge: Cambridge University Press, 2004.

[6] Burton, David. *Elementary Number Theory,* 7th ed. New York: Tata McGraw-Hill, 2012.

[7] Conrad, Brian. *Impossibility Theorems for Elementary Integration*. Available at math.stanford.edu/~conrad/papers/elemint.pdf

[8] Devaney, Robert L. *An Introduction to Chaotic Dynamical Systems*, 2nd ed. Boulder, CO: Westview Press, 2003.

[9] Derbyshire, John. *Unknown Quantity: A Real and Imaginary History of Algebra*. London: Atlantic Books, 2006.

[10] Devlin, Keith. *The Unfinished Game*. New York: Basic Books, 2010.

[11] Edgar, Gerald. *Measure, Topology and Fractal Geometry,* 2nd ed. New York: Springer Science+Business Media, 2009.

[12] Edwards, C. H., Jr. *The Historical Development of the Calculus*. New York: Springer-Verlag, 1979.

[13] Fenn, Roger. *Geometry*. London: Springer-Verlag, 2000.

[14] Gradshteyn, I. S. Ryzhik, I. M. *Tables of Integrals, Series and Products*, 7th ed. San Diego: Academic Press, 2007.

[15] Greenberg, Marvin Jay. *Euclidean and Non-Euclidean Geometries*, 4th ed. New York: W. H. Freeman, 2007.

[16] Guy, Richard K. The Strong Law of Small Numbers. *American Mathematical Monthly*, 95:8, 697–712, 1988.

[17] Guy, Richard K. The Second Strong Law of Small Numbers. *Mathematics Magazine*, 63:1, 3–20, 1990.

[18] Havil, Julian. *Gamma: Exploring Euler's Constant*. Princeton, NJ: Princeton University Press, 2009.

[19] Hellman, Hal. *Great Feuds in Mathematics: Ten of the Liveliest Disputes Ever*. Hoboken, NJ: John Wiley & Sons, 2006.

[20] James, Ioan. *Remarkable Mathematicians*. Cambridge: Cambridge University Press, 2003.

[21] Knuth, Donald. Fibonacci Multiplication. *Applied Mathematics Letters*, 2:1, 3–6, 1989.

[22] Livio, Mario. *The Equation That Couldn't Be Solved: How Mathematical Genius Discovered the Language of Symmetry*. London: Souvenir Press, 2007.

[23] Mahoney, Michael Sean. *The Mathematical Career of Pierre de Fermat*, 2nd ed. Princeton, NJ: Princeton University Press, 1994.

[24] Nahin, Paul J. *Inside Interesting Integrals*. New York: Springer Science+Business Media, 2014.

[25] Reid, Constance. *Julia: A Life in Mathematics*. Washington, DC: Mathematical Association of America, 1997.

[26] Roe, John. *Elementary Geometry*. Oxford: Oxford University Press, 1993.

[27] Rose, H. E. *A Course in Number Theory*, 2nd ed. Oxford: Oxford University Press, 1998.

[28] Silva, Tomás Oliveira e, Herzog, Siegfried, Pardi, Silvio. Empirical Verification of the Even Goldbach Conjecture and Computation of Prime Gaps up to 4×10^{18}. *Mathematics of Computation*, 83:288, 2033–2060, 2014.

Index

Printed in the United States
By Bookmasters